AF529860

Economic Mineralization

Economic Mineralization

A festschrift for
Professor Bhawani Shanker Paliwal

editor

K.L. SHRIVASTAVA

Professor of Economic Geology
Jai Narain Vyas University
Jodhpur, India

SCIENTIFIC PUBLISHERS (INDIA)

P.O. Box 91 JODHPUR (Raj.)

Published by:

Tanay Sharma
SCIENTIFIC PUBLISHERS (INDIA)
5-A, New Pali Road, P.O. Box 91
Jodhpur - 342 001 (India)
Tel.: +91-291-2433323
Fax: +91-291-2613449

E-mail: info@scientificpub.com
www.scientificpub.com

ISBN: 978-81-7233-572-4

Cover Illustration:
Lignite and Bentonite mine, Giral (Barmer), Rajasthan
Courtesy : Dr. S. C. Mathur

Printed in India

Hon'ble Reviewers

Prof. U. Aswathnarayana
Director, MICWM
B16, Santishikha
Somajiguda
HYDERABAD 500 082

Padam Sree
Dr. B.P. Radhakrishnan
Geological Society of India
63, 12th Cross, Basappa Layout
BANGALORE 560 019

Dr. S. Viswanathan
Former Director AMD
B 203 United Avenue Apt.
South end, 7-1-29, Ameerpet
HYDERABAD 500 001

Prof. S.K. Tandon
Pro-Vice Chancellor
Delhi University
DELHI 110 007

Prof. K.L. Rai
Former Head, Appl. Geol. ISM
45, Chatrapati Shivaji Nagar
Kolar Road
BHOPAL 462 016

Prof. A.B. Roy
Emeritus Professor
Deptt. of Geology
Preseidency College
KOLKATA 700 073

Prof. S.M. Naqvi
Former Director
Nat. Geophys. Res. Institute
Uppal Road
HYDERABAD 500 001

Prof. S.C. Sarkar
Emeritus Scientist
Deptt. of Geological Science
Jadavpur University
KOLKATA 700 032

Prof. B.K. Sahu
Emeritus Professor
Deptt. of Earth Sciences
I.I.T.B., Powai
MUMBAI 400 076

Padam Sree
Prof. K.S. Valdiya
Former Vice Chancellor
Jawaharlal Nehru Centre for
Adv. Sci. Res.
BANGALORE 560 064

Dr. A.V. Sankaran
10, P & T Colony
I Cross, II Block
R.T. Nagar
BANGALORE 560 032

Prof. M.K. Pandya
Emeritus Professor
Deptt. of Geology
MLS University
UDAIPUR 313 001

Cont.

Cont.

Prof. M.C. Khare
School of Studies in Earth Sci
Jiwaji University
GWALIOR 474 011

Prof. V.K. Vaya
Former Head
Geology Deptt. JNVU
L-7, Income Tax Colony
Tonk Road
JAIPUR 302 001

Prof. S.K. Lunkad
Founder Chairman
Department of Geology
Kurukshetra University
KURUKSHETRA 136 119

Prof. Nikhil Desai
Head
Department of Appl. Geol.
M.S. University of Baroda
VADODARA 390 002

Prof. P.K. Verma
Chairman
Department of Geology
Delhi University
DELHI 110 007

Prof. Ravindra Kumar
Chairman
Deptt. of Appl. Geol. and CAS
Panjab University
CHANDIGARH 160 014

Dr. Anil Bhardwaj
Head, Planetary Sci. Br.
Space Physics Laboratory
Vikram Sarabhai Space Centre
TRIVANDRUM 695 022

Prof. J.P. Srivastava
Deptt. of Geology
Delhi University
DELHI 110 007

Dr. Arun Kumar Shandilya
Department of Geology
Dungar College
Bikaner University
BIKANER 334 003

Prof. Arun Kumar
Head
Department of Earth Science
Manipur University
IMPHAL 795 003

Prof. R.K. Trivedi
Head, Deptt. of Appl. Geol.
Centre of Adv. Stud. in Geol.
Dr. H.S. Gaur University
SAGAR 470 003

Prof. R.P. Tiwari
Dean
Faculty of Forestry and Geol.
Mizoram University
AIZAWAL 796 009

Dr. D.K. Sinha
Scientist 'G'
Atomic Mineral Directorate
1-10-153-156 Begumpet
HYDERABAD 500 016

Prof. V.S. Palaria
Ex Head
Department of Mining Engg.
JNV University
JODHPUR 342 011

Prof. A.K. Sharma
Head
Department of Computer and Electrical Engineering
MPA Technical University
UDAIPUR 313 001

Prof. U.C. Singh
Chairman
School of Studies in Geology
Jiwaji University
GWALIOR 474 011

Prof. Mahshar Raza
Chairman
Department of Geology
Aligarh Muslim University
ALIGARH 202 002

Prof. A.S. Sheoran
Head
Department of Mining Engg.
JNV University
JODHPUR 342 011

Cont.

Prof. Madhumita Das
Head
Department of Geology
Utkal University
BHUBANESWAR 751 004

Prof. D.C. Gupta
Head
Department of Appl. Geol.
Barkatullah University
BHOPAL 462 010

Dr. S.S. Rathore
Head
Department of Mining Engg.
MPA Technical University
UDAIPUR 313 001

Prof. Hari B. Srivastava
Department of Geology
Banaras Hindu University
VARANASI 221 005

Mr. Alok Tripathi
Suptd. Geologist
O.N.G.C. Ltd., Camp.
JAISALMER 345 001

Prof. T.K. Biswal
Department of Earth Science
I.I.T.B. Powai
MUMBAI 400 076

Prof. Devendra Mohan
Department of Zoology
JNV University
JODHPUR 342 005

Prof. H.C. Verma
Department of Physics
I.I.T. Kanpur
KANPUR 208 010

Prof. Rajesh K. Srivastava
Department of Geology
Banaras Hindu University
VARANASI 221 005

Dr. N. Srivastava
Scientist
Physical Research Laboratory
Navrangpura
AHMEDABAD 380 009

Dr. S.C. Mathur
Department of Geology
JNV University
JODHPUR 342 005

Prof. L.S. Chamyal
Department of Geology
M.S. University of Baroda
VADODARA 390 002

Dr. K. Ramesh Kumar
Scientist 'G'
Atomic Mineral Directorate
Sanganer
JAIPUR 302 030

Dr. Deepak Dhingra
Scientist
Physical Research Laboratory
Navrangpura
AHMEDABAD 380 009

Dr. S.K. Trivedi
Department of Geology
JNV University
JODHPUR 342 005

Prof. P.K. Verma
Department of Geology
Vikram University
UJJAIN 456 010

Dr. Pankaj K. Srivastava
Department of Geology
University of Jammu
JAMMU 180 006

Dr. Rajesh Sharma
Senior Scientist
Wadia Inst. Himalayan Geol.
DEHRADUN 248 001

Dr. Krishan Kumar
School of Studies in Env. Sci.
Jawaharlal Nehru University
NEW DELHI 110 001

Mr. Hemant Kumar Jain
Sr. Ore Dressing Officer
Indian Bureau of Mines
AJMER 305 002

Dr. (Mrs.) Beena Tripathi
Department of Geology
JNV University
JODHPUR 342 005

Cont.

Preface

Economic Geology developed as a science, particularly in those times when it was closely linked with the pressing needs of human society. Even in prehistoric times, the development of material culture was inconceivable without the use of minerals. The neptunistic views, that water is the most important element on Earth and source of everything living or dead, and the plutonistic view that the basis of everything earthly was fire, have been advocated by two Greek thinkers, Thales (640 BC) and Xenon (340 to 270 BC), respectively. They have significantly influenced our understanding of the genesis of economic mineral deposits.

Perhaps, the first truly scientific approach was expressed by Georg Bauer (1494-1555), better known as Georgius Agricola, who considered that ore veins were formed by solutions circulating in the bowels of the Earth. Since ore was usually only in veins, the necessary richness and coarseness were achieved by the term, "Ore deposit", becoming virtually synonymous with vein fillings.

James Hutton (1726-1997), a leader of the plutonists, linked the formation of ore veins with internal (magmatic and igneous molten) processes within the Earth. A.G. Werner (1749-1817), a leader of the Neptunists, favoured that veins of ore were formed from aqueous solutions flowing downward through cracks in rocks, eroding them and precipitating ores in them. Interestingly, the two schools of contradictory views gradually subsided, but as late as in the first half of the nineteenth century, when the fact was better accepted that the ores can also be formed through the action of external geological processes, chiefly by sedimentation.

In those days, the bedded ores, accompanied by minor veins filling cracks within the adjacent sedimentary rock, attributed the vein sulphides to essentially the same source and process as those responsible for the bedded material. Breislak (1811), Boue (1822) and Scrope (1825) emphasized the likely importance of water in fluidity in magmas.

It is true that the development of ore genetic theory has been haphazard, repetitive and clearly shows an unclear pattern, yet some of these very old ideas have been revived as modern ones. Agricola (1546) was probably the first to put forth the idea of "lateral secretionist", Hutton (1788) sees to have been the first advocate of "dry ore magmas", Werner (1791) was the early proponent of "sedimentary" sulphide formation, Elie de Beaumont (1847) is probably the originator of the idea of igneous origin and aqueous transportation, well accepted by modern hydrothermalists and Sterry Hunt (1860) showed the important of surface and biological processes in collecting and localizing metals.

The first half of the twentieth century remained a period of reexamination. During this time, the contributions made by Fenner on pneumatolitic processes, Bowen on differentiation processes, Graton on hydrothermal depth zones, Emmons on lode associated with batholiths, and Buddington on selective mineralization in selective igneous rocks are worth mentioning. Schneiderhohn's genetic classification of ore deposits (1932) and Lindgren's volume on "hypogene deposits" remained two of the most valuable documents, dominantly considering epigenetic mineralization, and perhaps, the last to restrict the term "Mineralization" as a substitute for hydrothermal processes only.

Continuing in the second half of the twentieth century, Schneiderhohn suggested that the ore materials had been extracted by hot brines from underlying pre-existing rocks. He explained "framboidal texture", low temperature "gel" and lead isotopes to indicate that sulphide ores are older than the host rocks. He suggested a regenerative process and propounded the idea popularly called "secondary mineralization". Stanton (1955) suggested that some stratiform ores are reduced sediments of old volcanic sequences with iron sulphides of sedimentary origin. It is considered as a result of interplay between sedimentary facies and volcanic emission. Further, the idea of exhalative - sedimentary type was to cover many bedded deposits of iron and manganese. It led to the fact that the sulphide "mineralization" was of volcanic affiliation (Kinkel, 1962; Goodwin, 1965).

Also, the third quarter of the twentieth century, sulphide concentration formation from modern volcanic cones in Japan, deposition of sulphide in pipes discharging geothermal drilling water in Salton Sea and the most popular among 'live' formation of ores, the deposition of sulphide ores from springs on the floor of Red Sea, which gained a milestone status in the history of the evolution of the economic mineralization. The modern discoveries remained synchronous with the final opinion of Stanton (1972) who considered "Ores" as "Ordinary rocks," and hence, ore genesis as a part of petrogenesis.

The last quarter of the twentieth century, later, witnessed an electrifying conceptual revolution and information explosion in economic geology. It happened because of the following reasons.

First, the advent and full emergence of Plate Tectonic theory which revolutionalized the understanding of petrogenesis of igneous, metamorphic and sedimentary rocks during the time when ores were firmly recognized as rocks. Second, numerous discoveries of 'live' deposits, relating to collection, transport and deposition of ore/metals in a variety of 'habitats'. Third, application of scanning electron microscope to directly observe the fossilized micro life in ore deposit, greatly assisted by EPMA. Fourth, over the years, explosion in the invention and application in economic geology of geophysical and geochemical techniques; to include regional geophysical profiles, XRD, XRF, PIXE, LA-ICP-MS, Mossbauer spectroscopy and mass spectroscopy. Fifth, the mantle plume view which evolved parallel to the global tectonic theory, gave many logical explanations to the unanswered questions of ore genesis. Although not everyone believes in plumes, a vast amount of emerging data supported the idea, including seismic tomography of the Earth's mantle and path finder and Magellan Mission data on Mars and Venus. Sixth, high-speed computers allowed scientists to numerically model mineralization processes in a reasonable amount of time with increased accuracy. Application of computer-based data treatment and simulation practices became greatly significant as they led to more quantitative definition of some of the models.

This information explosion is greatly reflected in economic Geology. The most common textbooks suffered from either faster obsolescence or considerable revision. Many new journals exclusively devoted to economic Geology started. Accent on publication of thematic mostly multi-authored volumes appeared.

The present edited volume is also one of the result and upshots of the same race of economic mineralization against time. The purpose of this edited volume, at first, is to present new researches and to emphasize the many areas in which problems still exists with a view to contributing and encouraging further research. The second purpose is to present review articles summarizing our present knowledge authoritatively, and in so doing, demonstrate almost all aspects of the subject selected. 'Economic Mineralization'- the title has been preferred to express its modern meaning (specially post-Lindgren time) and broader perspective to cover the formation of almost all the economic minerals, epigenetic (metasomatic replacement or vein filling) and syngenetic (chemical precipitation, biogenic, volcanic exhalative), including mineral fuels.

The volume has been divided into five sections. The first section is on Crustal Evolution and Economic Mineralization. It includes eight papers. The second and third are on Economic Mineralization of Igneous Affiliation and Economic Mineralization of Sedimentary Affiliation, respectively. There are twelve papers in section two and thirteen papers in section three. Section four includes eight papers on prospecting and exploration. The fifth and the last section includes nine papers on Mining, Economics and Environment. All the fifty papers in the volume were reviewed by the editor and invariably scrutinized and approved by at least two external referees. The editor has regretfully been forced not to include as many as 27 papers submitted for the publication in this volume because, more than one referee advised against it.

The volume sets out to present various aspects of a very broad field of economic mineralization, at a time when the competitively growing economy of the nations, in other words, the pressing needs of the society, are compelling economic geology to grow, and pile of data is accumulating and opinions changing very rapidly. Although care has been taken by the editor, in a work of this nature, errors of omission and commission are inevitable. I unhesitatingly own responsibility for the same. I would not find it truly rewarding unless the volume delivered the desired goods to the specialized readers, both from academia and industry. I wish, hope and pray for the same.

30th October, 2008
Jodhpur

K.L. SHRIVASTAVA

Acknowledgements

It is an omni accepted truth that to make a beginning in an endeavour, one needs an impetus. Without overemphasizing the fact in the present case where the much derided inertia was shaken off and momentum imparted by many, from near ones including my colleagues and students in the university, to peer, professional colleagues and contemporaries positioned far away, I thank them from the innermost recesses of my heart.

I express my gratitude to all the authors for responding to my request and for contributing the papers to this special volume. The following authors and geoscientists of distinction have also extended their great support without which this initiative would have been difficult to sustain: Prof. K.L. Rai, Prof. U. Aswathanarayana, Prof. A.B. Roy, Prof. R.S. Sharma, Prof. B.K. Sahu, Prof. M.K. Pandya, Prof. S. Sinha Roy, Prof. G.S. Roonwal, Prof. N. Bhandari, Dr. R. Dhana Raju, Prof. Naresh Kochhar and Prof. O.P. Goel. The authors are also thanked for their tolerant attitude to the editorial revisions.

I place on record my sincere thanks to all the Hon'ble reviewers whose names are given separately, for giving their valuable time and expertise to review the papers critically and giving constructive suggestions which have helped in greatly improving the value of the contributions. I am specially grateful to Dr. B.P. Radhakrishna, Prof. S.C. Sarkar, Prof. K.S.Valdiya, Prof. S.M. Naqvi, Dr. A.V. Sankaran, Dr. S. Viswanathan, Prof. S.K. Tandon, Dr. S.K. Bhushan, Dr. M.N. Balasubrahhamanyan, Prof. V.K. Verma, Prof. A.K. Singhvi, Prof. I.B. Singh, Prof. T.V.V.G.R.K. Murty, Dr. M. Ramakrishnan, Prof. B.L. Sharma, Prof. P.K. Verma (Delhi), Prof. P.S. Ranawat, Dr. Fareeduddin, Prof. K.K. Singh, Prof. S.P. Singh, Prof. A.M. Pathan, Prof. P.S. Saklani, Prof. Ravindra Kumar and Prof. U.C. Singh.

I am greatly indebted to the following geoscientists who gave their friendly and valuable suggestions and help in preparation of this volume : Prof. V.K. Vaya, Prof. M.C. Khare, Prof. S.K. Lunkad, Prof. J.P. Shrivastava, Dr. A.K. Shandilya (Bikaner), Prof. Arun Kumar, Prof. P.K. Kathal, Prof. R.K. Trivedi, Prof. R.P. Tiwari, Dr. D.K. Sinha, Prof. D.C. Gupta, Prof. L.S. Chamyal, Prof. T.K. Biswal, Prof. H.B. Shrivastava, Prof. M. Raza, Prof. M.S. Sisodia, Prof. Madhumita Das, Dr. S.C. Mathur, Prof. A.S. Sheoran, Dr. K.K. Sharma, Dr. S.K. Trivedi, Dr. Beena Tripathi, Dr. Hema Achyuthan and Dr. Vimla Sheoran.

I am thankful to all my research students, present and former, for providing me editorial assistance. It includes Prof. Srikant Ojha, Dr. Virendra Gaur, Dr. Sameer Naval, Dr. Vijay Gaur, Dr. Narendra Singh, Dr. Jaimini Swadia, Dr. Yasmin Sayyed, Ms Veena Chaudhary, Mr Chandra Prakash Dadhich, Ms Mamta Chauhan, Ms Anju Mathur, Ms Tina Prajapat and Mr Prakhar Kumar. The assistance is gratefully acknowledged.

I express my deep appreciation to my wife, Dr. Raka Srivastava, and daughters Krati and Srasti, for the consideration and forbearance shown during the long hours this work kept me busy, and for cheerfully accepting the many holidays lost.

K.L. SHRIVASTAVA

Professor Bhawani Shanker Paliwal

Professor Bhawani Shanker Paliwal — An Appreciation

Born on 19th May 1946, Professor Bhawani Shanker Paliwal hails from the city of lakes, Udaipur. He completed his school education there and also his graduation from the erstwhile University of Udaipur (now Mohanlal Sukhadia University). He continued his academic growth by doing M.Sc. Geology in the year 1970 from the University of Rajasthan. As soon as he completed his post- graduation, he was appointed as Lecturer in Geology at Government Bangur College Didwana on 18th July 1970. He was awarded UGC Teacher Fellowship in 1977 which led to the Degree of Doctor of Philosophy in Science (Geology) from the University of Rajasthan in the year 1981 under the worthy supervision of Prof. A.B. Roy.

He was the man behind starting and establishing the second prestigious Postgraduate Department of Geology in the state at a remote place like Didwana in the year 1981, next to the one quite popular at Udaipur under the umbrella of University of Rajasthan, Jaipur. After that he moved to the Government Dungar College at Bikaner as Head of the Postgraduate Department of Geology in September 1994. In the same year, he was selected as Professor of Geology at Jai Narain Vyas University Jodhpur on 22nd April 1995. Prof. Paliwal became Head of the Department on 10th February 1999 and continued for more than seven years until his superannuation on 31st May 2006. He was appointed as Dean, Faculty of Science, at Jai Narain Vyas University, Jodhpur on 1st May 2000 and served for two consecutive terms of three years each. During this period Prof. Paliwal was Member of the Academic Council, Chairman of the Faculty of Science, Member of the Senate, Member of the Syndicate for two terms, Chairman of the University Library Board and Government Nominee in the Selection Committee. During this period he rendered his services with utmost sincerity, efficiency and wisdom. On superannuation from Jai Narain Vyas University, Jodhpur, Prof. Paliwal was appointed as U.G.C. Visiting Professor of Geology at his own *Alma Mater* Department of Geology, Mohanlal Sukhadia University, Udaipur (erstwhile Department of Geology, University of Rajasthan) on 1st June 2006.

Prof. Paliwal developed keen interest in research right in the beginning of his academic career. His first publication appeared in the Journal of the Geological Society of India in 1971, a paper which is considered as a land mark in the field of deformation of the Aravalli rocks; it was based on the findings of his M.Sc. Thesis. Subsequently, he concentrated on the region around and he has several papers to his credit on different aspects like the origin of salt in Rajasthan, sedimentary structures in the trans-Aravalli, Vindhyans, Strometolites and economic mineralization. In addition to study, the deformation of the Palaeoproterozoic Aravalli rocks, Prof. Paliwal also established tectonics of the Precambrian rocks in the region. He aptly handled the issues of deposition to deformation of the Aravalli rocks in south-central Rajasthan, deformation of the Delhi rocks around Ajmer, marble deposits of Makarana and scattered outcrops of the Precambrian rocks in the sands of the great Thar Desert. He also covered several other aspects of the region like the Malani Igneous Suite, and sediments of the Marwar Supergroup.

Prof. Paliwal's discovery of Elephant remains from the Quaternary gypsum deposits of Nagaur in the Thar Desert and Fossils of the Fish from the Palaeocene Palana Formation, received overwhelming appreciation from the geoscientific fraternity throughout the world. He investigated the problem of Fluoride in ground water of the Western Rajasthan and its impact on the human health. Prof. Paliwal is credited with the important meteorite falls like those of Didwana-Rajod and Bhawad. His discovery of life forming amino-acids like Phenylalanine, Tryptophan and Tyrosene from the 4.52 Ga. old. Didwana-Rajod meteorite raised curiosity in the scientific community. His ideas of "the extra-terrestrial origin of life and its transportation on the earth by meteorites and comets" and "the impact of extra-terrestrial bodies caused the evolution of life and its mass extinction on the earth" have impressed the geoscientists.

Prof. Paliwal had an opportunity to work with eminent geoscientists like Prof. H. Okada, Prof. M. Yoshida and Prof. N.N. Nakajima of Japan, Prof. A. Kroner and Prof. M.M. Raith of Germany, Prof. Abate of Italy, Prof. John W. Harbaugh and Prof. Nein of U.S.A. Prof. Rabban, Prof E.V. Sklyaro and Prof. Eugene Khain of Russia, Prof. R. Hutchison, Prof. S. Russell and Prof. I.A. Franchi of U.K. Prof. Alain Ploquin of France, Prof. Nan Junya of China, Prof. S.A. Wilde of Australia, Prof. Svetlana V. Bogdanova of Sweden, Prof. P.G. Cooray of Shri Lanka and Prof, K. Naha, Prof. K.S. Valdiya, Prof. A.B. Roy and Prof. N. Bhandari of India.

Prof. Paliwal was associated with UNESCO IGCP projects like IGCP-269-SEDBA and IGCP-440 Assembly and Fragmentation of Rodinia and Growth of Asia. He was appointed as the Chairman of the South Asia in the project IGCP-269-SEDBA for five years. Apart from this, Prof. Paliwal handled many research projects funded by the UGC, DST, and DRDO successfully.

Prof. Paliwal is a widely traveled person who visited several countries including USSR, Germany, U.K., Russia, China, France, Italy, Austria, Thailand, Australia, Hong Kong, Singapore, Sri Lanka, Bangladesh, Pakistan, Nepal, Spain, Egypt, Japan, Kenya, Somalia, U.A.E., Norway, Sweden and Switzerland in connection with research work. During these visits he has also undertaken field work in and around Siberia, Lake Baikal, Somali Plateau, Rift Valley, Tibetan Plateau, Alps, Himalaya and Atlas Mountain Ranges, Great Dividing Range, Rhine Valley, Deltas of Bhrahmputra and Nile Rivers, British Channel and coastal areas, of Spain, Eastern China and Japan.

Prof. Paliwal participated in a number of National and International Seminars held in India and abroad. He was invited to Chair Sessions in the International Geological Congress held in Moscow (USSR) in 1984 (27th IGC), Kyoto (Japan) in 1992 (29th IGC), and Florence (Italy) in 2004 (32nd IGC). "Because of his extensive work on Proterozoic sediments" Prof. Paliwal was invited to Chair the Session on Proterozoic Sediments of the 12th International Sedimentological Congress held in Canberra (Australia) in 1986. Prof. Paliwal also Chaired Sessions in the International Conferences like GEOSOM-87 held in Mogadishu (Somalia) in 1987, GEOSAS-I held in Islamabad (Pakistan) in 1992, GEOSAS-II held in Colombo (Sri Lanka) in 1995, Assembly and Fragmentation of Rodinia and Growth of Asia held in Irkutsk (Siberia, Russia) and Osaka (Japan) in 2001. Recently, Prof. Paliwal attended the 33rd International Geological Congress held in Oslo (Norway) during August 6 – 14, 2008 where he was the Convener of the General Symposium – Hydrogeology and Topical Symposium – Geohazards and Chaired as many as eight Sessions of this prestigious International Conference. Prof. B.S. Paliwal has been entrusted with the task of publishing the Proceedings of the Hydrogeology –HYH-2 Symposium of the 33^{Rd} International Geological Congress.

Prof. Paliwal edited two books: *The Indian Precambrian* (1998) and *Geological Evolution of Northwestern India* (1999). These books were well received and acclaimed by the geoscientists all over the world. His book, *The Indian Precambrian* was reviewed by renowned geologists like Prof. Brian Windley (Precambrian Research, Elsevier), Prof. Ravindra Kumar (Gondwana Research, Elsevier), Dr. M. Ramakrishnan (Journal, Geological Society of India), Dr. T.M. Mahadevan (DST News Letter), and Prof. Deepak C. Srivastava (IGC News Letter).

As a devoted teacher, Prof. Paliwal has produced a large number of geoscientists from Didwana, Bikaner and Jodhpur, presently working almost in all the prestigious organizations like ONGC, GSI, ISRO, SAC, GWD, DMG, CGWB, BITS, RAS, RPS and several universities and colleges. Many of his students are actively engaged in exploration and exploitation in mining, groundwater and mineral fuel activities. Some of them have received prestigious National Mineral Award of the Country. He has also supervised a dozen of students for the Doctorate Degree. Even after his superannuation at present, he is actively engaged in supervising eight students for the degree of Ph.D.

As a scholar of distinction, a dedicated researcher, and a devoted teacher, Prof. Paliwal has greatly influenced his surroundings. I was not an exception. It truely remained a matter of great honour for me to work as a colleague of Prof. B.S. Paliwal. When Prof. Paliwal joined this department as Professor in the year 1995, I was serving as Associate Professor, before becoming Professor of Geology in the year 2001. I feel proud to dedicate this book on 'Economic Mineralization', to Prof. B.S. Paliwal. By joining hands with the contributors of this volume, I wish him all the best for the future. A glimpse of his contributions over the years can be had from the following list of some of his publications.

K.L. SHRIVASTAVA

SELECTED PUBLICATIONS OF PROF. B.S. PALIWAL

1. Roy, A.B., Paliwal, B.S. and Goel, O.P. (1971). Superposed folding in the Aravalli rocks of the type area around Udaipur, Rajasthan. Journal, Geolo-gical Society of India, Vol.12, No.4, pp. 342-348.
2. Paliwal, B.S. (1975). Phosphatic algal stromatolites in the Trans- Aravalli Vindhyans of Badi Khatu, District Nagaur, Rajasthan, Current Science, Vol. 44, No. 5, p. 160-161.
3. Paliwal, B.S. (1976). Occurrence of Copper Mineralization around Tikali Dungari near Didwana, Rajasthan. Science and Culture. Vol. 42, No.4, p. 218-219.
4. Paliwal, B.S. (1976). Note on Layered Chromatographical Structures in the Vindhyan Sandstone of Badi Khatu, District Nagaur, Rajasthan, India. Bulletin Indian Geologists Association. Vol. 9, No. 1, p. 17-22.
5. Paliwal, B.S. (1978). A note on the Intraformational Recumbent Folds in the Jodhpur Sandstone of Ladnun, District Nagaur, Rajasthan. Current Science, Vol. 47, No. 19, p. 73.
6. Paliwal, B.S. (1980). Systematics of the Precambrian Stromatolites of Rajasthan A review in the light of tectonic deformation. Geological Survey of India, Miscellaneous Publication. No. 44, p. 68 74
7. Paliwal, B.S. (1980). Structural history of the Precambrian rocks southwest of Didwana, Rajasthan, India. Bulletin Indian Geologists Association. Vol. 13, No. 1, p. 7-14.
8. Roy, A.B. and Paliwal, B.S. (1981). Regional Stratigraphy and Palaeogeography of the stromatolitic rock phosphate bearing Aravalli rocks of Rajasthan. Fourth International Field Workshop and Seminar on Phosphorite, November 1981 at Udaipur. IGCP-166. Phosphorite. UNESCO and IGCP Project, Geological Survey of India, Western Region, Jaipur. p. 16.
9. Roy, A.B. and Paliwal, B.S. (1981). Evolution of the Lower Proterozoic Epicontinental deposits: Stromatolites bearing Aravalli rocks of Udaipur, Rajasthan, India. Precambrian Research, Vol. 14, p. 49- 74.
10. Paliwal, B.S. (1982). The Source of Salt in Rajasthan An investigation of the Salt Lake of Didwana. Proceedings of the Workshop on the Problems of Deserts in India. Geological Survey of India, Miscellaneous Publication. No. 49. p. 99- 104.
11. Paliwal, B.S. (1983). The Tectono-sedimentary history of the Lower Proterozoic rocks of Rajasthan, India: Palaeogeography, Depositional Environment and Structural Evolution. Geological Society of Zambia, Lusaka. April 1983, p. 23.
12. Paliwal, B.S. (1984). Evolution of the Trans- Aravalli Vindhyans of Rajasthan, India. 27th International Geological Congress, Moscow, USSR, August, 1984. Vol. II, Sec. 04, p. 160.
13. Roy, A.B., Paliwal, B.S. and Bejarniya, B.R. (1984). The Aravalli rocks: an evolutionary model and metallogenic trends. Indian Journal of Earth Sciences, CEISM Vol. p. 73- 83.
14. Paliwal, B.S. (1986). Source of salinity in soil, and surface and sub-surface waters and its bearing on the evolution of the Thar Desert, India. Transactions of XIII Congress of the International Society of Soil Science. Hamburg, August 1986. Commission VII A, Vol. IV, p. 1532.
15. Paliwal, B.S. (1986). Role of Tectonics in the Evolution of Palaeogeography, Depositional Environment and Sedimentation: Aravalli Super-group as an example. 12th International Sedimentological Congress, Canberra, Australia, August, 1986. p. 232-233.
16. Paliwal, B.S. (1986). Sedimentary structures of an Early Proterozoic fluvial sequence around Udaipur city, India. 12th International Sedimentological Congress, Canberra, Australia, August, 1986. p. 233.
17. Paliwal, B.S. (1987). Orientation of Geology Teaching in Developing Countries: a case study from Indian Universities and Colleges. GEOSOM -87, International Meeting on Geology of Somalia and Surrounding Region, November 1987, Somali National University, Mogadishu, Somalia. p. 107.
18. Paliwal, B.S. (1987). Problems in structural mapping and interpretation in a complexly deformed area. GEOSOM-87, International Meeting on Geology of Somalia and Surrounding Region, November 1987, Somali National University, Mogadishu, Somalia. p. 108.

19. Paliwal, B.S. (1987). Tectonic Evolution of the Crust in the Northwestern part of the Indian Peninsular Shield and its bearing on adjoining Regions. GEOSOM-87, International Meeting on Geology of Somalia and Surrounding Region, November 1987, Somali National University, Mogadishu, Somalia. p. 109.
20. Paliwal, B.S. (1988). Geoscience Research and Training in Developing Countries: a case study from the Research Institutions and Organizations in India. Geoscience in Development: An International Conference on the Application of Geology in Developing Countries. September, 1988. University of Nottingham, United Kingdom. p. 104.
21. Roy, A.B., Paliwal, B.S., Shekhawat, S.S., Nagori, D.K., Golani, P.R. and Bejarniya, B.R. (1988). Stratigraphy of the Aravalli Supergroup in the type area. Precambrian of Aravalli Mountain Rajasthan. Memoir Geological Society of India, No.7, p. 121-138.
22. Paliwal, B.S. (1988). Deformation pattern in the rocks of the Aravalli Supergroup around Udaipur City, Rajasthan. Precambrian of Aravalli Mountain, Rajasthan. Memoir Geological Society of India. No. 7, p. 153-168.
23. Paliwal, B.S. (1988). Excess fluoride content in the ground water of the central districts of Rajasthan State, India and it's bearing on the human health. Sixth IWRA World Congress on Water Resources, Ottawa (Canada). Vol. II, p. 303 A.
24. Paliwal, B.S. (1989). High Fluoride Content in Ground Water in Central Districts of Rajasthan, India and it's bearing on Human Health. 28th International Geological Congress, August, 1989, Washington, D.C., USA. Vol.2, p. 562.
25. Paliwal, B.S. (1989). Geoscience Research and Training in Developing Countries: Case Study from Research Institutes and Organizations in India.. 28th International Geological Congress, August, 1989, Washington, D.C., USA. Vol. 2, p. 562.
26. Paliwal, B.S. (1989). Precambrian Crustal Movement and Its Bearing on Tectonosedimentary Evolution of Rocks of Aravalli Mountain Belt in Northwestern Part of Indian Peninsular Shield. 28th International Geological Congress, August, 1989, Washington, D.C., USA. Vol.2, p. 562.- 563.
27. Paliwal, B.S. (1989). Fluvial sediments of the Aravalli Supergroup of India: Study of Depositional, Erosional and Penecontemporaneous Deformation Structures. Fourth International Conference on Fluvial Sedimentology, Sitges, Barcelona, Spain, October, 1989, p. 200.
28. Paliwal, B.S. (1989). Status of water resources in developing countries in the twenty first century and the need of spreading awareness in the society through education- a case study from the Republic of India. Proceeding of the International Seminar on Education and Training in Water Resources, Aurangabad, India, October, 1989, p. 347-353.
29. Paliwal, B.S. (1992). Tectonics of the post-Aravalli mountain building activity and its bearing on the accumulation of sediments along the western flank of the Aravalli Range, Rajasthan, India. GEOSAS-I: First South Asia Geological Congress, Islamabad, Pakistan. p. 31.
30. Paliwal, B.S. (1992). The Precambrian Stratigraphy of Rajasthan, India: a fresh approach. GEOSAS-I: First South Asia Geological Congress, Islamabad, Pakistan. p. 31-32.
31. Paliwal, B.S. and Shandilya, A. (1992). The Lower Proterozoic Aravalli rocks north of Udaipur city, Rajasthan: the deformation pattern. 29th International Geological Congress, Kyoto, Japan. II 2-2, 014 A, 294 p. 92 (299).
32. Paliwal, B.S. and Shandilya, A. (1992). The Lower Proterozoic Aravalli rocks north of Udaipur city, Rajasthan: the Tectono-Sedimentary Evolution. 29th International Geological Congress, Kyoto, Japan. II 6 2, 27 A, 429 p. 115 (469).
33. Paliwal, B.S. and Okada, H. (1993). Aravalli Supergroup of India: an example of the Lower Proterozoic Rift Tectonics and Sedimentation. The Journal of the Sedimentological Society of Japan. No. 39, p. 1-13.
34. Paliwal, B.S. (1993). Geology of Rajasthan a summarium. In: T.S. Chouhan (Ed.) Natural and Human Resources of Rajasthan. Scientific Publishers (India), Jodhpur. p. 1-22.
35. Paliwal, B.S. and Shakuntala (1993). Groundwater Conditions in Rajasthan. In: T.S. Chouhan (Ed.) Natural and Human Resources of Rajasthan. Scientific Publishers (India), Jodhpur. p.113-124.
36. Paliwal, B.S. and Shakuntala (1993). A network system of small interconnected dams: a proposal for the arid and semi arid regions of Developing Countries. International Conference on Environmentally Sound Water Resources Utilization. November, 1993. Asian Institute of Technology, Bangkok, Thailand.
37. Paliwal, B.S. and Shakuntala (1993). Change in the attitude of the society towards water resources through education: a basic need of Developing Countries. International Conference on Environ-mentally Sound Water Resources Utilization. November, 1993. Asian Institute of Technology, Bangkok, Thailand.
38. Roy, A.B., Sharma, B.L., Paliwal, B.S., Chauhan, N.K., Nagori, D.K., Golani, P.R., Bejarniya, B.R., Bhu, U. and Ali Sabah, M. (1993). Lithostratigraphy

and Tectonic Evolution of the Aravalli Supergroup - A Protogeosynclinal sequence. In: S.M. Cassyap (Ed.), Rift Basins and Aulacogens. Gyanodaya Prakashan, Nainital, India, p. 73 90.

39. Paliwal, B.S. (1994). Well preserved sedimentary structures in a complexly deformed fluvial sequence of the Precambrian period. International Symposium on Applications of Computers and Data Base to Sedimentology, Chengdu, China, July, 1994. p. 8.

40. Paliwal, B.S. (1994). Applications of SEDBA to Sedimentology: necessity of vertical profiles and sequences of sedimentary structures. International Symposium on Applications of Computers and Data Base to Sedimentology, Chengdu, China, July, 1994. p. 9.

41. Paliwal, B.S. and Shakuntala (1994). Deficiencies in Data Base in Sedimentary Petrology. International Symposium on Applications of Computers and Data Base to Sedimentology, Chengdu, China, July, 1994. p. 10.

42. Paliwal, B.S. (1994). Cratonization of the Indian Peninsular Shield along its Northwestern margin: Volcanism and Sedimentation. International Volcanology Congress, IAVCEI 94, Theme-1, Collision Related Magmatism, Ankara, Turkey. September, 1994, p. 6.

43. Paliwal, B.S. (1994). Tectonics of the post-Aravalli mountain building activity and its bearing on accumulation of sediments along the western flank of the Aravalli range, Rajasthan, India. In: R. Ahmad and A.M. Sheikh (Eds.), Geology in South Asia I, Proceedings of the First South Asia Geological Congress, GEOSAS-I. Hydrocarbon Development Institute of Pakistan, Islamabad. p. 52 60.

44. Paliwal, B.S. (1995). Application of SEDBA to Sedimentology: Deficiencies of information and need of vertical profiles of lithology and sequences of sedimentary structures. Journal of Sedimentary Facies and Palaeogeography. Chinese Academy of Geological Sciences. Vol. 15, No. 2. p. 82 89.

45. Paliwal, B.S. (1995). A Note on the Spots in Jodhpur Sandstone of Western Rajasthan. Rajasthan Mineral Bulletin. Directorate of Mines and Geology, Rajasthan, Udaipur. Vol. 16, No. 1-2, p.13 16.

46. Paliwal, B.S. and Shandilya, A.K. (1995). Tectono-sedimentary evolution and deformation pattern of the Lower Proterozoic Aravalli rocks north of Udaipur city, Rajasthan. GEOSAS-II, Second South Asia Geological Congress, Colombo, Sri Lanka, January 1995. p. 137 138.

47. Paliwal, B.S. and Pareek, U.S. (1995). Structural framework of the Makarana marble deposit and its bearing on the Precambrian Stratigraphy of Rajasthan, India. GEOSAS-II, Second South Asia Geological Congress, Colombo, Sri Lanka, January 1995. p. 135 136.

48. Paliwal, B.S. and Trivedi, S.K. (1995). Application of SEDBA in mineral based industries: a proposal. In: S. Bhandari and A. Tuteja (Eds.). Proceedings of National Seminar on Application of Computer in Mineral Industry. Scientific Publishers (India), Jodhpur. p. 155-159.

49. Paliwal, B.S., Chauhan, D.S. and Vaya, V.K. (1997). Didwana-Rajod Meteorite: An unusual Chondrite. Current Science Vol. 73 No. 6. p. 499-502.

50. Paliwal, B.S., Pareek, U.S. and Vyas, A. (1997). Structural framework of the Makarana marble deposit and its bearing on the Precambrian Stratigraphy of Rajasthan, India. In: N.P. Vijayananda, P.G. Cooray and Peter Mosley (Eds.). Geology in South Asia-II. Geological Survey and Mines Bureau, Sri Lanka. Professional Papers-7, p. 123-136.

51. Paliwal, B.S. (1998). Felsic volcanics interlayered with sediments of the Marwar Supergroup at Chhoti Khatu; district Nagaur, Rajasthan. Journal Geological Society of India. Vol.52: p. 81-86.

52. Paliwal, B.S. (1998). The Indian Precambrian, (Ed.). Scientific Publishers (India), Jodhpur. p. 1-556.

53. Paliwal, B.S. (1998). The Aravalli fluvial sequence of Udaipur city, India: an example of well preserved sedimentary structures in complexly deformed Precambrian metasediments. In: B.S. Paliwal (Ed.). The India Precambrian, Scientific Publishers (India), Jodhpur. p. 142-158.

54. Paliwal, B.S. and Vyas, A. (1998). The Precambrian Aravalli Supergroup of rocks south of Udaipur: deposition to deformation. In: B.S. Paliwal (Ed.). The Indian Precambrian, Scientific Publishers (India), Jodhpur. p. 219-232.

55. Paliwal, B.S. (1998). Contamination of ground-water and its effects on human health in developing countries. In: K.R. Chowdhury (Ed.). International Seminar on Geoscience and Urban Development, Bangladesh Geological Society (BGS) and Association of Geoscientists for International Development (AGID). Dhaka, Bangladesh. p. 22.

56. Paliwal, B.S., Vaya, V.K. Shukla, A.D. Chakraborty, S., Suthar, K.M., Dixit, M.H., Shukla, P.N. and Bhandari, N. (1998). Cosmogenic records in the Didwana-Rajod meteorite. 61st Meteoritical Society Meeting, Dublin, 1998, p. 5156, Meteoritics and Planetary Science. 33: A 120.

57. Paliwal, B.S. (1998). Fish skull from Palana Formation at Handla- Bhatiyan, District Bikaner, Rajasthan. Current Science Vol.76 No.12. p. 1536-1538.

58. Paliwal, B.S. (1999). Geological Evolution of North-western India (Editor). Scientific Publishers (India),

Jodhpur. p. 1-414.

59. Paliwal, B.S. and Vyas, A. (1999). Geology of the Middle Proterozoic Ras marble deposit: a potential reserve for the cement industry. In: B.S. Paliwal (Ed.). Geological Evolution of Northwestern India, Scientific Publishers (India), Jodhpur. p. 183-190.

60. Paliwal, B.S. (1999). High Fluoride content in groundwater: a curse to the human health in the central part of Rajasthan, India. In: B.S. Paliwal (Ed.). Geological Evolution of Northwestern India. Scientific Publishers (India), Jodhpur. p. 271 292.

61. Paliwal, B.S. (1999). Geology of the Region to the west of Aravalli Mountain: Some Observations. In: B.S. Paliwal (Ed.). Geological Evolution of Northwestern India., Scientific Publishers (India), Jodhpur. p. 387 414.

62. Paliwal, B.S., Tripathi, R.P., Verma, H.C. and Sharma, S.K. (2000). Classification of Didwana Rajod Meteorite: A Mossbauer spectroscopic study. Meteoritics and Planetary Science. Vol. 35, p. 639-642.

63. Paliwal, B.S. (2000). Discussion on 1. On the discovery of Middle Eocene larger Foraminifera from a limestone bed in Churu District, Rajasthan, by Raghav, K.S. Journal. Geological. Society of India, Vol. 55, p. 269-274. 2. Discovery of Forami-nifera from Bilara Group, Jodhpur District, Rajasthan by Raghav, K.S. Journal. Geological. Society of India, Vol. 55, p. 395-397. Comments, Journal Geological. Society of India, Vol. 56, p. 113-114.

64. Paliwal, B.S. (2000). Precambrian metasedimen- tary sequence exposed along the western flank of the Aravalli mountain range and its significance in the Precambrian stratigraphy of northwestern India. 31st International Geological Congress, Rio de Janeiro, Brazil, Section 9 (2).

65. Paliwal, B.S. (2000). Fascinating Glimpses of the Geology of the Udaipur Valley. In: P.S. Ranawat, (Ed.). Rajasthan Geology Education: Compendium "Smarika" 1950-2000, Department of Geology, Mohanlal Sukhadia University, Udaipur. p. 68-73.

66. Paliwal, B.S. and Rathore, P.S. (2000). Neoproterozoic volcanics and sedimentaries of Jodhpur: a reappraisal. In: K.C. Gyani and P. Kataria. Proceedings of the National Seminar on Tectonomagmatism, Geochemistry and Metamorphism of the Precambrian Terrains. Department of Geology, Mohanlal Sukhadia University, Udaipur. p. 75- 96.

67. Paliwal, B.S. (2000). Hazardous and environmentally degrading mining operation at Makarana. In: S.S. Rathore, and V. Lakshmi Narayana, (Eds.). Proceedings of National Workshop on Safety and Technology in Marble Mining and Prospecting in New Millennium. Department of Mining Engine-ering, CTAE, Udaipur and Directorate General of Mines Safety, Udaipur Region. p. 217-221.

68. Paliwal, B.S. (2000). Status of Research in Geosciences in India. 31st International Geological Congress, Rio de Janeiro, Brazil, Section 26, No. 3.

69. Paliwal, B.S. (2000). Declining interest of students in Geosciences in developing countries and its causes: a case study from India universities and colleges. 31st International Geological Congress, Rio de Janeiro, Brazil, Section. 26, No.3.

70. Paliwal, B.S. (2000). Proterozoic crustal growth along northwestern margin of the India Peninsular Shield. 31st International Geological Congress, Rio de Janeiro, Brazil, Section 26, No. 3.

71. Paliwal, B.S. (2001). Geology Teaching and Training in Developing Countries: A Case Study from India. In: Ramesh K. Kakar (Ed.) IGCP UNESCO National Seminar on Geoscience Education for Natural Resources Management, December, 2001. Centre for Advance Study in Geology, Punjab University, Chandigarh, India. p. 46- 50.

72. Paliwal, B.S. (2001). Assembly and Breakup of Rodinia: Signatures along the Western flank of the Aravalli Mountain, India. Field Workshop on the Assembly and Breakup of Rodinia Supercontinent: Evidence from South Siberia, July August 2001. In: E.V. Sklyarov, Guide Book and Abstract Volume, Workshop IGCP 440, Irkutsk, Russia, p.166.

73. Paliwal, B.S., Mahajan, R.R., Murty, S.V.S., Shukla, A.D., Shukla, P.N., Bhandari, N., Natarajan, R., Hutchison, R., Russell, S. and Franchi, I.A. (2001). Chemical and isotopic characteristics of the Didwana-Rajod (H5) Chondrite. Meteoritics and Planetary Science. Vol. 36, p. 1249-1256.

74. Paliwal, B.S., (2001). Assembly and breakup of Rodinia: signatures along the western flank of the Aravalli Mountain in India. Gondwana Research. Vol. 4, No. 4, p. 725-726.

75. Paliwal, B.S., and M. Khilnani (2001). Electrical Resistivity data to asses aquifer parameters for better interpretation of Hydrogeology: a study of the Nagaur region, Rajasthan. In: H. Bhu and V. Agrawal. (Eds.) Ground Water Resources: Evaluation and Management. IGC 2000 Volume I, XII IGC. Department of Geology, Mohanlal Sukhadia University, Udaipur, p. 48-67.

76. Tewari, V.C., Venkatakrishna, K. and Paliwal, B.S., (2001). Laser Raman Spectroscopy of Microbial Carbonates from the Lesser Himalaya and Didwana Rajod Meteorite, Rajasthan, India. Seminar on Contribution to Himalayan Geology, October, 2001. Wadia Institute of Himalayan Geology, Dehradun,

India. p. 19- 20.

77. Paliwal, B.S. and A. Vyas (2001). Hydrogeological studies of the Nagaur District, Rajasthan. In: H. Bhu and V. Agrawal (Eds.) Ground Water Resources: Evaluation and Management. IGC 2000 Vol. I, XII IGC. Department of Geology, Mohanlal Sukhadia University, Udaipur, p. 233-243.

78. Paliwal, B.S. (2002). Science Project- Solar Wheel. Science Reporter, National Institute of Science Communication (Council of Scientific and Industrial Research), New Delhi, India. Vol. 39, No. 2, p. 50.

79. Paliwal, B.S. (2002). Science Project- Solar Wheel. Science Reporter, National Institute of Science Communication (Council of Scientific and Industrial Research), New Delhi, India. Vol. 39, No. 2, p. 50. Comments by Ghosh, Samir Kumar and (2) Dhamangaonkar, P.R. and Reply, Science Reporter, National Institute of Science Communication (Council of Scientific and Industrial Research), New Delhi, India. Vol. 39, No. 6, p. 4.

80. Paliwal, B.S., Sisodia, M.S. and Tripathi, R.P. (2002). Bhawad L(LL)6 Chondrite: Petrology and Mossbauer study. *Current Science.* Vol. 83, No. 9, p. 1071 -1072

81. Paliwal, B.S. and Vyas, A. (2002). Ground Water Resources of Rajasthan. In: B.B.S. Kapoor, Ahmed Ali, S.K. Mathur and Satish Kaushik (Eds.). Advances in Resource Management of the Indian Desert. Madhu Publications, Bikaner, India. p. 61 78.

82. Tewari, V.C., Paliwal, B.S. and Venkatakrishna, K. (2002). Discovery of Amino Acids from Didwana - Rajod Meteorite and its Implication on Origin of Life. Journal, Geological Society of India. Vol. 60, No. 1, p. 107-110.

83. Tewari, V.C., Paliwal, B.S. and Venkatakrishna, K. (2002). Discovery of Amino Acids from Didwana - Rajod Meteorite and its Implication on Origin of Life. Journal, Geological Society of India. Vol. 60, p. 107-110. Reply of Comments by (1) Sukumaran, P.V. (2) Maurya, A.S. Journal Geological Society of India. Vol. 60, No. 5, p. 601 604.

84. Verma, H.C., Rawat, A., Paliwal, B.S. and Tripathi, R.P. (2002). Mossbauer Spectroscopic Studies of an Oxidized Ordinary Chondrite fallen at Itawa Bhopji, India. Hyperfine Interactions. Vol. 142, p. 643 -652.

85. Paliwal, B. S., Yadav, S.K., Rathi, G. and Baghela, A. (2003). Sandstone with Layered Chromatograp- hical Structures: A Prospective Decorative Stone. Sandstone Spectrum -03, Jodhpur, April, 2003, Department of Mines and Geology, Government of Rajasthan. Vol. I, p. 16 - 25.

86. Vyas, A., Paliwal, B.S. and Sharma Shishir, (2003). Paleoenvironment during Quaternary in Western Rajasthan, India: An Overview. International Geological Correlation Programme (IGCP-413), Chennai, India. p. 15-17.

87. Maanju, S.K., Vyas, A. Paliwal, B.S., Sinha, A. K., Mittal, G.S., Gaur, C.P. and Khan, H. (2003). Appraisal of Impact of High Fluoride Ground Water on Human Health: A Case Study. International Conference on Soil and Groundwater Contamination and Clean up in Arid Countries. January, 2003, Muscat, Sultanate of Oman. p. 29 30.

88. Vyas, A. and Paliwal, B.S., (2003). Jodhpur Sandstone: The Crowning Glory of Building Marvel. Sandstone Spectrum -03, Jodhpur, April, 2003, Department of Mines and Geology, Government of Rajasthan. Vol. II, p. 15 -20.

89. Paliwal, B.S. (2003). Fossilized Elephant bones in the Quaternary gypsum deposits at Bhadawasi, Nagaur district, Rajasthan. Current Science. Vol. 84, No. 9, p. 1188 1191.

90. Paliwal, B.S. (2004). Geology of Rajasthan (Northwest India) Precambrian to Recent by A. B. Roy and S.R. Jakhar, Scientific Publishers (India) Jodhpur, Book Review. Gondwana Research, (Gondwana Newsletter Section), Vol. 7, No. 2, p. 649 650.

91. Paliwal, B.S. (2004). Geology Teaching and Training in Developing Countries: A case Study from India. In: Ramesh K. Kakar, Naval Kishore and Naresh Tuli (Eds.). Proceedings of IGCP UNESCO National Workshop on Geoscience Education for Natural Resources Management. UNESCO, New Delhi and Centre of Advanced Study in Geology, Punjab University, Chandigarh, p. 151 165.

92. Vyas, A. and Paliwal, B.S., (2004). Mineral Resources of Rajasthan, India A Developing Scenario. Geological Survey of India, Special Publication, Vol.72, p. 13 25.

93. Paliwal, B.S. (2004). Abrupt rise in the ground water level in Jodhpur : an example of ground water interaction with the surface water in the Thar Desert of Western India. Proceedings: 32nd International Geological Congress, Florence, Italy, (32nd IGC August 2004). Oral Session: 98-13. Abstract Part -1, p. 459.

94. Paliwal, B.S. (2004). Origin and evolution of life and mass extinctions: possible extra-terrestrial causes. Proceedings: 32nd International Geological Congress, Florence, Italy, (32nd IGC August 2004). Poster Session: 335-21. Abstract Part -2, p. 1476.

95. Paliwal, B.S. (2004). Signatures of Neotectonics to the west of Aravalli Mountain in India: drainage reversal, desertification and formation of salt lakes. Proceedings: 32nd International Geological Congress, Florence, Italy, (32nd IGC August 2004). Poster Session: 198-37. Abstract Part -2, p. 937.

96. Paliwal, B.S. (2005). Can Habur limestone curdle milk- a myth or reality? Current Science, Vol.89, No.

10, p. 1647-1648.

97. Bhandari, N., Murty, S.V.S., Shukla, P.N., Mahajan, R.R., Shukla, A.D., Suthar, K.M., Parthasarthy, G. and Paliwal, B.S., (2005). Bhawad LL6 Chondrite: Chemistry, Petrology, Noble Gases, Nuclear Tracks and Cosmogenic Radio Nuclides. Meteoritics and Planetary Science, Vol. 40, No. 7, p. 1015- 1021.

98. Paliwal, B.S., (2005). Practical Work in Science Subjects. Current Science. Vol. 88, No. 11, p. 1715.

99. Kumar, K., Rana, R.S. and Paliwal, B.S., (2005). Osteoglosid and Lepisosteid Fish remains from the Paleocene Palana Formation, Rajasthan, India. Paleontology, Vol. 48, Pt. 6, p. 1187- 1209.

100. Paliwal, B.S., (2005). Geology Introduced as an Optional Subject in Schools in Rajasthan. Journal, Geological Society of India. Vol. 66, No. 5, p. 658.

101. Paliwal, B.S. and Baghela, A. (2005). Pre Tertiary Glacial events in the Northwestern part of the Indian Peninsula and their significance in Assembly and Breakup of Rodinia and Growth of Asia. Super continents and Earth Evolution Symposium, Fremantle, Western Australia, September, 2005. Tectonics Special Research Centre and the Geological Society of Australia (WA Division) & IGCP.

102. Paliwal, B.S. and Baghela, A. (2005). Tectonics of the North Western part of the Indian Peninsular Shield and its Significance. Super continents and Earth Evolution Symposium, Fremantle, Western Australia, September, 2005. Tectonics Special Research Centre and the Geological Society of Australia (WA Division) & IGCP.

103. Paliwal, B.S., (2006). Magnetic Spherules with Metallic Shine in the Recent Alluvium of Rajasthan. Journal, Geological Society of India. Vol. 67, No. 6, p. 709 714.

104. Paliwal, B.S., (2007). Magnetic Spherules with Metallic Shine in the Recent Alluvium of Rajasthan. Journal, Geological Society of India. Vol. 67, No. 6, p. 709-714. Discussion: Reply of Comments by (I)Kamal L. Prushet and Pramod Singh, Journal Geological Society of India, Vol. 68 No. 5, P. 194-195.

105. Paliwal, B.S., (2007). Magnetic Spherules with Metallic Shine in the Recent Alluvium of Rajasthan. Journal, Geological Society of India. Vol. 67, No. 6, p. 709-714. Discussion: Reply of Comments by (ii) Robin Westerman, Journal, Geological Society of India. Vol. 69, No. 1, p. 194-195.

106. Paliwal, B.S., Rathi, G., Sharma, T.S. and Baghela, A. (2007). Status of the Precambrian rocks exposed around Punagarh, District Pali, Rajasthan. In: P.C. Avadich and Harsh Bhu (Eds.) Proceedings of National Seminar on Emerging Trends of Research in Geology with Special Reference to Northwestern India, January, 2007. Department of Geology, Mohanlal Sukhadia University, Udaipur, p.15-27.

107. Paliwal, B.S. (2007). Some fossil like structures in the rocks of the Marwar Supergroup. In: P.C. Avadich and Harsh Bhu (Eds.) Proceedings of National Seminar on Emerging Trends of Research in Geology with Special Reference to Northwestern India January, 2007. Department of Geology, Mohanlal Sukhadia University, Udaipur, p. 57-64.

108. Paliwal, B.S. and Baghela, A. (2007). Alarming rise in groundwater levels beneath the city of Jodhpur: an example of ground and surface water interaction in the Thar Desert of western India. In: Ken W.F. Howard (Ed.) Urban Groundwater- Meeting the Challenge, Selected Papers from the 32nd International Geological Congress (IGC), Florence, Italy, August, 2004. Selected Papers on Hydrogeology-8, Taylor and Francis, New York, p. 229- 258.

109. Paliwal, B.S., Kaushik, S. and Yadav, S.K. (2008). Status of the deformed Proterozoic metasediments exposed along the western flank of the Aravalli Mountain Range in Rajasthan (India). Indian Journal of Geology, (in press).

110. Paliwal, B.S. (2008). Neoproterozoic Palaeomagnetic results of Jodhpur Sandstone, Marwar Supergroup, Western Rajasthan, by Pooranchandra Rao, G.V.S., Singh, S.B. and Prasanna Lakshmi, K.J. Journal Geological Society of India, v.69, no.5, 2007, pp. 901-908. Discussion: Comments. Journal Geological Society of India, vol. 71, No.1, p. 143-145.

Contents

Section II

Economic Mineralization of Igneous Affiliation

Section III

Economic Mineralization of Sedimentary Affiliation

Section IV

Prospecting and Exploration

Section V

Mining, Economics and Environment

Section I
Crustal Evolution and Economic Mineralization

Preamble

Crustal evolution and economic mineralization is an account of the apparent parallelism between the evolutionary pattern of the continental crust and genesis of the ore types that each continent contains. The perceived interrelationship between economic mineralization and that of evolutionary history of the earth's 'lithosphere- exosphere' is a field of our yet incomplete knowledge.

In the first paper, Prof. R.S.Sharma has discussed the Precambrian high- grade metamorphic granulite terrains of India and their potential for mineral deposits. The granulite facies terrains, a component of crustal evolution, were traditionally thought to be sterile (or poor) in economic mineralization. The author has discussed retrogression in the Dharwar craton following the granulite facies metamorphism and formation of the gold veins therewith, in the light of various genetic models available. The occurrence of graphite deposits at sheared contacts and ilmenite, Ti-magnetite as cross-cutting veins in the Eastern Ghat Mobile Belt have also been discussed. Considering Sandmata and Bhinai granulites of Rajasthan, it is proposed that during the Delhi Orogeny, the magma underplating has heated the already metamorphosed rocks of BGC, allowing the metals to get leached from their original setting, leaving the Rajasthan granulites sterile. Hon'ble reviewer, Dr. M. Ramakrishna, while reviewing the paper and respecting the opinion of the great author, has expressed a ray of hope for economic geologists by mentioning that there are other models of contradictory views existing.

The prevalence of several common geological features like stress patterns, rifts and magmatic events, coinciding with Malani Magmatism in NW India, Central Iran, Nubian-Arabian Shield, Somalia, Seychelles, has made Prof. Naresh Kochhar conceive the idea of a 'Malani Supercontinent', in the year 2000, by treating all these micro-continents as its constituting parts. He has also stressed this view in subsequent years in seminars and publications. In the present paper on " The Malani Supercontinent: Middle East connection during Late Proterozoic", he has further strengthened his earlier evidences and added new evidences to give ultimate justification for the 'Malani Supercontinent'. The paper discusses the similarities between the Trans-Aravalli Block (TAB) and the Arabian- Nubian Shield (ANS) in the configuration of the Malani Supercontinent to provide a new approach for the search of economic mineralization. The observations of Prof. N.

Kochhar suggest that there is a characteristic presence of alkali granites in TAB (732Ma) and ANS (700 Ma). The emplacement of these granites and associated volcanics is controlled by ring structures, a manifestation of plume activity and cauldron subsidence, an evidence of extensional tectonic environment. Both the terrains preserve remnants of Archaean crust, i.e. BGC of amphibolite facies (3.2 Ga) in TAB and trondjhemite (2.6 Ga) in ANS which acted as precursor to the younger granites marking the cratonization. There is a characteristic similarity in the occurrence of peralkaline and peraluminous granite-hosted mineralization of U,Th, Nb, Ta, REEs, Sn, W, Li and F in both terrains which make them akin to the younger mineralized granites of Nigeria. Both terrains show evidence of strutian glaciation and occurrence of evaporate-carbonate sequence of economic importance.

Prof. A.B. Roy has discussed "Precambrian Crustal Evolution and Metallogeny", the concepts related with the topic are in fact, continually changing as new data from a variety of disciplines make significant impact on geological opinions. The paper endeavours to present a balanced view of the current state of knowledge. Prof. Roy has discussed evolution of Earth prior to formation of the oldest recognized crust; during Archaean, Archaean-Proterozoic transition and Proterozoic. The paper outlines constraints and identifies assumptions which have led to arrive at various hypotheses for the early evolution of the lithosphere, atmosphere, hydrosphere and biosphere. The author has identified that the unidirectional crustal trends also find aggreable reflection in the Precambrian metallogeny, and presents a number of conclusions concerning Precambrian Earth history and metallogeny which have not been emphasized previously.

Prof. K.L. Rai has attempted to comprehend the inter-relationship among various episodes of tectonics, magmatism, sedimentation, and metamorphosm contributing to the crustal development and metallogenesis in different entities of the Baster craton during the Precambrian period. The craton contains important deposits of metals like copper, gold, uranium, tungsten, tin, niobium, tantalum, rare-earths, etc. besides diamond and other gemstones. Prof. Rai even feels that the craton has been studied and suffers numerous constraints arising from the inadequacy of requisite geological and reliable geochronological data and several unsolved controversies. The paper has analyzed and evolved an integrated picture of time-controlled crustal and metallogenic evolution of the Baster craton.

The compositional difference from that of the primitive solar nebula, the Earth, from Phobos and Demos and from the satellites of the outer planets call for a distinctive mode of origin for the moon which is depleted in the most volatile elements. It is suggested that in the moon, elements with condensation temperature in excess of about 1200 K (at 10" 3atm) must have recondensed without fractionation. This uncommon object may require an uncommon origin that includes (1) capture from an independent orbit; (2) fission from a rapidly rotating earth; (3) formation as a double planet; (4) disintegration of incoming planetesimals; and (5) earth impact by a mars-sized planetesimal. A reasonable possibility is a giant collision with the earth, at the right angle and velocity to project a jet of hot material from the impactor into circumterrestrial orbit. It possibly can produce the moon, a unique bone-dry refractory satellite with curious orbital properties.

These few and many other uncertainties compelled human beings to plan several missions for the exploration of the moon, including India's mission-'Chandrayan-l'. Prof. N. Bhandari, in his paper 'Mineral Exploration of the Moon,' has discussed the chemical origin of the moon, its economic mineral resources and the other resources which can be tapped in future by human habitation, either on Earth or Moon. The author has identified lunar polar regions of special interest as they may have hydrogen, water, methane, alcohol, ammonia and other hydrocarbons. Basalt regions are of interest for ilmenite and Ti-magnetite, being rich in titanium and oxygen. Solar power is another possibility because it occurs in abundance on the Moon. Importance of Helium-3 lies in the fact that it is possible to produce radioactivity-clean (unpolluted) energy by ^{3}He - ^{3}He fusion. The author feels that a few kg of ^{3}He can meet the annual energy requirement of the whole earth.

In the evolution of the Earth, micro-organisms were probably largely instrumental in the generation of an oxygenous atmosphere and have played important roles in the formation of various mineral deposits. The conditions of abiotic synthesis of living organisms, under which prebiotic compounds evolved into living systems, are extremely speculative. An extra-terrestrial source for the molecules which ultimately has given rise to life forms is an alternative over the possibility that mixtures of compounds which occur in contemporary biochemistry, can be chemically synthesized from mixtures of components in the Earth's primitive atmosphere /hydrosphere. Presence of amino acids in some meteorites and source material of life (e.g., ammonia, formaldehyde, hydrogen cyanide, alcohols, methane) in the molecular clouds in collapsing nebulae have been documented. These complex molecules exist widely in considerable quantities within the interstellar medium, are likely to be incorporated in the outer Earth during the formation of the Earth's primordial crust or by subsequent influx of organic molecules later, when the Earth became more hostile. The next paper of the section is entitled "Possible Role of Sulfur on the Early Diversification of Life on Earth: Astrobiological Implications" authored by Vinod C. Tiwari and Julian Chela Flores. The paper has reviewed the possible role of sulfur on the early diversification of life on Earth, which has greatly influenced biogenic mineralization. The significance of Chemoautotrophs has been discussed with special reference to Europa, accepting the fact that the presence of sulfur-reducing bacteria could be similar to the early stages of crustal evolution of the Earth.

Deep seismic reflection profiling and its implications to the metallogeny of the Aravalli-Delhi fold belt, NW India, have been presented by B. Rajendraa Prasad and V. Vijaya Rao in the next paper. It suggests that seismic images of the deep crustal reflection study of Aravalli-Delhi fold belts are able to recognize tectonic dislocation zones or mega-lineaments which are responsible for metallogeny in the craton, have long and complex history of crustal evolution involving multiphase rifting, collision and magmatism. The seismic study, assisted by geological and geophysical data, suggests that the subduction, arc formation and collision are responsible for the crustal evolution in the study area and two prominent metallogenic periods of Pb-Zn formation in ADFB, at -1800 Ma and -1000 Ma, coincide with global metallogenic events and formation of supercontinents, such as Columbia and Rodinia.

The last paper of the section, authored by Prof. S.Sinha Roy, is entitled "Tectonic polarity in plate tectonic- controlled interaction of Precambrian terrains of Rajasthan Cratons". The paper suggests that the Rajasthan Craton, which consists of four distinct terrains, namely, Bundelkhand, Mewar, Marwar and foldbelts (Aravalli and Delhi), clearly depicting Wilson cycles on the basis of tectono-thermal events and their manifestations allow for identification of plate tectonic signatures. It is further suggested that the crustal-scale deformation kinematics is controlled by westward push of the Bundelkhand on the Mewar terrains in Neoproterozoic. The Phulad Suture Zone of South Delhi fold belts, Moho-reaching Banas Dislocation Zone and the Great Boundary Fault are the resultant of Grenvillian, Pan-African and Caledonian events, respectively. The tectonic polarity and the disposition of sequential dislocation zones in Rajasthan seem to be Proterozoic analogues of Phanerozoic Himalayan collision tectonics.

Economic Mineralization
K.L. Shrivastava (Editor)
2009. pp. 4-14
Scientific Publishers (India), Jodhpur

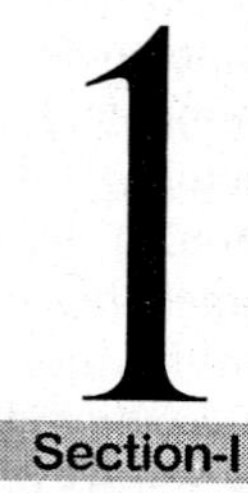

Section-I

Granulite Facies Terrains of India and Their Potential for mineral Deposits

RAM S. SHARMA

(Former Prof. and Head, Department of Geology, Banaras Hindu University)
Emeritus Professor, Department of Geology, University of Rajasthan, Jaipur 302 004
Email : sharma.r.sw@gmail.com

Abstract

The Precambrian high grade metamorphic granulite terrains, a component of crustal evolution, were traditionally thought to be sterile or poor in economic mineralization are able to provide insights into the effect of high grade metamorphism on the formation and or preservation of mineral deposits, and the relationship between thermotectonic activity in the middle to lower crust and the mineralization in the upper crust.

The Darwar Craton in the South India show retrogression following granulite facies metamorphism, which often result in the formation of gneissic terrain with incipient charnockitization or/and granulitization at several places. In the light of various genetic models available, the formation of the gold veins have been discussed. If the biotite hornblende gneiss are retrogressed enderbitic charnockitic granulites due to shearing during Proterozoic time, gold was leached from granulites and transported via fluids as carbonyl complexes to upper crustal levels and into the vein system. It is interpreted that mixing of low density gold bearing carbonic fluids of deep seated origin (derived from granulites) with low salinity aqueous fluids (derived from younger granites, has destabilized the gold bearing complexes, which were subjected to remobilization and deposition of gold in quartz veins.

The Eastern Ghat Mobile Belt contains charnockites, khondelites and migmatites with granitoids. Khondelites invariably contain graphite but workable deposits occur only at sheared contact between khondelite and charnockites, suggesting that the graphite is deposited from natural carbon bearing fluids. Also ilmenite, titaniferousmagnetite occur as cross cutting veins and lenses to represent residual liquid of anorthosite norite magma.

In a part of reconstructed basement of Banded Gneissic Complex in Rajasthan, the granutite are present mainly around Sandmata and Bhinai with a charnockite suit. It is proposed that during the Delhi Orogeny, the magma underplating have heated the already metamorphosed rocks of BGC (the precursors of Delhi and Aravalli Supergroups). During this high grade metamorphism the metals must have been leached from the original setting, leaving the Rajasthan granulites sterile.

Keywords : Granulites, Ore fluids, Southern Granulite Terrain, Sandmata Complex, Eastern Ghat Mobile Belt, Fluid Rock Interaction.

INTRODUCTION

Granulites are metamorphic rocks formed in the Earth's lower to middle crust at temperatures in the range of 750 to 950°C and pressures of 6 to 9 kilobars. Mineralogically, the granulites are anhydrous and are characterized by granoblastic (granular) texture, hence the name granulite. A common granulite rock is characterized by hypersthene (orthopyroxene) and plagioclase with accessory garnet, clinopyroxene, spinel, cordierite, sapphirine hornblende and oxides. This granulite assemblage points to reduced activity of water , i.e. $^{A}H_2O$ less than one. The mineralogy of granulites is controlled by the parent composition, and granulites vary in composition from basic to acidic, pelitic and quartzo-feldspathic. The pelitic granulites are recognized by garnet-sillimenite-k-feldspar ± graphite association, while quartzo-feldspathic granulites are characterized by hypersthene-plagioclase-Kfeldspar-quartz and are often called charnockites. The basic composition denoted by metamorphosed basalt contains orthopyroxene and clinopyroxene (garnet) and defines the granulite facies. The rocks of the granulite facies are of particular interest to geologists because most granulite samples represent lower- to middle-crust, mainly exposed in Precambrian cratons of the world.

Thermo-tectonic evolution of the Precambrian high-grade metamorphic terrains, particularly Archean granulite terrains, should throw light on the origin and evolution of the early crust. Also, research on mineral deposits in high-grade terrains, which were traditionally thought to be poor in economic mineralization, should provide insights into the effect of high-grade metamorphism on the formation and/or preservation of mineral deposits, and the relationship between thermo-tectonic activity in the middle-to-lower crust and mineralization in the upper crust.

OCCURRENCE OF GRANULITES

In almost all the shield areas, the granulite facies rocks are intercalated with amphibolite facies terrains. Some of these amphibolite facies terrains are considered to result from retrogression associated with amphibolitization of granulite's dry assemblages. Here, one would notice coronas around earlier minerals wherein hornblende and biotite are found bordering hypersthene and garnet. Contrary to this situation, geologists have also found a prograde transition of amphibolite facies to granulite facies. This feature is constrained by evidence of partial melting of metapelites as a result of dehydration of biotite and even hornblende. Orthogneisses of hornblende-biotite composition are also prograde to charnockites. Further increase of P and T are recorded by garnet- and sapphirine-forming reactions in metapelites which identify granulite grade metamorphism. In some terrains, peak metamorphic conditions are recorded by the formation of hecrynite (Fe-spinel) + quartz at the expense of cordierite and the formation of ossumulite in metapelites (Waters, 1991). These regionally metamorphosed granulites from orogenic belts show clockwise P-T path. Their origin is related to high heat flow during convergent plate tectonics, causing partial melting in rocks of suitable compositions, preferably metapelites and gneisses, in the mid-to-lower crust. The common dehydration melting reactions during granulite facies metamorphism involve biotite and hornblende (Biotite + quartz = hypersthene + Kfeldspar in melt + H_2O ± garnet ; Hornblende + quartz = hypersthene + clinopyroxene + plagioclase in melt + H_2O). The generated granitic melt and volatiles must have moved to higher crustal levels, leaving the dry granulites as residue of high-grade metamorphism attended with anatexis. The granulites evolved in this way are depleted in LREE, which get enriched in the associated acidic melts crystallized as granitoids in the vicinity of the upper amphibolite and granulite facies rock.

If cooling occurs at nearly constant pressures (cooling, IBC), these mid-crustal pelitic granulites would show overgrowth features, mainly of cordierite or sillimanite around hercynitic spinel and replacement of cordierite by garnet + sillimanite. The high heat flow in the mid-crust could also be attained by magmatic accretion that may have been caused by a thick basaltic underplate, about few kilometers below the present erosion surface. Some granulites also experience decompression from high temperatures in deeper levels of the crust to shallower crustal levels during which suitable compositions underwent decompression melting reactions stated above, giving rise to charnockite assemblages that are often characterized by symplectitic textures.

GRANULITES FROM DHARWAR CRATON

Retrogression following granulite facies metamorphism often results in the formation of gneissic terrains, as in Dharwar craton in southern India where incipient charnockitization/ granulitization is reported at several places. There are also views that the incipient charnockites are prograde and not due to retrogression following granulite facies

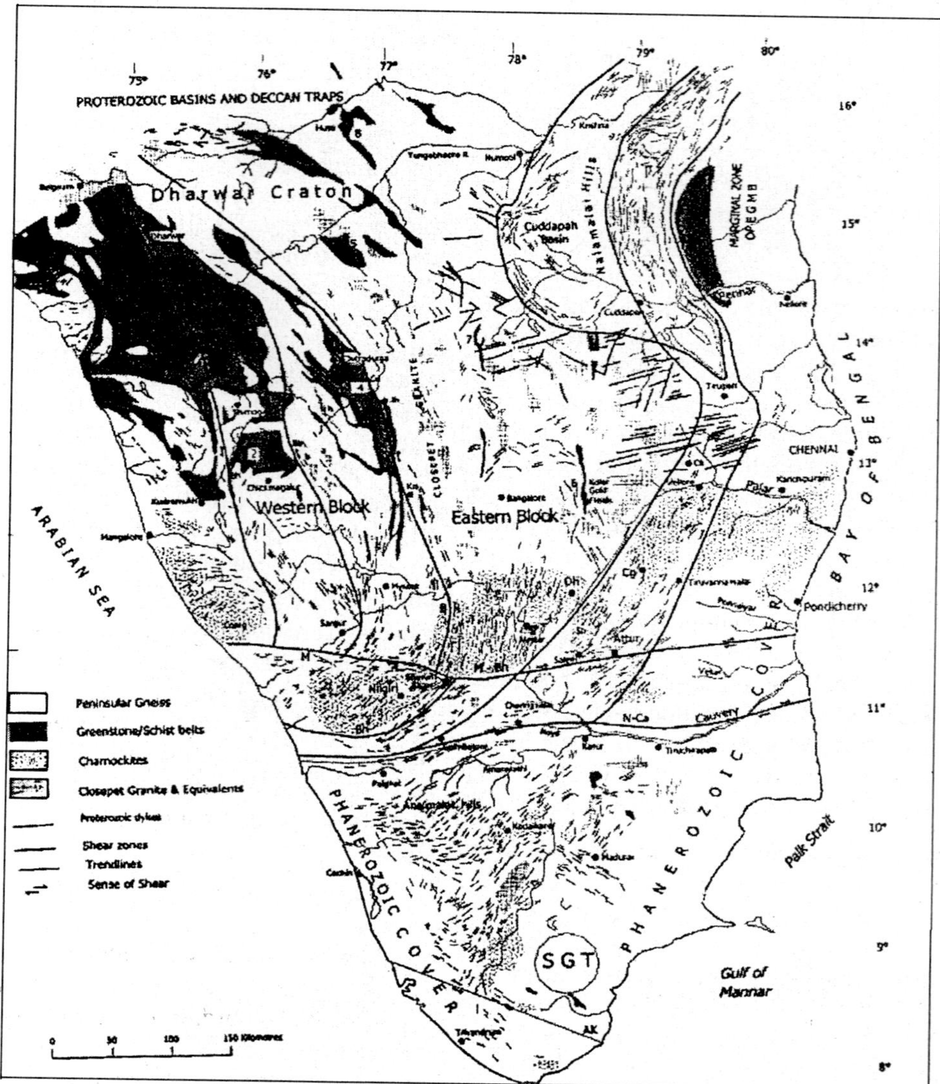

Fig. 1 Geology of South Indian Shield (After, Ramkrishnan, 2003; Naqvi, 2005) showing SGT= Southern Granulite Terrain (Pandyan Mobile Belt). CB = Cuddapah Basin, EGMB = Eastern Ghat Mobile Belt, BR = Biligiri Rangan Hills, Kn = Kunigal, Ch = Chitoor, Dh = Dharmapuri, Nk = Namakkal, Jh = Javanahalli, Sh = Shevaroy hills, Cg = Chengam. Greenstone/schist Belts : 1 = Shimoga, 2 = Bababudan, 3 = Western Ghats, 4 = Chitradurga, 5 = Sandur, 6 = Kolar, 7 = Ramagiri, 8 = Hutti, 9 = Nellore, 10 = Veligallu, Shear Zones : NCa = NoyilCauvery, M = Moyar, B = Bhavani, CB = Cuddapah boundary shear zone, M+ = Mettur shear zone linking up with Nallamalai, Ak = Achankovli, Bh = Balehornur, Bb = Bababudan, Cd = Chitradurga.

metamorphism as shown in Fig. 1 of geological map of South Indian Shield (After, Ramkrishnan, 2003; Naqvi, 2005). The crustal thickness map of South India is shown in Fig. 2 (After, Gupta et al. 2003; Naqvi, 2005) in which numbers denote moho depth. The crustal scale 'flower structure' is shown in a cross section of the Region in Fig. 3 (After, Chetty et al. 2003; Naqvi 2005). Fig. 4 showing major shear zones of South India (After, Jain et al. 2003; Naqvi, 2005). Zircon geochronology from these incipient charnockites (also called arrested charnockites) and associated granitoid bodies gave reliable age of 2500 Ma for the charnockitization in the region. The charnockite occurs along ductile shears traversing the amphibolite-facies gneisses as at Kabbaldurga, Southern India (Pichamuthu, 1960). Hypersthene is reported to have developed replacing biotite and/or amphibole along the fractures and shears that became the domains of $P_{H2O<<}P_{total}$. In view of the occurrence of hypersthene within structurally-controlled domains

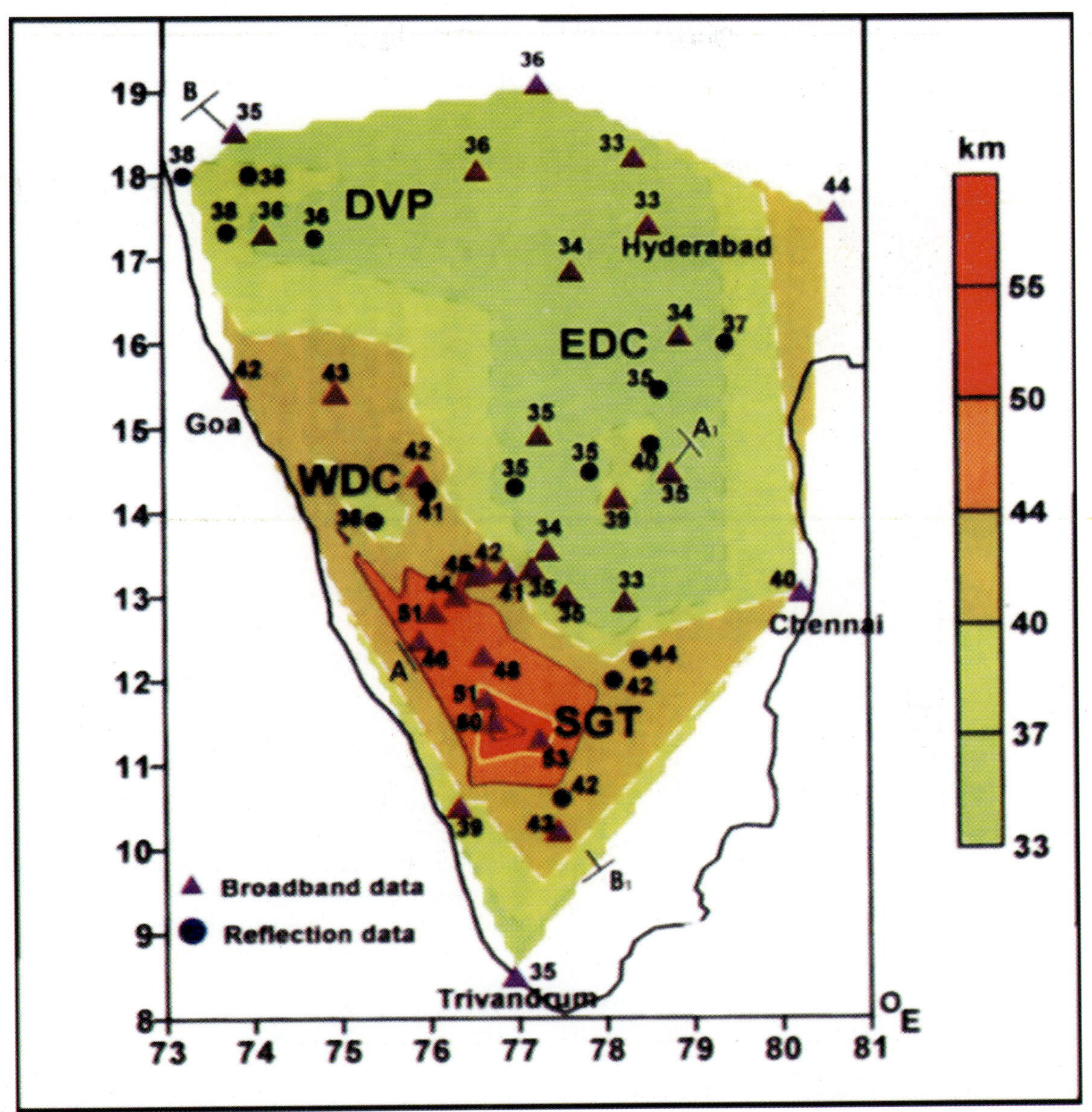

Fig. 2 Crustal thickness map for South India using data from broadband receiver function analysis and modeling of reflection sturies. Numbers denote Moho depth (After, Gupta et al., 2003; Naqvi, 2005).

within the gneisses, the reduced water activity was explained by infiltration of CO_2-rich fluids, originated from deep-seated sources (Newton et al., 1980; Janardhan et al., 1982). A variety of mechanisms have been proposed for the reduced water activity and formation of charnockites along structurally-controlled fabrics. These are: (i) infiltration of CO_2-rich fluids from the mantle (Newton et al., 1980); (ii) infiltration of CO_2-rich fluids from lower crustal granulites when subjected to deformation (Stahle et al., 1987); (iii) drop in fluid pressure, liberating the trapped fluid from the fractured minerals (Srikantappa et al., 1985); reaction of biotite in the presence of graphite whereby the activity of equilibrium water was reduced by mixing with generated gases, mainly carbon dioxide and methane ($2C + 2H_2O = CO_2 + CH_4$) (Hansen et al., 1987); vapour-absent dehydration melting where H_2O is partitioned into the melt (Burton and O'Nions, 1990); infiltration of concentrated chloride/carbonate brines (Hansen et al., 1995); and polyionic salt solutions, immiscible with CO_2, as feasible dehydrating fluids (Newton et al., 1998). All these mechanisms invoke mutual soluble CO_2-H_2O rich fluids streaming upwards through the

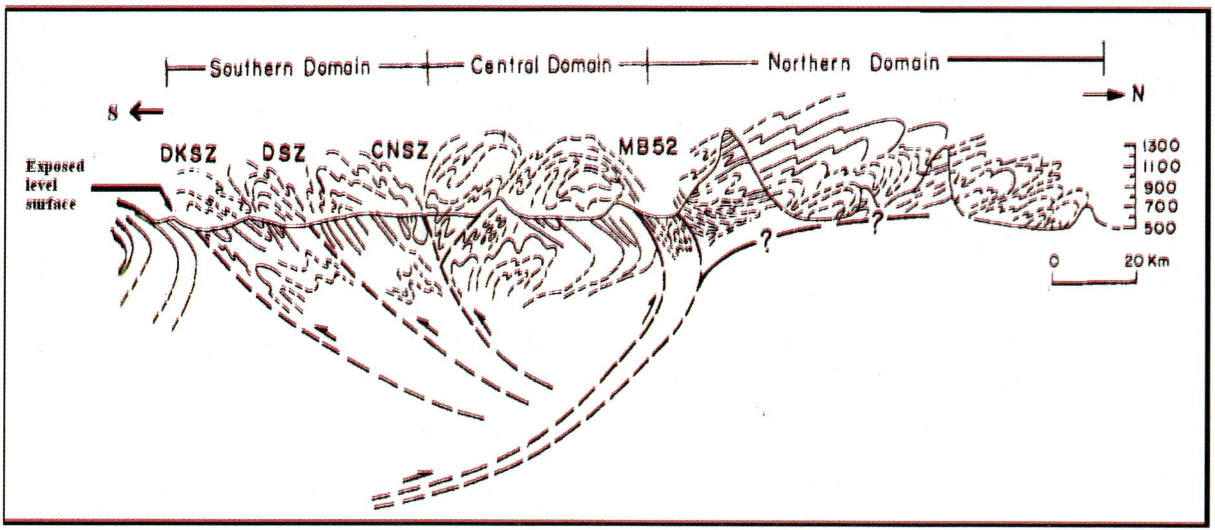

Fig. 3 A schematic representation of the inferred NS structural cross section and the crustal scale "flower structure" along the KuppamPalani corridor. Vertical scale is applicable to generalized topography (continuous lines). (After, Chetty et al., 2003; Naqvi, 2005).

crust, causing local reduction in volatile contents and accompanying mineralogical changes along the fluid flow paths.

In addition to the two aforesaid models for the origin of charnockites, viz. the dehydration melting and infiltration of CO_2-rich fluids in the deep crust, there are two more mechanisms for the genesis of the granulites and charnockites, both conforming to the reduced activity of water during their formation. First, if an already metamorphosed rock-suite undergoes high-grade metamorphism, the mineral reactions will take place under $P_{H2O} << P_{total}$ conditions, whereby replacement of biotite and amphiboles by hypersthene will result into anhydrous assemblages that characterize granulite facies rocks.. Such granulites will necessarily be polymetamorphic in nature in which paragenesis of two metamorphism can be inferred from the presence of two generations of minerals, specially Al-silicates, garnets and even Fe-Ti oxides. Age determination of zircons from different minerals from the granulites and associated high-grade rocks is likely to shed light on the two regional metamorphic events. In India and other parts of the globe, the Precambrian terrains appear to be polymetamorphic but their mineralogy is not yet properly worked out with the appreciation of relict mineralogy and/or textures. Second, the granulite facies rocks, viz. the charnockites are considered plutonic rocks crystallized from magmas in depth. The charnockites (acid, basic and ultrabasic) from Chennai (formerly Madras) have mineral assemblages corresponding with those of the granulite facies, but were considered plutonic rocks because their textures as well as chemical compositions recall those of corresponding plutonic rocks. From a comprehensive study of charnockites, Howie (1955, p.762) concluded that these rocks were originally a series of calc-alkaline plutonic rocks, although he was unable to find decisive criteria to distinguish products of granulite facies metamorphism from primary igneous rocks crystallizing in a similar deep-seated environment. Field and recent experimental and theoretical works on granitoid rocks showed that charnockite series can originate from magmatic crystallization under subcrustal $P_{H2O} << P_{total}$ conditions, and rocks from norite to charnockite can form in place of diorite to granite under reduced activity of water (see Isachesen, 1969; Clemens, 1990; Johannes and Holtz, 1996). In this context, the recently recognized Archaean age of 2800 to 3000 Ma for some charnockite-enderbite in southern India (Mahabaleswar et al., 1995; Buhl, 1987; Bhaskar Rao et al., 2003) possibly indicate a late Archaean crystallization of charnockite-enderbite plutonic rocks, instead of pointing to the recrystallization event of granulite facies metamorphism in late Archaean, specially in BR Hills and other local areas. Thus, it may imply that these igneous granulites, distinctly older than the 2.5 Ga old charnockitization, seen in form of arrested/incipient charnockite along structurally controlled fabric within the gneisses. Field observation at several places, notably in Bhavanisagar quarry suggests that the charnockites

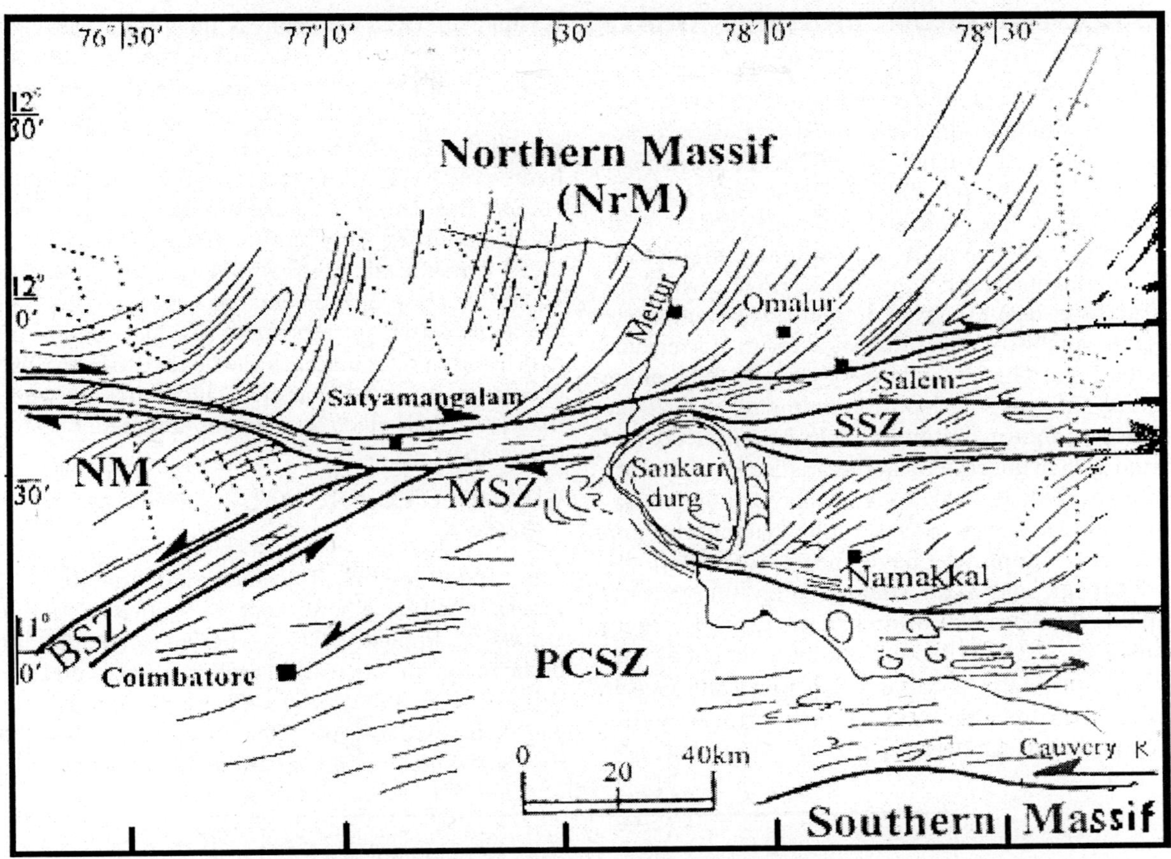

Fig. 4 Major shear zones of south India (after, Jain et al., 2003; Naqvi, 2005). MSZ = Moyar Shear Zone in the Moyar river satyamantalamSalemAttur area. BSZ = Bhavani Shear Zone along Bhavani River. PCSZ = PalghatCauvery Shear Zone in the PalghatCoimbatoreNamakkal area. SSZ = Sweta Nadi Shear Zone.

show intrusive relationship with the basic granulites (see Srikantappa, 2004 p.152). Massive charnockites are also reported from southern Kerala where they fringe the Khondalite Group of rocks, both on its north and south. Based on geochemistry, Chacko et al. (1996) emphasized that these charnockites, owing to their resemblance with the C-type magmas are of magmatic origin and were the dominant source for the metamorphism of the surrounding Kerala Khondalite Belt (Chacko et al., 1996). From metamorphic and geochronological considerations the gneisses from Southern Granulite Terrain (SGT) appear to be either retrograded charnockite-enderbite rocks (as magmatic or metamorphic) prior to 2.8 Ga or they are the Archaean basement Peninsular gneisses that were transformed into charnockite-enderbite by progressive metamorphism during Proterozoic event.

Numerous gold-quartz veins of varying dimensions are seen parallel as well as across the regional foliation in the biotite-hornblende gneiss. Here, gold is reported to occur as disseminated grains as well as invisible gold locked in pyrrotite (Malathi and Srikantappa, 2004, p.72). If the biotite-hornblende gneisses are retrogressed enderbitic-charnockitic granulites due to shearing during Proterozoic time, gold was leached from granulites and transported via fluids as carbonyl complexes to upper crustal levels and into the vein system. It seems that mixing of low density gold-bearing carbonic fluids of deep-seated origin (derived from granulites) with low salinity aqueous fluids (derived from younger granites, possibly Pan-African granites) has destabilized the gold-bearing complexes, which were subjected to remobilization and deposition of gold in quartz veins. This proposition finds support from fluid inclusion studies of auriferous quartz veins, which revealed the presence of low to moderate density carbonic inclusions mixed CO_2-H_2O inclusions and low salinity aqueous inclusions, in contrast to high density carbonic inclusions in enderbite-charnockite granulites.

GRANULITES FROM EASTERN GHATS MOBILE BELT

The Southern Granulite Terrain is bordered on the east by another granulite province called the Eastern Ghats Mobile Belt as shown in Fig. 5 (After, Ramakrishnan, et al., 1998; Naqvi, 2005). This belt contains charnockites, khondalites, and migmatites with granitoids from granite to granodiorite composition. The khondalites form high linear ridges while the charnockites occur in low-lying hills. A laterite cover is seen on top of the hills as a result of residual weathering. The khondalites invariably contain graphite but workable deposits commonly occur at the sheared contact between the khondalite and charnockite in the eastern Godavari district of Andhra Pradesh. In this shear zone, called Burugubanda-Tapasikonda shear zone, both syngenetic and epigenetic graphites have been identified by spectroscopic and X-ray studies, suggesting that the graphite is deposited from natural carbon-bearing fluids (CO_2, C, CH_4) (Parathasarathy et al., 2006). The role of fluids is also documented by the occurrence of zoned apatite-magnetite veins within which hyalophane and calciferous pyroxene is reported, as at Kasipatnam (Rao and Vijayakumar 1992). The veins are considered to be fracture filling metasomatic skarn deposits because of zoning and inward coarseness of mineral from margins of the veins. The mineral deposits are considered due to contamination of hydrothermal fluids by the calcareous material of the Khondalite Group. In the layered norite-gabbro-anothosite rock-suite from Kondapalle and other areas, ilmenite-titaniferous magnetite accumulations occur as cross-cutting veins and lenses. These veins are considered to represent residual liquid after anorthosite-norite magmas and hence can be exploited economically if geochemical anomalies as well as magnetic and electro-magnetic anomalies can be detected.

GRANULITES FROM RAJASTHAN TERRAIN

The granulites in Rajasthan are found in the reconstituted basement complex, called the Banded Gneissic Complex.(BGC) as shown in geological map of the region in Fig. 6 and a cross section of the region (Fig. 7) to illustrate various tectonostratigraphic units and their relationship (After, Roy and Jakhar, 2002; Roy et al. 2004). Main granulites occur at and around

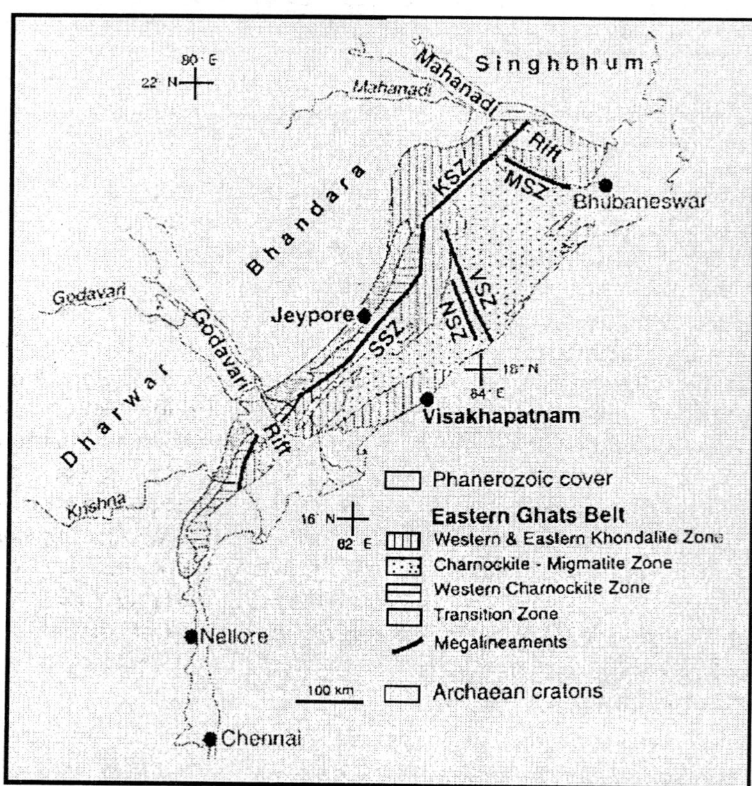

Fig. 5 Simplified Geological map of the Eastern Ghat and adjacent regions (After, Ramakrishnan et al., 1998; Naqvi, 2005).

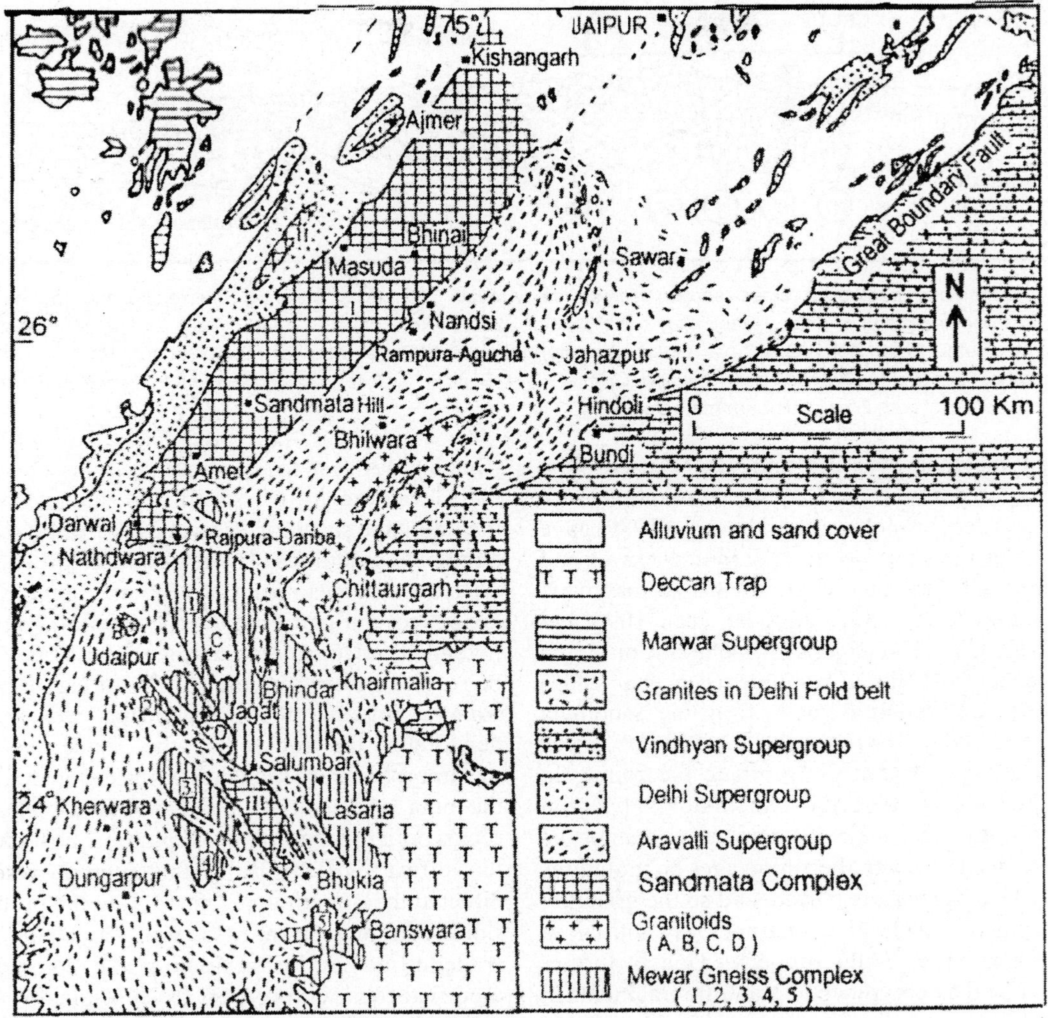

Fig. 6 Geological map of the Aravalli Mountains showing major lithostratigraphic units (after Roy and Jakhar, 2002; Roy et al., 2004).

Sandmata and also at Bhinai to the north, mostly as raised hills amidst the BGC. The rocks belong to charnockite suite, (charnockite to enderbite), mafic granulite and pelitic granulite. The rock-suite also contains metanorite which occurs as intrusion in the gneisses and the granulite-suite. The geothermo - barometry on the charnockitic assemblages gave pressure of 7-9 kbar and temperatures of 700 to 900 °C. (Sharma, 1988, 1999; Dasgupta et al., 1997). On the basis of P-T data and mineralogy, Sharma (1988) deduced anticlockwise P-T-t path for the Sandmata granulites and inferred their exhumation by thrusting along shear zones that are now seen at the granulite-BGC contact.

The anti- or counter-clockwise P-T path rules out crustal thickening by compressive plate tectonics, but clearly requires magmatic input early in their evolution. Hence crustal extension and thinning have been invoked as a precursor for the partial melting of the asthenosphere and upward movement of basaltic magmas into lower crust. It is difficult to increase pressure later except by secondary melting in the crust to form granitoid magmas. The separation of these magmas from their metamorphic residue (i.e. granulites), and their upward movement and emplacement as batholith-sized plutons would result in a greater thickness of the crust above the metamorphic rocks. In Sandmata area, the presence of late to post-tectonic Anjna Granite and its equivalent granitoids could represent the secondary melts that

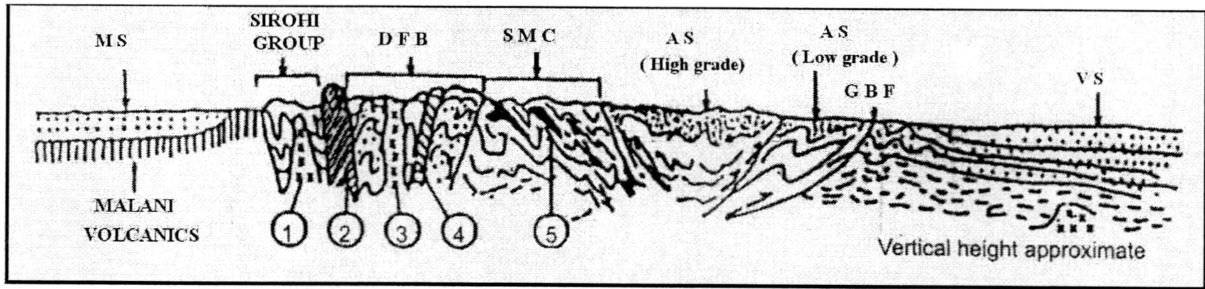

Fig.7 Schematic geological cross section across the Aravalli Mountains showing interrelationship between the major tectonostratigraphic units. (after Roy and Jakhar, 2002; Roy et al., 2004). A.S. = Aravalli Supergroup, D.F.B. = Delhi Fold Belt, G.B.F. = Great Boundary Fault. M.S. = Marwar Supergroup. V.S. = Vindhyan Supergroup. 1 = Balda Granite, 2 = Sewaria Granite, 3 = Sendra Granite, 4 = Basement inlier within D.F.B. 5 = Exhumed bodies of granulites in the Sandmata Complex.

increased the load pressure and also caused slow cooling of the granulites of Sandmata complex. From U-Pb zircon geochronology, Sarkar et al. (1989) gave 1723 Ma for the charnockite-enderbite rocks of and considered the Sandmata charnockites as magmatic rocks. Somewhat younger ages between 1690 and 1621 Ma are given by evaporation technique on single zircon crystal from Bhinai granulite (Roy et al. 2005) while associated tonalitic gneiss from the Sandmata gave 1692 Ma (Fareeduddin, 2004). Since evaporation ages do not discriminate the zoning in zircon, the ages are less reliable than the U-Pb zircon geochronology and their interpretation cannot be unambiguous. However, the zircon ages on granulites point to Mesoproterozoic period, and so the granulite metamorphism (including charnockitic intrusion) must be related to Delhi orogeny. This is further supported by the occurrence of granulite fragments in the Proterozoic cover rocks of the Delhi metasediemnts near Pilwa-Chinwali area to the north of Ajmer, and their total absence in the Palaeoproterozoic metasediments of the Aravalli Supergroup (Sharma, 1995; Sharma, 2003) has clearly shown that the protoliths of these granulites are not the sediments of the Delhi or Aravalli Supergroup but their precursors that were the already metamorphosed rocks of the BGC, which were heated by magma underplating during the Delhi orogeny. During this high-grade metamorphism the metals must have been leached from their original settings, leaving the Rajasthan granulite rocks sterile. In fact metamorphism and the prevailing fluids can significantly improve the value of a deposit through coarsening, recrystallization of ore minerals into purer end members. Literature survey indicates that metamorphism does not volatize sulphides. Both sulphide and oxide mineralizing hydrothermal (-magmatic) systems are commonly associated with the development of significant alteration zones, including the shear zones such as those abounding the Sandmata granulites. The ore-fluid might come from different sources, such as the metamorphic fluid generated during prograde dewatering, magmatic-juvenile fluid, or a combination of these two. This can be resolved from isotopic studies of ore components from a large number of deposits. Metamorphism dehydrates the rocks but the fluids lost during metamorphism are seen during mass transfer, altering the bulk composition and isotopic composition of rocks, both metamorphic and plutonic host rocks. It is this efficacy of the fluids that modern researchers directed their attention to for solving the problems of fluid-rock interaction. Mass transfer calculations are made to find out the variation of a given oxide in the modified rock with respect to its original rocks. But the fluids are fugitive and thus elusive, and their nature can be understood when they are trapped as inclusions, syngenetically formed with the host mineral. Thus fluid inclusions as microsamples are useful in revealing the nature of fluids. Again, it appears to be a broad spatial relationship between metal deposits and tectonic settings. (A.B. Roy, this volume). For example, a magmatic arc in zones of plate convergence is nearly always associated with porphyry-style Cu-Mo deposits and epithermal Au-Ag deposits. If so, one has to critically evaluate a proposed geodynamic model for the tectono-thermal evolution of the Indian granulite belts for specific deposits, especially when modern-style plate tectonics may not have operated in Archaean times.

References:

Bhaskar Rao, Y.J., Janardhan, A.S., Vijaya Kumar, T., Narayana, B.L., Dayal, A.M., Dayal, A.M., Taylor, P.N. and Chetty, T.R.K. (2003). Sm-Nd model ages and Rb-Sr isotopic systematics of charnockites and gneisses across the Cauvery shear zone, southern India: Implications for the Archaean-Neoproterozzoic terrane boundary in the southern Granulite Terrain. In: M. Ramakrishnan (ed.), Tectonics of Southern Granulite Terrain, Kuppam-Palani Geotransect. Mem Geol. Soc. India, No.50, pp.297-317.

Burton, K.W. and O'Nions, R.K. (1990). The time scale and mechanism of granulite formation at Kurunegala, Sri Lanka. Contrib. Mineral. Petrol., vol.106, pp.6-89.

Buhl, D. (1987). U-Pb and Rb-Sr Altersbestimmungen and Untersuchun-gen zum Strontium isotopenaustausch und Granuliten Studiens. Ph.D. thesis, University of Munster, Germany.

Chacko, T., Lamb, M. and Farquhar, J. (1996). P-T conditions in the Kerala Khondalite Belt. Revisited: evidence for ultra-high temperature metamorphism resulting from the intrusion of C-type magmas. Abstract, International Field workshop on the Proterozoic continental Crust of southern India, IGCP 368, Aug.20-27, 1996.

Chetty, T.R.K., Bhaskar Rao, Y.J. and Narayana, B.L. (2003). A Structural cross section along Krishnagiri Palani corridor, southern Granulite Terrain of India. In Tectonics of Southern Granulite Terrain. Ramakrishnan, M. (Ed.), Mem. Geol. Soc. India. vol. 50, pp. 255-278.

Clemens, J.D. (1990).The granulite-granite connection. In:D. Vielzeuf and Ph. Vidal (eds.), Granulites and Crustal Evolution, Kluwer, Dordrecht, pp.25-36.

Dasgupta, S., Guha, D., Sengupta, P., Miura, H. and Ehl, J. (1997). Pressure-temperature-fluid evolutionary history of the polymetamorphic Sandmata granulite complex, northwestern India. Precambrian Res., vol.83, pp.267-290.

Fareeduddin (2004). Granulite facies metamorphism and the Gondwanian correlations: Evidence from northwestern Indian shield. Abstract, International workshop on Tectonics and Evolution of the Precambrian Southern Granulite Terrain, India & Gondwanian Correlations. National Geophysical Res. Institute, Hyderabad, Feb.18-25, 2004, p.94.

Gupta, S., Rai, S.S., Prakasam, K.S., Srinagesh, D., Bansal, B.K., Chadha, R.K., Priestley, K. and Gaur, V.K. (2003). The nature of the crust in southern India : Implications for Precambrian crustal evolution. Geophys. Res. Lett., vol. 30 No. 8, pp. 1-4.

Hansen, E.C., Janardhan, A.S., Newton, R..C., Prame, W.K.B.N. and Ravindra Kumar, G.R. (1987). Arrested charnockite formation in south India and Sri Lanka. Contrib. Mineral. Petrol., vol.97, pp.225-244.

Howie, R.A. (1950). The geochemistry of the charnockite series of Madras, India. Trans. Roy. Soc. Edinburgh, vol.62, pt.3, pp.725-768.

Isachesen, Y.W. (1969). Origin of anorthosite and related rock. New York State Mus. Sci. Serv. Mem. 18.

Jain, A.K., Sandeep Singh and Manickavasagam, R.M., (2003). Intracratonic shear zones in the Southern Granulite Terrain : Their kinematics and evolution. In : Tectonics of Southern Granulite Terrain, Ramakrishnan, M. (Ed.) Mem. Geol. Soc. India. vol. 50, pp. 225-254.

Janardhan, A.S., Newton, R.C. and Hansen, E.C. (1982). The transformation of amphibolite facies gneiss to charnockite in southern Karnataka and northern Tamil Nadu, India. Contrib. Mineral. Petrol., vol.79, pp.130-149.

Johannes, W. and Holtz, F. (1996). Petrogenesis and experimental petrology of granitic rocks. Springer-Verlag, Berlin, p.335.

Mahabaleswar, B., Jayananda, M., Peucat, J.J. and Shadaksharswamy, N. (1995). Archaean high-grade gneiss complex from Satnur-Halgur-Sivasamudram areas, Karnataka, southern India. J. Geol. Soc. India, vol.45, pp.33-49.

Malathi, M.N. and Srikantappa, C. (2003). Composition and evolution of fluids and timing of gold mineralization in the Bhavani shear zone, Nilamur, Kerala, India. Abstract, In International workshop on Tectonics and Evolution of the Precambrian southern Granulite Terrain, India & Gondwanian correlations. National Geophys. Res.Institute, Hyderabad, Feb.18-25, 2004, p.72.

Naqvi, S.M. (2005). 'Geology and Evolution of the Indian Plate (From Haden to Holocene 4 Ga to 4 Ka). Capital Pub. Comp., New Delhi, India. 450 p.

Newton, R.C., Smith, J.V. and Windley, B.F. (1980). Carbonic metamorphism, granulites and crystal growth. Nature, vol.288, pp.45-50.

Newton, R.C., Vranovich, L.Ya., Hansen, E.C. and Vandenheuvel, B.A. (1998). Hypersaline fluids in Precambrian deep-crustal metamorphism. Precambrian Res., vol.91, pp.41-63.

Parthasarathy, G., Sreedhar, B. and Chetty, T.R.K. (2006). Spectroscopic and X-ray diffraction studies on fluid deposited rhombohedral graphite from the Eastern Ghats Mobile Belt. Current Sci., vol.90, No.7, pp.995-1000.

Pichamuthu, C.S. (1960). Charnockite in the making. Nature, vol.188, pp.135-136.

Ramkrishnan, M., (2003). Craton-mobile belt relations in southern Granulite Terrain in Tectonics of Southern Granulite Terrain. Ramkrishnan, M. (Ed.), Mem. Geol. Soc. India. vol. 50, pp. 1-24.

Ramakrishnan, M., Nanda, J.K. and Augustine, P.F. (1998). Geological evolution of the Proterozoic Eastern Ghat Mobile Belt. Proc. of workshop on 'Eastern Ghat Mobile Belt', Geol. Surv. India. Spec. Pub. Vol. 44. pp. 1-21.

Rao, A.T. and Vijaya kumar, V. (1992). Chemical petrology of charnockites from the Eastern Ghats granulite province, India. In: A. Barto Kyriakidis (ed.), High GradeMetamorphics, Theophrastus Publications, pp.217-235.

Roy, A.B. and Jakhar, S.R. (2002). Geology of Rajasthan (Northwest India) Precambrian to Recent. Scientific

Publisher (India), Jodhpur. 421 p.

Roy, A.B., Kumar, S., Laul, V. and Chauhan, N.K. (2004). Tectonostratigraphy of the Lead Zinc bearing Metasedimentary Rocks of the RampuraAgucha Mine and its Neighbourhood, District Bhilwara, Rajasthan : implications on metallogeny. In SedimentHosted LeadZinc Sulphide deposits. M. Deb and W.D. Good fellow (Eds). Narosa Pub. House, New Delhi, India.

Roy, A.B., Kroener, A., Bhattacharya, P.K. and Rathore, S. 2005). Metamorphic evolution and zircon geochronology of early Proterozoic granulites in the Aravalli Mountain of northwestern India. Geol. Mag., vol.143 (3), pp.287-302.

Sarkar, G., Ray Barman, T. and Corfu, F. (1989). Timing of continental arc magmatism in northwest India, evidence from U-Pb zircon geochronology. J. Geology, vol..97, pp.607-612.

Sharma, R.S. (1988). Patterns of metamorphism in the Precambrian rocks of the Aravalli Mountain belt. In: A.B. Roy (ed.), Precambrian of the Aravalli Mountain, Rajasthan, India, Mem. Geol. Soc. India, No.7, pp.33-75.

Sharma, R.S. (1999). Crustal development in Rajasthan craton. Indian J. Geology, vol.71, pp.65-80.

Sharma, R.S. (2003). Evolution of Proterozoic fold belts of India: a case of the Aravalli Mountain belt of Rajasthan, NW India. Mem. Geol. Soc. India, No.52, pp. 145-162.

Srikantappa, C. (2004). Bhavanisagar quarry (GPS: N 11°25'48"; E 77°08'18") - Retrogressed charnockite. Abstract in International workshop on Tectonics and Evolution of the Precambrian southern Granulite Terrain, India & Gondwanian correlations. National Geophys. Res.Institute, Hyderabad, Feb.18-25,2004, p152.

Srikantappa, C., Raith, M. and Spiering, B. (1985). Progressive charnockitization of a leptynite-khondalite suite in southern Kerala, IndiaEvidence for formation of charnockites through decrease of fluid pressure?. J. Geol. Soc. India, vol.26, pp. 859-872.

Stahle, H.J., Raith, M., Hoernes, S. and Delfs, A. (1987). Element mobility during incipient granulite formation at Kabbaldurga, southern India. J. Petrology, vol.28, pp. 803-834.

Waters, D. J. (1991). Hercynite-quartz granulites: phase relations and implications for Crustal processes. Eur. J. Mineral., vol.3, pp.367-386.

Hon'ble Reviewers

1. Prof. Ernst Hegner
   University of Munich
   Department of Earth Sciences
   Theresienstra Be 41 D80333 Muchen
   GERMANY

2. Dr. M. Ramkrishnan
   201, Skyline Surabhi Apartments
   Vidhyapeeth Road
   BSK III Stage
   BANGALORE 560 085 INDIA

Editorial Assistance

Ms. Mamta Chauhan

Economic Mineralization
K.L. Shrivastava (Editor)
2009. pp. 15-25
Scientific Publishers (India), Jodhpur

Section-I

The Malani Supercontinent : Middle East Connection During Late Proterozoic

NARESH KOCHHAR

Centre of Advanced Study in Geology, Panjab University, Chandigarh 160014, India
Email : nareshkochhar2003@yahoo.com

Abstract

The Trans Aravalli Block (TAB) is unique in the geological evolution of the Indian Shield as it marks a major period of anorogenic bimodal, high heat producing, Intra-Plate magmatism represented by the Malani Igneous Suite (50,000 Km^2, 732 Ma). The Arabian Nubian Shield (ANS) is an accreted terrain composed of crustal domains of island arc, oceanic and continental affinities overlain by post accretionary sediments and volcanics. The paper discusses the similarities between the two, in the configuration of the Malani Supercontinent, to provide a new approach for the search of economic mineralization.

Both, TAB and ANS are showing characteristic presence of alkali granites at 732 Ma and Ca. 700 Ma, respectively. The emplacement of these granites and associated volcanics is controlled by ring structures, a menifestation of plume activity and cauldron subsidence, an evidence of extensional tectonic environment. Both the terranes preserve remnants of archean crust i.e. BGC of amphibolite facies (3.2 Ga) in TAB and trondjhemite (2.6 Ga) in ANS. This older crustal protolith has acted as precursor to the younger alkali granites, which in both the terranes mark the cratonisation of the shield. Both the terranes show evidence of strutian glaciation.

In both the terranes, there is characteristic similarity in the occurrence of peralkaline and peraluminous granite hosted mineralization of U, Th, Nb, Ta, REEs, Sn, W, Li and F etc. In this respect they are akin to the younger mineralized granites of Nigeria. Also, both the terranes are characterized by the occurrence of evaporite carbonate sequence of economic importance.

Keywords : Malani, Trans Aravalli Block (TAB), Arabian Nubian Shield (ANS), Mineralization, Alkali granite, Strutian glaciation, Mantle Plume.

INTRODUCTION

The Indian subcontinent is composed of three main geotectonically different blocks or terrains, the South Indian Block (SIB), the Bundelkhand block (BB) and the Trans-Aravalli block (TAB) (excluding, Central Indian and Eastern Indian Blocks) which were juxtaposed and sutured during different periods of Earth's history. The TAB and BB are geologically less related to each other are separated by NE-SW trending 700 km long Proterozoic Aravalli-Delhi mobile belt. The paper discusses the similarities between the Arabian shield and the Trans-Aravalli block of the NW Indian shield in the configuration of the Malani

Supercontinent in terms of alkaline magmatism, ring structures and Precambrian glaciation including sedimentation at near equatorial position to provide a new approach for the search of economic mineralization in the light of new tectonic settings.

The Trans-Aravalli Block: The TAB (west of the Aravalli-Delhi mobile belt) is unique in the geological evolution of the Indian shield as it marks a major period of anorogenic (A-type) bimodal, high heat producing, "Intra-Plate magmatism" represented by the Malani igneous suite (50,000km^2; 732Ma) comprising peralkaline (Siwana), metaluminous to mildly peralkaline (Jalor) and peraluminous (Jhunjhunu and Tusham) granites with cogenetic carapace of acid volcanics (welded tuff, rhyolite, breccia, perlite etc.). The magmatism is characterized by volcano-plutonic ring structures and radial dykes (Fig.1). The suite is bimodal in nature with minor amounts of basalt flows, gabbro and dolerite dykes. The Siwana ring structure (25.6 km NS and 31 km EW) is the most spectacular feature of the Thar desert and it coincides with the low velocity anomaly centred around Sarnu-Dandali, and Mer-Mundwara, Rajasthan, representing fossil plume head (Kennett and Widiyantoro, 1999). The Malani magmatism is controlled by NE-SW trending lineaments of fundamental nature and owes its origin to the Malani Plume (Kochhar, 2000,2004 Bhushan, 2000). The representatives of Malani magmatism also occur at Kirana Hills, and at Nagarparkar in Pakistan. (Qasim Janet al., 1997)

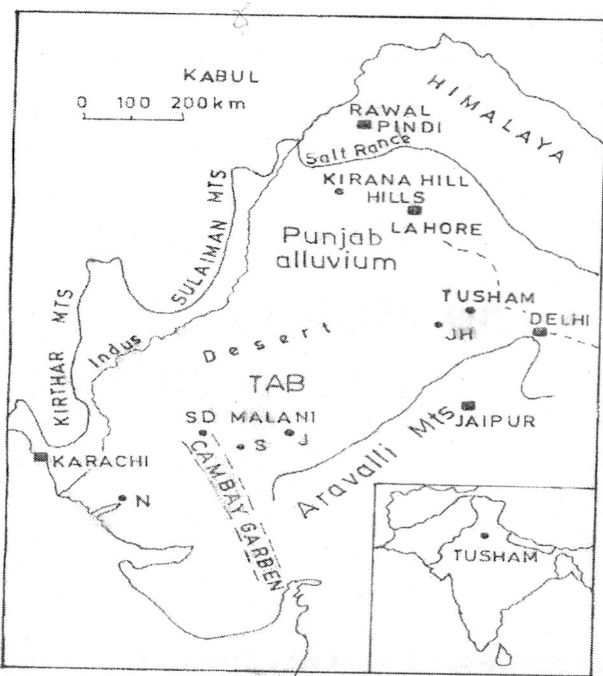

Fig. 1 Location map of the Malani igneous sutie . S: Siwana; J: Jalore, JH: Jhunjhunu; SD: Sarnu-Dandali, N: Nagarparkar; MM : Mer-Mundwara. TAB : Trans-Aravalli Block

The Arabian-Nubian Shield :

The Arabian Nubian shield (ANS) is an accreted terrane composed of crustal domains of island arc, oceanic and continental affinities overlain by post accretionary sediments and volcanics, and invaded by voluminous intrusions. The Saudi Arabia part of the Arabian-Nubian shield has been divided into five terranes separated by four suture zones. The three terranes i.e. Asir, Hijaz, Midyan west of the Nabitah suture zone are of ensimatic character, whereas, eastern two i.e. Afif and Ar-Rayan are of continental to marginal continental character (Fig. 2). The three ensimatic terranes are collectively referred to as Arabian-Nubian arc terranes (Stoesser, 1986).

Most of the shield was created in two distinct stages. These are (1) Island arc stage (between 660-900 Ma) in which subduction related processes operated for the generation of magma, and (2) a post accretionary stage (between 550-700 Ma) where magma was generated by fusion of continental protolith (Stoesser, 1986). According to Jackson (1986) the post accretionary stage is marked by the presence of granitic intrusions. According to Stern (2004) the evolution of the Arabian-Nubian shield occurred between 870 Ma and the end of Precambrian (ca.542Ma) and the crustal growth encompassed a time of dramatic climatic change.

In southern Yemen, the amalgamation of the two terranes ie. Abas (west) and Al-Mahfid (east), separated by Al Bayda island arc sequence, is marked by the 760 Ma magmatic activity (White house et al., 2004). According to Johnson and Kattan (2004) the Asir terrane is more closely linked structurally and tectonically with East African orogen, Madagascar, Ethiopia, Somalia, Yemen and Eritea than with the northern Shield, and that the southern and northern Arabian shields may be decoupled.

The remnant of the Archean crust as Proterozoic basement has been reported on the basis of lead isotope studies in the Afif terrane. A zircon from trondjhemite along the Al Amar fault has yielded 2067±74 Ma age there by indicating Late Proterozoic

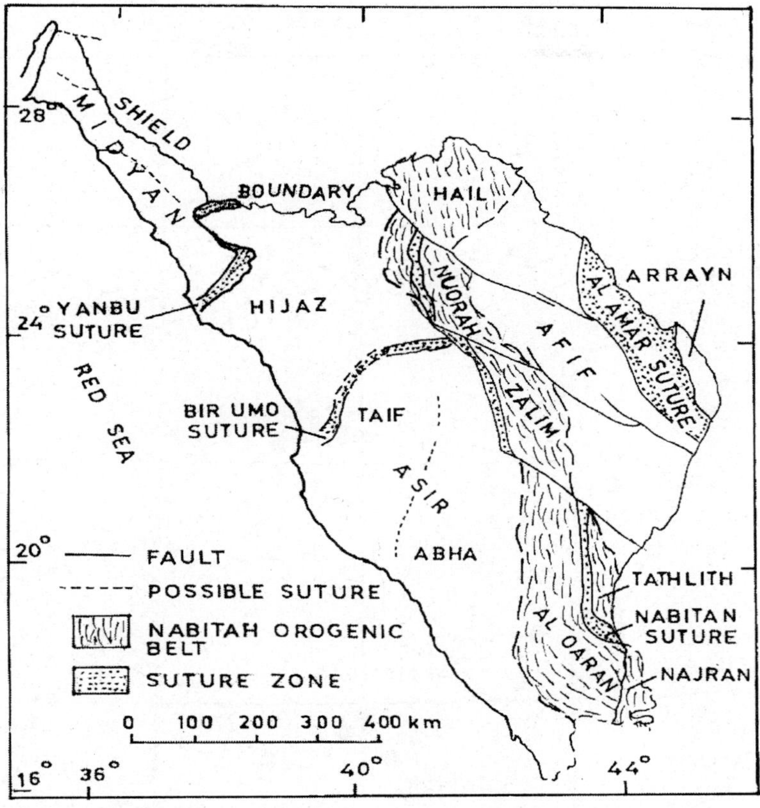

Fig. 2 Map of the Arabian shield showing terranes, provinces and Nabitah orogenic belt.

crust in the shield. Granodiorite gneiss from Jabal Khida in the easternmost part of the shield has been dated at 1629± 20 Ma (Stoesser, 1986).

According to Stoesser et al., (2004) there is no indication of crust formation between 1670-900 Ma as indicated by U/Pb and single zircon data. Between 760-740 Ma ago, the Khida terrane separated from the margin of Paleoproterozoic and Archean continetal block-the Arabian craton, presently concealed beneath the Phanerozoic cover in the east but exposed in Yemen.

Magmatism

The Arabian-Nubian shield contains one of the largest fields of alkali granites in the world. Forty nine major and more than a dozen minor alkali granite plutons occur in the Midyan and Hijaz terranes and in the Nabitah orogenic belt. The alkali granites do not occur in the SW and easternmost part of the shield. These alkali granites, emplaced over a span of 180 Ma, between 680-510 Ma ago, constitute the last phase of intrusives here. They are spatially associated with meta-aluminous subsolvus granites which form the core of the complex, and with a swarm of alkali rhyolite dykes in the northeast part of the shield (Stoesser and Ellion, 1985). A total of thirty three ring complexes and funnel shaped intrusions have been recognized related to the extensional teconic regime in the Arabian-Nubian shield. (Fig. 3) The magmatism (670-570 Ma) is mainly peralkaline to metaluminous with A-type affinity. The granites and the related rock types were emplaced by cauldron subsidence which operated in the final stage of cratonisation of the Arabian shield (Roobal and White, 1986).

The peraluminous and peralkaline granites with A-type geochemical characteristics are encountered at Jabal-Al Hasasin, Jabal-As Sawal, Jabal-Dahul (Stoesser et. al., 2004). According to Vail and Kuron (1978) high level ring complexes, ring dykes and cone sheets have been reported from the Red Sea Hills and Bayuda desert in NE Sudan. These complexes comprise granites, quartz-syenite, nepheline-syenite gabbro etc have been dated at 700 Ma. The Abas terrane is dominated by 760 Ma granitoids, (Whitehouse et. al., 2004), with Sm-Nd model age (2.3 Ga) indicating rework of a Paleoproterozoic or older crustal precursor.

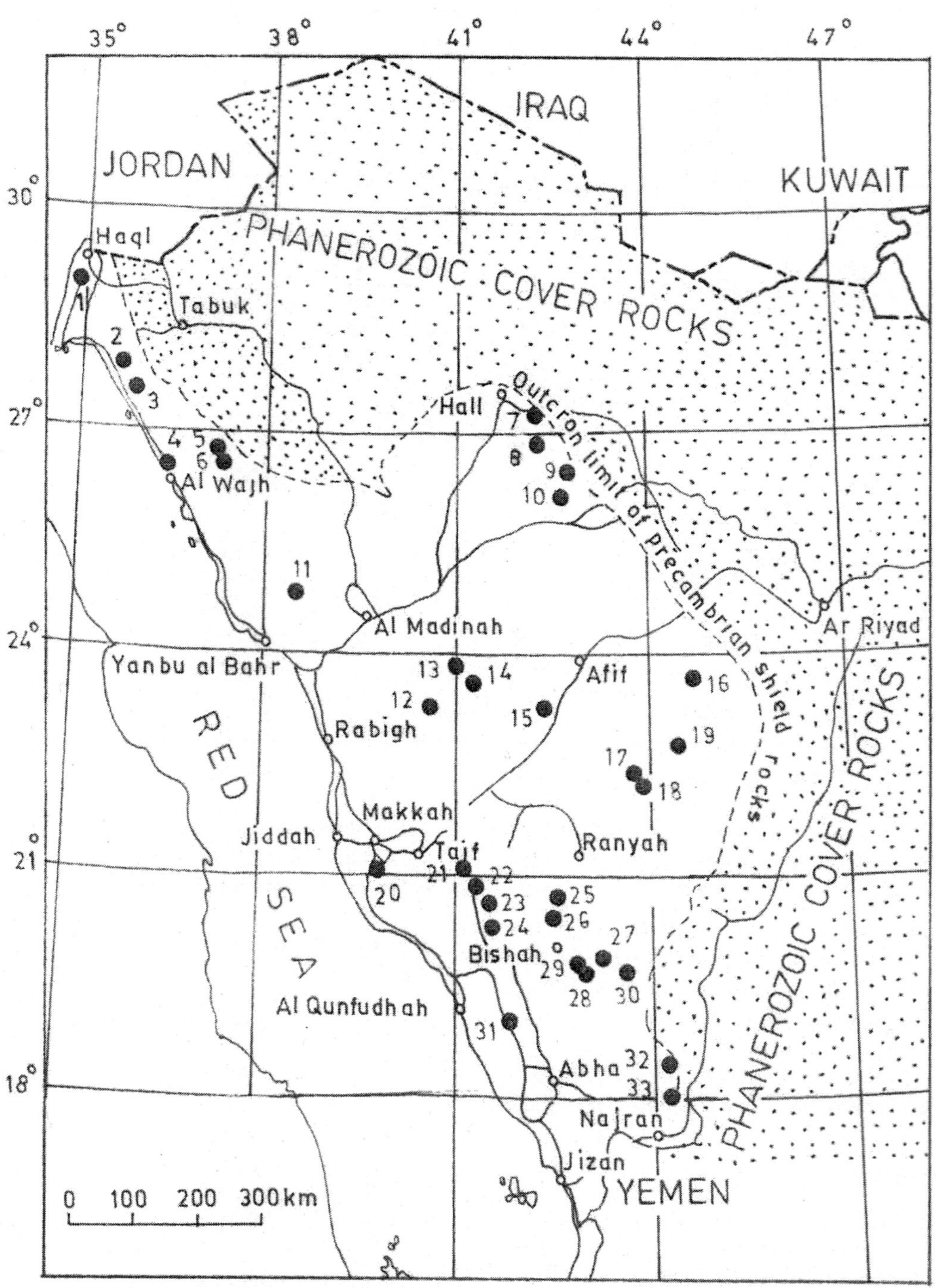

Fig. 3 Map Showing the location of Ring Complex (RC), Single Ring Structure (SR), Beljar Plutons (BJP) and Calderas (C) in the Arabian Shield. Number refers to the locations.

Table for Fig. 3

Sr.No.	Location	Type	Geochemical Affinity
1	Sawda Complex	RC	AA
2	Dabbagh Complex	RC	AA
3	Sulay Siyah Complex	RC	P
4	Liban pluton	BJP	P
5	Habd Complex	BJP	P
6	Warid Complex	RC	AA
7	Jabal Salma Complex	C	AA
8	Jabal AL-Makhrugah	SR	P
9	Jabal Raha	C	AA
10	Jabal Silrilah	SR	AA
11	Martabah Comples	RC	P
12	Jabal Ramram	C	P
13	Jabal Hadash Shara	RC	P
14	Jabal Hadbad Dayahm	RC	AA
15	Jabal as Safwah	RC	AA
16	Jabal Arawan	RC	AA
17	Jabal Kursh	RC	AA
18	Jabal Dahul Complex	RC	AA
19	Uyaijah ring dyke	RC	AA
20	Unnamed Complex	RC	P
21	Jabal Al-Qunnah	RC	P
22	Unnamed ring dyke	SR	AA
23	Khutamah ring dyke	SR	P
24	Aquq ring dyke	SR	P
25	Jabal Munirah	RC	P
26	Jabal Al Najiah	RC	P
27	Jabal Bani-Bisqan	RC	AA
28	Jabal Al-Hasser	RC	AA
29	Jabal Refdah	RC	P
30	Unnamed Complex	RC	P

Table for Fig. 3 Cont.

31	Lakathah Complex	RC	P
32	Wadi Idimah Complex	RC	AA
33	Jabal Ashirah Complex	RC	AA

AA : Alkaline affinity, Syenite, alkali granite with Riebeckite, aegirine etc. A-type. P Peraluminous / metaluminous affinity.

Mineralisation

The alkali granites of the Arabian-Nubian shield have been compared with those of Jos Plateu of Nigeria and Niger in terms of mineralisation. Alkali granites and microgranites and their hydrothermal derivatives are associated with disseminated and vein deposits of Nb, Ta, Sn, and REEs commonly accompanied by Th, U and F. Mineralisation is generally disseminated in porphyritic microgranitic stocks eg. Ghurayyah and Ummal Birak, or concentrated in apical or marginal layered pegmatitic-aplite complex eg. Jabbal said and Jabal Sumr al Ishar. Galena veins containing fluorite are also encountered at Jabal Hadb ad Dayahen (Jackson, 1986). The aluminous granites host rich deposits of Nb, Th, REEs, Li, Zr etc.

According to Ramsay (1986) there are four specialized felsic plutonic rocks in the Arabian-Nubian Shield i.e agpaitic, plumastic, calcic and miaskitic corresponding to alkali granite, alkali-feldspar granite, monzogranite and syenitic precursors. These granites host Nb-Zr-REE, Sn-W-Ta-Mo-Be and Mo-W-Bi-Pb, Zn-Ag respectively.

Malani Supercontinents

The Malani plume which gave rise to the Malani magmatism some 732 Ma B.P. was responsible for the separation of TAB from the East Gondwana and subsequent amalgamation of the Malani supercontinent comprising TAB of the Indian shield, Arabian Nubian, shield, Central Iran, Somalia, Seychelles and Madagascar and south China (Fig. 4) (Kochhar, 2001, 2004, 2006). It has been suggested that all these microcontinents were characterized by a common crustal stress pattern, rifting and thermal regime evidenced by the occurrence of widespread magmatism of alkali granites and comagamtic acid volcanics in the TAB of Indian shield (Kochhar, 2004), Central Iran (Forster, 1987), ANS (Kroner, et. al., 1989), Somalia, (Kroner et. al., 1989), Madagascar (Yoshida et. al., 1999), Seychelles (Hoshino, 1986), (Ashwal et. al., 2002, Torsvic et al., 2001, Li et. al., 2003).

Interestingly Forster (1987) compared the Infra-Cambrian and Cambrian rhyolites and the associated granites of Central Iran with the Malani rhyolites and the Pan-African granties of Arabian shield. The similarity has been attributed to the Gondwana connection based on paleomagnetic data. Central Iran held a position south of equator during Infra Cambrian and later drifted from the southern margin of the Gondwanaland to its present position as a microplate. According to Kroner et. al., (1990), the period between 850-750 Ma represents the most significant crust formation event in all parts of ANS including Somalia in the form of felsic intraplate magmatism. Rogers (1990) has also suggested that both Indian shield and Arabian-Nubian shields are similar in their development including production of alkali granites, subsidence of thick partly deformed basin on recently formed crust and ultimate development of plateform cover sediments.

Paleomagmatism and Sturtian glaciation :

The ubiquity of alkali magmatism in all the above Middle East countries, together with their comparable paleomagnetic data, and the common occurrence of carbonates, phosphorites and evaporites all strongly support an amalgamated landmass the Malani supercontinent (Fig. 4). The carbonate evaporite facies of the Hauqf Supergroup form a belt of evaporite basin and intervening carbonate plateform which can be traced in these microcontinents: Hormuz Series of Arabian Gulf and southern Iran, Infra - Cambrian carbonate plateform in Central Iran, Ghabar Group of South Yemen, India Ad Series of Somalia, Wadi Fatima Series and Abla Formation, South Arabia Saramuj Formation of Jordan, Salt Range Formation, Pakistan, and evaporites of Nagaur-Bikaner basin and subsurface evaporite and anhydrite sequence of SW Punjab and Haryana, India (Fig. 5) (Gorgin et. al., 1982, Kochhar, 2004, 2006). The Huaqf axis formed a barrier between evaporites an carbonate platform and or anhydrite plateform. The Oman-Ural megalineamnet controlled the distribution of evaporite-carbonate sequence.

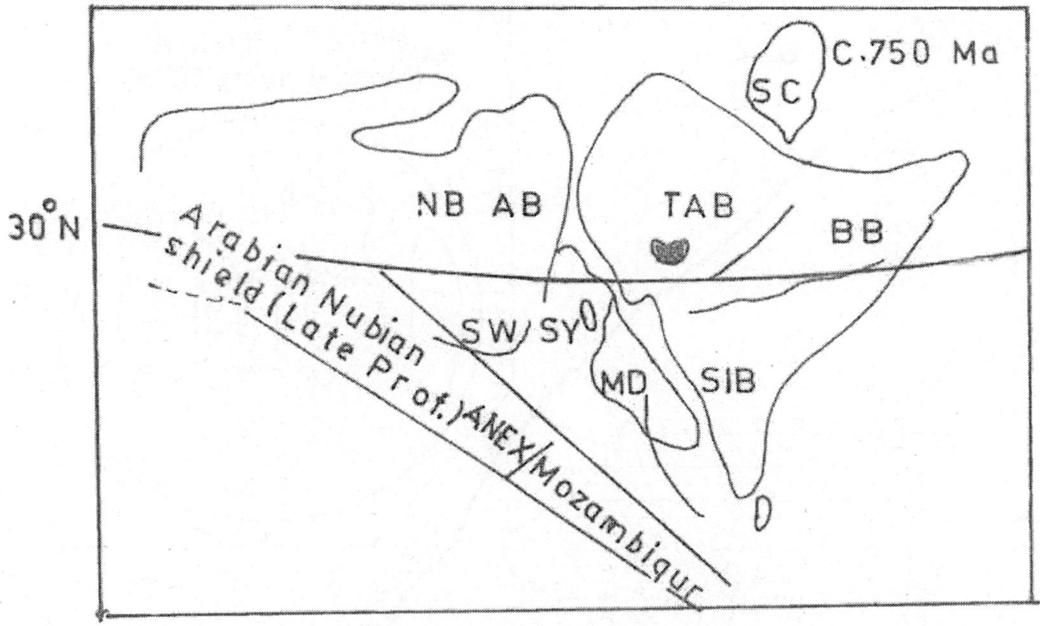

Fig. 4 Assembly of the Malani supercontinent.
NB : Nubian Arabian shield.MD : Madagascar, SM : Somalia, SC : South China, SY : Seychelles, BB : Bundelkhand block, SIB : South Indian block,TAB : Trans-Aravalli block

The APWP paleomagnetic pole at ca 800 Ma place South China at 55-70° N at par with high paleolatitudes of India during Malani rhyolite period (ca.750 Ma) (Li et al., 2004 Torsvic et al., 2001). The impact of the Precambrian glaciation which had covered vast areas of the globe including India, located at that time in the Mid to Low latitude equatorial zone, is evident in the occurrence of glacial boulders the Pokhran glacial boulder beds (Chauhan et al., 2001, Pareek, 1984, Paliwal and Rathore, 2000). The boulders of Malani granites also occur in the Salt Range. In the Yangtze craton, Linntua and Nantua deposits are also of glaciogenic origin. Kilner et al., (2005) have identified glacial and interglacial cyclicity within the Haqf supergroup, which has been attributed to low latitude glaciation. (Stern et al., 2004)

Around 700 Ma B.P. India, which still remained part of East Gondwana, was located in the tropical latitudes around 600 Ma (Powell et al., 1993). The combined 800-700 Ma polar wander path implies pole to equator rapid velocities of 20 cm per year for India and South China (Li et al., 2004). During this period the northern margin of India were subjected to desiccation. The carbonate mainly dolomite and phosphate deposits of Late Sinian-Late Precambrian age in Yangtze block corresponds with Hanseran evaporites in the Marwar basin of TAB, SW Punjab, and Bilara phosphorites in Rajasthan (Sinha-Roy et al., 1998). According to Pandit et al., (2001) a carbonate isotopic profile for the lower part of the Bilara carbonate rocks show marked oscillations and a broadly negative $\delta^{13}C$ as low as <-4.3% PDB in the lower unit of this Group and <-6.5% PDB in the overlying middle unit. The upper part of the profile shows a gradual positive shift. The ^{13}C shifts observed are related to the ^{12}C-dependent marine biota, whose population radically decreases during glacial period and proliferates with the onset of interglacial spells. This consequently leads to oscillations in the ^{13}C values corresponding to the changing uptake of ^{12}C by life present. Such shifts in the C-isotopes are well preserved in the marine carbonates of the period. For example, the extremely low ^{13}C values in Bilara carbonates indicate glacial related cold climatic conditions, with the positive shift in the carbon isotope values in the upper formation implies a warmer climatic conditions during the later period. The Bilara carbonate group carbon isotope profile have close correspondence with global carbon isotopic evolution curve in Haqf, Oman, Sibena platform, Mongolia and Morocco. According to Illyin (1990) the continental margin of Vendian-Early Precambrian supercontinent comprising South China, Kazakhstan, Mongolia were the sites of equatorial sedimentation in the form of shallow water

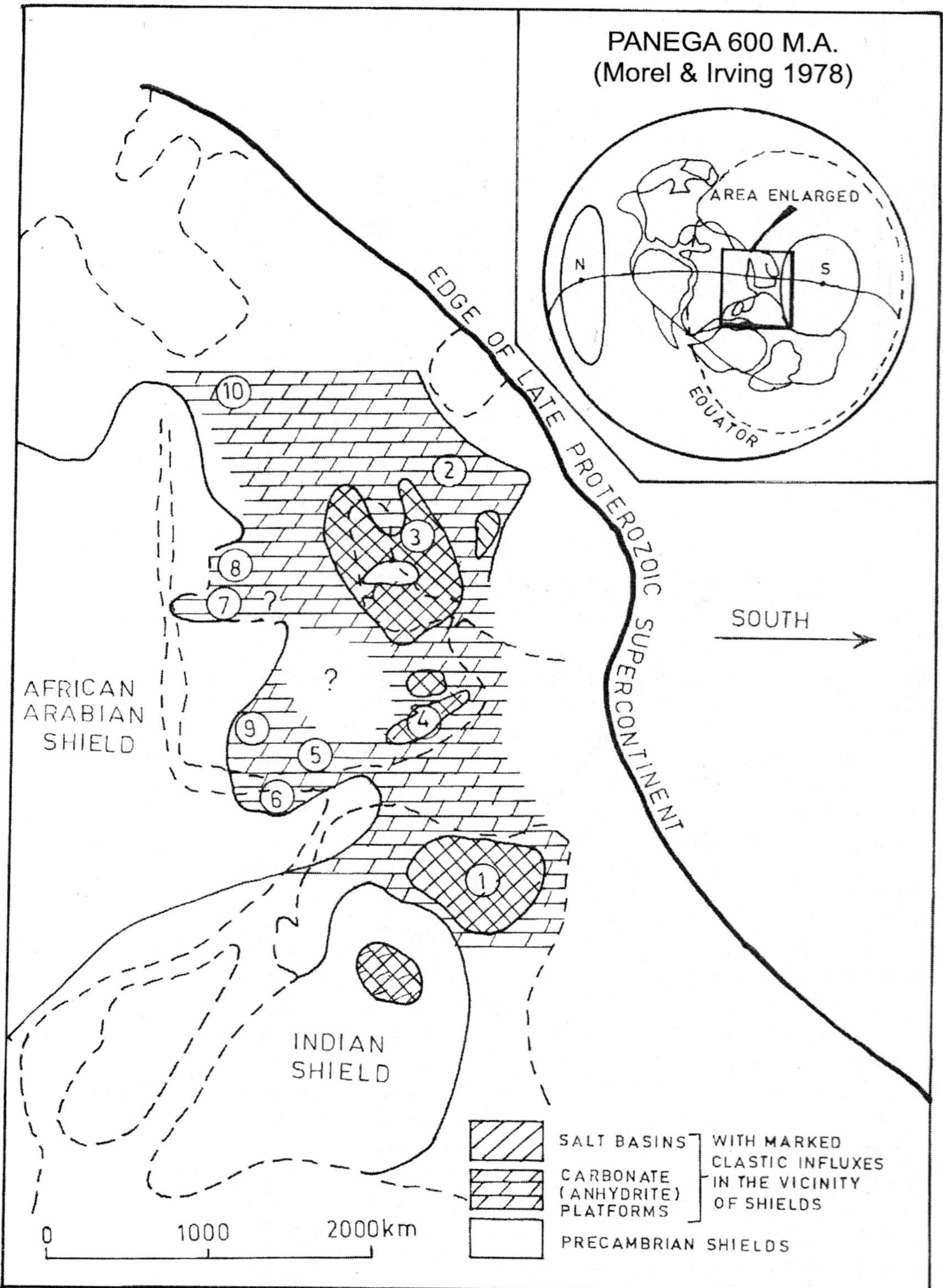

Fig. 5 Paleogeography of the Late Precambrian Early Cambrian sediments. Number refers to the locations.

carbonate and phosphorite deposits. Similar deposits also occur in the Lesser Himalaya (Banerjee and Mazumdar, 1999) (Kumar et al., 1997). Incidentally the Pokhran diamictite correlate with other

Precambrian diamictites in the Lesser Himalaya (Valdiya, 1995, Virdi, 1995, Tewari, 2003).

CONCLUSION

1. Both the terranes are characterized by the occurrence of alkali granites (peraluminous-peralkaline with A-type geochemical signatures) at 732 Ma and ca. 700 Ma respectively.
2. The emplacement of these granites and the associated acid volcanics was controlled by ring structures and cauldron subsidence. Ring complexes are the continetal manifestations of plume activity and are indicative of extensional tectonic environment.
3. Both the terranes preserve the remnants of Archean crust i.e. BGC of amphibolitic facies (3.2 Ga in the TAB (Kochhar, 2004) and trondjhemite of 2.6 Ga in the Afif terrane of ANS. This older crustal protolith has acted as precursor to the younger alkali granites.
4. In both the terranes the alkali granites mark the cratonisation of the shield.
5. Both the terranes are characterized by similar peralkaline and peraluminous granite hosted mineralisation of U, Th, Nb, Ta, REEs, Sn, W, Li and F etc. In this respect they are akin to the younger granites of Nigeria. (Kochhar,1998, 2000 and Vallinayagam, 2004, Somani 2006).
6. Both the terranes are characterized by the occurrence of evaporites-carbonate sequence and show evidence of Strutian glaciation.

The aforesaid similarities can only be explained if the Arabian-Nubian shield was attached to the Trans-Aravalli block of the Indian shield around 600-700 Ma B.P. in the configuration of the Malani Supercontinent.

ACKNOWLEDGEMENTS

I am thankful to Prof. K.L. Shrivastava for inviting me to contribute this paper to the Prof. B.S. Paliwal commemorative volume.

References:

Ashwal, L.D. Demaiffe, and Torsvic, J., 2002, Petrogenesis of Neoproterozoic granitoids and related rocks from the Seychelles, the case for an Andean-type arc origin, J. Pet., v. 43, no. 1, pp. 45-84.

Banerjee, D.M., and Mazumdar, A., 1999, On the Late Neoproterozoic- Early Cambrian transition events in part of east Gondwanaland. Gond. Res., v. 2, pp. 199-211.

Bhushan, N., 2000, Neoproterozoic magmatism of the Malani igneous suite, W. Rajasthan, India, Geol. Surv. Ind. Spl. Pub. No. 55, pp. 319-332

Chauhan, D.S., Mathur, K.M., Narayan Ram, 2001, Geological nature of the Pokaran Boulder Bed : Paleoenvironmental, paleoclimatic and stratigraphic implication Jour. Geol. Soc. Ind. v. 58, pp. 425-433.

Forster, H., 1987, Ignimbrite cauldrons, alkali granites and mineral deposits in fault block mountains. Geol. Rundsch., v. 76/2, pp. 373-388.

Gorin, G.E., Racz, L.G; and Walter, M.R, 1982, Late Precambrian-Cambrian sediments of Hugf Group. Sultanate of Oman. AAPG Bull. V. 68, pp.2609-2627.

Hoshimo, M., 1986, Amphiboles and coexisiting ferromagnesian silicates in granitic rocks in Mahe, Seychelles. Lithos, v. 9, pp. 11-21.

Ilyin, A.V., 1990, Proterozoic supercontinent, its latest Precambrian rifting, breakup, dispersal into smaller continents and subsidence of their margins: Evidence from Asia Geology, v. 18, pp. 1231-1235.

Jackson, N.J., 1986, Petrogenesis and evolution of Arabian felsic plutonic rocks, Jour. African Earth Sci., v. 4, pp.47-50.

Kennett, B.L.M. and Widiywantro, S., 1999, A low velocity wavespeed anomaly beneath northwestern India: A seismic signature of the Deccan plume? Earth Planet Sci. Lett., v. 165, No. 1, pp. 145-156.

Kilner, B., Macniocaill; and Brasier, M., 2005, Low latitude glaciation in the Neoproterozoic of Oman. Geology v. 33, pp. 413-416.

Kochhar, N., 1998, Rare metal potential of the A-type Malani Granties, NW peninsular India, Ind. Mineral, v. 52, pp. 271-276.

Kochhar, N., 2000, Attributes and significance of the A-type Malani magmatism, northwestern peninsular India. In M.Deb (Ed.) Crustal Evolution and Metallogeny in the northwestern Indian shield, Chapter 9, pp. 183-217, Narosa Publishing House, New Delhi.

Kochhar, N., 2001, Anorogenic magmatism, mantle plume and assembly of the Late Proterozoic sueprcontinent, NW Indian shield. GRG/CIGE Misc. Publ. No. 12, pp. 22-27.

Kochhar, N., 2004, Geological evolution of the Trans-Aravalli Block (TAB) of the NW Indian shield: constraints from the Malani igneous suite (MIS) and its Seychelles connection during Later Proterozoic. Geol. Surv. Ind. Spl. Pub. No. 84, pp. 247-264.

Kochhar, N.,2006, The Malani supercontinent : Middle East connection during Late Proterozoic. 6th International Conference on the Geology of Middle East. Al Ain, UAE, Abstract, p. 230.

Kochhar, N., 2007, was Yangtze craton, South China

attached to the Trans-Aravalli block of the NW Indian shield during Late Proterozoic ? Curr. Sci. vol. 92, No. 3, pp. 295-297.

Kroner, A., Eyal., M., and Eyal, Y., 1990, Early Pan-African evolution of the basement around Elat, Isreal, and the Sinai peninsula revealed by single zircon evaportion dating, and implications for crustal accretion rate. Geology, v. 18, pp. 545-548.

Kroner, A., Reischmann, T., Tojt, Y., Zimmer, M., Stern, R.J., Hussein, U., Mansour, M., Eyal, M., Eyal., Y., and Sassi, P. (1989) Timing mechanism and geochemical pattern of arc accretion in the Arabian-Nubian shield and its extension into Isreal and Somalia. Extended Abstr. 28th IGC, Washington, DC, USA, v. 2, p. 2/231.

Kumar, G., Shanker, R., Maithy, P.K., Mathur, V.K., Bhattacharya, S.K., and Jani, R.A., Terminal Proterozoic-Cambrian sequences in India: a review with special reference to Precambrian-Cambrian boundary. Paleobot., v. 46, pp. 19-33.

Li, Z.X., Evans, D.A.D., and Zhang, S., 2004. A 90° spin on Rodinia: possible casual links between the Neoproterozoic supercontinent. super plume, true polar wander and low latitude glaciation. Earth Planet Sci. Lett., v. 220, pp. 409-421.

Li, Z.X., Li, X.H., Kinny P.D., Wang, J. Zhang, S., Zhov, H., 2003, Geochronology of Neo-proterozoic syn-rift magmatism in the Yangtze craton, sourth China and correlation with other continents: evidence for a mantle super-plume that broke up Rodinia. Precamb. Res., v. 122, pp. 85-109.

Li, Z.X., Li, X.H., Kinny, P.D., and Wang, J., 1999, The breakup of Rodinia, Did it start with a mantle plume beneath south China. Earth Planet. Sci. Lett., v. 173, pp. 171-181.

Paliwal, B.S., and Rathore, P.S., 2000, Neoproterozoic volcanics and sedimentaries of Jodhpur, A reappraisal In K.C. Gyani and P.Kataria (Eds.) Tectonomagmatism, geochemistry and metamorphism of Precambrian terrain University Press, Udaipur, pp. 75-96.

Pandit, M.K., Sial, A.N., Jamrani, S.S., and Feroeira, V.P., 2001, Carbon isotopic profile across the Bilara group rocks of trans-Aravalli Marwar supergroup in Western India : Implications for Neoproterozoic Cambrian Transition Gond. Res., v. 4, pp. 387-391.

Pareek, H.S., 1984, Pre-Quaternary geology and mineral resources of north-western Rajasthan, Mem. Geol. Surv. Ind., v. 115, pp. 1-99.

Powell, C.Mc A., Li, Z.X., McElihiney, M.W., Meert, J.G. and Park, J.K., 1993, Paleomagnetic constraints on timing of the Neoproterozoic breakup of Rodinia and the Cambrian formation of Gondwana geology, v.2, pp. 889-892.

Qasim Jan, M., Lachari, A., and Asif Khan, M., 1997, Petrography of the Nagar Parkar igneous complex, Tharparkar, SE Sindh Geol. Bull. Univ. Peshwar, v. 30, 227-249.

Ramsay, C.R., 1986, Specialized felsic plutonic rocks of the Arabian shield and their precursors. Jour. African Earth Sci. v.4, pp. 153-168.

Rogers, J.J.W., 1990, Comparison of India and Nubian shield. In S.M.Naqvi (Ed.) Precambrian continental crust and its economic resources. Development in Precambrian Geology, 6, Elsevier, New York.

Roobal, M.J. and White, D.L., 1986, Caldera subsidence and Caldera above Arabian felsic plutons : a preliminary survey. Jour. African Earth Sci. v. 4, pp. 123-134.

Sinha-Roy, S., Malhotra, G., and Mohanty, M., 1998, Geology of Rajasthan, Geol. Soc. Ind. Publ. Bangalore. 278p.

Somani, R.L., 2006, Evolution of the tin-tungston and copper-bearing hydrothermal solutions associated with Tusham igneous compelx, Bhiwani, Haryana: a fluid inclusion study. Jour. Geol. Soc. Ind., v.67, pp.379-386.

Stern, R., Avigad, D. Beyth, M., Gebresilassie, S., Kuster, D., Mehare, K., and Miller, N., 2004, The Neoproterozoic Snowball earth hypothesis and the Arabian Nubian shield, 32nd IGC, Florence, Italy, Scientific sessions : Abstract (part2), p. 1191.

Stoesser, D., Whitehouse, M., and Stacey, J., 2004, Neoproterozoic evolution of the Khida terrane: Saudi Arabia : A detached microplate of the Arabian Shield, 32nd IGC, Florence, Italy, Scientific sessions : Abstract (part2), p1190.

Stoesser, D.B. 1986, Distribution and tectonic setting of plutonic rocks of the Arabian shield. Jour. African Earth Sci. v. 4, pp. 21-46.

Stoesser, D.B. and Elliott, J.E., 1985, Reconnaissance geology of the Al-Qasr Quadringle, sheet 27/41C Kingdom of Saudi Arabia, Open file Report USGS-OF-05/2, Scale 1:100,000

Tewari, V.C., 2003, Sedimentology, paleobiology and stable isotope chemostratigraphy of the terminal Neoproterozoic Buxa dolomite, Arunchal Pradesh, NE Lesser Himalaya, Him. Geol. v. 24, pp. 1-18.

Thomson, P.R., and Kattan, F.N., 2004, Geochemistry and structural reappraisal of the Asir Terrane assembly, S. Arabia Shield, 32nd IGC, Florence, Italy, Scientific sessions : Abstract (part2), p1190.

Torsvic, T.H., Carter, I.M., Ashwal, L.D. and Bhushan, S.K. Pandit, M.K. and L.D. Jamtveit, B., 2001. Rodinia refined or obscured : paleomagnetism of the Malani igneous suite, N.W. India Precamb. Res., v. 108, pp. 319-333.

Vail, J.R. and Kuron, J.L., 1978, High level igneous emplacement in the Red Sea Hills, Sudan Geol. Runds. v.

67, pp. 521-530

Valdiya, K.S., 1995, Proterozoic sedimentation and Pan-African geodynamic development in the Himalaya. Precamb. Res. v. 76, pp. 35-55.

Vallinayagam, G., 2004, A report on rare-metal and rare cartus in the Siwana ring complex, Rajasthan, Jour. Applied Geochem. v.6, pp. 387-391.

Virdi, N.S., 1999, Co-existing Late Proterozoic glacigene sediments and evaporites in the Lesser Himalaya and eastern Indian shield expression of contemporaneity of low latitude glaciation and tropical dessication. In B.S. Paliwal (Ed.) The Indian Precambrian, Scientific Publishers, Jodhpur, pp. 502-511.

Whitehouse, M., Flowerdew, M., Windley, B., Al-Khirbash, S., and Ba-Bttat, M., 2004, The Al Bayda arc, Yemen : A 200 Ma record of Neoproterozoic accretion at the eastern margin of the Arabina-Nubian shield. 32nd IGC, Florence, Italy, Scientific sessions : Abstract (part2), p1190.

Yoshida, M., Rajesh, H.M., and Santosh, M., 1999, Juxtaposition of India and Madagascar : a perspective, Gond. Res. v. 2, pp. 449-467.

Hon'ble Reviewers

1. Prof. S.C. Sarkar
   (Former Head)
   Emeritus Scientist
   Department of Geological Sciences
   Jadavpur University
   KOLKATA 700 032 INDIA

2. Dr. A.V. Sankaran
   10, P & T Colony
   I Cross, II Block
   R.T. Nagar
   BANGALORE 560 032 INDIA

Editorial Assistance

Dr. Virendra Gaur

Economic Mineralization
K.L. Shrivastava (Editor)
2009. pp. 26-34
Scientific Publishers (India), Jodhpur

Section-I

Precambrian Crustal Evolution and Metallogeny

A. B. ROY

Emeritus Professor, Department of Geology, Presidency College, Kolkata-700073
Email : abroy_g@yahoo.com

Abstract

The Evolution of the earth's most primitive crust followed an initial accretionary phase when the planet differentiated into a Fe-Ni rich inner core surrounded by an outer silicate Mantle. There is hardly any record to prove that the crust formed simultaneously with the core and mantle. There is also no evidence to conclude that the earth's early crust was as old as that of the Moon. Discovery of ca. 4200-4300 Ma old detrital zircons in the early Precambrian quartzites in western Australia along with finding of records of ca. 4000 Ma old Priscoan gneisses from the Slave Province of Canada can be interpreted as an evidence of very old sialic crust, which in all probability evolved at a number of isolated centres as "islands" surrounded by mafic-ultramafic primitive oceanic crust. The enlargement of these sialic crustal areas was attained through the addition of granitic materials, which formed due to the partial melting in the upper mantle. Extensive granite formations during the end-Archaean phase resulted in the thickening of the continental crust and its stabilisation, a process generally known as cratonisation.

The crust of the earth, which was considerably thickened by the end of Archaean became fragmented opening a number of intracratonic rift-basins or "protogeosynclines" comparable to those of the Palaeoproterozoic Aravalli basins. The Archaean-Proterozoic transition is also marked by a number of secular changes in terms of chemistry of atmosphere and hydrosphere, and the biotic evolution initiating prolific cyanobacterial activities. The earth also witnessed spectacular transformation in regard to the lithology and tectonic pattern. The onset of Mesoproterozoic is marked by a gradual transition in the tectonic processes from the early dominantly plume related tectonism to ' intra-plate-tectonism' with limited ocean opening. In the Neoproterozoic, the earth witnessed transition in the process of supercontinetal reconstitution through the break-up of Rodinia, and the subsequent re-assembly of different fragments to constitute the Pangea. The earth's Precambrian crustal history, therefore, is marked by records of unidirectional changes from very early-formed continental nucleus (appearing as small, isolated islands) to one or more frameworks of supercontinents. The unidirectional crustal trends also find agreeable reflection in the Precambrian metallogeny.

Keywords : Crustal Evolution, Precambrian Metallogeny, Tectonic Recycling, Cratonisation.

INTRODUCTION

The Precambrian as defined by the IUGS Subcommission of Stratigraphy (Plumb and James, 1986) constitutes about 8/9th of the Earth's total history, beginning almost from the origin of the planet to the onset of the Phanerozoic at 543 million years ago. However, there is hardly any preserved records of the very early part of the Earth's history since its inception. This period of unknown history is known as the Hadaean (Cloud, 1976). Initially, the Hadaean used to be described as the 'missing billion', implying a billion-year hiatus of any record of the earth's crust from 4500 Ma (when the possible first crust of the earth was formed) to about 3500 Ma (the oldest dated Crustal record known at that time). With the recent information about the presence of crustal materials as old as 4300 Ma, the Hadaean phase is now reduced only to about 200-300 Ma. The subdivisions of the Precambrian Aeon recognised today is shown in Table1.

The studies made during the last three and half decades witnessed remarkable progress in our understanding of the nature and evolution of the Precambrian crust due primarily to the interpretation of data from interdisciplinary subjects like isotope geology, geochemistry, geophysics, planetary as well as the other branches of geology. It is now known that the earth differentiated into a core and a mantle soon after its origin. There is no available inform to confirm that the earth's crust had also formed simultaneously with the core and the mantle. Even if we assume that the crust did form at the beginning, it must have been very thin and in all probability did not 'reside' on the surface for a long period. The formation and accumulation of the crustal materials was accomplished by partial melting of mantle through fractionation-differentiation process. Therefore, crust is a secondary feature on the earth's surface very much like the atmosphere and the hydrosphere.

It is now well known that the development of the earth's crust is linked with the internal heat source through the processes of heat transfer from the mantle (and perhaps core). The principal source of the earth's internal heat is due primarily due to the radiogenic decay with possible primordial supplement. The bulk earth heat production is diminishing with time. This is reflected not only in the crustal processes but also in

Table 1: Classification of the Precambrian Aeon

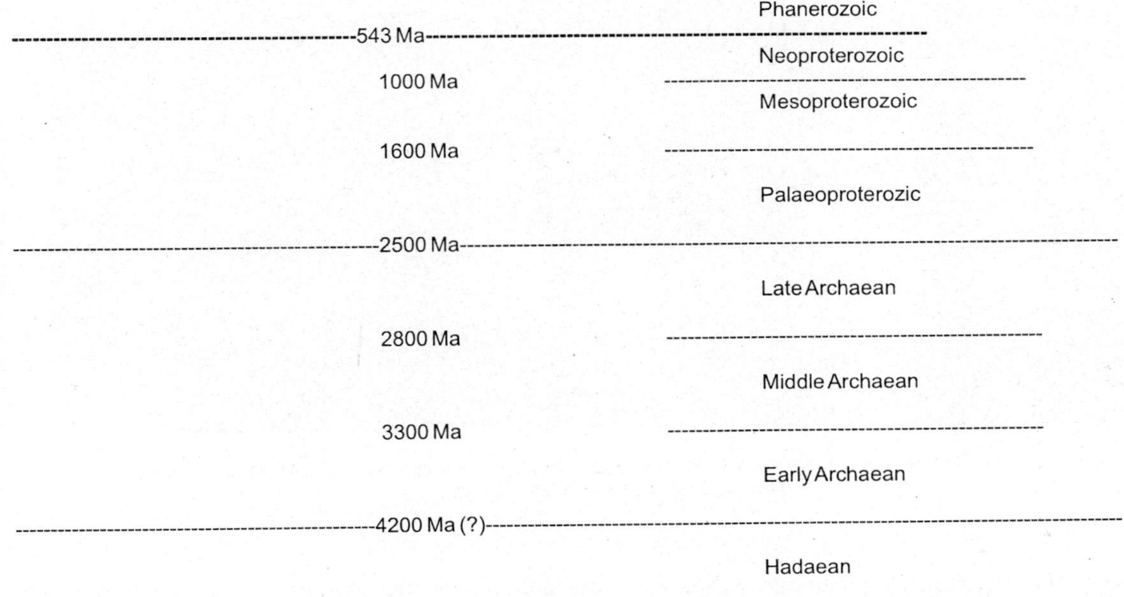

Age	Division
	Phanerozoic
543 Ma	
	Neoproterozoic
1000 Ma	
	Mesoproterozoic
1600 Ma	
	Palaeoproterozic
2500 Ma	
	Late Archaean
2800 Ma	
	Middle Archaean
3300 Ma	
	Early Archaean
4200 Ma (?)	
	Hadaean (Unknown phase of Crustal history)

the formation of the economically viable mineral deposits.

Nature and Origin of the Early Archaean Crust

The studies of lunar samples indicated that the moon's crust had formed as early as 4500 Ma. In the case of the earth, the discovery of ca. 4200-4300 Ma old zircon grains in quartzite of Jack Hills and Mount Narryer in western Australia provides an indirect proof of development of the earth's earliest crust. The only definitive record of the earth's earliest crust comes from the ca. 4000 Ma age of the Priscoan gneisses in the Slave province of Canada (Bowring et al. 1989). Thus till date the available information suggest that the oldest preserved crust on the earth is about 200 million years younger to that of the moon.

The oldest Archaean gneisses that occur in different parts of the world are typically grey gneisses showing well-developed banded character. Mainly because of the ubiquitous banded character, these ancient gneisses are described as the banded gneisses or the banded gneissic complex (Fig. 1). A vast majority of the early Archaean banded gneisses petrochemically belong to the category of tonalite-trondhjemite-granodiorite (TTG) granitoids. Isotope datings from the different Shield areas in the world indicate the period of formation of these TTG type banded gneisses between 4000 and 3000 Ma.

The first appearance of the 'TTG granitoids' marks a transition from the early simatic crust represented by mafic rocks like the hydrated basalts to a sialic Crust. Moorbath and Taylor (1981) explained the formation of the early sialic crust as a process of 'Accretion-Differentiation Superevents'. It is known that there was a massive generation of TTG granitoids during the Archaean. The most popular petrogenetic model for the evolution of the Archaean TTG granitoids involved fractional cystallisation from a melt having the composition of hydrated basalt. Such an explanation for the evolution of the Archaean crust might be considered a 'quasi-uniformitarian' idea that relied heavily on the modern type plate tectonics. However, the principal limitation in applying the

Fig. 1 A typical banded 3300 million years old TTG gneiss forming a component of the Archaean basement, Mewar Gneiss Complex, Rajasthan.

Phanerozoic type model is that there is an overwhelming dominance of trondhjemite-tonalite in the Archaean crust compared to the much subdued occurrence of this association in the younger oceanic-continental margin batholiths. This fact coupled with almost total lack of andesite in the in the Archaean continental crust favours adoption of models which are decidedly non-uniformatarian (Roy, 2001).

Irrespective of processes that generated the earliest sialic crust, it is almost certain that the accretion of the crust took place at isolated centres as sialic 'islands' surrounded by mafic-ultramafic materials. These initial islands were subsequently enlarged due to several accretionary processes.

Middle and Late Archaean Crustal growth history

By Middle Archaean (3500-3000 Ma) the earth's continental crust must have attained considerable thickness with predominance of orthogneisses of tonalite composition. Such Archaean terranes in different parts of the world are divided into high-grade gneissic and the greenstone belts. The high-grade terranes of which about 25% comprise granulite facies rocks, cover extensive areas in different shield areas of the world. The exposures of granulite complexes require exhumation of these bodies from deeper crustal levels to their present site of occurrence. The process actually involves upliftment of deep crustal blocks between 17 and 31 km. It is because of this the crustal sections containing granulites are considered the geodynamically most active component of the crust prior to their cratonisation. All the high-grade terranes expose three different types of rock associations:

1. Gneiss-migmatite-granulite complex;
2. Amphibolite facies assemblage; and
3. Layered intrusions of mafic and ultramafic rocks.

The Archaean high-grade terranes are characterised by extremely complex structural patterns, with the development of thrusts, repeated isoclinal folding and very intricate fold interference patterns.

The greenstone belts became the most characteristic feature during the last billion year of the Earth's Archaean history. Although there are reports of greenstone belts as old as 3700 Ma (contemporary to the Maria basalt magmatism on the Moon), or as young as 2100 Ma, the evolution of the majority greenstone belts are episodic evolving around 3000-2900 Ma, ~2800 Ma and 2600-2500 Ma (Fig. 2). In spite of extensive studies in different parts of the world, certain aspects of the greenstone belts continue to be enigmatic. Animated discussions are still continuing on the question of their basement and evolution. The importance of the Archaean greenstone belts lies in their association as the host of several vital mineral deposits in different parts of the world.

The end-Archaean scenario in the crustal evolution is marked by extensive granitic activities so much so that it is considered a major benchmark event in the Crustal history. The extensive granitoid magmatism during the last phase of Archaean caused thickening and considerable enlargement of crustal area, mainly through welding of smaller, isolated and disparate 'continental blocks' called the protocontinents. Talking about the Indian Shield, it has been suggested that the accretion of the Archaean Indian craton was through the process of amalgamation of a number of protocontinents (Roy, 2001). The accretion is by extensive granitic activites in different protocontinents at around 2500 Ma, which also marks the stabilisation of the early Crust (Wiedenbeck et al. 1996). An indirect evidence about the extensive development of the Archaean crust by 2500 Ma comes from the fact that a large majority of the Proterozoic fold belts all over the world have evolved on ensialic basements. The ensialic evolution of the Palaeoproterozoic Aravalli 'geosynclines' in the northwestern Indian Shield constitutes a classic example of such types of basin formation.

Proterozoic Crustal Evolution

Very much like the Hadaean-Archaean transition, the Archaean-Proterozoic boundary (APB) constitutes a major benchmark event in the Crustal evolutionary process. Constrained at 2500 Ma in the Standard Stratigraphic Scale, the APB is also characterised by a number of secular changes in terms of biotic evolution, transform of atmospheric character, composition of ocean water, lithological character of rocks and metallogeny. The major controlling factor for all these changes is the changing scenario in the crustal evolutionary process, from a more 'mobile' state to the stage of considerable stability.

The crust became cool and rigid prior to the onset of the Proterozoic with the resultant change in the geothermal gradient from 54°C/km in Archaean to 47°C/km in Proterozoic (Grambling, 1981). Such an abrupt lowering of geothermal gradient found expression in the composition of the mafic-ultramafic effusives. This is reflected in the dominance of Mg-

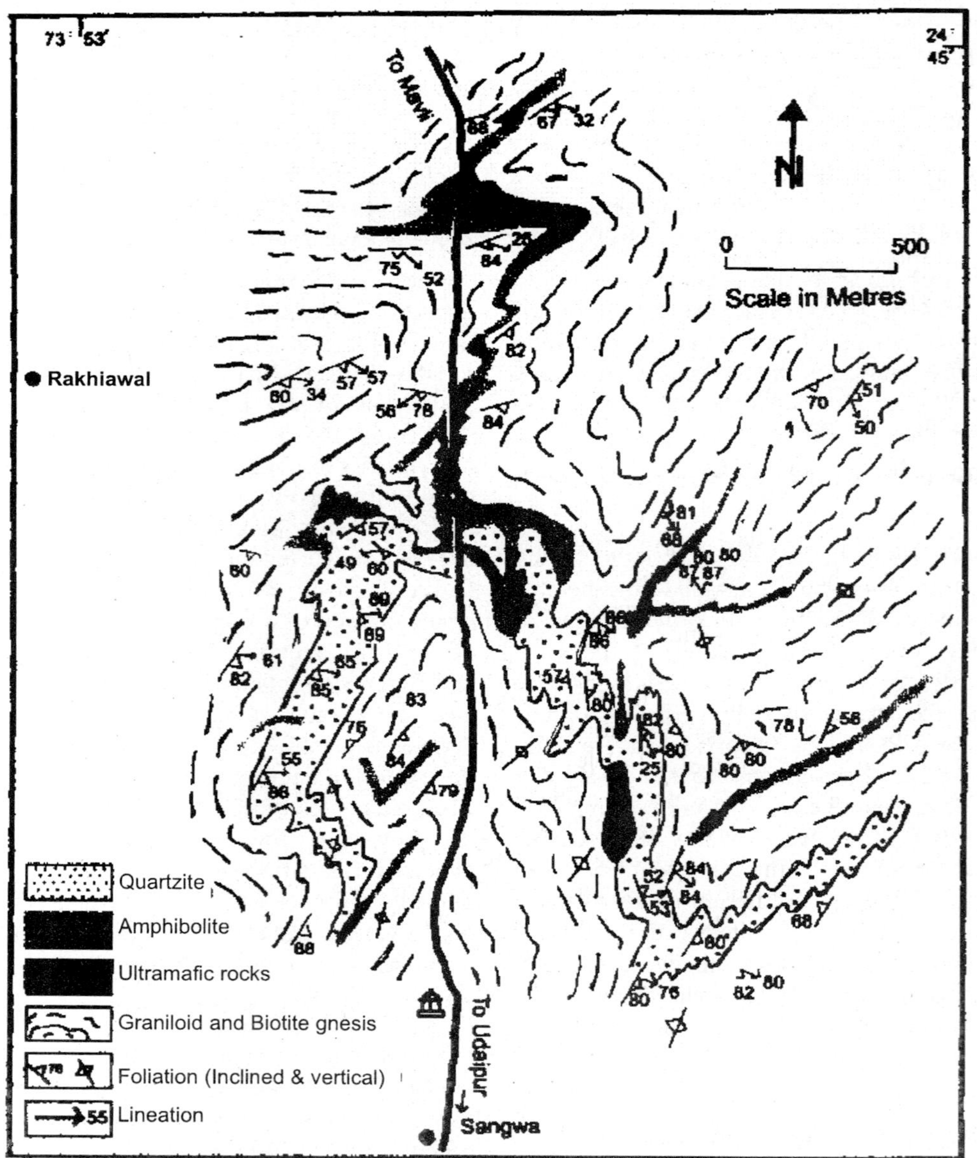

Fig. 2 Geological map of the Middle Archaean Greenstone belt from Rakhiawal, east of Udaipur, Rajasthan.

rich komatiitic lavas in Archaean to Fe-rich tholeiite basalts in the later period. The crustal stability at the end of Archaean not only changed the tectonic complexion at a later stage but also resulted in the explosion of cyanobacterial activities in the shallow shelf regions. The extensive growth of stromatolitic edifices (Fig. 3) in the Palaeoproterozoic carbonate as seen in the Aravalli Supergroup around Udaipur in Rajasthan (Roy and Paliwal, 1981) is the case in point. The biological explosion greatly enhanced the photosynthetically produced O_2 (Goodwin, 1981) in the atmosphere, which very significantly influenced the chemical weathering and deposition of siderophile elements, mainly iron.

The emplacement of dyke swarms (Great dyke of Zimbawe, for example) marks the first attempt to disrupt the cooled and the thickened crust. It is also indication of the prevailing extensional regime, which presumably resulted due to an increase in volume of the Earth as a result of the phase transformation in the mantle simulated by the accumulation of heat under the thermally insulated layer of the continental crust.

Fig. 3 Stromatolitic phosphorite occurring in the basal part of the Palaeoproterozoic Aravalli Supergroup at Jhamarkotra, southeast of Udaipur.

The extensional regime ultimately helped in opening up rift basins as 'protogeosynclines'. At many places, as in the case of the Palaeoproterozoic Aravalli basins, the rifting took place at 'triple junction' with one or two failed arms (Fig. 4A & B; Roy, 1991).

The Palaeoproterzoic crustal evolution has been viewed as a combination of intra-continent micro-oceans inside the partially disrupted supercontinent, and restricting the process of recycling of the oceanic lithosphere only to the domain of 'oceanic hemisphere'.

The onset of Mesoproterozoic is marked by the transitional stage when the prevailing supercontinental margins remained virtually passive, except in the southwestern Scandinavia and some parts of the modern China. The stage was set for the intraplate tectonics, which replaced the earlier process of 'plume-dominated tectonics'. The manifestation of intraplate tectonics was in the form of development of palaeo-rifts, aulacogens and intracratonic geosynclines, with varying degrees of extension and destruction of the continental crust. The closing of these basins resulted in the development of fold-thrust belts. Ophiolites obduction took place where new ocean crust developed during the stage of basin opening. Tectonically four different types of basins formed during this period:

1. Platformal basins;
2. 'Epi-cratonic' volcano-plutonic associations;
3. Aulacogens; and
4. Intracratonic geosynclines.

Many platformal basins were joined with aulacogens and geosynclines. In northwestern part of the India Shield, the Delhi basins forming under the different tectonic settings constitute a good example of such type of basin evolution.

An interesting development during the late Mesoproterozoic is the intrusions of anorthosite plutons in different Shield areas of the world. Most of the large anorthosite plutons are spatially linked with high-grade granulite terranes, which have undergone a phase of exhumation at ca. 1000 Ma. This tectono-thermal process is usually linked with the Grenville Orogenic cycle. The extension of this orogenic belt has now been traced beyond North America into East Antarctica and even to the Eastern Ghats of India as a reworked Archaean crust with the reset isotope ages. If the presently disparate belts are joined in a supercontinent framework, then the Grenville belt would turn out to be largest mountain belt in the reconstructed palaeo-world.

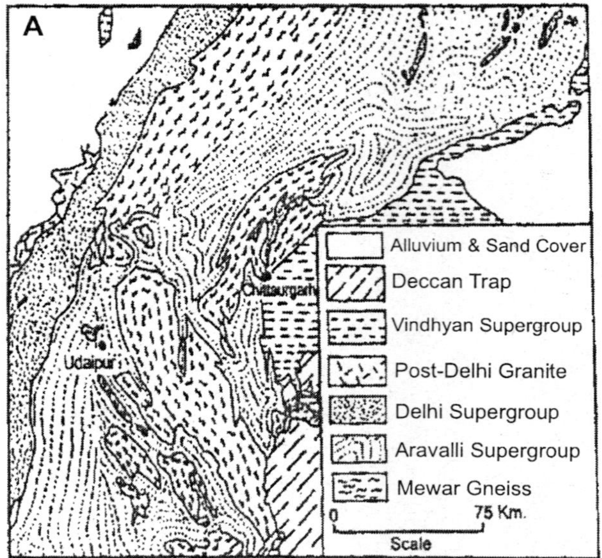

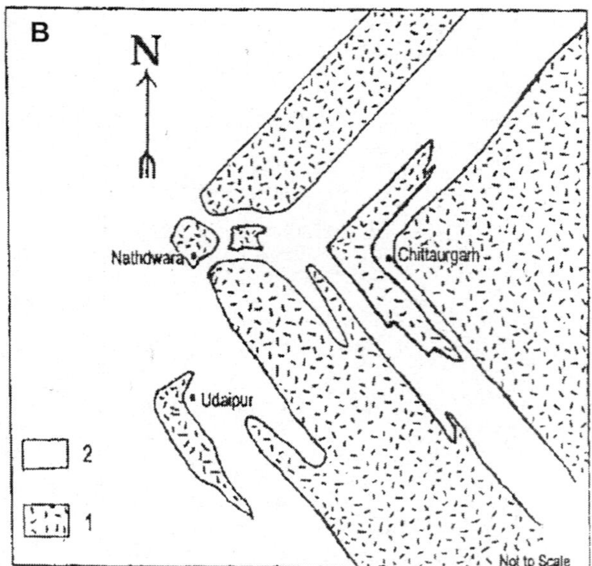

Fig. 4 A.Geological map showing the distribution of the Aravalli Supergroup in relation to the outcrops of the Archaean basement.(after Roy, 1991)
B. The model of evolution of the Aravalli basins along 'FRR' rifts.(after Roy, 1991) Legend: 1.: Landmasses of the basement rocks. 2.: Aravalli Basins.

The Neoproterozoic, which spanned between 1000 and 543 Ma represents a very important stage in the Crustal evolutionary history signalling the emergence of the modern plate tectonics. It is now believed that the conditions pre-Neoproterozoic phase did not allow operation of plate tectonics in full swing. Apart from drastic modifications in the global tectonic pattern, the Earth's atmosphere-hydrosphere systems also underwent significant changes in chemistry with the onset of the Neoproterozoic. The most significant global tectonic event in the Neoproterozoic is the break up of pre-Pangean supercontinent, the Rodinia, which resulted in a series of basin openings in intracratonic and ensiamtic settings. All these pointed towards intensification of rifting processes. A significant development during the terminal Neoproterozoic is the Pan-African tectono-thermal events with or without any new basin opening. In India, this tectonic cycle resulted in the exhumation of granulites in the Eastern Ghats and the Southern Granulite belts and also in the intrusion of predominantly alkaline granitoids all along the margins of the then 'Greater Indian Shield' areas (Roy, 1999; 2002).

Metallogenic trends in Precambrian

Very few mineral deposits are known in the Archaean gneisses and in the high-grade granulite terrains. The metamorphic/tectonic processes that operated under the prevailing conditions presumably destroyed the most of the economically viable minerals. By contrast, the Archaean greenstone belts and the associated granite terrains are found to host rich mineral deposits. The dominance of siderophile elements in the greenstone associations such as those of chromium, nickel, and asbestos deposits, are due to the prevalence of mafic and ultramafic rocks in these belts. Gold is another important mineral, which is associated with middle Archaean greenstone belts. The Kolar gold deposit in the Dharwar Craton constitutes one such example. The important iron occurrences world over are associated with banded iron formation (BIF), which formed an important component of the late Archaean greenstone belts. The oldest BIF occurs at the Isua Complex in Greenland.

The Archaean atmosphere was in all probability anoxic in character. Because of that the particles of unoxidised uranium mineral could deposit as placer minerals in a number of late Archaean to early Proterozoic epicontinental basins. The quartz pebble conglomerate, which is a typical end-Archaean formation, commonly shows the presence of uraninite. A close associate of placer uranium is the placer gold. These deposits are synchronous with or succeeded by extensive BIF accumulations, some of which formed the largest iron and manganese deposits of the world. Several Layered complexes, about 2.6

and 1.9 billion years old, respectively, contain the major nickel-copper and chromium and platinum resources in the world. This event marks a major shift in the nature of mineral deposits from siderophile to predominantly chalcophile metallogenic resources.

Prolific growth of cyanobacteria in the early Palaeoproterozoic not only helped to increase the oxygen content in the atmosphere through the process of photosynthesis, but also assisted in development of several rich deposits of stromatolitic phosphorite. Perhaps the best example of stromatolitic phosphorite deposit is that which occurs in the Aravalli succession (Aravalli Supergroup). The shale hosted lead-zinc (-copper) deposits are of major importance in the Palaeo- to Mesoproterozoic basins. One such important deposit occurs at the Rampura-Agucha in Rajasthan. There are also several occurrences of unconformity type uranium deposits in the interior basins of younger Proterozoic age. Certain Meso-/ Neoproterozoic stratabound copper deposits show unusual metal accumulations in stable basins dominanted by internal drainage systems. Large sediment-hosted manganese deposits occur in rocks of similar ages. Titaniferous magnetite deposits occur in some of the anorthosite massifs.

Summering, several distinctive metallogenic associations occur in the Precambrian rocks, which reflect the progressive growth and stabilisation of the continental crust. This indicates a change from the early mantle-dominated to later mantle-continental mixed activities and to the unidirectional change from prevailing anaerobic to aerobic ocean-atmosphere system.

Assuming that the plate tectonic processes were quite effective during the Precambrian, several important mineral deposits can be linked directly with the model as illustrated in the figure (Fig. 5). The Mid Oceanic Ridges, especially the central axial zones are the sites of hydrothermal vents, which locally appear as 'chimneys' discharging sulphides of iron, copper, and gold. Sulphides of elements like Mn, Li, Ca, Ba, and Si are not very uncommon associates of these vent-type deposits. The ancient oceanic crust, which occurs as ophiolite shows development of two major mineral deposits: Cyprustype Cu-Fe massive sulphide in the pillowed basalts, and podiform chromite especially in zones of layered gabbro. The Island Arc-Trench systems are characterised by the occurrences of hydrothermal vents in the volcanic rocks and in the plutonic granitic bodies. The common mineral deposits in the volcanic ensembles includes a major part of known base metals like Pb, Zn, Cu and Mo, precious metals like Au and Ag, and heavy metals like Sn, W, Sb and Hg. The granite plutons show important porphyry type deposits of Cu, Mo and Sn. Red bed association of V, U and Cu are reported from some foreland basins. On the other hand Kuroko-type massive sulphides of Cu, Pb and Zn form important deposits in the Back-Arc basins. In collision orogens, concentration of different minerals takes place through tectonic recycling aided by percolating fluids. Concentration of metallic minerals by this process of recycling turns them into economically viable deposits.

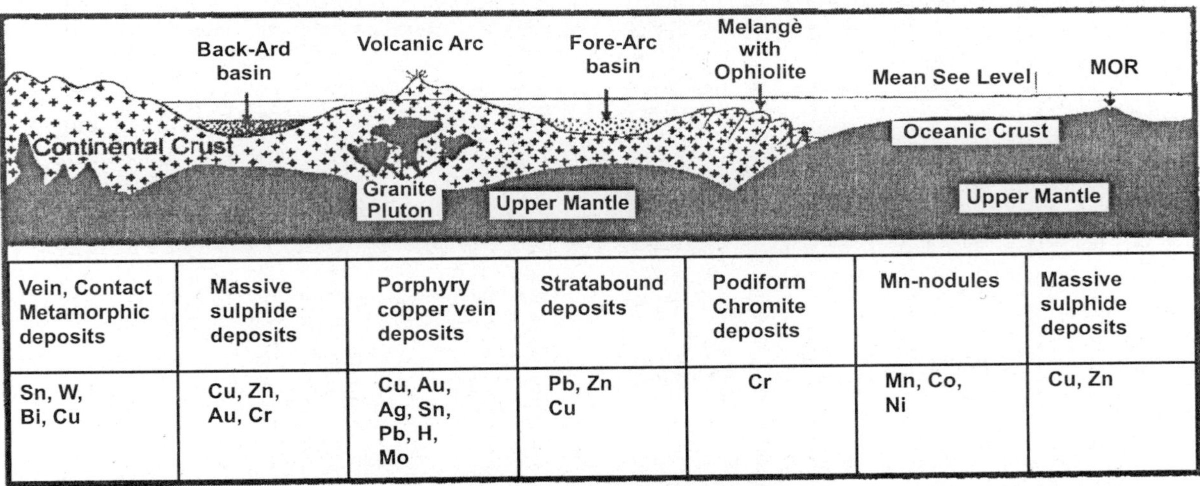

Vein, Contact Metamorphic deposits	Massive sulphide deposits	Porphyry copper vein deposits	Stratabound deposits	Podiform Chromite deposits	Mn-nodules	Massive sulphide deposits
Sn, W, Bi, Cu	Cu, Zn, Au, Cr	Cu, Au, Ag, Sn, Pb, H, Mo	Pb, Zn Cu	Cr	Mn, Co, Ni	Cu, Zn

Fig. 5 Model illustrating different types of mineral deposits under different Plate tectonic settings

ACKNOWLEDGEMENT

It gives me immense personal pleasure to be able to contribute this paper for the festschrift Volume honouring Dr. Bhawani Shanker Paliwal, who had been my first student; and at a later stage became my scientific collaborator. I fondly remember the days when we worked together looking for features that finally helped us in understanding the different aspects of the geology of the Aravalli Mountains.

References:

Bowring, S. A., Williams, I. S. and Compston, W. 1989. 3.96 Ga gneisses from the Slave Province , Northwest territories, Canada. Geology, 17, 971-975.

Cloud, P. 1976. Major features of crustal evolution. Geological Society of South Africa. Annex, 79, 1-33.

Goodwin, A. M. 1981. Archaean plate tectonics and greenstone belts.p. 105-135. In: A. Kröner (ed.) Precambrian Plate Tectonic. Elsevier, Amsterdam.

Grambling, J. 1981. Pressures and temperatures in Precambrian metamorphic rocks. Earth Planet Sci Letters, 53, 63-68.

Moorbath, S. and Taylor, P. N. 1981. Isotope evidence for continental growth in the Precambria. p. 491-526. In: A. Kröner (ed.) Precambrian Plate Tectonic. Elsevier, Amsterdam.

Plumb, K. L. and James, H. L. 1986. Dubdivision of Precambrian time: recommendation and suggestion by the Subcommision of Precambrian Stratigraphy. Precamb. Res. 32, 65-92.

Roy, A. B. 1991. Evolution of the early Proterozoic Aravalli depositional basin. In: S. K. Tandon, Charu Pant and S. M. Casshyap (eds). Sedimentary basins of India: Tectonic Context. Gyanodaya Prakashan, Nainital, 1-12.

Roy, A. B. 1999. Neoproteroozic Crustal evolution of Northwestern Indian Shield: Implications on Break up and Assembly of Superecontinents. Gondwana Res. 289-306.

Roy, A. B. 2001. Precambrian Crustal Growth: from the Earliest transient Crust to the Assembly and Break up of Supercontinents. Proc. Indian National Science Academy (Physical Sciences), 67, 439-460.

Roy, A. B. 2004. The Phanerozoic reconstruction of the Indian Shield as the aftermath of Break up of the Gondwanaland. Gondwana Res. 7, 387-406.

Roy, A. B. and Paliwal, B. S. 1981. Evolution of Proterozoic epicontinental deposits: stromatolite bearing Aravalli rocks of Udaipur, Rajasthan, India. Precambrian Res. 14,49-74.

Wiedenbeck, M., Goswami, J. N. and Roy, A. B. 1996. Stabilisation of the Aravalli Craton of northwestern India at 2.5 Ga: An ion-microprobe zircon study. Chem. Geol. 129, 325-340.

Hon'ble Reviewers

1. Padam Shree
   Dr. B.P. Radhakrishnan
   President Geological Society of India
   63, 12th Cross Basappa Layout
   Gavipuram Extension
   BANGALORE - 560 019 INDIA

2. Prof. Mike Katz,
   Key Centre for Mines
   University of New South Wales
   Sydney NSW 205
   AUSTRALIA

Editorial Assistance

Mr. Chandra Prakash Dadhich

Economic Mineralization
K.L. Shrivastava (Editor)
2009. pp. 35-46
Scientific Publishers (India), Jodhpur

4

Section-I

Precambrian Crustal Development and Metallogenic Evolution of Bastar Craton, Central India

K. L. RAI

(Formerly Prof. & Head, Department of Applied Geology, I.S.M. Dhanbad)
Prof. Emeritus, 45, Chhatrapati Shivaji Colony, Kolar Road, Bhopal - 462 016 (M.P.)
Email : rai_kishan@rediffmail.com

Abstract

The Bastar craton, constituting the northernmost peripheral part of the Dharwar or Deccan Protocontinent, just south of the Central Indian Tectonic Zone, is possibly the richest, yet inadequately studied and explored, metallogenic province of the Indian subcontinent. Existence of the rich endowments of several important metals like copper, gold, uranium, tungsten, tin, niobium, tantalum, rare-earths, etc. beside diamond and other gemstones in the Precambrian formations of different geological entities of this craton is already recognized and good scope exists for their further exploration and exploitation.

Despite numerous constraints arising from the inadequacy of requisite geological and reliable geochronological data and several controversies that remain yet unresolved, an attempt has been made in this paper to comprehend the inter-relationships among various episodes of tectonics, magmatism, sedimentation, metamorphism, etc. contributing to the crustal development, on the one hand, and metallogenesis on the other, in different geological entitites of the craton during the Precambrian period. Based upon this analysis, an integrated picture of era-wise crustal and metallogenic evolution of the craton has been visualized.

Keywords : Crustal development, Metallogenic evolution, Precambrian, Bastar craton, Central India.

GEOLOGICAL FRAMEWORK OF THE CRATON

The cratonic domain of Bastar, occupying large tracts of the newly-created state of Chhattisgarh and parts of adjoining states of Maharastra and Madhya Pradesh, is located in the northern marginal part of the Southern/ Dharwar Protocontinent (Yedekar et al., 1990) or the Deccan Province (Radhakrishna and Naqvi, 1986). It is bounded by the Central Indian Tectonic Zone (CITZ) including Satpura mobile belt in the north-west, the Mahanadi Gondwana Graben in north east, the Eastern Ghat Mobile belt in the east and south-east and the Godavari Gondwana Graben in west and south-west (Fig.1).

The Craton is a mosaic of different geological entities that include: (i) the oldest cratonic components of Bastar including Bengpal and Sukma gneiss- supracrustal-granite association of middle to

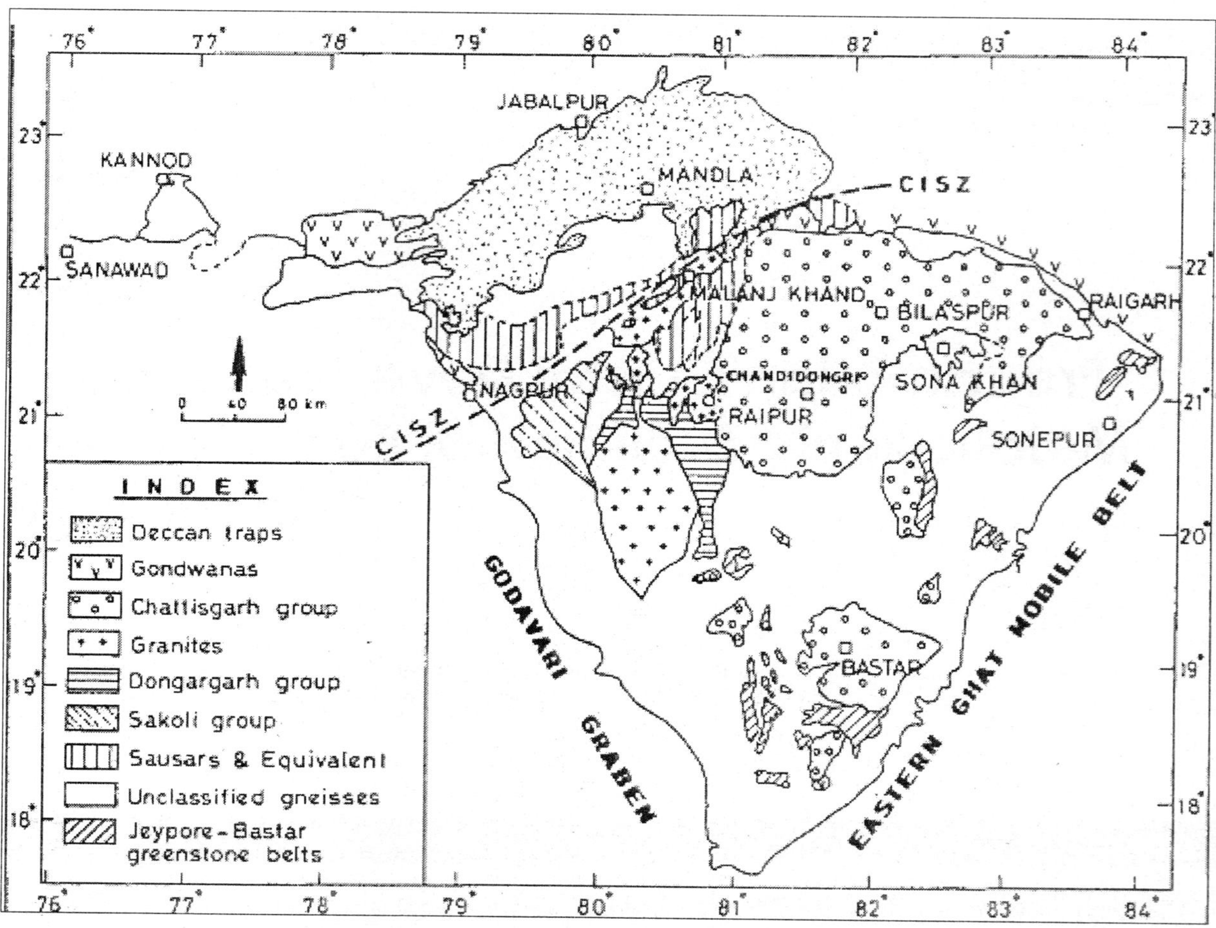

Fig. 1 Simplified Geological Map of Bastar Craton

late Archean age and (ii) the Proterozoics of Kotri-Dongargarh belt of bimodal volcanics-granite-supracrustal association; (iii) Sakoli belt with associated gneissic complexes and granitoids, and (iv) the cover sequences of the Chhattisgarh, Indravati, Albaka and Sullavai Supergroup/Group. These litho-associations span from middle Archean (3500-2900 Ma) to late Proterozoic (900-500 Ma) in age. Table-1 summarises their chronostratigraphic succession in generalised terms. It is to be noted that the tectonic and temporal relationships between different provinces and the chrono-stratigraphic successions within some of these segments have been the subjects of controversy for long and are still not fully resolved. Notwithstanding these constraints, an attempt has been made in this paper, to focus upon the broad relationships between the Precambrian crustal evolution, tectonics and metallogenic events in this, possibly the richest, yet metallogenically least studied, cratonic domain of the Indian subcontinent as follows:-

(I) MIDDLE TO LATE ARCHEAN STAGE (3.5 TO 2.6 GA)

Widespread mantle-fed volcanism in submarine or subaerial environment on a thin crust or directly on the proto-mantle, propelled by the vertical microplate tectonics, took place on global scale during this period. It was marked with high incidence of cyclically repetitive mafic-ultramfic, komatiitic-tholeiitic exhalative activity with least contaminated inputs from the upper mantle. Nickel was transferred to the crust through the Mg-rich, Al-poor Archean magmas in the form of immiscible sulphide droplets. According to Banerjee (1972), however, despite the known occurrences of favourable lithologies, workable concentrations of nickel have not been identified so far in the Indian subcontinent.

Heavy metals of Siderophile Group in Mafic-Ultramafic Complexes

In Bastar craton, sporadic occurrences of heavy metals of siderophile group e.g. Cr, Ti, V, etc.

Table-1 Generalised chronostratigraphic succession in the Bastar Craton (modified after Sarkar, 1957-58, 1994; Mishra et al., 1988; Ramakrishnan, 1990; and Ramachandra et al., 1998)

Era	Succession	
Neoproterozoic (900-500 Ma)	Chhattisgarh Supergroup	
	Indravati Group	
	---------- Unconformity ----------	
Mesoproterozoic (1600-900 Ma)	Intrusive granitoids	Dongargarh Malanjkhand
	Abhujmar Group	
	---------- Unconformity ----------	
Late Archean to Paleoproterozoic (2900-2500 Ma)	Sakoli/Kotri-Dongargarh Groups	
	Sonakhan Schist Belt	
	---------- Unconformity ----------	
Middle Archean (3500-2900 Ma)	Bailadila Group	
	---------- Tectonic contact/Unconformity ----------	
	Bengpal Group	
	---------- Tectonic contact/unconformity ----------	
	Sukma Metamorphic Suite	
	---------- Unconformity ----------	
	Gneisses with TTG suite of rocks	

separated in the early magmatic differentiation process, often forming layered cumulates in ultramafic-mafic complexes have been noted. The important ones among them include the chrome ores in dunite/ peridotite of Pauni - Taka - Bhisi (Bhandara and Chandrapur dists. Maharashtra) and Khondasanvali (Bastar, C.G.) and vanadiferous-titaniferous magnetite in norite - gabbro assemblage of Khursipar near Gondia (Maharastra) and Avapalli area (Bastar, Chhattisgarh) as per various G.S.I. Reports. However, no workable deposit of any of these metals, has been identified in the craton so far. The chrome-ore bearing ultramafics of eastern Maharashtra have been dated around 2600 Ma (mentioned in Bhoskar and Saha, 2001).

BIF of Sukma Group

The oldest supracrustals of the craton, represented by the Sukma Metamorphic suite comprise the cordierite- anthophyllite bearing rocks as a major unit with closely-associated and ubiquitous Banded iron formation. This association, according to Ramakrishnan (1990), is comparable to the well documented suite of rocks from Javanhalli in Dharwar craton and, by anology, denotes a passive continental margin or continental side of a back- arc basin. The Pb- isotope data hints at 3.4-3.5 Ga old crust in the area.

BIF of Bengpal Group

Relatively younger, mafic, platformal-type Bengpal Group rocks, essentially an amygdular basalt-pelite association, host poorly-developed Banded Iron Formation of Superior type. These rocks are exposed in approx 100 km long and 10-30 km. wide, WNW-ESE trending Chandenar-Tulsidongar mobile belt, east of the Bailadila ranges in southern Bastar. The short-spanned depositional history of the BIF in association with the basic volcanics and interlayered sediments in this belt, is indicative of the intracratonic rifting and basin opening akin to other mobile belts of the craton. The Paliam-Darba granites, that have intruded the Bengpals, set the upper limit of their age at 2409-148 Ma (Sarkar and Gupta, 1989).

Gold in Bengpal Group iron formation

Several ancient workings of gold in one of the BIF ridges of Bengpal Group in Sonadehi area near Bhanupratappur (Bastar district), mark possibly the most important site of gold mineralisation in the craton. Gold is associated here with a band of sheared ferruginous shale lying between quartzite and BHQ and extending over a strike-length of over 500 m. Gold values ranging from 0.1 ppm to as high as 20.4 ppm have been reported in the samples from the walls, roof and floor of the incline of an old working of this occurrence (Geol. Surv. Ind., 1994).

(II)ARCHEAN-PROTEROZOIC TRANSITION AND PALEOPROTEROZOIC STAGE (2.6 to 1.6 Ma)

The recurrent tectonothermal reactivations contributing to the intrusion of syn-to late-kinematic tonalite-tronjhemite-granodiorite (TTG) in Archean greenstone sequences, apparently ceased by the Late Archean/Early Proterozoic event. This marked the beginning of the cooling and progressive stabilization of the continental crust leading to the formation of the proto-Bastar craton. The oxygenation of the atmosphere and great progressive changes in tectonic patterns of this era from vertical microplate tectonics of the Archean to Wilson-cycle type, B subduction-related, macroplate tectonics of the Phanerozoic, led to the formation of new, unique types ore deposits, essentially by the recycling of material derived form the pre-existing rocks/ ores.

Rich occurrences of the ores of iron, manganese, gold, copper, uranium, etc. often in close spatial and temporal relationship with the huge volumes of basic volcanics of predominantly tholeiitic character which outpoured in and around the rift- generated basins, typically characterize the early part of Archean-Proterozoic transitional period. They indicate rifting and associated magmatism under extensional tectonic regime as the dominant process in Kotri-Dongargarh Belt and Sakoli Fold Belt (Fig.2) and analogous cases. Slow development of an oxygenated atmosphere following the above-stated widespread basic volcanism facilitated the extraction of iron, manganese and associated elements from the primordial oceans leading to large-scale deposition of the Lake Superior-type formations and, at later stages, of the manganiferous sediments at several places in and around the craton. Their typical examples include the BIF and associated rich iron ores of the Iron Ore Basin in Bailadila Ranges of Chhattisgarh and the manganiferous formations of Gondite (Sausar Group) type deposited in the Sausar basin bordering the Bastar craton all along north of the Central Indian suture. The reported presence of manganese-ore, with jasper, chert etc. in the form of pebbles in the typical polymictic conglomerate at the base of Upper Chilpi rocks, south-west of Malanjkhand (Sharma and Kumar, 1969), confirms the existence of manganiferous formations in older rocks of the Chilpi basin, south of the Central Indian suture, as well.

BIF of Bailadila Group

The Late Archean Bailadila Group, lying unconformably above the Bengpal and Sukma supracrustals, is exposed extensively in a north-south trending, northward plunging, synclinorium that extends from Rowghat through Surajgarh and Ari Dongri to Dalli-Rajhara in north. An intracontinental tectonic setting has been envisaged for its typical litho-association that comprises dominantly of BF with rich iron ore deposits, felspathic quartzites, phyllites and acid tuffs (Roy and Bandopadhyay, 1998).

Gold in Sonakhan Schist Belt

The other important type of metallogeny of this period that of gold-is associated with the, platform-phase greenstones in low extensional regime, post-dating much of the deformation and metamorphism, e.g. in Sonakhan schist belt of Raipur dist., Chhattisgarh, located close to the eastern margin of the craton. The bimodal volcanics of this belt have shown the presence of auriferous chert bands and BIF, beside a 200 m. wide zone of silica impregnations, within chert-argillite assemblage (Sarkar,2004). According to Das et al (1990), this belt of Late Archean age is comparable with the greenstone belts of other Precambrian shields of the world. Detailed geochemical study of basic volcanics of this Belt by Ray and Rai (2004) has revealed their tectonic setting as transitional between the typical island arc and N-MORB, the bimodal volcanism representing the aborted rifting of the basin due to rapidly falling geothermal gradient.

Gold, Copper and Uranium in Kotri-Dongargarh Belt

The volcano-sedimentary Kotri-Dongargarh belt, occupying the Kotri Rift zone extends along N-S strike for approx. 240 km. from the Abhujmar basin in the south to the southern truncated margin of the Sausar mobile belt in north. It is limited to the east by the Bailadila Group of rocks. The belt comprises older Nandgaon group bimodal acid and basic volcanics, Dongargarh and Malanjkhand granitoids and younger Khairagarh Group of volcano sedimentary sequence (Fig.2). The extensive association of andesites and rhyolites in this belt, suggests that this may either mark subduction zone of popular plate tectonic models or a rift-zone (Kroner, 1984). Neogi et al (1996) ruled out the possibility of island- arc setting on the basis of detailed geo-chemical study of the Dongargarh volcanics and interlayered sediments that, taken together points to a stable environment. Deshpande et al. (1990) have also concluded that these volcanies represent the upper horizons of a greenstone succession, while Krishnamurty et al. (1988) inferred dominantly a continental margin set-up with some oceanic affinities for them. Few others, however, have emphasized that the arc- like geochemical signatures of these volcanics have to be explained within a destructive plate margin tectonic scenario only (Yedekar et al 1990, Asthana et al., 1996). It is presumed from the available geo chronological evidences that this Belt predates the Sakoli Group of rocks exposed westwards (Roy and Bandopadhyay, 1998)

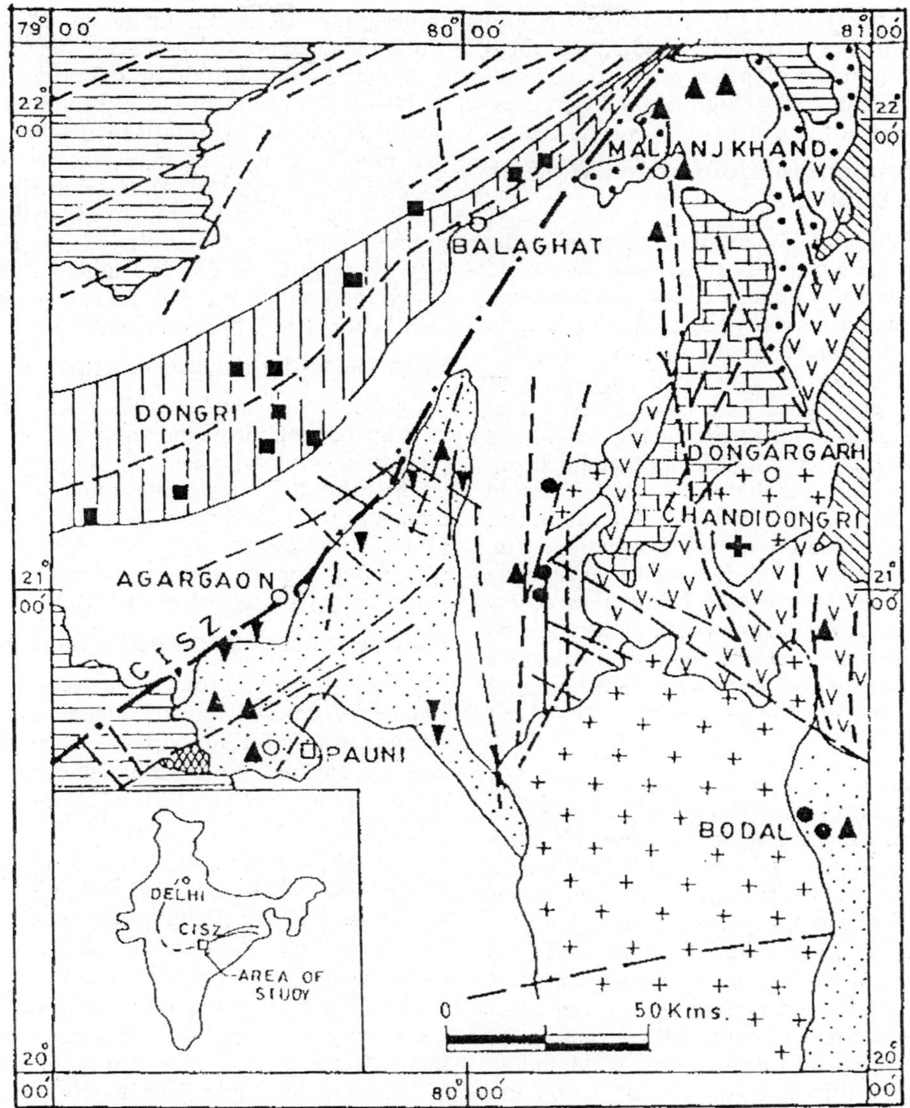

Fig. 2 Geological map of Kotri - Dongargarh and Sakoli Belts showing prominent locations of various types of mineral deposits (Modified after Raina 1980)

1. UNCLASSIFIED GRANITES AND GNEISSES, 2. SAKOLI GROUP,
3. SAUSAR GROUP, 4. NANDGAON GROUP ANDESITE, RHYOLITE,
5. DONGARGARH & RELATED GRANITES, 6. CHILPI GROUP,
7. KHAIRAGARH GROUP, 8. CHHATISGARH GROUP, 9. GONDWANA GROUP,
10. DECCAN TRAP, 11. SUTURE ZONE, 12. CENTRAL INDIA, 13. COPPER ORE,
14. URANIUM, 15. MANGANESE ORE, 16. TUNGSTEN ORE, 17. FLUORITE,
18. CHROMITE,

Gold in Nandgaon Group volcanics

Incidence of primary gold mineralisation in association with sulphides in a flow-laminated rhyolite band which has suffered weak brittle to ductile deformation, over a length of 2 km. in Gurwandi area of Bastar district. Ghosh et al. (1993) and Sarkar (2004). In a promising prospect here, two rich zones have been found at 150 m vertical depth from the surface. Possible presence of a gold deposit at depth is visualized here on the basis of the study of hydrothermal cycle (Sarkar, ibid).

Tholeiitic magmatism was sustained in Kotri and other coeval basins (like Sakoli) till Paleoproterozoic when a thickened crust gave rise to calc-alkaline type magmatism represented by Dongargarh and Malanjkhand- type intrusive batholiths. Both of them are comparable in their isotopic ages at around 2400 Ma (Sarkar and Gupta, 1989). Significant differences, however, characterize them in metallogenic respect.

Copper±molybdenum±gold association of Malanjkhand-type

The granitoid pluton of Malanjkhand, located at the northern end of the Kotri Rift zone, hosts the world famous, giant CuMoAu deposit of Malanjkhand and several other smaller copper-rich, gold-poor (e.g. at Taregaon, Pipardhar and Gidori); gold-rich, copper-poor (e.g. at Garhi Dongri and Khandapar) and poor copper and /or poor gold (.e.g at Shitalpani and Kundikasa) prospects or occurrences (Fig. 3). A good number of them, including the Malanjkhand deposit, are localized within parallely disposed, sheared quartz-veins that occupy shears and fractures, schistosity, faults, etc. along a N-S trending, 35 km. long, Malanjkhand-Nandora brittle ductile shear zone (Pal and Sindhupe, 2004; Raina, 1980). Few others like Garhi Dongri or Khandapar occur at the contacts of granitoids with older rocksthe Nandgaon volcanics or Amgaon gneisses or at places even within adjoining Amgaon gneisses e.g. at Kundikasa (Pal and Sindhupe, 2004).

The metallogeny of copper represented by the super-large copper deposit of Malanjkhand, has been a subject of fierce debate among the researchers in recent years. The proponents of one school of thought led by Sikka (1971,1989), Tripathi (1979) and Sarkar (1986) have advocated the porphyry model for its origin, while many others who have rejected this model on firm grounds of geological and geochemical evidences, have preferred to recognise it either as an epigenetic, mesothermal, fracture-filling vein/ lode-type deposit (Sharma and Kumar, 1969; Muktinath et al, 1976) that has primarily a volcanic affiliation, the host pluton being an integral part of the Kotri-Dongargarh rift environment (Rai and Venkatesh, 1990, 1993; Rai and Pandey, 2001;Rai et al.,2004;) or a fracture-controlled, quartz-reef hosted sulphide mineralisation from a low temperature ore-fluid of granitoid affiliation (Panigrahi and Mookherjee, 1997) or a fracture-filling, lode type deposit formed in an exhalative, SEDEX-type spreading environment (Vishwakarma, 2001) . Recently however, Stein et al. (2004) have reiterated that Malanjkhand is a subduction-related stockwork porphyry-style Cu-Mo-Au deposit, the entire mineralisation in the quartz reef as well as host granites being contemporaneous at 24908 Ma.

Uranium occurences/prospects of Bodal-type

Numerous occurrences of uranium have been reported from a variety of host-rocks of the Dongargarh Supergroup (Krishnamurthy et al., 1988). The workable deposits are, however, essentially confined to a metarhyolite-metabasic hybridized contact zone of Nandgaon group in close proximity to the BHQ members e.g. at Bodal in Rajnandgaon district or to the granite pluton borders e.g. at Bhandaritota. According to Sinha (1954), this occurrence is located in a volcano-plutonic setting in which the Dongargarh granites represent the intrusive phase. Mahadevan (1987) regards it as "a rather unique type of occurrence which may be related to the late phase fluids generated at the end of acid volcanism".

Fluoritelead-zinc association

A stock of sheared granites at Chandidongri, west of Rajnandgaon, hosts the unique fluorite-associated, lead-zinc mineralisation. A scrutiny of all such granite-hosted deposits and occurrences in Dongargarh supergroup has led Krishnamurthy (1996) to suggest that a crude zonality in terms of spatial deformation and type of mineralisation exists in the Kotri-Dongargarh Mobile Belt.

Polymetallic associations Sakoli Fold Belt

The Paleo-to Meso-Proterozoic triangular Sakoli Fold Belt of Eastern Maharashtra, located east of the Kotri-Dongargarh basin in north-western corner of the Bastar craton, constitutes another significant metallogenic province hosting several polymetallic associations of Zn, Cu, Au, Sn, W, Nb, Ta, etc. in various geological settings (Fig.4). The Belt has been considered to be of ensialic "Rift setting", peripheral to craton margin (Bandopadhyay et al, 1988) or "Island-arc type setting" (Yedekar et al., 1990). Bandopadhyay et al. (1988) believe that this Belt as well as the Kotri-Dongargarh Mobile Belt evolved from the same family of N-S trending intra-plate fundamental crustal zones of weakness characterised by aborted multiple linear rift systems, which controlled the crustal development of these Proterozoic supracrustal belts.

The Sakoli fold belt comprises argillite-dominated metasediments with phases of felsic-mafic bimodal volcanism, the ratio of volcanic to sedimentary rocks being 1:4 (Roy et al., 1995). Three distinct metallogenic modes that are represented in these rocks, according to Padhi et al. (2004) and Bhoskar and Saha (2001), include: (i) Syn-volcanogenic, stratiform zinc-scheelite mineralisation which is hosted in an exhalate-related rhtythmically-layered manganiferous chert interlayered sequence with basic volcancis (Padhi et al., 2004) e.g. in Bhaonri-Kolari area in SW part of the Belt, its host rock having been identified as managaniferous BIF; (ii) epigenetic, strata-bound or quartz-vein related, base-metal sulphidegoldscheelite mineralisation located dominantly within felsic volcanics/ meta-acid tuffs

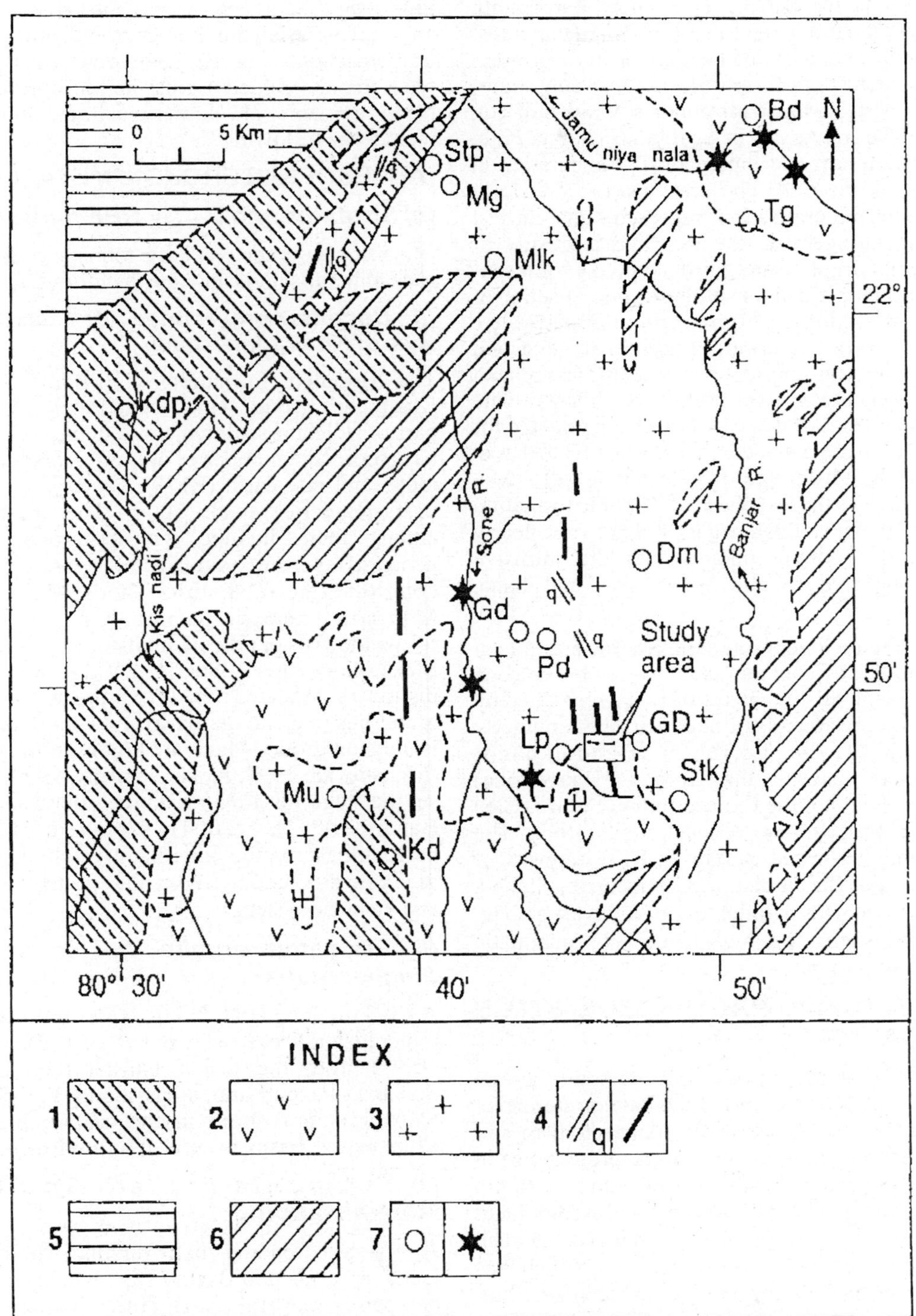

Fig. 3 Geological map of the Malanjkhand area : 1 - Amgaon schists and gneisses, 2 - Nandgaon volcanics (metabasalt), 3 - Grey and Pink Malanjkhand granitoids, 4 - Quartz veins/dolerite dykes, 5 - Sausar schists, 6 - Chilpi metasediments, 7 - Locality/gold panning location. Village locality : Bd - Bandhaniya, Tg - Taregaon, Stp - Shitalpani, Mg - Mohgaon, Mlk - Malanjkhand, Dm - Damoh, Kdp - Khandapar, Gd - Gidori, Pd - Pipardhar, GD - Garhi Dongri, Lp - Lalpur, Stk - Saletekri, Bm - Babamandi and Kd - Kundikasa (modified after Bhargava and Pal, 2000).

e.g. at Pular-Parsori, Khapri, Tuthanbori, Ranmangli, etc. and, to a smaller extent in adjoining metapelites/ mica schists; and (iii) metapelites-hosted, greisen, pegmatite-related wolframitescheelite, columbite-tentalite mineralisation with minor sulphides of Cu, Pb, Mo, Zn, Fe and As e.g. in Agargaon-Khobna-Kuhi belt in western part and Pipalgaon-Dahegaon belt in eastern part of the Belt. The first type i.e. stratiform zinc-scheelite mineralisation, represents the earliest phase of metallogeny in the Belt and is clearly of volcanogenic origin. The third- type i.e. greisen-related tungsten with minor sulphide mineralisation of nearly undisputed plutonogenicpneumatolytic hydrothermal origin. The mode of origin of the second type, i.e. vein-type gold-copper-scheelite associated ores, however, are not yet clearly established. The hydrothermal veins of this type have been targeted for platinum-group metals in the recent years (Bhoskar et al., 2003). The Pb-Pb age of galena as 1850 9 Ma, possibly indicates the last phase of sulphide formation in the Belt. However, this age data still does not clearly mark out the genetic linkage with the intrusive granitoids referred to above (Bhoskar and Saha, 2001).

The end of the Paleoproterozic era (i.e. around 1.6 Ga) was marked by another distinctive global episode of granitic magmatism that seems to have resulted by palingenesis presumably linked to the southerly subduction of the northern (Bundelkhand) protocontinent beneath the Deccan (Dharwarian) protocontinent along the Central Indian Suture (CIS) in the model proposed by Yedekar et al., (1990). The zoned pegmatite-hosted Sn-Nb-Ta-REE deposits of Govindpal type (Bastar dist., C.G.) occurring along a NW-SE trending belt parallel to the Godavari graben, typically represent the high-level acid magmatic activity of this period.

(III) MESO- TO NEO-PROTEROZOIC STAGE (1.6-0.5 GA)

After the terminal, 1.6 Ga old, tectono-thermal reactivation marking the end of the Paleoproterozoic, the changeover to Wilson-cycle like tectonism and crustal processes progressively gained prominance in further cratonisation of the Bastar craton. Rapid lowering of the geothermal gradient and a cooler and denser oceanic crust were further conducive for such global revolutions. The rift-displacement, which was essentially vertical during the Paleoproterozoic era, changed over progressively to mainly lateral during the Mesoproterozoic and by the Neoproterozoic time, evolved into the Phanerozoic type major lateral traslations and interactions of the oceanic and crustal plates. Cratonic basic magmatism that accompanied the rift displacement of this period (generally over 1.4-0.9 Ga) in rift-generated basins apparently played an important role in remobilising the pre-existing metals and concentrating them in new tectonically controlled fractures, faults, unconformities etc. The new deposit-types that developed during the Mesoproterozoic and Neoproterozoic era under such tectonic controls, include the following:

(A) MESOPROTEROZOIC METALLOGENY

(a) Strata-bound U. Th REE mineralisation

Notable low-grade, yet large tonnage endowments of such mineralisations include the following:

(i) Unconformity-related or clastic-hosted type of Khairagarh basin

The Bortalao formation of Middle Proterozoic Khairagarh group, which constitutes the lowermost part of volcano-sedimentary sequence of the group, has recorded a number of uranium, thorium and rare-metal and rare-earth (RMRE) occurrences in recent years (Banerjee et al., 2004). The promising ones among them include the strata-bound uranium mineralisation associated with polymicitc conglomerate-felspathic sandstone or grit at Mongarra (Ramesh Kumar et al., 1996) and in felspathic grits or arkosic sandstones of Malaharbodi-Makardhokra area (Banerjee, 1972) both in Bhandara district of Maharashtra. Proximity to large, K-rich, granitoids (e.g. Dongargarh) and Bijli rhyolites of the Nandgaon group, which are highly fertile in uranium (Mahadevan, 1995; Krishnamurthy, 1996) must have set the stage for a major source effect. According to Banerjee et al (2004) semi-acid depositional paleoclimate, on the other hand, might have played a vital role in retention of oxyphile elements, such as U and Th in the system.

(ii) Felspathic arenite- hosted uranium of Singhora-type

A 20 km. long and 5 km wide zone of uranium mineralisation has been reported by Bhairam et al. (1998) from the Kokabahal-Paraskol-Ghogra and Sonabala-Pandaritarai areas along a N-S to NNE-SSW structural grain in the eastern proximity of Chhattisgarh basin, in parts of Raigarh dist., M.P.

(b) Pegmatite-hosted Sn-Nb-Ta-Li Mineralisation

A NW-SE trending belt running parallel to the Godavari graben in Bastar/ Dantewadda dist. of C.G. comprises several pegmatitic bodies that host promising vein-type associations of tin and other granitophile elements like Nb, Ta, Be, Cs, Li and rare-earths. The outer zones of these pegmatites hold promise for bismuth-bearing minerals, specifically in the greisenised or skarnified portions of the wall-rocks. They are also well known as the ideal repositories of several gemstones (Deshpande, 1976;

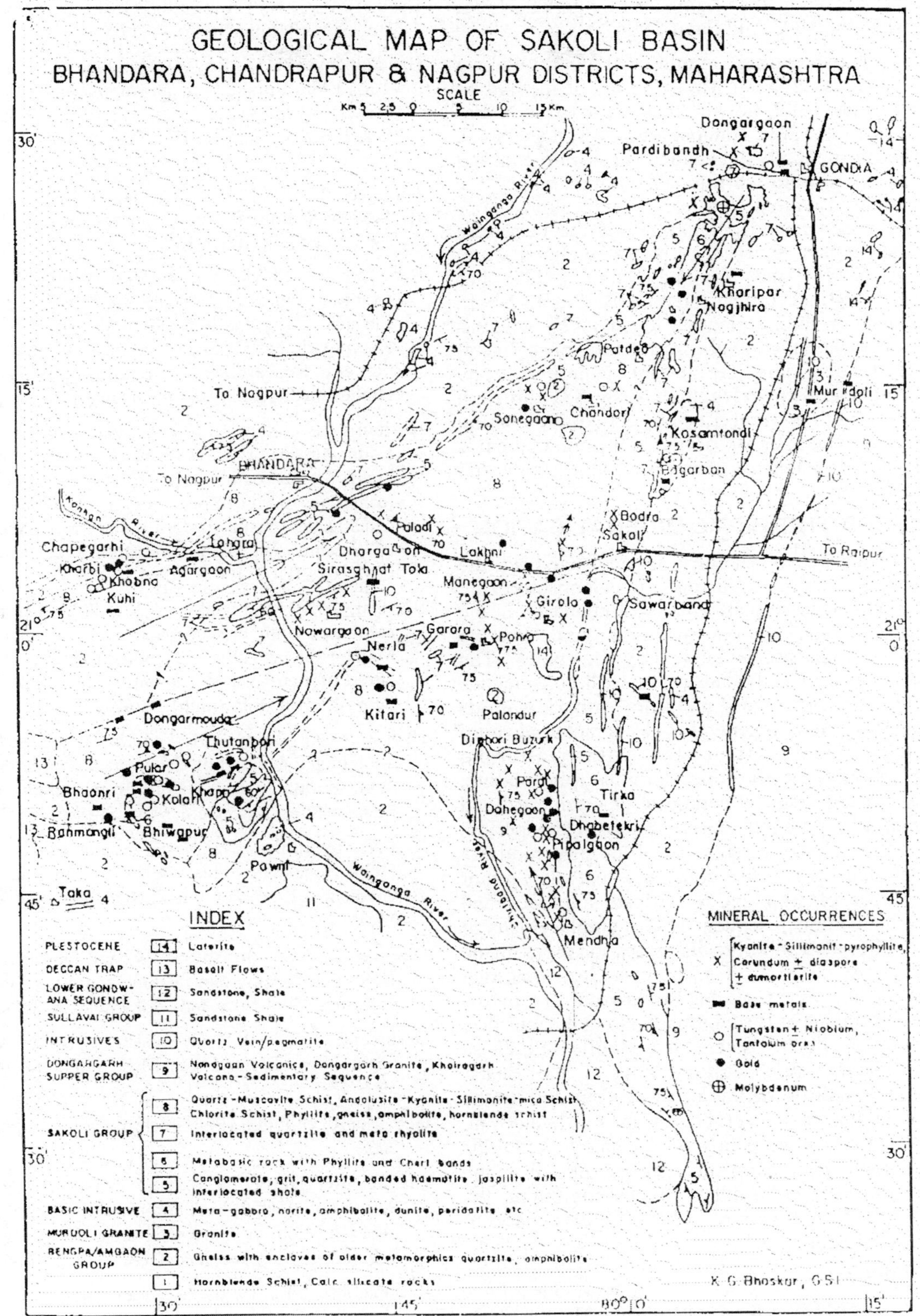

Fig. 4 Geological map of Sakoli basin showing locations of various types of mineral occurrence (After Bhoskar et al. 2003)

Ramachar et al., 1983; Lamba and Agarkar, 1988; Ramesh Babu, 1993; and Babu, 1994).

Tourmaline-bearing granites of Darba-Paliam type (dated as 2275±47 Ma using Rb-Sr method by Pandey et al., 1989) hosting these pegmatitic bodies (dated as 1528 Ma in Tongpal area by Sarkar, 1990), have been genetically linked to the tin, rare-metal, and rare-earth mineralisation associated with the pegmatities. Detailed study of several, almost non-metasomatised zoned pegmatities of the belt from Kawadgaon area in Bastar district by Lamba and Khanna (1981) and Yamuna Singh (1998) has revealed a distinct vertical zonation, Be and Be-Nb-Ta mineralisation taking place at lower levels whereas Sn-mineralisation and albitization are observed at higher levels. The Govindpal Mundval zone extending over 8 km in length in NNW-SSE direction is replete with several Li-rich, Sn-Nb-Ta bearing pegmatites. The intensity of mineralisation in them is related to the intensity of imposed albitisation and greisenisation.

(B)NEOPROTEROZOIC MINERALISATION

The prevalence of stability is envisaged in the continental regimes during the Neoproterozoic facilitated extensive platformal sedimentation in the shallow cratonic shelves of intracratonic marine basins of Chhattisgarh and Indravati types. The predominance of carbonate rocks with conspicuous absence or rarity of metallic mineralisations characterise them, in general. The characteristic non-metallic deposits associated with platformal sediments include limestone, dolomite, phosphorite, barite, gypsum, etc. reflecting an O_2- rich, arid and hypersaline depositional environment of this period. The black-shale facies of the intermediate inundational stage, by contrast, host the minor occurrences of iron and / or lead sulphides and uranium in thinner sequences.The emplacement of diamondiferous kimberlite diatremes and pipes, in the time span of 1.3 to 1.0 Ga, mark the last important mineralisation event of the Proterozoic era in the cratonic domains of Bastar mostly along the interface of the craton with the Eastern Ghat Mobile Belt. These emplacements imply cratonisation of the Bastar cratonic domain to depths of more than 200 kmthe source region of concerned magmas (Mahadevan, 1998). Petrochemical study of the mantle xenoliths (containing high calcium chromium pyrope) from the Kimberlite fields of Andhra Pradesh adjoining the Bastar craton, by Rao et al. (2001), lends support to such a conclusion.

References

Asthana, D., Khare, S.K., Pophare, Anil M., and Dash, Manas R., 1997. Comments on the paper by Neogi et. al. (1996). Precamb. Res. 84:105-107.

Babu, T.M., 1994. Tin in India. Geol. Soc. India. Mineral Resources of India Series, 7:217 p.

Bandopadhyay, B.K., Roy., A., Shukla, R.S., Guha Sarkar T.K., 1988. Bimodal volcanism, massive sulphides and associated Mn-garnet chert rocks in the Proterozoic Sakoli basin, Central India. Proc. IGCP-247 (A-3):27-28.

Banerjee, P.K., 1972. Geology and geochemistry of the Sukinda Ultramafic field, Cuttack Dist., Orrisa. Geol. Surv. Ind. Mem. 103:1-158.

Banerjee, R., Shivkumar, K. and Verma, S.C., 2004. Uranium mineralisation in the Middle Proterozoic Bortalao Formation in Malharbodi- Makardhokra area, Khairagarh basin, Bhandara dist., Central India. In 'Recent Researches in Econ. Geology (Vol.-2) Hindustan Publishing Corpn., New Delhi; 28-41.

Bhairam, C.L., Raju, B.V.S.N., Koteswar Rao, M. and Dattanarayana, T.A., 1998. Delineation of uraniferous horizons in cataclastic zones using integrated techniques in Precambrian basement in the environs of Chhattisgarh basin, Eastern India. Proc. Internat. Sem. on Precambrian crust in Eastern and Central India. UNESCO-IUGS-IGCP-368:155-159.

Bhoskar, K.G. and Saha, A.K., 2001. Metallisation associated with Precambraian magmatism in Eastern Maharashtra and adjoining parts of Madhya Pradesh. In "Precambrian Crustal Evolution and Mineralisation", Sem. Vol. (Ed. S.P. Singh), Patna: 365-375.

Bhoskar, K.G., Saha, A.K. and Mahapatra, K.C., 2003. Targeting for platinum Group metals in hydrothermal vein set-up: a case history from Sakoli fold Belt, Maharashtra. Gond. Geol. Mag., Spl. Vol., 7:421-429.

Das, N, Royburman, K.J., Vasta, U.S. and Mahurkar, Y.V., 1990. Sonakhan Schist BeltA Precambrian Granite-Greenstone Complex. Geol. Surv. Ind. Spl. Publ., 28: 118-132.

Deshpande, G.C., Mohabey, N.K. and Deshpande, M.S., 1990. Petrography and tectonic setting of Dongargarh volcanics. Geol. Surv. Ind., Spl. Publ., 28: 260-286.

Deshpande, M.L., 1976. The cassiterite-lepidolite bearing pegmatities, Bastar Dist., M.P. India. Ind. Minerals, 30:67-74.

Geol. Survey of India, 1994. Records, Vol. 127, Part 6: 202-203.

Ghosh, J.G.,Dutta, N.K. and Mishra, V.P., 1993. Petrographic study of deformed rhyolite and gold mineralisation in Gurwandi area, Kotri linear Belt, Bastar dist., M.P., Ind. Jour. Geol., 65(1): 25-34.

Krishnamurty, P., 1996. Are early-middle Proterozoic unconformity-related uranium deposits unique to Australia and Canada?: An appraisal with reference to India. Expl. Res. Atomic Min., 9:1-40.

Krishnamaurthy, P., Chaki., A., Sinha R.M. and Singh, S.N., 1988. Geology, geochemistry and genesis of metabasalts, metarhyolites and the associated uranium mineralisation at Bodal, Rajnandgaon dist., M.P., Expl. Res. Atomic Min. 1:13-39.

Kroner, A., 1984. Changes in plate-tectonic styles and crystal growth during the Precambrian. Bull. Geol. Soc. France., 16: 297-319.

Lamba, V.J.S. and Khanna, V.K.,1981. Characteristic features of tin-bearing rare-metal pegmatites of Konta Tehsil, Bastar Dist., M.P. Bull. Ind. Geol. Assoc. 14(2):151-154.

Lamba, V.J.S. and Agarkar, P.S., 1988. The tin potential of Precambrian rare-metal bearing pegmatites of Bastar, M.P., India. Mineral. Deposita, 23:218-221.

Mahadevan, T.M., 1987. Metallogeny concepts, constraints and prospects in India. Ind. Jour. Geol., 59(2):86-116.

Mahadevan, T.M., 1995. Identifying high-grade uranium deposits in the Proterozoic basins in India-A challenge to exploration. Expl. Res. Atomic Min., 8:49-60.

Mahadevan, T.M., 1998: Precambrian Lithospheric Evolution and chemical and thermal regimes of the mantle. M.S. Krishnan Centenary Commem. Vol., Geol. Surv. India:108-109.

Mishra, V.P., Singh P. and Dutta, N.K. 1988. Stratigraphy, structure and metamorphic history of Bastar Craton. Rec. Geol. Surv. Ind., 117 (3-8): 1-26.

Muktinath, Mathur, S.M., Sharma, R.S., Neelakantam, S., Narang, J.L. and Sonakia. A., 1976: Geology and Mineralization in Malanjkhand area, Balaghat district, M.P. Geol. Surv. Ind., Spl. Pub. 3:203-207.

Neogi, S., Miura, H. and Hariya, Y., 1996. Geochemistry of the Dongrargarh volcanic rocks, Central India: implications for the Precambrian mantle. Precamb. Res., 76: 77-91.

Padhi, R.N., Bhoskar, K.G., Saha, A.K., Mahapatra, K.C. and Raut, P.K., 2004. Gold mineralisation in Sakoli Group of rocks in Eastern Maharashtra and associated base metals and platinum group elements. Jour. Econ. Geol., & Geores. Mgmt.,1(1):103-109.

Pal, A.B. and Sindhupe, G.L., 2004. Gold investigations in Malanjkhand area with special reference to Garhi Dongri Prospect, Madhya Pradesh. Jour. Geol. Soc. Ind., 64:52-58.

Panigrahi, M.K. and Mookherjee, A., 1997. The Malanjkhand copper (molybdenum) deposit, India: Mineralisation from a low-temperature ore-fluid of granitoid affiliation. Mineral. Deposita., 32:133-148.

Radhakrishnan, B.P. and Naqvi, S.M., 1986. Precambrian continental crust of India and its evolution. J. Geol. 94:145-166.

Rai K.L. and Pandey, H.K. 2001. Comment on two papers about Malanjkhand copper deposit by M. Bhargava and A.B. Pal published in Jour. Geol. Soc. India, 53 (6) and 56(4), Jour Geol. Soc. India, 57:551-554.

Rai, K.L. and Venkatesh, A.S., 1990.Malanjkhand copper deposit- A pertrological and geochemical appraisal. Geol Surv. India, Spl. Pub., 28: 563-584.

Rai, K.L. and Venaktesh, A.S., 1993: Geological setting and nature of copper-molybdenum mineralisation in the intra-continental acid magmatic regime of Malanjkhand, Central India. Resources Geol (Japan). Spl. Issue, 15:285-297.

Raina, B.N., (1980): Landsat exploration of Himalayan and Peninsular regions (Remote Sensing and Mineral Exploration IGCP Project (143):Progress report of work done in India in "Advances in space Exploration" COSPAR Symp. Series, 10:59-67.

Ramachandra, H.M., Mishra, V.P., Roy, A. and Dutta, N.K., 1998. Evolution of the Bastar Craton: a critical review of gneiss-granitoid and supracrustal belts. M.S. Krishnan Centenary Vol., Geol, Surv. Indi. : 144-150.

Ramachar, T.M. Shivananda, S.R., Dwivedi, K.K. and Jayaram, K.M.V., 1983. The rare-metal and REE occurrences in South Bastar M.P., Geol. Surv. India. Spl. Publ. 13:104-107.

Ramakrishnan, M.,1990. Crustal development in Southern Bastar, Central Indian Craton. Geol. Surv. Ind., Spl. Publ., 28:44-66.

Ramesh Babu, P.V.,1993. Tin and rare-metal pegmatites of the Bastar- Koraput pegmatite belt, Madhya Pradesh and Orissa, IndiaCharacterisation and classification. Jour. Geol. Soc. Ind.,42:180-190.

Ramesh Kumar, K., Thangavel, S., Chavan, S.J., Parihar, P.S. and Sinha, D.K., 1996. Strata-bound Uranium mineralisaton in Proterozoic conglomerates at Mogarra, Bhandara Dist., Maharashtra, India. Expl. Res. Atomic Min., 9:55-60.

Rao, K.R.P., Rao, K.N., Dhakate, M.V. and Nayak, S.S., 2001. Petrology and mineralogy of mentle xenoliths of Wajrakarur and Narayanpet Kimberlite fields, A.P., Nat. Sem. "On Exploration and survey of noble metals and precious stones". Geol.Surv. Ind., Spl. Publ., 58:577-591.

Roy, R.K. and Rai, K.L., 2004. Geological setting and petrogenesis of the auriferous metavolcanic complex of Sonakhan, Raipur Dist., Chhattisgarh. Jour. Econ. Geol.& Geores Mgmt., 1(1):45-60.

Roy, Abhinaba and Bandopadhyay, B.K., 1998. Supracrustal Belts of Central India and their significance in the crustal evolution of Central Indian Shield. M.S. Krishnan Cent. Commem. Vol, Geol. Surv.India, Calcutta:160-166.

Roy, A., Bandopadhyay, B.K., Huin, A.K., Money, Charles,D., Saha, S.K., 1995. Geology of the Sakoli fold

Belt, Nagpur, Bhandara and Gadchiroli districts, Maharashtra, Central India. Unpubl. Report, Geol. Surv. India.

Sarkar, G. and Gupta S.N., 1989. Dating of early Precambrian granite complex of Bastar district, Madhya Pradesh. Rec. Geol. Soc. Ind., 122 (2): 22-30.

Sarkar, S.C., Kabiraj, S. Bhattacharya, S. and Pal, A.B., 1996. Nature, origin and evolution of the granitoid-hosted early Proterozoic copper-molybdenum mineralisation at Malanjkhand, Central India. Mineral. Deposita, 31:419-431.

Sarkar, S.K.,2004. Concept-oriented prognosis and prospecting for gold mineralisation in Kotri Belt, Bastar Craton. In 'Recent Researches in Econ.Geology, Vol. 2 (Ed. V.K. Khanna), Hindustan Publ. Corpn., New Delhi:93-103.

Sarkar, S.N., 1957-58. Stratigraphy and tectonics of the Dongargarh sytem: a new system in the Precambrian of Bhandara- Durg- Balaghat area, Bombay and M.P. India., Inst. Tec. Jour. Sci. Engr. Res. Kharagpur, 1:237-268, 2:145-160.

Sarkar, S.N., 1994. Chronostratigraphy and tectonics of the Dongargarh Supergroup Precambrain rocks in Bhandara-Durg region, Central Ind., Indi. Jour. Earth Sci., 21: 19-31.

Sharma R.K. and Kumar, H., 1969. Geology and copper occurrences of the Malanjkhand area, Balaghat Dist., M.P. Ind. Minerals, 23(2):23-39.

Sikka, D.B.,1971. Preliminary Report of Malanjkhand porphyry copper prospect, Dist. Balaghat, M.P. and Pular- Parsori copper prospect, Nagpur and Bhandara dists., Maharashtra, Hindustan Copper Ltd., Company Report quoted in Sikka et al., 1991:15 p.

Sikka, D.B., 1989. Malanjkhand Proterozoic Porphyry Copper Depsoit, M.P., India. Jour. Geol. Soc. Ind., 34:487-504.

Stein, H.J., Hannah, J.L., Zimmerman, A., Markey, R.J. and Sarkar, S.C., 2004. A 2.5 Ga porphyry Cu-Mo-Au deposit at Malanjkhand, Central India: implications for Late Archean continental assembly. Precam. Res., 134: 189-226.

Tripathi, C., 1979. the Malanjkhand-Taregaon Porphyry Copper deposit. Geol. Surv. Ind., Misc. Publ., 34 (II):161-168.

Vishwakarma, R.K., 2001. Isotopic characters of the Proterozoic Malanjkhand copper deposit, Central India: implications for exhalative activity. Jour. Appl.Geochem. 3(2):91-103.

Yamuna Singh, 1988. Early Proterozoic rare-metal and tin pegmatites near Kawadgaon, Bastar, M.P.An example of vertical pegmatite zonation. Jour. Geol.Soc. Ind., 51:175-182.

Yedekar, D.B., Jain, S.C., Nair, K.K.K. and Dutta, K.K.,1990. The Central Indian Collision Suture. Geol. Surv. Ind., Spl. Publ., 28:143.

Hon'ble Reviewers

1. Prof. B.K. Sahu
   Emeritus Professor
   Department of Earth Sciences
   Indian Institute of Technology
   Bombay POWAI
   MUMBAI 400 076 INDIA

2. Prof. M.K. Pandya
   Emeritus Professor
   Department of Geology
   MLS University
   UDAIPUR 313 001 INDIA

Editorial Assistance

Dr. (Mrs.) Yasmin Sayyed

Economic Mineralization
K.L. Shrivastava (Editor)
2009. pp. 47-52
Scientific Publishers (India), Jodhpur

5

Section-I

Mineral Exploration of the Moon

N. BHANDARI

Basic Sciences Research Institute, Ahmedabad 380009
Email : nnbhandari@yahoo.com

Abstract

Several missions for exploration of the Moon are planned during this decade starting with Kaguya-Selene (Japan), Chang'E (China), Chandrayaan-1 (India) and Lunar Reconnassaince orbiter (USA).These are all orbiting missions making remote sensing observations from a low orbit around the Moon, but they will be followed by more ambitious landing missions with a goal to establish a human outpost by 2015. Considerable efforts are being made to identify resources on the Moon which can support human activity there. Here we review possible resources on the Moon which require concentrated study for a quantitative assessment. Region of special interest include lunar polar regions which may have hydrogen, water, organic compounds and many other volatile deposits. Much effort is being made to map ilmenite which is rich in oxygen and titanium and also in ^{3}He, a possible fusion fuel of the future.

Key words : Moon, Lunar polar regions, Ilmenite, Fusion fuel, Chandrayaan

INTRODUCTION

Planetary exploration is emerging as a frontier area of research. It is motivated by two main objectives: (i) understanding the formation and evolution of the solar system and (ii) finding resources for human habitation on planets and for possible use on Earth. Of all the planets and satellites in the solar system, the Moon has been explored in great detail during the past half century by orbiter and lander missions and the samples brought back to Earth have been studied in detail for chemical, isotopic and mineral composition. Present attempt is to examine the possibility of using mineral resources of the Moon for human outposts or habitation on the Moon. Some resources may potentially be useful back here on the Earth.

CHEMICAL EVOLUTION OF THE MOON

Moon and Earth, having similar isotopic composition of oxygen, are believed to have been formed out of similar source material. The bulk composition of the Moon is however much different compared to that of the Earth, and in view of its low density (3.34 g/cc, compared to Earth's 5.25g/cc), it is believed that the bulk Moon is depleted in iron. In the "Giant Impact Hypothesis" for the origin of the Moon, the Moon is formed out of the material ejected from the Earth when a large asteroid impacted on the already differentiated Earth. The iron depletion of the Moon is accomplished on the Earth when the iron core separated in the formative stages of the Earth, before the asteroid impact. Thus Moon's bulk composition is largely similar to the Earth's mantle, with some component from the impactor. In the subsequent melting and chemical differentiation of the Moon which occurred in the beginning, about 4.5 billion years ago, ocean of magma was formed. Plagioclase containing refractory elements like calcium and aluminum, being lighter, floated on the Moon and formed its crust, as it cooled. Minerals containing iron and magnesium, being heavy, settled down at the bottom of the "magma ocean". Part of this magma, subsequently oozed out of the deep basins, formed in the crust which was punctured by large impacts between 3.1 and 3.8 billion years ago. The main surface units of the Moon, therefore, are anorthositic highland crust, covering about 80% of the Moon's surface and mare basalts which cover about 17%, mostly on the Earth facing side. After this primary differentiation, the Moon's chemical evolution stopped, although crystallization of basalt gave rise to some lithological units, like KREEP basalt, enriched in incompatible elements like K, REE, P, U, Th etc.

ECONOMIC MINERAL RESOURCES

As a consequence of this initial episode of melting and crystallization, the lunar surface contains four main lithological terraines (i) Feldspathic Highland terrane (FHT), (ii) Mare basalts, (iii) Procellarum Kreep Terrane (PKT), and (iv) South pole Aitken Terrane (SPAT). FHT consists of calcium-aluminum-plagioclase feldspar typically containing 11.3% Ca, 13% Al, 5% Fe and 4.2% Mg. Mare Basalts are rich in Fe, Ti, Mg etc. PKT covers about 2 to 3 % of the lunar surface and is high in Fe (15%), and low in Ca (6%) and Al (4%), and, as mentioned above, contains a high concentration of K, REE, P, U and Th. The SPAT occurs within the South Pole Aitken basin on the far side of the Moon, considered to be the largest basin of the solar system, about 2500 km. wide and 12 km deep. Unlike the basins occurring on the front side of the Moon, SPA basin was not filled by lava and hence still contains the excavated lower crust material exposed on the surface.

The major chemical composition of various litho-units of main highland rocks and mare basalts are compared with the bulk Moon composition in Table 1.The most abundant element is oxygen (40-45%) and others are common elements like Si, Al, Mg, Ti etc. which are of no economic consequence, except that if required for human base they can be locally extracted, and need not be transported from the Earth. The trace elements in the lunar soil contains sulphur (1800g/m^3) and phosphorus (1000 g/m^3), in addition to minor amounts of H, N, and noble gases like He, Ne, Ar, Kr and Xe, all implanted from solar wind (Schrunk et al,1999). Of these, hydrogen, if it can be extracted in sufficient quantity, will be very useful for producing energy and water and nitrogen and phosphorus for agriculture on the Moon. Although their concentrations are very small, carbon (~100 ppm) and hydrogen (~ 50 ppm) in lunar soil, occur in comparable amounts, appear promising for manufacture of methane which can be used as fuel or in propulsion. (Ruiz and Duke, 2004).

The chemical evolution of rocky planets is size dependent; the bigger the planet, the more evolved it is. In this respect the Moon has evolved to a very small degree, only a little more than asteroids but much less than Mercury, Mars, Earth, or Venus. Moon's evolution stopped about 3 billion year ago. After the primary differentiation, nothing much has happened on the Moon. The last isolated evidence of volcanic activity on the Moon goes as far back as 0.8 to 1 billion years ago. Precious metals, ores and minerals are formed after repeated differentiation and fractionation which is nearly absent on the Moon. Therefore it is unlikely that economic minerals will be found on the Moon, although there may be small pockets, rich in certain minerals but not yet discovered because of poor spatial resolution of lunar chemical maps. The mineral and chemical maps of the Moon are based on Clementine and Lunar Prospector observations. Japanese mission Kaguya- Selene (launched in Septemebr 2007), Chinese Chang'E launched October 2007 or India's Chandrayaan-1 orbiters are expected to provide high resolution data from their ~100 km altitude and it may be possible to identify small pockets of useful minerals. The best results are expected from the US-LRO mission expected to be launched in late 2008, because it will observe the Moon from lower altitude of 50 or 30 km.

One promising area on the Moon for some resources is the dark polar regions, which are never exposed to sun light and hence are very cold (about -230°C). Since day time temperature of the Moon may go up to 120°C, the Moon essentially acts as a distillation apparatus and volatiles, like water, mercury and some gases may get transported and deposited in polar regions, never to be out gassed. The other sources of volatiles may be comets, which have sporadically impacted on the inner planets in the past (Gehrels, 1994; Bhandari and Gehrels, 1999). There is evidence in terrestrial sediments that asteroids and comets have collided with the Earth over geologic times (Shukla and Bhandari,1997 and references therein). Comets are known to contain volatiles, specially hydrocarbons, ices etc. These impact events may produce transient atmosphere on the Moon, majority of volatiles either escaping from

Table-1 Model composition* of bulk moon and various typical lunar rocks

Element	Bulk Moon	Highland rocks						Mare basalts		
		Highland crust	KREEP basalt	Ferro-anorthosite	Norite	Gabbronorite	Troctolite	High Ti	Low Ti	High Al
Al	3.17	13.0	8.95	18.85	7.93	6.98	10.53	4.66	5.66	7.19
Mg	19.3	4.1	4.70	0.16	7.5	7.68	12	5.04	3.9	5.1
Ca	3.22	11.3	6.73	14.56	6.50	8.28	7.71	7.64	8.43	8.00
Si	20.3	21.0	24.47	20.74	23.81	23.63	20.04	17.6	21.57	21.62
Fe	10.6	5.1	7.02	0.163	7.69	7.70	3.89	15.3	15.46	13.05
Na	0.06	0.33	0.65	0.27	0.30	0.67	0.17	0.27	0.21	-
O	40.4	43.1	44.53	46.28	44.15	44.06	44.35	40.3	41.5	42.33
Ti	0.18	0.336	1.14	0.20	0.20	0.24	0.03	7.8	2.4	1.56
U	0.033	0.24	3.37	0.60	0.60	<0.6	0.054	0.13	0.22	0.59
K	0.0083	0.06	0.44	0.14	0.14	0.014	0.02	0.041	0.06	0.08
Th	0.125	0.90	10.3	1.84	1.84	0.19	0.19	0.34	0.84	2.1

* The bulk composition is based on the Model of Taylor (1992). All concentrations are in wt % except U and Th which are in ppm.

the Moon or eventually getting deposited at the cold poles. Therefore there is a good chance of finding some of the cometary volatiles deposited on the Moon's polar regions. These may include ammonia, water, methane, alcohol and other hydrocarbons.

Yet another source of gases on the Moon is the solar wind. Solar wind impinges on the uppermost soil grains lying on the surface of the Moon and get implanted. Solar wind is a perennial source of hydrogen with a flux of about 10^8 atoms per cm^2.s. Other than hydrogen, solar wind has helium and a number of elements in solar proportion. Indeed hydrogen enrichment has been found on the lunar poles by the Lunar Prospector neutron monitor data (Feldman et al, 2000; Crider and Vondrak, 2003). Helium is of special interest because it is rich in a rare isotope ^{3}He and will be discussed in some detail later.

OXYGEN AND WATER

The two chemicals essential for human survival on air-less Moon are oxygen and water. Efforts are being made to extract or manufacture them from lunar rocks which contain abundant oxygen. Ilmenite ($FeTiO_3$) containing oxygen is a mineral of special interest. Feldmann et al (2000) interpreted the Lunar Prospector neutron monitor data in terms of waterice and showed that several billion tons of water, mixed with lunar soil may be present at north and south poles of the Moon. There is some debate whether it is waterice or hydrogen which gave anomalous neutron fluxes (Crider and Vondrak, 2003). The occurrence of water on the Moon is thus yet to be established. Therefore much effort is being made by the currently planned missions to ascertain whether water is indeed present on the Moon. The Indian mission, Chandrayaan- 1 is designed to have a polar orbit and will observe both the lunar poles on every orbit. It will look for water in many ways. If water is present, it will result in subdued signal of rocky elements (like Si, Mg etc) and will be observed by the low energy (LEX) X-ray fluorescence spectrometer (Bhandari, 2003, 2005). Optical and near Infra red spectrometers (HySI, SIR-2 and Moon Mineral Mapper) may also be able to identify water if it is present on the surface from its characteristic absorption lines. If buried near the surface, radioactive gamma ray signal will be subdued and it can be identified by Chandrayaan-1-high energy (HEX) (Xray) spectrometer. Other instruments like Synthetic Aperture Radar (Mini SAR) will be able to identify surface or sub-surface water or ice by polarization of reflected signal and if water on the lunar surface gives rise to some neutrals in the lunar environment, it will be identified by SARA. These instruments and their objectives have been discussed in detail by Bhandari (2004, 2005 and references therein) Thus results of Chandrayaan-1 mission is expected to throw some light on this important question. Other missions like Kaguya and Chang'E are also well equipped for identifying water on the surface or even if it is present under the surface.

LUNAR TRANSIENT PHENOMENA

Lunar Transient Phenomena have been reported in the past from time to time. These are believed to be either features related to impacts of meteorites, asteroids and comets or due to degassing from the lunar interior. There is evidence from the Apollo alpha spectrometer that there are some craters, specially haloed dark craters around Mare Fecunditatis, which are emanating radon, which is produced in the lunar regolith from decay of uranium. If gases from the lunar interior are emanating from some regions of the Moon, they will be immensely useful for human habitation. Some of these LTP regions have been identified as areas of special observation by the forthcoming Moon orbiter missions.

OTHER LUNAR RESOURCES

Moon's resources have been discussed by Schrunk et al., (1999). Assessment and harnessing of resources are also required if human bases are to be established as are being planned on the Moon and Mars. We discuss here the potential of ^{3}He which is considered to be an ideal fuel for fusion and is of foremost interest although it is a very futuristic proposition.

HELIUM 3

Importance of Helium-3 lies in the fact that it is possible to produce unpolluting (or radioactively clean) energy by ^{3}He-D or ^{3}He-^{3}He fusion.

$$^3He + D = P + {}^4He + 18.4\ MeV$$

$$^3He + {}^3He = 2P + {}^4He + 12.9\ MeV$$

The large amount of energy release in these reactions make ^{3}He fusion an attractive proposition, since a few kg of ^{3}He can meet the annual energy requirement of the whole Earth.

Helium-3 is made in the Sun by thermonuclear reactions, when hydrogen is converted into helium. Helium-3 produced in the Sun is carried by solar wind, impinges on soil lying on the planetary surfaces e.g. of Mercury, Moon etc which have no atmosphere. The turnover of the soil cover (regolith) exposes grains in the top few meters to ^{3}He which eventually gets saturated to a level of few parts per billion. ^{3}He is very difficult to retain in grains since it easily escapes into space. However it has been found that in the Moon, there is an equilibrium amount of a few parts per billion (~5 to 10) in some minerals. The oxide of iron and titanium, Ilmenite ($FeTiO_3$) has relatively high retentivity for He-3. Mapping of iron and titanium on lunar surface can thus indicate the distribution of ^{3}He.One of the objectives of Chandrayaan-1 is to map the distribution of ilmenite on the lunar surface with high spatial resolution.

There are two problems associated with using ^{3}He (i) its extraction on the Moon for recovering reasonable amounts of ^{3}He involves heating billions of tons of soil to 600 to 800° C and (ii) development of suitable fusion reactors on the Earth.

SOLAR POWER

Solar power is another possibility because it occurs in abundance on the Moon. Not only it can be used to power the local requirements of transportation, industry and human habitation, but it can even be beamed to the Earth after suitable conversion to a more convenient form.

Transportation of material from the Earth to the Moon is very expensive, currently estimated at US$ 1000 per kg and therefore suitable technology has to be developed for local mining on the Moon for long term habitation. It is well known that transporting material from the Earth to Low Earth Orbit (LEO) is about 20 times more expensive than transporting the same material from the Moon to LEO. This is because of low gravity of the Moon which makes it cheaper to escape from the Moon and the high gravity of the Earth, which can be used to easily achieve LEO at considerable saving of fuel. In view of this consideration Blair et al (2004) have argued that it may be cheaper to transport space craft fuel from the Moon to LEO provided such an infrastructure is built on the Moon and its cost is discounted. Possibility of hydrogen and oxygen on the Moon makes manufacture of Liquid hydrogen and oxygen an attractive proposition.

ACKNOWLEDGEMENTS

It is an honour for me to write an article in the felicitation volume of Prof. B. S. Paliwal, with whom I had a long association. I record my appreciation to him for the exciting collaboration we had on study of meteorites which fell in Rajasthan centered around the J.N.V. University, Jodhpur with an amazingly regularity over the past decade. I thank Indian National Science Academy for Honorary Scientist Grant and D. Dhingra, N. Srivastava and A. Bhardwaj for useful comments.

References:

Bhandari N. and Gehrels,T., 1999 : Killer comets and asteroids over geologic history and their future hazards In Space Research in India: Accomplishments and prospects, (M.S. Narayanan et al, eds.), PRL Alumni Association, Ahmedabad.

Bhandari, N., 2005 : Chandrayaan-1: science goals, Jour. Earth System Sci., Vol. 114, pp. 699-710.

Bhandari, N., Adimurthy, V., Banerjee, D., Srivastava N. and Dhingra, D., 2004 : Chandrayaan-1 lunar polar orbiter : Science goals and payloads, Proc. International Lunar Conference 2003 (S.M. Durst et al, editors) American Astronautical Society, Science and Technology Series, Vol. 108, pp. 33-42.

Bhandari, N., 2004 : Scientific challenges of Chandrayaan-1: The Indian lunar polar orbiter mission, Curr. Sci., Vol. 86, pp. 1489-1498.

Blair, B. R., 2004 : An economic rationale for lunar mineral exploration, Proc. of International Lunar Conference 2003 (S.M. Durst, C.T. Bohannan, C.G. Thomason, M.R. Cerney and L. Yuen, Editors) American Astronautical Society, Science and Technology series, Vol. 108, pp. 237-258.

Crider D.H. and Vondrak, R.R., 2003 : The solar wind as a possible source of lunar polar hydrogen deposits, Jour. Geophys. Res, Vol. 105, pp. 26773-26782.

Diaz,J. and Duke M. B., (2004) Lunar polar ice propellant production plant study, Proc. of International Lunar Conference 2003 held at Hawaii (S.M. Durst, C.T. Bohannan, C.G. Thomason, M.R. Cerney and L. Yuen, Editors) American Astronautical Society, Science and Technology series, Vol. 108, pp. 309-313.

Feldman, W. C., Lawrence D. J., Elphic, R.C., Barraclough, B. et al, Polar hydrogen deposits on the Moon, 2000 : Jour. Geophys. Res., Vol. 105, pp. 4175-4195.

Gehrels, T., 1994 : Hazards due to comets and asteroids, University of Arizona Press, Tucson, USA, 1300 p.

Prettyman, T.H., Hagerty, J.J., Elphic, R.C., Feldman, W.C., Lawrence, D. J., McKinney, G. W., Vaniman, D. T., 2006 : Elemental composition of the lunar surface: Analysis of gamma ray spectroscopy data from Lunar Prospector Jour. Geophys. Res. (Planets), Vol. 111, No. E12,E1200710.1029/2005JE002656.

Ruiz, B. and Duke, M.B. (2004), Lunar methane production plant: CH_4/O_2 versus H_2/O_2 fuel production, Proc. of International Lunar Conference 2003 (S.M. Durst, C.T. Bohannan, C.G. Thomason, M.R.Cerney and L.Yuen,

Editors) American Astronautical Society, Science and Technology series, Vol. 108, pp. 303-307.

Schrunk D., Sharpe, B., Cooper, B., and Thangavelu, M., 1999 : The Moon: Resources, future Development and Colonisation, Wiley-Praxis, New York.

Schmitt H.H., 2006 : Return to the Moon: Exploration, Enterprize and energy in the human settlement of Space, Copernicus books.

Shukla P.N. and Bhandari N., 1997 : Physical, Chemical and biological changes at geological boundaries: Causes, Consequences and clues based on the study of Indian sections. Paleobotanist, Vol. 46, pp. 41-62.

Taylor, S.R., 1999 : Solar System Evolution, Cambridge University Press, Cambridge, 307 p.

Hon'ble Reviewers

1. Dr. Anil Bhardwaj
   Head, Planetary Sci. Br.
   Space Physics Laboratory
   Vikram Sarabhai Space Centre
   TRIVANDRUM 695 022 INDIA
2. Dr. N. Srivastava
   Scientist
   Physical Research Laboratory
   Navrangpura
   AHMEDABAD 380 009 INDIA
3. Dr. Deepak Dhingra
   Scientist
   Physical Research Laboratory
   Navrangpura
   AHMEDABAD 380009 INDIA

Editorial Assistance

Prof. S.K. Ojha

Economic Mineralization
K.L. Shrivastava (Editor)
2009. pp.53-56
Scientific Publishers (India), Jodhpur

6

Section-I

Possible Role of Sulfur on the Early Diversification of Life on Earth : Astrobiological Implications

VINOD C. TEWARI[1] AND JULIAN CHELA FLORES[2]

[1]Wadia Institute of Himalayan Geology, 33, General Mahadeo Singh Road, Dehradun -248001, Uttarakhand, India
[2]The Abdus Salam ICTP, Strada Costiera 11, 34014 Trieste, Italia and Instituto de Estudios Avanzados, IDEA, Caracas 1015A, República Bolivariana de Venezuela
Email : vinod1954@yahoo.co.in

Abstract

The purpose of the present paper to review the possible role of sulfur on the early diversification of life on Earth, which has also influenced greatly, biogenic mineralization. Recently the significance of chemoautotrophs (microorganisms using inorganic or organic substances as energy source rather than light) has been discussed with special reference to the Jupiter's satellite Europa. The search for presence of sulfur in the core of Europan ocean, its chemistry and sulfur reducing bacteria (SRB) could be similar to the earth, especially in its early stages of crustal evolution. They have emphasized that the sulfur patches on the icy surface of the Europa might contain biomarkers and should be aimed to study in future Europa mission for extraterrestrial life in the universe.

Key words : Early life, Astrobiology, Europa, Extraterrestrial life sulfur reducing bacteria.

INTRODUCTION

Our aim is to ascertain certain aspects of the evolution of sulfur reducing bacteria that might suggest hints as to which biomarkers would be likely signatures of life on Europa. Fortunately, the question of distinguishing between signatures of past life and signatures of nonlife in the sulfur cycle of the Earth has been carefully discussed (Machel et al, 1995, Machel, 2001). Distinguishing criteria of bacterial and **T**hermochemical **S**ulfate **R**eduction (TSR) have been considered.

THE ROLE OF SULFUR IN LIFE

Sulfur is an important element for all microorganisms, animals and plants on Earth. The main source of sulfur are bacterial sulfate reduction, plants and soils. Sulfur isotopic ratios provide valuable clues regarding the presence of sulfur-based metabolic activity on the early earth.

The redistribution of the primordial isotopic mixtures can be followed up in terms of the appropriate parameter, namely:

$$\text{delta}^{34}S = [(^{34}S/^{32}S)_{sa} / (^{34}S/^{32}S)_{st} - 1] \times 10^{3} [^{0}/_{00}, \text{CDT}]$$

For simplicity this function will be referred to as the delta ^{34}S parameter, or simply as the delta parameter. Its value is close to zero when the sample coincides with the corresponding value of the Canyon Diablo triolite (FeS) that is a meteorite, abbreviated as CDT.

This parameter allows a comparison of a sample (sa) with the standard (st) CDT. The relevant terms are the dominant sulfur isotope (^{32}S) and the next in abundance (^{34}S). In fact, $(^{34}S/^{32}S)st$ coincides with the average terrestrial fraction of the two most abundant isotopes of sulfur. We obtain positive values of the delta-parameter when by comparison we have a larger quantity of the less abundant isotope ^{34}S.

Sulfur isotopic values of delta ^{34}S for sulfide (pyrite) and sulfate (barite) minerals in the early Archaean display a relatively narrow spread around delta $^{34}S = 0 \pm 3\ ^0/_{00}$ for sulfides and delta $^{34}S = 4 \pm 1\ ^0/_{00}$ for sulfates (Strauss, 2003). Archaean oceans at 3.5 billion years ago (Ga) were sulfate rich and sulfides were formed by sulfate reducing bacteria (Ohmoto, 1992). The average sulfate content of the mantle is about 300- 400 ppm (Gehlen, 1992).

Sulfur isotope ratios in mantle sulfur are close to meteorites (delta $^{34}S = 0.5\ ^0/_{00}$). Sulfur isotopic compositions of sulfides are enriched in ^{32}S and may be of biogenic origin.

THE ANTIQUITY OF SULFUR REDUCING BACTERIA

The presence of pyrite in black shales, chert and phosphorite association in Proterozoic and Early Cambrian formations with delta $^{34}S > +4\ ^0/_{00}$ indicate that sulfate and sulfide reducing bacteria were present in these depositional environments (Skyring and Donnelly, 1982, Tewari, 1984, 1996, Krajewski et al, 1994).

Precambrian-Cambrian boundary black shale-pyrite, stromatolite and small shelly fauna indicate highly reducing palaeoenvironment in lagoonal facies, where sulfate reducing bacteria must have flourished (Tewari, 1984, 1994, 1996, Tewari and Qureshy, 1985). Neoproterozoic carbonates (1000 - 540 million years before the present) of the world are characterized by positive delta ^{13}C values (Tewari and Sial, 2007, Tewari, 2007), where is defined as follows:

$$\text{delta}\ ^{13}C=[(^{13}C/^{12}C)_{sa}/(^{13}C/^{12}C)_{st}-1]\times 10^3[^0/_{00}, \text{PDB}]$$

The value of delta ^{13}C is close to zero when the sample coincides with the PeeDee belemnite standard (PDB) in which $(^{13}C/^{12}C) = 88.99$ and delta ^{13}C is defined as equaling 0.00 $^0/_{00}$. This parameter can be used as a good biosignature. On the Earth biota, for instance, there is ample evidence that photosynthetic bacteria, algae and plants have typical significant deviations that yield values of up to - 30 and beyond, due to biological processes (Schidlowski et al., 1983).

But the main point that we have emphasized in the past is that negative values of the delta ^{13}C parameter do not arise exclusively from biogenic sources. For this reason we have mentioned in the present paper that sulfur is a better biomarker for the study of possible biosignatures.

THE ROLE OF SULFUR IN THE EXTINCTION OF LIFE

This could be related with global oceanic anoxia near the Precambrian/Cambrian boundary. This was also the period when soft-bodied Ediacaran metazoans declined (extinction?) on earth and another biological diversification of Cambrian life took place. It is interesting that small shelly fauna (protoconodonts and conodonts, phosphatised oncolites and stromatolites having pyritic and phosphatic microlaminae and algae) were restricted to the reducing (non photosynthetic) environment (Tewari, 1984, 1989, 1994, 1996, 2004, 2007, Tewari and Qureshy, 1985).

The global sulfur isotopic trends suggest a major increase in the importance of sulfate reducing bacteria with rising sulfate levels (Lambert and Donnelly, 1992). The decreasing delta ^{34}S values for sedimentary pyrite and increasing D34S for sea water sulfate may be the result of widespread reduction under conditions similar to those of modern oceans (Lambert and Donnelly, 1992).

DISCUSSION AND CONCLUSION

We would like to restate the main conclusion of the previous paper (Seckbach and Chela Flores, 2007). S isotope analysis is the most valuable for planetary exploration. In situ analysis of the Europan surficial patches of sulfur, together with carbon signatures could yield a clearer interpretation of biosignatures. In the present paper we have indicated that geological and biogeochemical data from many sources of the Precambrian demonstrate that pyrites and evaporates were formed biologically by dissimilatory sulfate reduction (Schidloswski et al, 1983, Konhauser, 2007). Rocks of Archean age [older than 2.5 Gyr BP] provide the best evidence of early metabolic processes. Their study allows reconstruction of the biogeochemical cycle for sulfur since the origin of life on Earth. The remarkable sulfur icy patches on the Europan surface will inevitably be targets for future space missions that are expected to return to Europa in the next decade.

With landers, or low-cost penetrators that could first of all be tried out on the Lunar surface (Smith and Gao, 2007), we would be in a position to test the redox state of the Europan ocean. Alternatively, the imprint of the possibly biogenic signature of the surficial sulfur would be retained in the dust cloud that surrounds this singular jovian satellite (Kruger *et al.*, 2003).

The arguments in the present paper continue to point towards mass spectrometry as the principal instrumentation for future probing of the Europan patches, either in orbit, or with penetrators of the Moon type.

ACKNOWLEDGEMENTS

We are grateful to the reviewers for greatly enhanc-ing quality of the first draft of the manuscript.

References

Chela-Flores, J.; 2006 : The sulphur dilemma: Are there biosignatures on Europa's icy and patchy surface ? Internat. Jour. of Astrobiol., Vol. 5, pp. 17-22. http://www.ictp.it/~chelaf/ss64.html

Chela-Flores, J.; 2007 : Testing the universality of biology. Internat. Jour. of Astrobiol., in press. Http://www.ictp.it/~chelaf/ss116.html

Gehlen, K.; 1992 : Sulfur in the Earth's mantle : A Review. in: Early Organic Evolution : Implications for Mineral and Energy Resources, In: M. Schidlowski (ed.), SpringerVerlag Berlin Heidelburg, pp. 359-366.

Konhauser, K.; 2007 : Introduction to Geomicrobiology, Blackwell Publishing, Malden MA, USA, pp. 320; 342-343.

Krajewski, K.P., Cappallen, P.V., Trichet, J., Kuhn, O., Lucus, J., Algarra, A.M., Prevot, L., Tewari, V.C., Knight, T, Lamboy , M.; 1994 : Biological processes and apatite formation in sedimentary environment. EclogaeGeol. Helv., Vol. 87, No. 3, pp. 701-745.

Kruger, H., Krivov, A.V., Sremcevi, M. and Grün, E.; 2003 : Impact-generated dust clouds surrounding the Galilean moons, Icarus, Vol. 164, pp. 170-187.

Lambert, I.B., and Donnelly, T.H.; 1992 : Global oxidation and a supercontinent in the Proterozoic : evidence from stable isotopic trends., in : Early Organic Evolution: Implications forMineral and Energy Resources. In : M. Schidlowski (ed.) Springer -Verlag Berlin Heidelburg, pp. 408414.

Machel, H.G.; 2001 : Bacterial and thermochemical sulfate reduction in diagenetic settingsold and new insights. Sediment. Geol. Vol. 140, pp. 143-175.

Machel, H.G., Krouse, H.R. and Sassen, R.; 1995 : Products and distinguishing criteria of bacterial and thermochemical sulfate reduction. Appl. Geochem. Vol. 10, pp. 373-389.

Ohmoto, H.; 1992 : Biogeochemistry of Sulfur and the mechanisms of Sulfide Sulfate mineralization In Archean oceans. in : Early Organic Evolution: Implications for Mineral and Energy Resources. In: M. Schidlowski (ed.) Springer Verlag Berlin Heidelburg, pp. 378-397.

Schidlowski, M., Hayes, J.M. and Kaplan, I. R.; 1983 : Isotopic Inferences of Ancient Biochemistries: Carbon, Sulfur, Hydrogen, and Nitrogen, in: Earth's Earliest Biosphere its Origin and Evolution, In: J. W. Schopf (ed.), Princeton University Press, Princeton, New Jersey, pp. 149-186.

Seckbach, J. and Chela-Flores, J.; 2007 : Extremophiles and Chemotrophs as Contributors to Astrobiological Signatures on Europa: A Review of Biomarkers of Sulfate-Reducers and Other Microorganisms, Invited Talk (6694-32) at Instruments Methods and Missions for Astrobiology X, 2007 SPIE OPTICS AND PHOTONICS SYMPOSIUM, San Diego, California U.S.A, August 2630, 2007, (to be published in the proceedings of the conference).

Smith, A., and Gao, Y.; 2007 : Concepts and instruments for low-cost lunar surface Missions, Geophysical Research Abstracts, Vol. 9, 07927, 2007, SRefID: 16077962/gra/EGU2007A07927, European Geosciences Union.

Strauss, H.; 2003 : Sulfur isotopes and the early Archaean sulfur cycle. Precamb. Res., Vol. 126, pp. 349-361.

Taylor, E.A., Ball, A.J., Barber, S.J., Miljkovic, K., McBride, N., Sheridan, S., Wright, I.P., Zarnecki, J.C., Hillier, J.K.; 2007 : A combined dust impact detector and ion trap mass spectrometer for a Europa orbiter, Geophysical Research Abstracts, Vol. 9, 10928, SRefID: 16077962/gra/EGU2007A10928, European Geosciences Union.

Tewari, V.C.; 1984 : Discovery of Lower Cambrian stromatolites from the Mussoorie Tal Phosphorites, India, Curr. Sci., Vol. 53 (6), pp. 319-321.

Tewari, V.C.; 1989 : Upper Proterozoic Lower Cambrian stromatolites and Indian stratigraphy . Himal. Geol., Vol. 13, pp. 143-180.

Tewari, V.C.; 1994 : Sedimentology of the rocks of Deoban basin, Dhuraphat area, Saryu valley, Eastern Kumaon Lesser Himalaya. Geosci. Jour., Vol. 15 (2), pp. 117-162.

Tewari, V.C.; 1996 : Controls of phosphorite formation superimposed on biological activity in the Lesser Himalaya, India. Geosci. Jour., Vol. 16 (2), pp. 135-153.

Tewari, V.C.; 2004 : Microbial diversity in Meso-Neoproterozoic Formations ,with particular reference to the Himalaya. in : Origins,In: J. Seckbach (ed.), Kluwer Academic Publishers, The Netherlands, pp. 515-528.

Tewari, V.C.; 2007 : The rise and decline of Ediacaran biota : Palaeobiological and stable isotope evidence from the NW and NE Lesser Himalaya, India. Geol. Soc. London, Spl. Publ. Vol. 237(1, 2), pp. 64-88.

Tewari, V.C. and Qureshy, M.F.; 1985 : Algal structures from the Upper Krol Lower Tal Formation of Garhwal and Mussoorie synclines and their palaeoenvironmental significance. Jour. Geol. Soc. India, Vol. 26, pp. 111-117.

Tewari, V.C. and Sial, A. N.; 2007 : Neoproterozpic Early Cambrian isotopic variation and chemostratigraphy of the Lesser Himalaya, India , Eastern Gondwana. Chem. Geol., Vol. 237, pp. 64-88.

Hon'ble Reviewers

1. Prof. J. Seckbach
   The Hebrew University of Jerusalem
   Jerusalem
   ISRAEL

2. Prof. N. Pugliese
   Department of Geological,
   Environmental and Marine Sciences
   University of Trieste
   ITALY

Editorial Assistance

Mr. Prakhar Kumar

Economic Mineralization
K.L. Shrivastava (Editor)
2009. pp.57-69
Scientific Publishers (India), Jodhpur

Section-I

Deep Seismic Reflection Profiling and its Implications to the Metallogeny of the Aravalli-Delhi Fold Belt, NW India

B. RAJENDRA PRASAD AND V. VIJAYA RAO

National Geophysical Research Institute, Uppal Road, Hyderabad-500007.
Email : vijayraov@yahoo.co.in ; rajbitragunta@yahoo.com

Abstract

The NW Indian shield with 3300 Ma old Archean basement is cratonised at 2500 Ma. The region had a long complex history of crustal evolution involving repeated rifting, collision and associated magmatism. It is represented by the Archean Banded Gneissic Complex, Paleoproterozoic Aravalli and Mesoproterozoic Delhi fold belts, intruded by various granitoids ranging in age from 3000-750 Ma. These tectonic activities developed a large number of dislocation zones / mega-lineaments along which major metallogenic activity is observed. They left several imprints which are analysed using geophysical probing, mainly seismic reflection profiling. Seismic images of the deep crustal reflection study delineated a number of thrust faults which are used to understand the metallogenesis of the Aravalli-Delhi fold belt region. The seismic study constrained by geological and other geophysical data suggests that the subduction, arc formation and collision are responsible for the crustal evolution of the fold belt region. Two prominent and distinct Pb-Zn mineralized zones formed at ~1800 Ma and ~1000 Ma seem to be closely related to the crustal evolution. The metallogenic periods of ADFB coincide with the global metallogenic activity and formation of supercontinents Columbia and Rodinia. The present study indicats that seismic images of fold belts when correlated with magnetotelluric data can effectively decipher metallogeny of a region. It also suggests a correlation between the subsurface structure, orogeny and metallogey in a plate tectonic framework.

Keywords: Deep Seismic Reflection Profiling, Metallogen, Aravalli-Delhi Fold Belt, Crustal evolution, Subduction, Collision.Belt, FluidRock Interaction.

INTRODUCTION

Knowledge of the deeper parts of the earth's crust and upper mantle is essential to understand the evolution of various structural features present on the surface. The shallow surface features of any region are in general controlled by the deep-seated geological processes that take place at depth. The distribution of structural forms within the earth's crust and mineral deposits on the surface has a close dependence on the geological processes occurring in the crust and upper mantle. Hence, it is very important to understand the significance of the crustal and sub-crustal lithospheric structure and their relationship with the surface features.

The main source of most of the metals in the Earth's crust is upper mantle. The crust-mantle interaction processes are generally responsible for their formation. Some of the metal deposits are found to be

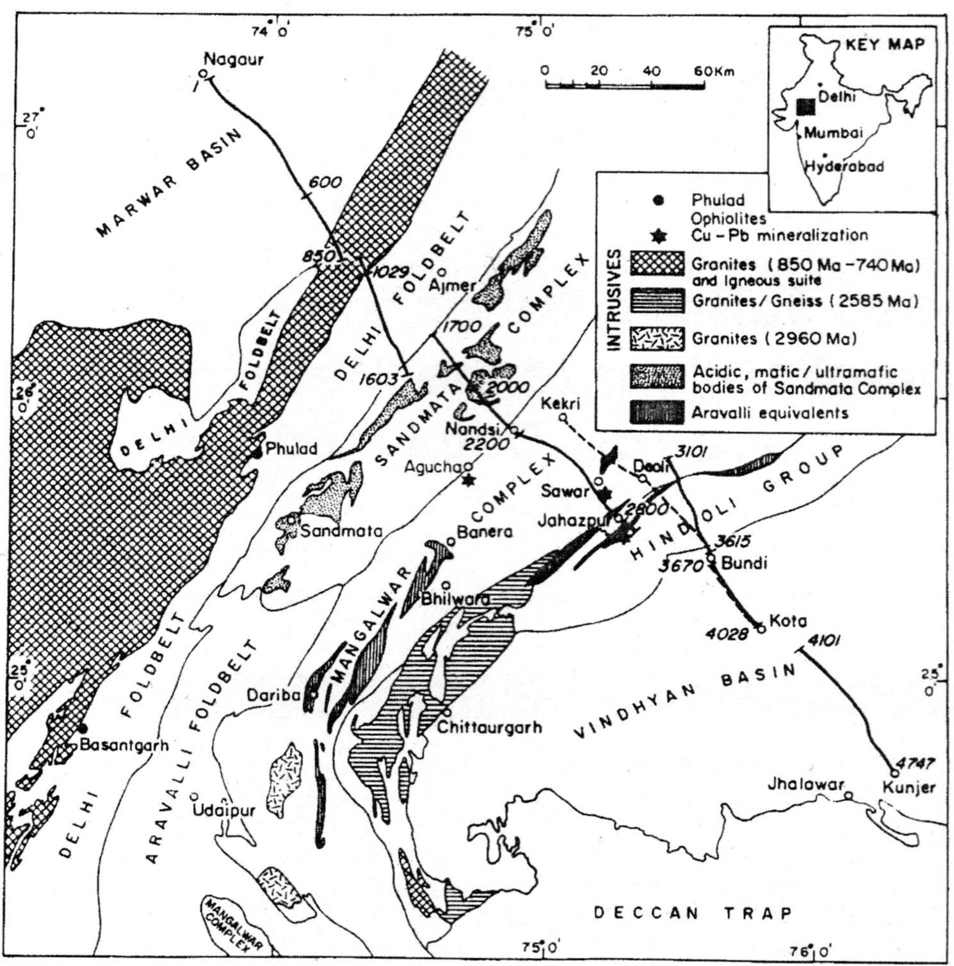

Fig. 1 Generalised regional geological map of the Aravalli-Delhi orogenic belt of the NW Indian shield along with the Nagaur-Jhalawar seismic transect.

time specific, following the secular time-bound changes in the evolution of the earth-system, namely the lithosphere, hydrosphere, biosphere and atmosphere. The present study mainly deals with the evolution of upper part of the lithosphere and its relationship to the metallogeny of the NW Indian shield. The NW Indian shield covering a major part of Rajasthan, and minor part of Gujarat and Haryana, is one of the most important metallogenic provinces of the country. It has the unique distinction of having the highest reserves of lead-zinc, phosphorites and evaporite minerals in the country. A large number of geological and geophysical studies are being carried out for a long time to the exploration of minerals by various organizations. Most of these studies were confined to the shallow part of the crust that are of economic importance. However, the structure of the deeper part of the crust and its relationship to metallogeny was poorly understood.

A deep seismic reflection profiling which provides the subsurface structural images of the crust and subcrustal lithosphere is carried out along a 400 km long Nagaur-Jhalawar transect (Fig. 1) to understand the crustal structure and tectonic evolution of the region (Tewari et al., 1998; Rajendra Prasad et al., 1998; Vijaya Rao et al., 2000). These seismic images reveal the deep tectonic processes and provide important clues essential to understand their surface manifestations. They unravel the interworking of the Earth's endogenic and exogenic processes and in turn the metallogenesis of the region. Even though there are various tectono-stratigraphic models available based on the geological data, there is no comprehensive understanding of the subsurface structure derived from the geophysical data with the metallogeny of the region. Present paper is an attempt in that direction with emphasis in understanding the structure, crustal evolution and metallogeny. The present paper deals with the correlation of the seismic structure of the Aravalli-Delhi fold belt with the metallogeny in a plate tectonic framework.

GEOLOGY AND TECTONICS

The NW Indian shield is the only region in India where records of all the geological events starting from the early Archean (~3.3 Ga) to Neoproterozoic (~550 Ma) are preserved. It is regarded as one of the most important Precambrian terrains of India because of the presence of a number of fold belts, ductile shear zones, metallogenic provinces and paleo-sutures. The lithostratigraphy of the NW Indian shield is mostly due to the pioneering work of Heron (1953), who identified the geological evolution of the region through three major orogenic cycles namely, the Archean Banded Gneissic Complex (BGC), the

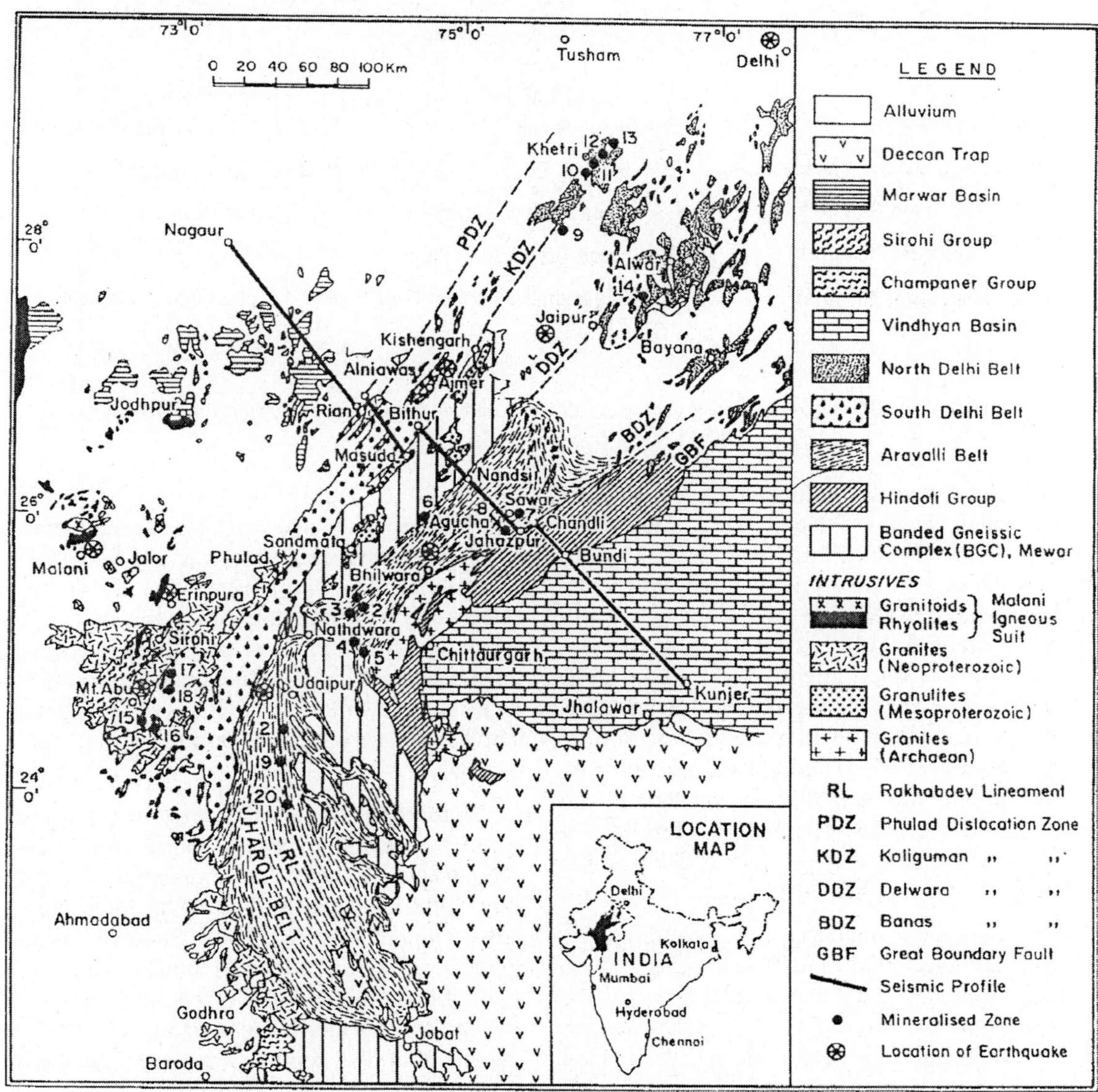

Fig. 2 Tectonic map of the Aravalli-Delhi orogenic belt of NW India (modified after Roy, 1988 and Deb et al., 1989) showing the locations of mineralised zones (Pb-Zn, Cu, W, Sn). 1) Bethumni 2) Sindeswar Kalan 3) Rajpura 4) Dariba 5) Wari 6) Rampura-Agucha 7) Sawar 8) Jahazpur 9) Saladipura 10) Akwali 11) Chandmari 12) Kolihan 13) Madan Kudan 14) Kho-Dariba 15) Deri 16) Ambaji 17) Basantgarh, 18) Golia, 19) Zawarmala 20) Rakhabdev 21) Umra. Uranium Deposits : Phulad, Kishangarh.

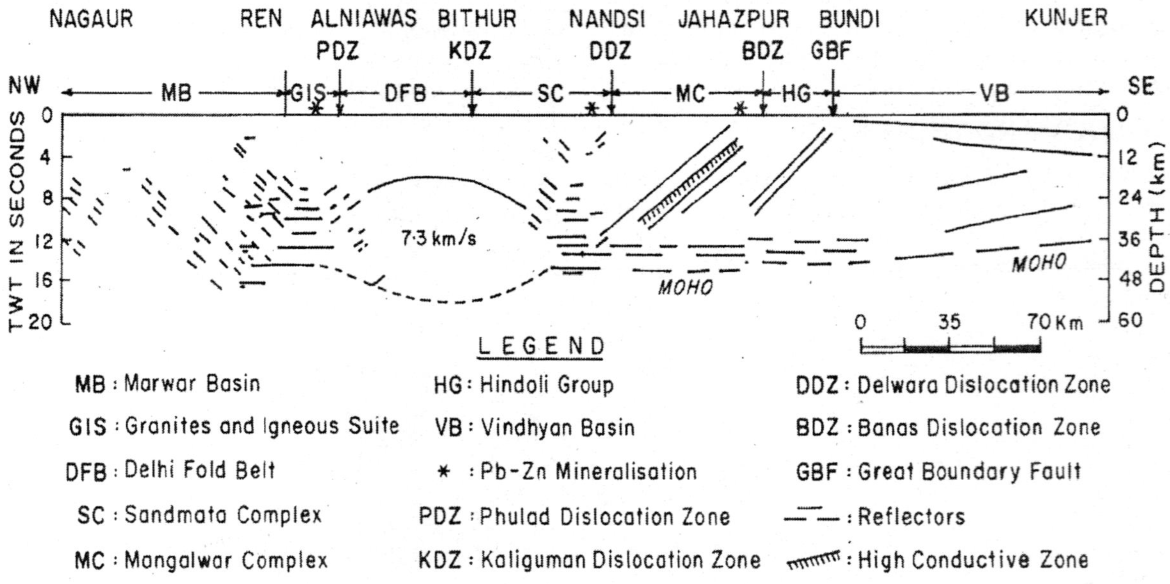

Fig. 3 Line drawing representing the crustal structure derived from the deep seismic reflection study.

Paleoproterozoic Aravalli fold belt (AFB) and the Mesoproterozoic Delhi fold belt (DFB). The DFB consisting of the Khetri, Alwar and Bayana-Lalsot basins located to north of Ajmer is referred to as the North Delhi Fold Belt (NDFB) and the one to its south as the South Delhi Fold Belt (SDFB). In the present study the SDFB is intermittently referred as DFB. Other important events of the region are the post-Delhi magmatic activity witnessed in the form of Erinpura granites (850 Ma) and Malani igneous suite (750 Ma) in the Marwar Terrain, located on the western side of the DFB. The Marwar Terrain is a distinct Neoproterozoic anorogonic crustal block that separates from orogenic ADFB in the east.

The oldest cratonic nucleus (Mewar craton) of the NW India exhibits 3300 Ma age (Gopalan et al., 1990; Roy and Kroner, 1996) and usually referred to as the Banded Gneissic Complex (BGC). It is the Archean basement for the subsequent tectonic activities of the region. The Paleoproterozoic Aravalli and Mesoproterozoic Delhi sediments were deposited over this basement. Subsequently, they were deformed, metamorphosed and folded and presently represented by a 700 km long mountain chain (Aravalli-Delhi fold belts) extending from Palanpur in the SW to Delhi in the NE (Fig.2). The unmetamorphosed and undeformed sediments of the Meso-Neoproterozoic Vindhyan and Neoproterozoic-Eocambrian Marwar basins were located in the eastern and western flanks of this fold belt region. The region exhibits variety of metamorphic activity viz. the greenschist, amphibolite, granulite and blue schist facies. It is reworked by multiple recurrent magmatic activities of varying ages during its long geological history.

Tectonic evolution of the region is suggested by several research workers using geological and geophysical data in a plate tectonic framework (Sinha-Roy, 1988; Deb and Sarkar, 1990; Sinha-Roy et al., 1995; Vijaya Rao et al., 2000). Some of these models suggest rifting of a crustal block, whereas others suggest involvement of two separate cratonic blocks viz. the Mewar and Bundelkhand. All these models suggest that the region went through an orderly Wilson cycle operations during the Proterozoic in a manner similar to that of late Phanerozoic orogenic belts. Rifting, sedimentation, subduction and collision processes were found to be responsible for the crustal evolution of the Proterozoic Aravalli-Delhi fold belt region. (also see, Sinha Roy, This Volume)

Table : Major mineral deposits of Aravalli-Delhi fold belt region

S.No	Tectonic Domain	App. Age (in Ma)	Metal Deposit	Name of the Locality
1.	AFB	1800	Zinc-Lead	Rajpura-Dariba, Rampura -Augucha, Pur -Banera, Jahazpur, Zawar, Bethumni, Sindeshwar Kalan, Wari, Mochia Magra
			Uranium	Balaria Zawarmala, Baroi Magra Umra-Udaisar (near Udaipur), Devpura, Amalda, Chainpura, Kherpura, Putholi
			Phosphorites	Jamarkotra, Kanpur,
2.	NDFB	1800-1500		Matoon (all near Udaipur)
			Copper	Khetri, Alwar and Bayana-Lalsot basins. Kho-Dariba, Madan Kudan, Kolihan, Chandmari, Akwali, Satkui-Dhanaota
			Pyrite-Pyrrhotite	Saladipura
			Tin	Tosham Khetri & Alwar areas : Ghateswar, Diara,
			Uranium	Kerpura, Maonda, Kushlawala ki Dhani, Dhanota, Antri, Bairat
3.	SDFB	1100	ZincLead	Ambaji-Deri, Basantgarh-Pipela-Golia, Birantya, Khurd, Kalabar, Chitar
4.	Eranpura (Post-Delhi) Granites	850	Tungston	Degana, Balda, Phalwadi-Positara, Sewariya
5.	Malani Igneous Suite	732	Tin and Tungston Fluorite	Tosham, Mungeria (Both, only accurrence) Jalerakhurd, Karara
6.	Marwar Supergroup	570	Phosphorites	Birmania
			Evaporites	Nagaur, Didwana

METALLOGENY

The tectonic activities developed a large number of maga-lineaments / dislocation zones separating different lithotectonic domains along which variety of metallogenic activity is observed. The Proterozoic ADFB hosts a large number of economically important strata bound basemetal sulphide deposits and a few others such as tungsten, tin, uranium, phosphorites fluorite etc. Majority of these deposits are confined to the tectonic boundaries, whereas others are located in the interior parts. The basemetal deposits are basically divided into Paleoproterozoic sediment-hosted highly deformed and metamorphosed Aravalli sequence (~1800 Ma) and Mesoproterozoic Volcanic-associated Massive Sulphide (VMS) type of the SDFB (~1000 Ma). Various metal deposits identified in different tectonic domains of the region are shown in the table. But, no attempt has been made to show the exhaustive list of all metal deposits of the region.

SEISMIC STUDY

A deep crustal seismic reflection profiling was carried out across the Aravalli-Delhi fold belt in the NW-SE direction covering most of the important tectonic domains of the region. It traverses the Marwar Terrain, Delhi fold belt, Sandmata Complex (Banded Gneissic Complex), Mangalwar Complex, Hindoli group and finally the Vindhyan basin. The multifold reflection data were acquired using 35 kg explosives loaded in 20 m depth shot holes and the resultant seismic energy is recorded with 120 channel, DFS-V, digital seismic equipment. The data were processed using the standard reflection processing sequence (Rajendra Prasad et al., 1998; Vijaya Rao et al., 2000).

DISCUSSION

The reflection study has delineated a number of thrust faults viz. the Jahazpur thrust (Rajendra Prasad et al., 1998), South Delhi thrust (Vijaya Rao et al., 2000) and the Great Boundary thrust fault (Rajendra Prasad and Vijaya Rao, 2006). Some of these thrust faults developed counter thrusts in the opposite direction. Most of these thrust faults extend from shallow part of the crust to the crust-mantle boundary located at various depth levels from 36 to 46 km. The line drawing derived from the seismic sections indicating these faults is shown in figure 3. Based on the deep crustal seismic data, the evolution of these thrust faults are explained within a plate tectonic framework involving subduction, generation of its related volcanic arcs and their subsequent collision with the interaction of the Mewar and Bundelkhand cratons (Vijay Rao et al., 2000). Similar models are also proposed based on the geological and geochemical data (Sinha-Roy, 1988; Deb and Sarkar, 1990). Presence of metal deposits along the tectonic boundaries and their structure derived from seismic images suggest the necessity for a proper understanding between crustal evolution and metallogeny.

The line drawing of the seismic reflection study (Fig. 3) shows a 30 km wide prominent thrust fault with 80 km of lateral extension and is referred as the Jahazpur thrust. It extends almost from the surface to crust-mantle boundary. It represents the paleo-subduction / suture zone formed in a compressional environment. The most striking evidence for the subduction of sediments together with interstitial brines has come from magnetotelluric soundings (Fig. 4a) made across the Jahazpur thrust (Gokaran et al., 1995; Rajendra Prasad et al., 1999). The thrust fault shows high conductive anomaly of 50 ohm.m. A large wavelength Bouguer gravity high coincides with the Jahazpur thrust (Figs. 5 and 6). The geophysical data along with geological information suggest presence of paleo-island arc signatures in the AFB region (Vijaya Rao et al., 2000). The sediments initially deposited under extensional regime were later subjected to compression during the evolution of the Aravalli orogeny. Subduction, development of island arc and subsequent collision were found to be responsible for the tectonic evolution of the AFB at ~1800 Ma (Deb and Sarkar, 1990; Sinha-Roy et al., 1995; Vijaya Rao et al., 2000). During this period extensive orogenic activity took place globally evolving a large number of fold belts namely, the Trans Hudson, Wopmay, Penokean, Sveco-Fennian, Capricorn orogeny etc. with the formation of supercontinent Columbia (Vijaya Rao and Reddy, 2002). The Pb-Zn deposits (~1800 Ma) of Rajpura-Dariba, Rampura-Augucha, Pur-Banera, Jahazpur, Zawar (refer Table) etc are located at the tectonic boundaries, mainly the Banas and Delwara dislocation zones of this fold belt (Fig.2). These evidences suggest a close relationship between orogeny and metallogeny.

The basemetal deposits of the Paleoproterozoic AFB region are regarded as sediment-hosted massive sulphides which were formed by convective seawater circulation in zones of crustal extension at ~1800 Ma (Deb and Sarkar, 1990). Further, metal content of brines are found to indicate prolific biological activity at the time of their formation. However, the subsurface structure derived from deep seismic reflection study suggests that the major metal deposits are located at the locations of thrust faults developed during the subduction and collision process. The geochemical data from the metavolcanics of AFB resemble low-k tholeiites to calc-alkaline basalts suggesting their origin from mantle source and is consistent with their emplacement in an ocean-floor setting (Deb et al., 1989). The Aravalli belt near Udaipur is charecterised by shelf and deep-sea facies lithologies (Roy and Paliwal, 1981) with ultamafic beccia and pillowed basalts occurring as a series of composite slivers in between them, referred as the Rakhabdev suture (Fig. 2). These ultramafic and associated rocks represent ophiolitic mélange (Mohanty et al., 1993) and possibly the tectonic slices of obducted Jharol oceanic crust emplaced during the evolution of AFB (Sinha-Roy, 1988). These evidences further substantiate the inferences drawn from the geophysical data.

The seismic reflection image observed in the form of southeast dipping reflection band on the western boundary of DFB (Fig. 3) represents the South Delhi thrust fault. This thrust fault extends upto the Moho and beyond. World wide such signatures are generally observed at the paleo-subduction zones (Steer et al., 1998). The profuse basic volcanics and tectonised mafic / ultramafic rocks of the DFB, collectively named as the Phulad ophiolites (Fig. 1), are regarded as the dismembered fragments of oceanic crust (Sinha-Roy and Mahonty, 1988). The geochemical characteristics of this ophiolitic suite are similar to low-k tholeiites to calc-alkaline basalts occurring close to trenches in island arcs (Bhattacharya and Mukherjee, 1984). Further, Volpe and MacDougall (1990) based on geochemical and isotopic studies from the rocks of DFB concluded that they are the

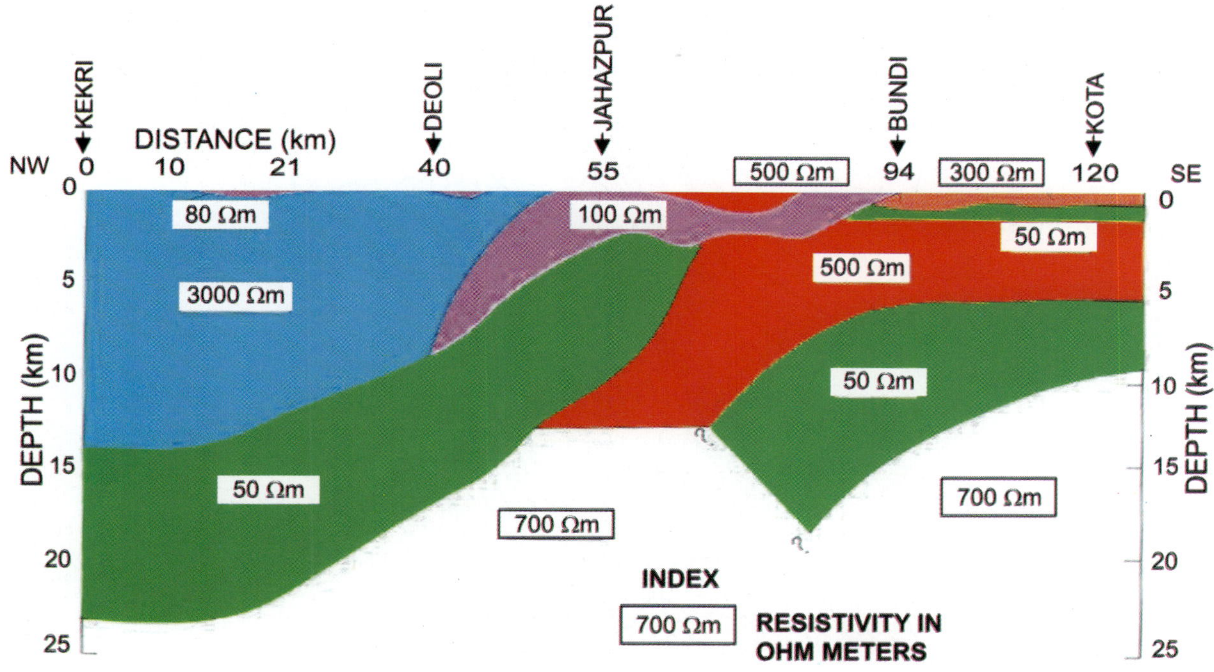

Fig. 4 .a

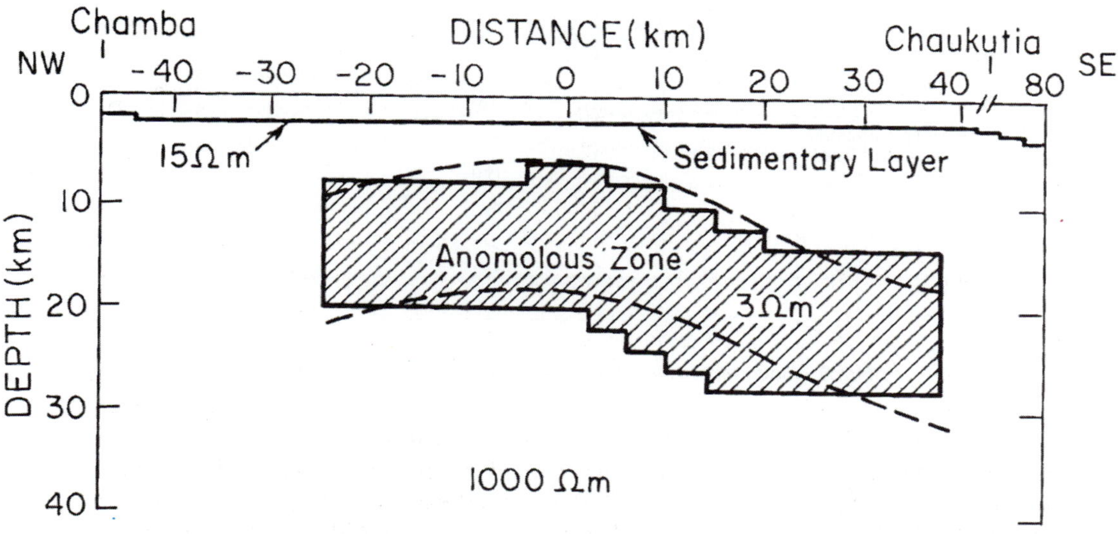

Fig. 4 :b

Fig. 4 : Magnetotelluric signature across **a)** Aravalli Fold Belt **b)** the northern part of Delhi Fold Belt.

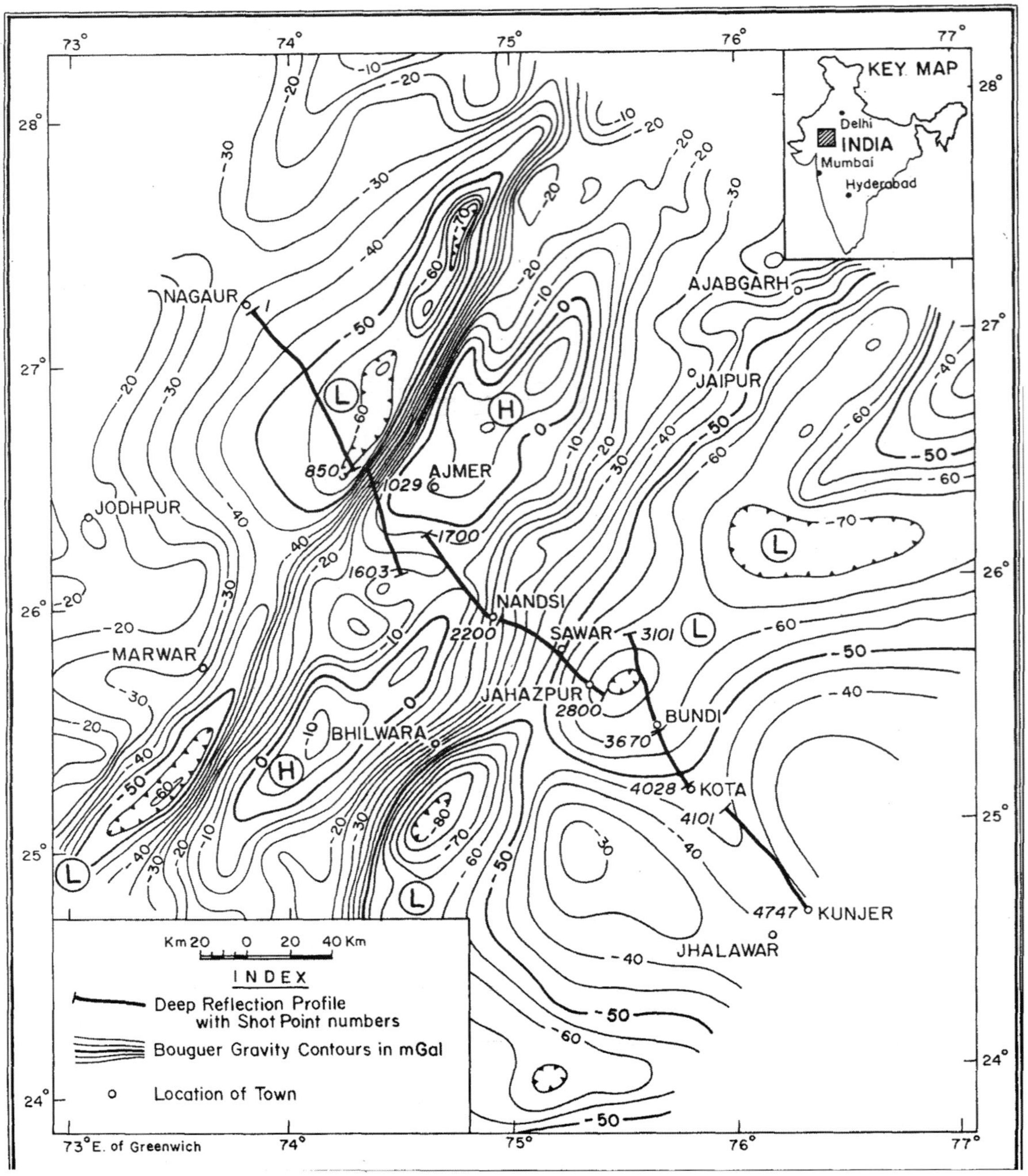

Fig. 5 Bouguer gravity map of the ADFB region (after NGRI, 1978)

fragments of a Proterozoic island arc complex formed at ~1000 Ma. These evidences suggest presence of a paleo-arc at the western boundary of DFB. The Bouguer gravity map of the region (NGRI, 1978) shows gravity high of the order of ~70 mGal with a steep gradient all along the western boundary of DFB (Figs. 5 and 6). It coincides with the South Delhi thrust fault. Such a regional feature must be closely related to the crustal evolution. The magnetometric array studies carried out across the northern part of DFB (Lilley et al., 1981) show high conductive anomaly of 3 ohm.m (Fig. 4b). The subsurface seismic images at the SDFB with extensive mafic volcanism, ophiolitic melange, and island-arc-trench signatures suggest subduction-arc formation and collision during 1100-1000 Ma is mainly responsible for the tectonic

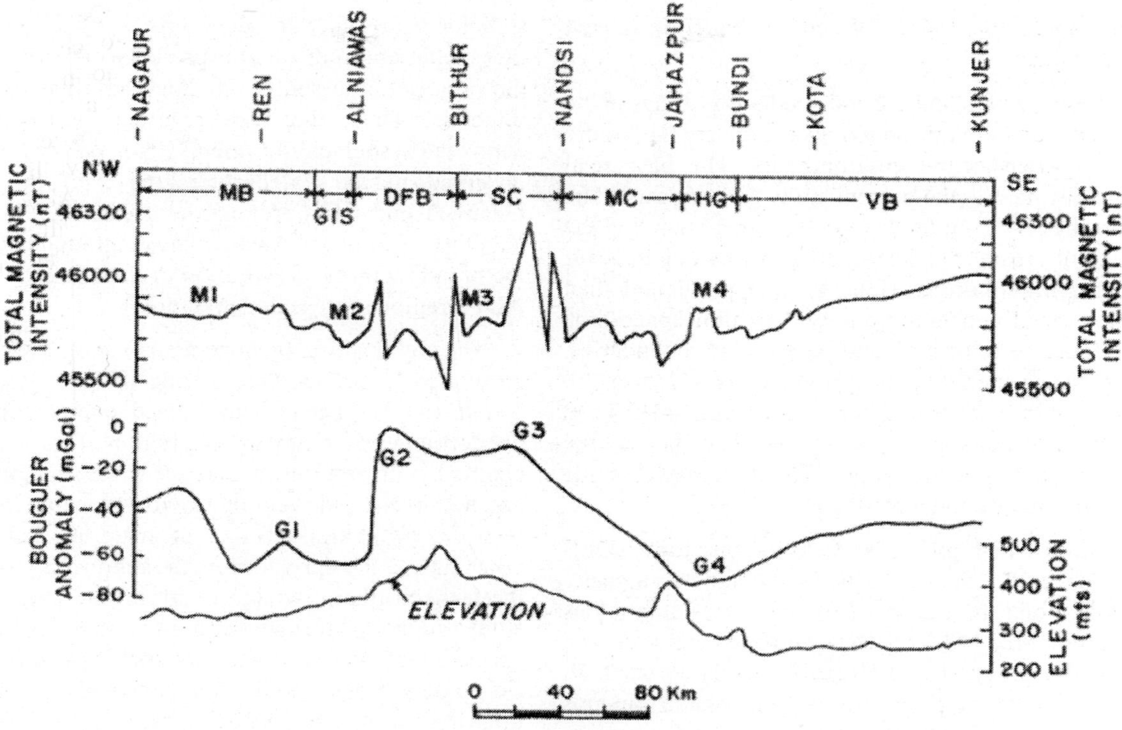

Fig. 6 a) Magnetic b) Gravity c) Elevation profiles along the Nagaur-Jhalawar transect (after Mishra et al., 1995).

evolution of the region. During this period, large scale global orogenic activity is observed forming the Grenvillion, Sveco-Norwegian, Sunsas, Albany-Fraser, Pinjara, Mozambique orogeny etc with the evolution of Rodinia supercontinent (Hoffmann, 1989). The Pb-Zn (~1000 Ma) deposits of Ambaji-Deri, Basatgarh-Pipela-Golia, Birantiya, Kalabar, Chitar (refer table) etc are located along the western boundary of DFB / Phulad dislocation zone (Fig. 2). It is observed from the geological studies that these deposits are characterized as VMS type. Thus, it can be inferred that the mineralization observed along the DFB must be related to the crustal evolution and orogeny.

Metallogeny is intimately related to the magmatic activity of a region. The magmatic processes in the lower crust / mantle act as a source for the ore body and also provide necessary heat which sets the deep water in motion and generates a convective hydrothermal system. The source for water could be meteoric, magmatic, metamorphic or sea water. The water in contact with high heat flow absorbs heat and rises up being lighter. These hot solutions leach metals from the rocks through which they pass by progressive fluid-rock interaction and transport the derived ore fluids via a weak fault zone to ultimately precipitate them near the surface (Sakins, 1990; Mills et al., 1981).

The processes of crustal extension, back-arc spreading or compression at the subduction zone of the study area allowed the sea water to percolate downwards producing convective geothermal systems that extend progressively to greater and greater depths with time and brought the ore fluids to surface. The deep-seated fault zones, the low angle thrusts and counter thrusts namely, the Jahazpur thrust, South Delhi thrust fault etc. of the subduction zones acted as channels to ore-generating system. Thus, metal deposits are formed where convective hydrothermal systems operated with appropriate transport mechanism to carry ore fluids from the deeper depths to surface. Such regions of the world posses a notable concentration of deposits of a certain metals and these regions are known as metallogenic provinces. The convective hydrothermal system developed during the subduction and collisional episodes must have enhanced the process of metallogenesis and probably indicate multiple sources atlest for some of the metals in the region. Thus, it is inferred that the metal deposits are atypical of massive sulphide ores formed in subduction-related environments.

Presence of island arc and subduction zones in the present study area provided all characteristics of ore forming-transporting environments. The basemetal mineralization in the present study area appears to be controlled by deep faults / geofractures bounding the different crustal blocks which serve as conduits for ore-bearing solutions. Thus, it can be predictable that major metal concentrations will exhibit space-time relationships to present and paleo- plate boundaries. Large-scale metallogenic provinces world over are associated with subduction zones (Sawkins, 1984 and 1990). The geological and geophysical data of the region suggest a probable correlation between the crustal structure and metallogeny.

The gravity high is observed over the entire ADFB region (Fig. 5). Similarly, the total magnetic intensity profile along the transect (Fig. 6) shows anomalous features (M2, M3 and M4) confined between the South Delhi thrust and Great Boundary thrust fault (Fig. 3) indicating concentration of magnetic minerals in this region. The observed gravity and magnetic anomalies seems to be closely related to the basemetal deposits of this region. On the other hand, magnetic signature is more or less smooth in the Marwar and Vindhyan basins with a steady rise of magnetic values in the SE direction. The rise in these values is attributed to the magnetic basement in the region.

Subsequent to the Delhi orogeny, the Marwar Terrain located to the west of DFB experienced granitic (Erinpura) and felsic (Malani) magmatism during 850-750 Ma. The Malani magmatism with characteristic ring structures and radial dykes is related to the mantle plume activity (Kochaar, 1984). Malani volcanics represent third largest felsic magmatism in the world and largest in India (Preek, 1981) and wide spread up to Tusham in the northeast and Kirana hills in Pakistan in the northwest. Presence of major Sn-W deposits in the Marwar Terrain suggests its relationship with the Malani igneous suite. The period of Malani volcanism (732 Ma) coincides with the breakup of Rodinia supercontinent. Thus, it can be concluded that in a plate tectonic framework these metallic deposits can be related to the rifting of Rodinia supercontinent.

Very large deposits of Phasphorite and evaporite (salt and gypsum) sequences of Neoproterozoic-Eocambrian (750-450 Ma; cf. 570 Ma) ages are observed overlying the Malani volcanics of the Marwar Supergroup. They are referred as the sediments of Marwar super group. Occurrence of these sediments indicates the presence of ocean, as it is the only possible source for huge amounts of these minerals. Thus, they are regarded as sediments deposited in shelf environments. Phosphorites are also observed in the Paleoproterozoic Aravallis (near Udaipur) and Mesoproterozoic Delhi (Alwar) and Vindhyan sediments. A schematic diagram illustrating the broad features of evolution of metallogeny in the ADFB region is presented in figure 7.

The composition (geochemistry) of ores can often be related to the tectonic setting under which they were formed and hence can be used to understand the ore-forming environment of a region. This evidence coupled with the subsurface structure, magmatism and tectonics of the region provide probable insights into the orogenesis. Bulks of material that build volcanic arcs are derived from the subducting slab and the asthenospheric wedge overlying it. The present study suggests formation of specific deposits at specific tectonic environments such as sulphides of Zn-Pb deposit near subduction zone and arc, Sn-W in granites due to mantle plume activity, phosphorites and evaporites near shelf area and Cu mostly in less deformed rift basins etc. Thus, it can be inferred that the ore deposit geology indicate paleo-environment and probable tectonic setting. It is generally observed that Zn, Pb and Ag are the most characteristic metals of the back arc deposits (Sakins, 1990). Tin deposits are associated with anorogenic granites which may be generated due to mantle plume activity. The present study suggests a close relationship between the subsurface structure, crustal evolution and metallogeny. The geological data also suggest that plate tectonics control the Precambrian deformation and metallogeny of the NW Indian shield (Sychanthavong and Desai, 1977).

Lineaments represent fundamental flaws in the continental crust and are the major factors in the control of metallogenesis. They help in understanding spatial distribution of major metal deposits. They have controlled geological processes such as rise of magma and fluid flow. But, lineaments simply do not provide adequate models for understanding the space-time distribution of metal deposits. Conversely, models based on plate tectonic theory permit the division of most metal deposits into subtypes that are distinctive in tectonic, lithologic and genetic terms. Many ore-generating systems can be viewed as the culminations of normal geological processes. Such systems might have undoubtedly repeated themselves through

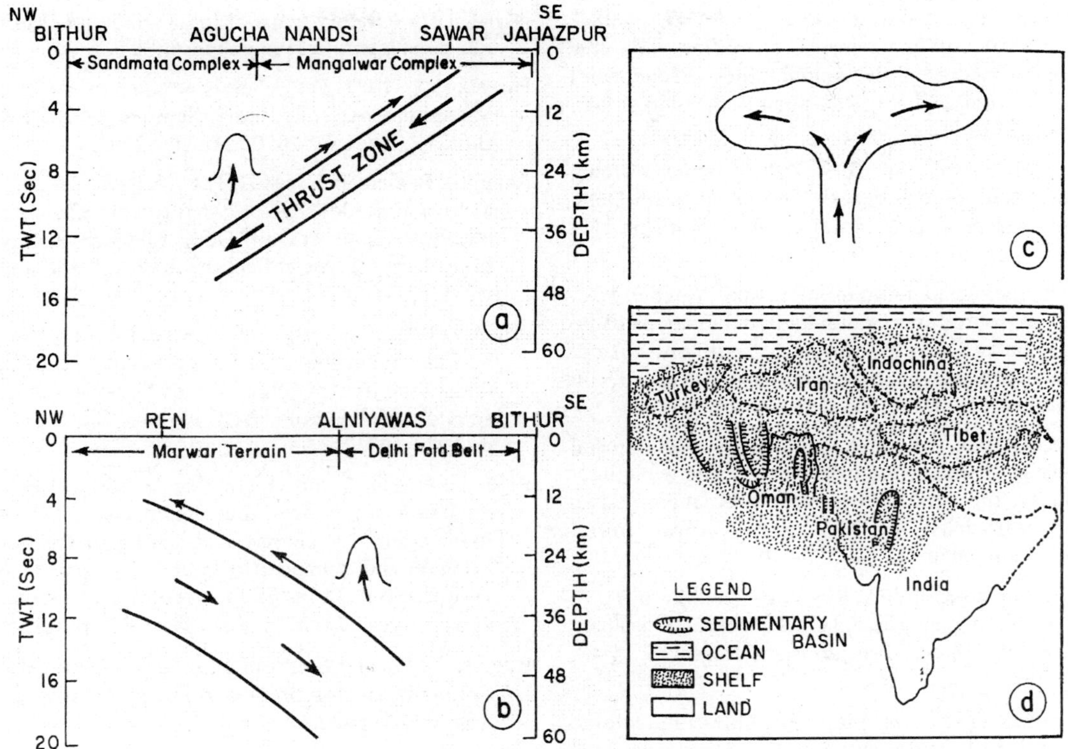

Fig. 7 Schematic diagram illustrating the broad features of evolution of metallogeny in the ADFB region. a) Aravalli fold belt (~1800 Ma), b) Delhi fold belt (~1100 Ma) c) Malani mantle plume magmatism (~750 Ma), d) Paleo-shelf of NW India (~550-400 Ma)

geologic time. As plate tectonics have operated since late Archean, then a variety of environments favorable for ore generation must have repeated themselves through geological time as observed in the present study area. The period of Zn-Pb deposits of the present study area show two distinctly separate ages one around 1800 Ma and another ~1100 Ma. They indicate repetition of subduction, arc formation and collision tectonics for their formation with a gap of ~700 Ma. These metallogenic periods coincide with the end of Paleo- and Mesoproterozoic periods and formation of supercontinents Columbia and Rodinia (Hoffman, 1989; Rogers and Santosh, 2002; Vijaya Rao and Reddy, 2002). These evidences clearly indicate operation of plate tectonics since Paleoproterozoic with a Wilson cycle period of ~700 Ma. The deposition period of these metallogenic provinces are close in time to the global orogenic activity. Various types of mineralization are correlatable with large scale crustal structures. Plate convergence, spreading centres, mantle plumes, rifting and collision tectonics are the important features that control the large-scale metallogenic province of a region.

CONCLUSIONS

* Metallogeny of the Aravalli-Delhi fold belt region seems to be intimately related to the crustal evolution. A close relationship exists between the crustal structure, tectonics and mineralization of the region.
* Basemetal deposits of the region are confined to the paleo-subduction and collision zones.
* Plate tectonic framework provides the relevant models to understand the ore genesis and space-time distribution of large-scale metallogenic provinces. We strongly believe that large-mineralised provinces necessarily be explained within plate tectonic framework.

ACKNOWLEDGEMENTS

We thank Prof. K.L. Shrivasthava for inviting us to contribute a paper to the special volume in honour of Prof. Paliwal and Dr.V.P.Dimri, Director NGRI for permission to publish this work. We also thank Sri M.Shankaraiah for the drawings.

References

Bhattacharya, P.K. and Mukherjee, A.D., 1984. Petrochemistry of metamorphosed pillow and the geochemical status of the amphibolites (Proterozoic) from the Sirohi district, Rajasthan India. Geo.Mag., v.121, pp.465-473.

Deb, M. and Sarkar, S.C.,1990. Proterozoic tectonic evolution and metallogenesis in the Aravalli-Delhi orogenic complex, NW India. Pre.Res., v.46, pp.115-137.

Deb, M., Thorpe, R.I., Cumming, G.L. and Wagner, P.A., 1989. Age, source and stratigraphic implications of Pb isotope data for conformable, sediment-hosted, basemetal deposits in the Proterozoic Aravalli Delhi orogenic belt, NW India. Pre. Res, v.43, pp.1-22.

Gokaran, S.G., Rao, C.K. and Singh, B.P., 1995. Crustal structure in southeast Rajasthan using magnetotelluric techniques. In: S.Sinha Roy and K.R.Gupta (Eds.), Continental crust of NW and Central India, Geo.Soc.India, Mem. 31, pp. 373-381.

Gopalan, K., Macdougall, J.D., Roy, A.B. and Murali, A.V., 1990. Sm-Nd evidence for 3.3 Ga old rocks in Rajasthan, northwestern India. Precambrian Research, v.48, pp.287-297.

Guilbert, J.M., 1981. A plate tectonic-lithotectonic classification of ore deposits. In : W.R.Dickinson, and W.D.Payne (Eds.), Relations of tectonics to ore deposits in the southern Cordillera. AR Geol. Soc Digest, v.14, pp1-10.

Heron, A.M., 1953. The geology of Central Rajputana: Mem.Geol.Surv.India, v.79, pp.389.

Hoffman, P.F., 1989. Precambrian geology and tectonic history of North America, In: A.W.Bally and A.R.Palmer (Eds.), The geology of North America - An overview. Geo. Soc. Amer., Boulder Co., pp. 447-512.

Kochhar, N. 1984. Malani Igneous Suite : hotspot magmatism and cratonisation of the northern part of the Indian shield. Jour.Geol.Soc.India, v. 25, pp.155-161.

Lilley, F.E.M., Singh, B.P. Arora, B.R., Srivastava, B.J., Prasad, S.N. and Sloane, M.N., 1981. A magnetometer array study in NW India. Phys.Earth Plant.Int., v.25, pp. 232-240.

Mishra, D.C., Laxman, G., Rao, M.B.S.V. and Gupta, S.B., 1995. Analysis of gravity-magnetic data around Nagaur-Jhalawar geotransect. In: S.Sinha Roy and K.R.Gupta (Eds.), Continental crust of NW and Central India, Geo.Soc.India, Mem. 31, pp. 345-352.

Mills, H., Halliday, A.N., Ashton, J.H., Anderson, I.K. and Russell, M.J., 1987. Origin of a giant ore body at Navan, Ireland. Nature, v.327, pp. 223-226.

Mohanty, M., Gupta, G.P. and Salu, R.L., 1993. Occurrence of ultramafic breccia in the Aravalli Fold Belt, Rajasthan. Curr. Sci., v.64, pp. 679-681.

NGRI, 1978. Gravity maps of India. National Geophysical Research Institute, Hyderabad, pp. 1-5.

Pareek, H.S., 1981. Petrochemistry and petrogenesis of the Malani igneous suite, India: Summary, Geol. Soc. Am. Bull., **92**, Part I, 67-70; Part II, 206-273.

Rajendra Prasad, B., Tewari, H.C., Vijaya Rao, V., Dixit, M.M. and Reddy, P.R., 1998. Structure and tectonics of the Proterozoic Aravalli-Delhi fold belt in NW India from deep seismic reflection studies. Tectonophysics, v.288, pp.31-41.

Rajendra Prasad, B., Vijaya Rao, V. and Reddy, P.R., 1999. Seismic and magnetotelluric studies over a crustal scale fault zone for imaging a metallogenic province of Aravalli Delhi Fold Belt region. Curr. Sci., v.76, pp.1027-1031.

Rajendra Prasad, B. and Vijaya Rao, V., 2006. Deep seismic reflection study over the Vindhyans of Rajasthan : Implications for geophysical setting of the basin. In: J.S.Ray and Chandan Chakraborty (eds.), Special volume on "Vindhyan Geology : Status and Perspectives", J.Ear. Syst. Sci., v.115(1), pp.135-147.

Rogers, J.J.W. and Santosh, M., 2002. Configuration of Columbia, a Mesopropterozoic supercontinent. Gond. Res., v.5, pp. 5-22.

Roy, A.B. and Paliwal, B.S., 1981. Evolution of lower Proterozoic epicontinental deposit : Stromatolite-bearing Aravalli rocks of Udaipur, Rajasthan, India. Precamb. Res., v.14, pp.49-74.

Roy, A.B. and Kroner, A., 1996. Single zircon evaporation ages constraining the growth of the Archean Aravalli craton, NW Indian shield. Geol.Mag., v.133, pp. 333-342.

Sawkins, J.F., 1984. Metal deposits in relation to plate tectonics. Springer-Verlag, Berlin-Heidelberg, 325p.

Sawkins, J.F., 1990. Integrated tectonic-genetic model for volcanic-hosted massive sulfide deposits. Geology, v.18, pp. 1061-1064.

Sinha-Roy, S., 1988. Proterozoic Wilson cycles in Rajasthan. In: A.B.Roy (Ed.), Precambrian of the Aravalli mountain, Rajasthan, India. Geo.Soc.India, Memoir 7, pp.95-108.

Sinha-Roy, S. and Mohanty, M., 1988. Blueschist facies metamorphism in the ophiolitic melange of the late Proterozoic Delhi fold belt, Rajasthan, India. Precamb. Res., v.42, pp. 97-105.

Sinha-Roy, S., Malhotra, G. and Guha, D.B., 1995. A transect across Rajasthan Precambrian terrain in relation to geology, tectonics and crustal evolution of south-central Rajasthan, In: S.Sinha-Roy and K.R.Gupta (Eds.), Continental crust of NW and Central India. Geol.Soc.India, Mem. 31, pp.63-89.

Steer, D.N., Knapp, J.H. and Brown, L.D., 1998. Super deep reflection profiling : exploring the continental mantle lid. Tectonophysics, v.286, pp.111-121.

Sychanthavong, S.P.H. and Desai, S.D., 1977. Protoplate tectonics controlling the Precambrian deformations and metallogenic epochs of NW Peninsular India: Mineral Sci.Engg., v.9, pp.218-237.

Tewari, H.C., Divakar Rao.V., Narayana.B.L., Dixit, M.M., Madhava Rao, N., Murthy, A.S.N., Rajendra Prasad,B., Reddy, P.R., Venkateswarlu, N., Vijaya Rao, V., Mishra, D.C. and Gupta, S.B. (1998). Nagaur - Jhalawar geotransect across the Delhi/ Aravalli fold belt in NW India. J. Geol. Soc. India, v. 52, pp. 153-161.

Vijaya Rao, V. and Reddy, P.R., 2002. A Mesoproterozoic Supercontinent : Evidence from the Indian Shield, In : J.J.W.Rogers and M.Santosh (Eds.), Special volume on Mesoproterozoic Supercontinent. Gondwana Research, v.5, pp.63-74.

Vijaya Rao, V., Rajendra Prasad, B., Reddy, P.R. and Tewari, H.C., 2000. Evolution of Proterozoic Aravalli Delhi fold belt in the northwestern Indian shield from seismic studies. Tectonophysics, v.327, pp.109-130.

Volpe, A.M. and Macdougall, J.D., 1990. Geochemistry and isotope characteristics of mafic (Phulad Ophiolite) and related rocks in the Delhi supergroup, Rajasthan, India: Implications for rifting in the Proterozoic. Pre.Res., v.48, pp.167-191.

Hon'ble Reviewers

1. Padam Shree Prof. K.S. Valdiya
   (Former Vice Chancellor Kumaon University)
   Jawahar Lal Nehru Centre for
   Advanced Scientific Research
   Jakkur Campus
   BANGALORE 560 064 INDIA
2. Prof. J.P. Srivastava
   Department of Geology
   Delhi University
   DELHI 110 007 INDIA

Editorial Assistance

Dr. Jaimini Swadia

Economic Mineralization
K.L. Shrivastava (Editor)
2009. pp.70-89
Scientific Publishers (India), Jodhpur

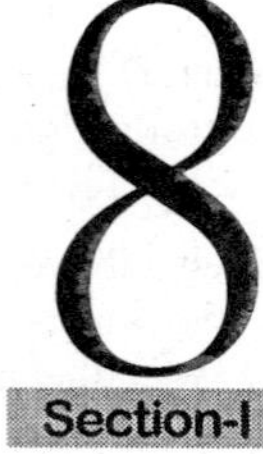

Section-I

Tectonic Polarity in Plate Tectonics-Controlled Interactions of Precambrian Terrains of Rajasthan Craton

S. SINHA-ROY

Birla Institute of Scientific Research, Statue Circle, Jaipur 302001.
Email : ssinharoy@yahoo.com

Abstract

Four distinct terrains, Bundelkhand, Mewar, Marwar and Foldbelts (Aravalli and Delhi) are recognized in Rajasthan craton. The Greater Bundelkhand terrain was rifted in Neoarchaean to early Palaeoproterzoic time at Hindoli (= Mahakoshal) secondary greenstone belt to produce a greater Mewar terrain on the west. The Mewar terrain experienced repeated rifting and basin formation with generation of oceanic/transitional crusts. These ensialic/ensimatic basins were crushed first in the Aravalli foldbelt at ca. 1.5 Ga and later in the Delhi foldbelt at ca. 1.0 Ga with the production of ophiolitic mélange and suture zones. Subsequently, the deformation sequence migrated eastward (present coordinates) forming a number of crustal-scale dislocation zones, and depicting an east-directed tectonic polarity. The suture zone (Phulad suture) of the South Delhi foldbelt is the product of the Grenvillian event while Moho-reaching Banas Dislocation Zone is Pan-African. The last dislocation to form in this sequence is the Great Boundary Fault, representing probably the Caledonian event.

These tectonothermal events and their responses allow for identification of plate tectonics signatures in Rajasthan craton, depicting Wilson cycles. The far-field deformation kinematics is controlled by the westward push of the Bundelkhand on the Mewar terrain in Neoproterzoic. The tectonic polarity and the disposition of sequential dislocation zones in Rajasthan seem Proterozoic analog of Phanerozoic Himalayan collision tectonics. The mineralization types as recorded in the different terranes can be accommodated in a plate interaction model for the region.

Keywords: Rajasthan Craton, Precambrian Terrains, Metallogeny, Tectonic Polarity, Plate Tectonics.

INTRODUCTION

Precambrian continental crust has evolved through accretion of mobile belts formed generally by inversion of rift systems that fragment crustal segments, and by subsequent welding of these disparate terrains. The dimension of these terrains may be variable, and they may or may not share similar tectonothermal history, but on a crustal scale these terrains are distinct because their boundaries are defined by deep-seated dislocation or suture zones. These inter-terrain interaction zones commonly contain signatures of plate tectonic processes that operate in a sequence of events having spatial progression and variability in response characteristics. The latter feature imparts a tectonic polarity to craton development, signifying a secularity and spatial dimension of tectonothermal processes in terrain evolution.

Rajasthan, a part of northwestern Indian craton, has a protracted geological history spanning ca. 3.5 0.5 Ga. During this period a number of fold-thrust belts, crustal-scale dislocations were formed, and repeated metamorphic-magmatic events took place. These features help define different terrains in the region whose Precambrian tectonic development shows a polarity similar to that of Phanerozoic orogenic belts. The purpose of the paper is to establish this tectonic polarity, and to show that the tectonic sequence in space and time was caused by plate tectonics-controlled crustal interactions between different terrains. Which will help in explaining the metallogeny of the craton.

Terrain characters

The geologic and tectonic setting of Rajasthan allows for identification of four major terrains (Fig. 1) (Sinha-Roy, 2000a). The position of different mineralization types in the tectonostratigraphy of the Rajasthan terranes have been shown in Fig. 2 (after, Sinha-Roy 2004b).

The Palaeo-Mesoproterozoic Aravalli fold belt and the Meso-Neoproterozoic Delhi fold belt separate two complex terranes in Rajasthan, namely, the Mewar terrane in the east and the Marwar terrain in the west. The Mewar terrain containing Archaean basement rocks is an old cratonic block while the status of the Marwar terrain as having an Archaean or Palaeoproterozoic basement is yet to be demonstrated although several authors have speculated on its old cratonic characters. For, example, Fareeduddin (2000) reported Pb-Pb age-date of 1434 Ma for the granulites of the eastern part of the Marwar terrain, west of the South Delhi foldbelt (SDFB)

Apart from the west-directed structural sequencing in the fold belt development, particularly in the SDFB, the prominent Neoptroterozoic magmatic activity of the Malani Igneous Suite and the Precambrian-Cambrian boundary setting are located in the western block, i.e., the Marwar terrain. Moreover, the Phanerozoic sequences have developed

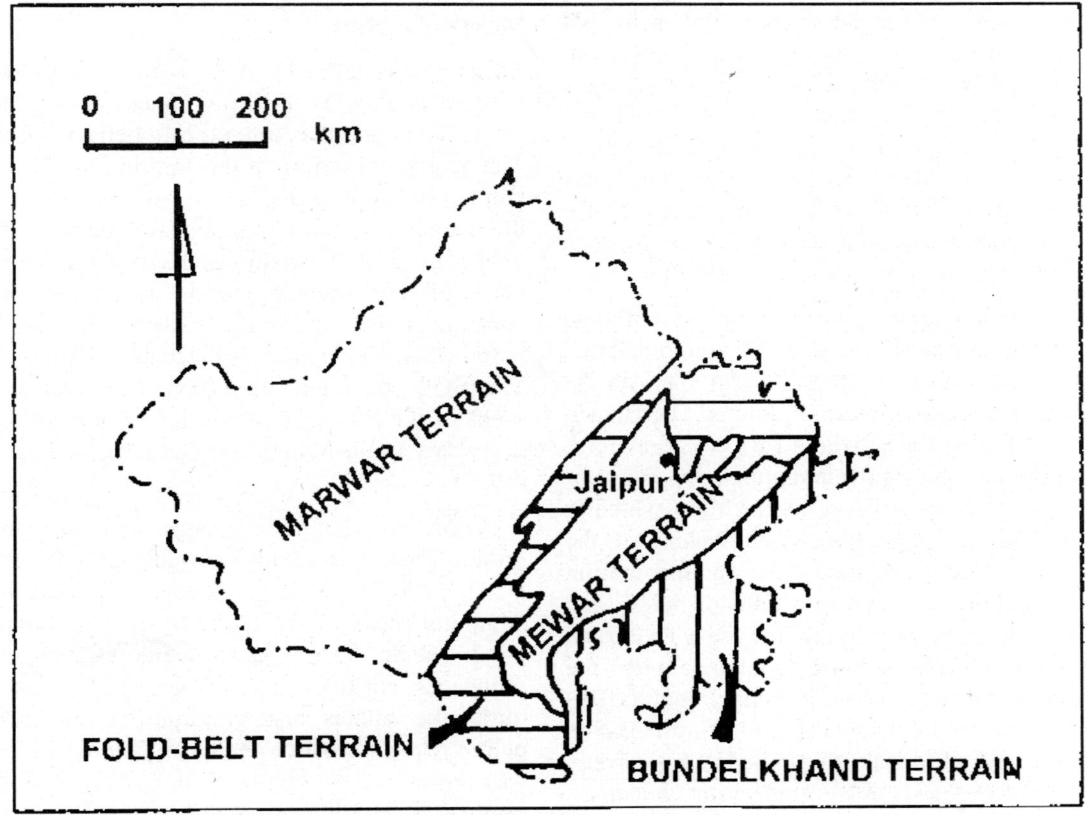

Fig. 1 Major Precambrian terrains in Rajasthan craton (after Sinha-Roy, 2000a).

only in the Marwar terrain.

The terrain boundaries are defined by crustal-scale dislocations, ductile shear and suture zones and, in some, by suture zones. The Mewar terrain supports Proterozoic cover sequences contained in the Aravalli and Delhi fold belts. These fold belts constitute the Fold-belt terrain, and each belt has a history of repeated ensialic rifting, followed by collision tectonics. The Marwar terrain is a possible fragment of the Greater Mewar block that got rifted during Mesoproterozoic time to form the SDFB oceanic trough, and subsequently got welded in Neoproterozoic time with the fragmented Mewar terrain along the SDFB suture zone. The easternmost terrain is the Bundelkhand terrain, which is an extension of the Greater Bundelkhand block that lies to the north of the central Indian Bastar craton.

The principal characters of these terrains are discussed below.

Bundelkhand Terrain

This terrain in Rajasthan forms the western part of the Bundelkhand craton that is largely covered by the paratectonic Vindhyan sequence. In Rajasthan region, apart from the Vindhyan rocks, the principal components of this terrain are end-Archaean Berach granite, the Hindoli metasediments and the Jahazpur cover sequence. This terrane is separated from the adjoining Mewar terrain in the west by a prominent crustal-scale dislocation zone, termed as the Banas Dislocation Zone (BDZ) (Sinha-Roy et al, 1993), imaged as mantle-reaching reflector zone in deep seismic reflection profile (Tewari et al, 1995).

An Archaean granite-greenstone assemblage forms the oldest rocks of the Bundelkhand block that has been cratonised at around 2.5 2.4 Ga with the emplacement of large granitoid plutons. The Berach granite (2528 Ma, Rb-Sr whole rock, cf. Crawford, 1970) and its possible equivalent pluton, the Jahazpur granite, have intruded the Hindoli Group which is most probably stratigraphically equivalent of the Mahakoshal Group, developed at the southern margin of the Bundelkhand craton in Madhya Pradesh and Uttar Pradesh. Both the Hindoli and the Mahakoshal Group rocks have tectonic contacts with the Vindhyans, namely, the Great Boundary Fault (GBF) in the case of the Hindoli and the North Narmada-Son Fault in the case of Mahakoshal. The Hindoli Group, comprising greywacke, felsic and mafic volcanics, volcaniclastics and BIF, is thrust in the east and southeast against the Vindhyan sequence along the Great Boundary Fault (GBF), and in the west, it is overthrust by the migmatites of the Mewar terrane along the BDZ. The status of the Hindoli Group is controversial. Gupta (1934) and Heron (1953) suggested that it is equivalent to the Proterozoic Aravalli sequence. On the contrary, Sinha-Roy (1985) and Gupta et al. (1997) assigned an early Palaeoproterozoic to late Archaean age to the Hindoli Group because the Berach (ca. 2.5 Ga) and Jahazpur granites have intruded the Hindoli sequence causing Buchan type contact-metamorphism. From the presence of a turbidite sequence and bimodal volcanics, a secondary greenstone affinity for the Hindoli sequence has been suggested (Sinha-Roy, 1988).

Mesoproterozoic Jahazpur Group that unconformably overlies the Hindoli Group, is probably equivalent of the Jungel Group of the Greater Bundelkhand block. The unconformity at the Hindoli-Jahazpur Group contact is a folded one Sinha-Roy, 1984a). The Jahazpur Group, a dolomite, orthoquartzite, BIF sequence, developed in narrow rift-related pull-apart basins at the margin of the Bundelkhand terrain.

Mewar Terrain

This terrain is defined in the E and SE by the BDZ, and in the W and SW by the thrust zones marking the base of the Delhi and the Aravalli fold belts, respectively. The geological setting of this terrain is controversial. Gupta (1934) proposed a two-fold classification for the Banded Gneissic Complex (BG rocks. According to him, the BGC-I constitutes the southern and central parts of the Mewar terrane while the BGC-II constitutes the northern Mewar terrain where the basementcover relations are debatable (Bose, 2000). The BGC rocks and the Aravalli equivalent cover rocks of the Bhilwara region have been collectively designated as the Mangalwar Complex by Gupta et al. (1997).

While the basement status and an Archaean protolith age for BGC-I is generally accepted (Bose et. al, 1996; Roy, 2001), the status of BGC-II and the adjoining rocks of the Bhilwara belt is a matter of much debate. The western segment of these rock sequences consists of deformed TTG, mafic and ultramafic bodies and granulite-facies mafic and pelitic rocks for which the term Sandmata Complex was introduced by Gupta et al, (1980). The eastern segment in the Bhilwara belt consists of amphibolite-facies gneisses and migmatites that contain

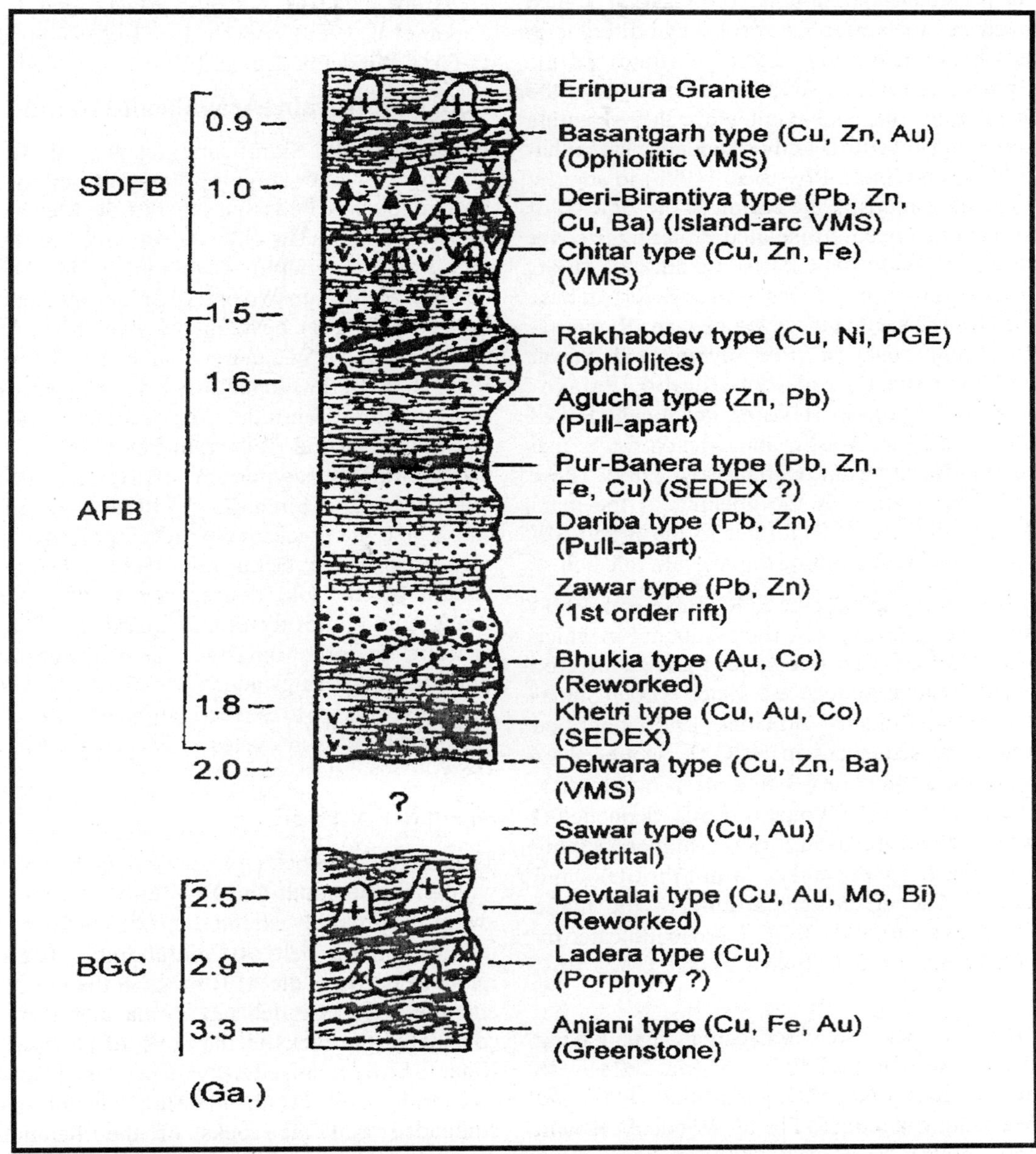

Fig. 2 Position of different mineralization types in the tectonostratigraphy of the Rajasthan terraines (after Sinha-Roy, 2004). BGC = Banded Gneissic Complex, AFB = Aravalli Fold Belt, SDFB = South Delhi Fold Belt.

greenschist to lower amphibolite-facies metasedimentary enclaves. Parts of these two segments of the northeastern part of the Mewar terrane, particularly the Bhilwara belt, is considered variably metamorphosed and migmatised equivalents of the Aravalli rocks (Naha and Roy, 1983), an idea originally mooted by Heron (1953). On the contrary, discordant structural relations, presence of folded unconformity, metamorphic breaks and lithofacies contrasts between the migmatitic substrate and the sedimentary cover, and also intrusive relations of the Berach granite (ca. 2.5 Ga) with the substrate migmatites and not with the cover sequence led Sinha-Roy (1984b) and Sinha-Roy et al. (1998) to suggest that the enclave rocks in the region of the BGC-II of the Bhilwara belt occur as isolated Proterozoic cover sequences in synformal keels within the BGC gneisses. This cover sequences, the repository of base metal deposits, are well developed in Rampura-Agucha. Rajpura-Dariba. Pur-Banera, and Sawar belts. These belts are considered Palaeo-Mesoproterozoic pull-apart basins, genetically linked with the regional strike-slip and wrench faults that transect the Mewar terrane (Sinha-Roy, 1989). These faults are responsible for the opening of the main Aravalli ensialic rift basin farther to the S and SW (present coordinates), forming the Aravalli fold belt.

The lithostratigraphy of the BGC (Bose et al, 1996) suggests that it is a composite rock sequence in which Sinha-Roy (1985) first recognised disemembered granite- greenstone sequences. Later work (Sinha-Roy et al, 1992; Mohanty and Guha, 1995) identified two greenstone sequences in BGC-II. Sequence-I (Sawadri Group) contains komatiites, ultramafics, chert, basic tuff and carbonates while Sequence-II (Tanawan Group), occurring at a higher tectonic level, contains greywacke, amphibolites and carbonates. The greenstone affinity of the mafic components of a part of the BGC-I is also indicated by their geochemistry (Upadhyaya et al, 1992; Roy, 2001).

The BGC-I rocks are Archaean in age, having yielded such age-dates as 3307+/- 65 Ma, 2828+/- 46 Ma (whole-rock Sm-Nd, cf.Gopalan et al. 1990), and 3230 Ma (single zircon U-Pb, cf. Wiedenbeck and Goswami, 1994), and 3230 Ma (single zircon evaporation age, Roy and Kroner, 1996). The age of the BGC-I is not well-constrained, except that the whole or a segment of it in the eastern part is older than the intrusive Berach granite and its age-equivalent plutons that have yielded an ages of 2450+/- 8 Ma (single zircon U-Pb, cf. Wiedenbeck et al, 1996), 2585 Ma (whole-rock Rb-Sr, cf. Crawford, 1970). The granulite-facies metamorphism of the Sandmata Complex, bounded by ductile shear zones, seems to have taken place between 1.7-1.6 Ga because the tectonised bodies of enderbite-charnockite in Gyangarh area and the granites in Amet and Anjana areas are dated at 1723+/-7 Ma (zircon U-Pb, cf. Sarkar et al, 1989), and 1641 +/-14 Ma (single zircon U-Pb cf. Wiedenbeck et al, 1996), respectively.

Fold Belt Terrain (Aravalli and Delhi)

A geologically significant feature of Rajasthan geology is the development of cross-cutting Proterozoic fold belts that separate the Mewar and the Marwar terrains. The older Aravalli fold belt records a multistage stratigraphic development. The Delhi fold belt is divisible into two parts, namely, the South Delhi Fold Belt (SDFB), developed in Ajmer-Mt. Abu tract extending into NW Gujarat, and the North Delhi Fold Belt (NDFB), developed in Alwar-Ajabgarh-Bayana areas in North Rajasthan. The terminology of NDFB and SDFB for the Delhi fold belt (used in place of main Delhi Synclynorium of Heron, 1953) was introduced by Sinha-Roy (1984b) in order to emphasize the diachronous development of the two segments of the Delhi fold belt. A characteristic feature of the fold belt terrain is that it records stratigraphic and tectonic signatures of repeated ensialic rifting, narrow ocean opening, and subsequent closing and formation of collisional foldbelts. This led to the recognition of a history where Proterozoic Wilson cycles in Rajasthan (Sinha-Roy, 1988).

Aravalli Fold Belt

The Aravalli fold belt (AFB) tapers in the north and it widens in the south and SE, from where it takes a swing and probably extends beneath the Deccan Traps into the Sausar belt of Central India. The contact relations between the AFB rocks in the northern part and the BGC are debated. Naha and Roy (1983) considered the rocks to the north of Nathdwara and Katar to be migmatised Aravalli sequence. Sen (1983) and Sinha-Roy et al, (1993a), on the contrary, contended that the rocks of the 'hammer-head' syncline of Heron (1953) are indeed parts of BGC-II, and that the contact between the Aravalli rocks in the south and the BGC-II rocks in the north is an E-W trending dextral strike-slip fault that delimits the SW extension of the BDZ (Sinha-Roy, 2000b).

The BDZ is a significant terrrain boundary in Rajasthan, and it exhibits tectonic and tectonothermal characteristics of the Himalayan Main Central Thrust (MCT), a crustal-scale dislocation of continental collision (Sinha-Roy, 2000b). Interestingly, the strike-slip fault in the northern part of the AFB does not cross-cut the adjoining Delhi rocks. It is likely that this post-Aravalli strike-slip fault played role in the extensional opening of the South Delhi oceanic trough. Similar E-W trending strike-slip fault (Sambhar-Jaipur-Dausa Fault, cf. Sinha-Roy, 1988) also played a significant role in the opening of the Delhi trough.

The stratigraphy of the Aravalli sequence, suggested by Heron (1953), has undergone many changes with the recognition of a number of rock groups and formations. A two-fold classification (Lower and Upper Aravalli Groups) with a younger deep-water carbonate-free sequence has been proposed by Roy (1988). Sinha-Roy et al. (1993b, 1998) suggested a three-fold classification, namely, Delwara Group at the base, Debari Group in the middle and the Jharol Group at the top. The group terminology was originally proposed by Gupta et al. (1980) who proposed a detailed sector-wise stratigraphy of the Aravalli foldbelt.

The Aravalli basal sequence (Delwara Group) marks the first stage of ensialic rifting of the Aravalli basin. The Delwara Group contains high-Mg and high-Fe tholeiites, quartz-pebble conglomerate, chert, BIF and minor dolomite. Raza and Khan (1993) interpreted the sequence as Palaeoproterozoic greenstone-type lithologic assemblage. The inversion of the Delwara basin is associated with granitic activity. After a prominent hiatus the Debari Group, a platformal sequence, marking the second stage of Aravalli rifting. A pronounced boulder bed and conglomerate sequence marks the erosional unconformity between the Debari and Delwara sequences. The second stage of rifting also produced elongated horst-graben structures within the Aravalli basin, the horsts being represented by the Sarara and other basement inliers while the grabens are the repository of carbonate-hosted base metal mineralization (e.g. Zawar) and phosphate deposition (e.g. Jhamarkotra, Matun, Kanpur etc.). The Jharol Group developed as a deep-water turbidite facies of the Aravalli sequence in the western part of the Aravalli fold belt, and it is separated from the Debari-Delwara horst-graben facies by a prominent tectonic line (Rakhabdev lineament), located at the median part of the AFB. The Rakhabdev lineament is defined by dismembered bodies of serpentinite, metagabbro, metabasalt and metachert etc., and it is interpreted as an ophiolitic melange zone depicting the Aravalli suture (Sinha-Roy, 1999).

The exact timing of AFB formation is not clearly known. Since the oldest Aravalli rocks unconformably overlie ca. 2.5 Ga granites (Ahar, Jhamarkotra etc., cf. Roy, 2001), the oldest age of the Aravalli sequence is younger than 2.5 Ga. The Darwal granite, dated at 1990±60 Ma (whole-rock Rb-Sr, cf. Chaudhary et al, 1984), and its possible equivalent granitoids (the younger components of Udaisagar, Jaisamand, Lakapa and Ahar granitoids) within the AFB represent post-Delwara but pre-Debari granitic activity (Sinha-Roy et al, 1993b). It is inferred that the oldest age of the first stage of Aravalli rifting (Delwara rifting) is older that ca. 1.9 Ga., but younger than 2.5 Ga. The Delwara basic volcanics have yielded galena Pb age-date of 2150 Ma (M.Deb, R.Thorpe and others, personal communication, quoted in Roy, 2001). Thus, the oldest age of the Aravalli sequence seems to be between 1.9 Ga. and 2.1 Ga.

The timing of the closure of the Aravalli basin and the formation of the Rakhabdev suture zone are also not known. The Amet and Anjana granites have been dated at 1641±4 Ma (single zircon ion-probe, cf. Wiedenbeck et al, 1996) while the Tekan pink granite has been dated at 1.55 Ga. (Sarkar et al, 1992). Considering that all these granite bodies that occur within the BGC at the fringe of the AFB are the products of crustal melting caused by NE-directed Aravalli ocean floor subduction beneath Mewar terrain, and are the tectonothermal response of the Aravalli basin closure (Sinha-Roy, 1984b, 1988), the youngest age of the AFB seems to be between 1.6 Ga. and 1.5 Ga. The timing of the Aravalli basin closure event corresponds to the timing (ca. 1645 Ma. cf. Roy, 2001) of the first phase of exhumation of the Sandmata granulites that occur adjacent to the AFB. The opening of the Debari basin is constrained by 1.6 Ga galena U-Pb model age of the Zawar mineralization (Deb et al. 1989). It can thus be concluded that the AFB spans 2.0-1.5 Ga. of the Precambrian history of Rajasthan.

Delhi Fold Belt

This fold belt has two segments, namely, the South Delhi Fold Belt (SDFB) and the North Delhi Fold Belt (NDFB). The SDFB is a narrow and linear NE-SW trending belt that fans out in the SW and NE. The NDFB from where the terminology of Alwar and Ajabgarh stratigraphies has been derived contains a folded sequence of graben-filled clastics, volcaniclastics and carbonates. The relationship between and the continuity of the NDFB rocks into the SDFB, proposed by many workers, are problematic because the gross stratigraphy of the SDFB do not match with that of the NDFB. This feature forms one of the bases of the concept of "suspect terrain" in the Delhi fold belt, originally proposed by Sinha-Roy (2001), and on the basis of tectonic consideration expounded further by Sinha-Roy (2004a) to suggest that the Khetri belt in the NDFB is Aravalli-equivalent transpressional "suspect terrain". The rocks of the Khetri belt are traditionally correlated with the Ajabgarh sequence, but from the regional tectonic setting, and lithological and geochronological considerations these are suggested to be equivalent of the Raialo Group (=Delwara group) (Sinha-Roy et al, 1998; Sinha-Roy 2000a). The principal factor, namely, the presence of 1.83 Ga (Rb-Sr) old felsic volcanics in the North Khetri belt (Gupta et al, 1998) and 1.78 Ga (Pb-model) age of Saladipura sulphide mineralization (Deb et al, 1989) indicate that the Khetri sequence is older than the Delhi stratigraphy. Sen (1980) recognized five tectonic zones within the SDFB of which the eastern two zones he considered to belong to the Aravalli fold belt. Thus the contention of the NDFB rocks extending into the SDFB appears suspect.

The substrate of the Alwar-Ajabgarh sequence is the Raialo Group whose stratigraphic status is problematic. While Roy and Kataria (1999), following Heron (1953), consider it to be the basal part of the Delhi Supergroup, Sinha-Roy (1988) and Sinha-Roy et al. (1995) correlated the Raialo sequence with the Delwara Group of the Aravalli Supergroup because the unconformity between the Alwar and the Raialo Groups is very profound representing a strong hiatus. Moreover, apart from the lithologic similarities between the Delwara and the Raialo sequences, the feature of the NDFB granitoids (e.g., Bairat), whose ages are around 1.8 1.6 Ga (Chaudhary et al, 1984), intruding the Raialo rocks and not the overlying Alwar rocks, is an indication that the Raialo Group has a separate identity, much older than the NDFB stratigraphy, and hence, it does not belong to the Delhi sequence.

The rocks of the Khetri belt are traditionally correlated with the Ajabgarh sequence, but from the regional tectonic setting and lithologic consideration these are suggested to be equivalent of the Raialo Group (=Delwara Group) (Sinha-Roy et al, 1998; Sinha-Roy, 2000a). This implies thereby a break in the continuity of the NDFB stratigraphy into the SDFB via the Khetri belt. This runs counter to the idea generally held by earlier workers. The rocks of the Khetri belt and those farther south, including the BGC-type inlier within the SDFB, SW of Ajmer, thus do not appear to be *in-situ* and would be candidate for "suspect terrane" in the Delhi fold belt (Sinha-Roy 2001).

The diachronous nature of the Delhi fold belt in which the NDFB pre-dated the SDFB, originally proposed by Sinha-Roy (1984b), found support in the later work (Ray, 1992, Gupta et al, 1991; Bose, 1989). Gupta et al.(1991) recognised the distinctive lithostratigraphic packages in the SDFB that are quite different from those of the NDFB. The clastic wedge of the Gogunda Group at the SDFB-AFB interface decreases in thickness in the NE, and in Ajmer sector, the interface is marked by a tectonised unconformity defined by deformed and detached conglomerate and grit beds (Srinagar Conglomerate). The Sendra Group, occurring to the west of this interface, contains rift-related mafic volcanics that are high-Mg at places (Bose et al, 1990). In the western part of the Sendra Group occurs the Barotiya Group whose important volcanic components are mafic and felsic volcanics, felsic tuff, rhyodacite and minor andesite (Bose et al, 1990). On the basis of the difference in the composition of the volcanics in the Sendra and Barotiya sequences Sinha-Roy (1996, 1999) differentiated an eastern ensialic rift-related sequence (Sendra Group) deposited in a basin that prograded into a deeper trough floored by oceanic or transitional crust, and a western mature to immature island arc sequence (Barotiya Group). These sequences are not uniformly developed in the SDFB such that in the south the basic volcanics-dominant Sendra sequence reduces in thickness where the platformal Bhim sequence of essentally calc-schist becomes the dominant lithology in the eastern part of the SDFB. The island arc sequence (Barotiya Group) extends from west of Ajmer in the north through Ranakpur to Deri-Ambaji in south Rajasthan and into Balaram-Mawal area in north Gujarat.

The SDFB stratigraphy at its western margin, south of Phulad, through Pindwara, Basantgarh, and Abu Road and beyond, is different from the adjacent Barotiya Group. The former sequence (Basantgarh Group, cf. Sinha-Roy and Mohanty, 1988), designated as the Phulad ophiolite (Gupta et al, 1997), comprises tectonised and dismembered bodies of metagraywacke, mafic meta-tuff, pillow-basalt, pillow-breccia, gabbro, ultramafics and plagiogranite. Apart from the above lithologic association, the presence of HP-LT metamorphic assemblages such as grospydite and talc-chlorite-kyanite (Sychanthavong and Merh, 1984) and relict blueschist-facies mineralogy in metagraywacke (Sinha-Roy and Mohanty, 1988) suggest that the Phulad-Basantgarh rocks are a part of an obducted ophiolite sequence. Sinha-Roy (1988) interpreted the Phulad ophiolite as having been derived from SDFB back-arc oceanic setting whose accompanying island arc was the Barotiya sequence.

The available geochronology roughly constrains the geotectonic evolution of the SDFB. The initial rifting and the deposition of the Sendra Group took place at ca. 1.5 Ga. as a consequence of and complementary to the compressional tectonics related to the closure of the Aravalli basin at that time. The nepheline syenite, dated at 1490+/- 15 Ma. (whole-rock Rb-Sr, cf. Crawford, 1970) within the probable Delwara-equivalent rocks in Kishengarh area, adjoining the northeastern margin of the SDFB, is the distensional response of the initial South Delhi rifting (Sinha-Roy, 1984b). The mafic volcanics of the Sendra Group, not yet dated, are likely to yield similar age-dates. The same event is also responsible for the emplacement of the Saladipura, Udaipurwati, and Seoli etc. granitoids that yielded age-dates around 1400 Ma. (Chaudhary et al, 1984). The timing of the Birantiya-Deri-Ambaji island arc formation is given by the age of the Ranakpur magmatic complex, the diorite component of which has yielded an age of 1012+/- 78 Ma. (whole-rock Sm-Nd, cf. Volpe and Macdougall, 1990). This age-date matches with the Pb-Pb model ages between 1087-1102 Ma. (Deb et al, 1989) given by the base metal sulfides of Deri-Ambaji area. From zircon U-Pb ages of the rocks of the Ambaji-Sendra belt Deb et al, (2001) suggested the presence of 1.0 Ga arc terrane in the SDFB. The timing of the back-arc spreading which is contemporaneous to or younger than the timing of island arc formation is given by Pb-Pb model age of 990 Ma. (Deb, 1996) of the base metal sulfides associated with the Phulad ophiolite in Basantgarh area. The timing of the Phulad ophiolite obduction and the closure of the SDFB basin, marking the end-phase of the "Delhi orogeny", is given by the extensive Erinpura granitic activity at ca. 900-800 Ma. (Chaudhary et al, 1984; Tobisch et al, 1994).

Marwar Terrain

The geology and tectonics of the trans-Aravalli Marwar terrain is poorly known because of extensive and thick sand cover of the Thar desert. The principal Proterozoic rock sequences comprise the Erinpura granitoids adjacent to the SDFB, Neoproterozoic metasedimentary and metavolcanic packages (Sirohi, Punagarh, Sindreth Ras, Makrana belts etc.), the Malani Igneous Suite (MIS), and the end-Neoproterozoic to Eocambrian Marwar Group of platform sedimentary sequence with evaporates.

The basement rocks in the Marwar terrane are poorly exposed. These rocks adjacent to the SDFB include gneisses that have been correlated by Heron (1953) with the BGC of the Mewar terrane. A contrary view considers these gneisses as migmatised South Delhi rocks (Sen, 1983; Sychanthavong and Merh, 1984). However, a pre-Delhi and even pre-Aravalli status of the Marwar terrane basement rocks appears a reasonable proposition (Sinha-Roy, 1988; Bose, 1989; Gupta et al, 1991). Tectonic slices of pelitic granulites and charnockite occur within the migmatitic gneisses in the NW of Ajmer and west of the SDFB (Fareeduddin, 1995). These rocks might represent the western counterparts of the Sandmata granulites representing the exhumed lower crustal basement rocks of the Marwar terrane.

Multiphase Erinpura grantoids (900-800 Ma, cf. Chaudhary et al, 1984), are generally S-type (Gangopdhyay and Lahiri, 1984). These granitoids occur in a narrow belt to the west of the SDFB. This magmatic belt, probably marking either a Cordilleran sub-terrain or an continental arc, developed at the edge of the Marwar hinterland on the leading edge of the SDFB subduction system. The specialized granitoids of this belt containing W,Sn, Li mineralisations are peraluminous showing calc-alkaline differentiation trends (Banerji and Pandit, 1995).

The Sirohi Group occurring in isolated basin to the SW of SDFB, considered earlier (Coulson 1933) the western counterpart of the Aravalli rocks, are now regarded as post-Delhi in age (Sen, 1980; Gupta et al. 1997). The Sirohi Group of low-grade phyllites, quartzite and carbonates and its equivalent sequences in Ras and Makrana areas in the NE developed in linear grabens to the west of the SDFB at ca. 830 Ma (Sharma and Roy, 1997). The isolated basins (Punagarh, Sindreth) that contain conglomerate, quartzite, phyllite and synsedimentary mafic flows indicate a younger phase of rifting in the Marwar terrane. The second phase of rifting event appears to represent a precursor to the MIS magmatic episode between ca. 750-730 Ma (Crawford and Compston, 1970) or between 779-681 Ma (Rathore et al. 1999; Roy, 2001).

The MIS, extensively developed in the Marwar terrane, represents 'within-plate' A-type anorogenic magmatism, and is divisible into three main events.

These are an initial extrusive phase of dominantly felsic volcanics, followed by an intrusive phase of meta-aluminous to peralkaline granites (Siwana, Jalore, Tosham) and ending with a major mafic and minor felsic dykes swarm. The Malani volcanism was of central to fissure type eruption (Bhushan 1984), related probably to hot-spot activity (Kochhar, 1984) or to plume-related magmatism (Roy, 2001).

The youngest Neoproterozoic sedimentary sequence of the Marwar terrane is the Marwar Group rocks that extend from near Jodhpur in the south to Haryana and Punjab in the NE, and beyond to the Salt Range in Pakistan. The Marwar stratigraphy comprises a basal arenite unit (Jodhpur) resting unconformably on the rocks of the MIS, followed by a carbonate formation and salt bearing evaporate sequence (Bilara-Hansaran) and ending with a arenite-phyllite sequence (Nagaur), (Das Gupta et al. 1988; Kumar, 1999). The Nagaur rocks have been correlated with the Purple Sandstone of early Cambrian age of the Salt Range, Pakistan (Jones 1970). The Marwar basin is the likely response of Pan-African tectonism in the Marwar terrain.

Crustal-scale dislocations

The boundaries of the major crustal terrains of Rajasthan craton, described above, are important dislocation zones. Apart from these zones, individual terrains contain polyphase deformation structures, dislocations and suture zones of different generations (Fig. 3). The suture zones of the Aravalli foldbelt (Rakhabdev suture) and South Delhi foldbelt (Phulad suture) represent important crustal-scale dislocation zones.

The Aravalli fold belt has a tectonic contact with the basement gneisses of the BGC, which in fact is a tectonised unconformity, marked prominently in the north where the Delwara sequence of the basal Aravalli is tectonically juxtaposed against the BGC along a reworked unconformity (Sinha-Roy et al. 1995). This tectonic zone, the Delwara Dislocation Zone (DDZ), extends south and SE along the Aravalli-basement interface, and truncates a number of lithofacies. The DDZ is offset by the strike-slip

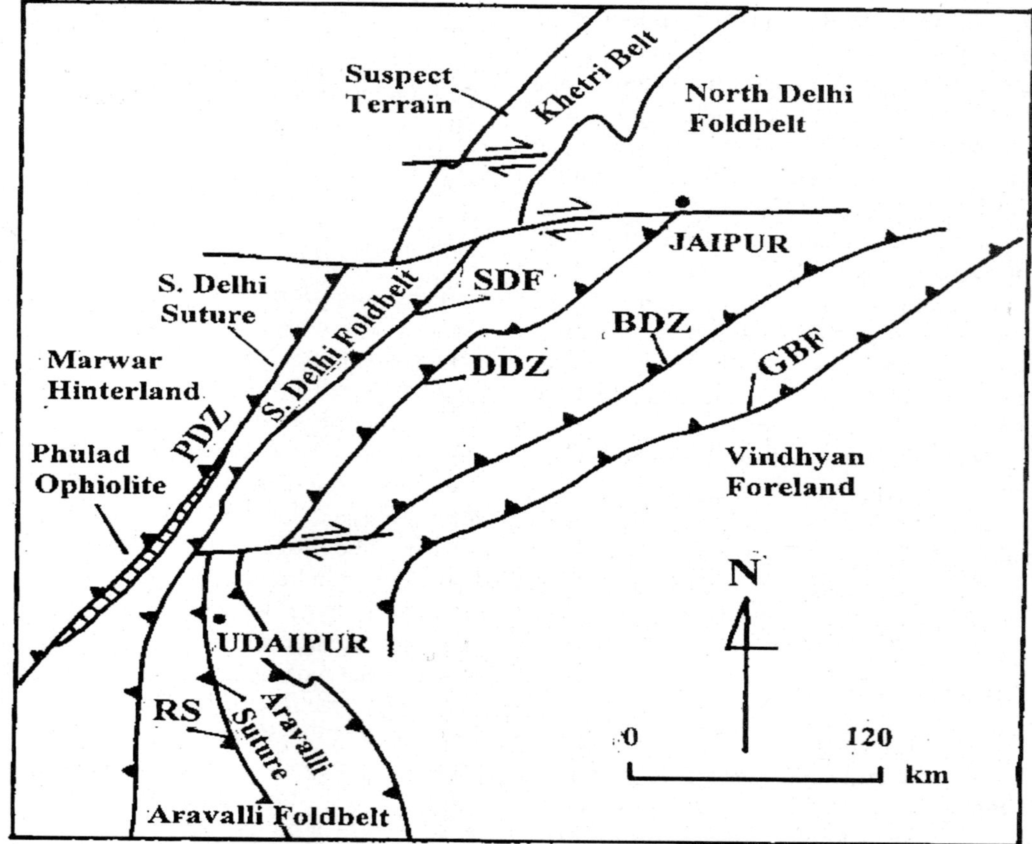

Fig. 3 Major crustal-scale dislocation and suture zones in Rajasthan craton (after Sinha-Roy, 2004a).

segment of the BDZ in Nathdwara and adjacent areas, and it extends northeastward up to the Sambhar-Jaipur-Dausa strike-fault. The DDZ in this segment marks the tectonic boundary of the Sandmata granulites and the associated amphibolite-facies gneisses against the reworked BGC of the Bhilwara belt.

Another important tectonic zone of the Aravalli fold belt is the Rakhabdev Suture Zone (RS) that divides the fold belt into two contrasting segments, namely, the platform carbonate-arenite sequence in the east and deep-water turbidite facies (Jharol) in the west. The Banas Dislocation Zone (BDZ) is an important crustal-scale dislocation showing oblique thrusting in the northern segment and dextral strike-slip movement in the southern segment. The latter segment truncates and delimits the Aravalli fold belt in the north (Sinha-Roy et al, 1993) while the thrust segment defines boundary between the Mewar and Bundelkhand terranes.

Apart from the DSZs related to the ophiolite zone, the South Delhi fold belt contains an upthrust basement wedge, flanked by two prominent shear zones in the Ajmer-Bewar sector. This basement wedge is probably attached to the Khetri "suspect terrain". The Phulad Dislocation Zone (PDZ) at the western boundary of the south Delhi fold belt defines the terrain boundary between the Marwar and the fold belt terrains. The South Delhi fault (SDF) at the eastern margin of the South Delhi fold belt marks the boundary between the Mewar and the fold belt terrains. Apart from these dislocation zones the South Delhi fold belt contains a number of shear zones that are the result of transpressional tectonics (Gupta et al. 1991; Sinha-Roy, 2001).

The Great Boundary Fault (GBF) appears to be a relatively young and thin-skinned tectonic zone that separates the Vindhyan cover sequence and the pre-Vindhyan basement rocks of the Bundelkhand terrane. Although its exact age is difficult to ascertain it is likely that the GBF has been reactivated during Tertiary collision tectonics of the Indian shield in the Himalayan region.

The tectonic features of Rajasthan craton have been brought out by the deep seismic profiles studies (Tewari et al. 1995). The crustal structures show the following characteristics: (i) moderate reflectivity of the Marwar terrain, (ii) strong lower crustal reflections and their continuity among various Moho bands beneath the SDFB, (iii) steeply dipping reflectors in the upper crust near both the margins of the SDFB, (iv) poor reflectivity of the BGC crust in the Mewar terrain, (v) dip reversals at mid-crustal levels with dome-shaped reflections in the BGC, and (vi) dipping Moho reflections from both sides of the BGC crust, coinciding with the base of the SDFB.

An interpretation of the deep continental structures of Rajasthan craton and an intracontinental underthrusting model as proposed by Sinha-Roy (2000b, 2004b) are shown in Figure 6 and Figure 7, respectively. The salient features are as follows: (1) The lower and upper crusts are dissected by a number of dislocation zones which correspond to important tectonic features such that Phulad Dislocation Zone (PDZ), South Delhi Fault (SDF) and Banas Dislocation Zone (BDZ). These are all prominent seismic reflection zones. These zones bound several tectonic domains. (2) The SDFB is fairly well defined with a deep root-zone. (3) The dome-shaped reflection beneath the SDFB and the Sandmata Complex, is a

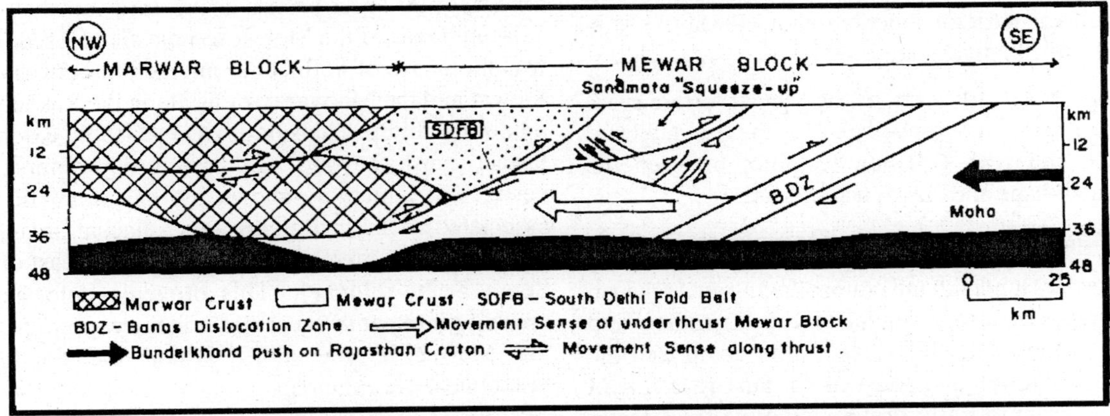

Fig. 4. Tectonic model of Rajasthan craton interpreted from seismic reflection profile. For explanation see text. (after Sinha-Roy, 2000b).

composite structure of reflectors of more than one generations such that SE-dipping reflector probably represents an AFB structure, while NW-dipping reflector is a SDFB structure. (4) The Sandmata Complex occurs in a tectonic slice bounded by SDF and DDZ, and is likely to have been tectonically emplaced "squeeze-up" imbricate sliver from the BGC lower crust. This structure is responsible for the gravity high feature of this region (5) The SDF seems to be an oblique intracratonic-underthrusting zone (A-subduction), having large-scale strike-slip movement component, and it separates the fold-belt and Mewar terrains.

An analysis of the disposition of the reflectors in the seismic reflection profile and its correlation with geological and tectonic setting (Fig. 4) indicate the following.

1. The Moho beneath Mewar and Marwar terrains is undulating which would mean involvement of sub-continental Moho in tectonic development of the region.
2. The Marwar continental crust is split at mid-crustal level by a sub-horizontal detachment that merges with the Phulad suture zone. This detachment could correlate with the mid-crustal electrical conductor and the low-velocity channel.
3. The SDF extends up to the lower crustal level in the Marwar terrain.
4. The SDF bifurcates beneath the Sandmata Complex that contains a number of antithetic thrust faults, and the Sandmata "squeeze-up" is caused by a duplex propagation at DDZ ramp.
5. The BDZ extends up to Moho depths, and it may act as a lower crustal detachment zone separating the Bundelkhand and the Mewar terrains, and act as a decollement for the indentation of Mewar block by the Bundelkhand block.
6. Deep crustal structures of the Mewar terrain seem consistent with westward push by the Bundelkhand block and with collision tectonics between the Bundelkhand and the Mewar blocks.

Plate tectonics-controlled terrain evolution

The Greater Bundelkhand craton, having an Archaean TTG-greenstone core, included the Rajasthan and Bastar cratons. The Rajasthan craton comprised the combined Mewar and Marwar terrains (MMT). In Late Archaean to Early Proterozoic time the Greater Bundelkhand was rifted to form the Mahakoshal trough that separated the Baster craton and the Hindoli trough that separated the MMT from the Bundelkhand. These troughs containing secondary greenstone sequences closed at Neoarchaean to Early Palaeoproterozoic time, and a major cratonisation event took place in the region at ca. 2.5-2.3 Ga with the emplacement of K-rich granitoids. This event welded and the end-Archaean granitoids stitched the Bundelkhand terrain with the MMT. The stitching zone later formed the site of intraconinental A-subduction along the Banas Dislocation Zone.

A plate tectonic model of orogen development and major dislocation zone formation in Rajasthan craton is shown in Figure 5.

After cratonisation event the MMT was rifted at ca. 2.0 Ga to initiate the formation of the Aravalli basin (Delwara - Raialo trough) (Aravalli Rifting Stage-I). The trough extended across the MMT, dividing it into the eastern Mewar terrain and the western Marwar terrain. Aravalli Rifting Stage-II produced the Debari trough in the south, and probably also the Alwar-Ajabgarh troughs in the north at ca. 1.7 Ga. Large-scale strike-slip faults within the Mewar and Bundelkhand terranes during the Aravalli Rifting Stage II produced linear and narrow metalliferous (Pb-Zn) transpressional and transtensional pull-apart basins in Rajpura-Dariba-Bhindar, Pur-Banera, Jahazpur, Rampura-Agucha, Sawar etc areas. The Aravalli basin eventually opened into a deep oceanic trough (Jharol ocean), floored by oceanic/transitional crust. A compressional regime, starting at ca. 1.5 Ga closed this basin and crushed the older platformal (Debari) and rift (Delwara) sequences through a east-directed subduction of the oceanic/transitional crust beneath the Mewar terrane. This formed the Aravalli fold belt as a collisional orogen. At about the same time the pull-apart basins in the Bhilwara belt and adjacent areas of the Mewar terrane also closed. This was the phase of collisional interaction between the Mewar and the Marwar terrains along the Rakhabdev ophiolitic suture. The suture zone probably extended farther north beyond the present-day limits, the vestiges of which are shown by the ultramafic and associated rocks in Rajgarh and adjacent areas and also within the pre-Delhi basement rocks, east of the South Delhi fold belt. This orogenic deformation created a number of shear zones in the western part of the Mewar terrane along which the Sandmata granulites were exhumed.

East-directed Aravalli A- and B-subductions caused extensional tectonics at ca. 1.5 Ga in the

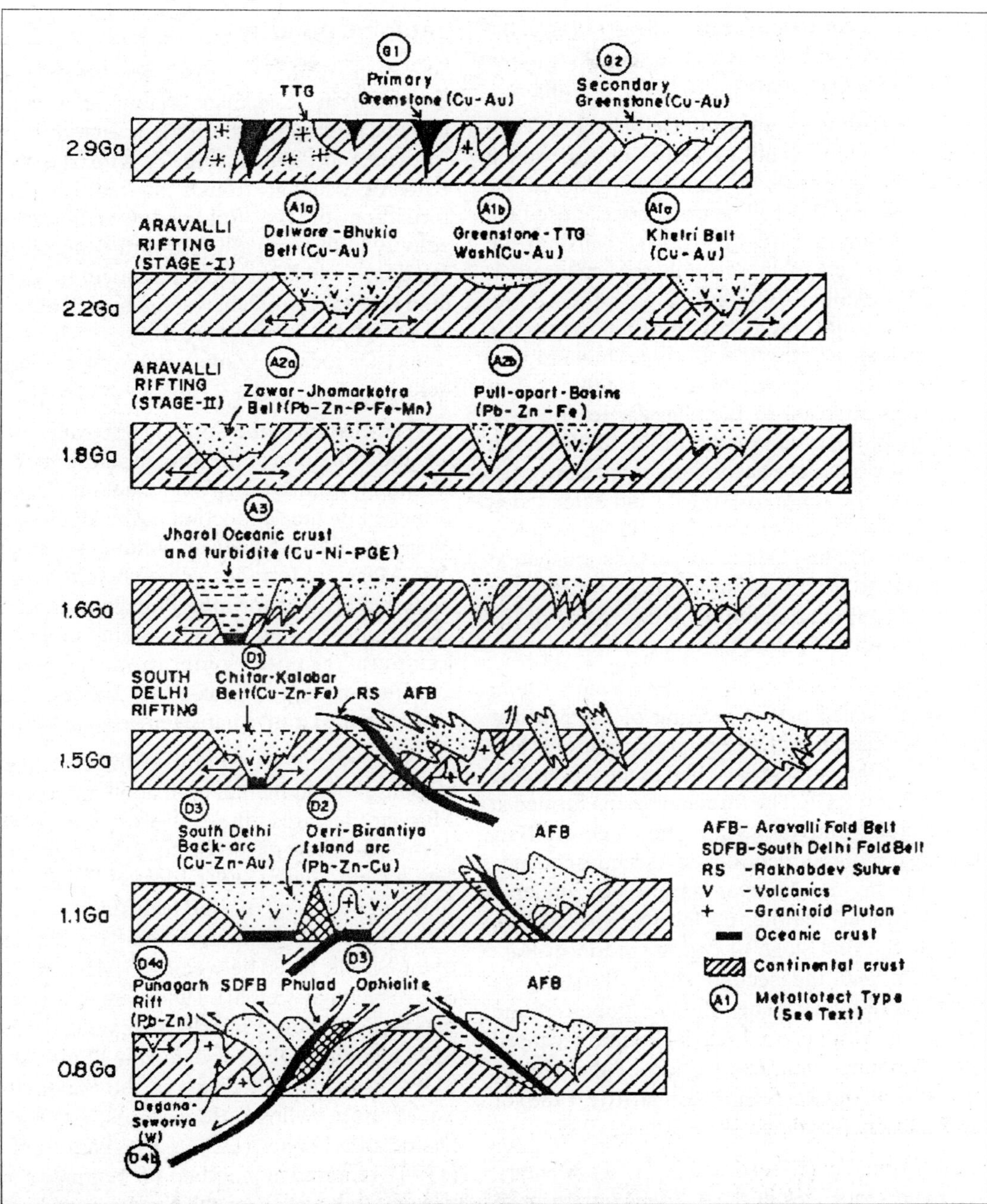

Fig. 5 Plate tectonic model for evolution of Proterozopic foldbelts and related metallogeny in Rajasthan. (after Sinha-Roy, 2000c)

Mewar terrane. Judging from the linearity of the narrow South Delhi fold belt and the rapid lithofacies variations it seems likely that the South Delhi rift basin (Gogunda, Bhim, Sendra troughs) opened as transtensional foredeep troughs at the western margin of the Mewar terrain in response to Aravalli collisional tectonics. The North Delhi graben system except for the Khetri belt developed during almost the same period of time. The difference in the geology between the NDFB and SDFB, emphasized by Sinha-Roy (1984b, 1988), arose from the difference in their tectonic settings and kinematics of deformation for these developed in two different terrains. The SDFB developed as an inter-terrain orogen between Mewar and Marwar terrains while the NDFB developed within the Mewar terrain. It is also likely that the difference is also due to the fact that the geology of the NDFB correlates with the Aravalli sequence rather than with the Delhi sequence.

Post-1.5 Ga rifting in the South Delhi basin opened an oceanic trough Sendra-Kalabar-bhim), much similar to the present-day Red Sea, floored by oceanic/transitional crust. This trough was closed via westward subduction of this crust, causing the formation of a chain of island arcs (Barotiya-Ranakpur-Deri-Ambaji-Balaram) at ca. 1.0 Ga. (Fig. 5). This arc contained distinctive magmatic components and Kuroko/Besshi type metallogeny (Cu-Pb-Zn-Ba)(Fig. 6).(Sinha-Roy, 2000c). The island arc was attached to the Marwar terrain as a continetal arc in the northern part of the SDFB, north of Phulad, but a back-arc basin with an oceanic crust floor opened to the south of Phulad and extended at least down SW to Abu Road through Basantgarh and probably beyond. This oceanic basin was crushed at ca. 0.9 Ga due to westward subduction, causing the emplacement of Phulad ophiolite and suturing of the Delhi orogen with the Mawar terrain.

The westward subduction of the South Delhi oceanic crust caused partial melting of the Marwar terrane crust, generating the Eerinpura granitoids and forming a Cordilleran magmatic belt to the west of the SDFB at ca. 0.9-0.8 Ga. The foreland basins formed at this period in the Marwar terrain contained the Barr and associated molassic sediments. As a consequence of this tectonism the Marwar terrain was under a strong extensional regime during Neoproterozoic time causing the first stage of rifting in Sirohi-Ras-Makrana belt, and the second stage of rifting in Punagarh-Sindreth belt. The Malani Igneous Suite was formed in response to partial melting caused either by intracontinental subduction of the Mewar crust beneath the Marwar terrain (Sinha-Roy, 1988) or by hotspot activity (Kochhar, 1984; Raval, 1995).

The extensional event recorded in the sedimentary and magmatic sequences in the Marwar terrain from ca. 870 Ma to the end of Neoproterozoic, namely, Sirohi-Punagarh-Malani-Marwar sequences in western Rajasthan, are linked with the break-up of the Rodinia supercontinent and assembly of the East Gondwanaland, the boundary between the West and the East Gondwanaland being located to the west of the Marwar terrain. It is possible to draw an analogy in the tectonic, magmatic and sedimentary characters between Neoproterozoic Marwar terrain and Cenozoic Tibet (Sinha-Roy, 1988, 2000a).

Tectonic polarity

The plate tectonics model for Precambrian terrain evolution in Rajasthan craton, enumerated above, depicts migration of orogenic activity toward east (present coordinates) (Fig. 7). After the closure of the Aravalli oceanic trough at ca. 1.5 Ga and the formation of the Rakhabdev suture the tectonic activity in the Aravalli fold-thrust belt shifted toward NE due to which the BGC basement rocks were thrust onto the Delwara-Debari sequences. This event seems to be related to the formation of the crustal-scale Banas Dislocation Zone (BDZ) at ca. 550 Ma (see below).

The closure of the South Delhi trough at ca. 1.0 Ga and the formation of the Phulad suture zone caused the Marwar terrain to thrust over the South Delhi Foldbelt, followed by the subduction of the Delhi oceanic crust attached with the Mewar continental block beneath the Marwar terrane. This resulted in retroarc hinterland rifting forming Punagarh, Sindreth, Sirohi etc. basins, Malani Igneous Suite emplacement, and Erinpura granite generation at South Delhi Cordilleran belt (Sinha-Roy, 2004b). This would represent the Grenvillian event in Rajasthan.

Subsequent to South Delhi suturing, and in response to continental collision between Mewar and Marwar terrains the deformation zone shifted eastward which gave rise to a detachment zone, the South Delhi Fault (SDF), located at the SE margin of SDFB. The timing of SDF is difficult to ascertain, but if the Sendra, Ranakpur, Ambaji and similar granites of the SDFB, dated between 765 Ma and 830 Ma (Rb-Sr) (Chaudhary et al, 1984) are considered to be magmatic response of crustal-scale underthrusting at SDF, then the SDF post-dated the Phulad suture zone. SDF forms the roof thrust of the Sandmata tectonic 'squeeze-up' whose floor thrust is the Delwara Dislocation Zone (DDZ). Nagpaul and Nagpaul (1974) reported a number of pegmatite muscovite fission track ages near the Sandmata Complex of the Bhilwara belt at around 800 Ma. Although the granulite-facies M1 metamorphism of the Sandmata Complex rocks appears to have taken place at 1723 Ma (Sarkar et al, 1989; Sinha-Roy et al, 1992), M2 metamorphism and tectonic exhumation of the granulite-facies rocks aided by SDF and DDZ seem to have occurred at ca. 1000 Ma (Fareeduddin, 2000). Another response of DDZ detachment was the creation of extensional fracture system at the footwall of DDZ where albitite was emplaced in nearly 150 km long belt (Ray, 1990). Ca. 800 Ma event in and around

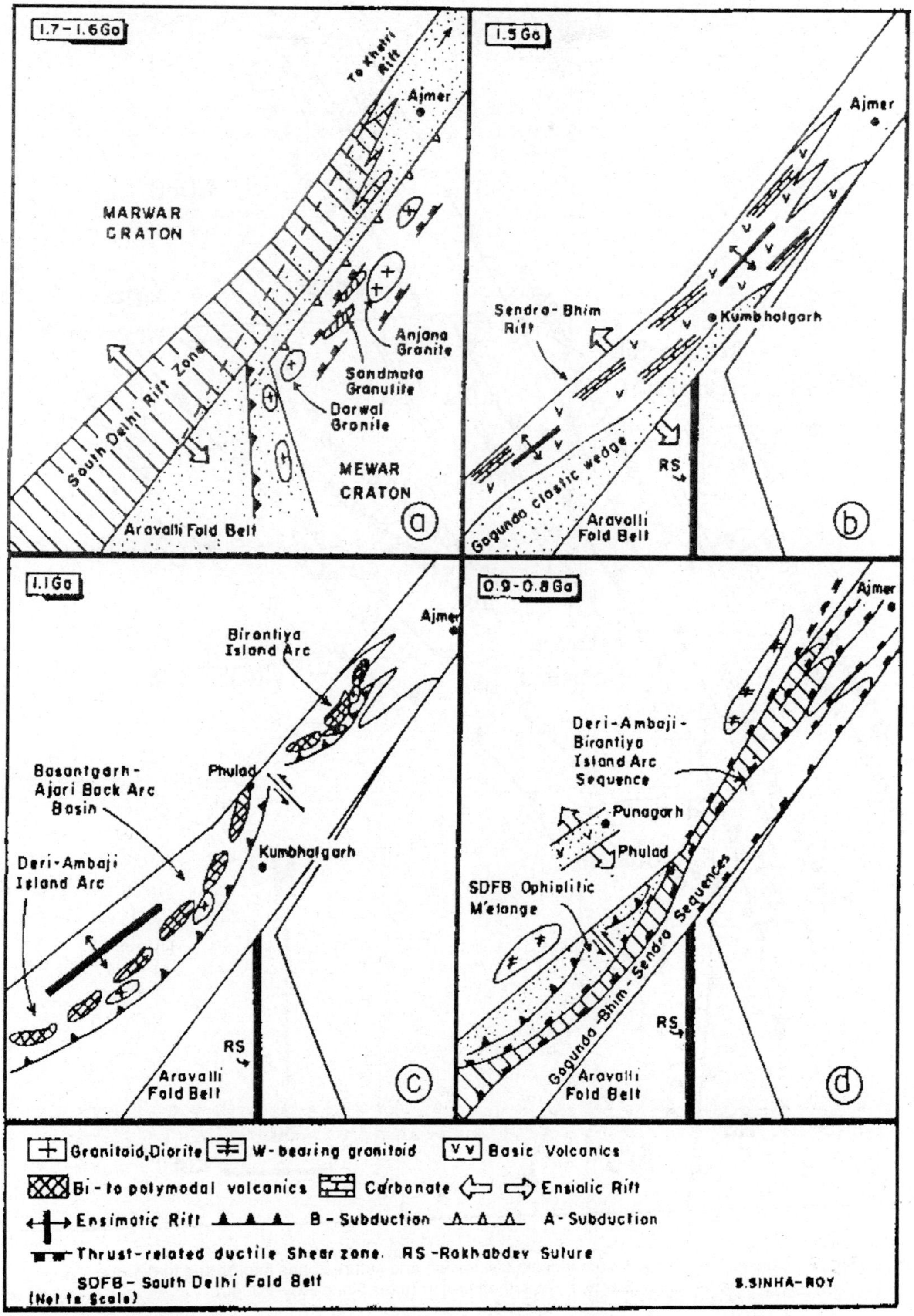

Fig. 6 Plate tectonic evolution of the South Delhi foldbelt (after Sinha-Roy, 2000c).

the Sandmata Complex of the Mewar terrane marks the beginning of the Pan-African regime in Rajasthan craton.

The subsequent tectonic event migrated eastward

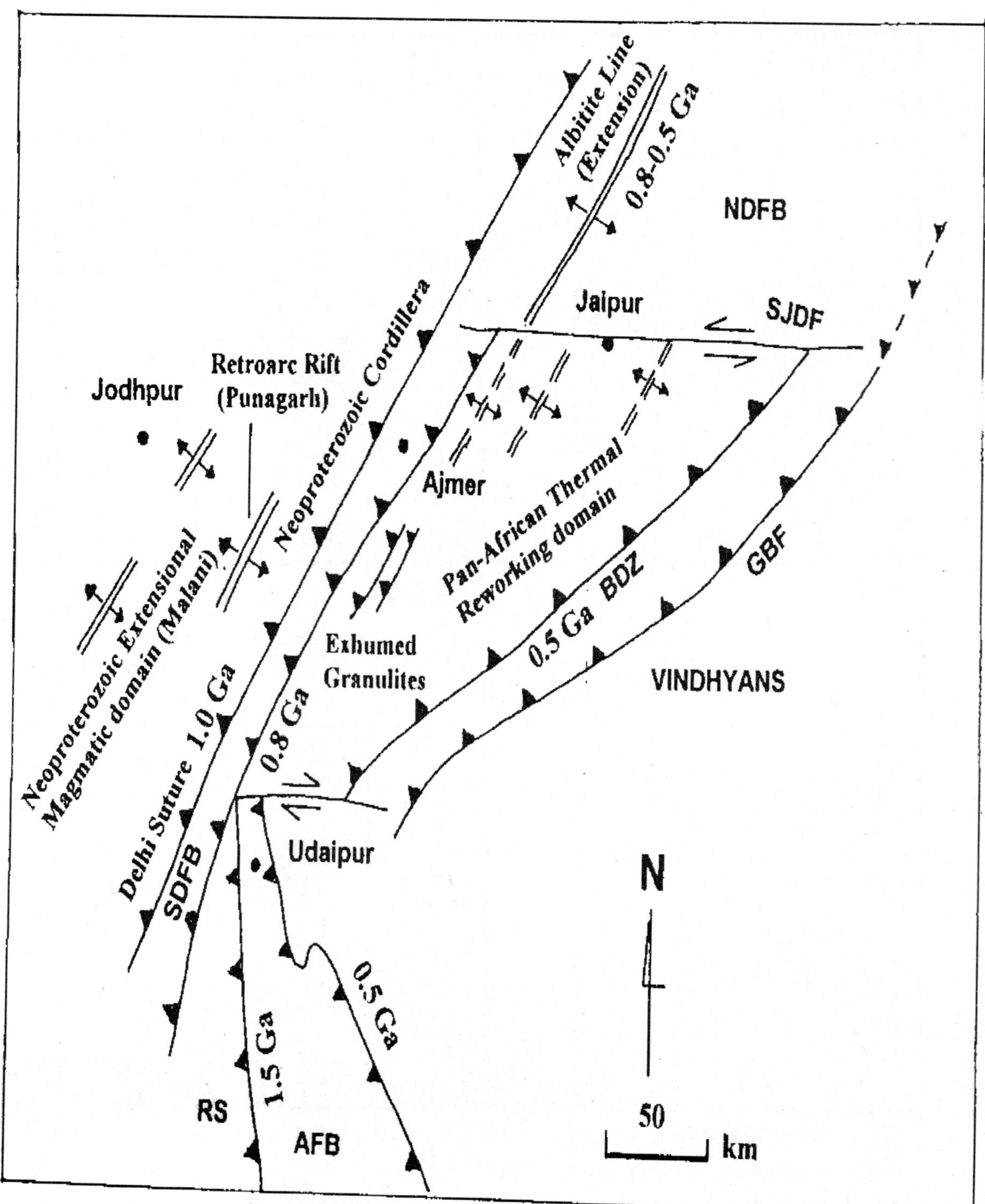

Fig. 7 Disposition of major crustal-scale dislocation and suture zones interpreted to deduce the tectonic polarity in Rajasthan craton (after Sinha-Roy, 2004c).

and formed the Banas Dislocation Zone (BDZ), marking the boundary between the Mewar and Bundelklhand terrains. The timing of its formation is difficult to ascertain, but judging from the fact that the hangingwall slab (Mewar terrain) experienced pervasive thermal reworking evidenced by fission track age-dates of biotite and apatite in pegmatites of Bhilwara belt, ranging from 530 Ma to 600 Ma

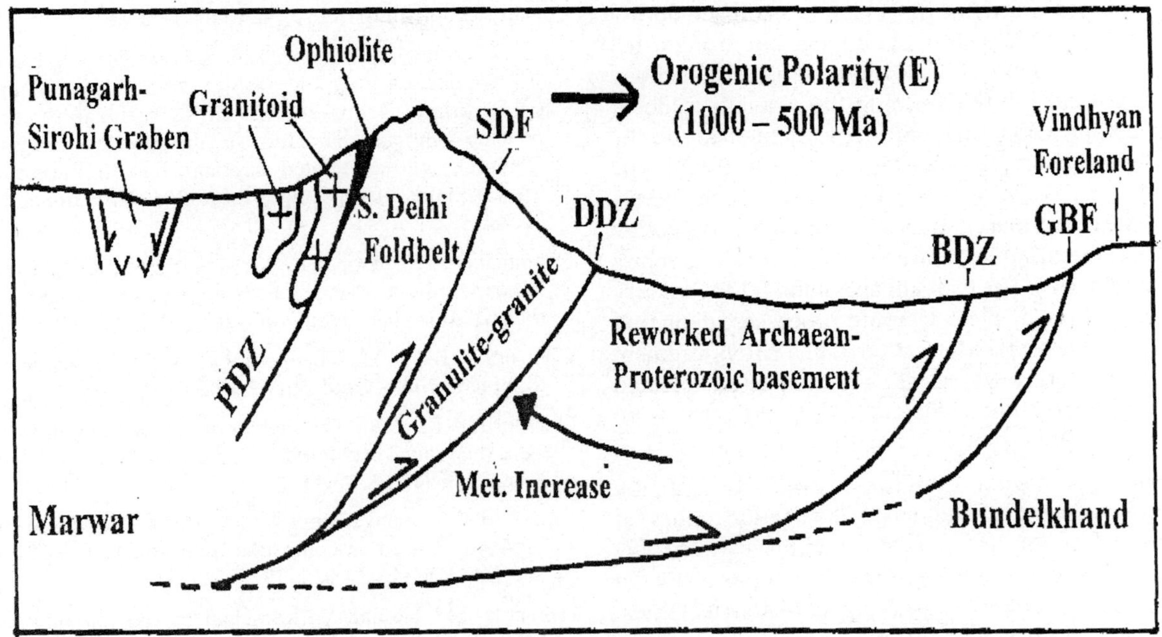

Fig. 8 Crustal section across Rajasthan craton showing the sequential development of major dislocations and suture zones, during 1000-500 Ma period. Note the decollement nature of the BDZ and metamorphic inversion in the hangingwall Mewar block consequent upon Bundelkhand indentation (after Sinha-Roy, 2006).

(Sharma et al, 1975), and at 580 Ma (Aswathanaryana , 1964), the BDZ seems to represent the culmination of Pan-African tectonothermal event in Rajasthan craton at ca. 550 Ma.

In conformity with east-directed tectonic polarity the Great Boundary Fault (GBF) is the youngest dislocation zone of the region. The youngest Vindhyan rock sequence, namely, the Bhander Group, dated at ca. 550 Ma (Crawford and Crompton, 1970) is affected by the GBF. The GBF, therefore, seems to be Eo-Cambrian in age, presumably representing Caledonian tectonism (ca. 400 Ma) (Sinha-Roy, 2004c) in Rajasthan.

The Himalayan tectogene records classical example of N-S tectonic polarity from the Indus suture zone through MCT to MBF and HFT toward the foreland due to progressive India-Tibet collision tectonics. The tectonic polarity and the disposition of the major dislocation zone, foldbelts and metamorphic characteristics etc., in Rajasthan craton (Fig. 7) have a spatial similarity with those of the Himalayan region (Sinha-Roy, 2006). This would suggest that Cenozoic Himalayan collision tectonics has an analog in Precambrian terrain evolution in Rajasthan craton.

CONCLUSIONS

Proterozoic tectonothermal events have produced at least four distinct terrains in Rajasthan, namely, Bundelkhand, Mewar, Marwar and Foldbelts (Aravalli and Delhi). These terrains interacted with one another at different times producing polyphase deformation structures, and more importantly, deep-seated crustal-scale dislocation and suture zones. It also produced the "suspect terrain" of the Khetri belt and "squeeze-up" tectonic wedges of lower crustal granulite-facies rocks in Sandmata Complex area. The spatial disposition of these tectonic zones and their possible timing of formation allow for working out a tectonic polarity of terrain interaction events. Neoproterozoic tectonic polarity was directed toward east (present coordinates).

The Rajasthan craton comprising the Greater Mewar block separated from the Greater Bundelkhand craton in Neoarchaean to Early Palaeoproterozoic time, the two being welded at the Hindoli secondary greenstone belt. The Greater

Mewar block was rifted first at the Aravalli belt during ca. 2.0 to 1.8 Ga to produce an oceanic trough that closed at an ophiolitic mélange suture (Rakhabdev suture) at ca. 1.5 Ga, forming the Aravalli foldbelt. This compressional tectonics led to extension in the foreland that created the South Delhi Red Sea type trough at ca. 1.5 Ga. This trough closed at ca. 1.0 Ga via island arc-marginal basin generating ophiolitic suture (Phulad suture) and a Cordilleran belt (Erinpira magmatic belt). The Aravalli and Delhi foldbelts carry the signatures of plate tectonic processes, depicting Wilson cycles. The Marwar terrain, a rifted fragment of the Greater Mewar block, has Neoproterozoic and Phanerozoic geologic history totally different from that of the Mewar block.

The east-directed tectonic polarity and the geometry of the crustal-scale dislocation zones in Rajasthan craton are consistent with westward far-field push and indentation of the Mewar terrain by the Bundelkhand craton, signifying collision tectonics. The terrain tectonics of Rajasthan craton is a Proterozoic analog of Phanerozoic Himalayan collision tectonics.

References

Aswathanaryana, U. 1964. Age determination of rocks and geochronology of India. Preoceedings International Geol. Congress, New Delhi, Pt. 10, 12 p.

Ave Lallemant, H.G., and Oldow, J.S. 1988. Early Mesozoic southward migration of Cordilleran transpressional terranes, Terctonics, 7(5), 1057-1075.

Banerji, S. and Pandit, M.K. 1995. Lithium and tungsten mineralisation in Sewariya pluton, South Delhi Fold belt : evidences for preferential host rock affinity. Curr. Sci., 69(3), 252-256.

Bhattacharya, T. Sengupta, S., and Mukhopadhay, D. 1995. Tectonic status of the Delhi-pre-Delhi contacts. In : Continental Crust in Northwestern and Central India, (Eds. : S. Sinha-Roy and K.R. Gupta), Mem. Geol. Soc. India, 31, 217-230.

Bhusan, S.K. 1984. Classification of the Malani Igneous Suite, Special Publ. Geol. Surv. India, 12, 199-205.

Bose, U. 1989. Correlation problems of the Proterozoic stratigraphy of Rajasthan. Indian Minerals. 43, 183-193.

Bose, U, Fareeduddin and Reddy, A.B. 1990. Polymodal volcanism in parts of the South Delhi Fold Belt, Rajasthan. Jour. Geol. Soc. India, 36, 263-276.

Bose, U. Fareeduddin and Rdddy, A.B. 2000. The nature, origin and evolution of the Banded Gneissic Complex basement rocks of Rajasthan. In : Crustal Evolution and Metallogeny in the NE Indian Shield (Ed. : M Deb), Narosa Publishing House, New Delhi, 73-86.

Bose, U. Mathur, A,K, Sahoo, K.C. Bhattacharya, S. Dutta, K. Kumar, A. Sarkar, S.S., Chawdhury, S and Chawdhury, I. 1996. Event stratigraphy and physicochemical characters of the Banded Gneissic Complex and associated supracrustals in the south Mewar plains of Rajasthan, Jour. Geol. Soc. India, 47, 325-338.

Chaudhary, A.K., Gopalan, K., and Sastry, C.A. 1984. Present studies on the geochronology of the Precambrian rocks of Rajasthan. Tectonophysics, 105, 131-140.

Coulson, A.K. 1933. The geology of the Sirohi State, Rajputana. Mem. Geol. Surv. India, 63, pp. 469.

Crawford, A.R. 1970. The Precambrian geochronology of Rajasthan and Bundelkhand, northern India. Canad. Jour. Earth Sci. 125, 91-110.

Crawford, A.R., and Compston, W. 1970. The age of the Vindhyan System of Peninsular India. Quart. Jour. Geol. Soc. London, 125(1), 351-372.

Dasgupta, S.P. Kumar, V., Ramchandra and Jairam, M.S. 1988. A framework of Nagaur-Ganganagar evaporite basin, Rajasthan. Indian Minerals, 42(1), 57-64.

Deb, M. 1996. VMS deposits : a contemporary genetic model and review of database of these deposits in the southern Aravalli range (Abstract). In : National Seminar on Mineralisation in the Western Indian Craton, (Eds. : M. Deb and D.M. Banerjee), Delhi University, New Delhi, 26-30.

Deb, M. 2000. VMS deposits : geological characteristics, genetic models and a review of their metallogenesis in the Aravalli range, NW India. In : Crustal Evolution and Metallogeny in the NW Indian Shield, (Ed. : M. Deb), Narosa Publishing House, New Delhi, 329-363.

Deb, B., Thorpe, R.I., Cumming, G.L. and Wagner, P.A. 1989. Age, source and stratigraphic implications for conformable sediment-hosted base metal deposits in the Proterozoic Aravalli-Delhi orogenic belt, NW India, Precambrian Res., 108, 1-22.

Deb, M. Thorpe, R.I., Karstic, D., Corfu, F., and Davis, D.W. 2001. Zircon, U-Pb and galena Pb isotopic evidence for an approximate 1.0 Ga terrane constituting the western margin of the Aravalli-Delhi orogenic belt, NW India. Precambrian Res., 108, 195-213.

Fareeduddin. 1995. Field setting, petrochemistry and P-T regime of the deep crustal rocks to the northwest of the Aravalli-Delhi mobile belt, North-central Rajasthan. In : Continental Crust of Northwestern and Central India, (Eds. : S. Sinha-Roy and K.R. Gupta), Mem. Geol. Soc. India, 31, 117-140.

Fareeduddin, 2000. Polyphase granulite facies metamorphism in Rajasthan : tectonometamorphic evolution and isotopic constraints. Geol. Surv. India, Spl. Publ. 55, 307-318.

Gangopadhya, P.K. and Lahiri, A. 1988. Anjana granite and associated rocks of Devgarh, Udaipur district, Rajasthan, In : The Precambrians of the Aravalli Mountains (Ed. : A.B. Roy). Mem. Geol. Soc. India, 7, 307-316.

Gopalan, K., MacDaugall, J.D., Roy, A.B., and Murly, A.V. 1990. Sm-Nd evidence for 3.3 Ga old rocks in Rajasthan, NW India, Precambrian Res. 48, 287-297.

Gupta, B.C. 1934. The geology of central Mewar. Mem. Geol. Surv. India, 65, 107-168.

Gupta, P. Fareeduddin, Mukhopadhya, A.K. and Reddy, M.S. 1991. Structural style in the rocks of the Delhi Supergroup along the eastern part of the Aravalli range in central Rajasthan. Indian Jour. Geol., 63(3), 190-205.

Gupta, P., Chattopadhyay, B., and Guha, D.B. 1998. Basement-cover relationship in the Khetri Copper Belt and the emplacement mechanism of the granite massif, Rajasthan, India, Jour. Geol. Soc. India, 52, 417-432.

Gupta, S.N. Arora, Y.K. Mathur, R.K. Iqbaluddin., Prasad, B. Sahai, T.N., and Sharma, S.B. 1980. Lithostratigraphic Map of the Aravalli region, southern Rajasthan and northeastern Gujarat. Geol. Surv. India. Jaipur.

Gupta, S.N. Arora, Y.K. Mathur, R.K. Iqbaluddin., Prasad, B. Sahai, T.N. and Sharma, S.B. 1997. Precambrian geology of the Aravalli region, southern Rajasthan and northeastern Gujarat, India, Mem. Geol. Surv. India, 123, 262.

Heron, A.M.1953. The geology of central Rajputana, Mem. Geol. Surv. India, 78, pp. 389.

Jones, C.L. 1970. Potash in halite evaporite salt , Western Pakistan. U.S.G.S. Prof. Paper, 25, 140-145.

Kochhar, N. 1984. Malani Igneous Suite : hotspot magamatism and cratonisation of the northern part of the Indian shield. Jour. Geol. Soc. India, 25, 155-161.

Kumar, V. 1999. Evolution and geological set-up of the Nagaur-Ganganagar basin, NW Rajasthan. In : Geological Evolution of Northwestern India, (Ed. : B.S. Paliwal), Scientific Publishers, Jodhpur, 34-60.

Mohanty. M, and Guha, B.B. 1995. Lithotectonic stratigraphy of dismembered greenstone sequence of the Mangalwar Complex around Lawa-Sardagrah and Parasoli areas, Rajsamand district, Rajasthan. In : Continental Crust of Northwestern and Central India, (Eds, : S. Sinha-Roy and K.R. Gupta), Mem Geol. Soc. India, 31, 141-162.

Nagpaul, M.K. and Nagpaul, K.K. 1974. Thermal history of Rajasthan pegmatites of India as revealed by Fission Track studies. Canad. Jour. Earth Sci., 11, 519-528.

Naha, K. and Roy, A.B. 1983. The problem of Precambrian basement in Rajasthan. Precambrian Res., 19, 217-223.

Rathore, S.S, Venkatesan, nd Srivastava, R.K. 1999. Rb/Sr isotope dating of Neoproterozoic (Malani Group) magmatism from southwest Rajasthan, India Evidence of younger Pan-African thermal event by 40Ar/39Ar studies. In : Neoproterozoic Crustal Evolution and India-Gondwana Linkage, (Guest Ed. : A.B. Roy), Gondwana Res. 2(2), 271-286.

Raval, U. 1995. Geodynamics of the tectomagmatic and geophysicsl signatures within mobile parts of the transect, In : Continental Crust of Northwestern and Central India, (Eds.: S. Sinha_Roy and K.R. Gupta), Mem. Geol. Soc. India, 31, 37-61.

Ray, D.K. 1992. The bearing of crustal evolution on the metallogeny in Precambrian Shield of India, Indian. Jour. Geol. 64(2), 161-186.

Ray, S. K. 1990. The albitite line of northern Rajasthan a fossil intracontinental rift zone. Jour. Geol. Soc. India, 36, 413-423.

Raza, M., and Khan, M.S. 1993. Geochemical characterization of Delwara volcanics : evidence for the origin of the Aravalli basin by extensional rifting. In : Rifted Basins and Aualcogens, (Ed. : S.M. Cassyap), Gyanodaya Publications, Nainital, 119-129.

Roy, A.B. 1988. Stratigraphic and tectonic framework of the Aravalli mountain range, In : Precambrians of the Aravalli Mountain Range, (Ed. : A.B. Roy), Mem. Geol. Soc. India, 7, 3-31.

Roy, A.B. 2001. Neoproterozoic crustal evolution of NW Indian shield : implications on break-up and assembly of supercontinents, Gondwana Res., 4(3), 289-306.

Roy, A.B., and Kataria, P. 1999. Precambrian geology of the Aravalli mountains and neighbourhood : analytical update of recent studies. In : Geology of Rajasthan : Status and Perspective. (Ed. : P. Kataria), M.L. Sukhadia University, Udaipur, 1-56.

Roy, A.B., and Kroner, A. 1996. Single zircon evaporation ages constraining the growth of Archaean Aravalli craton, NW Indian shield. Geol. Mag., 133(3), 332-342.

Sarkar, G. Burman, T. and Corfu, F. 1989. Timing of continental arc type magmatism in NW India : evidence from Pb zircon geochronology. Jour. Geol. 97, 607-612.

Sarkar, G. Bishui, P.K. Chattopadhyay, B. Chawdhury, S. Chawdhury, I. Saha, K.C. and Kumar,A. 1992. Geochronology of granites and felsic volcanic rocks of Delhi Fold Belt. Rec. Geol. Surv. India, 125(2), 21-23.

Sen, S. 1980. Precambrian stratigraphic sequence in a part of the Aravalli range : a re-evaluation. Quart. Jour. Geol. Min. Met. Soc. India, 52(2), 63-76.

Sen, S. 1981. Proterozoic palaeotectonics in the evolution of crust and location of metalliferous deposits, Rajasthan. Quart. Jour. Geol. Min. Met. Soc. India, 53, 162-185.

Sen, S. 1983. Stratigraphy of the Precambrians of central and northern Rajasthan : a review. In : Structure and Tectonics of the Precambrian Rocks of India, (Ed. : S.

Sinha-Roy), Hindustan Publishing Corp., New Delhi, 10, 26-39.

Sharma, K.K., and Roy, A.B. 1997. Neoproterozoic geology and evolution of the trans-Aravalli crust, In : Precambrian'97, (Eds. : K.C. Gyani and P. Kataria), M.L. Sukhadia University, Udaipur, (Abstract), 15-16.

Sharma, K.K., Nagpaul, M.K., Nagpaul. K.K. and Jhingran, A.G. 1975. Fission Track geochronology of early Precambrian, Rajasthan, India, Chayanika Geol., 1, 196-209.

Sinha-Roy, S. 1984a. Folded unconformity in Rajasthan. Curr. Sci., 53, 1205-1207.

Sinha-Roy, S. 1984b. Precambrian crustal interactions in Rajasthan, NW India, In : Crustal Evolution of the Indian Shield and its Bearing on Metallogeny. Indian Jour. Earth Sci. Seminar Vol., 84-91.

Sinha-Roy, S. 1985. Granite-greenstone sequence and geotectonic development of SE Rajasthan. In : Megastructures and Plate Tectonics and their Role as Guide to Ore Mineralisation. Bull. Geol. Min. Met. Soc. India, 53, 115-123.

Sinha-Roy, S. 1988. Proterozoic Wilson cycles in Rajasthan, In : Precambrians of Aravalli Mountain Range, (Ed. : A.B. Roy), Mem. Geol. Soc. India, 7, 95-108.

Sinha-Roy, S. 1989. Strike-slip faults and pull-apart basins in Proterozoic fold belts development in Rajasthan, Indian Minerals, 43 (3-4), 226-240.

Sinha-Roy, S. 1996. Precambrian crustal evolution and metallogeny in Rajasthan (Abstract), In : Mineralisation in the Western Indian Craton, (Eds. : M. Deb and D.M. Banerjee), Delhi University, New Delhi, 24-25.

Sinha-Roy, S. 1999. Proterozoic sutures in Rajasthan, In : Geology of Rajasthan : Status and Perspective, (Ed. : P. Kataria), M.L. Sukhadia University, Udaipur, 87-100.

Sinha-Roy, S. 2000a. Precambrian terrane evolution in Rajasthan, Geol. Surv. India, Spl. Publ. 55, 275-286.

Sinha-Roy, S. 2000b. Plate tectonic evolution of crustal structures in Rajasthan. In : Research Highlights in Earth Systems Science, Dept. of Science and Technology, Govt. of India (Eds. : O.P. Verma and T.M. Mahadevan). Indian Geol. Congress, Roorkee, 17-25.

Sinha-Roy, S. 2000c. Precambrian metallotects and mineralisation types in Rajasthan : their relation with crustal evolution, In : Crustal Evolution and Metallogeny in the NW Indian Shield. (Ed. : M. Deb), Narosa Publishing House, New Delhi, 217-239.

Sinha-Roy, S. 2001. Plate tectonics controlled Precambrian metallotects in Rajasthan and their role in predictive metallogeny. Special Publ. Geol. Surv. India, 72, 11-34.

Sinha-Roy, S. 2004a. Intersecting Proterozoic transpressional orogens, major crustal and suspect terrains in Rajasthan craton : a plate tectonic perspective. Geol. Surv. India, Spl. Publ., 84, 207-226

Sinha-Roy, S. 2004b. Precambrian terranes of Rajasthan, India and their linkages with plate tectonics controlled mineralisation types and metallogeny. In : Sediment-hosted Lead-zinc Sulphide Deposits, (Eds : M. Deb and W.D Goodfellow), Narosa Publishing House, New Delhi, 222-245.

Sinha-Roy, S. 2004c. Proterozoic rifting and major unconformities in Rajasthan and their implications for uranium mineralisation. Exploration and Research for Atomic Minerals, 15, 69-97.

Sinha-Roy, S. 2006. Proterozoic suture, suspect terrane and tectonic polarity in rajasthan : a Himalayan analog. In : Active and Fossil Suture Zones, Dehradun, Abstract Vol., 17-18.

Sinha-Roy, S. and Mohanty, M. 1988. Blueschist-facies metamorphism in the ophiolite mélange of the late Proterozoic Delhi fold belt, Rajasthan, India, Precambrian Res., 42, 97-105.

Sinha-Roy, S., Guha, D.B., and Bhattacharya, A.K. 1992. Polymetamorphic granulite-facies pelitic gneisses in Precambrian Sandmata Complex, Rajasthan, Indian Minerals, 46(1), 1-12.

Sinha-Roy, S. Mohanty, M. and Guha, D.B. 1993a. Banas Dislocation Zone in Nathdwara-Khamnor area, Udaipur district, Rajasthan and its significance on basement-cover relations in the Aravalli fold belt. Curr. Sci., 65(1), 68-72.

Sinha-Roy, S., Mohanty, M., Malhotra, G., Sharma, V.P., and Joshi, D.W. 1993b. Conglomerate horizons in south-central Rajasthan and their significance on Proterozoic stratigraphy and tectonics of the Aravalli and the Delhi fold belts, Jour. Geol. Soc. India, 41, 331-350.

Sinha-Roy, S. Malhotra, G. and Guha, D.B. 1995. A transect across Rajasthan Precambrian terrane in relation to geology, tectonics and crustal evolution in south-central Rajasthan, In : Continental Crust in Northwestern and Central India, (Eds. : S. Sinha-Roy and K.R. Gupta), Mem. Geol. Soc. India, 31, 63-90.

Sinha-Roy, S. Malhotra, G., and Mohanty, M. 1998. Geology of Rajasthan. Geol. Soc. India, Bangalore, pp. 278.

Sychanthavong, S.P. and Merh, S.S. 1984. Proto-plate tectonics : the energetic model for the structural and igneous evolution of the Precambrian rocks, NW Peninsular India, Special Publ. Geol. Surv. India, 12, 419-458.

Tewari, H.C. Rao, V.V., Dixit, M.M., Prasad, R.B., Rao, M.N., Venkateshwaralu, N., Khare, P., Rao, K.G., Raju, S. and Kaila, K.L. 1995. Deep crustal reflection studies across the Delhi-Aravalli fold belt : results from NE parts, In : Continental Crust in Northwestern and Central

India, (Eds. : S. Sinha-Roy and K.R. Gupta), Mem. Geol. Soc. India, 31, 383-402.

Tobisch, O.T., Collerson, K.D., Bhattacharya, T., and Mukhopadhay, D. 1994. Structural relationships and Sr-Nd isotope systematics of polymetamorphic granite gneisses and granitic rocks from central Aravalli craton. Precambrian Res., 65, 319-339.

Upadhyaya, R., Sharma, B.L., Sharma, B.L (Jr)., and Roy, A.B. 1992. Remnants of greenstone sequence from the Archaean rocks of Rajasthan. Curr. Sci., 63(2), 87-92.

Volpe, A.M., and MacDougall, J.D. 1990. Geochemistry and isotope characteristics of mafic (Phulad ophiolite) and related rocks in the Delhi Supergroup : implications for rifting in the Proterozoic. Precambrian Res., 48, 167-191.

Weidenbecke, M. and Goswami, J.N. 1994. An ion-probe single zircon 207Pb/206Pb age of Mewar gneiss at Jhamarkotra, Rajasthan, Geochim Cosmochim Acta, 58, 2135-2141.

Wiedenbecke, M. Goswami, J.N. and Roy, A.B. 1996. Stabilisation of the Aravalli craton of NW India at ca. 2.5 Ga : an ion-probe zircon study. Chemical Geol. 129, 325-340.

Windley, B.F. 1983. The Evolving Continents. John Wiley & Sons, New York, pp. 399.

Hon'ble Reviewers

1. Prof. S.M. Naqvi
   Emeritus Scientist
   National Geophysical Research Institute
   Uppal Road
   HYDERABAD 500 001
2. Dr. Fareeduddin
   Editor
   Geological Society of India
   No. 63, 12th Cross Road
   Gavipuram
   P.B. No. 1922
   BANGALORE 560 019

Editorial Assistance

Ms Anju Mathur

Section II
Economic Mineralization of Igneous Affiliation

Preamble

The section on economic mineralization of igneous affiliation has been examined in twelve papers by various authors after obtaining the direct and indirect systematic information which is required in different scales of observations from ore deposits and its surrounding host rock and structures to synthesize the findings, giving due considerations to the principles, constraints, limitations and uncertainties involved.

The first paper of the section is on 'An ore genetic model for the Gogi hydrothermal type uranium deposit in Neoproterozoic Bhima basin, Southern India', authored by Dr. S.A.Pandit and Dr. R. Dhana Raju. The newly-discovered Gogi uranium deposit is about to gain the status of a first-order uranium deposits of the country. The authors have observed that the deposit is a poly-phasic, poly-episodic and poly-metallic, litho-structurally controlled, epithermal vein- type having an average grade of 0.19 percent U_3O_8 of almost three thousand tonnes. In this exhaustive study, almost every aspect of the deposit including the conceptual evolution, geological setting, host rocks, structures and the uranium mineralization have been covered, justifiably. The paper has proposed an eight-stage evolutionary genetic model of the deposit.

"Volcanogenic-pneumato-hydrothermal intrusive and collapse breccia deposit of fluorite at Jalera-Khurd in Malani Igneous Suite, northwest India" is the title of the second paper of the section, authored by Prof. K.L.Shrivastava and Dr. Virendra Gaur. The paper incorporates field studies, with production of the first geological map of the deposit, detailed petrology and geochemistry. The complex data obtained have been synthesized to suggest a genetic model of the deposit comprising twelve stages to satisfy various features observed in the deposit. The Jalera-Khurd deposit is not only the first deposit of considerable size in the Malani Igneous Suite, which is conventionally being treated as sterile, but also displays excellent features of both hydrothermal intrusive breccias and hydrothermal collapse breccias.

The next paper of the section by Dr. Brajesh Singh and Prof. Santosh Kumar is entitled "Tungsten Mineralization associated with felsic magmatism in Purbani district of Kinnaur district, Himachal Higher Himalayas. Three types of granites of the region, namely, Neoproterozoic granite gneiss (GGn), Early Proterozoic granitoids (EPG), and tourmaline-bearing leucogranite (TLg) have been studied geochemically for tungsten content and for incompatible elements such as K, Rb, Sr, Ba and La. At and around Purbani region, the GGn commonly hosts shear-zoned skarn which has scheelite mineralization. Distinctly higher values of W in EPG and CaO in skarn rock compelled the authors to draw a conclusion that skarn-forming solutions were most likely derived from EPG melt enriched in boron, fluorine and tungsten.

The fourth paper of the section authored by Dr. Sukh Chain and Dr. P.K. Shrivastava is entitled "Huebnerite/Ferberite(H/F) ratio and depositional conditions for wolframite : an overview". Two extreme members, heubnerite($MnW0_4$) and Ferberite($FeW0_4$) of the solid solution series of wolframite always show their variation in the regional studies and also in single crystal. The authors have utilized various techniques like EPMA, XRD, isotopic and fluid inclusion to know factors like temperature, pressure, oxygen fugacity, pH, which possibly may be responsible for such chemical variations. Analyzing data for the various wolframite deposits, the paper has drawn a very important conclusion to suggest that the changing temperature of formation has the direct effect on MnO/FeO ratio of the wolframites.

Dr. Kamal Kant Sharma has authored the next paper of the section entitled "Tungsten metallogeny in the Sirohi group, Rajasthan". Balda, Nana, Sewariya and Degana linear belt of granites represents a tungsten metallogeny and closing of the Sirohi cycle in the ensialic Sirohi basin during Neoproterozoic. Detailed chemical analysis of basement granites, metasediments of Sirohi group and Balda granite have given enough evidences for the complex yet characteristic geochemical nature of the mineralizers and of the hosts during the tungsten metallogeny. The mineralization is mainly associated with quartz and greisenised veins present along the shear zones. The late magmatic fluids, boron, fluorine and lithium, caused the mobility and concentration of wolframite along the shear zones.

"Ophiolite suite of rocks in Manipur-Nagaland as hosts for chromites and related economic minerals: a review" is the next paper of the section which has been authored by Prof. O.P. Goel and Prof. Arun Kumar. The paper has presented economic aspects of platinum-bearing potential of chromite in Ukhrul, Moreh and Chandel areas of Manipur district, hosted in the ophiolite suite exposed along Nagaland-Manipur in Indo-Myanmar belt. The fact is that genetically and spatially associated economic minerals with the chromite mineralization in similar host rocks the world over have produced various genetic models. The authors have added these studies as review in the paper. The outline of the paper has suggested that the mechanism of ore formation, according to vapour-liquid crystal crystallization, reflects the main characteristics of the compositions, structures and relations between individuals in aggregates containing RPGE minerals.

The seventh paper of the section authored by Prof. K.L.Shrivastava, Dr. Yasmin Sayyed, Ms Mamta Chauhan and Prof. Ernst Hegner is entitled "Experimental, petrological and field constraints on the petrogenesis of economic carbonatites with special reference to the Sarnu-Dandali area, Northwestern India". The paper has extensively examined experimental, petrological and field constraints to understand the petrogenesis of economic carbonatites, one of the most discussed controversies of economic geology. The studied example, Sarnu-Dandali carbonatite demonstrates its evolution by fractional crystallization of an alkali melt following a magma series of ijolite-melanephelinite-phonolite-carbonatite in petrological and field study. Also, it has shown coherent and typically mantle C- and O- isotopic ratio. The authors have suggested that the Sarnu-Dandali carbonatite is in fact, an ideal natural example showing field and petrological data well in accordance with the extensive experimental data now available.

Dr. G. Vallinayagam has authored the eighth paper of the section and is entitled "Occurrence of Zinc rich A-type granites in the trans Aravalli anorogenic ring complexes, Northwestern India". The Neoproterozoic Malani Igneous Suite covers A-type anorogenic granites of Tosham, Jhunjhunu, Siwana, Kundal and Jalore associated with ring complexes. These granites have interesting characteristics including sub-volcanic nature, hypersolvus to subsolvus, peralkaline to peralminous and HHP type. These have high silica, total alkalies, iron, HFSE and LILE and low Ca, Mg, Sr, Ba, Eu, Cr, Co, Ni and V. The author has conducted a search for Zn granites. He has further suggested that high Zn may be attributed to sodic metasomatism by circulation of post-magmatic fluids; and must have high Zn content in the source magma. The findings are important for further exploration of Zn in the granites.

The paper entitled "Structural setting of the nepheline syenite pluton, west of Khariar, eastern ghat mobile belt, Orissa, India" is the ninth paper of the section authored by Prof. T.K.Biswal and Dr. Harish Ahuja. The nepheline syenite pluton which is subject of the present study is emplaced in a very specific tectonic setting, within a fold thrust of mobile belt that developed on its NW margin. The fold thrust belt consists of a basal decollement and a stack of nappe sheet structurally resting over the Baster Craton. The structural features of Mylonite and the magmatic fabric of the pluton have been analyzed critically and a phenomenon called coupling has been identified to suggest that the magma has been emplaced under a non-coaxial strain. Further, a compressional heating and consequence of melting of the lithosphere are proposed for the origin and travel of alkaline magma through mantle-deep fractures during thrusting.

The tenth paper of the section authored by Prof. R.K.Trivedi and Dr. L.K. Soni is entitled "Stanniferous pegmatites on the eastern flank of the Bailadila iron range, Dantewara district, Chhattisgarh". The paper has discussed the geological set-up to include field studies, structural, petrological and geochemical aspects of the stanniferous pegmatites and the host rocks. The authors have also identified and marked a few stanniferous pegmatites for the first time in the eastern flank of Bailadila iron range on the western extremity of the Baster- Koraput pegmatite belt.

Dr. Pooja Shrivastava and Prof. V.K.Khanna have authored the next paper entitled "A note on significance of high rubidium in the lithiumn mica from Govindpal area of Baster district, Chhattisgarh". The geochemistry of lithium mica, as determined and suggested by the authors, can be utilized as geothermometer for mineralizars. The paper has expressed the possibility that the pegmatites of Govindpal area of Baster seems promising to explore for Li, Rb and Cs contents.

There are some soapstone deposits formed in the vicinity of Sarada inlier due to impact of metamorphism and associated tectonic activity. The last paper of the section entitled "Geological characteristics and genesis of the soapstone deport around Sarada inlier, Rajasthan" is authored by Dr. A.K. Shandilya, Dr. N. Srivastave, Dr. M.K.Singh and Dr. S.K.Gehlot. The paper has attempted to understand the genesis of soapstone using field and laboratory evidences. Temperature, pressure and geochemical variations occurred due to metamorphism. These have been identified as keys to reveal the physio-chemical environment of the hydrothermal solutions and finally to establish the genetic model.

Economic Mineralization
K.L. Shrivastava (Editor)
2009. pp.94-104
Scientific Publishers (India), Jodhpur

Section-II

An Ore-genetic Model for the Gogi Hydrothermal-Type Uranium Deposit in the Neoproterozoic Bhima Basin, Southern India

S.A. PANDIT[1] AND R. DHANA RAJU[2]
[1]Atomic Minerals Directorate for Exploration & Research, Bangalore-560 056
[2]6-3-124, Hastinapuri, Sainikpuri P.O. Hyderabad 500 094
Email : rdhanaraju@yahoo.co.in

Abstract

The Bhima basin, the youngest (~0.625-0.55 Ga) and the smallest (~5,300 sq km in area) amongst the intracratonic 'Purana basins' (Meso-/Neo-proterozoic) of India, has become a first-order target for uranium (U) exploration, on the analogy of similar basins in Canada and Australia being the hosts for many large-tonnage and medium- to high-grade unconformity-type U-deposits, as well as the Cuddapah basin in southern India becoming a U-province with diverse types of U-mineralisation, and two of them being of deposit-grade. Multi-disciplinary exploration by the Atomic Minerals Directorate (AMD) for Exploration and Research in the Bhima basin and its environs since 1988 has led to the identification of U-mineralisation in four diverse lithounits, viz., phosphatic limestone near Ukinal, shale at Kasturipalle, non-phosphatic limestone and basement granitoid in the Gogi area, with the last two being potentially significant and constituting the Gogi U-deposit.

The Gogi U-deposit is a poly-phasic, poly-episodic and poly-metallic, lithostructurally-controlled, hydro (epi)-thermal vein-type having ~3,000te of an average grade of ~0.19% U3O8. Salient aspects of this deposit, like conceptual evolution, geologic setting, host rocks, structure and characteristic features of U-mineralisation, viz., structure, ore minerals, loci, textural aspects, mineral-chemistry, chemical ages of discrete U-phases, mineral-paragenesis, nature, wall-rock alterations, physico-chemical conditions of mineralisation, possible recovery of high-value by-products of Ag, Au, Co and Ni during extraction of the main product, U, are presented. Based on these, an 8-stage evolutionary genetic model of the deposit is proposed.

Keywords : Gogi Uranium Deposit, Uranium mineralization characteristics, Bhima basin, Uranium genetic model.

INTRODUCTION

Meso-/Neo-proterozoic intracratonic basins in Canada and Australia, represented, respectively, by the Athabasca basin region, Saskatchewan and the Pine Creek Geosyncline, Northern Territory, became one of the most attractive exploration targets for uranium mineralisation, as they host many large-tonnage and medium- to high-grade unconformity-type uranium deposits. These are structurally-controlled by faults or fracture zones, and mostly occur at the base of Mesoproterozoic sandstone in unconformable contact with Palaeoproterozoic graphitic/ pyritic metapelites, e.g., monometallic Rabbit Lake (15,769t with average grade of 0.27% U), P2 North (100,000t of 4.22%), Eagle Point (24,720t of 1.27%), Dominique-Peter (11,587t of 0.66%) and polymetallic Cigar Lake (110,000t of 12.2%), Key Lake (70,000t of 1.99%), Midwest (19,300t of 1.6%) and Collins Bay B (11,400t of 0.38%) in Canada; and monometallic Jabiluka 1 and 2 (2,883 t of 0.21% and 172,992t of 0.33%), Koongara (11,278t of 0.228%), Nabarlek (8,700t of 1.56%) and Ranger 1 and 3 (32,915t of 0.273% and 72,123t of 0.17%) in Australia (Ruzicka, 1993). On the same analogy, similar basins, popularly known as the 'Purana basins', and their basements, have become the first order targets for uranium exploration in India (see papers in EARFAM, v. 8, 1995). Integrated exploration program of the Atomic Minerals Directorate (AMD) for Exploration and Research, comprising surveys by airborne-remote sensing, hydrogeochemistry, aero- and ground-geophysics, ground radiometry and drilling, together with relevant multi-disciplinary laboratory studies, led to the identification of diverse types of uranium mineralisation in one such basin, namely, the Mesoproterozoic Cuddapah basin, especially along its margins, in Andhra Pradesh. These include: (i) stratabound-type in phosphatic siliceous dolostone of the Vempalle formation in the Tummalapalle-Gadankipalle-Rachakuntapalle area in Cuddapah district in the southwestern part of the basin (Vasudeva Rao et al., 1989; Dhana Raju et al., 1993; Rai et al., 2002); (ii) unconformity-proximal type, hosted by both the basement granitoid and its overlying Srisailam and Banganapalle quartzite, respectively, below and above the unconformity, in the Lambapur-Peddagattu-Chitrial area in Nalgonda district along the northern margin of the basin (Sinha et al., 1995); (iii) fracture-controlled hydrothermal-type, hosted by the deformed granitoids around Lakkireddipalle, Cuddapah district in the southern environs of the basin (Zakaulla et al., 1998; Dhana Raju et al., 2002); (iv) hydrothermal-type hosted by the Gulcheru quartzite in the Gandi area, Cuddapah district in the southwestern part of the basin (Umamaheswar et al., 2001); and (v) shear-controlled type in the schistose rocks in close proximity to the eastern margin of the basin at Kasturigattu, Gudarukoppu and Kullru in the Nellore district (Veerabhaskar et al., 1991; Rai et al., 1995); of these, the first two are established as deposits.

Similar exploration efforts in the Neoproterozoic Bhima basin in Karnataka since 1988 have brought to light uranium mineralisation in diverse lithounits, viz., phoshpatic limestone and phosphorite near Ukinal (Achar et al., 1997; Dhana Raju et al., 2002), non-phosphatic limestone and basement granitoid in the Gogi area (Pandit et al., 1999 and 2002; Pandit, 2002; Dhana Raju et al., 2002); and shale at Kasturipalle (Pandit, 2002; Pandit et al., 2002). Of these, the one at Gogi is established by subsurface exploration as a uranium deposit of medium-grade (av. ~0.19% U_3O_8) and low-tonnage (~3,000 te). This deposit is unique in many respects, viz., (i) being of the richest grade so far established in India; (ii) located within the smallest (5,300 sq km in area) and the youngest (Neoproterozoic) Purana basin of India; (iii) hosted by two markedly diverse lithounits of Shahabad limestone and its basement biotite granitoid (2.5 Ga old; Sastry et al., 1999), respectively, much above and below the unconformity; (iv) polymetallic deposit with a distinct possibility of recovering precious metals of Ag and Au (?) as well as other high-value metals of Ni, Co and As, during extraction of the main metal, U; and (v) somewhat unusual type since, generally, limestone is considered as an unsuitable host for U-mineralisation due to transport of U in soluble form as uranyl bi- or tri-carbonate complex. In the light of these, we present here salient aspects of this hydrothermal-type U-deposit, based on which its evolutionary genetic model is proposed.

HISTORY OF EVOLUTION OF THE CONCEPT

Reconnoitory radiometric traverses in the Bhima basin and its crystalline environs by P.S. Naidu during the field season, 1988-89 revealed a few radioactive anomalies during hydrogeochemical surveys near Wadi (with U up to 110 ppb) and a few low-order U-values in the basement granite near Ladlapur. Later, a few spot values of high radioactivity due to U were reported in the phosphate-rich Halkal shale. Integrated exploration, involving field techniques of ground radiometry, study of satellite imagery, jeep-borne and hydrogeochemical surveys led Gangadharan and Achar to identify U-mineralisation hosted by brecciated phosphatic limestone near Ukinal. This is marked by high positive correlation between P and U, with sources of U being mainly collophane and to a minor extent clays, thereby resulting in poor leachability of U, due to which further exploration in this area was not undertaken. (Achar et al., 1997; Dhana Raju et al., 2002). Continued reconnoitory radiometric traverses in the Bhima basin, mainly guided by local geological setting, led to recording of low-order radioactivity by Patnaik and Ganesh Bhat in a few isolated, small outcrops of P-poor limestone in the soil-covered areas in the village, Gogi. By semi-detailed exploration, involving digging of a few trenches, followed by shielded probe logging, the U-mineralisation was traced over 400 m, extending farther into the village. The surface expression of radioactivity in these trenches is poor, except for a few samples assaying 0.04-0.05% eU_3O_8. One such sample assayed 0.07% eU_3O_8, 0.12% U_3O_8 and <0.01% ThO_2. The highest value of P_2O_5 in this radioactive zone is 0.28%. Mary K. Kumar and Dhana Raju, while studying a sample from this zone by ore microscopy, have identified hydrothermal vein-type U-mineralisation, in the form of coffinite and pitchblende. To trace this low-order radioactivity further, there were many constraints in the village, like numerous houses and a lake on one side, and a burial ground on the other. At this stage, gamma-ray logging of drinking water borewells in the village was carried out, which led to the discovery of better grade mineralisation with values of 0.136% U_3O_8 x 6.4 m and 0.052% U_3O_8 x 12.3m, intercepted in two borewells. However, it was not certain whether the rock hosting mineralisation is limestone or granite. Hence, sub-surface exploration by core drilling was commenced during May, 1997, with the first borehole (GGL-1) located just adjacent to the bore-well (BWG-2) in which the highest radioactivity was intercepted. Drilling of a few more vertical boreholes helped in establishing the continuity of U-mineralisation, its variable apparent thickness as well as the amount and attitude of major fault and cross-faults. Later, inclined boreholes, commencing from granite towards the basin, were drilled so as to intercept the ore zones for estimating their true thickness. With over 100 coring boreholes (both vertical and inclined) and a few down-the-hole (DTH) non-coring boreholes, over 2 km strike length of U-mineralisation was established, with both ends in the strike direction being still open. The mineralisation is within ~50-70 m wide tectonized-zone that partly encompasses the fractured basement granite and its overlying sediments of the Bhima Group.

GEOLOGIC SETTING

The Bhima basin, in the form of a reverse sigmoid on the northeastern margin of the Dharwar craton, is sandwiched between the basement of Peninsular Gneissic Complex (PGC) and its associated younger granitoids to its south, and Deccan traps to its north (Fig. 1). A large part of its ~5,300 sq km-area lies in the districts of Bagalkot, Bijapur and Gulbarga in Karnataka, with a small NE part extending into the Rangareddy and Mahaboobnagar districts of Andhra Pradesh. Geologically, the sediments of the basin are devoid of major imprints of metamorphism and tectonism (except block faults), with the result that, most of them are almost horizontally disposed with low-dips, retaining their original sedimentary fabric and giving rise to stratiplains, complains or sediplains, which indicate conformity of the landforms with sub-surface disposition of the rocks. This scenario, however, changes along the major fault zones that cut across the basin and, at a few places, its basement granitoids too. These major faults, dissecting the sediments, include the E-W-trending Kurlagare-Gogi (KG), Thirth-Tintini and Rabbanapali, and NW-SE-trending Wadi-Bhimanahalli. Due to strike-slip movement, rhomboid- to parallelogram-shaped sedimentary blocks are arranged in an en-echelon pattern. Moderate to intense folding is observed in the vicinity of these faults, resulting in brecciation, dragging and, at places, isoclinal, overturned and recumbent folding. The width of the tectonized zones along these faults varies widely with a maximum of about 500 m along KG fault. River Bhima and its major tributary, Kagna, drain the basin, with the latter joining the former almost in the middle of the basin, and the former, in turn, joins the major river, Krishna. The soil profile varies from almost nil to 4 m deep (along the KG fault), with thickness and alluvium, being slightly more along river courses; the soil is of black-cotton type. In the basin, there is practically no forest-cover and only abundant thorny bushes. Except along fault zones and fractures, groundwater is fairly deep.

Stratigraphically, the Bhima Group of sediments, with an aggregate thickness of ~300 m, unconformbaly overlies the crystalline basement and underlies the basaltic flows (Deccan Traps) and inter-trappean sediments (Table 1, after Jayaprakash, 1999). From the base to the top, these sediments are

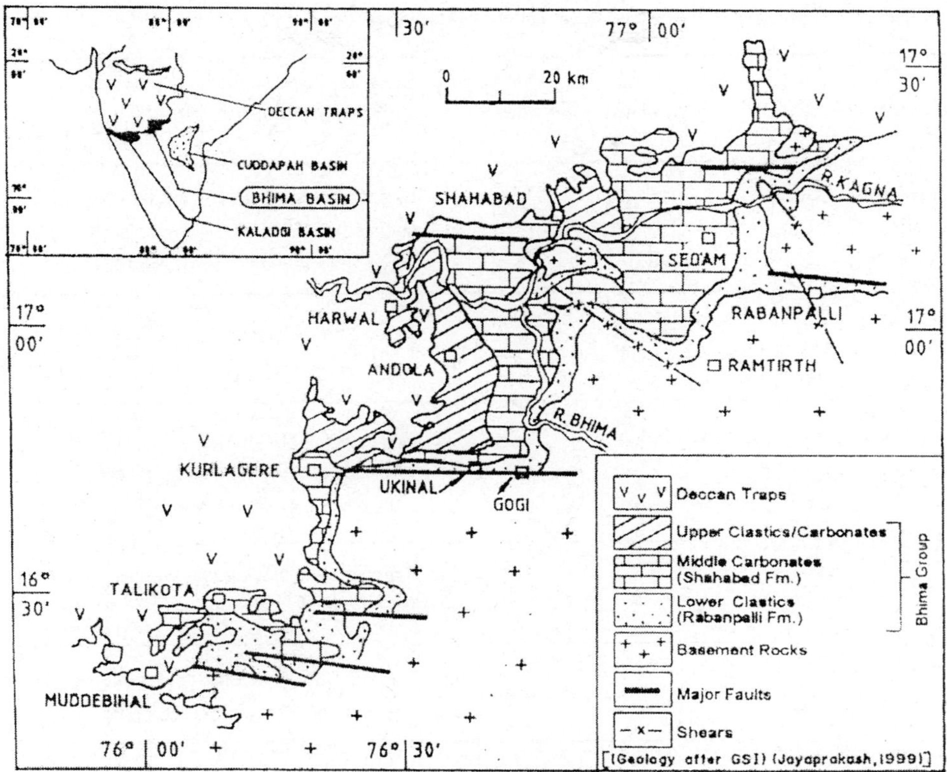

Fig. 1 Geological map of the Bhima basin with location of the uranium deposit at Gogi and uranium occurrence near Ukinal along the Kuragare-Gogi fault.

divided into the formations of Rabanpalli (Lower Clastics, mostly of conglomerate, arenite and shale), Shahabad (Middle Carbonates, viz., limestone flaggy, bedded, stylolitic and variegated in different shades), Hulkal (shale-silt), Katamadevarahalle (limestone) and Harwal (shale), with the last three constituting the Upper Clastics/ Carbonates (Fig. 1). Of these, the Shahabad Formation is the thickest (~200 m) and, from the point of economic potentiality, is the most important in the stratigraphic column of Bhima sediments, as it comprises voluminous quantity of limestone that supports cement industry, besides hosting the potential U-deposit in the Gogi area. The basement crystalline rocks comprise greenstone/schist belt (Mangalur) of the Hutti Group (3.1-3.4 Ga), present as enclaves in PGC (~3 Ga) that occupies lower contours with peneplained topography and intruded by pink and gray granitoid (2.5 Ga, Sastry et al., 1999; Closepet equivalent), which rims the basin as ridges, hills and mounds. Many basic dykes traverse the crystalline terrain in the environs of the basin. The age of the Bhima sediments, as deduced from biota and litho-stratigraphy, points to Vendian/ Neoproterozoic (<625 - >550 Ma). One of the most striking features of the Bhima basin is the absence of stromatolites, unlike in the nearby Kaladgi and Cuddapah basins, and this may be attributed to the emergence of Metazoa that scavenged stromatolite-producing algal biota in shallow marine environment (Awarmik, 1990), in which the sediments were deposited in inter-tidal to sub-tidal environment.

HOST ROCKS OF MINERALISATION

The host rocks for uranium mineralisation in the Gogi area are: (i) Shahabad Limestone, and (ii) the basement biotite granitoid.

Uranium-mineralized limestone is light dark gray-colored, fine-grained, compact, brecciated (Fig. 2). It contains dominant carbonate minerals with lesser chert, organic or carbonaceous matter, clays (illite and smectite), sulfides and limonite, together with accessory glauconite and barite, at places. The predominant carbonate mineral is ferroan calcite that is both primary and secondary. Primary calcite is the major constituent (~60-70 vol.%) that is fractured; it is mainly micritic and subordinate sparry type, and is turbid due to presence of opaque impurities. The

Table-1 Stratigraphic sequence of the Bhima basin (after Jayaprakash, 1999)

Eon/Epoch	Group	Formation	Member	Lithology (thickness in m)
Cretaceous *to Eocene*	Deccan Trap			Basaltic flows with intertr appean sediments
		Harwal		Brown, pink to vermillion shale (45)
		Katamde-varahalli		Deep gray, occasionally stylolitic, flaggy limestone (40)
		Hulkal		Gray, blackish buff, dull & pale pink shale, occasionally with fine grained, thin, silty beds at the base (30)
			Mulkod Limestone	Deep gray to black flaggy lime - stone (10)
			Gudur Limestone	Akin to Wadi limestone, yet slightly inferior chemical composition (20)
	BHIMA GROUP	Shahabad	Sedam Limestone	Variegate d medium to thickly bedd-ed, siliceous limestone (60)
Eocambrian			Wadi Limestone	Thickly bedded, stylolitic, relatively superior, cement-grade limestone (15)
			Ravoor Limestone	Flaggy limestone with prominent fissility (Shahabad slabs) (10)
			Korla Shale	Fine silty base, g rades into green shale, followed by chocolate brown Shale with prominent parting(50)
		Rabanpalli	Kundrapalle - Sandstone	Fine grained quartz arenite, sub - felaspathic arenite, ferruginous -cemented, medium-grained quartz arenite (15)
			Muddebihal Conglome-rate	Pebbly orthoconglomerate, lo-cally or at the top matrix-sup-ported and also granular (2)
		Unconformity		
Proterozoic to Archaean	BASEMENT CRYSTALLINES			Younger Granites, Eastern Block Greenstone Belts, Peninsular Gneiss

Fig. 2 Three-centimeter-thick U-mineralised vein (black; containing pitchblende and coffinite) traversing the non-phosphatic limestone at Gogi.

secondary type occurs as large anhedral and subhedral grains along the fractures of the primary calcite. A little rhomb-shaped dolomite occurs cutting across primary calcite. Chert is usually less than calcite and in some samples, it exceeds the latter, when they are designated as 'brecciated calcitic chert'. Chert contains scattered rhombs of dolomite, veins of calcite, clay and a little detrital quartz. It is bordered by chalcedony, at places. Organic matter (OM) and sulfides are in major amount, megascopically conspicuous (Fig. 2) and are post-calcite in formation. Chemical data on the U-mineralized carbonate rock from Gogi (see Dhana Raju et al., 2002 for detailed data) point out that the ore is essentially non-phosphatic (0.04-0.42% P_2O_5), with major impurity of silica (7-19%), manifest as chert, and minor impurities of alumina (1.2-3.6%), potash (0.4-1.1%), soda (0.2-0.8%) and iron, with ferric (0.3-4.3% Fe_2O_3) dominant over ferrous (0.3-0.7% FeO), and these are mineralogically expressed in the form of clays, limonite and Fe-sulfides. The U-ore (0.013-1.68% U_3O_8) is generally characterized by high contents of Co (up to 129 ppm), Ni (up to 77 ppm), Pb (20->1000 ppm; mostly radiogenic), V (58-1143 ppm), Mo (up to 721 ppm) and Ag (up to 62 ppm), with a general positive correlation of the first three with U.

The U-mineralized basement granitoid is a deformed pink and gray biotite granite-granodiorite (s.s.) with cataclastic texture. It comprises major quartz, sodic plagioclase, microcline and microcline perthite (replaced by sodic plagioclase along margins in some places), with minor to major amounts of chloritized and epidotized biotite, minor hornblende and accessory zircon, apatite, allanite, OM and sulfides, besides anatase, ilmenite and limonite. Its chemical data (see Dhana Raju et al., for data) point out that it is highly silicic (with silica up to 76 wt.%), low-Ca type (0.21-1.1% CaO), peraluminous with A/CNK of 1.0-1.37, both sodic and potassic; as for trace elements, it has relatively higher Rb (142-258 ppm), Ba (up to 1072 ppm), Y (14-37 ppm) and Pb (27-77 ppm) and lower Sr (84-177 ppm), Cr (27-39 ppm), V (up to 20 ppm), Co (9-10 ppm), Ni (up to 11 ppm) and Zn (16-50 ppm). The sodic type, compared to the potassic type, has higher U, Fe, Mg, Mn, Cr V and Ni, possibly attesting to some degree of Na-metasomatism and alterations like chloritisation and epidotitsation on an original potassic (biotite) granite leading to enrichment of ferromagnesian elements with which U also got remobilized and concentrated.

REGIONAL STRUCTURE IN SELECTION OF TARGETS FOR U-MINERALISATION

Regional structural mapping, using remote sensing data, revealed 3 sets of major lineaments over PGC. These are: (i) NNW-SSE to NW-SE; (ii) NE-SW to ENE-WSW; and (iii) E-W to WNW-ESE, with the first set appearing to be the oldest, and best represented by trends of the major greenstone belts (Achar et al., 1997). This set is superimposed or cut across by the NE-SW and E-W lineaments that are believed to have developed in response to N-S extensional tectonics (Drury, 1984; Chetty, 1995). This synoptic view suggests that the impress of these three lineaments is best etched over PGC in the south of the central parts of the Bhima basin. Incidentally, this sector also has a large proportion of younger granitoid (~2.5 Ga; Closepet equivalent) (Fig. 1) that constitutes a fertile provenance for uranium, demonstrated by its U content (20-24 ppm), in contrast to the barren PGC (with 3-4 ppm U). NW-SE and NE-SW to ENE-WSW shear systems in the region are known for the occurrence of epigenetic vein-type U-mineralisation in the Raichur granitoid (Chopra and Jagannadha Rao, unpublished report, SR, AMD, 1989). U-mineralisation, controlled by fracture and fault system, is also known in the eastern and western cratonic blocks of Karnataka [(Bhairam and Pandit, unpublished rep., SR, AMD, 1995) and (Swarnkar et al., 1998)]. Probably, these major shear systems could have acted as conduits for migration of U-bearing solutions that originated from the fertile younger granitoid (of Closepet eq.). Areas of such older lineaments and their intersection with younger lineaments/shear sets, especially at the sediment-basement contact are, thus, identified as ideal sites for the concentration of U under suitable physico-chemical conditions. Accordingly, the following three zones were identified as the first-order targets in the central Bhima basin (Fig. 1):

(i) E-W trending KG fault that is orthogonal or near orthogonal to the major trends over PGC, with many structural features, like shear zones, fractures and faults extending from the basement into the basin, besides a number of cross-faults along it;

(ii) Intersection of ENE-WSW lineament and the Wadi fault between Alur and Bhimanahalli; and

(iii) E-W trending Hunsigi-Devathkal fault with intense shearing and spatially close to the younger granitoid, with major ENE-WSW and NNW-SSE linear trends. There is also hot spring activity at Wajhar along these faults.

URANIUM MINERALISATION

Structural Aspects of the Ore Body

The ore body is in the form of intricate network of veins, veinlets and fracture-fillings within the brecciated limestone and tectonized granitoid. Two distinct zones hangwall and footwall are identified, with the former being prominent in the central portion of the mineralised strike length and located after the marker bed, viz., hangwall shale, as seen in most of the transverse sections and in the fence diagram (Fig. 3; Pandit, 2002). Often, the hangwall lodes coalesce and attain locally huge (tens of meters) thickness. The footwall zone is more consistent and better correlatable than the hangwall zone, along both the strike and dip directions. Examination of radiometric beta-gamma data on numerous drill-core samples points out that nearly 75% samples analyzed indicate disequilibrium (av. 50%) in favor of parent and 25% in favor of daughters, with only a few showing equilibrium (Pandit et al., 2002).

Minerals, Loci and Textures of Ores

In the non-phosphatic carbonate U-ore, uranium is accounted mainly by coffinite and pitchblende, with minor contribution from OM, anatase and adsorbed U associated with clays and chert; a little labile U is also present along the grain boundaries of minerals. The discrete U-phases of both coffinite and pitchblende are intimately associated with OM and sulfides, and occur as veins, veinlets and fracture-fillings. Pitchblende is of two generations, and the older one is always rimmed by coffinite, with patches of galena occurring in the vicinity. The younger generation of pitchblende replaces coffinite within OM as well as along the fractures of coarse-grained pyrite. OM is a low-rank, migratory (labile)-type bituminous variety and occurs as large globular aggregates with variable reflectivity that is a little higher than that of coffinite. It contains pockets of clay, chert and U-phases. Among the sulfides, pyrite is the most dominant and occurs in three modes, viz., euhedral to subhedral coarse grains (0.5-1 cm) that are fractured and brecciated at places, framboids and in reticulate pattern along the grain boundaries of calcite. Coarse-grained pyrite is replaced along its margins by OM that also fills its

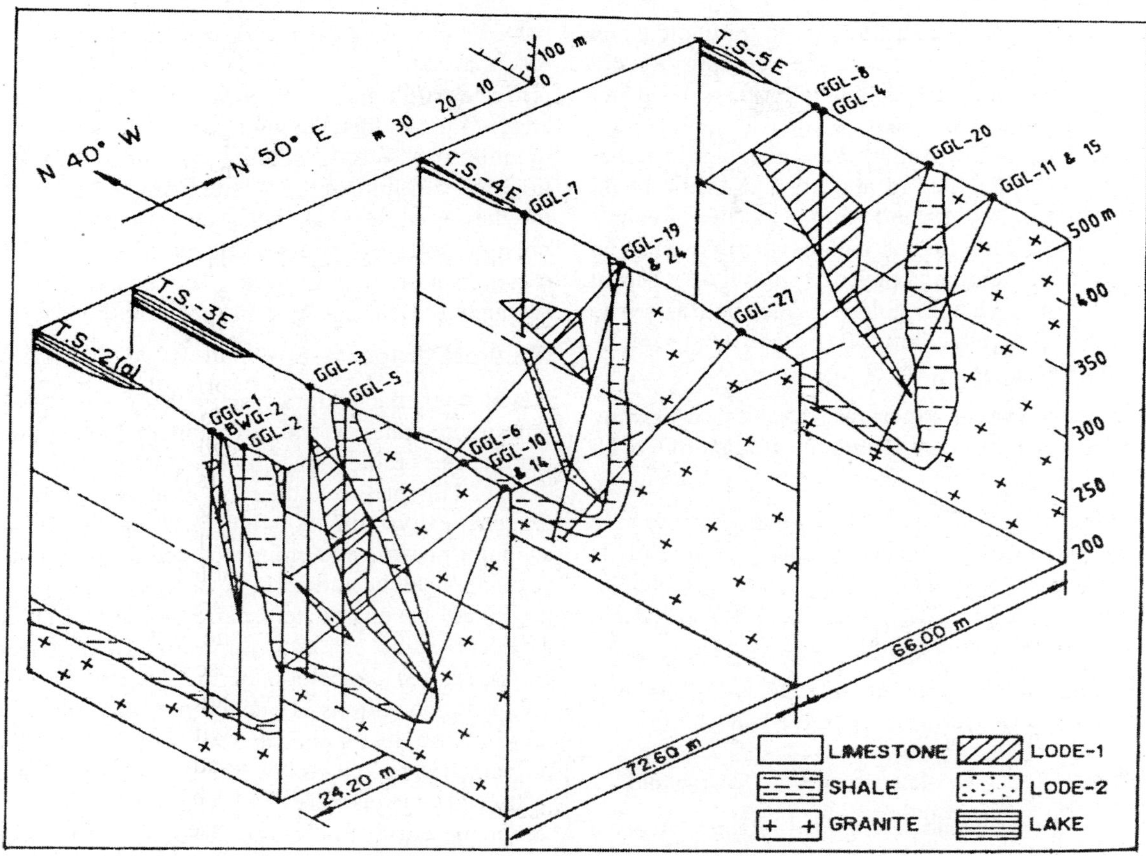

Fig. 3 Fence diagram of the main uranium mineralised zone in the Gogi area.

fractures. Other sulfides in minor amounts are marcasite, chalcopyrite and galena (Dhana Raju et al., 2002).

In the basement granitic U-ore, uranium is mainly due to coffnite and minor pitchblende, with lesser contribution from U-Ti complex; some adsorbed U is associated with sericite and chlorite. The discrete U-phases occur as veins, veinlets and fracture-fillings, with pitchblende being replaced by coffinite, and both U-Ti complex due to 'Pronto' reaction. Similarly, sulfides, comprising mainly pyrite and minor marcasite, arsenopyrite, cahlcopyrite and galena, occur as veins traversing the granitoid (Dhana Raju et al., 2002).

Mineral Chemistry

Electron-probe microanalysis of the radioactive mineral phases in the U-ores of brecciated limestone and granitoid points out the following features (see Dhana Raju et al., 2002 for data):

(i) Pitchblende in limestone, compared to that in granitoid, is marked by higher contents of UO_2 (av. 84%) and PbO (av. 6.3%), and lower contents of silica (av. 2.73%) and Ln_2O_3 (av. 0.5 %), with ThO_2 being negligible (av. 0.04%) in both.

(ii) Silica content in pitchblende of granitoid (5.4-10.2%) indicates that it is a coffinitized pitchblende, which corroborates the textural relationship of coffinization of pitchblende, mentioned earlier;

(iii) Coffinite from limestone, compared to that from granitoid, is marked by higher contents of UO_2 (av. 77.9%) and PbO (av. 3.2%), and lower contents of silica (av. 10.3%) and Ln_2O_3 (av. 0.28%), with negligible Th in both; these patterns almost mimic those of pitchblende given above.

(iv) In limestone, the framboidal pyrite, in contrast to that of euhedral-subhedral one, is marked by slightly lesser contents of major elements of Fe and S, and this is compensated by its higher contents of minor/trace elements, like Co, Ni, Cu,

Pb, As and Ag. In granitoid, both framboidal and euhedral-subhedral pyrite are compositionally much similar in major element contents, but differ in the trace element contents, with latter, compared to the former, containing a little higher amount of Co, Ni, Sn and Se, and a little lower amount of As, Ag and Au. Of the two precious metals, silver is higher in both types of pyrites (up to 0.09%) and chalcopyrite (up to 1.18%) from limestone, whereas gold is higher both in pyrite and galena from granitoids (up to 0.19%), with As mimicking the pattern of Ag.

(V) Chalcopyrite in limestone, like its co-existing framboidal pyrite, has tangible amount of Co, Ni and Ag.

(vi) OM is compositionally heterogeneous, as demonstrated by the wide range of some of its analyzed constituents, like UO_2 (0.0-14.9%), silica (0.06-14.6%) and CaO (0.01-5.4%); Th and Zr contents are almost nil.

(vii) Radioactive clay has notable UO_2 (4.5-7.9% and PbO (3.9-4.5%), and is free of Th and Zr.

U-Mineral	Limestone		Granitoid	
	Range	Average	Range	Average
Pitchblende	334-862	533 (n=9)	9-798	226 (n=14)
Coffinite	62-1146	300 (n=16)	1-480	126 (n=20)
U-Ti Complex	-	-	1-12	7 (n=6)

Chemical Ages of U-minerals

The range and average of chemical ages (in Ma) of U-minerals, determined from their respective contents of U, Th and Pb, in both limestone and granitoid, as per the equation of Montel et al., (1994) are as follows:

The above data suggest the following: (i) The average chemical age of pitch-blende is relatively older than that of coffinite in the ores of both carbonate and granitoid, whereas that of U-Ti complex, present only in the latter, being the least, implying poly-cyclic U-mineralisation and conforming their textural relationship. (ii) The range of UO_2 in pitchblende (~82%) and coffinite (~75%) is relatively less, whereas, that of PbO, considering its much lower absolute content, is much large (0-12%) in both types of ores, which attests to the higher degree of redistribution of radiogenic Pb due to remobilization leading to large spread in chemical ages. (iii) The wide range of chemical ages of discrete U-phases implies polyphase remobilistion and precipitation of U due to repeated structural dislocations, including faulting and folding; in this, the younger chemical ages of the uranium phases in granitoid, as compared to that in limestone, in spite of the former being the source of U for the latter, suggest that such remobilization and precipitation of U continued even after cessation of mineralisation in limestone. The post-ore stage is accounted essentially by vein-type calcite and quartz.

Mineral Paragenesis

Based upon the textural relationships between and within the gangue and ore minerals, the mineral paragenesis (Table 2) is established (Dhana Raju et al., 2002). The pre-ore stage in limestone comprises diagenetic crystallization of primary calcite and chert as major phases, with minor to accessory contents of secondary calcite and barite, whereas in granitoid it is accounted by quartz, feldspars, biotite and chlorite. The ore stage in both limestone and granitoid comprises (a) deposition of the discrete U-phases, viz., two generations of pitchblende with intervening coffinite; (b) lesser contributors of radioactivity, viz., organic matter and clay, the formation of which started prior to discrete U-phases; and (c) sulfides that are mainly prior to the formation of U-minerals and include predominant pyrite with lesser chalcopyrite and galena (with its Pb being radiogenic). The pre-ore stage minerals were affected by brittle and ductile deformation, possibly as a consequence of faults that resulted in intense brecciation, macro-fracturing and microfracturing, thereby providing suitable structural backdrop for mineralisation.

Nature of U-mineralisation

The uranium mineralisation in both the carbonate and granitoid ores occurs mainly as veins, veinlets and fracture-fillings, and confined to the structurally weak zones of faults, folds and brecciation. The mode of occurrence of the major U-mineral assemblage of coffinite and pitchblende, the textural relationship between them and with OM and pyrite, their Th-free mineral chemistry and presence in different host rocks of limestone, chert and granitoid cumulatively indicate that the Uranium mineralisation in the study Gogi area is essentially hydro (epi)-thermal vein-type. Generally, this type of mineralisation is accompanied by wall-rock alterations. These include chloritisation of biotite in the granitoid and carbonatization (secondary coarse-grained calcite and minor dolomite) and argillitisation in the formation of

Table-2 Paragenesis of Ore and Gangue Minerals in the Gogi Uranium deposit,Gulbarga district, Karnataka, southern India (after Dhana Raju et al., 2002)

Mineral	Pre-Ore Stage	Ore Stage	Post-Ore Stage
Gangue			
In Limestone			
Calcite (primary)	—		
-do-(secondary)	—		—
Dolomite	—		
Barite	—		
Chert	—	—	
In Granite			
Quartz (primary)	—		
-do-(secondary)			—
Feldspars	—		
Biotite	—		
Chlorite	—		
Sulfides			
Pyrite			
Idiomorphic		—	
Reticulate		—	
Framboidal		—	
Galena		—	
Chalcopyrite		—	
Uranium Sources			
Coffinite		—	
Pitchblende		— —	
U-Ti Complex		—	
Organic Matter		—	
Clays		—	

smectite and illite in limestone and chert. The source for the uranium mineralisation is the fertile basement biotite granitoid, labile U from which was remobilized during ductile and brittle deformation and transported most probably as uranyl tri-carbonate (UTC) complex and precipitated in the zones of reduction, rich in organic carbon and pyrite. The first phase is pitchblende that, with increasing silica-activity got coffinitized resulting in the formation of dominant coffinite. This coffinite, in turn, was replaced by pitchblende on reduction of silica-activity. Thus, the U-mineralisation is of more than one episode that is also corroborated by wide range of chemical ages of both pitchblende and coffinite. Furthermore, the presence of notable contents of sulfides, mainly pyrite, containing appreciable amounts of As, Ag, Co, Ni and occasionally Au, with minor barite in limestone and accessory fluorite in the granitoid makes this hydrothermal uranium mineralisation mineralogocially complex, with U-minerals precipitated after the formation of much of reductants, viz., sulfides and OM. Thus, the U-mineralisation in the Gogi area is poly-phasic, poly-episodic and polymetallic hydro (epi)-thermal vein-type. Comparing the mineral assemblage with the experimental results of Krumbein and Garrels (1952), the most probable range of Eh and pH for U-mineralisation is 0.2 to 0.3V and 7 to 8, respectively. The mineral assemblage is typical of low-temperature of formation (~<200°C), and this is also corroborated by the presence of framboidal pyrite, occasional marcasite, low rank, bituminous type OM and Th-free/poor pitchblende and coffinite. The association of pitchblende, OM and pyrite vis-à-vis experimental studies (cited from Rich et al., 1978) indicate log f_{O2} of 62 to 68, log f_{CO2} of ~ -2, log f_{S2} of 14 to 18, with transportation of U, most probably as UTC at a pH of 7.2 to 8.

Controls for U-Mineralisation

The following are the main controls for the U-mineralisation in the study area:

(i) Fertile basement biotite granitoid (Closepet equivalent) containing U up to ~15 times that of crustal average;

(ii) Post-sedimentary, structurally-disturbed zones, marked by major and minor faults like the KG fault and its other intersecting faults, folds and brecciation;

(iii) Presence of weak and permeable zones, like unconformity between the basement and sediments of the Bhima basin, fractures and zones of brecciation, which facilitated the easy mobility for U-bearing solutions;

(iv) Presence of strong reductants, viz., OM and sulfides, prior to the precipitation of discrete U-minerals;

(v) Impervious cover rocks of reactive limestone and shale for preservation and better grade of the ore; and

(vi) Dyke (dolerite) activity in the basement granitoid as source for heat and S, besides creating necessary geothermal gradients to remobilize U of the basement.

By-Products

The presence of precious and other high-value elements, like Ag, Au, Ni, Co and As, present almost entirely in sulfides makes them as potential bye-products during extraction of U from carbonate gangue by alkali route or from silicate gangue by acid route. The sulfides can be easily concentrated and separated from the rest by flotation.

AN ORE-GENETIC MODEL FOR THE GOGI URANIUM DEPOSIT

Synthesizing the geological observations of the intracratonic Neoproterozoic Bhima basin and its crystalline environs as well as ore-forming processes in the study area, a conceptual model for the Gogi uranium deposit, involving the following sequential stages, is proposed (Pandit, 2002) (Fig. 4).

1. In the first stage, the lithological framework of basement, comprising the Archaean PGC (~3Ga), intruded by the late Archaean-early Paleoproterozoic (2.5 Ga) biotite granitoid (equivalent to the ~2.5 Gaold Closepet Granite) was formed. The introduction of U into the crust apparently occurred in Na-rich granitic magmas resulting in PGC (of TTG composition) early in the development of the crust. According to Glikson (1979), these magmas were probably produced by partial melting of upper mantle mafic rocks and had U content of 1-3 ppm and Th content of 4-10 ppm. The first significant concentration of U, however, took place during the formation of K-rich felsic magma that developed due to partial melting of PGC, resulting in ~2.5 Ga-old Closepet equivalent fertile biotite granitoid with ~15-20 ppm U (Dhana Raju, 2006). The crystalline basement was intruded by minor dykes (~1.8 Ga?) that could have been the source for S.

2. The second stage involves the formation of the Neoproterozoic (1-0.8 Ga) Bhima basin within and over the eroded parts of the above cratonic basement. For the basinal arenaceous and argillaceous sediments, the source is the crystalline basement granitoids, whereas, the carbonates were formed by chemical precipitation, possibly in a marginal shelf environment, with plenty of reductants of organic matter and sulfides.

3. During the third stage of post-sedimentation (~0.6 Ga), major structural disturbances affected the basin, manifest as major faults, like the Kurlagare-Gogi and others, which affected both the basinal sediments near the basal unconformity and crystalline basement.

4. During the fourth stage, precipitation of OM and pyrite-dominant sulfides took place in the deformed basement granitoid and brecciated

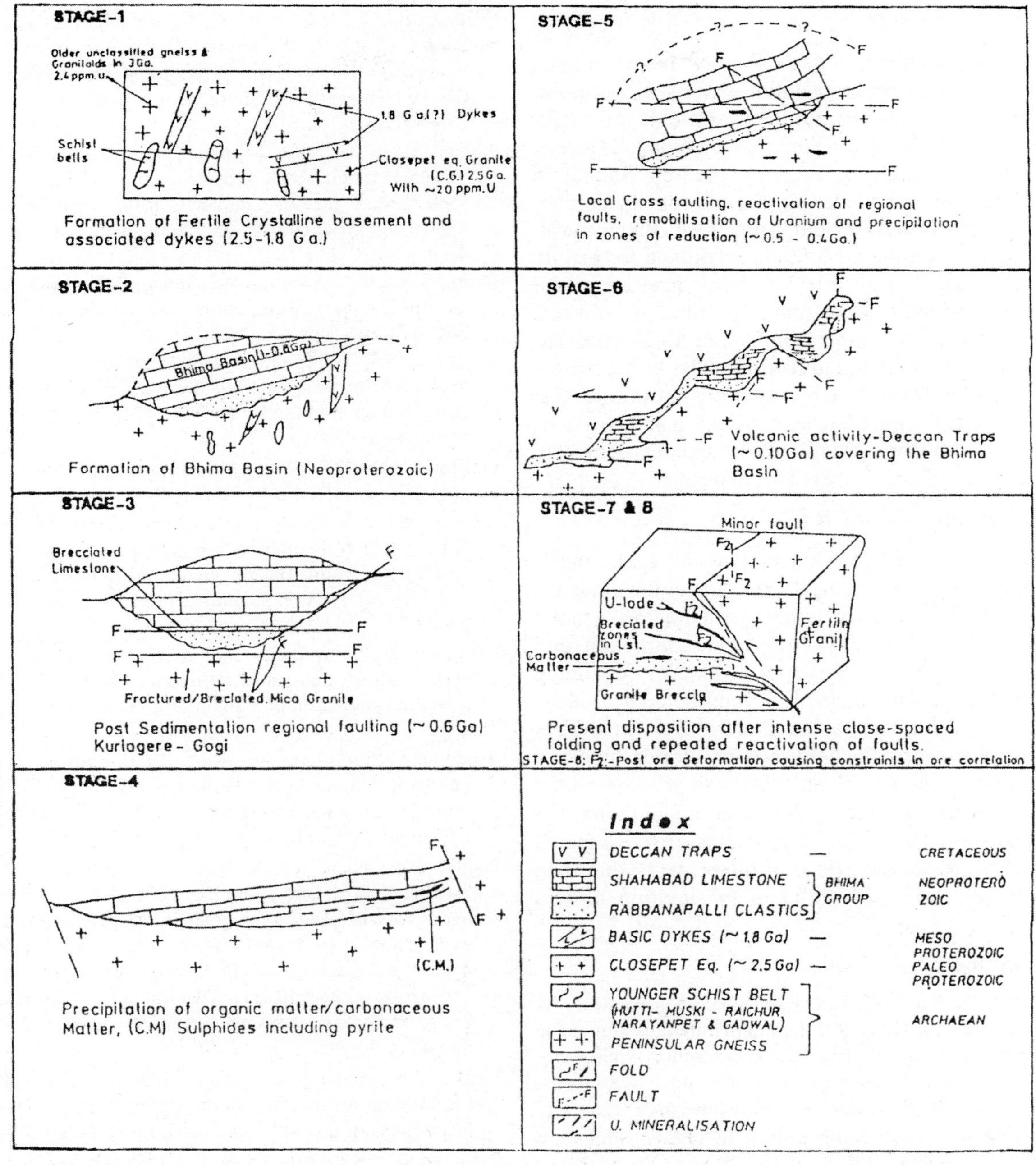

Fig. 4 Evolutionary genetic model for the Gogi uranium deposit in the Bhima basin.

limestone.

5. In the fifth stage, local cross-faulting and reactivation of regional faults resulted in remobilization - transportation (as uranyl tri-carbonate complex) precipitation of U in the zones of reduction, viz., OM- and sulfide-rich zones.

6. During the sixth stage (~0.1 Ga), large-scale volcanic activity, in the form of Deccan traps, took place covering the basinal sediments.

7. During the seventh stage, intense close-spaced folding and repeated reactivation of faults took place and this could be the reason for poly-episodic U-mineralisation.

8. The eighth stage is marked by post-ore deformation effects.

CONCLUSIONS

The Gogi uranium deposit in the Neoproterozoic Bhima basin in southern India is poly-phasic, poly-episodic and poly-metallic, lithostructurally-controlled, hydro- (epi)-thermal vein type. It is hosted by the brecciated, non-phosphatic, Shahabad limestone and tectonized basement biotite granitoid, and occurs as bands in both the hangwall and footwall zones. The U-mineralisation is accounted for mainly by discrete U-minerals of two generations of pitchblende and a generation of coffinite in between, with some contribution from minor U-Ti complex, organic matter (OM) and clays. It is intimately associated with OM and pyrite-dominant sulfides that host high-value metals, like Ag, Au, Co and Ni. An 8-stage evolutionary genetic model is consistent with the characteristic features of the deposit.

ACKNOWLEGEMENTS

The present work forms a part of the Ph.D. thesis (Bangalore University) of Pandit, under the guidance of Dhana Raju. They thank the Director, Atomic Minerals Directorate (AMD) for Exploration and Research of the Dept. of Atomic Energy for timely support and encouragement during the course of the work. Grateful thanks are due to their numerous colleagues, both in the field and laboratories, for many fruitful discussions and multi-disciplinary data, which played a crucial role in formulating ideas for the proposed model. Dhana Raju thanks Prof. Kanhaiya Lal Shrivastava, the Editor of the present Volume in honour of Prof. B.S. Paliwal, for inviting him to contribute the paper.

References

Achar, K.K., Pandit, S.A., Natarajan, V., Kumar, Mary K. and Dwivedy, K.K. 1997. Uranium mineralisation in the Neoproterozoic Bhima Basin, Karnataka, India. In: Recent Developments in Uranium Resources, Production and Demand, IAEA, Vienna, pp. 1-22.

Awarmik, S.M. 1990. Stromatolites. In: D.E.G. Briggs and P.R. Crowther (Eds.), Paleobiology: A synthesis. Blackwell Sci. Publ., London, pp. 336-341.

Chetty, T.R.K. 1995. Significance of the block rotation model in tectonics and mineralisation in the Precambrian terrain an example from the South Indian shield. Jour. Geodynamics, v. 20 (3), pp. 255-266.

Dhana Raju, R. 2006. Geochemistry of the Peninsular Gneiss (PG) and Closepet (Eq.) Granite in the Rayachoti area, Andhra Pradesh, India: Its implication on the concepts of age-zonation of PG in the Dhrwar Craton and 'Dharwar Batholith'. Jour. Applied Geochem., v. 8, no. 2A, pp. 558-577.

Dhana Raju, R., Kumar, Mary K, Babu, E.V.S.S.K. and Pandit, S.A. 2002. Uranium mineralisation in the Neoproterozoic Bhima basin at Gogi and near Ukinal: An Ore Petrological Study. Jour. Geol. Soc. India, v. 59, pp. 299-321.

Dhana Raju, R., Roy, Minati, Roy, Madhuparna and Vasudeva, S.G. 1993. Uranium mineralisation in the southwestern part of Cuddapah basin: a petromineralogical and geochemical study. Jour. Geol. Soc. India, v. 42, pp. 135-149.

Drury, S.A. 1984. A Proterozoic intercratonic basin, dyke swarms and thermal evolution in south India. Jour. Geol. Soc. India, v. 25, pp.437-444.

Earfam 1995. Exploration and Research for Atomic Minerals. Spl. Issue on 'First order uranium exploration target selection in the Proterozoic of India'. v. 8, AMD, Hyderabad-500 016, 213 p.

Glikson, A.Y. 1979. Early Precambrian tonalite-trondjemite sialic nuclei. Earth-Science Rev., v. 15, pp. 1-73.

Jayaprakash, A.V. 1999. Evolutionary history of Bhima basin. Field Workshop on 'Integrated Evaluation of the Kaladgi and Bhima Basins', Geol. Soc. India, Bangalore, Abstract vol. (V.S. Kale and R.H. Sawkar, Compilers), pp. 23-28.

Krumbein, H.C. and Garrels, R.M. 1952. Origin and classification of chemical sediments in terms of pH and oxidation-reduction potentials. Jour. Geol., v. 60, pp. 1-33.

Pandit, S.A. 2002. Uranium Mineralisation in the Neoproterozoic Bhima basin, Gulbarga district, Karnataka, India Geology and Exploration. Ph.D. thesis, Bangalore University (Unpublished), 278 p.

Pandit, S.A., Natarajan, V., Bhat, G., Dhana Raju, R., Swarnkar, B.M. and Banerjee, D.C. 1999. Neoproterozoic Bhima basin as a potential target for uranium exploration a case study from Gogi area, Gulbarga district, Karnataka, India. Field Workshop on 'Integrated Evaluation of the Kaladgi and Bhima Basins', Geol. Soc. India, Bangalore, Abstract vol. (V.S. Kale and R.H. Sawkar, Compilers), pp. 58-60.

Pandit, S.A., Natarajan, V. and Dhana Raju, R. 2002. Exploration for uranium in the Bhima basin in parts of Karnataka, India. Expl. Res. At. Miner., v. 14, pp. 59-78.

Rai, A.K., Sharma, S.K., Umamaheswar, K., Nagabhushana, J.C. and Vasudeva Rao, M. 1995. Uranium mineralisation in the eastern margin of Cuddapah basin. Abstract

Rai, A.K., Zakaulla, S., Jeyagopal, A.V., Vasudeva Rao, M., J.C. Nagabhushana, J.C. and Varadaraju, H.N. 2002. Uranium mineralisation in the southwestern part of the Cuddapah basin, Andhra Pradesh. Expl. Res. At. Miner., v. 14, pp. 79-94.

Rich, R.A., Holland, H.D. and Petrson, U. 1978. *Hydrothermal Uranium Deposits.* Elsevier, Amsterdam, 264 p.

Ruzicka, V. 1993. Unconformity-type Uranium Deposits. In: R.V. Kirkham, W.D. Sinclair, R.I. Thorpe and J.M. Duke (Eds.), *Mineral Deposit Modeling*. Geol. Assoc. Canada, Spl. Paper 40, pp. 125-149.

Sastry, D.V.L.N., Krishna, V., Sharma, U.P., Bhatt, G., Patnaik, J.K., Mumar, Mary, K. and Chabria, T. 1999. Rb-Sr isochron ages on granites from basement of Bhima Basin, Gulbarga Dist., Karnataka. Proc. Eight ISMAS Symp. on 'Mass Spectrometry', Indian Institute of Chemical Technology, Hyderabad, pp. 638-641.

Sinha, R.M., Shrivastava, V.K., Sarma, G.V.G. and Parthasarathy, T.N. 1995. Geological favourability for unconformity-related uranium deposits in northern parts of Cuddapah basin: evidences from Lambapur uranium occurrence, Andhra Pradesh, India. Expl. Res. At. Miner., v. 8, pp. 111-126.

Swarnkar, B.M., Dhana Raju, R., Pandit, S.A. and Umamaheswar, K. 1998. Eastern Dharwar craton as a potential uranium province. Extended Abstract, Ind. Miner., v. 32 (1), pp. 90-91.

Umamaheswar, K., Basu, H., Patnaik, J.K., Ali, M.A. and Banerjee, D.C. 2001. Uranium mineralisation in the Mesoproterozoic quartzites of Cuddapah Basin in Gandi area, Cuddapah district, Andhra Pradesh: A new exploration target for uranium. Jour. Geol. Soc. India, v.57, pp. 405-409.

Vasudeva Rao, M., Nagabhushana, J.C. and Jeyagopal, A.V. 1989. Uranium mineralisation in the middle Proterozoic carbonate rock of the Cuddapah Sukpergroup, southern Peninsular India. Expl. Res. At. Miner., v. 2, pp. 29-38.

Veerabhaskar, D., Sarkar, M., Thimmaiah, M., Sharma, M. and Dhana Raju, R. 1991. Uranium mineralisation at Kastuigattu, Nellore district, Andhra Pradesh, India: An example of shear-conttrolled, syngenetic-remobilized-type in Proterozoic granitic and low-grade metamorphic rocks. Expl. Res. At. Miner., v. 4, pp. 27-37.

Zakaulla, S., Rai, A.K., Tripathi, B.K., Thirupathi, P.V., Umamaheswar, K. and Dhana Raju, R. 1998. Alterations in the basement fracture zones and their bearing on uranium mineralisation in southwestern environs of the Cuddapah basin, Andhra Pradesh. Expl. Res. At. Miner., v. 11, pp. 13-24.

Hon'ble Reviewers

1. Dr. S. Sinha Roy
   (Former Dy. Director General
   Geological Survey of India)
   Emeritus Scientist
   Birla Institute of Scientific Research
   Statue Circle
   JAIPUR - 302 001

2. Dr. S. Viswanathan
   (Former Director, AMD)
   Flat B-203, Block-B
   United Avenue Apartment
   South End, 7-I-29
   Ameerpet
   HYDERABAD - 500 016 INDIA

Editorial Assistance

Dr. Vijay Gaur

Economic Mineralization
K.L. Shrivastava (Editor)
2009. pp.108-138
Scientific Publishers (India), Jodhpur

Section-II

Volcanogenic PneumatoHydrothermal Intrusive and Collapse Breccia Deposit of Fluorite at Jalera Khurd in Malani Igneous Suite, Northwest India

K.L. SHRIVASTAVA AND VIRENDRA GAUR
Department of Geology, J.N.V. University, Jodhpur 342 005 INDIA
Email : klsgeology@yahoo.co.in

Abstract

The present investigations of the fluorite deposit at Jalera Khurd is the study of an economic deposit of considerable size in the Malani Igneous Suite (745±10Ma) which is the largest suite of anorogenic acid volcanics in India and the third largest acid volcanics of the globe. The fluorite deposit is situated one kilometre northwest of village Jalera Khurd, (Lat. 72°08'69"N and Long. 24°48'00"E) about 15 kilometre northwest from Raniwara, Bhinmal tehsil, Jalore district, Rajasthan, India. Bimodal volcanics of the Malani Igneous Suite (Neoproterozoic) consisting of lower porphyritic rhyolite and upper porphyritic trachyte are the host rocks. Plume tectonics, circular intrusions and collapse caldera conditions have been explained by earlier workers for the regional geology. Out of three prominent Joint sets, at the fluorite deposit site, only one, named Joint I, has been mineralized with its strike N43°E-S43°W and dip 85°W. The main ore body constituting four veins of fragmental mass collectively constituting a tabular body, with its average thickness, length and depth 20m, 500m and 100m, respectively, within an elongated stockwork. There are few parallel mineralized veins available nearby. Detailed geochemical and petrological studies favour that the host rock magma must have originated relatively at deeper parts of the earth. Crustal assimilation and fractional crystallization are the proposed possible mechanisms. The range of geological features displayed by fluorite breccia indicate that disruptive intrusion by a turbulent fluidized system was the dominant process, initially. The intrusive aspects of fluorite breccia show angular killas fragments encompassing a wide range of sizes all cemented by microcrystalline fluorite itself. This is followed by two episodes of collapse, a major, and a minor are resulting in enclosure of early-formed fluorite fragments into later-formed fluorite matrix. These episodes were followed by pervasive silicification and metasomatic alterations. A twelve stage genetic model is suggested for the fluorite ore body, which combines the features of both hydrothermal intrusive breccia and hydrothermal collapse breccia to form the economic deposit.

Key words : Fluorite Mineralization, Hydrothermal Intrusive Breccia, Hydrothermal Collapse Breccia, Jalera Khurd, Malani Igneous Suite.

INTRODCTION

Malani volcanism, chiefly acidic in nature, is a late Proterozoic (745±10 Ma) tectonomagmatic event in western and southwestern Rajasthan. Scattered outcrops of Malani cover an area of 50,000 sq. km. from south of Sirohi to north of Pokaran and east of Jodhpur to the edge of Thar desert of India where exposures are covered by sand dunes.

Three stages of igneous activity have been recognized on the basis of field evidences. Volcanics of the first stage are mostly basalt with occasional andesite or trachybasalts followed by a second stage of the voluminous outpouring of peralkaline and peraluminous rhyolite, basalt, dacite and trachyte flows covering the former. The third stage ceased with the outburst of ash flow deposits. The dominant felsic volcanics are rhyolites and rhyodacites spread over an area of about 31,000 km^2. The other rock types associated with the rhyolites are trachyte, dacite, pitchstone, welded tuff, vitric, lithic and crystal ash, ignimbrite, obsidian, pyroclastic slates, agglomerate, volcanic breccia and volcanic conglomerates. The majority of the acid volcanics are high potassic and a few are calcalkaline or lowpotassic in composition.

Though mineralization in Malani volcanics is expected and pointed out by earlier works (Kochhar, 1984, Shrivastava et al. 2000). However, the present investigation of the fluorite deposit at Jalera Khurd forms the first scientific/technical study (not discovery) of this huge deposit in the Malani volcanics, the largest anorogenic acid volcanic suite in India and the third largest acidic volcanism in the world.

The Jalera Khurd fluorite deposit, (Lat. 72°8'69"N and Long. 24°48'00"E) is situated one kilometre northwest of village Jalera Khurd, which is situated about 15 kilometres northwest from Raniwara, on Raniwara-Sanchore road and is part of a Bhinmal tehsil, Jalore district, Rajasthan in Northwest India (Fig. 1). The study area forms part of toposheet number 45 D/1 of Survey of India. The nearest railway station is Raniwara of Northern Railway of Indian Railways. The Jalera Khurd is approachable by metalled road from Jalore, Bhinmal and Raniwara cities of northwestern India.

The terrain conditions vary widely, ranging from ergs or typical dune fields with interdunal plains to rocky and gravel pediments mainly. Older dunes are generally established with alluviated calcium carbonate horizon within it. Still flanks and crests of a few are reactivated. Many active dunes of younger alluvium are also present. The rock outcrops are available as residual hill tors, inselbergs and scattered hummocks. Some thirty percent of the Jalera Khurd fluorite deposit area is blanketed by wind blown sand and sand dune.

Previous work :

La Touche (1902) named the volcanics as "Malani series", whereas Coulson (1933) called them as "Malani System", Pascoe (1960) described them as "Malani granite and volcanic suite" and grouped them under Trans-Aravalli Vindhyan. Paliwal (1992) referred them as "Malani Series", Malani Volcanic Series", Malani beds; Murthy et al. (1961) grouped the rhyolites and associated granites as Malani Igneous Suite, Chore and Mohanty (1998) called these rocks as "Malani Igneous Province", while Bhushan and Khullar (1998) named them as "Malani Igneous Complex". The authors of this paper prefer to use the term 'Malani Igneous Suite'. It is more as a conventional name, a noun, rather than in its true geological meaning.

Several other attempts have been made to study different aspects of these rocks by La Touche, 1902; Heron, 1932; Pascoe, 1959; Kochhar, 1983, 1984, 1998; Murthy, 1979; Pareek, 1984; Bhushan, 1985, 1989, 1995, 2000; Vallinayagam, 1998; Vallinayagam and Kochhar, 1998; Srivastava, 1988; Srivastava et al., 1989a and b; Rathore et al., 1996 a and b; Basu, 1982; Paliwal 1992, 1998, 1999; Pandit et al., 1999; Shrivastava et al. 2000; Paliwal and Rathore 2000 and Torsvik et al., 2001. La Touche (1902) discussed the physical, petrographic features and nature of eruption. Most of his field observations and interpretations continue to remain undisputed. Pascoe (1960) attempted a generalized classification based on previous data. His stratigraphic order however, did not include any basic volcanics, which are now firmly believed to prevail in many areas below the rhyolites. Murthy et al. (1961) studied the petrographic characters of sodic and peraluminous rhyolites of Barmer. The acid volcanics of the same area have also been investigated by Venkataraman et al. (1968), Krishnan (1968), Mukherjee (1962) and Mukherjee and Roy (1981). Pareek (1981a, b) attempted petrochemistry of Malani Igneous Suite (MIS). Kochhar (1984) and Eby and Kochhar (1990) studied the volcano-plutonic association and type of magmatism. Srivastava (op.cit.) elucidated the major and trace element geochemistry of felsic volcanics from Pali district. Maheshwari et al. (1995) described

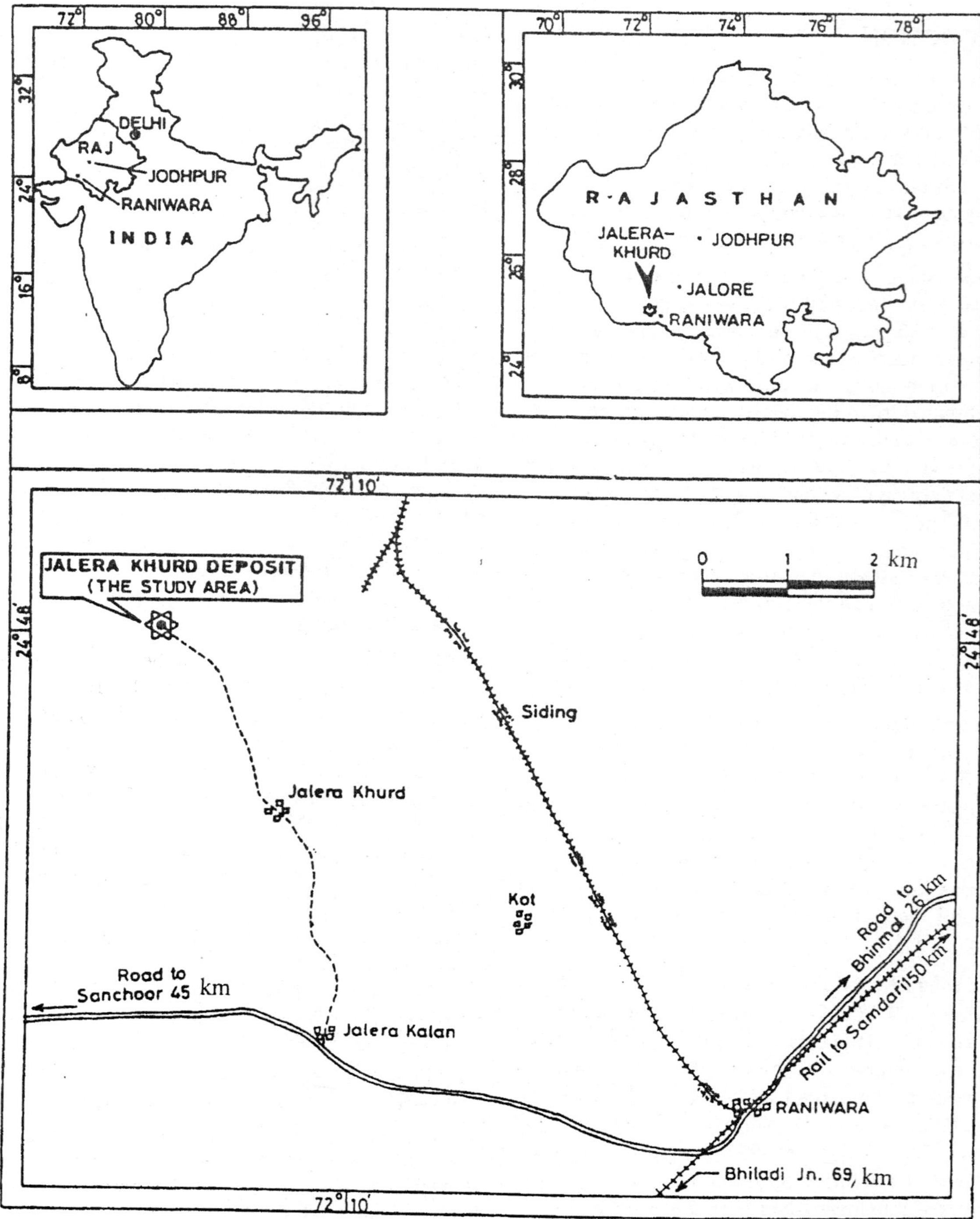

Fig. 1 Location of the Study Area : Jalera Khurd Deposit

the crustal influence in the petrogenesis of rhyolites based on oxygen isotope studies. Dhar et al. (1996), on the basis of Sr, Pb and Nd isotope studies concluded that the Siwana alkaline magma for the granites as well as the rhyolites is mantle derived. Their Sr-isotopic data (Sr^{87}/Sr^{86} 0.7062 ± 0.0020) have indicated some crustal contamination and presence of Archaean crust in the area. Bidwai and Krishnamurthy (1997) illustrated some of the volcanic landforms from parts of the Siwana Caldera sequence. Barring some marginal attempts, a comprehensive genetic model pertaining to generation of both basic and acid magma, ascent of the fluids and emplacement of the lava piles and their physico-chemical condition of crystallization are yet to be clearly made out.

In fact, intrusive breccias of hydrothermal association from various mineralized provinces in the world are known (Locke, 1926; Bryner, 1961; Kents, 1964; Johnston and Lowell, 1961; Llambias and Malvicini, 1969; Sawkins, 1969; Sillitoe and Sawkins, 1971; Fletcher, 1977; Knutson et al., 1979 and Gultekin, 2005). It has now become clear that such bodies constitute one of the characteristic features of hydrothermal systems generated around plutons of acids to intermediate composition emplaced at shallow depths (<4-5 km) in the crust. The frequent occurrence of tourmaline-rich matrices in these intrusive breccias (Kents, 1964; Sillitoe and Sawkins, 1971; Fletcher, 1977; Grant et al., 1980) suggests that boron is strongly enriched in the residual phase during the final stages of crystallization of certain granodioritic and granitic plutons. Tourmaline breccia pipes are notably well developed in the copper-bearing porphyry instrusives in the Western Cordillera of the Andes (Kents, 1964; Sillitoe and Sawkins, 1971) and similar breccias are invariably found to be an important feature of the sub volcanic tin deposits of the Eastern Cordillera of the Central Andes in Bolivia (Grant et al., 1980). The Cornubian province of the Hercynian tourmaline within hydrothermal vein structures and as a major constituent of those granites and killas country hydrothermal solutions emanating from the granites (Dines, 1956; Exley, 1959; Charoy, 1979). In recent years it has been noted that more dynamic expressions of the activity of boron-rich fluids are also present in the Cornubian province suggesting that it is a chemical and mechanical sense there are certain similarities between the behaviors of the eruptives in the Cornubian region and those of the Bolivian tin belt (Halls et al. 1977; Goode and Taylor, 1980). Sillitoe and Sawkins (1971) have given an account of certain tourmaline bearing breccia pipes in Chile and decided, on the basis of geological observation, that the process chiefly responsible for the formation of the Chilean pipes was one of chemical solution and in situ chemical brecciation (Sawkins, 1969) followed by collapse under the influence of gravitational forces and cementation by the residual boron-rich hydrothermal solution. An epithermal fluorite-barite-REE deposits of vein type in rhyolite-dacite-andesite-trachyte-basalt sequence at Kizilcaoren, Turkey have been evaluated by Gultekin (2005).

Regional Geology

The Neoproterozoic geology of Rajasthan begins with opening of shallow rift basins to deposit non volcanic sediments described as the Sirohi Group. Before that, there are diorite intrusion in South Delhi Fold Belt (DFB) (ca. 1000 Ma, Volpe and Mac daugall, 1990) and exhumation of granulites around 1128 Ma (Fareeduddin and Kroner, 1998). West of DFB, the Sirohi Group allowed deposition of fine clastics and chemogenic sediments which were deformed and metamorphosed concomitantly with the emplacement of A-type granitoids (ca. 850 Ma) termed Erinpura Granites. Culmination of orogenic activities is marked by closing of Sirohi basins and concomitant emplacement of Erinpura Granites.

Subsequently evolved a huge cratonic platform of Rajasthan Shelf, started geological activities in western Rajasthan with anorogenic, plumerelated magmatism along with the development of ephemeral basins. The change of geological scenario from orogenic to typical platformal activities is observed at Sirohi, where an undeformed and unmetamorphosed volcano-sedimentary ensemble, termed the Sindreth Group, occurs unconformably overlying the deformed and metamorphosed sediments of the Sirohi Group. Similar relationships have also been observed in the hills of Diri, Gurapratap Singh and Manihari in the Pali District. The volcanic succession in these areas is dominantly bimodal representing both mafic and felsic volcanics. Associated with these volcanics are sedimentary conglomerates, grits and arkosic sandstones with minor shale beds. These dominantly volcanic sequences, which constitute linear N-S trending outcrops are the oldest beds of lateNeoproterozoic volcanic/plutonic and (minor) sedimentary formations of the Malani Group (Roy, 1998), popularly called 'Malani Igneous Suite', though not in strict geological sense.

Neoproterozoic Malani rocks of western Rajasthan, occur as unconformity bounded sequence of dominantly bimodal volcanics with minor interbedded sediments and peralkaline meta-aluminous to per-aluminous granites. Outcrops of rocks belonging to the Malani Igneous Suite (MIS) occur over widely scattered areas of western Rajasthan and Sindh (Pakistan).

Stratigraphically, Malani overlies Sirohi Group (ca. 835 Ma) unconformably. The upper age limit of the Malani is provided by the Marwar Supergroup

(PrePalaeozoic, Pareek, 1984) which overlies unconformably the granites and lava flows of the Malani (745 Ma to 680 Ma).

Lithostratigraphy of MIS, according to Bhushan (2000) is :

Phase III	Dykes of basic, acid, trachyte porphyry and andesite. Porphyry dykes, aplite and dioritic plugs.
Phase II	Intrusion of discordant peraluminous and peralkaline granites as plutons, ring dykes and bosses.
Phase I	Eruption of basic flows followed by voluminous acid flows, and culminating with eruption of ash flow deposits.

There are few uncertainties about Phase III of the lithostratigraphy of MIS. Roy and Jakhar (2002) questioned the characteristics of dykes of Phase III as there has also been KTB (CretaceousTertiary Boundary) underplating activity in the region (Basu et al., 1993). (See also Bhushan, 1984). Also there is possibility of post-granitic, volcanic activity (equivalent to Phase III). A second phase of flow post dating granites is inferred from the Rb-Sr isochron age of ultrapotassic rhyolite by Rathore et al. (1999).

The Malani area was reactivated during Tertiary times to produce a few younger volcanics. These are considered as manifestations of the Reunion plume (Basu et al. 1993, Roy, 2003). Such as those at Sarnu-Dandali in Barmer District (Basu et. al. 1993, Bhushan and Chandrashekar, 2002), Tavidar in Jalore District (Rathore, 1995) and Mundwara in Sirohi District (Subrahmanyan and Rao, 1972; Rathore et al. 1996). Some of the good reviews on Malani and associated rocks, that appeared in the literature are Pareek, 1984; Bhushan, 2000; Roy and Jakhar, 2002; Bhushan and Chandrashekhar, 2002; and Sharma, 2004.

The Geology of the area

The geological map of the area is presented in Fig. 2. Jalera Khurd deposit, the local geology (Fig. 2) shows that here, the host rocks consist of two flows a lower rhyolite and an upper trachyte. There are No pyroclastic materials separates the two flows but the presence of xenoliths and bombs in the lower part of the trachyte flow are evidences for the initiation of the trachytic explosion above rhyolite. The rhyolites the trachytes are porphyritic in nature. Rhyolites are distinctly leucocratic, varying from cream colour in the pink and purple tints, while the trachytes are meso to melanocratic exhibiting shades of green which in same exposures are darker.

The geological map and cross sections (Fig. 2) of the area show that the breccia body is distinctly elongated in the N43°E-S43°W direction. Only the southwest limit of the body can be defined with certainty; the northeastern extensions of the breccia are still unexposed even on the lower topographic reaches. The known extent of the outcrop measures over 500 m in the main strike direction.

The fluorite mineralization is later to the porphyritic trachyte flow, as it shows both intrusive and collapse breccia features of the stockwork in which the broadly defined veins are available.

Four fluorite veins constitute the main fluorite ore body, all of them collectively 20 m thick each vein averaging 5 m and slightly separated. Two veins, among them, present at both the lateral sides carry low grade fluorite while the other present centrally are of medium grade on the west and high grade towards the east. Apart from this main body, one prominent and continuous vein of low-grade fluorite are also present in the eastern side of the main fluorite body almost at 20 m apart, with a thickness of 5 m. Also one more low grade fluorite vein is also present at further east of this single vein. This later vein (5 m thickness) and is from 20 m distance from the single vein. This later single vein has only been traced in the northern part of the deposit. The western side of the main fluorite body, traces of three more low-grade fluorite veins, each about 2-3 m thick and 100-50 m length are present. These veins are better studied with the help of trenching and pitting.

As a rule, all the fluorite veins follow the trend of prominent Joint I showing N43°E-S43°W strike and dip 85°W. Though mineralization in the area is joint controlled, it has occurred mostly along the first set and not the later, Joint II and Joint III sets. However, all of them have fragmented the main mineralized body. Apparently, no post mineralization tectonic as fragmented mass. But along joints activity has occurred as neither shifting of the blocks nor crushing of the available minerals are seen along the joints.

As the second and third sets of joints are unaffected by the mineralization, there is no bifurcation of veins (or the ore body, in general). It is clear from the plan of the breccia veins constituting stockwork that the shape of the ore body has been closely controlled only by the first set of joints, the formation of which presumably immediately pre-dates the emplacement of the breccia body itself.

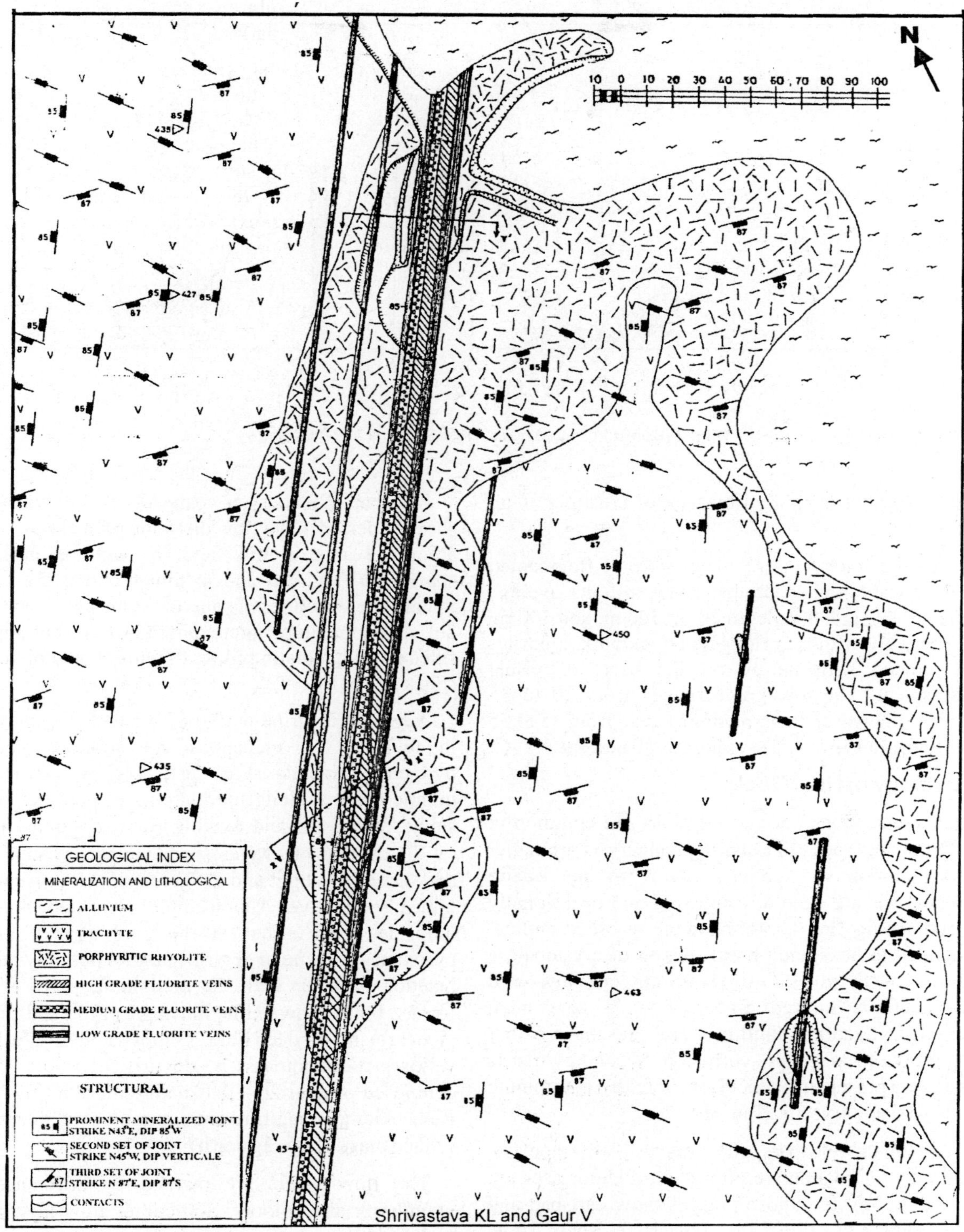

Fig. 2 Geological Map of Jalera Khurd Fluorite Deposit, District : Jalore, Rajasthan, India

Trends of mineralized veins within the stockwork are almost sub-parallel to parallel, suggesting joint formation by tensional forces. Qualling apart of rock perpendicular to the trend of the joint (mineralized

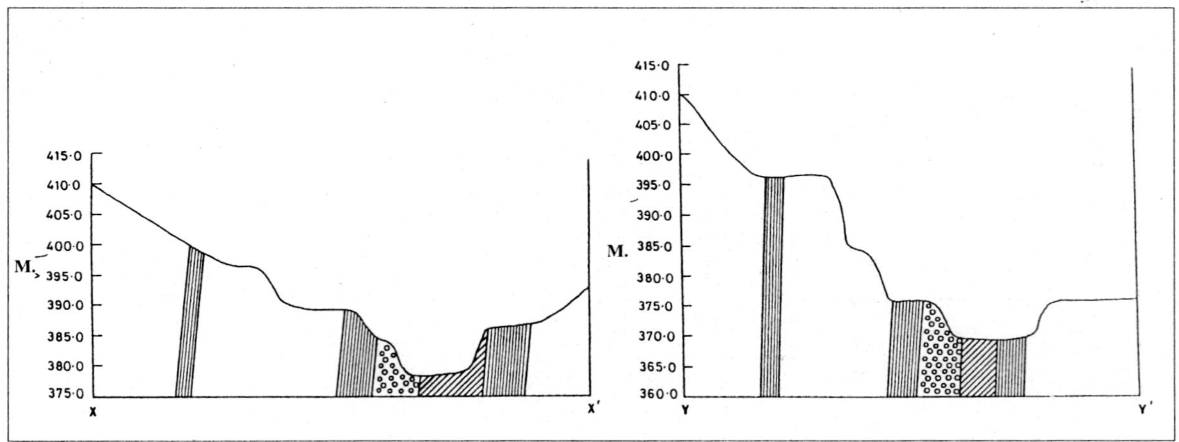

Fig. 2 (Contd.) : Geological cross sections along xx' and yy'

one) is further supported by lack of crushing along veins.

All the four veins of main body of fragmented mass constituting a tabular body with its average thickness, length and depth 20 m, 500 m, and 100 m, respectively, within an elongated stockwork. Fluorites are graded here purely on the basis of visual assessment. The low grade ranges from 20 to 35 percent fluorite content, medium grade from 35 to 55 and high grade from 55 to as high as 80 percent.

Petrology of Host Rocks :

Phenocrysts of quartz and feldspars are common in both the host rocks but quartz phenocrysts are rather less common in trachytes. Most of the quartz phenocrysts are generally anhedral and occasionally euhedral (Fig. 3a), is also showing resorbed outline. Rounded to hexagonal basal section of crystoballite grains with dark interference colours and characteristic curved fractures have also been recorded. Resorbed outlines are present possibly suggestive of magmatic corrosion. Secondary quartz grains are present in phenocrysts of feldspar as minute and eye shape crystals. (Fig. 3b)

Among the feldspars, phenocrysts of anorthoclase, sanidine, microcline are prominent. Plagioclases are rare (Fig. 3e), few those present show carlsbad and albite twinning (Fig. 3e). Potash feldspars show simple twinning and microcline grains show cross twinning (Fig. 3e). Feldspar minerals are string with aegirine inclusion, indicating earlier crystallization of the latter. (Fig. 3d)

Feldspar phenocrysts particularly in trachytes show preferred in the flow direction of the lava. These are observable in the field, as well as in hand specimens (Fig. 3a) and in thin sections (Fig. 3h). Spherulitic structures or spherulites are also seen (Fig. 3e), and sometimes microspherulites present appear filled up with radiating fibres of feldspar and/or quartz (Figs. 3e and h).

Aegirine, aegirine-augite of braided mesoperthite in feldspars as inclusions are common among pyroxene minerals (Fig. 3g). Aegirine occurs as minute needles along the margins of perthites or as subhedral prisms and exhibit low extinction angles (5^0-7^0) on 010 cleavages, strong pleochroism (X = dark green, Y= light and Z= greenish yellow) and are optically negative. Alkali amphiboles, riebeckite or diopside occur either at the periphery of alkali pyroxene (aegirine or aegirine augite) possibly due to reaction of latter with the melt in the core of the former. It has a low extinction angle (1°-5°) and strong pleochroism (X=deep blue, Y=green and, Z=greenish yellow). Groundmass is devitrified acidic glass composed of quartzo - feldspathic mosaic (Fig. 3h). Riebeckite and aegirine are common constituents of ground mass. (Figs. 3g and h)

The flow layers are marked by difference in granularity and colours. Sometimes flow layers are replaced by iron minerals, calcite or quartz (Figs. 4b, e, f and g). Glass shards are seen. Broken grains of feldspars and quartz are embedded in a dense vitric mass. (Fig. 4j) Boundaries of these angular fragments are commonly fractured and displaced (Figs. 4c, h and

Table-1 Geochemistry of host rocks of fluorite deposit at Jalera Khurd

Major oxides (in Percent)	Rhyolite samples					Trachyte samples										Xenolith sample	Average	
	Jkh/v$_2$	Jkh/v$_4$	Jkh/v$_7$	Jkh/v$_8$	Jkh/v$_{15}$	Jkh/v$_1$	Jkh/v$_3$	Jkh/v$_5$	Jkh/v$_6$	Jkh/v$_9$	Jkh/v$_{10}$	Jkh/v$_{11}$	Jkh/v$_{12}$	Jkh/v$_{14}$	Jkh/v$_{16}$	Jkh/v$_{13}$	Rhyolite	Trachyte
SiO_2	69.87	72.06	69.38	74.84	73.83	64.03	64.77	62.30	61.48	67.91	67.66	67.34	68.07	67.01	62.97	45.82	71.99	65.35
Al_2O_3	12.33	12.63	13.64	12.63	12.68	16.00	15.82	15.28	15.60	14.63	12.88	13.65	13.60	13.15	14.59	18.82	12.78	14.52
Fe_2O_3	5.64	5.34	6.15	4.61	5.46	5.21	4.46	5.21	5.00	5.20	5.72	5.53	5.67	5.92	7.20	8.99	5.44	5.51
TiO_2	0.46	0.47	0.80	0.49	0.45	1.06	0.83	0.84	0.80	0.67	0.67	0.76	0.75	0.82	0.97	2.83	0.53	0.81
MnO	0.11	0.12	0.21	0.21	0.09	0.03	0.03	0.20	0.18	0.12	0.25	0.14	0.25	0.16	0.17	0.11	0.14	0.15
P_2O_5	0.10	0.08	0.14	0.21	0.17	0.27	0.32	0.27	0.29	0.19	0.26	0.46	0.20	0.19	0.26	0.76	0.14	0.27
CaO	3.37	1.47	1.28	0.36	0.35	2.33	2.74	1.56	1.84	0.42	2.88	1.43	1.15	1.76	1.60	3.95	1.36	1.77
MgO	0.67	0.58	0.81	0.52	0.46	0.59	0.51	0.69	0.74	0.62	0.76	0.72	0.81	0.96	0.83	3.78	0.60	0.72
Na_2O	3.76	4.26	3.77	2.31	4.41	4.36	5.29	6.86	5.80	5.54	5.48	4.87	4.90	4.91	5.75	2.74	3.70	5.37
K_2O	5.19	5.00	6.58	5.75	5.04	5.83	5.36	4.89	5.80	4.52	4.42	5.22	4.78	4.75	5.33	3.08	5.51	5.09
Trace Element (in ppm)																		
Ba	248	256	209	326	268	298	541	911	1401	254	279	447	455	408	593	572	261.4	558.7
Cr	280	427	263	20	321	117	137	39	12	20	394	26	19	41	496	489	262.2	130.1
Ni	10.2	7.9	9.7	5.0	5.2	2.1	5.0	9.9	4.6	4.2	25.8	5.3	5.4	12.9	40.5	129.5	7.6	11.57
CU	24.2	15.5	25.8	36.3	8.8	12.3	28.8	20.2	20.7	12.0	25.1	13.2	19.7	14.5	25.1	261.5	22.12	19.16
Zn	135.9	141.5	211.0	267.6	162.9	77.6	124.1	111.7	102.0	115.1	165.4	120.6	141.1	169.2	248.8	366.7	183.66	137.56
Th	15.5	21.2	12.1	14.7	25.2	11.6	1.7	6.9	2.4	14.9	11.4	12.0	15.0	17.4	13.1	9.9	17.74	10.64
Rb	186	173	206	192	171	212	162	149	126	161	135	151	143	200	174	418	185.6	161.3
U	5.7	5.7	5.8	6.5	6.2	6.4	4.6	3.4	4.2	7.6	5.1	5.7	4.4	5.9	5.3	8.5	5.98	5.26
Y	61	73	50	69	82	55	32	36	34	59	51	57	56	68	48	18	67	49.6
Zr	699	796	537	640	861	606	273	337	271	658	495	570	564	688	560	282	706.6	502.2
Nb	110	128	115	102	137	112	68	76	69	120	91	101	99	123	103	71	118.4	96.2

Table-1 (Contd.) :

CIPW Norm																		
Q	25.27	28.11	21.93	37.04	30.36	12.96	11.49	6.14	5.91	19.18	18.40	19.50	21.13	19.67	9.90	6.24	28.54	14.42
Or	30.69	29.52	38.98	33.97	29.28	34.47	31.69	28.91	34.30	26.68	26.13	30.85	28.24	28.07	31.52	18.18	32.48	30.08
Ab	24.89	26.14	22.84	19.49	27.24	36.83	39.24	39.45	35.78	38.72	31.59	30.49	32.75	30.91	33.95	23.10	24.12	34.97
An	–	–	–	1.77	–	0.278	–	–	–	–	–	–	–	–	–	15.17	1.02	0.278
Apa	0.23	0.168	0.30	0.47	0.36	0.638	0.73	0.63	0.67	0.43	0.50	1.07	0.47	0.43	0.60	1.78	0.30	0.62
Hem	3.53	2.32	3.39	4.60	2.4	5.20	2.78	–	0.96	0.27	0.91	2.24	2.96	2.70	2.73	8.97	3.24	2.30
Dio	8.39	4.31	4.35	1.3	1.48	6.17	6.17	4.72	4.95	1.96	7.23	3.64	3.91	5.59	4.74	9.45	3.96	4.90
Acm	6.05	8.68	7.94	–	8.82	–	4.80	15.01	11.64	7.11	12.93	9.37	7.62	9.28	12.88	–	7.87	10.07
Namt Sil	–	–	–	–	–	–	–	0.34	–	–	–	–	–	–	–	–	–	0.34
Ratios																		
Salic / Femic	4.44	5.41	5.24	18.19	6.69	7.03	1.16	3.67	4.17	8.65	3.51	4.93	5.48	4.36	3.59	3.10	7.99	4.65
Q or L / F	0.45	0.50	0.35	6.67	0.53	0.18	0.16	0.08	0.08	0.29	0.31	0.31	0.34	0.33	0.15	0.11	.5	0.22
$(Na_2O+K_2O)1.92$ / CaO	4.64	5.73	15.35	0.20	3.18	2.91	5.84	4.73	18.30	2.63	5.25	6.33	4.12	5.24	1.08	5.56	5.85	
K_2O / Na_2O	0.91	0.77	1.15	1.64	0.75	0.88	0.66	0.47	0.65	0.53	.53	0.70	0.64	0.19	0.61	0.74	1.04	1.01

i). Calcite, quartz and fluorite veins have penetrated groundmass and the fractured grains. (Figs. 4d and e)

Vugs are present and internally lined mainly by cryptocrystalline quartz. (Fig. 4g) Internal lines are rarely of clay minerals, sericite or calcite specify.

Geochemistry of Host Rocks:

Detailed geochemical studies of the host rocks were undertaken to understand petrogenesis and its role in the genesis of the fluorite mineralization. Chiefly, there are two flows in the area; lower rhyolite and upper trachyte. Representative samples covering lateral and vertical dimensions of five porphyritic rhyolites and ten trachytes were collected for the geochemical analysis by X-ray fluorescence technique. Major oxides such as SiO_2, Al_2O_3, Fe_2O_3, TiO_2, MnO, P_2O_5 CaO, MgO, Na_2O and K_2O and minor and trace elements Ba, Cr, Ni, Cu, Zn, Th, Rb, U, Y, Zr, Nb have been determined. Based on the data available CIPW norm have been calculated and presented in Table 1. This Table also shows ratios of salic/femic $(Na_2O+K_2O)/CaO$ and K_2O/Na_2O and the average of all the data are also given. The normative value shows that the content of quartz in rhyolite (28.54) is much higher then in trachyte (14.42). Feldspar in both the rocks are essentially orthoclase and albite, respectively 32.48 and 24.12 percent in rhyolite and 30.08 and 34.97 percent in trachyte. Among other normative minerals, acmite, diopside and hematite are of interest.

Using appropriate plots of major element oxides (mostly alkalies and silica), the volcanic rocks of the study area could be classified whether (a) belonging to the rhyolite or trachyte families (b) they are high-potassic rhyolites, potassic-trachytes or potassic dacites (c) they could be classified as meta-aluminous or per-aluminous or per-alkaline (d) belonging to which of the four major alkaline groups like nephelinites, agpaites, miaskites, alkali basalts or sub-alkalines (e) nature of source magma and (f) influence of hydrothermal activity.

Results of above plottings indicate that

(1) when plotted as per Total Alkali Silica diagram of Le Bas et. al. (1986) rock samples numbered 2,4,7,8 and 15 collected from lower flow fall in the field of rhyolite (Fig. 5). Rest of the rock samples from upper flow are trachytes and not andesite, shown in the published Geological map of Abu quadrangle.

(2) Fig 6, showing plot of K_2O versus SiO_2 (wt%) by Peccerillo and Taylor (1978) (Fig. 6) show that among the volcanic rocks of Jalera Khurd fluorite deposit rhyolite samples 4, 8 and 15 from lower flow fall in high K-rhyolite area though two samples (number 2 and 7) slightly exceed. Remaining samples of upper flow fall in the K-trachyte of high K-dacite fields. However, when plotted by this method of Le Maitre (1989), all the host samples from this deposit fall in the field of high potash volcanic rocks. (Fig. 7)

(3) The relationship of molar $Al_2O_3/(CaO+Na_2O+K_2O)$ proposed by Maniar and Piccoli (1989) shows that all the host rock samples clearly fall in the field of peraluminous rocks, and not in metaaluminous or in peralkaline fields (Fig. 8).

(4) All the host rock samples belong either to alkali basalt family or to subalkaline rocks based on Alkali-Silica diagram prepared for alkaline rocks by Hyndman (1985, p. 377) and shown in Fig. 9.

(5) Fig. 10 showing Na_2O versus Al_2O_3 diagram of comendites and pantellerites of Shrivastava (1990). Position of rhyolites and trachytes from Jalera Khurd is normal in the diagram Fig. 11 of SiO_2 versus Al$_2$O3 of peralkaline silicic volcanic rocks showing that Jalera Khurd rhyolite could have been derived from less silicic soda trachyte magma.

(6) SiO_2 values of all the rhyolite samples, (2,4,7,8 and 15) plotted against P_2O_5 (Fig. 12) show more silica percentage in comparison with trachytes. Also, all the trachyte samples are slightly higher in P_2O_5 than rhyolites. These are largely attributed to hydrothermal redistribution or introduction of SiO_2 or both. The deposition of large amount of quartz in the microfractures and groundmass might be expected under condition of decreasing temperature from limited volume of SiO_2 saturated fluids.

(7) Plot of P_2O_5 versus TiO_2 (Fig. 13) of the host rocks reveal that percentage. All the rhyolites from Jalera Khurd is low in TiO_2 content than the trachytes. The origin of rhyolite and trachyte by crustal fractionation of a basalt (average less than 5 Wt%) is too small to produce K_2O enriched liquid, like rhyolites (Edward, 1985).

(8) Jalera Khurd rhyolites show (a) less TiO_2 and high SiO_2 (Fig. 14). (b) higher Zr (550900 ppm) than in trachytes (250-700 ppm) (Fig. 15). (c) and cluster toward higher Zr and low P_2O_5 (Fig. 16). These trends are attributed to hydro thermal redistribution or introduction of SiO_2 or both, and the occurrence of quartz in microcracks and groundmass suggest

conditions of decreasing temperature and limited volume of silica saturated fluids (Edwards and Hodder, 1991).

(9) The relationship Zr versus TiO_2 and Zr versus Cr are shown in Figs. 17 and 18. In the former plot trachytes exhibit slightly low Zr and higher TiO_2, while no pattern is seen in the latter plot.

(10) Variations among the common trace elements present in the host rocks has been determined. The variations for Zr, Th, Zn, Cr, U, Ni, Rb, Nb, Ba, Cu and Y the trace element composition, are considered to be the consequence of variable degrees of diffusive exchange between felsic and more mafic magmas during slow cooling (Waight et al, 2001).

THE FLUORITE MINERALIZATION

(i) Significance of regional tectonic setting:

The Jalera Khurd fluorite deposit is located on the outer circular ring fractures developed during collapse of caldera, in this region of Malani Igneous Suite. These fractures show only linear nature, locally. It is also expected that Because of collapse of caldera deeper towards north, gases (like fluorine), alongwith magma (basaltic) carrying mantle xenoliths have surfaced during petrogenesis and mineralization the latter more towards south.

In relation with the global tectonic the nature and origin of magmas of different compositions and their role in the genesis of mineral deposits is a field of general interest currently, (Abdel Rahman and Elkibbi, 2001; Drury et. al. 2001; Kabeto et. al. 2001; Mocek, 2001; Okay et al. 2001; Phillips and Woods, 2001; Scambelluri and Philippot, 2001; Seghedi et al. 2001; Sheppard et al. 2001; Spath et. al. 2001; Waight et al. 2001; Wang et al. 2001 and Zhu et. al. 2001).

Role of magma chambers in the generation of felsic and basaltic magmas, crustal assimilation, contamination and their fractional crystallization, remain subject of studies by many workers (Aguirre Diaz, 2001; Blake and Bruno, 2001; Borisova et al. 2001; Kabeto et al. 2001; Kuritani 2001, Lindsay et al. 2001; Rabinowicz et al. 2001; Riley et al. 2001; Stevenson et al. 2001; Tatsumi 2001 and Verma 2001). The Jalera Khurd parent magmas for the rhyolites must have depths and fractionated during the ascent to surface and assimilated crustal fragments. The emplacement of the magma must have been forceful and led to the development of regional Caldera having tens of kilometer diameter. Occurrence of similar Caldera structures elsewhere have been reported by many (Aquirre Diaz, 2001; Chabiron, et al. 2001; Fiske, et al. 2001; Lindsay, et al. 2001; Riller, et al. 2001 and Verma, 2001). Occurrences of xenoliths from deeper parts of crust (or from mantle?) have also been discussed by a few (Andersen and Neumann, 2001; Chabiron et al. 2001; Vandenkerkhof and Thiery, 2001). Presence of CO_2 and H_2O in the magma and their possible role in the mineralization, have been studied by Diamond (2001) Edmonds et al. (2001) Evans et al. (2001) and Tamic et al. (2001).

(ii) Significance of Shape of Fluorite body:

The contacts of the mineralized breccia are generally very sharp and subvertical, and the overall geometry of the body is therefore tabular and subvertical. The contacts tend NESW and the southeastern footwall contact and northwestern hanging wall contact are controlled only by the same joint set and not controlled or influenced by other sets of joints. Breccia apophyses and veinlets are totally absent in the host rocks neither in the trachytes at higher reaches nor in the rhyolites at lower reaches. The apophyses measure 0.1 to 0.5 mm and are identifiable only under microscope (Fig. 4f). These tiny apophyses, to begin with, are narrow along strike and terminate in a system of an astonishing fluorite filled joints. This aspect of the breccia contact is indicative of the hydraulic rupture of the hard trachyte or the host rhyolite wall rock by fluids under pressure. In the outer limits of this zone of hydraulic disruption, negligible disorientation of the intervening host has occurred but, as the zone is followed inwards, rocks are progressively displaced and tumbled into the disorder (Fig. 4a and b), in the stockwork mineralization of fluorite. These form

Fig. 3 (a) porphyritic rhyolite, a typical host rock, showing phenocrysts of feldspar laths. (b) Photomicrograph of rhyolite showing filling of secondary quartz in a phenocrysts of feldspar. (c) Photomicrograph of rhyolite showing alteration patterns of phenocrysts. (d) Photomicrograph of a volcanic bomb showing bleb-shaped core of dark glass, surrounded by pyroxene-feldspar mass with spinifex texture. (e) Photomicrograph showing many phenocrysts of plagioclase (PL) which have got accumulated in the fine-grained groundmass of quartz and feldspar crystals. (f) Field photograph covering 2m exposed surface of trachyte rock showing flow oriented feldspar (whitish pink) phenocrysts. Volcanic bombs or their pseudomorphs are visible. (g) Photomicrograph of trachyte rock showing albitization of phenocrysts of feldspar. (h) Photomicrograph of groundmass of trachyte rock showing quartz-feldspar mosaic with few isotropic minerals.

Fig. 3

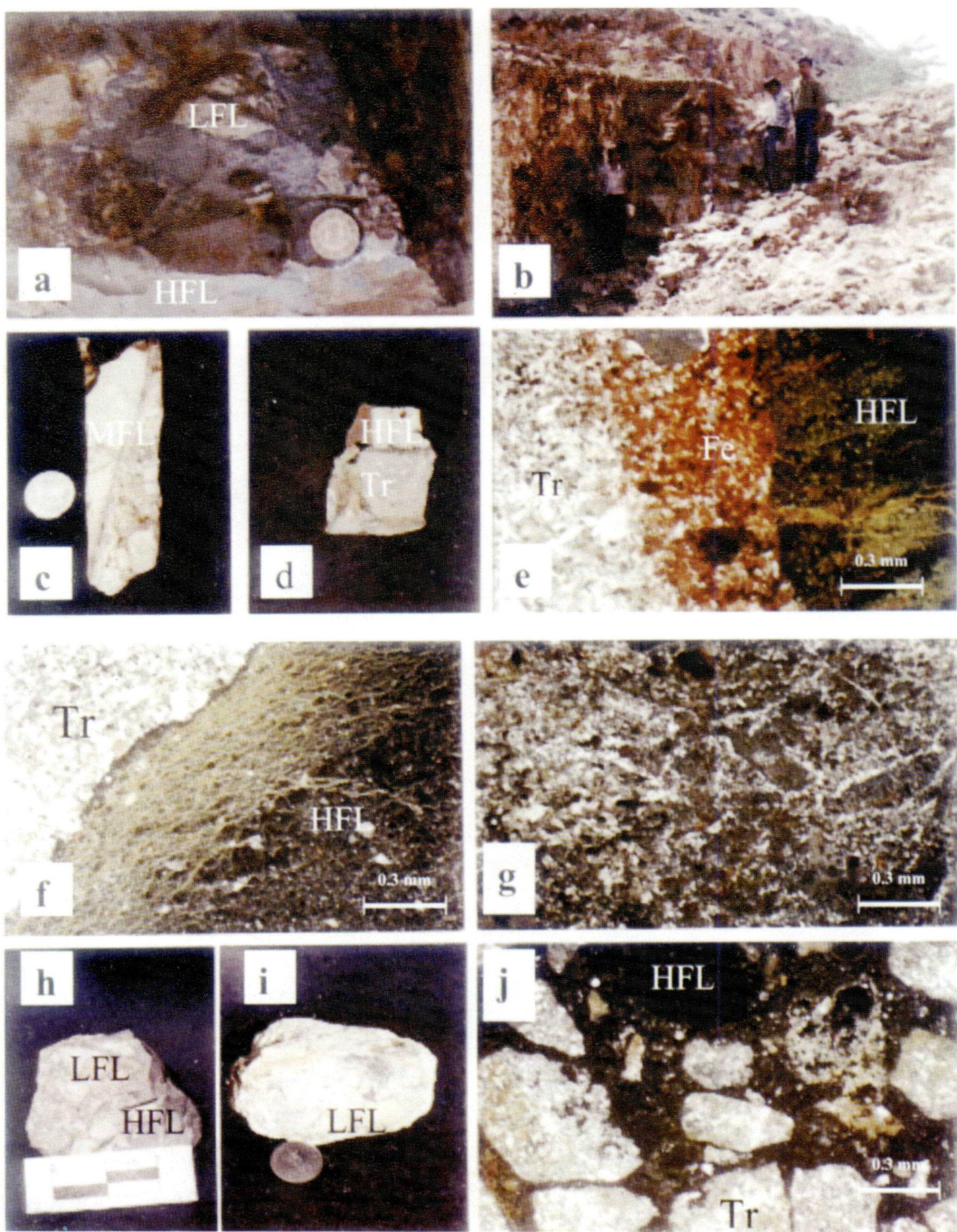
a
LFL
HFL
b
c
MFL
d
HFL
Tr
e
Tr
Fe
HFL
0.3 mm
f
Tr
HFL
0.3 mm
g
0.3 mm
h
LFL
HFL
i
LFL
j
HFL
0.3 mm
Tr

Fig. 4

good criteria to recognize local zones of brecciation *in situ*, which are transitional to the intrusive body in which the included rock fragments are completely disrupted and undergone varying degrees of lateral and vertical transport within the breccia system.

(iii) Composition and texture of Fluorite Breccia

Mineralization mostly occurs as filling of open spaces with permissive to forceful penetration. Texture suggesting filling of cavities, and to a lesser extent, replacement by digestion of fragments of host breccia certainly formed earlier. Open spaces along the joint planes have provided channel ways for fluorite breccia and mineralizing solutions and sites for the deposition of the fluorite.

The central high grade and lateral low and medium grade fluorite veins are collectively termed the main stage breccia. This stage is evidently the first, and volumetrically the main expression, of the violently disruptive release of fluorine-rich fluid from the host magma.

Morphology, composition and texture of the fragments formed in the main stage breccia range from angular to sub-angular shapes and also high proportion of comminuted host rock fragments in the matrix of fluorite itself is evident. The matrix of breccia stage has formed chiefly by a process of precipitation into the spaces dynamically opened during explosive vaporization of the fluorite and fluorine-rich fluid during decompression. The central fluorite fill has also formed partly by replacement in a second episode of dilation fluid intrusion. The volume of metasomatically replaced material contributes much more to the substance of the structure than does the central fill. Such fractures are typical fractures, which were formed in a stage prior to the main episode of brecciation.

The characteristics of fluorite breccia are the following: -

(i) Closed framework of angular clasts.

(ii) Compositionally homogeneous to heterogeneous (basaltic to rhyolitic and fluorite fragments cemented into fluorite itself).

(iii) Disorganized fluorite breccia.

(iv) Massive to diffusely layered fluorite mass and its fragments.

(v) The breccia is typical of large scale sector (up to hundreds of meters) generally known to form during collapse.

(vi) Preservation potential is moderate.

(vii) Recognition potential is moderate to good.

(viii) Clasts are angular to sub-angular fragments.

(ix) 'Jigsaw Puzzle' fit of clasts is preserved, where there has been no redistribution from site of fragmentation by turbulent mixing.

A general observation regarding the volume ratio of fragments to matrix and the ratio of killas fragments to host rock fragments in various parts of the mineralized stockwork breccia have been noted. The general conclusion which can be drawn from these observations is that the lateral veins of breccia consisting predominantly of host fragments and that for the such parts the volume proportion of fragments lies between 75 to 35 percent. In fact, it finally defines the various grades of fluorite.

By projecting the contacts along reasonable lines, based on measured dips at the present level of exposure, it is inferred that the eroded roof of the intrusion should possibly lie within 200 m vertically above the highest point now observed. The abundance of killas fragments in the deposit even indicate that the projected contact must lies even closure.

The breccia consists essentially host rock fragments and fluorite and lesser amounts of quartz. The fragments are cemented by fine grained fluorite

Fig. 4 (a) Field photograph of exposed high grade fluorite showing large fragments of high grade fluorite (HFL) and of low grade fluorite (LFL) breccia. (b) Multiple fractured rhyolite in the old working area hosting fluorite veins and irregularly constructed benches. At lower reaches, like this one, more metasomatic alteration by coatings of iron oxides are present. (c) Hand specimen of medium grade (MFL) fluorite breccia showing great variations in the size of the fragments. (d) High grade fluorite (HFL) vein (white) is attached with host trachyte rock (Tr) to form a sharp contact with iron colourations. (e) Photomicrograph of the same specimen (4d), showing sharp contact of high grade fluorite with host trachyte rock. The contact showing iron oxides (Fe). (f) Photomicrograph showing sharp contact of fluorite (HFL) (dark green) with the host (Trachyte) rock (Tr). (g) Photomicrograph showing deposition of silica in the form of quartz grains over low grade fluorite fragments by filling the available space in host trachyte. (h) Hand specimen of high grade fluorite (HFL) enclosing angular fragments of low grade fluorite (LFL). (i) Typical fluorite breccia sample, outer surface altered to kaoline. Low-grade fluorite fragments (LFL) are enclosed in high-grade fluorite matrix to form breccia. (j) Photomicrograph showing angular fragments of host trachyte rock (Tr) embedded in high grade fluorite (HFL) (dark green).

and quartz (Fig. 3a). With in the population of host rock fragments which all carry either lesser or greater amount of fluorite, there is a large range of textures. For example, the degree of rounding of fragments is quite variable and a good percentage of the constituting fragments of the variable sizes ranging from very large to moderate, retains an angular form convincingly. A typical exposure of the breccia is shown in Fig. 4b, c, h, i and j.

The main breccia shows that within the population of the host rock fragments, there are many textural types of fluorite available. Fluorites themselves often carry still smaller bites of fluorite, mostly very fine grained and sometimes with lesser amounts of banded type and small segments of host within an aphanitic matrix.

Alteration of boundaries, moderate to intense are possibly the characteristic features for the identification of the killas fragments. This alteration has become intense in some cases. The rather uniform distribution of killas fragments within the breccia shows that considerable differential movement of fragment took place during the formation of main mineralized body.

The hydraulic fracturing and disruption evident on the contact of the mineralized body indicates this body must have formed from a fluid under high pressure. As a consequence of this at lower levels of exposure, these fragments appear to have moved laterally or downward over almost two hundred meters.

(iv) Pneumato-Hydraulic action in mineralization

Field and microscopic evidences show that a fluid system in subsurface must have developed a pressure possibly exceeding the tensile strength of the enclosing (host) rock and the minimum principal stress component, then that fluid must have extended its capacity resulting to propagate the fracture and fracture system, to open it in a tensile manner. Such fracture propagation and openings are always expected to lead a kind of the weaker forces produced narrow openings in the brittle rocks. It is, therefore, for this reason that above-mentioned host rocks carry numerous stringers and veinlets of fluorite in them. Such fracture-filled veinlets are better studied in rhyolite and under microscope (Plate 4 Fig. d and e). They are usually parallel to one another and are seen traversing the host rocks or the earlier formed minerals. These veins are characterized by nearly matching walls, though complete matching is not found. Bastin has observed that "Perfect matching is not always found, even in undoubted fracture fillings, since one wall may have moved in the direction inclined to the plane of the section available for study, and fragments torn from the wall in the fracturing may also have moved out of that plane".

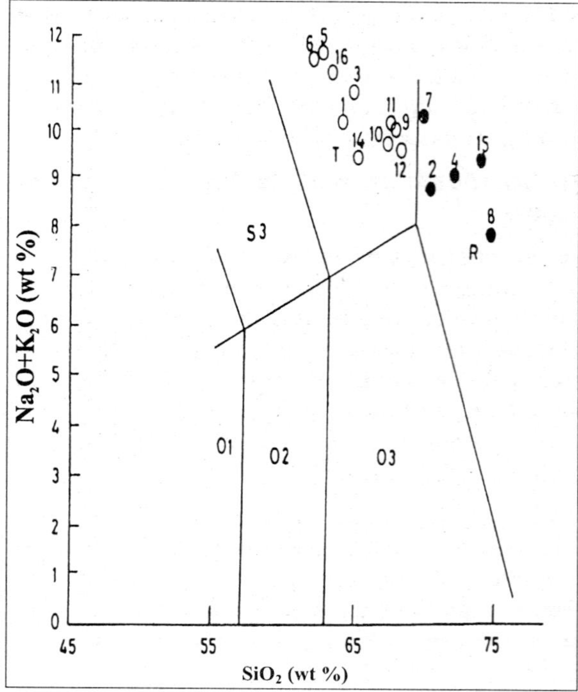

Fig. 5 Total Alkali Silica diagram of the host rocks from Jalera Khurd fluorite deposit.

The rock is essentially shattered in tensile fashion by the overpressed fluids working through the host

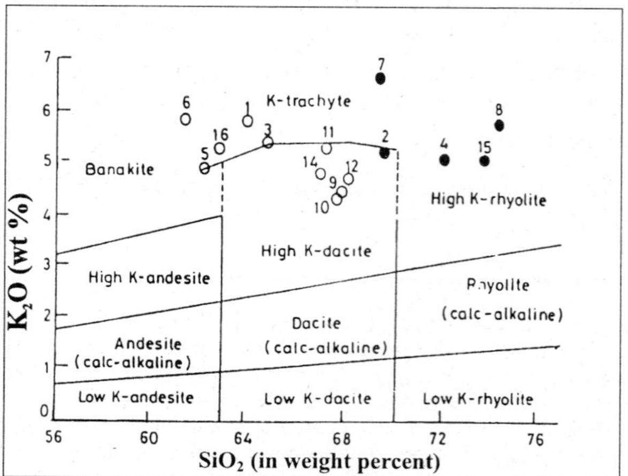

Fig. 6 Peccerillo and Taylor (1978) (K_2O vs SiO_2) diagram of the host rocks from Jalera Khurd fluorite deposit.

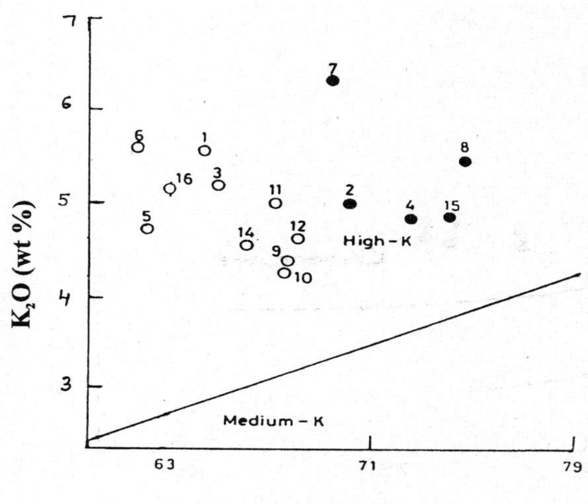

Fig. 7 Le Maitre (1989) diagram of plot of SiO_2 vs K_2O of the host rocks from Jalera Khurd fluorite deposit.

rock. Fracturing caused by this mechanism favour that the resultant fracture need not be involved in shear motion and have the breccia fragments and the associated wall rocks need not be pervasively foliated or sheared (Phillips, 1972). This fluid-induced fracturing and brecciation, called Pneumato-hydraulic fracturing, have effectively operated at Jalera Khurd

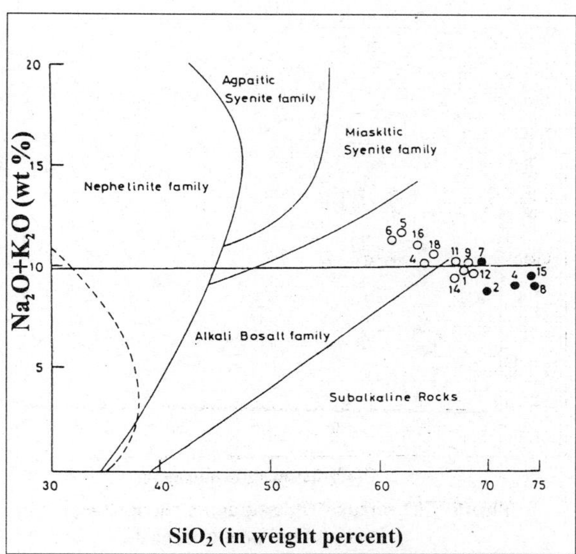

Fig. 9 Alkali-Silica diagram (Hyndman, 1985) of the host rocks from Jalera Khurd fluorite deposit.

deposit.

The manner of successive depositions of two or

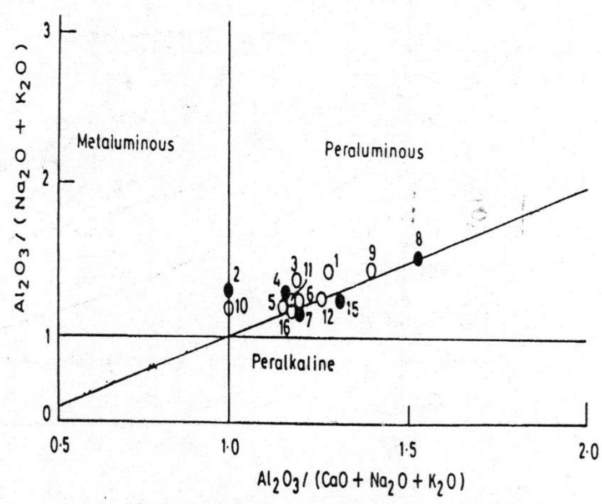

Fig.8 Maniar and Piccoli (1989), $Al_2O_3/(CaO + Na_2O + K_2O)$ versus $Al_2O_3/(Na_2O+K_2O)$ diagram of the host rocks from Jalera Khurd fluorite deposit.

more mineral starting from the walls of the openings giv

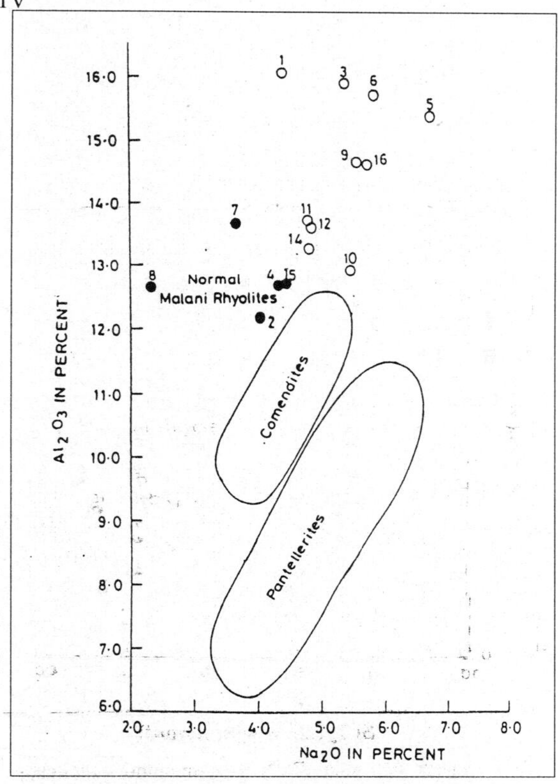

Fig.10 Na_2O versus Al_2O_3 diagram of the host rocks from Jalera Khurd fluorite deposit.

e crustified veins. The veins of this type provide

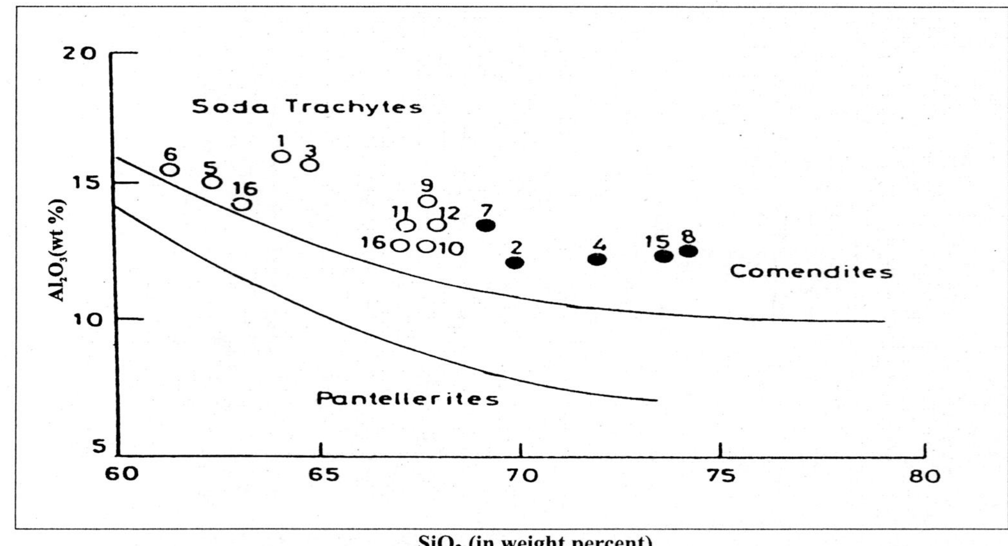

Fig.11 SiO_2 versus Al_2O_3 diagram of the host rocks from Jalera Khurd fluorite deposit.

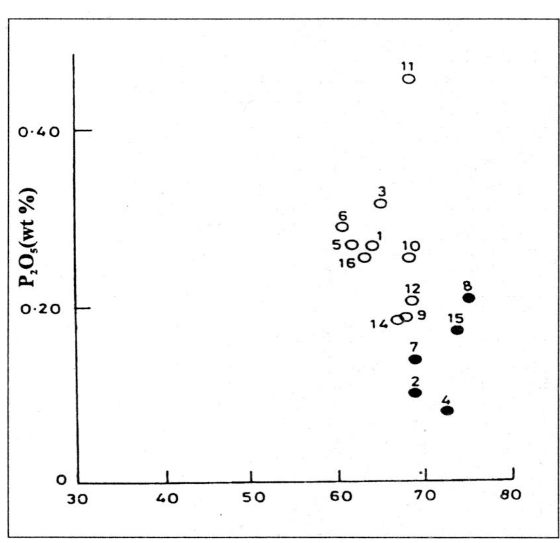

Fig.12 SiO_2 versus P_2O_5 diagram of the host rocks from Jalera Khurd fluorite deposit.

information on paragenesis and nature of mineralizing solutions. The minerals is present next to the vein wall are the earliest formed followed by the minerals towards the center of the cavity. The crustification in all the cases was found to be of symmetrical type i.e., nature and sequence of minerals on both the walls is similar. Calcite grew along the length of the vein while crystallization of quartz commenced from the walls only, thus producing comb quartz. Not in the main fluorite veins but in other small veins, fluorite is followed by quartz crystals with their

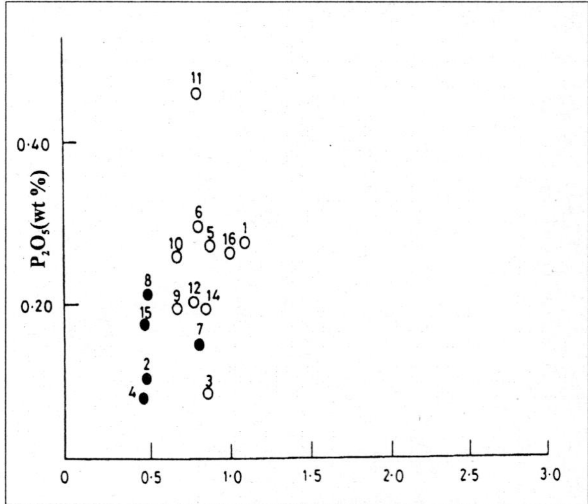

Fig.13 TiO_2 versus P_2O_5 diagram of the host rocks from Jalera Khurd fluorite deposit.

long axis perpendicular to the vein wall thus producing comb structures. In these cases, however, it is rare that successive mineral crusts have deposited in perfect parallelism for any appreciable distance. It is possibly because of the initial irregularity of the surface of

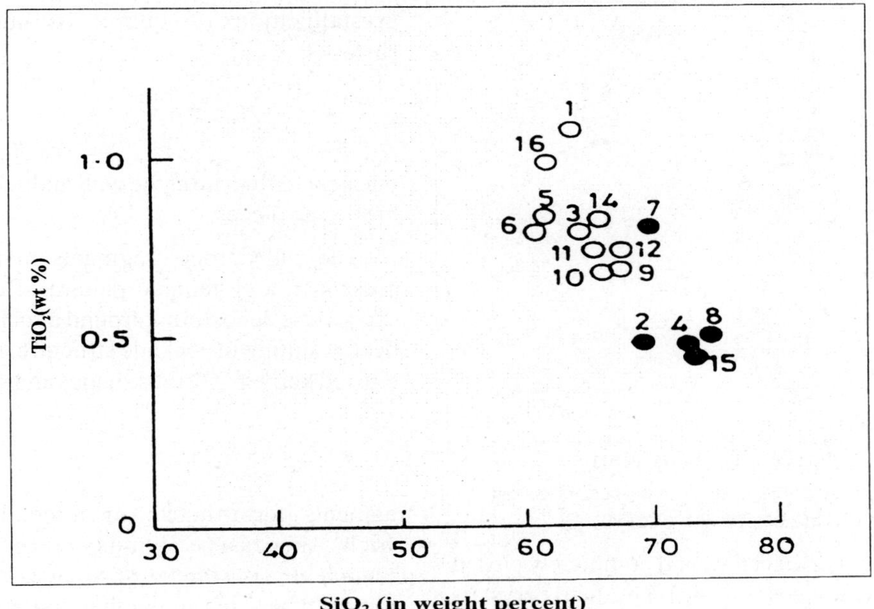

Fig.14 TiO_2 versus SiO_2 diagram of the host rocks from Jalera Khurd fluorite deposit.

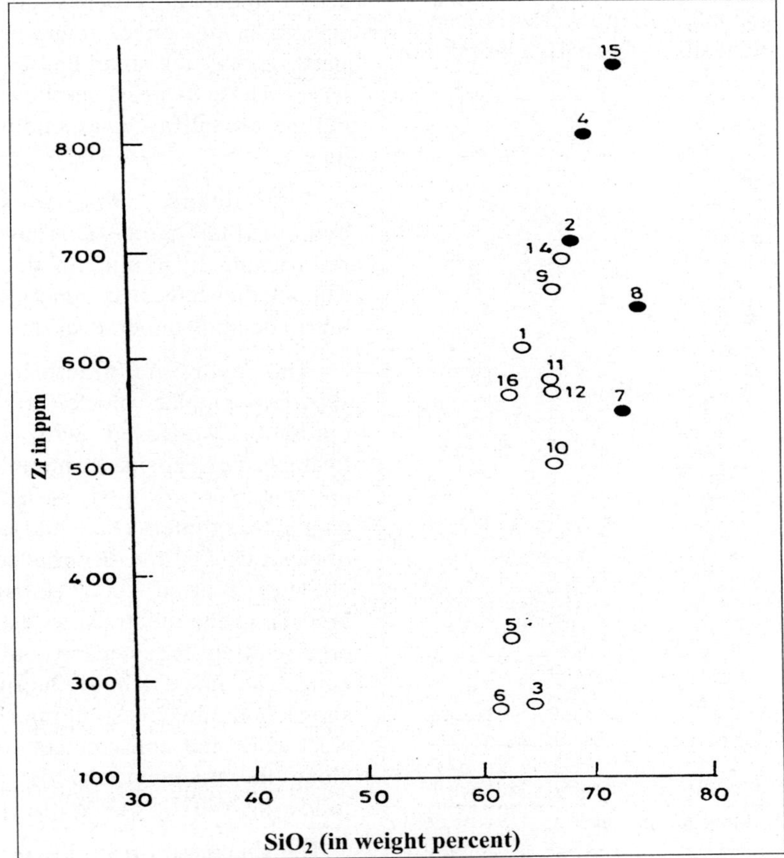

Fig.15 Zr versus SiO_2 diagram of the host rocks from Jalera Khurd fluorite deposit.

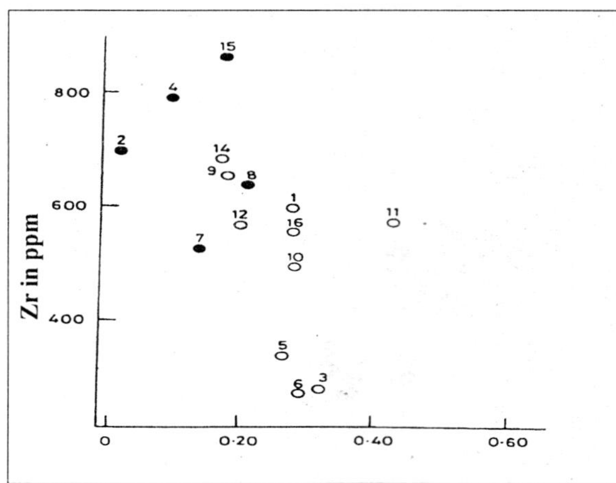

P_2O_5 (in weight percent)

Fig.16 Zr versus P_2O_5 diagram of the host rocks from Jalera Khurd fluorite deposit.

deposition or even rate of growth of crystals.

The quartz crystallized last and completely closed the cavity on all the sides (Fig. 4g). The quartz crystals are oriented normal to the torn pieces. This shows that fluorite which was deposited along the walls was broken before solutions started depositing quartz. Hence, the quartz crystallized afterwards is seen to line the fluorite at the wall. In this particular portion crystallization of quartz commenced at two surfaces, one next to fluorite at the wall and second around the broken up pieces.

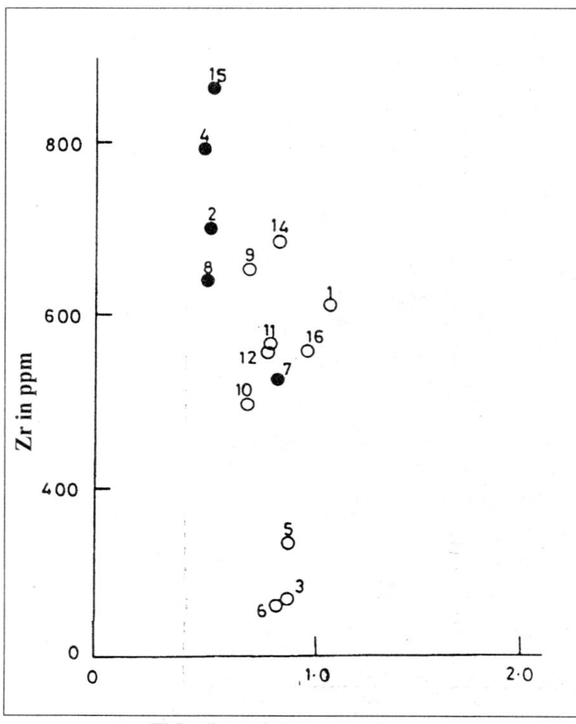

TiO_2 (in weight percent)

Fig.17 Zr versus TiO_2 diagram of the host rocks from Jalera Khurd fluorite deposit.

The rolled rock fragments in the deposit are present in a systematic pattern of deposition. The successive depositions around a rock fragment have been resulting in cockade structure, ring structure or crusted kernels. Various shapes and sizes of rock fragments form nuclei of mineralization around which successive bands were formed. The phenomenon is reverse of what is seen in circular cavities, where the mineralization starts at wall and proceeds towards the centre of the cavity. In contrast, the mineralization in cockade ores starts around the rock fragments and proceeds outward from the centre thus producing convex fractures. The earlier formed quartz crystals are small but the later ones are much larger. These depressions show polygonal or square outline around a corresponding zoning typical of quartz.

In volcanic conditions, the pneumato-hydrothermal fluid systems have enormous potential for causing hydraulic fracturing and brecciation. Though they have been reported from several places but association mineralization is not common.

The hydraulic fracturing and brecciation showing angular, blocky to splintery fragments could be confused with quench-fragmented hyaloclastic deposits. Pneumatic-hydraulic breccias are characterized by (i) variably altered clasts (ii) angular to splintery clasts and (iii) 'jig saw puzzle' fit of clasts where little disturbance or transportation of clast has occurred. The Pneumato-hydraulic fracture breccia, in the outcrops, is distinguishable by their cross-cutting character and confinement to a fracture zone. The Jalera Khurd fluorite deposit of course shows both, the cross-cutting characters to produce stockwork and confinement of the mineralization along the fracture zone developed and oriented following N43°E-N43°W direction.

The thickness of the hydraulic fracture zone at Jalera Khurd has been observed almost 20 metre. It is in the order with the concordant view that such zones

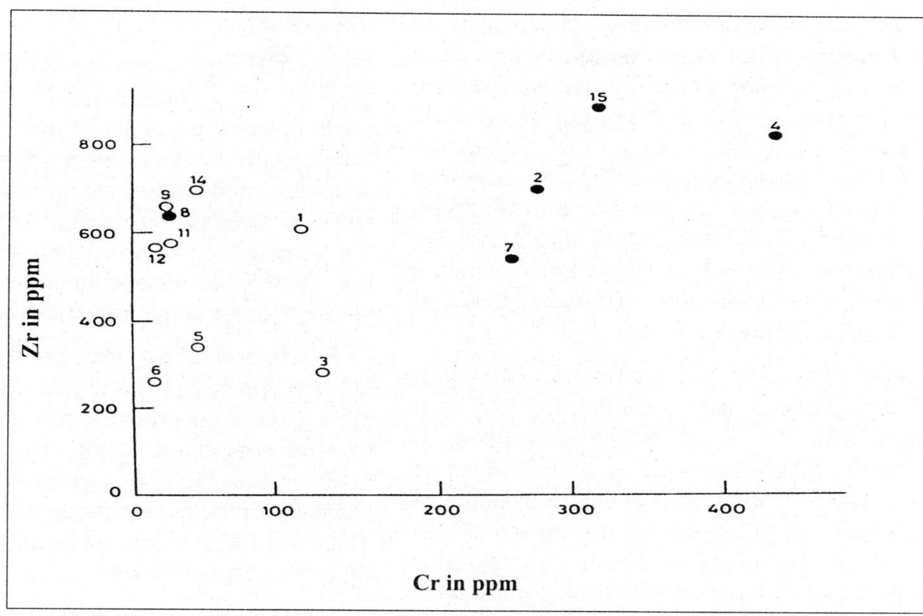

Fig.18 Zr versus Cr diagram of the host rocks from Jalera Khurd fluorite deposit.

may vary from several centimetres to many metres (Cas and Wright, 1987, p. 421). Jalera Khurd fluorite breccia also shows incipiently brecciated zones (Fig. 4a). The breccia clasts show 'jigsaw puzzle' like fit. In the mineralized stockwork deposit in the field, the angular clasts show their movement relative to each other. Some of them have rotated significantly and even transported a little through the system, leading to heterogeneous mixture of clast lithologies and various generations of fluorite. Some small scale occurrences of the process have also been observed (Figs. 4a, c and h).

The hydraulic fracturing has also produced sites for considerable hydrothermal alterations. These alteration processes, have produced significant weakening of the rock. Such zones may preferentially absorb iron and other stains (Figs. 4b, d and e) and become suitable site for subsequent hydrothermal action. The third and the last phase of fluorite formation has also occurred in the fractured brecciated sites and has cemented fragments of low grade and medium grade earlier formed fluorites, of pneumatic and hydraulic action.

The features observed in the ore breccia could best be explained in terms of the 'fluidization' model first applied to geological process by Reynolds (1954). Fluidized systems are those in which a dynamic state of suspension of solid particles by a flow of gas or vapour prevails, although a condensed liquid of lower density than the suspended fragments may also fluidize a system. Reynolds (1954) pointed out that the solid and gas (vapour) systems are characterized by a condition of turbulent flow within which great agitation and mixing of suspended fragments occurs. These results in abraded and well-rounded rock fragments, which may not have been appreciably transported away from their source and which, upon the exhaustion of the system, settle with lack of grading. Considerable amounts of kinetic energy would be absorbed by the multiple impacts of one fragment against another and it would be natural to expect that such systems should contain a significant proportion of well rounded fragments, the cause of which would be the physical process of attritive rounding by impact. The degree to which such rounding developed would depend upon the durability of the rocks, the viscosity of the suspending fluid phase, the kinetic energy of the system, and the duration of the fluidized condition. The possible corrosive chemical effects of the suspending fluid would also have to be taken into account (Sawkins, 1969; Sillitoe and Sawkins, 1971).

The mineralization of fluorite breccia at Jalera Khurd does not show high population of well rounded fragments that might be expected if mechanical attritive effects had been of long duration Examination of the contacts of the well rounded fragments also shows that the rounded shape is product of concentric exfoliation, rather than of attrition. Evidence of the operation of this process is shown in which the concentrically spalled layers are apparently frozen in the initial stages of disaggregation within the matrix. The explanation chosen for this phenomenon is that of explosive rupture following an instantaneous drop in confining pressure. For this reason, it is termed 'decompressive shock exfoliation'. It is evidently this process, which is responsible for the population of larger spheroidal fragments within the breccia body.

A number of textural features observed at Wheal Remfry by Sillitoe and Sawkins (1971) have similarities with the Jalera Khurd deposit. It is considered by Allman-ward et al. (1982) that the principal cause of disruption was the propagation of tensile fractures initiated during the escape of the boron-rich fluid from confinement at depth and that the disorientation of fragments depended more upon the turbulent passage of the fluids as they passed down a pressure gradient guided by the fracture system, than upon passive collapse under the influence of gravity into space created by chemical solution. The breccia thus displays features suggesting that, at the time of its formation, it consisted of a suspension of solid rock fragments in a stream of boron-rich fluid within which differential movements took place in response to the differences in hydraulic equivalence of fragments of different mass, shape, and density at different levels within the system. The fluid component was, no doubt, a boron-rich hydrous vapour carrying significant amounts of silicon, aluminium, and alkalis (Pichavant, 1979).

(v) Collapse Fluorite Breccia:

The presence of killas mixed within the fragments at all levels of the body is conclusive evidence of the turbulent mixing of fragments within a fluidized system at the time of formation. The impression conveyed by many of the textures of partial disaggregation is that the dynamic processes were frozen in situ by the rapid crystallization of the supporting matrix of fluorite. This inference is supported by the fact that the fluorite matrix is essentially microcrystalline and may have crystallized nearly instantaneously by a process of solute dumping as the fluids responded to a rapid change in physical conditions, in all probability a catastrophic fall in confining pressure. Although the breccia was certainly initiated as a fluidized suspension it is suggested that this condition was of comparatively short duration, in geological terms, perhaps, instantaneous.

The dynamic intrusive process of the fluorite breccia is different. The fluorite breccia is predominantly made up of angular killas fragments encompassing a wide range of sizes packed together and cemented again by high-grade fluorite. The constituent killas fragments are either partially or fully fluoritized. The matrix is made up chiefly of fluorite, but quartz and crystal-lined vugs are commonly observed in the larger interstices between fragments.

It is important to note here that predominance of killas fragments in the mineralized body some 200-300 m below the probable position of the porphyritic rhyolite (with granite at still lower parts) contact taken together with the purely siliceous and locally vuggy character of the matrix suggest an origin by collapse. Thus, this stage of breccia is, therefore, designated as fluorite collapse breccia.

The fluorite collapse breccia evidently formed at a later stage the intrusive breccia. The matrix shows the same textural geometry as the fluidized matrix of the main stage. The comminuted rock fragments and all the larger clasts have suffered complete fluoritization so that the rock shows much less visual contrast between clasts and matrix and has a generally grey appearance. In the inner part of this elongated halo of fluoritization the host rock clasts are completely pseudomorphed by fluorite within which felted patches of fluorite and cavities lined with small terminated fluorite crystals mark the site of pre-existing feldspar phenocrysts. Towards the outer limits of the fluoritized zone some fragments, which have suffered only partial fluoritization display distinctive alteration haloes.

(vi) Relationship between intrusive and collapse fluorite breccia

The precise definition of contacts between the main breccia and the collapse breccia is not made easily in the field because of the marked effects produced by the zone of fluoritization which evidently spread from the centre (in the elongated form) of collapse. There can, however, be little doubt that the collapse breccia formed, and was subject to the effects of pervasive fluoritization after the emplacement of the main body as a fluidized mass. Clearly the later stages of the formation of the breccia body were marked by collapse of the killas roof rocks into space created at a lower level by processes of physical displacement or chemical dissolution caused during the earlier stage of explosive release of fluids rich in fluorie. At the same time the composition of the active hydrothermal fluids changed from being fluorine-rich to being silica-rich.

The range of geological features displayed by the different parts of the Jalera Khurd fluorite breccia body show that the process of disruptive intrusion by a turbulent fluidized system was the dominant process at the initial stages of formation. This was followed by a stage of collapse and that this took place in at least two episodes, one major and one minor. Resultantly, the early formed fluorite in the form of angular fragment included into the later formed rich fluorite matrix. It was followed by pervasive silicification. The fluorite ore body, therefore, combines the features of both hydrothermal intrusive breccias and hydrothermal collapse breccias. The precise role of chemical processes in producing the brecciation of the rocks is difficult to assess but the balance of evidence suggests that hydraulic and mechanical forces were probably dominant.

(vii) Metasomatic alteration and mineralization

Potash metasomatism is main resultant of hydrothermal action of fluid and gases at Jalera Khurd fluorite deposit and also in the nearby areas. Mineralogically, this is characterized by microclinization due to the partial or complete replacement of sodium by potassium in the feldspars, accompanied by haematitic reddening. There is development of new mica in the compositional range annite to ferrous siderophyllite. Chemically, there is an increase in K and a concomitant loss of Na.

Acid metasomatism (hydrogen ion metasomatism) resulted into progressive breakdown of rhyolitic and trachytic minerals in response to change K^+/H^+ ratio in the fluids. Perthite or microcline feldspar is destroyed and replaced by siderophyllite/ zinnwaldite and accompanied by fluorite and sericite. Chemically, hydrogen ion metamorphism is characterized by a marked decrease in potash and alumina due to breakdown of feldspar and an increase of Fe_2O_3 and SiO_2 content.

Silica metasomatism resulting in quartz vugs and replacement of early-formed minerals, especially in trachytes. The process of silicification is responsible in increasing total silica balance in the host rocks.

Argillic alteration have been involved in the formation of clays at the expense of feldspars. As the main ore body is dipping somewhat westward, this alteration has accumulated kaoline at the hanging wall of the ore main body, at its centre.

The hydrothermal system responsible for kaolinization was clearly later than breccia formation (fluoritization). Towards the northwest of the main veins, (hanging wall of fluorite mineralization), the host rocks are completely kaolinized. Except in extreme fluoritized and silicified rocks, the breccia body retained permeability sufficient to allow access for solutions for kaolinization of feldspars. Hematitic alteration is prominently seen in the extreme northeast end of the deposits at the lower reaches the host rocks of the stockwork. Within these areas there are fractures carrying pyrite surrounded by oxidized iron oxides. This relationship is also not uncommon under microscopic observations for pyrite and hematite and other iron oxides. It is probable that the haematization reflects a process of flushing of iron from the kaolinized parts of the system or from intersecting fracture systems but the precise relationship between the two processes requires further study before a coherent explanation can be given. The fluorite breccia body itself posses no cogenetic mineralizations of economic significance.

THE GENETIC MODEL

It is clear from textural and mineralogical considerations that the emplacement of the main stage of the fluorite breccia body took place as a consequence of disruption and fluidization by a fluorine rich fluid which escaped down a pressure gradient guided by a set of joint, subvertical in nature, in the apical part of the host rocks.

From further consideration of the fluorite-rich facies of the host rocks (two feldspar) and the ubiquitous occurrence of fluorite and or fluorine within the mineralized hydrothermal structures of Jalera Khurd, it is clear that the fluorine-rich fluid was generated by differentiation and crystallization of the parent magma of the host rocks from subvolcanic to volcanic levels.

Presence of gases like fluorine, boron, CO_2 and H_2O in the magma and their vital role in the resultant processes including possible mineralization have been studied by Diamond (2001), Edmonds et al. (2001), Evans et al. (2001) and Tamic et al. (2001). Absence of gases and dry melt in Qz-Ab-Or system has also been evaluated by Hess et al. (2001). It is known that the presence of boron in significant concentration in the system also leads to enhanced solubility of H_2O in the silicate melt and the solidus temperature suffers a significant depression so that the hydrous phase may coexist with silicate melt at temperatures of the order of 575-600°C (Chorlton and Martin, 1978; Pichavant, 1979). It shows that boron plays a very important role in decreasing the miscibility gap between coexisting silicate and hydrous phases in water-saturated magmas and also lowers the temperatures at which silicate melts can survive in coexistence with the hydrous fraction. Whether fluorine play a similar role or not is still in question awaiting future experimental studies.

Considering the geological evidences including the field relationships observed at Jalera Khurd, and the specimens, a genetic model is proposed for explaining the phenomenon of the rapid channeled escape of the fluorine-rich fluids from the apical part of the igneous rocks. Burnham (1979) has presented a model describing the emplacement and solidification of bodies of hydrous acid magma. This model has also been found satisfactory in the case of tourmaline rich breccia in the western part of the St Austell, England (Allman Ward et al. 1982). This type of model is considered to form a good basis for explaining the petrogenesis and mineralization events at Jalera Khurd. The as many as 12 distinct stages have been recognized. They are :

(i) Formation of Carapace above the magma chamber

The first stage of the model involves the rise of the magma under the influence of telluric pressure differences to a level in the crust between 5 ± 2 km depth under confining pressures of 2 ± 1 kbar. Charoy (1979) has estimated that around 4 ± 1 weight percent of initial water content of magma, the granite will cross the saturated solidus boundary at a depth almost of three kilometres (Harris et al., 1970). A solid crystalline crust will develop at the top of the magma chamber as soon as solidus-liquidus barrier is crossed (Bumham, 1979) seals the upward progress of the rest of the magma melt (740°C ± 40°C) (Fig. 19) temporarily,

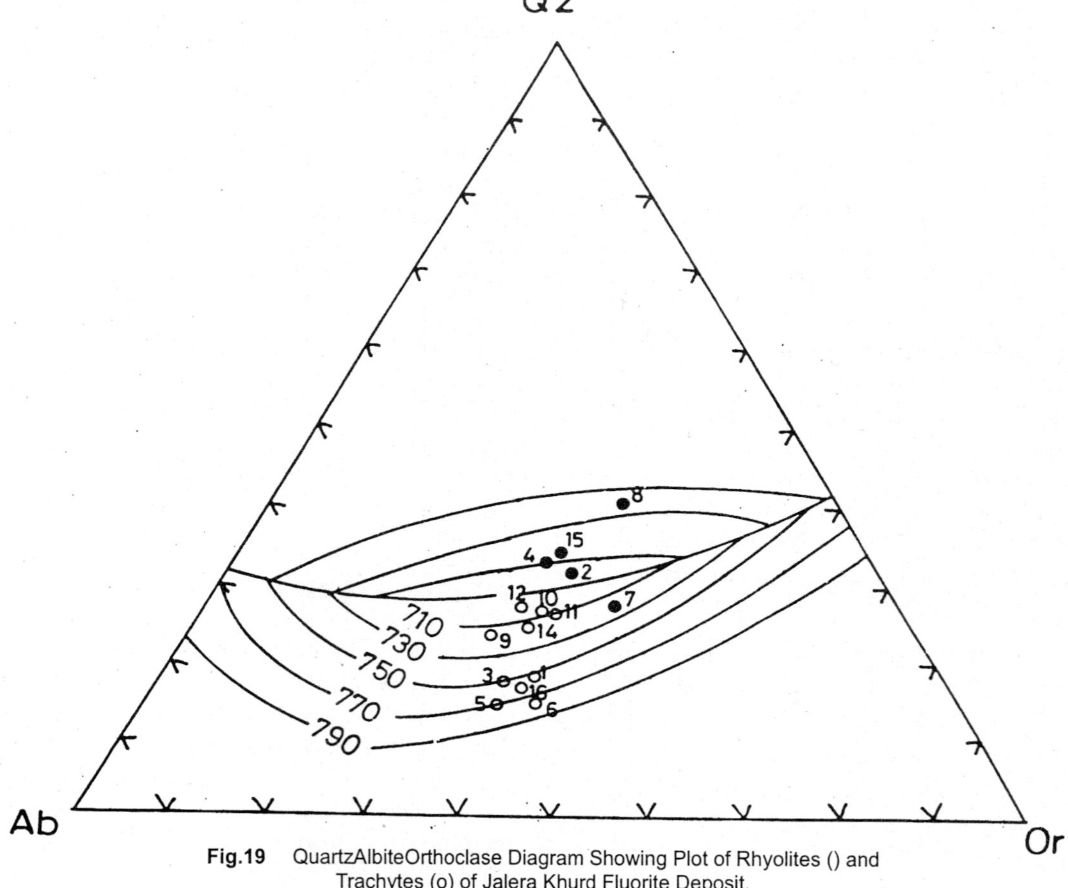

Fig.19 QuartzAlbiteOrthoclase Diagram Showing Plot of Rhyolites () and Trachytes (o) of Jalera Khurd Fluorite Deposit.

though the blocked pressure get released before development of the much solid carapace.

(ii) Initiation of Crystallization and differentiation

The next stage is represented by the crystallization and differentiation in confinement begins within the rest of the magma beneath the solid crystalline carapace above the halted magma. It is assumed that before the development of solid crystalline envelope above the magma chamber, all the volatiles generated during solidification may have escaped into the surrounding country rocks. Differentiation due to progressive crystallization in the static magma would now take place leading to a saturated condition with respect to the volatile aqueous component. The latter exsolves to form a separate phase, and rise to get concentrated in the apex of the elongated balloon. Given the experimental results obtained by Pichavant (1979), a hydrous phase enriched in fluorine could possibly coexist with the 600°C silicate melt in the top of such balloon.

(iii) Formation of collapse breccia

Interestingly in the next stage, because of increase of the internal pressure, the above hard newly formed envelope would fail by fracturing. In fact, from the saturated liquidus boundary to the saturated solidus boundary change at this stage is represented by water saturated melt crystals and the hydrous phase. Following the opinion of Pichavant (1979), it would also allow the partition of fluorine in favour of hydrous fraction in a ratio of 2:1.

It is known that the volume covered by a unit mass of crystals and hydrous phase is always greater than the same unit mass of water saturated silicate melt. Thus, when the crystallization proceeds, the volume expansion cannot be accommodated by progressive plastic deformation and expansion owing to the effects of firstly, the wall rocks, secondly, the rigidity of the hard apex (recently crystallized) and lastly the confining lithostatic pressure. It is now clear that as the resultant of these factors, during crystallization and cooling, the internal pressure in the apex of the magmatic body will keep increasing gradually.

A possible answer to adjust this additional internal pressure could be a certain finite compression of the contiguous magma system. But it is too little to absorb the internal additional pressure generated during the crystallization and cooling.

According to Burnham (1979), within pockets of water saturated magma undergoing crystallization, an internal pressure of almost five kilobar could be generated. If so, at depths corresponding to internal pressures of one kilobar, the carapace developed at the apex of the body would fail much earlier before pressures of this magnitude were achieved, and produce the first set of collapse and brecciation, a scenario promoting formation of both fluorite crystals and the host rocks.

(iv) Formation of Joints

The crystallizing magma would fail at its top by fracturing the newly formed crust when the internal pressure achieved a value to the minimum principal stress and the tensile strength value of the newly-formed envelope of the rocks. This is a critical condition for the enlargement of the effective volume of the system by tensile fracturing. Presuming that this process in the case of Jalera Khurd took place within a field of regional tectonic stress, the principal mechanical effect of increasing internal overpressure would be the creation of sets of subvertical tensile joints with an orientation normal to the trajectory of principal stress in the brittle envelope recently developed at the top of the host rocks of the fluorite deposit. This model is proposed following the similar sequence of stages explained by Phillips (1973).

(v) Formation and filling of outer mineralized veins

With the development of joints as described in stage iv the internal overpressure built up is relieved temporarily, as they enable the escape of portions of the hydrous fluids and gases from the elongated apical zone. At Jalera Khurd, apart from the main central mineralized set of veins, there are poorly mineralized veins available, parallel to the main veins at both sides, almost symmetrical to the main body. This is attributed to the continued progressive buildup of internal overpressure arising from further crystallization of magma leading to the and the critical conditions once again followed by renewed fracturing producing the outer mineralized veins.

(vi) Richness of mineralized veins

In course of time, this repetition of the sequence of failure, renewed pressure build up in the apex, result in over pressure so large to increase number of fractures which will finally connect fractures open already and connected to the surface. This will sharply result in an instantaneous drop in the confining pressure. Corresponding to the depth of initiation, this value must almost be from lithostatic value to hydrostatic value. By this time, the fluids would have boiled and a large volume of vapour would have been generated. This turbulent fluidized system with explosive energy would be channeled down the controlled joint system within which this sudden change will lead to deposition of a large quantities of fluorite, silica and iron to finally produce the complicated mineral relationship now observed in the mineralized body.

(vii) Pneumatic or buoyant plume action

The last stage can also be explained following a classical theory of 'buoyant plumes' suggested by Phillips and Woods (2001). CO_2 is relatively insoluble in basaltic magma at low crustal pressures, and therefore, exists as a gas phase in the form of bubbles in shallow crustal reservoirs. Over time, these bubbles may separate gravitationally from the magma in the chamber. As a result, any fresh magma which recharges the chamber from the depth will be rich in such bubbles, creating a turbulent bubble plume within the chamber. A relatively deep bubbly layer of magma may then form below a thin foam layer at the roof. As their eruption proceeds, there will be continuing influx at the base of the chamber, followed by a gradual increase of the bubble content of the bubbly layer. This may ultimately lead to a transition from lava flow activity to one more explosion. The foam layer at the top of the chamber may provide a flux for the continual outgassing from the flanks of the volcano.

(viii) Intrusion of Breccia

The emplacement of breccia evidently marked the climax of the sequence of events in which fluorine-rich fluids escaped from the apical zone of the crystallizing magma. Though the conditions necessary for the creation of such brecciated bodies are reported elsewhere e.g., local development in Cornubian province (Goode and Taylor, 1980) or regional as in Yukon, Canada (Thorkelson et al. 2001) nowhere they are so well developed or dramatically exposed as at Jalera Khurd deposit.

(ix) Explosive expansion

In the model proposed for the generation of the breccia, there is no implicit need for the breccia itself to have reached surface. But, there are indications that it has been emplaced in the roots of a volcanic system at Jalera Khurd. The mechanism proposed does, however require that an instantaneous change in the confining pressure of the system took place and that explosive expansion of the hydrous vapour could occur. Thus, the vapour under hydrostatic confining conditions probably escaped into the meteoric system, having dispersed its energy during the process of fluidization, hydraulic fracturing, and expansion in the guiding structures closest to the focus of internal overpressure generated in the elongated apex of the host rocks.

(x) Formation of Stockwork (more fractures)

The type of mechanism explained in Stage ix, operating under the influence of the prevailing tectonic stress field would explain the extended paragenetic and structural sequence observed in the pre-breccia fracture systems. Progressive stages in the relief of internal stresses by fracturing resulted in the development of intersecting sets of fractures changed in response to the jacking or wedging effect produced as the veins were filled under pressure. The interplay of this effect combined with slight changes in the regional stress field could explain in the principal sets of fractures (Joint I), which are developed in the Jalera Khurd area.

(xi) Repetition of Processes to form multigenerational deposit

The same processes possibly have been repeated atleast thrice to produce two generations of fluorite fragment in the breccia. The first set of fragments is one which is enclosed by later second set, and still later, the third set is represented, chiefly by fine grained fluorite which is either cementing the grains of earlier fluorites and/or host rocks or making somewhat layers of the fluorite itself.

(xii) Metasomatism

Post-mineralization metasomatism and alterations are present as kaolinization, pyritization and haematitization. The central body of fluorite breccia showing good deposition of kaoline at the hanging wall, originated by metasomatic fluid action at the expense of feldspars. Pyrite generally represent the last activity of the volcanic action. It has a coat of hematite, showing iron oxide as a result of still later action of metasomatism. Rusty colours of iron oxides have also penetrated deep inside the host rock and the fluorite breccia and preferencial occurrence is at the contact of the two.

CONCLUSION

The Jalera Khurd fluorite deposit in Malani Igneous Suite, locally showing bimodal, lower porphyritic rhyolite and upper porphyritic trachyte as host rocks for the joint-controlled parallel mineralized veins constituting almost subvertical tabular body with its average thickness, length and depth 20 m, 500 m and 100 m respectively within an elongated stockwork. Detailed geochemical and petrological studies of the host rocks favour that the parent magma must have originated relatively at deeper parts of the earth. Crustal assimilation and fractional crystallization are the possible mechanisms.

The Jalera Khurd fluorite deposit is an excellent example of volcanogenic pneumatohydrothermal intrusive and collapse breccia deposit of fluorite. The range of geological features displayed by fluorite breccia shows that the process of disruptive intrusion by a turbulent fluidized system was the dominant process, initially. The intrusive aspects of fluorite breccia show angular killas fragments encompassing a wide range of sizes packed together and cemented again by microcrystalline fluorite itself. This was followed by a stage of collapse atleast in two episodes, one major, another minor. It resulted in inclusions of earlyformed fluorite fragments into later-formed rich fluorite matrix. These episodes were further followed by pervasive silicification and metasomatic alterations. The genetic model suggested is comprise of twelve stages. The fluorite ore body, combines the features of both hydrothermal intrusive breccia and hydrothermal collapse breccia to form the economic deposit.

ACKNOWLEDGEMENTS

Authors are grateful to Dr. S. Viswanathan and Dr. A.V. Sankaran for greatly improving the first draft of the manuscript. One of the authors (KLS) is also grateful to DST for providing the Special Assistance under SAP, to the Department of Geology, JNV University, Jodhpur.

References

Abdel Rahman, A.F.M., Elkibbi, M.M., 2001: "Anorogenic Magmatism: Chemical evolution of the Mount El-Sibai A-type complex (Egypt), and implications for the origin of within-plate felsic magmas". Geol. Mag., Vol. 138, No. 1, pp. 67-85.

Aguirre Diaz, G.J., 2001: "Recurrent magma mingling in successive ignimbrites from Amealco Caldera, Central, Mexico". Bull. Volcano., Vol. 63, No. 4, pp. 238-251.

Allman-Ward, P., Halls, C., Rankin, A., and Bristow, C.M., 1982 : "An Intrusive Hydrothermal Breccia Body at Wheal Remfry in the Western Part of the St Austell Granite pluton, Cornwall, England". In "Metallization Associated with Acid Magmatism" (Ed. A.M. Evans), John Wiley and Sons Ltd., pp. 1-28.

Andersen, T., Neumann, E.R., 2001 : "Fluid inclusions in mantle xenoliths". Lithos, Vol. 55, No. 1-4, pp. 301-320.

Aswathanarayan, U., 1964 : "Age determination of rocks and Geochronology of India". Brochure XXII Ind. Geol. Cong., New Delhi. p. 23.

Basu, A.R., Renne, P.R., Dasgupta, D.K., Teichmann, F., and Poreda, R.J., 1993 : Early and late alkali igneous pulses and a high ^{3}He plume for Deccan flood basalts. Science Vol. 261, pp. 902-906.

Basu, S.K., 1982 : "Phases of acid igneous activity in Malani Suite of rocks around Jhunjhunu, Rajasthan". Rec. Geol. Surv. India, Vol. 12(7), pp. 89-94.

Best M.G.; 1986 :"Igneous and Metamorphic Petrology"CBS Pub. and Dist. Delhi. 630 p.

Bevins, R.E., Lees, G.J. and Roach, R.A., 1991 : "Ordovician bimodal volcanism in SW Wales : Geochemical evidence for Petrogenesis of silicic rocks". Jour. Geol. Soc. London, Vol. 148 No. 4, pp. 719-730.

Bhushan, S.K. and Chandrashekhran, V.; 2002 : Geology and Geochemistry of the magmatic rocks of Malani Igneous Suite and Tertiary Alkaline Province of western Rajasthan. Mem. Geol. Surv. Ind. Vol. 126, 181 p.

Bhushan, S.K. and Khullar, V.K., 1998 : "Geochemistry and tectonic significance of dyke swarm in Malani Igneous Complex around Sankara, district Jaisalmer, Rajasthan". In : B.S. Paliwal (ed.). The Indian Precambrian, Scientific Publishers (India), Jodhpur, pp. 482-491.

Bhushan, S.K., 1984 : Classification of Malani Igneous Suite. Symposium on three decades of development in Petrology, Mineralogy and Petrochemistry in India. Sp. Publ., Geol. Surv. Ind. Vol. 12, pp. 199-205.

Bhushan, S.K., 1985 : "Malani volcanics in Western Rajasthan". Ind. Jour. Earth Sci., Vol. 12, pp. 58-71.

Bhushan, S.K., 1988 : "Stratigraphic position of Marwar Supergroup, western Rajasthan". Abs. National seminar on "Stratigraphic Boundary Problems in India", Jammu, pp. 5-6.

Bhushan, S.K., 1989 : "Mineral Chemistry and Petrogenetic aspects of Malani Volcanics, Western Rajasthan". Ind. Min., Vol. 43, No. 3-4, pp. 325-338.

Bhushan, S.K., 1991 : "Granitoids of Malani Igneous Complex, Western Rajasthan". India. Jour. Earth Sci., Vol. 18, No. 3-4, pp. 184-194.

Bhushan, S.K., 1995 : "Late Proterozoic Continental growth : Implications from geochemistry of acid magmatic events of West Indian Craton, Rajasthan". Mem. Geol. Soc. India, Vol. 34, pp. 339-355.

Bhushan, S.K., 2000 : "Malani rhyolites - A review". Gondwana Res., Vol. 3, No. 1, pp. 65-77.

Bidwai, R. and Krishnamurthy, P., 1997 : "Possible lava channels and tubes from the Siwana caldera sequence, Siwana, Barmer district, Rajasthan". Jour. Geol. Soc. India, Vol. 50, No. 2, pp. 103-115.

Blake, S., Bruno, B.C., 2001 : "Modelling the emplacement of compound lava flows". Earth and Planetary Science Letters, Vol. 184, No. 1, pp. 181-197.

Borisova, A.V., Belyatsky, B.V., Portnyagin, M.V.,

Sushchevskaya, N.M., 2001 : "Petrogenesis of olivine-phyric basalts from the Aphanasey Nikitin Rise : Evidence for contamination by cratonic lower continental crust". Jour. Petrol., vol. 42, No. 2, pp. 277-319.

Bryner, L., 1961 : "Breccia and Pebble Columns associated with epigenetic ore deposits" Econ. Geol., Vol. 56, pp. 488-508.

Burnham, W.C., 1979 : "Magmas and hydrothermal fluids", in 'Geochemistry of Hydrothermal Ore deposits', H.L. Barnes (Ed.) Wiley, New York, pp. 71-136.

Cas R.A.F. and Wright J.V.; 1987 : "Volcanic successions : Modern and Ancient". Allen and Unwin. London. 528 p.

Chabiron, A., Alyoshin, A.P., Chuney, M., Deloule, E., Goluber, V.N., Velitchkin, V.I., Poty, B., 2001 : "Geochemistry of the rhyolitic magmas from the Streltsovka caldera (Transbaikalia, Russia) : a melt inclusion study" Chem. Geol., Vol. 175, No. 3-4, pp. 273-290.

Charoy, B., 1979 : "Definition et importance des phenomenes deuteriques et des fluides associes dans les granites, consequences metallogeniques", Sci. Terre., Mem. No. 37.

Chore, S.A. and Mohanty, M., 1998 : "Stratigraphy and tectonic setting of the Trans-Aravalli Neoproterozoic Volcanosedimentary sequences in Rajasthan". Jour. Geol. Soc. India, Vol. 51, pp. 57-68.

Chorlton, L.B., and Martin, R.F., 1978 : "The effect of boron on the granite solidus", Can. Mineral. Vol. 16 pp. 239-244.

Choudhary, A.K., Gopalan, K. and Shastry, A., 1984 : "Present status of the Geochronology of the Precambrian rocks of Rajasthan" Tectonophys., Vol. pp. 131-140.

Chouhan, D.S., 1999 : "Tectonic and sedimentary evolution of the Marwar Basin : A Neoproterozoic - Early Cambrian Intracratonic Sag Basin". Proceedings of seminar on Geology of Rajasthan (Ed. P. Kataria) Geol. Deptt. MLSU, Udaipur, pp. 111-125.

Condie, K.C., 1982 : "Early and middle proterozoic successions and their tectonic settings". Am. Jour. Sci. Vol. 282, pp. 341-357.

Cosgrove, M.E., and Elliott, M.H., 1976 : "Supra batholithic volcanism of the south west England granites". Proc. Ussher Soc., Vol. 5, pp. 76-80.

Coulson, A.L., 1933 : "Geology of Sirohi State". Mem. Geol. Surv. India, Vol. 63, No. 1, pp. 1-166.

Crawford, A.R. and Compston, W., 1970 : "The age of the Vindhyan system of Peninsular India". Quart. Jour. Geol. Soc. London, Vol. 125, No. 1, pp. 351-372.

Crecraft, H.R., Nash, W.P. and Evans, S.H., 1981 : "Late Cenozoic Volcanism of Twin Peaks, Utah, Geology and Petrology". Jour. Geophys. Res., Vol. 86, p. 10320.

Dana, E.S. 1955 : "Text Book of Mineralogy" John Wiley and Sons, New York 851 p.

Deer, W.A., Howie, R.A. and Zussman, J. 1962 : "Rock forming minerals : Vol. 1,2,3,4, Ortho- and ring silicates". 333 p., Longmans.

Dhar, S., Frei, R., Kramers, J.D., Nagler, T.F. and Kochhar, N., 1996 : "Sr, Pb and Nd. Isotope studies and their bearing on the Petrogenesis of the Jalore and Siwana Complexes, Rajasthan, India". Jour. Geol. Soc. India. Vol. 48, No. 2, pp. 151-160.

Diamond, L.W., 2001 : "Review of the systematics of CO_2-H_2O fluid inclusions". Lithos, Vol. 55, No. 1-4, pp. 69-99.

Dines, H.G., 1956 : "The Metalliferous Mining region of South-west England". Mem. Geol. Surv. Gt. Britain.

Drury, M.R., Van Roermund, H.L.M., Carswell, D.A., DeSmet, J.H., VandenBerg, A.P., Vlaar, N.J., 2001 : "Emplacement of deep upper-mantle rocks into cratonic lithosphere by convection and diapiric upwelling". Jour. Petrol., Vol. 42, No. 1, pp. 131-140.

Dunhan, K.C., 1952 : "Fluorspar, Great-Britain" Geol. Survey, Mem., Spec. Rept. Min. Resources, Vol. 4, pp. 143.

Eby, G.N. and Kochhar, N., 1990 : "Geochemistry and Petrogenesis of the Malani Igneous Suite, North India". Jour. Geol. Soc. India. Vol. 36, pp. 109-130.

Edmonds, M., Pyle, D., Oppenheimer, C., 2001 : "A model for degassing at the Soubriere Hills volcano, Montserrat, West Indies, based on geochemical data". Earth and Planetary Science Letters, Vol. 186, No. 2, pp. 159-173.

Evans, W.C., Sarey, M.L., Kennedy, B.M., Stonestrom, D.A., Rogie, J.D., Shuster, D.L., 2001 : "High CO2 emissions through porous media : transport mechanism and implications for flux measurement and fractionation". Chem. Geol. Vol. 177, No. 1-2, pp. 15-29.

Ewart, A., 1979 : "A review of the mineralogy and chemistry of Tertiary - Recent dacitic, latitic, rhyolite and related salic volcanic rocks". In : Trondhjemites, dacites and related rocks. Barker, F. (Ed.), Elsevier Pub., Amsterdam, pp. 13-121.

Exley, C.S., 1959 : "Magmatic differentiation and alteration in the St. Austell granite". Jour. Geol. Soc. Lond., Vol. 114, pp. 197-230.

Fareeduddin and Kroner A.; 1998 : Single Zircon age constrains on the evolution of Rajasthan granulite in the Precambrians (B.S. Paliwal, Editor). Sci. Pub., Jodhpur. Pp. 547-556.

Fareeduddin and Kroner, A. 1998 : Single Zircon age constrains on the evolution of Rajasthan granulite in The Precambrians (B.S. Paliwal, Editor), Sci. Pub., Jodhpur, pp. 547-556.

Fiske, R.S., Naka, J., Iizasa, K., Yuasa, M., Klaus, A., 2001 : "Submarine silicic caldera at the front of the Izu-Bonin arc, Japan : Voluminous sea floor eruptions of rhyolite pumice". Geol. Soc. Am. Bull., Vol. 113, No. 7, 813-824.

Fletcher, C.J.N., 1977 : "The geology, mineralization and alteration of Ilkwang Mines, Republic of Korea A Cu-W-bearing tourmaline breccia pipe". Econ. Geol., Vol. 72, pp. 753-768.

Garuti, G., Bea, F., Zaccarini, F., Montero, P., 2001 : "Age, geochemistry and Petrogenesis of the ultramafic Pipes in the Jvrea Zone, NW Itlay". Jour. Petrol., Vol. 42, No. 2, pp. 433-457.

Geological Survey of India; 1998 : "Geological map of Abu quadrangle" (Published).

Goode, A.J.J., and Taylor, R.T., 1980 : "Intrusive and pneumatolytic breccias in south-west England". Inst. Geol. Sci. Rep. No. 80/2.

Grant, J.N., Halls, C., Sheppard, M.F., and Avila, W., 1980 : "Evolution of the porphyry tin deposits of Bolivia". Min. Geol. (Japan), Special issue, vol. 8, pp. 151-173.

Grapes, R., Yogi, K. and Okumura, K., 1979 : "Aenigmatite, Sodic Pyroxene, arfvedsonite and associated minerals in syenites from Morotu, Sakhalin". Contrib. Mineral Petrol., Vol. 69, pp. 97-103.

Gultekin, A.H. 2005 : "Epithermal Kizilcaoren fluorite Barite REE deposit, Eskisehir, Turkey : Preliminary Results". Jour. Geol. Soc. Ind. Vol. 66, pp. 105 116.

Halls, C., Rankin, A., Ferriday, I.L., Blain, C.F., Bristow, C.M. and Gronow, C., 1977 : "A tourmalinized hydrothermal intrusion breccia at Wheal Remfry in the western part of the St. Austell pluton, Cornwall". Trans. Inst. Min. Metall., Vol. 86(B), 100-163.

Harris, P.E., Kennedy, W.Q., and Scarfe, C.M., 1970 : "Volcanism versus plutonism - the effect of chemical composition". In G. Newall and N. Rast (Eds.) Mechanism of Igneous Intrusion : Geol. Jour., Special issue, Vol. 2, pp. 187-200.

Hawkes, J.R., 1974 : "Volcanism and metallogenesis : the tin province of south-west England". Bull. Volcanol., Vol. 38, pp. 1125-1146.

Heron, A.M., 1932 : "The Vindhyans of Western Rajputana". Rec. Geol. Surv. India, Vol. 65, No. 4, pp. 457-489.

Hess, K.U., Dingwell, D.B., Gennaro, C., Mincione, V., 2001 : "Viscosity-temperature behaviour of dry melt in the Qz-Ab-Or system". Chem. Geol., Vol. 174, No. 1-3, pp. 133-142.

Hyndman, D.W., 1985 : "Petrology of Igneous and Metamorphic rocks. McGraw Hill, New York, p. 786.

Johnston, W.P., and Lowell, J.D., 1961 : "Geology and Origin of Mineralized breccia pipes in Copper Basin, Arizona". Econ. Geol., Vol. 56, pp. 916-940.

Kabeto, K., Sawada, Y., Iizumi, S., Wakatsuki, T., 2001 : "Mantle sources and magma-crust intractions in volcanic rocks from the northern Kenya rift : geochemical evidence". Lithos, Vol. 56, No. 2-3, pp. 111-139.

Kents, P., 1964 : "Special breccias associated with hydrothermal developments in the Andes". Econ. Geol., Vol. 59, pp. 1551-1563.

Knutson, J., Ferguson, J., Roberts, W.M.B., Donnelly, T.H., and Lambert, I.B., 1979 : "Petrogenesis of the Copper-bearing breccia pipes, Redbank, Northern Territory, Australia". Econ. Geol., Vol. 74, pp. 814-826.

Kochhar, N., 1983 : "Tusham ring Complex, Bhiwani, India". Proc. Indian Nat. Sci. Acad., No. 4, pp. 459-490.

Kochhar, N., 1984 : "Malani Igneous Suite : Hot Spot Magmatism and Cratonization of the northern Part of the Indian Shield". Jour. Geol. Soc. India, Vol. 25 No. 3, pp. 155-161.

Kochhar, N., 1998 : "Malani Igneous Suites of rocks". Jour. Geol. Soc. India, Vol. 51, pp. 120.

Kovalenko, V.I., Kuz'min, M.I., Antipin, V.S., and Petrov, L.L., 1971: "Topaz bearing quartz keratophyre (Ongonite), a new variety of subvolcanic igneous vein rock", Ookl. Acad. Sci. USSR, Earth Sci. Sect., Vol. 199, pp. 132-134.

Krishnan, T.K., 1968 : "Morphology of zircons from Jasai granite, Barmer district, Rajasthan". Geol. Surv. Ind., Misc. Pub. 9, pp. 70-76.

Kuritani, T., 2001 : "Replenishment of a mafic magma in a zoned felsic magma chamber beneath Rishiri volcano, Japan". Bull. Volcano., Vol. 62, No. 8, pp. 533-548.

Kuscu, G.G., Floyd, P.A., 2001 : "Mineral compositional and textural evidence for magma mingling in the saraykent volcanics". Lithos, Vol. 56, No. 2-3, pp. 207-230.

La Touche, T.H.D., 1902 : Geology of Western Rajputana. Mem. Geol. Surv. India, Vol. 35, No. 1, pp. 1-116.

Le Bas, M.J., Le Maitre, R.W., Streckeisen, A. and Zanettin, B., 1986 : "A chemical classification of volcanic rocks based on the total alkali-silica diagram". Jour. Petrol., Vol. 27, pp. 745-750.

Le Maitre, R.W. (Ed.), 1989 : "A classification of Igneous rocks and glossary of terms". Recommendations of the IUGS subcommission on the systematics of Igneous rocks. Blackwell Oxford.

Lindsay, J.M., Schmitt, A.K., Trumbull, R.B., DeSilva, S.L., Siebel, W., Emmermann, R., 2001 : "Magmatic evolution of the La Pacana Caldera system, Central Andes, Chile : Compositional variation of two cogenetic, large-volume felsic ignimbrites". Jour. Petrol., Vol. 42, No. 3, pp. 459-486.

Llambias, E.J., and Malvicini, L., 1969 : "The geology and genesis of the Bi-Cu mineralised breccia-pipe, San Francisco de los Andes., San Juan, Argentina". Econ. Geol., Vol. 64, pp. 271-286.

Locke, A., 1926 : "The Formation of certain ore bodies by mineralization stoping". Econ. Geol., Vol. 21, pp. 431-453.

Maheshwari, A., Coltori, M. and Sial, A.N., 1995 : "Petrogenesis and tectonic setting of Malani rhyolites : Evidence by trace elements and Oxygen isotopic composition". Curr. Sci., Vol. 68 No. 9, pp. 954-956.

Maheshwari, A., Coltori, M. Sial, A.N. and Mariano, G., 1996 : "Crustal influences in the petrogenesis of the Malani rhyolites, South-western Rajasthan : Combined trace-element and oxygen isotope constrains". Jour. Geol. Soc. India, Vol. 47, No. 5, pp. 611-619.

Maniar, P.D. and Piccoli, P.M., 1989 : "Tectonic discrimination of granitoids". Geol. Soc. Am. Bull., vol. 101, pp. 635-643.

Manning, D.A.C., 1979 : "An experimental study of the effect of fluorine on the crystallisation of granitic melts, with application natural fluorine-rich granitic rocks". Ph.D. Thesis, University of Manchester.

Miyashiro, A., 1978 : "Nature of alkalic volcanic rock series". Contrib. Mineral. Petrol., Vol. 66, pp. 91-104.

Mocek, B., 2001 : "Geochemical evidence for arc-type volcanism in the Aegean Sea : the blueschist unit of Siphons, Cyclades (Greece)". Lithos, Vol. 57, No. 4, pp. 263-289.

Moore, J. McM., 1975 : "A mechanical interpretation of the vein and dyke systems of the S.W. England Ore field". Mineral : Deposita, Vol. 10, pp. 374-388.

Mukherjee, A.B. and Roy, A., 1981 : "Cooling conditions of the high level Precambrian granites of Siwana : Evidence of experimental melting behaviour and the sodic amphibole reaction relation". Ind. Jour. Earth Sci., Vol. 8, No. 2, pp. 99-108.

Mukherjee, A.B., 1962 : "Pantellaritic assamblages from western Rajasthan". Quart. Jour. Geol. Min. and Metall. Soc. India, Vol. 34, No. 2-3, pp. 149.

Mukherjee, P.K., Gupta, A.K. and Chatterjee, V.P., 1985 : "Viscosity determination of some rhyolites and a pitchstone of variable temperature under atmospheric pressure". Ind. Jour. Earth Sci., Vol. 12, No. 2, pp. 165-172.

Murthy, M.V.N., 1979 : "Some aspects of mineralization associated with acid magmatism in Rajasthan. Proc. Workshop MAWAM Geol. Surv. India Spel. Publ., Vol. 13, pp. 15-24.

Murthy, M.V.N., Siddiqui, H.N. and Venkataraman, P.K., 1961 : "Riebeckite - aegirine rhyolite and granites in the Malani suite of igneous rocks, Western Rajasthan". Ind. Min., Vol. 15, No. 4, pp. 427-428.

Okay, A.I., Tansel, I., Tuysuz, O., 2001 : "Obduction, subduction and collision as reflected in the upper cretaceous - Lower Eocene sedimentary record of Western Turkey". Geol. Maga., Vol. 138, No. 2, pp. 117-142.

Paliwal, B.S. and Rathore, P.S., 2000 : "Neoproterozoic Volcanics and Sedimentaries of Jodhpur : A Reappraisal". In Tectonomagmatism, Geochemistry and Metamorphism of Precambrian terrains. K.C. Gyani and P. Kataria (Eds.) pp. 75-96. MSU Udaipur. 438 p.

Paliwal, B.S., 1992 : "Tectonics of the Post-Aravali mountain building activity and its bearing on the accumulation of sediments along the western flank of the Aravalli Range, Rajasthan, India". In : Geology in south Asia-I Proceedings of GEOSAS-I Islamabad, Pakistan, R. Ahmed and A.M. Seikh (eds.), February 23-27, 1992. Hydrocarbon Development Institute of Pakistan, 1994, pp. 52-60.

Paliwal, B.S., 1998 : "Felsic volcanics interlayered with sediments of the Marwar Supergroup at Chhoti Khatu". Jour. Geol. Soc. India, Vol. 52 pp. 81-86.

Paliwal, B.S., 1999 : "Geology of the region to the west of Aravalli mountain : some observations". In : Geological Evolution of Northeastern India. B.S. Paliwal (ed.) Scientific Publishers (India), Jodhpur, pp. 387-414.

Pandit, M.K., Shekhawat, L.S., Ferreira, V.P., Sial, A.N. and Bohra, S.K., 1999 : "Trondhjemite and granodiorite assemblages from west of Barmer : probable basement for Malani magmatism in western India". Jour. Geol. Soc. India, Vol. 53 pp. 89-96.

Pareek, H.S., 1981(a) : "Petochemistry and Petrogenesis of the Malani Igneous Suite, India". Geol. Soc. America Bull. Vol. 92, Part I, pp. 67-70.

Pareek, H.S., 1981(b) : "Petrochemistry and Petrogenesis of the Malani Igneous Suite, India". Geol. Soc. America Bull., Vol. 92, Part II, pp. 206-273.

Pareek, H.S., 1984 : "Pre-Quaternary geology and mineral resources of north-western Rajasthan". Mem. Geol. Surv. India, Vol. 115, pp. 1-99.

Pascoe, E.H., 1959 : "Manual of the Geology of India and Burma". Mem. Geol. Surv. India, Vol. 2 pp. 485-1343.

Pascoe, E.H., 1960 : "Manual of the geology of India and Burma", Geol. Surv. Ind., Vol. 2, pp. 553-560.

Peccerillo, A., and Taylor, S.R., 1976 : "Geochemistry of some calc-alkaline volcanic rocks from the Kastamonu area, northern Turkey". Contrib. Mineral. Petrol., Vol. 58, pp. 63-81.

Phillips, J.C., Woods, A.W., 2001 : "Bubble plumes generated during recharge of basaltic magma reservoirs". Earth and Planet. Sci. Lett., Vol. 186, No. 2, pp. 297-309.

Phillips, W.J., 1972 : "Hydraulic Fracturing and Mineralisation". Jour. Geol. Soc. Lond., Vol. 128, pp. 337-359.

Phillips, W.J., 1973 : "Mechanical effects of retrograde boiling and its probable importance in the formation of some porphyry ore deposits".Trans. Inst. Min. Metall., Vol. 82(B), pp. 90-98.

Pichavant, M., 1979 : "Etude experimantale a haute temperature et 1 kbar du role du bore dans quelques systemes silicates. Interet Petrologique et metallogenique". Ph.D. Thesis, Institute National Polytechnique de Lorraine, CRPG-ENSG.

Rabinowicz, M., Genthon, P., Ceuleneer, G., Hillairet, M., 2001 : Compaction in a mantle mush with high melt concentrations and the generation of magma chambers". Earth and Planet. Sci. Lett., Vol. 188, No. 3-4, pp. 313-328.

Rathore, S.S. and Venkatesan, T.R., 1999 : "Rb-Sr age of peralkaline granites and associated peralkaline volcanics of Barmer district, Rajasthan". In : Geological evolution of north-western India. B.S. Paliwal (Ed), Scientific Publishers (India), Jodhpur, pp. 61-68.

Rathore, S.S., Venkatesan, T.R. and Srivastava, R.K., 1996 : "Rb-Sr and Ar-Ar systematics of Malani Volcanic rocks of south-western Rajasthan : Evidence for a younger post-Crystallization thermal event". Proc. Ind. Acad. Sci. (Earth Planet Sci.), Vol. 105, pp. 131-141.

Reynolds, D.L., 1954 : "Fluidisation as a geological Process and its bearing on the problem of intrusive granites". Am. Jour. Sci., Vol. 252, pp. 577-614.

Riley, T.R., Leat, P.T., Pankhurst, R.J., Harris, C., 2001 : "Origins of large volume rhyolitic volcanism in the Antarctic Pennisula and Patagonia by crustal melting". Jour. Petrol., Vol. 42, No. 6, pp. 1043-1065.

Riller, U., Petrinovic, I., Ramelow, J., Strecker, M., Oncken, O., 2001 : "Late Cenozoic tectonism, Collapse Caldera and Plateau formation in the Central Andes". Earth and Planet. Sci. Lett., Vol. 188, No. 3-4, pp. 299-311.

Roy, A.B., 1998 : Malani Igneous Suite of rocks (correspondence). Jour. Geol. Soc. India, Vol. 51, No. 4, pp. 556-557.

Roy, A.B., and Jakhar, S.R.; 2002 : Geology of Rajasthan (Northwest India) Precambrian to Recent. Sci. Pub., Jodhpur. 421 p.

Sawkins, F.J., 1969 :"Chemical brecciation, an unrecognized mechanism for breccia formation ?" Econ. Geol., Vol. 64, pp. 613-617.

Scambelluri, M., Philippot, P., 2001 : "Deep fluids in subduction zones". Lithos, Vol. 55, No. 1-4, pp. 213-227.

Seghedi, L., Downes, H., Pecskay, Z., Thirlwall, M.F., Szakacs, A., Prychodko, M., Mattey, D., 2001 : "Magmagenesis in a subduction-related post-collisional volcanic arc segment : the Ukrainian Carpathians". Lithos, Vol. 57, No. 4, pp. 237-262

Sharma, K.K., 2004 : The Neoproterozoic Malani Magmatism of the northwestern Indian Shield : Implications for crust building processes. Proc. Ind. Acad. Sci. (Earth Planet. Sci.) Vol. 113, No. 4, pp. 795-807.

Sheppard, S., Griffin, T.J., Tyler, I.M., Page, R.W., 2001 : "High and Low K granites and adakites at a Palaeoproterozoic plate boundary in northwestern Australia". Jour. Geol. Soc., Vol. 153, No. 3, pp. 547-560.

Shrivastava, K.L., Gaur, V. and Gaur. S., 2000 : "A Search of possible target Loci of Mineralization in Malani Volcanic Rocks, Northwestern India" (Abstract). International Association of volcanology and Chemistry of the Earth's Interior. General Assambly at Bali, Indonesia (July 18-22, 2000).

Sillitoe, R.H., and Sawkins, F.J., 1971 : "Geologic, Mineralogic and Fluid inclusion studies relating to the origin of copper-bearing tourmaline breccia pipes, Chile". Econ. Geol., Vol. 66, pp. 1028-1041.

Smith, P. and Parrons, I., 1975 : "The alkali-feldspar solvus at 1 kilobar water-vapour pressure". Mineralog. Mag. London, Vol. 39, pp. 747-767.

Spath, A., LeRoex, A.P., OpiyoAkech, N., 2001 : "Plume-lithosphere intraction and the origin of continental rift-related alkaline volcanism the chyulu Hills volcanic province, southern Kenya". Jour. Petrol., Vol. 42, No. 4 pp. 765-785.

Srivastava, R.K., 1988 : "Alkaline and Peralkaline rocks of Rajasthan". Mem. Geol. Soc. India., Vol. 15, pp. 3-24.

Srivastava, R.K., Maheshwari, A. and Upadhyaya, R., 1989(a) : "Geochemistry of felsic volcanics from Gurupratap Singh and Diri, Pali district, Rajasthan. (Pt. I, Major elements)" Jour. Geol. Soc. India, Vol. 34, pp. 467-485.

Srivastava, R.K., Maheshwari, A. and Upadhyaya, R., 1989(b) : "Geochemistry of felsic volcanics from Gurupratap Singh and Diri, Pali district, Rajasthan (Pt. II, trace element)". Jour. Geol. Soc. India, Vol. 34, pp. 617-631.

Stemprok, M., 1971 : "Petrochemical features of tin-bearing granites in the Krushe Hory Mts., Czechoslovakia". Soc. Mining Geol. Japan Spec. issue, Vol. 2, pp. 112-118.

Stevenson, R.J., Dingwell, D.B., Bagdassarov, N.S., Manley, C.R., 2001 : "Measurement and implication of "effective" viscosity for rhyolite flowemplacement". Bull. Volcano., Vol. 63, No. 4, pp. 227-237.

Subrahmanyan, N.P. and Rao, G.V.U.; 1972 : Age of the Mundwara Igneous Complex, Rajasthan. Curr. Sci. Vol. 41. pp. 388-391.

Tamic, N., Behrens, H., Holtz, F., 2001 : "The solubility of H2O and CO2 in rhyolitic melts in equilibrium with a mixed CO2-H2O fluid phase". Chem. Geol., Vol. 174, No. 1-3, pp. 333-347.

Tatsumi, Y., 2001 : "Geochemical modeling of partial melting of subducting sediments and subsequent melt-mentle interaction : Generation of high - Mg andesites in the Setouchi volcanic belt, southwest Japan". Geol., Vol. 29, No. 4, pp. 323-326.

Torsvik, T.H., Carter, L.M., Ashwal, L.D., Bhushan, S.K., Pandit, M.K., Jantveit, B., 2001 : "Rodinia refined or obscured palaeomagnetism of the Malani Igneous Suite (NW India)". Precam. Res. Vol. 108, No. 3-4, pp. 319-333.

Tuttle, O.F. and Bowen, N.L., 1958 : "Origin of granite in the light of experimental studies in the system $NaAl_3O_8$-$KAlSi_3O_8$-SiO_2-H_2O". Mem. Geol. Soc. Am. No. 74.

Vallinayagam, G. and Kochhar, N., 1998 : "Geochemical characterization and petrogenesis of A-Type Granites and the associated acid volcanics of Siwana Ring Complex, North Peninsular India". In : The Indian Precambrian, B.S. Paliwal (Ed.), Scientific Publishers (India), Jodhpur, pp. 460-481.

Vallinayagam, G., 1998 : "Rare metal bearing and barren

distinction of alkali granites, Piplun area, Siwana Ring Complex, Rajasthan, India : petrological and geochemical evidence". In : The Indian Precambrian, B.S. Paliwal (Ed.), Scientific Publishers (India), Jodhpur, pp. 492-501.

Vallinayagam, G., 1999 : "Nb, Zr, Y, REE Rich acid dyke rocks from the Piplun area of Siwana ring complex, western Rajasthan, India". In : Geological evolution of north-western India, B.S. Paliwal (Ed.), Scientific Publishers (India), Jodhpur, pp. 94-102.

Vandenkerkhof, A., Thiery, R., 2001 : "Carbonic inclusions". Lithos, Vol. 55, No. 1-4, pp. 49-68.

Venkataraman, P.K., Sensharma, R.M. and Xavier, J., 1968 : "Petrology of some peralkaline granites of Barmer district, Rajasthan". Geol. Surv. India, Misc. Pub., Vol. 9, pp. 57-72.

Verma, S.P., 2001 : "Geochemical and Sr-Nd-Pb isotopic evidence for a combined assimilation and fractional crystallisation process for volcanic rocks from the Huichapan Caldera, Hidalgo, Mexico". Lithos, Vol. 56, No. 2-3, pp. 141-164.

Volpe, A.M., and Macdaugall, J.D.; 1990 : Geochemistry and isotopic characteristic of mafic (Phulad ophiolite) and related rocks in the Delhi Supergroup, Rajasthan, India : Implications for rifting in the Proterozoic. Precamb. Res. Vol. 48. pp. 167-191.

Waight, T.E., Maas, R., and Nicholls, I.A., 2001 : "Geochemical investigations of microgranitoid enclaves in the S-type cowra granodiorite, Lachlan Fold Belt, SE Australia" : Lithos. Vol. 56. pp. 165-186.

Wang, J.H., Yin, A., Harrison, T.M., Grove, M., Zhang, Y.Q., Xie, G.H., 2001 : "A tectonic model for cenozoic igneous activities in the eastern Indo-Asian Collision Zone". Earth and Planet. Sci. Lett., Vol. 188, No. 1-2, pp. 123-133.

Whalen, J.B., Jenner, G.A., Long Staffe, F.J., Robert, F. and Goriepy, C., 1996 : "Geochemical and isotopic (O, Nd, Pb & Sr), constraints on 'A' type granite petrogenesis based on top soils igneous suite, New land Applachains". Jour. Petrol., Vol. 37, pp. 1463-1489.

Zhu, Y.F. Sun, S.H., Gu, L.B., Ogasawara, Y., Jiang, N., Honma, H., 2001 : "Permian Volcanism in the Mongolian Orogenic Zone, North east China". Geol. Mag., Vol. 138, No. 2, pp. 101-115.

Hon'ble Reviewers

1. Dr. A.V. Sankaran
   10, P & T Colony
   I Cross, II Block
   R.T. Nagar

Economic Mineralization
K.L. Shrivastava (Editor)
2009. pp. 139-148
Scientific Publishers (India), Jodhpur

Section-II

Tungsten Mineralization Associated with Felsic Magmatisms in Purbani Region of Kinnaur District, Himachal Higher Himalaya

BRAJESH SINGH AND SANTOSH KUMAR
Department of Geology, Kumaun University, Nainital 263002, Uttranchal, INDIA
Email : bsinghng69@yahoo.co.in

Abstract

The Higher Himalayan Crystallines of Vaikrita Group in Purbani area are mainly represented by Neoproterozoic granite gneiss (GGn), metasediments, and Early Palaeozoic granitoids (EPG), which are intruded by Tertiary tourmaline bearing leucogranite (TLg). High molar $Al_2O_3/CaO+Na_2O+K_2O$ (= 1.1) and low magnetic susceptibility value (< 2×10^{-3} SI) of GGn, EPG and TLg suggest their derivation by anatexis of fusible crustal components forming peraluminous (S-type), ilmenite series (reduced type) granitoid suites. The GGn in and around Purbani region commonly hosts shear zoned skarn rocks, which endure tungsten (scheelite) mineralization. X-ray diffraction patterns of ore mineral samples suggest the presence of two principal ore groups viz. base metal sulphides consisting of chalcopyrite and bornite associated with pyrite and pyrrhotite, and oxide of tungsten (scheelite) ore mineral. Scheelite occurs as anhedral to rhombohedral grains, gray in colour showing distinct anisotropism masked by white internal reflection, and exhibits weak to strong diffraction peaks (2.2360 ? - 3.1617 ?) suggesting lean part of mineralization. The EPG are relatively enriched in tungsten content (av. 47.58 ppm) as compared to GGn (av. 11.30 ppm) and TLg (av. 31.72 ppm) whereas CaO content is higher in skarn rocks (11.90 wt%) as compared to EPG (1.02 wt%), GGn (0.97 wt%) and TLg (0.16 wt%), which indicate that EPG has served likely source for tungsten mineralization. The observed higher content of incompatible elements such as K, Rb, Sr, Ba, and La in most EPG samples as compared to that of GGn and TLg suggests that skarn forming scheelite solutions were most likely derived from EPG melt enriched in boron, fluorine and tungsten.

Key words : Tungsten Mineralization, Scheelite, Skarn, Purbani Region, Higher Himalaya.

INTRODUCTION

The study area forms a part of the Higher Himalayan Crystallines (HHC) of the Vaikrita Group constituted of a variety of gneissic rocks such as granite gneiss (GGn), migmatites and intercalated metasediments, which are spatially associated with Early Palaeozoic granitoids (EPG) and tourmalinebearing leucogranite (TLg) of the Haimanta Group. These lithounits are best exposed along the Satluj river valley and its tributaries in and around Akpa, Ribba and Purbani regions of Kinnaur district in Himachal Pradesh. In and around Purbani the exposed rocks show a rugged topography with broad glaciated valleys and narrow scarp faced ridges. General topographic slope of the area varies between 25-35° being steeper towards the riverbed. The GGn unit near Purbani hosts a shear zoned skarn mineralization. Iyenger (1949) carried out a survey on mineral resources of Kinnaur district. Many workers have described the geology of different parts of Kinnaur basin (e.g. Sharma, 1977, 1983; Kakar, 1988; Srikantia and Bhargava, 1998; Kwatra et al., 1999; Islam and Gururajan, 2003). More recently Singh and Kumar (2005) have discussed petrogenetic appraisal of EPG of Kinnaur region. Kumar and Singh (2005) have studied biotite compositions from granitoids of Kinnaur district and have discussed the oxybarometer and nature of host felsic magmas. Exploration for polymetallic mineralization in the Kinnaur district has been carried out by Geological Survey of India (1991-93). Kumar and Jain (1993) and Sharma (1995) have reported the occurrence of scheelite, pyrite, chalcopyrite, pyrrhotite and bornite in Purbani region of Kinnaur district. Also, tungsten mineralization as scheelite has been reported from the adjoining regions (Dorka and Gupta, 1993; Ruby and Singh, 1994). In this paper mineralogy and geochemistry of scheelite and associated felsic magmatic rocks have been studied in order to assess the nature and source of tungsten mineralization.

GEOLOGY AND FIELD RELATIONS

The country rocks of the area comprise of metamorphosed lithounits (meta-sediments with well developed gneissic character) of the Vaikrita Group (Sharma, 1977; Kumar and Jain, 1993) and are represented by variety of gneisses (psammitic gneiss, banded gneiss, granite gneiss, coarse psammitic gneiss), collectively referred to as granite gneiss (GGn), migmatites, calcareous bands and pegmatites (Fig. 1). Early Palaeozoic granitoids (EPG) have sharp to gradational contacts with the coarse variety of GGn of the Vaikrita Group, north of Ribba village and

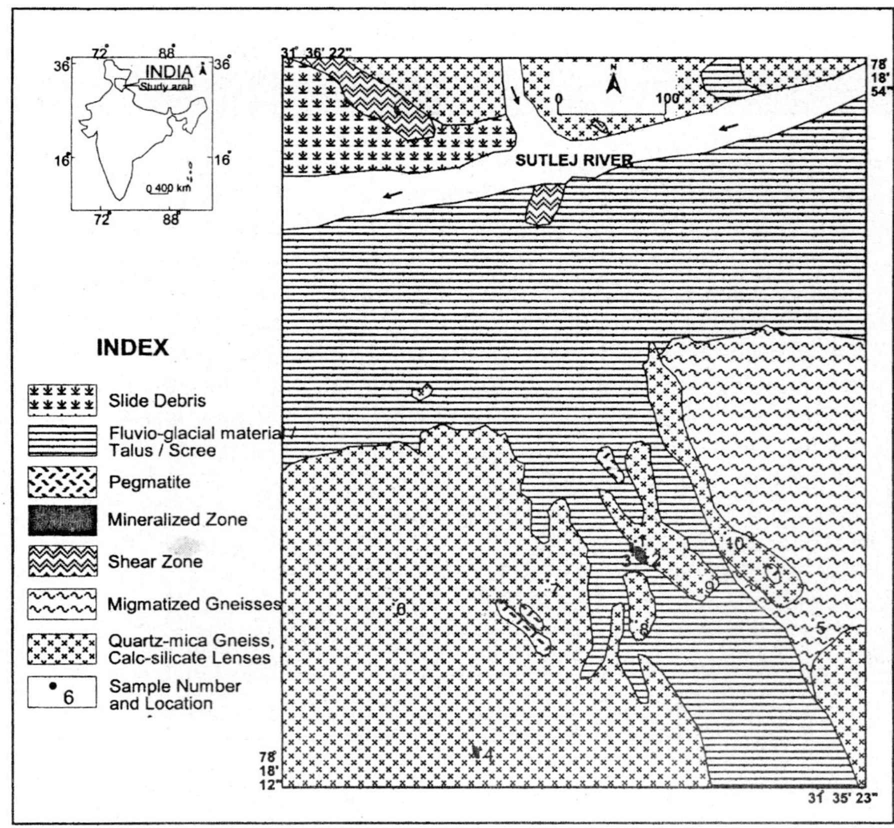

Fig.1 Geological map of the area (after Kumar and Jain, 1993)

between Sangla and Rakcham in the south. The EPG are leucocratic, fine- to coarse-grained and mostly porphyritic two-mica (muscovite and biotite) granites (*sensu stricto*), and show non-foliated, well-developed joints and features of spheroidal weathering. The EPG have compositional and grain size variations, but equigranular variety of EPG is more common. Enclaves (mostly as xenoliths) of variable shapes and sizes are commonly hosted in EPG, and are represented by phyllite, biotite schist, fine-grained to porphyritic, mesocratic biotite granite, occurring more frequently and larger in size near the marginal parts. Siliceous pods and patches were also observed.

Skarns are developed in calcareous bands along the shear zones occurring within the GGn unit. The skarns are fine- to medium-grained, hard, compact and light gray to grayish black. They are mainly calc-silicates (plagioclase, pyroxene, garnet etc.), and act as host for tungsten mineralization. Tungsten mineralization of scheelite is associated with these skarns. Pyrrhotite, pyrite with some chalcopyrite and bornite are also associated with these skarns. Migmatites, as metatexite (palaeosome structure) and diatexite (melt flow structure) similar to those described by Brown (1973) in Sawyer (1996), are associated with the GGn and are localized mainly in the vicinity of shear zone. These migmatites might have formed due to the remobilization of metamorphics and granitoids along the foliation and other planes of weakness of the protoliths.

The tourmaline-bearing leucogranite (TLg) is medium- to coarse-grained rock, generally found intruding the EPG and GGn. The TLg as crudely to un-foliated unit occurs in the form of veins, pockets and lenses within AG. It commonly contains radiating needle-shaped to well developed tourmaline crystals, which at places appear broken and rotated. Pegmatite and quartz veins of various dimensions profusely intrude the AG and the GGn, and are sometimes cross-cutting each other. Few pegmatite veins contain very large grains of quartz and feldspar. These pegmatites vary in size from a meter to several meters in length and 5cm to 30 cm in width. Pegmatites are also observed in close vicinity of the mineralized zone, and a few of them are also traversing it. The pegmatites traversing the mineralized zone are devoid of mineralization (Sharma, 1995).

ANALYTICAL TECHNIQUES AND METHODS

X-ray diffraction (XRD) investigation of ore samples was carried out using fine powder (1-50 μm) and XRD machine (PW 1140/09) at Institute Instrumentation Centre, Indian Institute of Technology, Roorkee. The XRD machine has scanning range of 5° - 90° at operating current of 20mA. Chart speed was fixed at 1cm/min and goniometer speed at 1°/min. The CaO wt % from ore samples was analyzed on pressed powder pellets using XRF (SIEMENS-SRS 3000 M) in Wadia Institute of Himalayan Geology, Dehradun. Tungsten (WO_3 wt%) was analyzed at Indian Bureau of Mines, Nagpur with the help of XRF (PW 1400, PHILIPS make) using pressed pellets mixed with 1ml 5% polyvinyl acetate. The results of XRD measurements, whole rock CaO and WO_3analyses are given in tables 1a,b and 2.

PETROGRAPHY

Granite Gneiss, Early Palaeozoic granitoids and tourmaline-bearing leucogranites

The Ggn exhibits variable textures ranging from schistose, through gneissose to porphyritic textures with large grains of quartz and feldspars with preferred orientations of flaky biotite and muscovite along the schistosity plane. Accessories are represented by garnet, apatite, zircon, and iron oxides. Plagioclase occurs as subhedral to euhedral laths and most of them have corroded margins due to replacement by perthitic microcline. Primary biotite is flaky and yellowish green, whereas the secondary biotite is anhedral and dark brown. Zoned muscovite of secondary origin is also found replacing the plagioclase. K-feldspar is represented by subhedral to anhedral grains of orthoclase and microcline, and varies in size from 1mm to 5mm.

The EPG are fine- to coarse-grained texture and are constituted of quartz, potash feldspars, sodic plagioclase, biotite, muscovite, accessories include tourmaline as coarse grains, apatite and occasional Fe-Ti oxides and zircon. Other minerals as retrograde sericite and rare residual garnet. Size of most minerals varies from 3mm to 10mm. Plagioclase occurs as subhedral to euhedral laths, and some have corroded margins due to replacement by anti perthitic microcline. Primary biotite, as randomly oriented flakes, and secondary biotites occur associated with zoned muscovite. The K-feldspar occurs as large and small grains. The larger ones contain inclusions of biotite, muscovite and plagioclase indicating

Table-1a Sample No. OPL 97/213

2 θ	d (Å)	Visual intensity	Specific Mineral
26.2	3.3984	Vs	Bornite
27.8	3.2063	Vs	Bornite
28.2	3.1617	Vs	Scheelite
28.5	3.1291	Vs	Scheelite
30.3	2.9472	Vs	Pyrrhotite
31.6	2.8289	Vs	Scheelie
33.0	2.7120	Vs	Bornite
33.8	2.6496	Vs	Pyrrhotite
34.4	2.6048	Vs	Scheelite
36.6	2.4531	S	Quartz
39.5	2.2794	S	Quartz
40.3	2.2360	S	Quartz

magmatic growth, and some are perthitic and partially develop microcline twinning; whereas smaller grains of K-feldspar are anhedral to subhedral, perthitic and contain well-developed microcline.

Quartz, K-feldspar, plagioclase, biotite, muscovite, apatite, zircon, tourmaline and rare iron oxides constitute the TLg, and are coarse-grained. Perthite, myrmekite and graphic intergrowths are common. Plagioclase occurs as subhedral laths showing lamellar twinning, and contains inclusions of mica, and is altered to saussuritechlorite at grain boundaries. Biotite and muscovite occur as subhedral to euhedral and subhedral to anhedral flakes respectively. K-feldspar is represented by orthoclase and microcline. Microcline shows welldeveloped, cross-hatched twinning. Tourmaline is zoned and fractured, and shows poikilitic relation with feldspar and quartz.

Skarns and Mineralization

The skarns reveal the presence of ores of scheelite, pyrite, chalcopyrite, pyrrhotite and bornite. Scheelite occurs as subhedral to anhedral grains showing dark gray to gray white colours and exhibits distinct anisotropism masked by white internal reflections. Pyrite has a more-or-less idiomorphic shape and is yellowish white, and found associated with chalcopyrite, which occurs as medium- to coarse-grained anhedral aggregates showing yellow to brass-yellow colours. Pyrrhotite as anhedral granular mass shows a creamy pinkish brown colour and is strongly anisotropic from yellowish gray to grayish blue. Bornite occurs as irregular polycrystalline aggregates often intergrown with chalcopyrite and shows weak anisotropism from pinkish brown to orange.

Quartz, K-feldspar, plagioclase, biotite, muscovite and pyroxenes also constitute parts of skarns. Sphene, apatite, zircon and iron oxides are present as accessories. Pyroxenes as subhedral grains of augite are enclosed in a matrix of quartz and feldspar and may contain inclusions of other minerals. Subhedral plagioclase laths show lamellar twinning and is altered to Saussurite along the grain boundaries Biotite occurs as subhedral flakes and is generally interlocked with muscovite, and also found as inclusions in quartz, feldspar etc. Muscovite is found as subhedral to anhedral grains in close association

Table-1b Sample No. OPL 97/211

2 θ	d (Å)	Visual intensity	Specific Mineral
26.8	3.3237	Vs	Bornite
26.9	3.3115	Vs	Bornite
27.9	3.1951	Vs	Bornite
28.1	3.1728	Vs	Bornite
28.6	3.1184	Vs	Sphalerite
31.6	2.8289	Vs	Scheelite
33.1	2.7040	Vs	Pyrite
36.6	2.4531	S	Quartz
39.5	2.2794	S	Scheelite
40.3	2.2360	S	Scheelite
42.2	2.1396	S	Quartz
42.4	2.1309	S	Quartz

with biotite and feldspar. K-feldspar as subhedral to anhedral orthoclase and is often perthitic. Sphene occurs as pleochroic euhedral rhombic grains associated with opaques, and shows partings. Apatite occurs as inclusions in quartz, feldspars etc., and contains inclusions of tiny zircon crystals. Zircon also occurs as euhedral and zoned crystals included within quartz, feldspar and apatite. Iron oxides generally occur as disseminated grains.

X-RAY DIFFRACTION STUDIES

From the diffraction patterns (Figs. 2a,b) obtained for the ore samples, the position of all the interference peaks were noted with respect to their angle (2θ). The visual intensity of the peaks was expressed by the following abbreviations: Vs Very strong; S- Strong; Ms Medium strong and W Weak. The corresponding interplanar spacings (d 001) with respect to 2θ angle for each of the peaks were determined and such 'd' values were compared with those of the probable minerals given in the ASTM cards (American Standard for Testing and Materials). Consideration of at least three strongest lines has formed the basis for confirming the identity of any mineral. But sometimes when only one or two strongest lines are obtained without having no overlapping values, the identification of that particular mineral can also be confirmed. All the peaks of diffraction patterns have not been considered because of their overlapping nature.

GEOCHEMISTRY

Tungsten deposits are commonly associated with granites and pegmatites. Tungsten mineralization in Purbani area occurs as scheelite ($CaWO_4$) in the skarns hosted in the GGn unit. The geochemistry of scheelite has been studied in terms of CaO (wt%) and W (ppm) contents, and has been compared to those of proximal GGn, EPG and TLg lithounits in order to examine probable genetic relation. From the Table 2 it is observed that the average contents of CaO wt% is higher in the skarn rocks (11.90 wt%) as compared to the EPG (1.02 wt%), Ggn (0.97 wt%) and TLg (0.16 wt%). Average W content of skarns is 204.34ppm, for EPG it is 47.52ppm, for TLg it is 31.72ppm and for GGn it is 10.94ppm. Observed high W contents of EPG and TLg suggest likely genetic relationships of W with these lithounits.

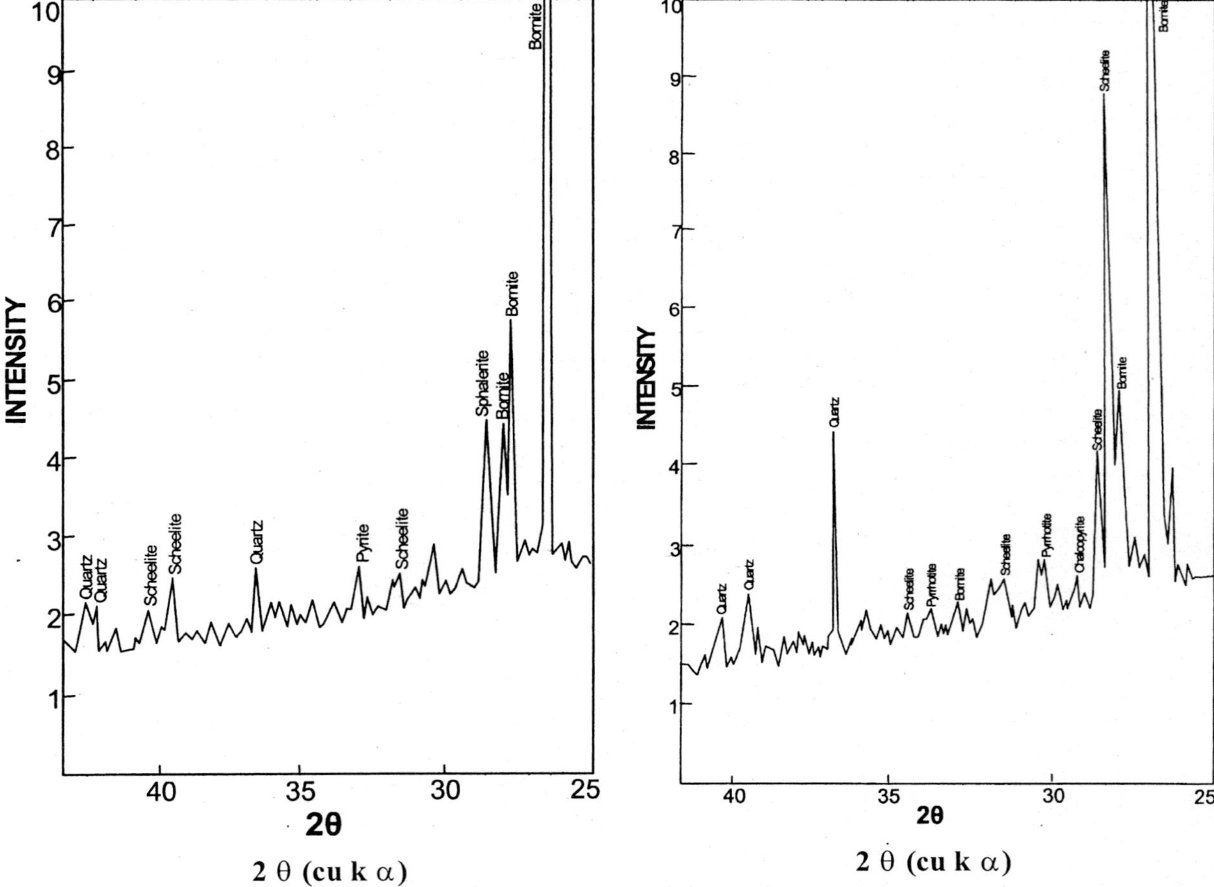

Fig.2a XRD Pattern for sample OPL 97/211

Fig.2b XRD Pattern for sample OPL 97/213

SCHEELITE ENRICHMENT IN SKARNS

Enrichment of scheelite as tungsten ore mineral is associated with the development of calcareous skarns along shear zones intruding granite gneisses under post-magmatic conditions (e.g. Shabynin, 1961; Zharikov, 1968). Calcareous skarns are formed at contacts between granitoids and limestone, the principal skarn minerals being pyroxenes coexisting with garnet. Formation of scheelite in such skarns is largely due to the acid leaching stage (Ushakov, 1955), where scheelite appears more readily among those minerals whose formation was accompanied by the liberation of calcium. Granite-hosted tungsten mineralization is formed at an early post-magmatic stage, and the fluids concerned may have been derived atleast in part by exsolution from magma during the end stage of crystallization (Manning and Henderson, 1984). For W mineralization in the Purbani area, it is therefore proposed that W was concentrated into volatile rich hydrothermal fluids during the late phase of EPG melt differentiation.

The tungsten (W) mineralization in the study area occurs as scheelite in the skarns along shear zone (trending N35°W-S35°E and dipping 60-70° towards NE) developed in calcareous bands within GGn. The Wmineralization occurs sporadically as specks and stringers in the skarns forming lenticular zones. Shear zones hosting skarns are usually reported near contacts between granitoids and limestones formed under post-magmatic conditions (e.g. Korzhinshiy, 1953; Zharikov, 1959; Pertsev, 1974), and skarn forming solutions are essentially derived from granitic magmas (e.g. Shimazaki, 1980), which are generally enriched in incompatible elements (e.g. Swanson et al., 1990), consistent with the observed higher values of K, Rb, Sr, Ba and La in most EPG compared to that of GGn and TLg. Average distribution of W and Nb contents of GGn, EPG and TLg is shown (Fig. 3) and compared to that of granite reported elsewhere (Turekian and Wedepohl, 1961). The W content of GGn (av. 11.30ppm) is substantially lower than that reported for EPG (av. 47.58ppm) and TLg (31.72ppm). However,

Table-2 WO_3 (wt%), CaO (wt%) and W (ppm) contents of skarns, EPG, GGn, and TLg

Sample Number	WO_3 (wt%)	CaO (wt%)	W (ppm)
Skarn rocks			
OPL 97/211	0.0421	9.08	333.90
OPL 97/213	0.0430	9.51	341.03
OPL 99/250	0.0110	14.39	47.24
OPL 99/261	0.0120	14.65	95.17
***Early Palaeozoic granitoids (RG)**			
OPL 97/175	0.0060	1.03	47.58
OPL 97/187	0.0060	1.00	47.58
Granite gneiss (GGn)			
OPL 97/205	0.0010	1.41	7.93
OPL 97/209	0.0020	0.78	15.86
OPL 99/262	0.0013	0.74	9.52
OPL 99/270	0.0014	0.96	11.20
OPL 99/301	0.0017	0.96	10.20
Tourmaline bearing leucogranite (TLg)			
OPL 97/190	0.0040	0.16	31.72

*Kumar and Singh (2005), GGn and TLg samples are from the mineralization area, other than the samples used for petrogenetic studies.

Nb-content of GGn (11.2 ppm) is slightly higher than EPG (6.75 ppm) and TLg (4.5 ppm). It is likely that enrichment of tungsten and other incompatible elements in TLg might have occurred due to remobilization of EPG during the Himalayan orogeny. The genetic relationship of Wmineralization either with EPG or TLg appears enigmatic because both are W-enriched. Since the skarns in the GGn occur near the vicinity of W-enriched (upto 47.58 ppm) EPG, therefore, it is quite likely that magmatic solutions responsible for tungsten mineralization were related to EPG, consistent with the opinion of Sharma (1995). The carbonates of country rocks might have neutralized the acid halides of hydrothermal solutions, exsolved during the crystallization of granitic magma, causing precipitation of W together with other ore elements (e.g. Kelly and Rye, 1979; Plimer, 1987).

DISCUSSION AND CONCLUSION

The Ggn, EPG and TLg are mostly characterized by high molar $Al_2O_3/CaO+Na_2O+K_2O$ (mol A/CNK) and magnetic susceptibility values lesser than 2×10^{-3} SI (Singh and Kumar, 2005). The observed field relations, petrography and geochemical features of these lithounits suggest derivation by anatexis of

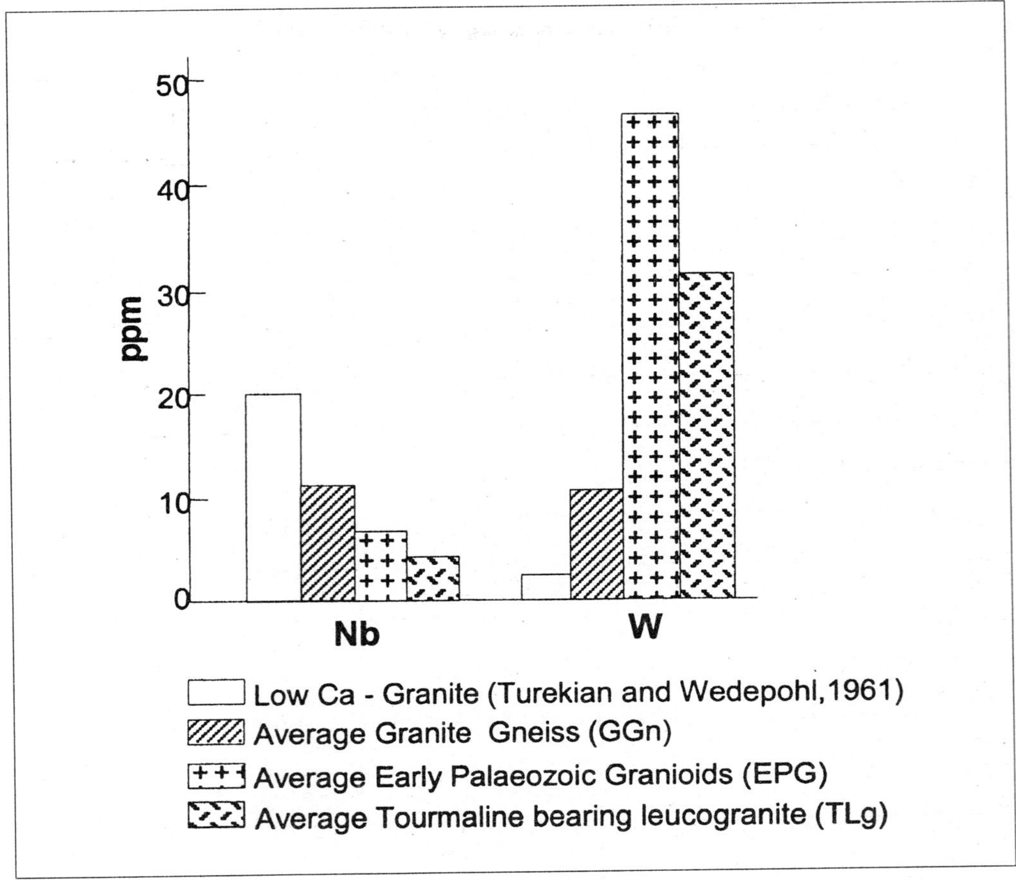

Fig.3 Bar diagrams showing average Nb and W distribution in EPG GGn and TLg compared with lowCa granites

crustal (metasediments) source rocks, evolving to a typical S-type (peraluminous), ilmenite series granitoids (Chappell and White, 1974; Clemens and Wall, 1981; Ishihara, 1977).

The XRD patterns of the ore samples from the Purbani area clearly indicated that there were two groups of ore minerals. One constitutes the base metal sulphides, viz. chalcopyrite and bornite occurring together with pyrite and pyrrhotite belonging to the iron-ore group, whereas the other group comprises the metallic oxide of tungsten, identified as scheelite. Among the iron sulphides sufficient quantity of pyrrhotite is present almost in every samples, which shows a very strong diffraction peak at d= 2.9472Å. Pyrite also exhibits strong diffraction peak at d= 2.7040Å. Copper ore bornite shows a strong diffraction peak at d=3.3984Å. However, with respect to the ore mineral scheelite, the weak and medium interference peaks suggest its occurrence in limited amounts, i.e., lean mineralization. Scheelite exhibits strongest peak at d=3.1617Å.

The relationship of tungsten mineralization to the rocks of various phases and subphases suggests that tungsten mineralization occurred after the formation of concluding phase of felsic magmatism, particularly when residual melt became saturated with volatile components and ore elements. The most probable scene for the generation is exsolution of tungstenbearing fluids from the main mass of the granitic melt when its upper parts were already solidified (e.g. Kelly and Rye, 1979; Plimer, 1987). During the formation of tungstenbearing complex, the felsic magma became excessively more acidic and richer in potassium, volatiles and certain ore elements, particularly fluorine and tungsten (Rub et al., 1970).

Higher values of W in EPG support the concept that EPG served as potential source for W mineralization in the area, although TLg also shows high W value. The average value of W for these lithounits is substantially higher than those recognized for specialized granites associated with acid magmatic rocks (Tischendorf, 1977). From the average distribution of W for these lithounits, it is observed that the W content is higher in EPG as compared to the GGn and TLg. Tauson and Kozlov (1973) suggested that leucogranites derived from water-rich granite melt contain high concentrations of F, Li, Sn and W, granite magmas formed by partial melting of meta-sedimentary rocks are associated with tungsten deposits. Presence of primary tourmaline in granites may also support its derivation from pelitic metasediments. The presence of primary tourmaline in granitic rock is an evidence of boron saturation in the felsic melt (Chorlton, 1973). Field and geochemical features of EPG suggest that source rock responsible for the formation of EPG magma was rich in B, F, W and water vapour leading to the development of water, boron, fluorine and tungsten rich magma. The presence of quartz and pegmatite veins in EPG is indicative of an aqueous phase involved in the formation of EPG melt. For W mineralization in the Purbani area, it is proposed that tungsten (scheelite) was concentrated into volatile-rich hydrothermal fluids during the late phase of EPG melt differentiation.

ACKNOWLEDGEMENTS

We extend our sincere thanks to the Head, Department of Geology, Banaras Hindu University, for providing necessary facilities to carry out the research works. Support from Prof. B.R. Singh (now deceased) is highly acknowledged. The authors thank the Directors, IIC, IIT, Roorkee, WIHG, Dehradun and IBM, Nagpur for providing XRF and ICP-AES analytical facilities respectively. Earlier version of manuscript has been greatly improved by the generous comments of Dr. N.C. Pant and Dr. P.K. Srivastava.

References

Brown, M. 1973. Definition of metatexite, diatexite and migmatite. Proc. Geol. Ass., v.84, pp. 371-382.

Chappell, B.W. and White, A.J.R. 1974. Two contrasting granite types. Pacific Geology, v.8, pp.173-174.

Clemens, J.D. and Wall, V.J. 1981. Origin and crystallization of some peraluminous (S-type) granitic magmas. Canadian Min., v.19, pp.111-131.

Chorlton, L.B. 1973. The effect of boron on phase relations in the granite-water system. M.Sc. Dissertation. McGill University, Montreal, Canada, pp.78-81.

Dorka, B.S. and Gupta, S.K. 1993. Study of granites and pegmatites of Himachal Pradesh and search for tin-tungsten and lithium mineralisation in Jeori-Wangtu Gneissic Complex, Kinnaur District, Himachal Pradesh. Geol. Surv. India Records 126(8), pp.112-114.

Ishihara, S. 1977. The magnetite-series and ilmenite-series granitic rocks. Mining Geology (Tokyo), v.27, pp.293-305.

Islam, R. and Gururajan, N.S. 2003. Geochemistry, petrogenesis and tectonic setting of Akpa-Rakcham granites of Sutlej valley, Himachal Pradesh, India. Him. Geol. v.24(2), pp.63-76.

Iyengar, B.R.C. 1949. Report on mineral investigations in Bushahr State, H.P. (Unpublished report) Geol. Sur. India, pp.1-7.

Kackkar, R.K. 1988. Geology and tectonic setting of Central Crystalline rocks of southern part of Higher Himachal Himalaya. Jour. Geol. Soc. India, v.31, pp.243-250.

Kelly, W.C. and Rye, R.O. 1979. Geologic fluid inclusion and stable isotope studies of the tin-tungsten deposits of Panasqueira, Portugal. Econ. Geol., v.74, pp.1721-1822.

Korzhinskiy, D.S. 1953. Review of metasomatic processes. In: Fundamental problems in the study of magmatogenic ore deposits. Izd-vo AN SSSR, Moscow.

Kumar, Prem and Jain, A.K. (1993) Detailed investigation of the polymetallic mineralisation at Purbani, Kinnaur District, Himachal Pradesh. Geol. Surv. India Records, v.126(8), pp.109-111.

Kwatra, S.K., Sandeep Singh, Singh, V.P., Sharma, R.K., Bimal Rai and Naval Kishor, 1999. Geochemical and geochronological characteristics of the Early Paleozoic granitoids from Sutlej-Baspa Valleys, Himachal Himalayas. In: Jain, A.K., Manickavasagam, R.M. (Eds.), Geodynamics of the NW Himalaya. Gondwana Research Group Memoir Number 6, pp.45-158.

Kumar, Santosh and Singh, B. 2005. Biotite from Neoproterozoic granite gneiss, Cambro-Ordovician granitoids and Neogene leucogranite of Kinnaur district, Himachal Pradesh. In: Saklani, P.S. (Ed.) Himalaya (Geological Aspects), v.3, Satish Serial Publishing House, Delhi, pp.79-99.

Manning, D.A.C. and Henderson, P. 1984. The effect of fluid composition on tungsten partitioning between granitic melt and vapour. Mineral. Soc. Bull., v. 61, 2 (Abs.).

Pertsev, N.N. 1974. Skarns as magmatic and postmagmatic formations. Internat. Geol. Rev., v.16(5), pp.572-582.

PLIMER, I.R. 1987. Fundamental parameters for the formation of granite related tin deposits. Geol. Rund., v.76, pp.23-40.

Rub, M.G., Toksubayeva, G.P. and Chernov, B.S. 1970. Composiotion and origin of a tungsten-bearing magmatic complex in an area of Soviet Maritime Province. Internat. Geol. Rev., v.12(3), pp.313-326.

Ruby, S. Balvinder and Singh, J. 1994. Detailed investigation for polymetallic mineralisation at Purbani, Kinnaur District, Himachal Pradesh. Geol. Surv. India

Rec., v.127(8), pp.88-90.

Sawyer, E.W. 1996. Melt segregation and magma flow in migmatites: Implications for the generation of granite magmas. Trans. Royal Soc. Edinburg, Earth Science, v.87, pp.85-94.

Shabynin, L.I. 1961. Some features of formation of ore-bearing skarns at dolomite contacts. Geol. Rudnik. Mestorozh., v.1, pp.3-18.

Sharma, K.K. 1977. A contribution to the geology of Satluj valley, Kinnaur, Himachal Pradesh, India. In: Colloq. Inter. CNRS: Himalaya Science de la Terre 268, CNRS, Paris, pp.369-378.

Sharma, K.K. 1983. Granitoid belts of the Himalayas. In: Shams, F.A. (Ed.), Granites of Himalayas, Karakorum and Hindu Kush, Punjab University, Lahore, pp.341-354.

Sharma, V.K. 1995. Tungsten mineralisation in parts of Himachal Pradesh. Symposium on recent advances in geological studies on the NW Himalayas and the foredeep. Geological Survey of India, Lucknow, 21-23 Feb.'95, pp.218-219.

Shimazaki, H. 1980. Characteristics of skarn deposits and related acid magmatism in Japan. Econ. Geol., v.75, pp.173-183.

Singh, B. and Kumar Santosh 2005. Petrogenetic appraisal of Early Palaeozoic granitoids of Kinnaur district, Higher Himachal Himalaya, India. Gond. Res., v. 8 (1), pp.67-76.

Srikantia, S.V. and Bhargava, O.N. (1998) Geology of Himachal Pradesh. Geological Society of India, Text Book Series 9, Bangalore, pp.1-406.

Swanson, S.E., Newberry, R.J., Coulter, G.A. and Dyehouse, T.M. 1990. Mineralogical variations as a guide to the petrogenesis of the tin granites and related skarns, Seward Peninsula, Alaska. Geol. Soc. America Special Pap. 246, pp.143-159.

Tauson, L.V. and Kozlov, V.D. 1973. Distribution, function and ratios of trace element concentrations as estimators of ore-bearing potential of granites. Geochemical Exploration, 1972, Institute of Mineralogy and Metallurgy, London, pp.37-44.

Turekian, K.K. and Wedepohl, K.H. 1961. Distribution of elements in some major units of the earths' crust. Geol. Soc. America Bull. 72, pp.175-192.

Ushakov, N.D. 1955. Comments on Kh. M. Abdullayev's book: genetic relation between mineralization and granitoid intrusions. AN SSSR Izvestiya, Ser. Geol. 3.

Zharikov, V.A. 1959. Geology and metasomatic phenomena of the skarn-polymetallic deposits of west Karamazar. IGEM AN SSSR Trudy, 14.

Zharikov, V.A. 1968. Physico-chemical petrology of skarn deposits. Doct. Diss., IGEM AN SSSR.

Hon'ble Reviewers

1. Prof. O.P. Goel
   Emeritus Professor
   Department of Earth Sciences
   Manipur University
   Canchipur
   IMPHAL 795 003 MANIPUR

2. Prof. S.K. Lunkad
   Founder Chairman
   Department of Geology
   Kurukshetra University
   KURUKSHETRA 136 119 HARYANA

Editorial Assistance

Dr. Narendra Singh

Economic Mineralization
K.L. Shrivastava (Editor)
2009. pp.149-154
Scientific Publishers (India), Jodhpur

12

Section-II

Huebnerite/Ferberite (H/F) ratio and Depositional Conditions for Wolframite : An Overview

SUKH CHAIN AND P.K. SRIVASTAVA
Department of Geology, University of Jammu, Jammu
Email : sharmageol_ju@rediffmail.com

Abstract

The Wolframite (Mn, Fe) WO_4 with its two end members huebnerite and ferbirite shows a great variations in MnO/FeO co-efficient on a regional as well as within a single crystal. Various techniques i.e. EPMA, X-ray analysis, isotopic and fluid inclusions study have been applied to know factors like temperature, pressure, oxygen fugacity, ph condition responsible for such chemical variations. The findings of the various workers from the different regions have different views regarding the chemical environments conducive for the chemical zoning within the wolframite. The variation of the Mn/Fe ratios on the local and regional scale have been to the effect of temperature, pH, MnO/FeO ratio of the mineralizing solution and whole rock composition of the source material. The case studies from the wolframite deposits of the different regions reveal that the changing temperature of formation has the direct effect on MnO/FeO ratio of the wolframites.

Keywords : Wolframite, Huebnerite, Ferberite, Huebnerite/Ferberite ratio, Hydrothermal.

INTRODUCTION

Wolframite (Mn, Fe) WO_4 forms solid solution series with huebnerite ($MnWO_4$) and ferberite ($FeWO_4$) two end members. A great variation in Mn/Fe ratio has been observed within the single crystal as well as from the different lodes of the particular tungsten deposit. In general, the different ore and gangue minerals are deposited as the chemistry of an ore solution evolves along its path with the changes in parameters like pressure, temperature, pH, oxygen fugacity, depth of ore formation and chemical exchange with wall rocks. Similarly, the minerals which form the solid solution series also vary in response to the changing environmental conditions in which they are formed. These changes in the environmental condition resulted in the compositional zoning of the many minerals which form the solid solution series. Such changes in the chemical zonation of the wolframite have been encountered in many deposits at the regional as well as local scale. The physico-chemical parameters responsible for such types of zoning are determined by using the various modern techniques. The temperature variation within the vein systems of the same deposits has been determined with the help of fluid inclusion studies. The studies have also been attempted to correlate solid solution zoning with bulk metal ratio zoning.

The concept of compositional variation in wolframite with the depositional condition gains general acceptance in Germany, the use of caution become apparent when Boulden (1954) in Taylor and Hosking (1970) examine the tungsten deposits in Vogt land, he discovered an H/F coefficient which decreased in depth, indicating unusual genesis with the temperature of deposition decreases with increase in depth. With the development of X-ray analysis technique (1959) for obtaining the H/F ratio (Starke, 1959) as the smaller samples of the individual crystals were able to analyze studies of isotopic received some momentum.

The purpose of this paper is to give an overview of the compositional variations of wolframite with the depositional conditions by using the various parameters.

HUEBNERITE/FERBERITE RATIO

Concept of Regional compositional variation

Wolframite (Mn, Fe) WO_4 often shows a great variations in its Mn/Fe ratio in a single ore deposit. The studies of the many workers revealed that the huebnerite is found mainly in the higher temperature parts of the wolframite veins and Mn/Fe ratio increases with the increase in the temperature of formation (Oelsner, 1952; Leutwein, 1952). The regional as well as local variation in the wolframite composition has been suggested by the number of workers (Soeda et al., 1979; Taylor and Hasking, 1970). Sasaki (1961), Barabanov (1971), Groves and Baker (1972) and Soeda et al. (1979) suggested that the Mn/Fe ratio of wolframite at regional scale is also effected by the composition of source material and surrounding rocks. Grovers and Baker (1972) have reported that the regional variation of wolframite composition in Tasmania tungsten deposit is controlled by the Fe/Mn ratio of the mineralizing solutions. Hasking and Palkinghorne, (1954) carried out a study in Cornwall, where they examined a suits of wolframite from various geological environments. The wolframites from the pegmatite were poorest in Fe, whereas in general those from the normal hypothermal lodes were the most Fe- rich, whereas wolframites from greisen-bordered veins tended to occupy an intermediate position. The perusal of the literature on the compositional variation of the wolframite suggest that their exists compositional differences in the wolframites with the depositional temperature. These observations are based on few assumptions :

(1) The temperature of hydrothermal fluids decreases away from the sources.

(2) Pegmatite veins formed at higher temperatures than the normal hydrothermal veins.

(3) Outermost zone of a single crystal is formed relatively at lower temperature.

(4) Within an individual ore body higher temperature prevails at greater depth. With few exception based on the above assumption indicated a general decrease in MnO/FeO ratio in wolframite away from the source rock and high values of MnO/FeO in wolframite from pegmatite.

The regional and local scale variations in the composition of wolframite series associated with granitoids in the inner zone of the tungsten deposit of NW Japan was investigated using X-ray diffraction and microprobe analysis. While studying the depositional condition of the wolframite in the area, Tsusue and Ishihara (1974) suggested that the magnetite series of granitic magmas were more oxidized than those of the illeminite series of the granitic magmas in the area. Such regional differences in the oxidation state may have an influence on the wolframite composition. However, the thermodynamic calculation made by Shibue, (1982) suggested that the relatively Mn-rich wolframites associated with the magnetite series may be explained by the relatively high oxidation state of the related magmas. They concluded that the regional variation in the wolframite series may also be affected by the differences in oxidation state of the related magmas.

Mn/Fe ratio of wolframite series may vary according to the depositional conditions in particular with temperature was pointed out by number of earlier workers especially from Germany (Leutwein, 1952; Oelsner, 1944; 1952, 1954; Boldman, 1958) who studied the huebnerite/feberite ratios (H/F coefficients) of wolframites from various source led them to conclude that huebnerite ($MnWO_4$) was deposited at higher temperature than ferberite ($FeWO_4$). Lower the coefficient lowers the temperature of deposition.

Belgium and French workers (Magnee, 1961; Varlamoff, 1959) have noted a related phenomenon at Ruanda-Urundi, where from the deepest zones of mineralization in granite, towards the exterior zones there is a progressive decrease in MnO content, corresponding to lower temperature of formation.

Concept of Local compositional variation

The local variation in the Mn/Fe ratio of the wolframite may occur within the single vein system as well as within a crystal. The tendencies of the wolframite zoning as found in natural occurrences depends upon the various functions. Many of the parameters will define adequate conditions of wolframite depositions and zonation, the pH value, W concentration and oxygen fugacities of the solution will control the stability and the occurrences of wolframite. The reason for the compositional zoning with in the single wolframite has been a controversial subject for many years.

Regarding the empirical concept that the Mn/Fe ratios of the wolframite series may vary according to the depositional conditions, many geologists initially argued that the high Mn/Fe ratio indicate high temperature of deposition, whereas other workers (Moore and Howie, 1978) believe low Mn/Fe ratio as indicator of a high temperature of wolframite deposition. Most of the recent workers have determined the chemical composition of the wolframite with the help of electron microprobe analyzer and found that the wolframite crystals are often chemically zoned (Moore and Howie, 1978; Nakashima et al., 1986 and Campbell and Peterson, 1988).

The contribution from America (Berman and Capbell, 1957), England (Hosking and Palkughjorne, 1954) and Australia (Edwards and Lyon, 1957; Lawerence, 1962) are very meager regarding the depositional condition of the wolframite. However, the European workers have considered MnO/FeO coefficient of wolframite as indicative of their temperature of deposition. Where as, the result from the USSR indicates exactly the reverse and this anomaly between the workers of the two regions is drawn to general attention. Srivastava and Sukhchain (2006) studied the wolframite composition within the single crystal as well as from the different lodes of the Degana tungsten deposit. The study revealed that there was not much difference in the composition of the wolframite from the different lodes whereas, significant single crystal variation occurs from the wolframite crystal. The chemical variation within the single crystal of the Degana tungsten deposit is shown in Fig. 1.

DISCUSSION AND CONCLUSION

The workers from the different regions differ in their opinion which does not reveals any mutual specific conclusion regarding the compositional variations of wolframite at the regional as well as local scale. The compositional variation of the wolframite series can not be used to evaluate physico-chemical variables such as oxygen fugacity, sulfur fugacity, pressure and temperature during ore formation (Hsu, 1976). On comparing the depositional conditions of ferberite with huebnerite, Horner (1979) emphasized the effects of pH and temperature of ore forming solutions on the wolframite compositions and suggested that the variation of oxygen fugacity was of secondary importance compared to the temperature and pH values. The variation of MnO/FeO and FeO/MnO ratio with changing depositional environments according to depth has been used by the number of

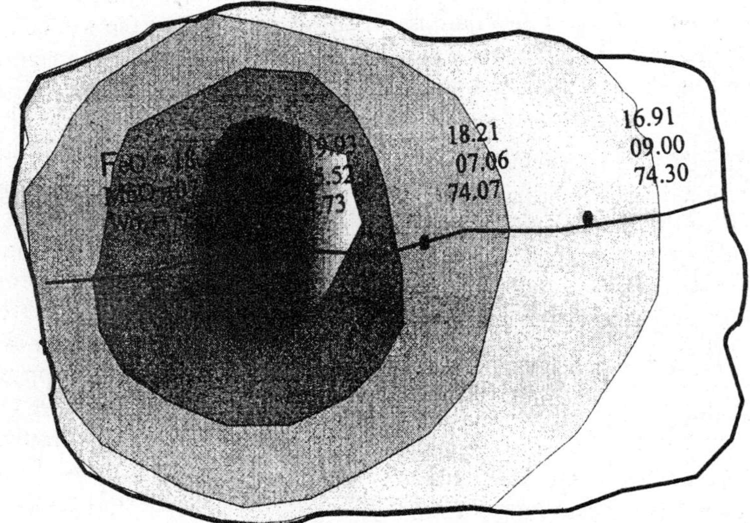

Fig. 1 Compositional variation in a single wolframite crystal of Degana tungsten deposit

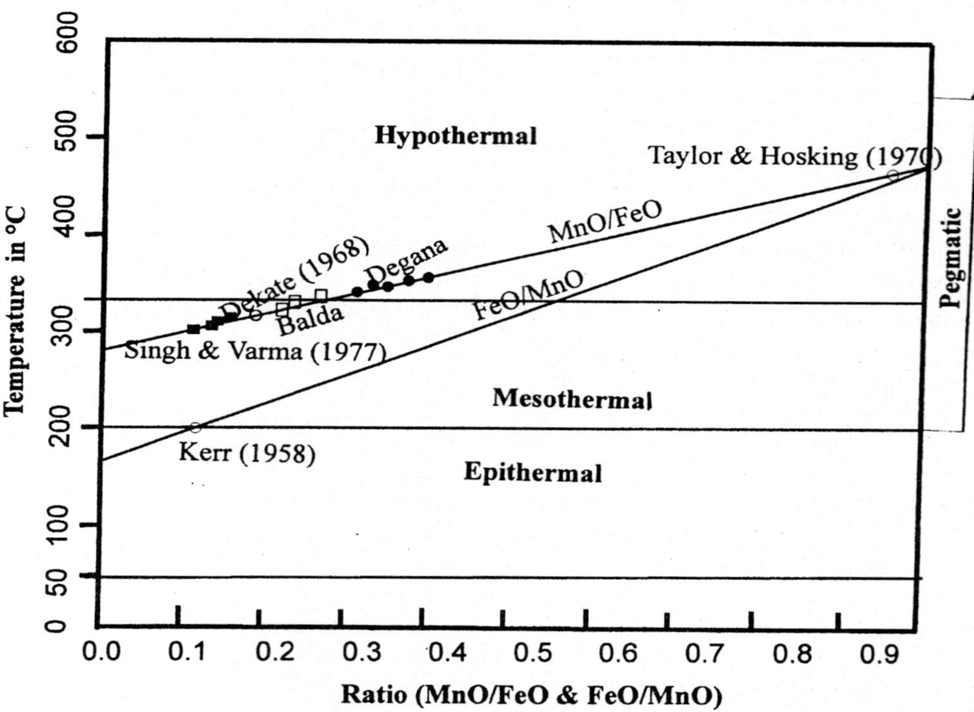

Fig. 2 Plot of MnO/FeO and FeO/MnO ratios vs. depositional temperatures of the wolframite series from different areas

workers (Dekate, 1968; Singh and Varma, 1974). The MnO/FeO and FeO/MnO has bee plotted against depositional temperature (Fig. 2) and the data has been used to draw the relationship between solute/solvent ratios and temperature of formation. Dekate (1968) observed that Ferberite from Agargaon is having low MnO/FeO ratio of 0.18 has deposited at 310° C temperature. The wolframite from South Crofty Mine shows MnO/FeO ratio of 0.91 deposited at higher temperature 450° C. At Degana the MnO/FeO ratio is 0.40 which on plotting this data against the depositional temperatures it shows the temperature range of 375° C.

To know the compositional variations in wolframites from different lodes as well as within the single wolframite crystal of the Degana tungsten the samples were runs for the EPMA analysis. The data suggest that the H/F ratio was very low and different lodes do not shows any much variation in Mn/Fe ratios, whereas within the single crystals compositional zoning was found and significant variation in the Mn/Fe ratios is present. This study reveals that the Fe-rich wolframites are formed relatively at lower temperature. In the wolframite from Balda tungsten deposit H/F ratios ranges from 0.060.15 at temperature of 350-400° C.

Srivastava and Sukhchain, (2006) have reported a oxygen fugacity between the range of 10^{-25} to 10^{-35} bars, and the pH value between 5.0 and 6.2 for the Degana tungsten which is well within the pH range of 5 to 7 for the crystallization of Fe-rich wolframite from magmatic hydrothermal solution as suggested by Shalton (1984) and So et al., (1981). High pH is required for the deposition of Mn rich wolframite. The whole rock composition from the Degana tungsten deposit suggest that the Mn content was very low in the initial crystallizing magma from which the hydrothermal fluid was separated which was responsible for the formation of tungsten at Degana. Moreover, the temperature of tungsten deposition at Degana is suggested to be ranging from to 335 to 487° C by the fluid inclusion studies (Sukhchain, 2004). This reveals that the composition of the host rock and high temperature of the wolframite formation are the main controlling factors for the ferbiritic nature of the wolframite. This ratio increase from core to rim of the crystals. This view is also supported by Shibu (1982) and Schrocke et al., (1984). The change of composition of wolframite from high FeO to low FeO may be attributed to the fluctuation in the pH conditions. Shibu (1982) suggested that the Fe-rich wolframite are formed under relatively reducing conditions and are stable under strongly acidic to weakly alkaline conditions. As pyrite and pyrrhotite are present with the wolframite in Degana, Shibu's observations holds true for the Degana wolframite also.

References

Barabanov, V.F., 1971. Geochemistry of tungsten. Intn. Geol. Rev., v. 13, pp. 332-344.

Berman, T., and Campbell, W.J., 1957. Relationships of composition to thermal stability in the Huebnerite-Ferberite series of tngstates : U. S. Bur. Mines. Rept. Invest., v. 53, 14 pp.

Boldman, H. 1958. uenetische Untersuchung der Wolframitlagerstatte Peehtelsgrun (Vogtl.) unterbesonderer Berucksichtigung der Verteilung des H/F = Koeffizienten and der Spurenalemente Niob und Tantal im Wolframite : Freiberg Forschungsh, C. 10, pp. 46-61. (English abstr. Annot. Bib. Econ. Geol., XXXI (1), 32, No. 224).

Campbell, A. and Peterson, U. 1988. Chemical zoning in wolframite from San Cristobal, Peru. Mineral. Deposita. V. 23, pp. 132-137.

Dekate, Y.G. 1968. Geology and Genesis of Agargaon tungsten deposit, Nagpur district Maharashtra. Mineral Research. V. 1, pp. 9-16.

Edwards, A.B. and Lyon, P.J. 1957. Mineralization at Aberfoyle Tin mine, Rossarden, Tasmania : Proc. A.I.M.M., v. 181, pp. 93-145.

Groves, D.I. and Baker, W.E. 1972. The regional variation in composition of wolframites from Tasmania. Econ. Geol., v. 67, pp. 362-368.

Horner, C. 1979. Solubility and hydrolysis of $FeWO_4$ and $MnWO_4$ in the 250-300 C ranges and zonation of wolframite. Chem. Geol., v. 27, pp. 85-97.

Hosking, K.F.G. and Palkughjorne, J.P.R., 1954. The significance of the variable iron content of the Wolframites of the west of England : Trans. R. Geol. Soc. Corn., v. 18.

Hsu, L.C. 1976. The stability relationships of the wolframite series. Am. Mineral, v. 61, pp. 944-955.

Kerr, P.F., 1958. Tungsten mineralization in the United States. Mem. Geol. Soc. Am., v. 15, 241 p.

Lawerence, L.J., 1962. Crystal habit of wolframiteas an indication of relative temperature of formation : Neues jahrbuch fur Mineralogie, Stuttgart, 1962. Monatshefte : pp. 241-247.

Leutwein, F., 1952. Die chemische Zusammensetzung der wolframite und ihre lagerstattenkundliche Bedeutung. Acta Geol. Akad. Sci. Hungary, v. 1, pp. 133-141.

Magnee, I.De., 1961. Contribution a la connaissance du tungsten-belt ruandais : Acad. Roy. Sci. Ontre-Mer. Cl. Sci. nat., Mem. In 8, n.s.t. 11.f.7, p. 56.

Morre, F. and Howie, R.A. 1978. On the application of the Hubnerite : Ferberite ratio as geothermometer. Mineral. Deposita, v. 13, pp. 391-399.

Nakashima, K., Watanabe, M. and Soeda, A. 1986. Regional and local variations in the composition of the wolframite series from NW Japan and possible factors controlling compositional variations. Mineral. Deposita. V. 21, pp. 200-206.

Olsner, O. 1944. Uber Erzgebirgische wolframite. Ber. Freib. Geol. Ges., v. 20, pp. 44-49.

Olsner, O. 1952. Die pegmatitisch-pneumatolytischen Lagerstatten, des Erzgebirges mit Husriahme der Kontaklagerstatten : Freib. Forschungsh, C. 10, pp. 1-80.

Olsner, O. 1954. Bemerkungen uber die Anwendbarkeit des H/F Koefizienten zur Dendung der Genese von Wolframitten : Freib. Forschungsgh, C.10, pp. 62-67.

Sasaki, A., 1961. Some problem on the wolframite series. Bull. Geol. Surv. Japan. V. 12, pp. 89 (Abstr.).

Schrocke, H.H., Trumm, A. and Hochleitner, R. 1984. Uber den transport von wolfram und den absatz von wolfram-doppel-oxiden in fluiden Wabrigen Losungen. Geochem. Cosmochim. Acta., v. 48, pp. 1791.

Shelton, K.L. 1984. Interpretation of characteristic mineral paragenesis in the tungsten molybdenum quartz vein type deposits : Neues Jahrbuch fur Mineralogue Monatssheft., v. 84, pp. 455-460.

Shibue, Y. 1982. Applications of thermodynamic data of the stability fields of ferberite, huebnerite, and ferric tungstate. N. Jb. Miner. Mh. V. 3, pp. 125-132.

Singh, V.K. and Varma, O.P. 1977. Iron-Manganese ratios in wolframites as geological thermometers. J. Geol. Soc., Ind., v. 8 (3), pp. 125-128.

So, C.S., Shelton, K.L., Chi, S.J. and Yun, S.T., 1981. Geochemical studies of the Gyeongehang W-Mo mine, republic of Korea : Progressive meteoric water inundation of a magmatic hydrothermal system; Econ. Geol., v. 86, pp. 750-767.

Starke, R., 1959. Rontgenographische Bestimmung des Hubnerite-Ferberite-Verhaltnisses in wolframiten : Bergakademie. V. 11, pp. 23-25.

Soeda, A., Takeno, Watanabe, M., 1979. Mineralogical study on the wolframite series from the Chugoku district, Southwest Japan (I). Relationships between lattice parameters and chemical compositions. J. Japan. Assoc. Min. Petr., Econ. Geol., v. 74, pp. 357-375.

Srivastava, P.K. and Sukhchain (In Press) Thermodynamic condition for the tungsten deposition at Degana: Evidence from the compositional variation of Wolframite. Ind. J. Geochemistry.

Sukh Chain, 2004. Metallogenetic modeling of the Degana tungsten deposit, Nagaur district, Rajasthan (India). Ph.D. thesis (Unpub.), University of Jammu, 149 p.

Taylor, R.G. and Hosking, K.F.G. 1970. Manganese, iron ratios in wolframites, South Crofty Mine, Cornwall, Econ. Geol., v. 65, pp. 47-53.

Tsusue, A. Ishihara, S. 1974. The iron-titanium oxides in the granitic rocks of Southwest Japan. Min. Geol., v. 24, pp. 13-30.

Varlamoff, N. 1959. Les gisements de tungstene au Congo belge et au Ruanda-Urundi, materiaux pour Petude de leur geologie et de leur classification : n.s., T.8.F2, : 73.

Hon'ble Reviewers

1. Prof. S.P. Singh
   (Former, Vice Chancellor
   VKS University, Ara)
   IISM Campus, Pundag
   RANCHI 834 004 JHARKHAND

2. Mr. Prakhar Kumar
   Scientist 'G'
   Atomic Mineral Directorate, DAE
   Pratap Nagar, Sector V Ext.
   Sanganer
   JAIPUR 302 030

Editorial Assistance

Ms. Veena Chaudhary

Economic Mineralization
K.L. Shrivastava (Editor)
2009. pp.155-165
Scientific Publishers (India), Jodhpur

Section-II

Tungsten Metallogeny of the Sirohi Group, Rajasthan

KAMAL KANT SHARMA
Department of Geology, Government Post Graduate College, Sirohi 307 001 Rajasthan
Email : sharmasirohi@yahoo.com

Abstract

The Sirohi basin developed during Neoproterozoic time along Sirohi-Sewaria-Ras-Degana linear belt in the western Rajasthan. The sedimentation took place in the ensialic basin, which began with the deposition of fine argillites followed by carbonates and carbonaceous shale. The completion of the orogenic event is marked by the intrusion of acid magmatism at Balda, Nana, Bera, Sewaria and Degana with tungsten mineralization in the Aravalli craton. The wolframite is concentrated in the recycled Neoproterozoic crust of Aravalli craton. The geological setting indicates that tectonic features and host rock chemistry control the tungsten mineralization. The basement rocks of this region are granitoids and granitic gneisses, which show tectonised contact with the cover rocks. The mineralization is mainly associated with quartz and greisenised veins situated along the shear zones. The late magmatic fluids boron, fluorine and lithium caused the mobility and concentration of wolframite along the shear zones.

Keywords : Tungsten Metallogeny, Neoproterozoic,Granite, Sirohi Group, Mineralization

IINTRODUCTION

The Neoproterozoic acid magmatism in the western Rajasthan contributed tungsten mineralization in the Aravalli Mountain belt. After end of Delhi orogenic cycle, the Sirohi group of rocks deposited Sirohi-Sewaria-Ras-Degana linear basin in western Rajasthan (Fig. 1) and the closing of the Sirohi cycle marked with intrusion of granites rich in W, Sn, Li, B and F (Table 1). The mineralization took place along shear zones and fractures. The wolfram metallogeny took place at Balda, Sewariya and Degana region. All these place located along a linear zone can be named as Tungsten Line. All the deposits exhibit similarity in the wall rock alteration, mineralization, and geochemistry. There is total absence of wolfram mineralization in the Aravalli and Delhi rocks. The recycling of the Aravalli crust in western Rajasthan caused concentration of wolfram in the basement rocks. The present contribution describes the geology and tectonic setting of the Sirohi Group type area, and associated tungsten mineralization.

GEOLOGICAL SETTING

The Banded gneissic complex (BGC) includes the oldest supracrustal rocks of Archaean age in the western Indian shield. Several mobile belts viz. Aravalli, Delhi and Sirohi mobile belts traverse this region. The beginning of the Neoproterozoic is marked by the intrusion of gabbro-diorite at Ranakpur (Volpe and Macdougall, 1990). This magmatic event is part of the development of a narrow linear basin from Ras-Degana-Sewaria to Sirohi, named as the Sirohi basin (Roy and Sharma, 1999). The Sirohi

Table-1 Startigraphic succession of rocks present in the Sirohi region (After Sharma, 1996)

Malani Magmatism	Hypabyssal intrusion	Dhanta Porphyry and dolerite Dykes
	Plutonism	Mirpur Granite
	Volcanism and associated sedimentary units	Angor conglomerate Arkose Volcanic silicic tuff Angor rhyolite Pamta rhyolite (Dellenite) Basalt and pyroclasts Sindreth conglomerate
Angular Unconformity		
Sirohi Group	Balda granite Carbonaceous phyllite (local) Calc-silicates and marble Mica-schist	
Basement Rocks	Granitoids, granitic-gneiss, and meta-gabbro	

Group contains rocks of fine shale-carbonate facies. There is a total absence of rocks with larger clasts (conglomerate, arkose and quartz arenite). The inversion tectonics took place under compressional regime at the end of the orogenic cycle, which resulted in the development of strong penetrative deformation and intrusion of younger leucocratic granite in the axial zone of the Sirohi-Ras basin. The development of andualsite, garnet and biotite as hornfelsic minerals in the mica schist around granite indicates a perfect thermal metamorphic aureole. The acid magmatism resulted in a wolframite metallogeny deposit in the Sirohi Group of rocks. The tungsten metallogeny and one phase of regional deformation of green schist facies make it different from the metasediments of Delhi mobile belt and unmetamorphosed Sindreth-Punagarh rocks (Sharma, 1996).

The basement rocks in this region are granitoids, and other gneisses. The cover rocks represents fine argillites. The mineralizing granites are well-foliated medium grained in nature characterized by the presence of K-feldspar, quartz and muscovite. The intruding granite exhibits wall rock alteration; boron metasomatism and intrusive relationship with cover rocks. Tungsten, Lithium and Boron concentrated in the magma after melting of much-evolved Neoproterozoic crust in the western Rajasthan. Boron, Lithium along with fluorine caused high mobility of the wolframite, which enriched in the late magmatic residual fluids. These fluids injected along shear zones, fracture and schistosity plane of the host rock in form of quartz, pegmatite, greisen and mineral veins. Besides wolframite, high fluorine content in residual fluids gave rise to fluorite deposits at Balda. The cover rock chemistry played a significant role for the localization of the wolframite. The wolframite deposits occur in the phyllite, mica schist host rock, whereas scheelite developed in carbonate rocks.

The Degana wolframite deposit is associated with high silica magmatism emplaced during the Neoproterozoic time in a post tectonic setting. The granites are homogeneous in composition and exhibits textural heterogeneity from coarse-grained hypidiomorphic to fine-grained porphyritic to hypabyssal granite porphyry. The region is affected by hydrothermal and pneumatolytic activities. The differentiation of melts resulted enrichment of Li, Rb, and W. concentration (Chattopadhyay et al., 1994). The Vein type tungsten mineralization is genetically associated with the Degana Granite. This is characterized by wall rock alteration around the mineralized veins. Three different alteration zones, greisen, silicification and potassic zones are present at Degana. Plagioclases, alkali feldspar, K-feldspar, quartz, zinnwaldite, muscovite, topaz and fluorite minerals are present along with wolframite

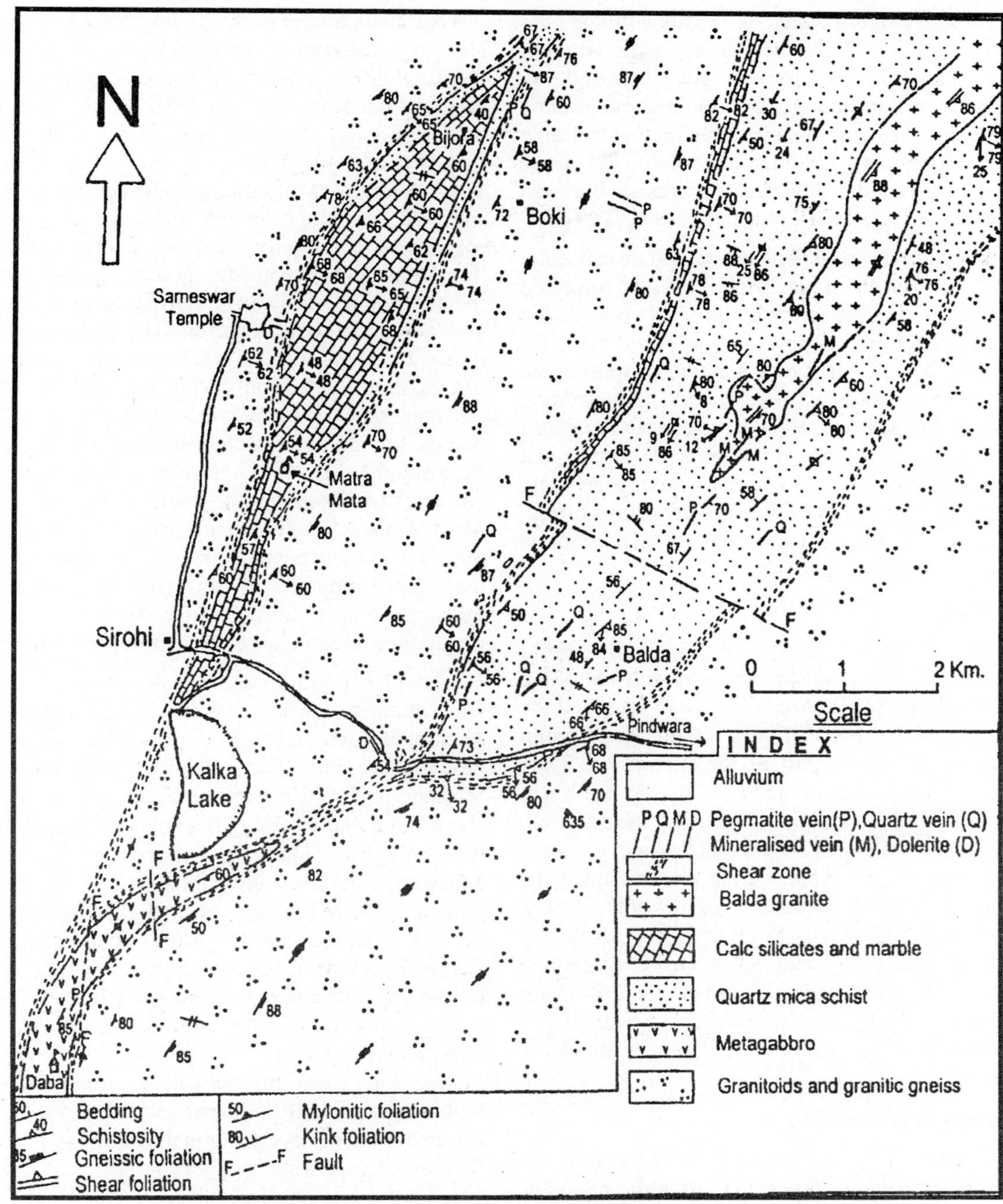

Fig. 1 Geology of Sirohi region (Sharma, 1996)

mineralized veins. The topaz and fluorite occur as primary and secondary minerals and fluorine-bearing micas suggest fluorine partitioning in substantial amount between granitic melt and coexisting aqueous fluid phase. (Srivastava and Sukhchain, 2005).

The Geological Survey of India reported wolframite mineralization in quartz veins from Motiya, Pipaliya, Richmaliyan, Kotariya and Bijathal in Sewariya-Govindgarh area (Jain and Bhattacharjee, 1992). The tungsten-mineralized quartz veins trend N-S to NNE-SSW and occur along the sheared contact between mica schist and granite. The Sewariya granite shows gneissic foliation. These veins consist of quartz, tourmaline, muscovite and wolframite. The

Sewariya-Govindgarh area is located along the western margin of Delhi Fold Belt in the Aravalli craton. The metasediments are intruded by leucocratic granite named as Govindgarh granite. This leucogranite has close resemblance with Balda and Degana granite. The tungsten-mineralized quartz veins are restricted along sheared contact between mica-schist and granite (Bhattacharjee et al., 1993).

Mukhopadhyay (1979) discovered wolframite deposit at Balda near Sirohi. The tungsten mineralization is associated with leucocratic Balda granite in the form of quartz veins, greisen veins, along shear zones. The veins are rich in tourmaline and fluorite. Besides this, topaz, beryl, zinnwaldite, muscovite, feldspar etc. are also present in the veins. The tungsten metallogeny of Degana, Sewariya and Balda have close resemblance. To explain the tectonic and geological setting of Neoproterozoic wolfram metallogeny of western Rajasthan, the discussion is emphasized on the Balda deposit.

Lithology :

Coulson (1933) classified the geology of the Sirohi region. Gupta et al. (1980) proposed a revision of the lithostratigraphy of the Aravalli region, included the Sirohi rocks as part of the Delhi Supergroup constituting the upper part of the succession. They recognized two groups in the Sirohi region; the lower as the Sirohi Group which underlies the Sindreth Group. A time span from 2000 to 700 Ma was assigned for the entire Delhi cycle, which also included the Sirohi and Sindreth Groups. Roy (1988, 1990), Sharma (1996), Roy and Sharma (1999), Roy and Jakhar (2002) considered the Sirohi Group as separate tectono-stratigraphic units younger to the Delhi Supergroup. Different lithological associations, which occur in this region (Fig. 1), are grouped as following:

a. Basement granitoids

The basement rocks outcropping in the region are granitoids of various types viz: granitic gneisses, banded gneisses, porphyritic gneisses etc., popularly described as Erinpura granites (Coulson, 1933). These rocks have tectonised contact to overlying metasediments and nowhere intrusive relationship is observed. These granitoids indicate all deformational patterns of the overlying cover rocks. The term 'Erinpura granite' is used as a basket term in geological literature to describe various Precambrian granites of Rajasthan. Looking to this the term 'basement granitoids' is used here.

Overall character of the basement rock is gneissic, showing alteration of quartz-feldspar and mafic rich bands. Folding, contortion and transposition of the gneissic banding is a common feature notable in different outcrops of the granite. In coarse grained varieties, feldspar megacrysts (sensu Lepedes, 1978) occur as large oval or tabular shaped grains measuring over 10 cm along length in some cases. These megacrysts commonly occur parallel to the foliation. The contact between the granite gneiss and the metasediments of the Sirohi Group generally marked as a zone of ductile shearing. The granite gneiss in these zones shows is development of mylonitic foliation and lineation. With the development of feldspar megacrysts between the mylonite foliation planes, the granite has changed into augen gneisses. Petrographically, the granite gneiss ranges in composition from granodiorite to normal granite. Most of the rocks of the region are characterized by the presence of large megacrysts of K-feldspar. The other important minerals in the rock are plagioclase feldspar, quartz and biotite. Muscovite, garnet, magnetite and chlorite occur as minor constituents.

K-feldspar grains are of variable sizes ranging from about a mm to over 10 cm in length. In thin sections, the winged regions are seen to be filled with biotite along with quartz and muscovite. Plagioclase feldspar is fine to coarse grained, with subhederal shape. Lamellar twinning is a characteristic feature of the plagioclase. Quartz grains are of variable shapes and sizes. Polygonal and strain-free grains are quite common. There are also few stretched grains, which occur as ribbons, parallel to the mylonitic foliation. Large irregular shaped grains showing patchy polygonization are not uncommon. Biotite is the only mafic mineral in the granites and occurs in considerable amount. Muscovite occurs as small flakes along with biotite and as small inclusions within large grains of feldspars. Small grains of garnet (a mm to a cm in size) occur in the biotite rich bands of the banded gneisses.

Five samples of basement granites which have been analysed, out of which three (sample no. KK-37, KK-38 and KK-39) are from the mylonite zone, and two (KK-40 and KK-41) are of granite gneiss (Table 2). High SiO2; approximately 72 % in gneissic rocks, and 75 % in the mylonitic types characterizes the samples of Erinpura granite. Alumina is around 13 %, which is a little higher in gneisses. The increase in Fe_2O_3 and decrease in FeO in mylonites may be due to oxidation during shear deformation.

Table-2 Chemical Composition, CIPW norms and important ratios of basement granites of Sirohi region.

Sample No.	KK-37	KK-38	KK-39	KK-40	KK-41
SiO_2	73.59	74.55	76.74	72.29	72.06
TiO_2	0.1	0.05	0.06	0.29	0.27
Al_2O_3	13.09	12.98	13.03	13.89	13.59
Fe_2O_3	2.35	2.16	1.43	1.63	2.04
FeO	0.4	0.4	0.48	2.52	2.4
MnO	0.01	0.02	0.02	0.06	0.06
MgO	0.06	0.32	0.05	0.82	0.93
CaO	0.47	0.29	0.24	1.34	1.4
Na_2O	3.23	3.43	3.21	2.22	2.14
K_2O	6.75	5.8	5.58	5.51	5.68
P_2O_5	0.15	0.14	0.14	0.14	0.17
Total	100.2	100.14	100.98	100.71	100.74
(ppm)					
Sr	25	16	22	44	48
Rb	373	252	439	243	250
Ni	27	23	16	37	29
Co	8	6	7	10	14
Cr	17	0	0	5	16
Li	19	36	15	41	42
Ba	122	82	36	443	544
Cu	11	10	9	15	16
Zn	32	31	14	47	50
Q	28.31	31.7	36.52	33.26	32.87
or	39.89	34.28	32.98	32.56	33.57
ab	27.33	29.02	27.16	18.79	18.11
an	1.28	0.52	0.28	5.73	5.83
c	0	0.87	1.61	2.17	1.78
di	0.05	0	0	0	0
hy	0.13	0.8	0.12	4.96	4.71

Table-2 Contd.

mt	1.03	1.21	1.44	2.36	2.96
il	0.19	0.09	0.11	0.55	0.51
hem	1.64	1.33	0.44	0	0
ap	0.35	0.32	0.32	0.32	0.39
K/Rb	150.22	191.05	105.51	188.22	188.6
Ca/Sr	134.34	129.52	77.96	217.63	208.42
Rb/Sr	14.92	15.75	19.95	5.52	5.21
Cr/Ni	0.63	0	0	0.14	0.55
D.I.	95.53	95	96.66	84.61	84.55
A/CNK (Mol)	0.97	1.04	1.11	1.15	1.11
A/NK(Mol)	1.04	1.09	1.15	1.44	1.4
A.I.	0.97	0.92	0.87	0.69	0.71
alk	9.98	9.23	8.79	7.73	7.82
FeO*	2.75	2.56	1.91	4.15	4.44

b. Sirohi Group

The metasediments of the Sirohi Group include argillites along with carbonates. The major rock units include phyllite, mica schist, marble and calc silicate rocks. Mica schist is the most common lithological unit in the Balda region. Carbonaceous phyllite occurs locally in pockets.

Mica-Schist

The mica-schist is a silvery grey coloured rock, showing development of pervasive schistosity planes defined by prisms of muscovite and biotite, and flattened grains of quartz. It is composed of quartz, muscovite and biotite as major minerals. Garnet and andalusite occur as porphyroblasts. Andalusite has developed as hornfelsic mineral in the rocks, which surround the Balda granite. Epidote, chlorite, microcline, tourmaline and some opaque minerals occur as accessory mineral. Garnet occurs as tiny skeletal grains (approximately upto 1 mm size) in the quartz mica schist. Quartz occurs in different forms, as flattened grains defining schistosity, or as polygonal shaped fine grains. There are also some large anhedral grains showing irregular grain boundaries. The last described grains also show undulose extinction. Biotite flakes are aligned parallel to schistosity along with muscovite. Geochemical data of four samples (KK-47, KK-48, KK-49 and KK-50) of mica schist is presented in Table 3. The rocks show high silica content and comparatively low alumina. Rubidium and lithium show high values, which indicate fluid activity related to the mineralization.

Calc-silicate

Two linear outcrops of calc-silicate and marble occur in the study area.Calc-silicate occurs at the contact of the mica schist or the granitoids is mylonitised. This is light green coloured and fine grained. Intense shearing has destroyed all the primary features turning the rock into mylonite gneiss which show development of paper thin laminations and down dip lineation.

The calc-silicate shows presence of fine-grained calcite along with diopside, talc and wollastonite. Tremolite-actinolite schist/ gneiss occurs in the central part of the calc-silicate outcrop. It is banded in character showing thin alternate bands rich in carbonate and silicate minerals. The rock may thus be described as a calc-gneiss. At places, the calc-silicate

Table-3 Chemical Composition of Sirohi Metasediments

Sample No.	KK-47	KK-48	KK-49	KK-50	KK-29	KK-30
Locallity	BD	BD	BD	BD	SRO	SRO
Rock Type	MS	MS	MS	MS	CS	CS
SiO_2	76.76	75.98	69.67	71.46	24.83	27.73
TiO_2	0.33	0.37	0.46	0.44	0.24	0.06
Al_2O_3	11.5	11.53	15.01	12.36	2.48	2.5
Fe_2O_3	2.05	2.23	5.84	4.19	2.34	0.3
FeO	2.04	3.4	1.72	3.24	1.6	0.46
MnO	0.5	0.06	0.14	0.07	0.06	0.05
MgO	1.12	1.36	2.05	2.17	6.67	8
CaO	0.16	0.09	0.22	0.41	29.96	28.72
Na_2O	0.28	0.23	0.13	0.81	0.57	0.42
K_2O	4.68	3.93	4.27	4.2	1.96	0.01
P_2O_5	0.17	0.19	0.29	0.19	0.16	0.04
Total	99.59	99.37	99.8	99.54	70.87	68.29
(ppm)						
Sr	13	4	27	26	202	211
Rb	512	529	230	170	94	96
Ni	24	26	51	53	14	8
Co	8	10	22	9	8	8
Cr	32	51	27	21	16	14
Li	159	136	32	35	13	12
Ba	221	292	480	257	540	468
Cu	7	7	21	37	22	19
Zn	28	41	92	105	69	11

BD = Balda, SRO = Sirohi, MS = Mica-Schist, CS= Calc Silicate

shows schistose character defined by the presence of parallelly oriented actinolite, tremolite and mica. Dolomite is the dominant mineral of the rocks, which occurs along with mica, quartz, wollastonite and garnet. All these minerals except actinolite and tremolite, occur in subordinate amount. In some thin sections, carbonaceous material is also observed which contributes black colour to the rock. Dolomite occurs as fine to medium sized flattened grains parallel to foliation plane. Tremollite and actinolite occur as needles and prism-shaped grains also defining the foliation planes. The former grains predominate over the latter ones. The geochemistry of two samples (KK-29 and KK-30) is shown in Table 3.

c. Balda granite

It is a medium grained, well foliated, leucocratic granite in which biotite is either absent or occurs rarely. The granite is highly homogeneous in composition. There are several shear zones along NNE-SSW direction in the Balda granite, which are occupied by greisenised veins and quartz-feldspar-rich pegmatite veins. The foliation in the granite is a result of shearing and recrystallisation. The attitude of foliation planes in the Balda granite suggests that the shear zones are either vertical or have very steep dip towards east. Mineralogically, the granite contains quartz, K-feldspar, plagioclase and muscovite as major constituents. Epidote, and opaques occur in minor amount. Quartz, which shows extreme variation in size and shape, constitutes nearly one third of the total mineral constituents of the rock.

The Balda granite shows effects of cataclasis subsequent to its emplacement. This is evident from solid-state deformation features like mylonitic foliation and spaced cleavage. Recrystallisation of minerals, particularly of muscovite parallel to foliation (spaced cleavage) has transformed the granite into mica gneiss. The last mentioned variety is a typical example of mylonite gneiss.

The analytical data of five Balda granite samples are presented in Table 4. The data show 73.24 to 75.92 % of silica and approximately 13% of alumina. The Balda granite has higher Na_2O (around 4%) in comparison to the basement granites (around 3%). The total iron content varies from 1.79% to 2.55% whereas the basement granites show 1.91% to 4.44%. It shows considerably higher Rb (766 to 1050 ppm) and Li (136 to 164 ppm) in comparison to basement granites Rb (250 to 439 ppm), Li (15 to 42 ppm), which indicate higher fractionation of the melt. The Sr and Ba are compatible to CaO and show lower values relative to those in the basement granite gneisses. Balda granite shows higher normative albite than othoclase and negligible anorthite. The D.I. is quite high ranging from 94.48 to 97.02 percent. The average A.I. value is less than one. The normative corrundum is low which suggests that the Balda granite belongs to metaluminous to peraluminous type.

Tungsten Metallogeny :

In the Balda region, the metasediment have been intruded by a linear granite body, which extends along the axial zone of the regional antiform. The granite is a medium grained, well foliated and characterized by the presence of muscovite. There is also abundant evidence of pneumatolysis in the granites as well as in the surrounding contact rocks. The wall rock alteration of metapelites at the contact of the granite and development of pneumatolytic veins are commonly observed. A common evidence of wall rock alteration is the growth of fine needles of tourmaline and induration of the schists. A variety of minerals have been formed in the altered wall rocks which include tourmaline, appetite, beryl, topaz, epidote, muscovite, fluorite, zinwaldite and chlorite. Tourmalinasation and silicification of mica-schist are intensive along the contact of the granite body.

The mineralization is associated with pneumatoloytic and greisenised pegmatite veins which intrude the Balda granite as well as schists along a number of shear zones. Banerjea et al. (1979, 1981) reported six mineralised veins in the Balda region showing strike length of 1.5 km, 1 km, 750 m, 500 m, 300 m, and 200 m; with an average width being 4 m. The major ore mineral is wolframite of ferberite variety (Jamthe and Dwivedi, 1979), which is associated with a fair amount of scheelite. The ore mineral occurs as stringers and veins (approximately one cm thick), comprising coarse to fine crystals. The main gangue minerals of the ore are quartz and tourmaline. Other associated minerals are muscovite, fluorite, topaz, lepidolite, pyrite, rutile and apatite. In some cases fluorite, becomes a major mineral in the pneumatolyitc quartz veins forming linear bodies of about 20 m length and 1 m thickness. The amount of fluorite in the mineralized veins varies from 70 to almost 100 percent. This resulted in workable deposits in the region.

CONCLUSION

The tungsten metallogeny in Rajasthan is genetically associated with Neoproterozoic granites. These granites are intruded in the Sirohi group of rocks along Sirohi-Sewariya-Degana belt. The intruding granite exhibits wall rock alteration; boron metasomatism and intrusive relationship with cover rocks. Tungsten, lithium and boron concentrated in the magma after melting of much-evolved Neoproterozoic crust in the western Rajasthan (Sharma, 1996; Krishanmurthi and Prabhakar, 2000). Boron, lithium along with fluorine caused high mobility of wolframite, which enriched in the late magmatic residual fluids. The intrusive granites are enriched in SiO_2, total alkalis and are

Table-4 Chemical Composition, CIPW norms and important ratios of Balda granite

Sample No.	**KK-42**	**KK-43**	**KK-44**	**KK-45**	**KK-46**
SiO_2	75.41	74.91	73.24	75.92	75.02
TiO_2	0.02	0.03	0.03	0.04	0.04
Al_2O_3	12.85	13.33	13.23	12.86	12.76
Fe_2O_3	1.57	1.17	1.93	1.77	1.19
FeO	0.56	0.62	0.62	0.48	0.6
MnO	0.01	0.01	0.05	0.01	0.03
MgO	0.01	0.02	0.01	0.01	0.02
CaO	0.09	0.56	0.73	0.43	0.55
Na_2O	4.4	4.72	4.38	3.3	4.26
K_2O	5.56	4.88	5.66	4.74	4.65
P_2O_5	0.16	0.32	0.33	0.32	0.36
Total	100.64	100.57	100.21	99.88	99.48
(ppm)					
Sr	14	13	14	11	9
Rb	920	902	1050	907	766
Ni	26	7	25	17	16
Co	11	11	8	6	9
Cr	10	10	12	14	15
Li	136	164	164	156	150
Ba	92	67	54	48	44
Cu	6	6	6	7	7
Zn	9	16	11	7	12
Q	29.02	28.39	25.89	38.55	32.2
or	32.86	28.84	33.45	28.01	27.48
ab	35.14	39.94	36.54	27.92	36.05
an	0	0.69	0	0.04	0.38
c	0	0.03	0	2.29	0.58
di	0	0	0.05	0	0
hy	0.27	0.19	0	0.02	0.16

Table-4 Contd.

mt	1.35	1.7	2.07	1.46	1.73
il	0.04	0.06	0.06	0.08	0.08
hem	0	0	0.34	0.76	0
ap	0.37	0.74	0.76	0.74	0.83
ac	1.84	0	0.46	0	0
wo	0	0	0.58	0	0
K/Rb	50.17	44.91	44.75	43.38	50.39
Ca/Sr	45.94	307.83	372.61	279.34	436.7
Rb/Sr	65.71	69.38	75	82.45	85.11
Cr/Ni	0.38	1.43	0.48	0.82	0.94
D.I.	97.02	97.17	95.88	94.48	95.73
A/CNK (Mol)	0.95	0.94	0.89	1.12	0.97
A/NK(Mol)	0.96	1.01	0.98	1.2	1.05
A.I.	1.03	0.98	1.01	0.82	0.94
alk	9.96	9.6	10.04	8.04	8.91
FeO*	2.13	1.79	2.55	2.25	1.79

depleted in CaO, FeO, TiO_2 and MgO. They are higher Rb, Li, Sn, B, Be and are highly differentiated. The presence of tourmaline needles at Balda, Sewariya and Degana indicates boron rich fluid activity at the time of wolframite mineralization, Boron and fluorine enhanced the wolframite enrichment in late magamtic fluids. These fluids injected along shear zones, fracture and schistosity plane of the host rock in the form of quartz, pegmatite, and greisenised and mineral veins. The cover rock chemistry played a significant role for the localization of wolframite.

ACKNOWLEDGEMENT

I am glad to contribute this paper for the special volume in the honour of Professor B.S. Paliwal, who has contributed a lot to the geology of Rajasthan for more than three decades. I am extremely thankful to Professor K. L. Shrivastava for extending help during the preparation of manuscript.

References

Banerjea, G.M., Mukhopadhyay, A.K. and Singhai, R.K., 1979. Tungsten mineralization of Sirohi district, Rajasthan. Workshop on mineralization associated with acid magmatism, IGCP-26 Nagpur.

Banerjea, G.M., Mukhopadyay, A.K. and Snghai, R.K., 1981. Tungsten mineralization of Sirohi, district Sirohi, Rajasthan. G.S.I. Sp. Pub., Vol. 13, pp. 130-133.

Bhattacharjee, J., Fareeduddin and Jain, S. S., 1993. Tectonic setting, petrochemistry and tungsten metallogeny of the Sewariya granite in the South Delhi Fold Belt, Rajasthan. J. Geol. Soc. India, v.42, pp.316.

Chattopadhyay, B., Chattopadhyay, S. and Bapna, V. S., 1994. Geology and geochemistry of the Degana plutonA Proterozoic rapakivi granite in Rajasthan, India. Mineralogy and Petrology.v50(1-30 pp. 69-82.

Coulson, A.L., 1933. The Geology of Sirohi State, Rajputana. Mem. Geol Surv., Vol. 63 part-I, p. 166.

Gupta, S.N., Arora, Y.K., Mathur, R.K., Iqbuluddin, Prasad, B. and Sahai, T.N., Sharma, S.B., 1980. Litho-stratigraphic map of Aravalli region, southern Rajasthan and northern Gujrat. Geol. Surv. Ind. Calcutta.

Jain, S. S. and Bhattacharjee, J., 1992. A note on the wolframite prospects associated with the Sewariya granite pluton, Rajasthan. IndianMiner., v 46, pp. 159-164.

Jamthe, N.M. and Dwivedi, G.L., 1979. X-ray studies of some rare minerals from the tungsten mineralised zone of Sirohi pegmatites in Rajasthan. Proc. workshop on MWAM GSI, pp. 153-154.

Krishanmurthi,R., Prabhakar, T. 2000. The nature, fluid characteristics and genesis of tungsten mineralization in Rajasthan: A brief review. . In : Deb, M. (Editor) Crustal Evolution and Metallogeny in the Northwestern Indian shield. Narosa Publication, pp. 417-427.

Lepedes, D.N., (Ed.), 1978. McGraw hill Encyclopedia of the Geological Sciences. McGraw Hills, 625 p.

Mukhopadhyay, A.K., 1979. Geochemical reconnaince around Sirohi and the Tungsten-Fluorite-Topaz mineralisation, NE Sirohi, district Sirohi, Rajasthan. Progress Report, Geol. Surv. Ind. 1977-78 (unpub.).

Roy A.B. and Sharma K.K. 1999. Geology of the region around Sirohi town, western Rajasthan-Story of Neo-proterozoic evolution of the Trans Aravalli crust. In: B.S. Paliwal (Ed.) Geological Evolution of Northwestern India, Scientific Publishers (India) Jodhpur, pp. 19-33.

Roy, A.B. and Jakhar, S.R. 2002. Geology of Rajasthan (Northwest India) Precambrian to Recent. Scientific Publishers, Jodhpur. 421 p.

Roy, A.B., 1988. Stratigraphic and tectonic framework of the Aravalli Mountain range. Geol. Soc. Ind. Mem., Vol.7, pp. 3-32.

Roy, A.B., 1990. Evolution of the Precambrian crust of the Aravalli Mountain range. Developments in Precambrian Geology, Elsvier Pub., Vol. 8, pp. 327-348.

Sharma, K.K. 1996. Stratigraphy, structure and tectonic evolution of the metasediments and associated rocks of the Sirohi region. Unpublished Ph.D. thesis, M.L. Sukhadia University, Udaipur, 103 p.

Srivastava, P. K. and Sukhchain, 2005. Petrographic Characteristics and Alteration Geochemistry of Granite-hosted Tungsten Mineralization at Degana, NW India. Resource Geology, v 55(4), pp. 373-384.

Hon'ble Reviewers

1. Prof. A.M. Pathan
   Vice Chancellor
   Mollana Abdul Kalam Azad
   Urdu University
   HYDRABAD 500 001 INDIA

2. Prof. R.K. Trivedi
   Head
   Department of Applied Geology
   Centre of Adv. Studies in Geology
   Dr. Hari Singh Gaur Viswavidyalaya
   SAGAR 470 003 INDIA

Editorial Assistance

Ms. Tina Prajapat

Economic Mineralization
K.L. Shrivastava (Editor)
2009. pp.166-179
Scientific Publishers (India), Jodhpur

14

Section-II

Ophiolite Suite of Rocks in Manipur-Nagaland as Hosts for Chromites and Related Economic Minerals : A Review

O.P. GOEL AND ARUN KUMAR
Department of Earth Sciences, Manipur University, Imphal 795 003 Manipur
Email : arun_kumar610@yahoo.com

Abstract

The present article reviews some economic aspects of the platinum-bearing potential of chromite ores in the Ukhrul and Moreh, Chandel, districts Manipur, hosted in ultramafic-mafic members of ophiolite suite of the rocks, exposed along Nagaland-Manipur in Indo-Myanmar Belt. Besides occurrences of chromite and PGE elements in these rocks, a general survey of available literature has been carried in order to understand the crystallization of chromite and other related economic minerals, for example, magnetite-chromespinels. One of these aspects lies in the concept that the main factor of the RPGE mineralization is fluid mobilization, concentration, and deposition of RPGE minerals. Such a process obviously accompanies the chromites ore formation, creating the prerequisites for the RPGE mobilization from a large substrate volume, possibly containing RPGE in a dispersed state. The subsequent RPGE behavior by their concentrating together with other components that do not participate in the main crystallization process, as well as by the successive deposition of RPGE minerals from highly saturated supercritical fluids. We suggest that the mechanism of ore formation according to the vapor-liquid-crystal crystallization scheme reflects the main characteristics of the compositions, structures, and relations between individuals in aggregates containing RPGE minerals. There are a number of economic minerals which are genetically and spatially associated with chromites mineralization in similar host rocks. Thus the various models of chromites and associated economic mineralization processes are outlined as a review in the present study.

Keywords : Ophiolite, Ultra-mafic, Chromite Mineralization, PGE, RPGE

INTRODUCTION

Ophiolites are pieces of oceanic crust that have been thrusted (obducted) onto the edge of the continental plate. They provide models for processes at Mid-Oceanic Ridges (MOR).The entire sequence ranges in composition from meta-peridotites to ultramafic-mafic cumulate, mafic volcanic sequence, volcanogenic sediments associated cherts, limestones and ore-deposits, especially of chromite are found exposed throughout, but they are dismembered. Generally, the ophiolites occur as fault contacted slices within ocean pelagic shale-sandstone association. The contact between the ophiolites and the enclosing sedimentaries are mostly sharp. Ophiolites are sections of the oceanic crust that have been uplifted or emplaced to be exposed at the land

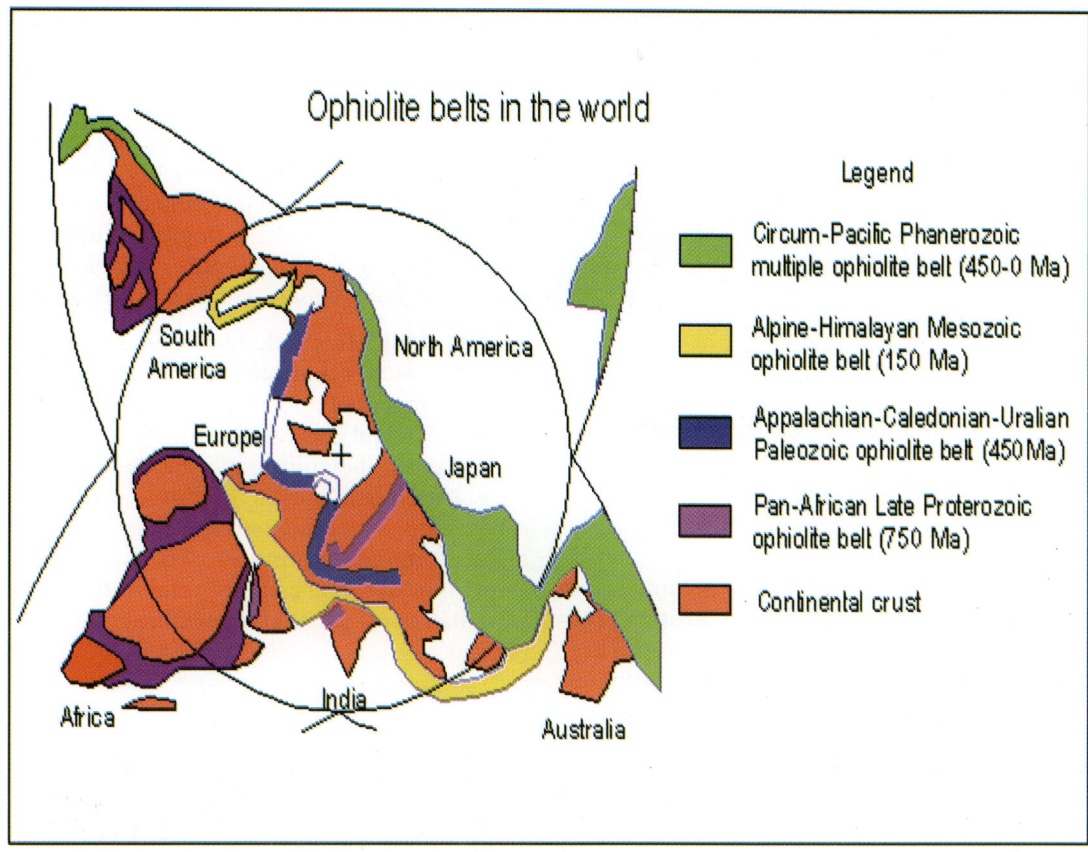

Fig. 1 Ophiolite belts in the world (after Ishiwatari, 2001)

surface. Examples include: the Troodos Ophiolite in Cyprus, the Oman Ophiolite in UAE, and Lizard Point in Cornwall, U.K. The stratigraphic sequence observed in ophiolites corresponds to the crust-forming processes at Mid- Oceanic Ridges: Mudstones and cherts deposited since the crust formed, Pillow lavas show magma/ seawater contact. Vertical, parallel fractures, which fed pillows above gabbros, indicative of fractionated magma chamber cumulate sequence. Dunite-rich layers of unfractionated magma mantle sequence comprise: Harzburgite-lherzolite-wehrlite-rich mantle rocks. The circulation of hydrothermal fluids through young oceanic crust causes alteration of the minerals observed: for example, chlorite, in the sheeted dykes. Often, ore bodies such as epidosites (Fe-rich epidosites) are found- evidence of (the now relict) black smokers that continue to dominate the seafloor setting today. Thus there are reasons to believe that ophiolites are indeed oceanic crust, however, certain problems arise when looking closer.

THE INDO-MAYANMAR BORDER OPHIOLITE BELT

The ophiolite belt, which is exposed along the Indo-Myanmar border in Manipur and to the north in Nagaland, forms a part of the Naga-Patkai, Chin Hills and Arakan-Yoma flysch trough of Cretaceous to Middle Miocene age. The Manipur- Nagaland ophiolite extends for about 200 Km from Phokpur in Nagaland to Moreh southeast of Manipur (Fig 2). The average width of the belt is about 15 Km. The general trend of the Ophiolite belt is NNE-SSW to NW-SE dipping steeply toward SSE/ SW along the extreme border of the country.

Field Observations

The study of ophiolites provides information regarding the tectonic history of the orogenic belts that contain them and leads to a better understanding of the genesis and the evolution of the oceanic lithosphere. As per the Plate Tectonics model developed during 1970s; it was accepted that ophiolites are formed by ocean-floor spreading, followed by tectonic emplacement onto continents. Thus, their study assumes importance in a better understanding the genesis, economic potential of metallic and non-metallic mineral deposits and evolution of oceanic lithosphere in the Chandel-Moreh and Ukhrul areas, Manipur.

The various litho units that characterize the ophiolite suite of rocks in Manipur include (Venkataramana et al., 1984):

1) Sediments belonging to Disang and Barail groups.
2) Ultramafic and mafic magmatic rocks
3) Mafic dykes and volcanic rocks- latter showing Pillow structure
4) Oceanic Pelagic sediments and
5) Tectonic mélanges.

More or less continuous outcrops of ophiolites are found as allochthonous bodies in the area of study. The present ophiolites are commonly dismembered. They are generally composed of upper mantle lithologies including dunites, serpentinites, peridotites and rarely gabbros and volcanic basalts- showing pillow structures. The latter are exposed near Silk Research Centre to the east of Ukhrul town.These lack sheeted dykes and prolific pillow lavas in the present study area. However, they are invariably associated with tectonic mélanges- petrograhicallly these show presence of contrasted lithologies_ ultramafic-mafic in one part and almost pure pelagic sediments- composed of sandstones, shales and carbonates.

The ultramafics including dunites and peridtotites are highly altered and deformed. These are manifested in development of microfolds, micro faults and cross-cutting different generations of serpentinite veins. The harzburgites, lherzolites and the dunites of the area have undergone plastic to brittle deformation, as shown by olivines and pyroxenes that are elongated in the direction of deformation. The dunites are mostly serpentinized and contain relict olivine crystals. Cumulus and adcumulus textures are commonly present.

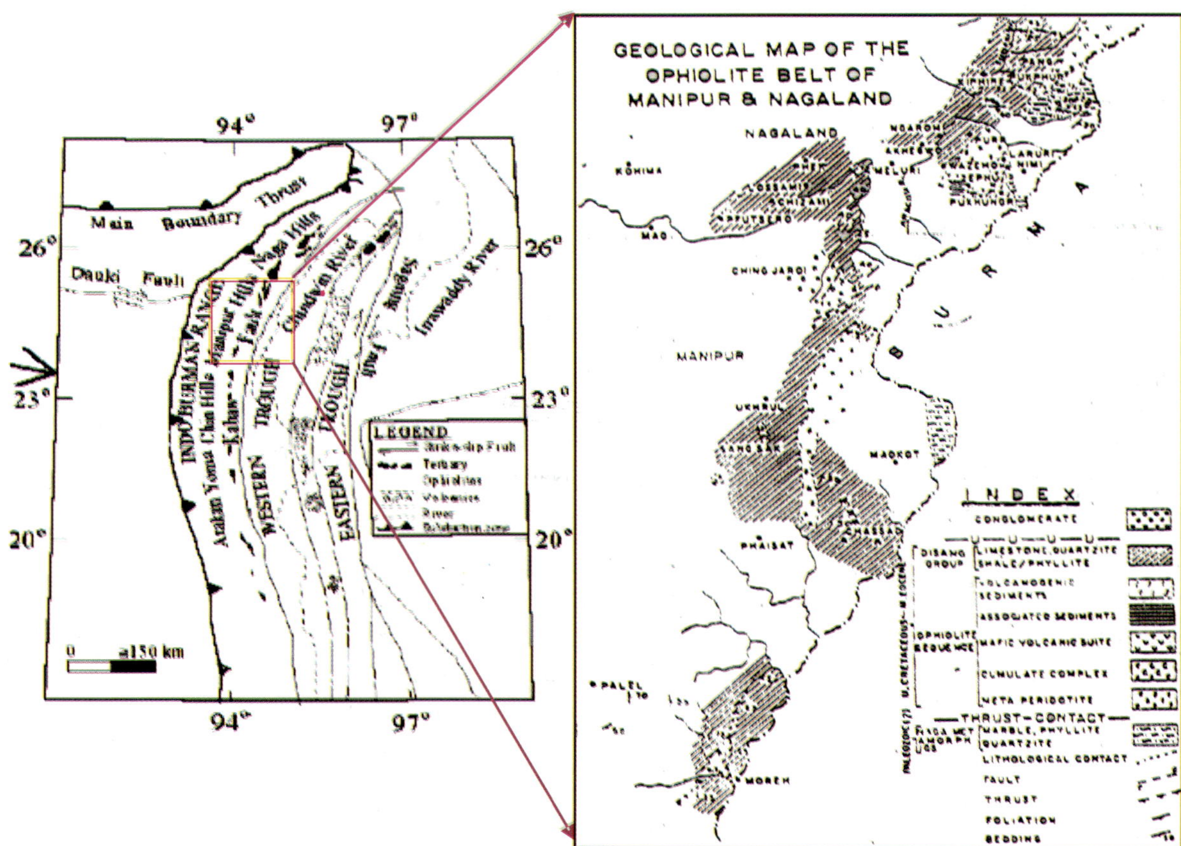

Fig. 2 Ophiolites in Manipur-Nagaland(After GSI 1984)

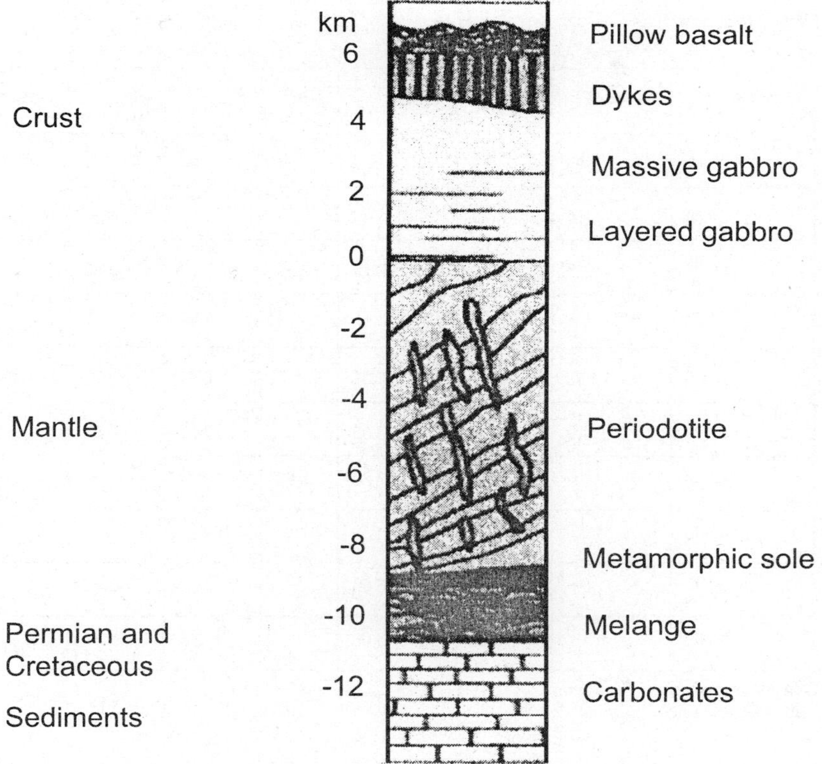

Fig. 3 Section of an Ophiolite Suite

By early 1980s, it was recognized that many ophiolites exhibit a geochemical component of subduction zone magmatic process that could not be explained by spreading at Mid-Oceanic Ridges (MOR). According to Pearce et al., 1984 and Bloomer et al., 1995), these ophiolites, known as supra subduction zone ophiolites (SSZ) may constitute a large portion of the ophiolites in the world. In subduction zones, rifted island arcs, back-arc basins and fore-arc basins have been suggested as sites of SSZ ophiolite zones. In addition, some ophiolites exhibit both MOR and SSZ geochemical signatures (e.g. Semail, Betts, Coe and Pindos, Pearce et al., 1984).

According to King (1982), well developed marginal-basin ophiolites should be commonly overlain by deep-sea sediments that contain a volcaniclastic component and tuffs derived from an adjacent volcanic arc, whereas, MOR ophiolites would be overlain by deep-sea pelagic sediments (Moos and Zeback, 1982). However, the well-known example of SSZ ophiolites of the southern branch of Neothethys in the eastern Mediterranieans is overlain by pelagic sediments with no significant volcanogenic components (see Pearce op cit.). The ideal SSZ ophiolites could be overlain by pelagic sediments, which in turn may be overlain by thick oceanic arc units (Hawkins et al., 1984; Robertson, 1994). Pearce et al., (1984) note that South Anatolian ophiolites constitute a complete sequence, with harzburgitic depleted mantle, sheeted dykes and island arc tholeiitic-type volcanics, locally including calc-alkaline plutonics and volcaniclastics characteristics of the SSZ ophiolites.

Recent field and petrographical studies indicate that the Late Cretaceous ultramafic magmatic rocks of Chandel-Moreh-Ukhrul area are composed of partially dismembered ophiolitic bodies associated with pillow lavas and volcano- sedimentary rocks that are intruded by gabbros and wehrlites0. This sequence indicates that the ophiolites of the area were formed in a SSZ setting during Late Cretacceous time. This can be very well related to Alpine orogeny and is the youngest ophiolite suite in the Indian subcontinent.

Ophiolites are characterized by a classic sequence of rocks. This sequence is well exposed in and around Ukhrul, Manipur and also at the Samail Ophiolites (Fig.3). (Coleman, 1977, 1981).

Table-1 Modal Analysis of Thin Sections

Sl No.	SampleNo	Olivine	Serp.	Opx	Cpx	Chr	Sp	Mt	Remarks
1.	MS-2/1	61.30	5.80	16.6	14.0		2.0		Lherzolite
2.	MS-4/1	61.30	11.4	24.70	1.0			1.60	Harzburgite
3.	M-8/1	88.32	5.15	0.40	2.40	0.15	0.60	3.40	Dunite
4.	MS-8/2	57.00		37.00			6.00		Harzburgite
5.	M-9	86.05	4.13	3.02	1.04		2.04	3.72	Dunite
6.	MSA-3/1	77.10	11.9	4.20	3.50		2.70	0.60	Lherzolite
7.	GM-1/A	48.40		32.80	13.20	5.60			Lherzolite
8.	GW-5/1B	82.00		8.30	4.80		4.10		Lherzolite
9.	GW-3/3	58.80		4.80	33.80			2.60	Wehrlite
10.	GW-4/2	73.00		17.20	3.00		3.4	3.4	Harzburgite

Opx = Orthopyroxene; **Cpx** = Clino-pyroxene;
Chr = Chromite; **Sp** = Spinel; **Mt** = Magnetite.

Based on the petrological and metallogenic features, the sequence in Manipur-Nagaland belongs totally to the ophiolites (of the harzburgitic type) of the second type lherzolite with the ortho- and clino-pyroxene sequence, formed in the close relationships with the formation of the ophiolites of the northeastern type (Nagaland), during the Cretaceous, with the younger sediments at the uppermost levels.

Petrological Studies

The generalized sequence of the ultramafic rocks of the area, from bottom to top, is as follows:

a) Mantle series (m), Ultramafic tectonites, mainly fresh harzburgites (m1) and Serpentinized dunites,

b) Intercalation of harzburgite tectonites with the thin dunite tectonite bands

The tectonite-harzburgites has banded to gneissic texture showing the foliation elements derived from the plastic deformation, with the typical cataclastic and porphyro-clastic texture with neoblasts (Fig 4). They consist of magnesian olivine Fo = 90-94 and enstatite with ex-solutional needles of clinopyroxene with the chrome spinel disseminations. Modal analyses of a selected periditites from Manipur are presented in Table 1

Recent petrological and geochemical studies of such rocks from different parts of the world have revealed the presence of another very important group of economic and strategic minerals, viz., Platinoid Group of Elements (PGE), other chrome-spinels and Magnetite-Ilmenite etc. There are reports of occurrences of gold in mafic rocks, e.g. in South Africa and Brazil.

REVIEW OF CHROMITE, PGE AND RPGE IN OPHIOLITES

In view of the above, we present here, as review the various models of chromite, PGE and associated economic minerals' mineralization, as proposed by a number of workers. First, the general processes of chromite and chrome-spinels will be discussed, to be followed by other PGE mineralization under different conditions of mineralization. World wide chromite bearing ophiolite and layered intrusions are shown in Fig. 5.

Chromite Mineralization in Ladakh, India:

The chromite in the lowermost part of ultramafic zone occurs as pods and disseminated in dunite and harzburgite host rock in Nidar Ophiolitic sequence of Indus Suture Zone. The compositions of ore forming chromite in depleted mantle rock is uniform, showing high Cr-number (100 Cr/(Cr+Al)= 78-81) and Mg-

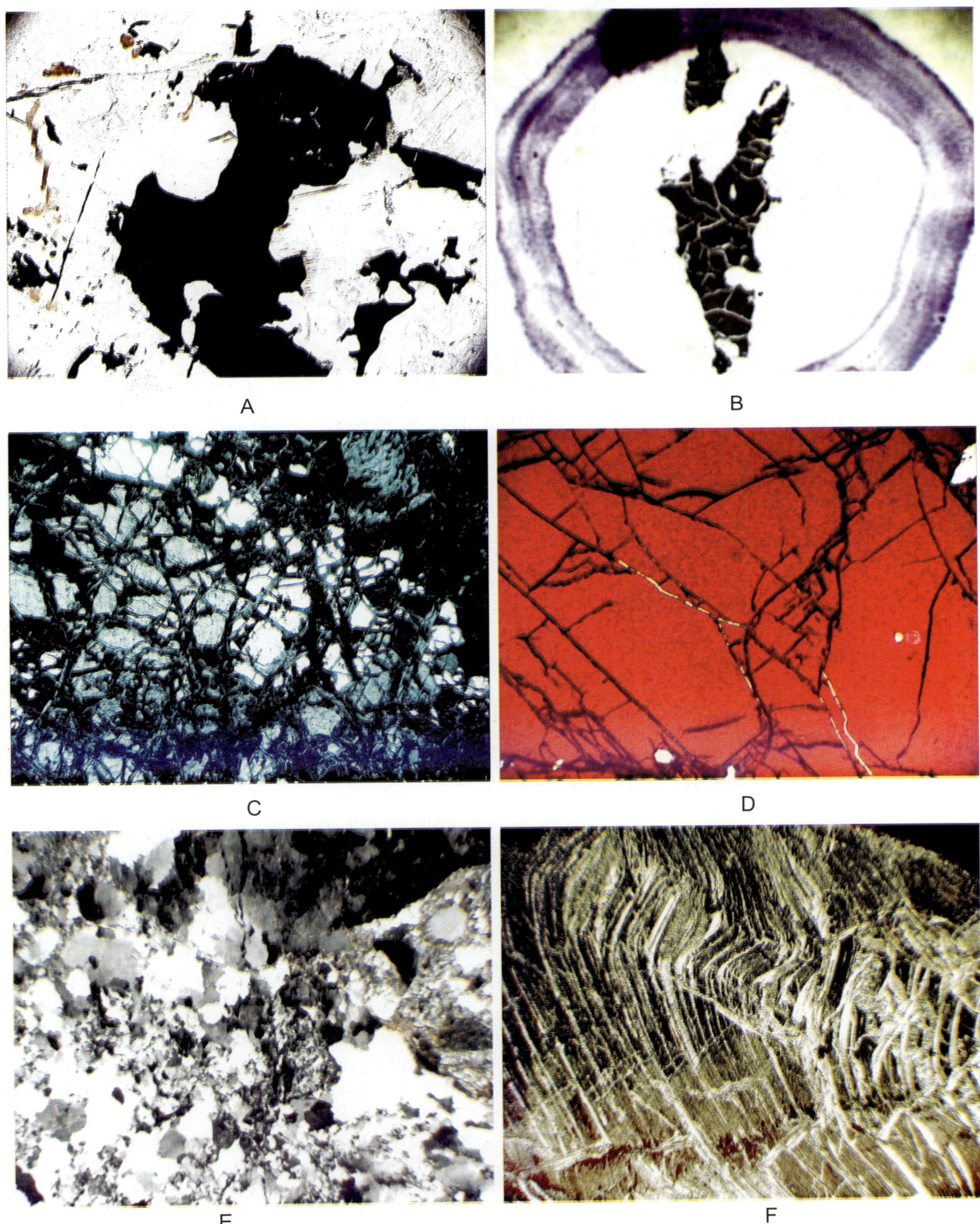

Fig. 4 Photomicrographs of samples (A) Sample No. GM-1/A, Photomicrograph of Magnetite present within ultramafics (Lherzolite) (B) Sample No.GW-5/1B,Photomicrograph of Spinel Present within Lherzolite (C) Sample No.-GW-3/3, Photomicrograph of serpentinised olivine grain within wehrlite. (D) Sample No. GW-1 Photomicrograph of Chromite (E) Sample No. MS-15 Quartz grains associated with ultramafics (Mélange) (F) Sample No. MS-4/2 Folded Serpentinite

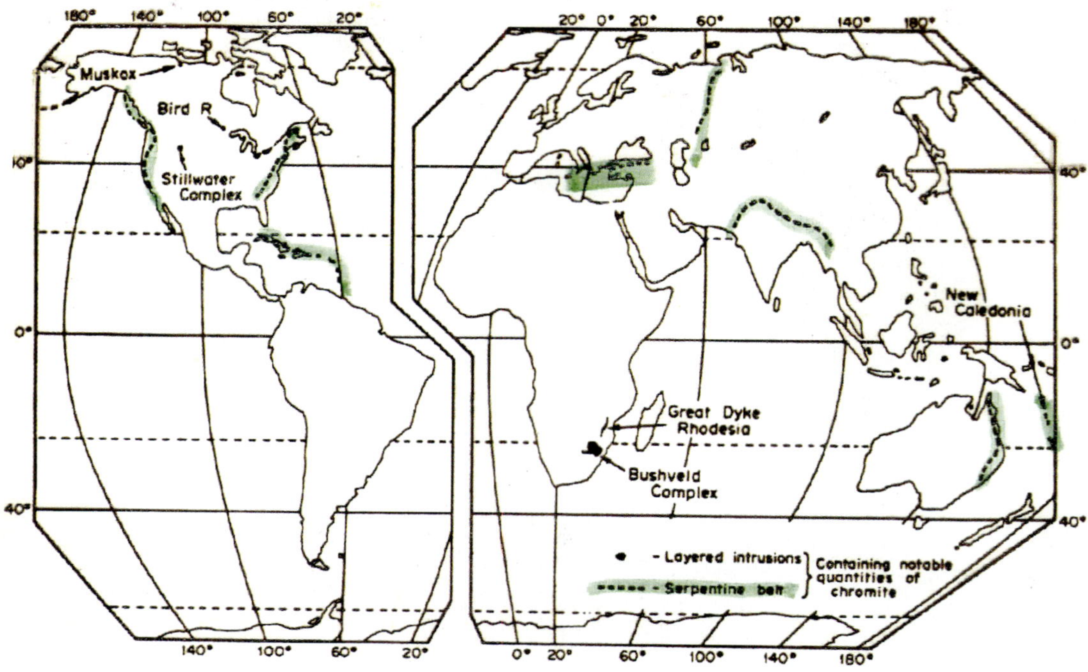

Fig 5 : Worldwide Chromite-bearing Ophiolite and Layered Intrusions

number (100 Mg/(Mg+Fe)= 65-78) values. The primary silicate inclusions generally have high Mg number (>88-98) and are dominated by pargasitic amphibole, olivine, diopside and enstatite suggesting that the melt responsible for crystallization of chromite had volatile rich fluid phase. The calculated parental melt compositions for chromite crystallization have Al203 10-07-10.75 (Avg. 10.42), while FeO/MgO liquid ratio 0.4-0.98 (Avg. 0.64). This parental melt composition indicates the chromite crystallizing melt was of boninitic in nature (WIHG Report, 2005).

Chromite Mineralization in British Columbia, Canada (Hancock, 1991)

Deposits of massive chromitite occur as pods, lenses or layers within ophiolitic ultramafic rocks in British Columbia, Canada. They are obducted fragments of oceanic, lower crustal and upper mantle ultramafic rocks within accreted oceanic terranes. They are formed as a primary magmatic differentiate during early olivine and chrome-spinel crystal fractionation of basaltic liquid at an oceanic spreading centre; (1) as massive to disseminated pods and lenses of chrome-spinel surrounded by a dunite envelope within depleted mantle harzburgite; or (2) as massive to disseminated cumulate layers in dunite at the base of the crustal plutonic section. Age of mineralization is Mesozoic and younger. The main host/associated rock types are serpentinized peridotite; residual mantle harzburgite; cumulate dunite.

Early fractional crystallization of chromite from a basaltic liquid either (1) just below the crust-mantle transition (syn. petrological MOHO) in small magma pockets or possibly conduits within the residual mantle harzburgite; or (2) immediately above the crust-mantle transition as cumulate layers within dunite at the base of the axial magma chamber. Pods and lenses in harzburgite obtain their diagonistic shape as a result of subsolidus to hypersolidus ductile deformation due to mantle convection.

Ophiolites of suprasubduction zone (SSZ) affinity with harzburgite mantle sections appear to be the only ophiolite type to host economic deposits of podiform chromite. A lack of any sizable chromite occurrence in British Columbia may reflect the fact that most ophiolitic complexes in the province are of mid- ocean ridge affinity. Occurrences of podiform chromite are found in ophiolitic ultramafic rocks in the Slide Mountain, Cache.

Recent studies have shown the existence of many Late Proterozoic Alpine type ultramafic massifs, as part of dismembered ophiolites along highly strained suture zones of the Pan-African Arabian Nubian Shield. The mantle sequences of these ophiolites contain numerous pods of chromite, which at some localities furnish economic deposits.

Chromite and Nickel in Sudan (Linnebacher, 1991)

Podiform chromite occurrences were first reported in Sudan. There are many Late Proterozoic Alpine type ultramafic massifs, as part of dismembered ophiolites along highly strained suture zones of the Pan-African Arabian Nubian Shield. The mantle sequences of these ophiolites contain numerous pods of chromite, which at some localities furnish economic deposits.

Nickel deposits are classified as either magmatic stratiform or lateritic deposits. So far the magmatic deposits have not been reported in Sudan. However, lateritic horizons with anomalous nickel values have been found associated with late type ores as in Shkalla deposit. Chromite mineralization of the transitory zone consists of the lensy-stratified, banded and disseminated bodies, concordant to subconcordant and more seldom, lensy-massive ores. The banded chromitite mineralization is associated with the pentlandite disseminations and the PGE mineralizations. (Beshkul et al., In Press). Nickel-bearing ferruginous laterite derived from the underlying ultramafic rocks covers close to ninety percent of the island. It is dark reddish-brown in colour generally darker in shade in the lower portions. It is quite variable in thickness with some areas having as much as 20 meters mantle though this is relatively rare. Generally, however, areas with steep slopes have a thin veneer of laterite whereas those with gentle slope have a thick cover (Arco, 1962).

Chromitite in kamchatsky MYS Peninsula (E. Kamchatka) (Melcher et al., 1994)

The chromite mineralization belongs to the podiform type, is located in lens-shaped dunite bodies, and is represented by several structural-textural varieties. It is found in the ultramafic rock complex of the Valizhgen Peninsula. The ore and accessory chrome-spinels have some specific compositional features that differ in the rocks of either series: chrome spinels of the cumulative dunite-orthopyroxenite-chromitite series have elevated Cr, Ti and Fe^{3+} contents, while spinels from residual harzburgites are typically low in Ti and Fe^{2+}, Fe^{3+} and variable in Cr and Mg contents. The specific location of chromitites in mafic-ultramafic massifs of the peninsula and the composition of ore and accessory chrome-spinels from ultramafics can be accounted for by interaction of primitive mantle magmas with host harzburgites. The calculations using mineral thermobarometry indicate relatively low pressure and high temperature and oxygen fugacity during the formation of chromite mineralization. The chemical and Re-Os isotopic compositions are also studied for PGE minerals from placers of the Valizhgen Peninsula. It is suggested that chromitites of the mafic-ultramafic massifs are sources of these placers.

Chromite in Kempirsai Hyperbasite Massif: (Melcher et al., 1997)

The Kempirsai massif is a part of the South-Ural hyperbasite belt, which is a continuation of the basite - hyperbasite platinum-bearing belt of the Middle and Northern Urals. The Kempirsai massif is an integral part of the large Sakmarian allochthone, which is a relict of the Early Proterozoic oceanic crust, overthrust on the Proterozoic ophiolitic mafic-ultramafic rocks in some investigated areas like Ingessana Hills, Jebel Rahib, and the Nuba Mountains.

The Ingessana massif is the largest mafic-ultramafic complex in the Sudan. It represents the main and central part of the Ingessana-Kurmuk Ophiolite Belt. The massif comprises a highly serpentinized dunite-harzburgite lower facies overlain by peridotite, minor pyroxenite, layered gabbro and dolerite. There are more than a hundred chromite occurrences in the Ingessana massif hosted by the lower dunite-harzburgite facies. The ore is generally massive, compact and consists of chrome- spinel sometimes ferruginous, with minor chrome-chlorite and chrome-garnet. The eastern part of the Nuba Mountains in the Central Sudan is predominantly covered by low-grade volcano-sedimentary sequence with dismembered ophiolitic bodies. The Red Sea Hills region of Eastern Sudan is an integral part of the Late Proterozoic Pan-African Arabian Nubian Shield. It comprises distinct terrains of related arc volcano-sedimentary sequences, and plutonic equivalents, all separated by ophiolite-decorated suture zones. The Jebel Rahib area is located in Northern Darfur State. It is dominated by a low-grade sequence of rift-type sediments with a complete, but dismembered Late Proterozoic ophiolite complex. The ore is very compact, massive and coarse-grained. Average Cr_2O_3 contents about 55%, the ore reserves, though promising, are not yet determined. In the Southern Sudan, chromite mineralization has been reported from two different areas: Juba area and Kapoeta-

Nagishot area. In the River Kit area, to the southeast of Juba town, a chromite occurrence was discovered on the footpath between River Kit and River Afoyi to the east.

Chromite in Eastern Albanides (Aleksander 2006)

The metallogeny of the Bulqiza ultramafic massif is characterized by the wide spread of the chromitite mineralization, represented by the unical deposits regarding the dimensions, morpho-structural features and the highest chrome- bearing potential in the eastern belt of Albanides. The chromitite mineralization is mainly of the metallurgic type. The refractory type mineralization occurs in the southwestern part, localized within the mantle sequence and the transitory zone.

Chromite mineralization of mantle section is localized throughout the ultramafic tectonite sequence, within which, some morpho-structural types are distinguished: folded concordant, pseudo-tabular ore bodies and/or subconcordant ones as in Lugu Gjate, Qafe Buall, Ternove and other deposits and discordant pipe eastern margin of the East-European paleocontinent in Middle Upper Paleozoic time. The Sm - Nd age (Edwards and Wasserburg, 1985; Melcher *et al.*, 1994) determined for unaltered rocks of the mafite - ultramafite complex is 397 Ma with ε (T) = +8.68. A somewhat younger age (359 Ma) was obtained by the isochrone for chromite ores that contain hydrated phases of silicates.

In the northern and central parts, the massif is made up mainly of harzburgites and enstatite dunites with subordinate pyroxenites, cumulative plagiodunites, wehrlites, and plagiowehrlites. The peripheral parts of the massif are composed of banded and taxite gabbro, gabbronorites, and olivine gabbro. The southern and south-eastern parts of the massif have a peculiar structure and compositions. Two submeridional zones are made up of large bodies of massive dunites among rocks of the schlieren-banded complex, enstatite dunites and harzburgites with subordinate wehrlites and lherzolites (Pavlov and Grigor'eva, 1997). The submeridional zones of massive dunites contain the main chromite deposits, which form the Main ore field of the Kempirsai massif. It should be noted that individual chromite ore bodies are known in the central and northern parts of the massif; however, they are considerably smaller in size and have ores of lower grade. Altogether, about 160 deposits and occurrences are known within the Kempirsai massif.

Chromites in Cuban ophiolites (Proenza et al., 2001)

Cuban Ophiolitic Belt hosts chromite mineralisation in its easternmost part. Genesis of sulphide-rich chromite ores by the interaction between chromitite and pegmatitic olivine-norite dikes in the deposit. Chromite mineralization occurs within the mantle-crust transition zone. Two events of magma intrusion overprint the chromitite bodies: one gave rise to the crystallization of pegmatitic olivine- norite dikes and the other produced pegmatitic gabbro dikes. sulphide-poor chromite ores, brecciated chromite ores. sulphide -rich chromite ores can be distinguished in the different chromitite bodies. sulphide -poor ores represent more than 80 vol% of the chromitites. This type occurs far from the zones intruded by pegmatitic gabbro dikes and shows petrographic and chemical features similar to other chromitite bodies described in the Moa-Baracoa massif. Brecciated chromite ores occur within pegmatitic gabbro dikes. In this type, chromite crystals occur included within chromian diopside and plagioclase. These silicates often contain droplet-like sulphide aggregates.

Sulphide -rich ores are spatially associated to the contacts between sulphde-poor chromite and pegmatitic olivine- norite dikes. These ores mainly consist of recrystallized (coarse) chromite with interstitial pyrrhotite. pentlandite, cubanite and chalcopyrite. The sulphide-rich ores are PGE-rich (up to 1,113 ppb of total PGE), and show nearly flat chondrite-normalized PGE patterns, slightly above 0.1 times chondritic values. The interaction between pre-existing sulphide-poor chromite ores and the intruding volatile-rich silicate melts produced strong brecciation, partial dissolution, and recrystallization (coarsening) of chromite. The sulphide assemblage formed by fractionation of the immiscible sulphide melt segregated from the volatile-rich silicate melt that generated the pegmatitic olivine-norite. The segregation of the sulphide melt can be interpreted as the consequence of chemical interaction between intruding melts and the host chromite. The variable extent of this interaction produced chromite ores with variable sulphide ratios. The magmatic nature of the sulphide mineralization is supported by sulphur isotope data, which range from -0.4 to +0.9 parts per thousand. Sulphide melt collected incompatible PGE (Rh, Pt, Pd) to produce the typical flat chondrite-normalized pattern of sulfide-rich chromite ores.

The host rock for the Scottie Creek prospect is an upper Paleozoic serpentinite wedge in the eastern facies of the Carboniferous to Jurassic Cache Creek Complex. This consists of a Late Triassic accretionary prism/subduction complex associated with the Nicola volcanic arc. The mélange contains Pennsylvanian and Early Permian limestones, chert, basalt and ultramafic rocks in a matrix of Permo-Jurassic chert and argillite.

Chromite, Ni-Cu and PGE mineralization in Manitoba, Canada (Theyer *et al.*, 2001)

The Bird River Sill has been explored for chromite, Ni-Cu and platinum-group elements (PGEs). Past studies focussed on the Dumbarton-Maskwa Ni-Cu mine and the well-exposed and accessible Chrome property. Three PGE-bearing stratabound layers, two of them intimately associated with chromitite layers, were defined on the Chrome property in previous studies. A sulphide-bearing layer associated with chromitite layers, thought to be equivalent to the 'Disrupted Chromitite' layer on the Chrome property, was defined this summer on the Page property. This approximately 6 m thick sulphide-bearing layer, recognized on most of the Page property, is a conspicuous, easily traced marker horizon containing disrupted chromitite layers and a vast number of chromitite pebbles. These pebbles are thought to be the product of episodic disruption and fragmentation of solid chromitite layers. The chromitite fragments were abraded and rounded during transport and eventually deposited on a semisolid substratum of olivine crystals.

This model requires a magma chamber that is at least periodically dynamic and characterized by magmatic surges of sufficient momentum to fragment solid chromitite layers and transport the fragments over an undetermined distance that was sufficient to abrade and round the fragments. Such surges would generate turbulence in at least part of the magma. Periodic magmatic turbulence during emplacement of the Bird River Sill is highly significant for guiding the search of PGE-bearing layers. It is widely believed that many 'reef-type' PGE deposits were formed as a result of magmatic interaction due to the injection of a batch of sulphide-rich primitive magma into a chamber containing PGE-bearing magma. The ensuing turbulence of the magmatic reservoir is assumed to facilitate thorough mixing of both magmas, resulting in the efficient scavenging and concentration of chalcophile elements, including the PGEs. Resulting sulphide droplets are thought to concentrate in discrete layers or reefs. (Theyer et al., 2001). Most geological studies of the Bird River Sill focussed on the well-exposed and readily accessible Chrome property. The Bird River greenstone belt and the Bird River Sill were mapped by Theyer (1982,) Trueman (1971, 1981). Bannatyne and Trueman (1982) and Cerny *et al.* (1981) summarized these studies. Geological investigations in the 1940s and early 1950s concentrated on chromite (Bateman, 1943). From the 1950s on, interest switched to the exploration for Ni-Cu mineralization (Karup-Møller and Brummer, 1971), but the focus reverted to chromite in the early 1980s.

Primary Platinum Mineralization in Urals, Russia (Edwards and Wasserburg, 1985)

Genetic Model for PGE Concentration in Chromite-Rich Zones in the Nizhny Tagil and Kachkanar Ultramafic Complexes, Urals, Russia:

The larger concentrations of chromite include brecciated and net-textured ore or massive veins that reflect the dynamic accumulation of chromite crystals in cavities along magma conduits inside the dunite body. These cavities are analogous to small magma chambers within the consolidated part of the dunite body and, similar to podiform chromitites in ophiolites, were created by continuous feeding of magma to the system. Chromite mineralization is restricted to the dunitic parts of the serpentinite. Chromite occurs as disseminations, small stringers and massive lenses. The observed chromite and associated Platinum mineralization can result from the normal magmatic evolution of the complex, provided that a structural (hydrodynamic) trap is available to concentrate the chromite.

PGE Mineralization in Bushveld Complex, South Africa: (Wager and Brown, 1967):

The concentrations of platinum metals in the Bushveld intrusion reflect their multistage accumulation both in the course of primary basite-hyperbasite layering of the magmatic chamber in the mantle (with isolation of dolerite-picrite metals which selectively extract sulfur) and at subsequent massif layering with isolation of ore-concentrating sulphide - bearing horizons (sulphide-depleted chromitic and pyroxenitic ones) and extremely iron-rich (hortonolitic) dunites. The Bushveld intrusion is unique in having such ferruginous rocks as hortonolitic dunites formed in it. This is due to its isolation from ascending transmagmatic fluid fluxes under the action of which extremely iron-rich melts, corresponding to them, must lose stability and be subject to sulphurization. Such interaction may result in the formation of complex (heterogeneous) ores whose metallic composition is due to two processes, i.e., magmatic differentiation of the Bushveld Type and metals introduction by transmagmatic fluids in connection with sulphurization of iron-rich melts. The combination of these processes may endow the mineralization with an extremely large scope leading to the formation of gigantic deposits, for example, the copper-nickel Talnakh deposit in the Norilsk region.

PLATINUM MINERALIZATION IN CHROMITE ORES

The process of the formation of platinum mineralization in complexes of the Alpinotype hyperbasites is an integral part of the general process of the formation of chromite ores and accompanying dunites, which make up a peculiar wall rock rim. The authors hold the viewpoint that chromite ores were formed as a result of a fractionation in the magmatic melt of an immiscible chromite liquid, the stability of which was defined, in particular, by a high concentration of a fluid dissolved in the magma. In addition, the magmatic fluid defined an overcooled state of the melt, as whole and correspondingly relatively low-temperature conditions of the crystallization of ore-localizing dunites. There are numerous data on the realization of such a mechanism, for example, during the formation of the chromite-olivine association in layered magmatic complexes. Thus, we believe that the mechanism of a tectonic localization of the chromite-dunite complex in a secondary (contemporary) occurrence of this complex is far from being of principal significance for the process of rock and ore formation. The distribution of PGE in various chromite Alpinotype hyperbasite complexes in the Urals, Cuba, Central Siberia show that the size of chromite ore bodies in hyperbasite massifs indicates not only the scale (reserves) of the PGE mineralization but the average PGE contents as well. In other words, the scale (extensiveness) of the chromite mineralization defines the degree of the mobilization or involvement of platinum metals in the ore-forming process.

Typical Evolutionary Succession in Formation of RPGE Mineral Phases

Mineral intergrowths in inclusions of the same chromospinelide crystals can considerably differ in quantities and ages of mineral phases. The results of the detailed study of phase relations allow us to reveal the most typical evolutionary succession in the formation of RPGE mineral phases and nonmetalliferous minerals, paragenetic to these phases. The MSS of the osmiride-iridosmine series and Ru disulfides of the laurite-erlichmanite series in equilibrium with them are obviously the earliest. Of nonmetalliferous minerals, forsterite, garnet, and some amphiboles can be included into this paragenesis, although their intergrowths with RPGE minerals are rarely observed. The main mass of the considered solid solutions and disulfides crystallizes without accompanied nonmetalliferous minerals. The main facts for understanding of the conditions of the crystallization and mechanisms of the growth of RPGE minerals are as follows:

RPGE minerals and coexisting mineral phases were formed mainly in isolated volumes as inclusions in the chromite matrix; the phases in inclusions crystallized after the crystallization of the matrix itself, i.e., the crystallization temperatures of phases are lower than those of the chromite matrix; volumetrically, RPGE minerals prevail in the compositions of mineral aggregates;

1. The leading form of the MSS crystallization is their primary unidirectional platy dendrite growth with a subsequent growth of dendrites and the creation of earlier dense solid solutions of the Os-Ir-Ru system or later porous RPGE solid solutions with base metals; the change of both types of MSS takes place without corrosion phenomena, only by a successive growth;
2. Dendrite relicts as evidence of a primary dendrite growth might be preserved at terminal facets of dense crystals of osmiride-iridosmine;
3. There exists a typical succession in the change of RPGE chemical compounds (metallic solid solutions-sulfides-sulfoarsenides-arsenides); however, it might be interrupted due to the absence of some compounds.

Based on these data, we suggest that the primary composition of a substance in isolated inclusions of chromospinelides was represented by a highly concentrated (relative to RPGE) supercritical predominantly gas fluid, the liquid phase of which appeared after the crystallization of, at least, iridosmine-osmiride and Ru-Os disulfide in equilibrium with it. In this case, the main role during the formation of the earlier RPGE minerals should be played by gas-transport reactions, whereas the deposition of RPGE minerals could take place according to a growth mechanism in the systems: vapor-crystal (VC mechanism) or vapor-liquid-crystal (VLC mechanism) (Givargizov, 1977). The dendritic growth of crystals and predominant formation of metallic forms of RPGE and their epitaxial aggregates with other ore or nonmetalliferous minerals are typical for both mechanisms. However, according to a number of other characteristics, the VLC mechanism (Givargizov, 1977) seems to be the most acceptable; in the field of precise chemical technologies, methods based on the VLC mechanism are the main ones used in obtaining metal monocrystals, including those of noble metals. The potential occurrences of RPGE are found in Keupirssai Russia and Big Trout Property, Ontario, which is analogous to the Bushveld complex. The Big Trout Lake Igneous Complex is analogous to other major chromium- rich intrusions, which contain very large stratabound PGE deposits.

DISCUSSIONS

The present study was taken up to review the chromite, associated economic minerals and PGE mineralization in ultramafic-mafic member hosted in ophiolite suite of rocks and allied rocks from selected parts of the world. Majority of chromites are of Podiform in ophiolitic suite of rocks of Cretaceous to Middle Miocene age. However, there are certain rocks with similar petrological and geochemical characters but differ in age from almost Greenstone Belts to Proterozoic, for example, Bushveld Complex, South Africa and Big Trout property, Canada. Chromite mineralization has been observed to have been formed under varied conditions, but majority of the host rocks are ultramafic-dunites, peridotites to olivinenorite etc. Further, the process of sulphurization has also been accounted for chromite and associated economic minerals mineralization. Also a comparative study of ophiolite sequences from different parts of the world also indicates that the most common host rocks for chromite mineralizations are ultramafic dunites and peridotites, viz. harzburgites, lherzolites and wehrlites.

Besides, these ultramafic-mafic intrusive rocks, which are very abundant in the study area, the volcanic constituents are rather sporadic to rare in the field. Also volcanic breccia and agglomerates have been observed from Lungphor village, near Tussar Silk Farm, Ukhrul, Manipur. Such a varied lithological character indicates that the material was expelled in pulses. The pillow structure indicates the pouring of basaltic magma over the marine surface, while the agglomerates and volcanic breccias suggest their violent explosion as well. The ideal suprasubduction zone (SSZ) ophiolites could be overlain by pelagic sediments, which in turn may be overlain by thick oceanic arc units (Hawkins et al., 1984, Robertson, 1994. Pearce et al., (1984) note that South Anatolian ophioltes constitute a complete sequence, with harzburgitic depleted mantle, sheeted dykes and island arc tholeiitic-type volcanics, locally including calc-alkaline plutonics and volcaniclastics characteristics of the SSZ ophioltes as follows:

(1) The SSZ-type ophioltes have gabbros, which differ petrographically from gabbros formed at Mid Oceanic Spreading Centres. Gabbros from SSZ have Cpx and Opx crystallizing before plagioclase (Bloomer et al., 1995).

(2) Elevated concentration of LILE relative to HFSE in SSZ-type magmas originate from fluids and/or siliceous melts derived from subduction oceanic slab. These slab-derived fluids carry high concentrations of LILE, while the HFSE are retained in the slab (Sajona et al., 2000). The slab fluids metasomatize the overlying mantle-wedge and trigger melting.

(3) The trace element signature of subduction-related magma is derived from slab-derived fluids and the overlying mantle-wedge (Pearce 1982, Aculus and Powell 1986, Yodogodzinski et al., 1993).

In view of the above discussion, it is concluded that more than 90% of the present study samples recorded equilibrium with compositionally indistinguishable parent melt, similar to Mid Oceanic Ridge Basalt (MORB) and the melt that formed the crust in the area.

Occurrences of variants of peridotites harzburgites, lherzolites and wehrlites- in the study area may also represent polybaric mantle derived melts. A segregation of polybaric melts may occur at the base of the lithosphere, because of the accumulation of melt beneath a porosity barrier at the base of the conductive boundary layer (Sparks and Parmentier, 1991). It is interesting to note that in the present study, though the dunites are the most common ultramafic rocks, but lacks completely in olivine-rich troctolites. This clearly supports the present observation that there is absence of plagioclase significantly and the majority of the rocks formed under deep mantle conditions. The absence of plagioclase is also evident from common occurrence of chrome-spinel in all the variants of peridotites.

CONCLUSIONS

The ultramafic rocks of the study area form the basal part of the ophiolte sequence, but the presence of wehrlites may indicate shallowing with rise of the parental magma, which may also suggest a polybaric differentiation from upper mantle up to crust-mantle Transition Zone. It is concluded that the more than 90% of the present study samples recorded equilibrium with compositionally indistinguishable parent melt, similar to Mid Oceanic Ridge Basalt (MORB) and the melt that formed the crust in the area. The absence of plagioclase significantly and the majority of the rocks formed under deep mantle conditions These basal constituents of Manipur-Nagaland Ophiolite suite were obducted during the subduction of the Indian Plate under the Myanmar plate during Cretaceous to Middle Miocene time.

ACKNOWLEDGEMENTS

Authors are grateful to the Department of Science & Technology for providing the funds under DST-FIST Assistance to the department and Drs. S. Manichandra and R.K. Chingkhei for improving the first draft of the manuscript.

References

Aculus, R.J. and P. Owell, R. 1986. Source component mixing in the regions of arc magma generation . J. Geophys. Res. v. 91, pp. 5913-5926.

Albers, J.P. 1986. Descriptive Model of podiform chromite: in Mineral Deposit Models, Cox, D.P. and Singer, D.A. (eds.) U.S. Geological Survey Bulletin, 1963. 34 p.

Aleksander Çina 2006. Chromites of Albanian ophiolite complex: a comparison with other northern and southern ophiolites. (In Press)

Andrada E.T., Sosa, L.A. et al., 1980. Memorandum report on the investigation for mineral resources in certain land areas in Carrascal, Cantillan and Madrid, Surigao Del Sur and Claver, Surigao del Norte, MGB Internal Report:SR-1990. 10 p.

Arco, R. C., 1962. Geologic reconnaissance and mineral resources of Bohol province, Philippines, MGB Internal Report: BO394, pp 1-68

Argaño, W. P., 1980. Geological verification of Benguet Corporation. Formerly (BCI) Mineral Claims Sta. Cruz, Zambales, MGB Internal Report: ZB-2071, 12 p.

Ash, Chris, 1996. Podiform chromite, in Selected British Columbia Mineral Deposit Profiles. V.2- Metallic Deposits, Lefebure, D.V. and Hoy, T. (eds.), British Columbia Ministry of Employment and Investment, Open File:1996-13, pp.1090112.

Bannatyne, B.B. and Trueman, D.L. 1982. Chromite reserves and geology of the Bird River Sill, Manitoba; Manitoba Energy and Mines, Mineral Resources Division, Open File Report OF82-1, 73 p.

Bateman, J.D. 1943. Bird River chromite deposits, Manitoba: Transactions of the Canadian Institute of Mining and Metallurgy, v. 46, p. 154-183.

Beshkul, H., Dobil, A., Gjoni V. and Pirdeni, A. (In press) Chrome bearing prospective of Mali lopes-10 Korribu-Theban-Ternove region (Bulquiza ultramafic Massif).

Bloomer, S.H., Taylor, B., Macleod, C.J., Stern, R.J., Fryer, P., Hawkins, J.W. and Johnson, L., 1995. Early arc volcanism and the ophiolite problem: A perspective from drilling in the western Pacific. In: B.Taylor and J. Natland (Eds.) Active Margin and Marginal Basins of the Western Pacific: Washington, D.C., Amer. Geophysics Union, pp.1-30.

Cerny, P., Trueman, D.L., Ziehlke, D.V., Goad, B.E. and Paul, B.J. 1981. The Cat Lake Winnipeg River and the Wekusko Lake pegmatite fields, Manitoba: Manitoba Energy and Mines, Mineral Resources Division, Economic Geology Report ER80-1, 215 p.

Coleman, R.C. 1981. Journal of Geophysical Research, v. 86, pp. 2497-2508.

Coleman, R.G. 1977. Ophiolites, Springer-Verlag, New York. 229 p.

Christiansen, F.G. 1986. Structural classification of ophiolitic chromite Deposits: In Metallogeny of Basic and Ultrabasic Rocks, In: Gallaghar, M.J., Ixer, R.A., Neary, C.R. and Pichard, H.M. (Eds.). The Institution of Mining and Metallurgy, pp. 279-289.

Edwards, R.L. and Wasserburg, G.J. 1985. The age and emplacement of obducted oceanic crust in the Urals from Sm-Nd and Rb-Sr Systematizes, Earth Planet. Sci. Lett., 1985, vol.72, pp. 389-404.

Givargizov, E.I., Rost Nitevidnyl'h i Plastinchatykh Kristallov 1977. In Para (The Growth of Filliform and Lamellar Crystals from Vapor), Moscow: Nauka, 1977.

Hancock, Kirk D. 1991. Ultramafic associated chromite and nickel occurrences in British Columbia, Canada. Ministry of Energy, Mines and petroleum Resources, Geological Survey Branch, pp. 1-68.

Hawkins, J.W., Bloomer, S.H., Evans, C.A. and Melchior, J.T. 1984. Himalaya, Him. Geol. V.10, pp. 412-439.

Karup-Miller, S. and Brummer, J.J. 1971. Geology and sulphide deposits of the Bird River claim group, southeastern Manitoba; Geological Association Canada, Special Paper 9, pp. 143-154.

Linnebacher, P., Stern, R.J., Reischmann, T., Manton, W. & Hussein, I.M. 1991. Evolution of Pan-African island arc assemblages in the southern Red Sea Hills, Sudan, and in southwestern Arabia as exemplified by geochemistry and geochronology. Precamb. Res. Vol. 53, pp. 99118.

Melcher, F., Grum, W., Simon, G., Thalhammer, T.V., and Stumpfl, E.F. 1997. Petrogenesis of the Ophiolitic Giant Chromite deposits of Kempirsai, Kazakhstan: A study of solid and fluid inclusions in chromite, J. Petrology, 1997, Vol. 38, No. 10, pp. 1419-1458.

Melcher, F., Stump, A.E.F. and Distler, V.V. 1994. Chromite deposits of the Kempirsai Massif, Southern Urals. Kazakhstan, London: Transactions of the Institution of Mining and Metallurgy (Section B.), vol. 103, pp. 107-

120.

Moos, D. and Zoback, M. 1982. In situ studies of seismic velocity in fractured crystalline rocks: Jour. Geophys. Res. vol. 88 pp. 2345-2358.

Pallister, J.S. & Fleck. R.J. 1992a. Age of initial oceanic magmatism in the late-Proterozoic Arabian Shield. Geology. Vol. 20, pp. 803-806.

Pearce, J.A., Harris, N.B.W. and Tindle, A.G. 1984. Trace element discrimination diagrams for tectonic interpretations of granitic rocks. J. Petrol. Vol. 25, pp. 956-983.

Proenza, J.A., Gervilla, F., Melgarejo, J.C., Vera, O., Alfonso, P. and Fallick, A. 2001. Genesis of sulfide-rich chromite ores by the interaction between chromitite and pegmatitic olivine-norite dikes in the Potosi Mine (Moa-Baracoa ophiolitic massif, eastern Cuba). Mineralium Deposita. V. 36(7), pp. 658-669.

Roberts, S. 1988. Ophiolitic Chromitite Formation: A marginal basin phenomenon? Economic Geology, v 83, 1034-1036.

Robesten, A.H.F. 1994. Role of tectonic fabrics concept in orogenic analysis and its application to Tethys in Eastern Miditernean region, Earth Sciences Reviews v.37, pp.139-213.

Sajona, F.G., Maury, R.C., Pubellier, M., Leterrier, J., Bellon, H. and Cotten, J. 2000. Magmatic source enrichment by slab-derived melts in a young post-collision setting, central Midanao (Philippines). Lithos, v. 54, pp 173-206.

Singer, D.A, Page, N.J. and Lipin, B.R. 1986. Grade and Tonnage Model of Major Podiform chromite; In Mineral Deposit Models, Cox, D.P. and Singer, D.A.(Eds.) U.S.Geological Survey, Bulletin 1693, pp. 38-44.

Sparks, D. and Parmentier, E. 1991. Melt extraction from the mantle beneath Mid Oceanic Ridges, Earth Planet. Sci. Lett. Vol.105, pp. 368-377.

Stowe, C.W. 1987. Evolution of Chromium Ore Fields: Van Nostrund Reinhold Co., New York, 340 p.

Thayer, T.P. 1964. Principal features and origin of podiform chromite deposits, and some observations on the Guleman-Soridag District, Turkey; Economic Geology, Vol. 59, pp. 1497-1524.

Theyer, P., Bruni, E. and Sundell, C. 2001. (Part of NTS 52L/5; In Report of Activities 2001, Manitoba Industry, Trade and Mines, Manitoba Geological Survey, pp. 126-132.

Trueman, D.L. 1971 Petrological, structural and magnetic studies of a layered basic intrusion Bird River Sill, Manitoba; M.Sc. thesis, University of Manitoba, 67 p.

Trueman, D.L. 1981. Stratigraphy, structure and metamorphic petrology of the Archean greenstone belt at Bird River, Manitoba; Ph.D. thesis, University of Manitoba, 150 p.

Venkataramana, P. and Bhattacharyya, S. 1989. Mode of occurrence and origin of chromite, magnetite and Ni-bearing laterite in the Naga Hills ophiolite belt, N.E. India. In: Ghose, N.C. (Ed.) Phanerozoic Ophiolites of India, Sumna Pub. & Distributors, Patna, pp. 213-234.

Wager, L.R. and Brown, G.M. 1967. Layered Igneous Complexes. Oliver & Boyd, Edinfurgh, 599 p.

Yodogodzinski, G.M., Rubenstone, J.L., Kay, S.M. and Kay, R.W. 1993. Mag-matic and tectonic development of the western Aleutians; an oceanic arcin a strike-slip setting: Journal of Geophysical Research, Vol. 98, pp. 11,807-11,824.

Hon'ble Reviewers

1. Prof. Naresh Kochhar
   Centre of Advanced Study in Geology
   Department of Geology
   Panjab University
   CHANDIGARH 160 014 INDIA

2. Prof. U.C. Singh
   Chairman
   School of Studies in Geology
   Jiwaji University
   GWALIOR 474 011 INDIA

Editorial Assistance

Mrs. Mamta Chauhan

Economic Mineralization
K.L. Shrivastava (Editor)
2009. pp.180-204
Scientific Publishers (India), Jodhpur

15

Section-II

Experimental, Petrological and Field Constraints on the Petrogenesis of Economic Carbonatites with Special Reference to the Sarnu-Dandali Area, Northwestern India

K.L.SHRIVASTAVA[1], YASMIN SAYYED[1], MAMTA CHAUHAN[1] AND ERNST HEGNER[2]

[1]Department of Geology, JNV University, Jodhpur 342005, INDIA
[2]Department of Earth Sciences, University of Munich, D80333 Munchen, GERMANY
Email : klsgeology@yahoo.co.in

Abstract

Carbonatites, are endogeneous accumulations of calcite, dolomite and other carbonate minerals. They are spatially and genetically associated with ultramafic, alkaline complexes and represent the sole economic source of Nb and the principal source of Rare Earth elements (REEs). In addition they constitute some of the largest sources of igneous phosphates and a rare source of strontianite. In the present paper experimental, petrological and field constraints have been reviewed to understand the petrogenesis of economic carbonatites, one of the most discussed controversy of economic geology. The largest carbonatite occurance located on the Kola Peninsula(Russia), the carbonatite of Mount Pass (California) showing the highest REE abundances, the Palabora carbonatite(Transvaal) showing the widest spectrum of economic deposits,as well as the ongoing eruptions of carbonatite of Oldoinyo Lengai (Africa) all have been well described with respect to field and petrological constraints. Yet it is realized that the Sarnu-Dandali carbonatite (Northwestern, India) is possibly providing the ideal evidence to satisfy the extensive experimental data now available.

There is now a general consensus among geoscientists that the parental magmas of carbonatites are mantle derived, and must be enriched in CO_2 ('carbonated'). Most carbonatite magmas are derived from carbonated silicate parental magmas by one of the three crystallization paths: (a) cooling of a magma to a liquidus-solidus terminal point where it precipitates silicates and CO_2 vapor, (b) cooling of a magma to a silicate-carbonate field boundary where silicates and carbonates are co-precipitated, or (c) cooling of a magma to a miscibility gap field boundary where a carbonate-rich magma is exsolved without the co-precipitation of calcite, except along the limiting field boundary where the two major liquidus surfaces intersect. Leaving aside many experimental data, the pseudoquaternary system CaO-(MgO +FeO (total Fe as Fe^{2+})-($Na_2O + K_2O$)-($SiO_2 + Al_2O_3 + TiO_2$) with CO_2 provides a framework for evaluating possible magmatic processes. Experimentally determined phase boundaries indicate that immiscible carbonate-rich melts, separated from melts similar to many natural silicate magmas tend to be concentrated near calciocarbonatite compositions, with a maximum $CaCO_3$ of being ~75-80 wt %, and $(Na, K)_2CO_3$ contents near 15 wt %.

The Sarnu-Dandali carbonatites have coherent and typically mantle C- and O- isotopic ratios ($\delta^{13}C_{v\text{-}PDB}$ -5.00 and -5.06; $\delta^{18}O_{v\text{-}SMOW}$ +19.06 and +18.92). The evolution of the carbonatite, by fractional crystallization of an alkaline melt following a magma series of ijolite melanephelinite- phonolite-carbonatite is well documented by petrological data and the field observations. Modeling the mineral associations determined, with the experimental data available, it has been suggested that the carbonatites of Sarnu-Dandali area must have been formed within a narrow temperature range of 925°C to 935°C. Associated mildly alkaline igneous rock were probably produced by different degrees of melting of a common mantle source.

Key words: Carbonatite, Sarnu-Dandali, Mantle Source, C-O isotopes, REE.

INTRODUCTION

Endogeneous accumulations of calcite, dolomite and other carbonates closely and genetically associated with complex ultrabasic alkali intrusives are known as carbonatites. In general, they represent the late phases in the development of alkaline complexes, particularly during the later stage during which CO_2 has been injected or volcanic CO_2 has remained trapped. The term carbonatite was introduced by Brogger (1921), while he was describing the Fen deposit (Norway).

Carbonatites constitute a distinctive group of rocks representing a distinctive ore type and environment of its formation. They are granular igneous rocks consisting primarily of calcite, dolomite, ankerite and other rock forming carbonate as principal constituents. Subordinate are apatite, magnetite, silicates and other accessory phases. Apparently, they crystallize from a carbonate rich melt and intrude into the rocks in which they are found in close association with igneous silicate rocks having low silica and high alkali and alkaline metal contents. They appear to be confined to zones of rifting in old stable shields of the continents. Carbonatites typically comprise of 80 99wt.% of carbonate and calcite sovite. Dolomitic, ankeritic and siderite manganiferous carbonatites show an increasing order of rareness. The combination of brecciation, multiple intrusions, assimilation and fenitization produces a very complicated overprinted fabric.

Carbonatites are the sole commercial source of Nb, the principal source of REEs, the largest source of igneous phosphates and a rare source of strontianite. Large deposits of magnetite are also known (e.g. Palabora). Due to the fact that the magnetites are titaniferous, too rich in Ti to constitute a good iron ore and too rich in Fe to constitute a valuable Ti ore. Additionally, a myriad of other metals and industrial minerals, such as vermiculite, barite, flourite, ZrO_2, U and Ni-sulphates, U-Ta and Th-Nb rich pyrochlore can be extracted as economically valuable by products. Thus carbonatites are potentially one of the valuable single rock type with such a diverse spectrum of mineralization.

Carbonatites began to be exploited for minerals since the second quarter of the last century when it was realized that they have considerable reserves of rare earth, apatite and elevated concentration of zirconium, tantalium and strontium. In places magnetite, phlogopite and vermiculite are of importance. The delayed exploitation of the carbonatites was because of the overdue realization of the usefulness of the elements, they contain. Most of the studies on carbonatites were carried out since the end of the second world war. Later, the discovery and investigation appears to have been accelerating steadily. About 150 massifs or groups of massifs of deep seated ultramafic alkali rocks containing carbonatites are known to date. Carbonatites came into geological history of India in 1960's with the discovery of flourite and conspicuous radioactivity at Amba Dungar in Gujarat. Radioactivity due to pyrochlore and apatite mineralisation led to Newania discovery (Dar, 1964). Like most enigmas they have aroused great curiosity and have provoked the development of a number of hypotheses. Carbonatites and primitive silicate volcanic rocks are mantle derived. On the basis of phase equilibrium experiment carbonatitic melts can be generated by (a) primary mantle melting, (b) liquid immiscibility, or (c) crystal fractionation.

The petrogenetic issues of carbonatites outlines the

most outstanding current problems in carbonatite research and raise several basic question. As suggested by various scientists (cf. Bell, 1989; Bell et al., 1998) these can be summarized as (1) what is the parental magma for alkali carbonatite suites as a whole or is there more than one parental magma type? (2) Is the parental magma primary or a derivative of. Is it derived from the lithospheric mantle or the asthenosphere, or does it represent mixture of both? (3) How is it derived from the ultimate source rock and from one another ? (4) How easily can parental magmas migrate through the mantle? How do such magmas effect the upper mantle chemically and mineralogically. (5) What are the spatial and temporal relations between silicate rocks such as melilites, nephelinite, ijolite, syenite and phonolite and associated carbonatite? (6) How do carbonate magmas fractionate and evolve?

The aim of the paper is to review the present knowledge on carbonatites covering almost all gamuts of controversies, chiefly, experimental, petrological and field constraints on the petrogenesis of economically valuable carbonatites. The study of carbonatites from the Sarnu- Dandali area, which demonstrates an ideal condition, have been added as a special reference.

GENERAL CHARACTERISTICS OF CARBONATITES

1. Form and Structures

Carbonatites occur as small plugs, stocks, ringdykes, conesheets and carbonate breccia in association with alkali rocks. At places they also occur in the form of large irregular or sheet like bodies, sills or dyke swarms, adjacent to the alkali rock complexes or as an independent composite complex.

The typical pipe like bodies have sub-circular or elliptical cross sections and are 34 km in diameter. Magmatic mineralisation within the pipe-like carbonatites is commonly found in crescent shaped and steeply dipping zones. Metasomatic mineralisations occur as irregular forms or veins.

Carbonatites are associated with ultramafic alkali intrusions of a central type occurring exclusively on platforms. The major faults of the platform, which control the ultramafic alkali intrusions including carbonatite deposits, can be found generally along the boundaries of trough and rises. The early-formed faults have been rejuvenated in the course of magma intrusions.

Extrusive carbonatites are rarely found as lava flows as found at Oldoinyo Lengai volcanoe (Tanzania), as pyroclastic tuffs (Hopi Butes, Arizona) or coarse-grained bombs (Laacher See, Germany).The form of the lava appears to be essentially similar to that of normal silicate lavas. Carbonatitic volcanic activity is characterized by the abundance of volatiles and ranges from coarse agglomerates to fine tuff.

The geological structures that govern the position and morphology of carbonatite bodies have a single source of deforming stresses but morphologically they are of two types:

(i) Central stocks they occur in cylindrical explosion vents formed by an eruption of gas along the axis of volcanic structures before carbonatite formation and open the conduits for the ascent of agents that create carbonatite plugs. The explosion leads to brecciation of rocks, breached by the force of gas pressure.

(ii) Dykes they occur in a circular structure as radial, ring and conical dykes.

Ring and conical dykes (shear structure) are caused by increasing upward pressure in an abyssal magma source and appear as shear fractures along the surface of maximum strain. Radial dykes are rupture structures. Dyke structures are most typical of deeply eroded parts of the carbonatite deposits. A sudden increase in the pressure of the gaseous phase may result in a magma, creating cylindrical vents for an explosion, which is characteristic of the near- surface conditions.

The character of tectonic deformation also changes in accordance with the scheme of evolution of geological structure within carbonatite deposits. Biaxial strains are typical of the early stages of development of geological structures, prior to the injection of mineral-bearing melts. During the injection of the melt and the formation of carbonatites, the deformation becomes three dimensional. The post-carbonatite disturbances that shift the ore bodies, are also caused by the plane strains.

2. Mineralogy and Texture

Carbonatite rarely occurs alone. A typical association includes carbonatite nephelinite (olivine poor), alkali peridotite-pyroxenite, ijolite- syenite and carbonatite-syenite complexes. The varieties include calcium carbonate (sovite), dolomitic carbonatite (beforsite), ferroan or ankeritic calcite-rich carbonate (ferro-carbonatite), magnesite-olivine-apatite± phlogopite rocks, nephelinite, syenite, pyroxenite, peridotite and phonolite. Country rocks may be of various lithologies and of different metamorphic grade. Single or varied types of carbonatites such as calcitic to magnesio and ferro-carbonatites usually differentiated with decreasing age, are commonly present. Heterogeneity in structure and texture, due to mineral banding and clotting, is very common.

Ore mineralogy (principal and subordinate) in carbonatite is represented by different ore minerals in various genetic classes:

In magmatic types occur: bastnaesite, pyrochlore, apatite, zircon, baddeleyite, magnetite, monazite, parasite and fersmite.

In replacement/veins occur: flourite, vermiculite, bornite, chalcopyrite and other sulphides and haematite.

In residual types occur: anatase, pyrochlore, apatite. Locally a crandallite group of minerals containing REE.

Gangue mineralogy is represented by calcite, dolomite, siderite, ferroan calcite, ankerite, haematite, biotite, titanite, olivine and quartz.

There is a definite sequence, observed in carbonatite for the formation of minerals. Calcite is followed by dolomite and then ankerite. Other minerals (over 70 in number) are virtually accessory in quantity. Quite specific ones are those accessories that are regarded as hypomorphic and they include phlogopite and apatite and less often the following minerals:

(1) Pyrochlore $(Na, Ca, Ce)_2$, $(Nb, Ti, Ta)_2 O_6 (OH, F)$.

(2) Baddeleyite or brazilite (ZrO_2)

(3) Perovskite-knopite-dysanalyte (Ca, (Ce) (Ti, Fe^{3+}, Nb) O_2

(4) Carbonatite minerals of rare earth like synchisite Ca (Ce, Ba) $(CO_3)_2$ F, parasite Ca (Ce, La)$_2$ $(CO_3)_3$ F_2 and bastanaesite (Ce, La) (CO_3) F.

Most carbonatites are massive in texture but occasionally they are banded, spotty or craggy. The latter three types are due to aggregation of dark accessory minerals that form linear chains or thin layers in carbonatite bodies. The carbonatite is normally granular with difference in size of their constituent minerals.

REE minerals form pockets and fill fractures within ferro-carbonatite bodies. Pyrochlore is dissiminated, apatite can be dissiminated to semi-massive, bastnaesite occurs as dissiminated to patchy accumulations, fluorite forms veins and massive portions, haematite occurs as semi-massive dissimination and chalcopyrite and bornite are found in veinlets.

3. Tectonic Setting and Age

Carbonatites occur mainly in a continental environment and rarely in oceanic environments as on the Canary Islands. They are generally confined to large scale, intra-plate fractures, graben or rifts that correlate with the periods of extension, and may be associated with a broad zone of epiorogenic uplift. Ranging in age from the Proterozoic to Recent, carbonatites are widespread in space and time but are most conspicuous in East Africa during the last 150 Ma. They appear to be increasingly abundant with decreasing age. The grand period of carbonatite development includes the Kaiserstuhl occurrence (Tertiary), the Tanzania-Uganda-Kenya-Nyasaland province of E. Africa (Mesozoic-Tertiary-Recent), Sao Paulo Minas Gerais- Santa Caterina belt of Brazil (Early Jurassic) and the Monteregian province of Eastern Canada (Mesozoic).

In India, four periods of carbonatite activity have been recorded, namely the Middle Proterozoic, Late Proterozoic, Late Mesozoic and the Cenozoic.

4. Classification

(i) Petrological Classification

(1) Sovite:Coarse grained calcite rich carbonatite with > 90% calcite and containing upto 10% dolomite.

(2) Rauhaugite: Dolomite or magnesian carbonatite containing 10% calcite.

(3) Alvikite and beforsite : Hypabyassal equivalent of sovite and rauhaugite.

(4) Silico-carbonatite : Containing a larger proportion of the silicate Phase than carbonates.

(5) Manganeferrous carbonatite or ferro-carbonatite: Carbonatite enriched in Fe and Mn or Ankerite and Siderite.

(6) Metacarbonatite: Carbonatite subjected to metamorphism.

(ii) Economic Classification

(a) Apatite-magnetite type:

It is most widespread. It has a content of niobium

and titanium minerals. Reserves of iron ore are occasionally high.

(b) Rare-metal rare earth type:

It is subordinate. Occurance of pyrochlore, bastanaesite, parisite, baddeleyite characterizes this type.

(c) Phlogopite type:

It is distinguished by mica in the outer margins of the carbonatite stocks.

(iii) Genetic Classification

C_1 (Sovite): Usually emplaced in an envelop or explosive brecciated rock. Form: penetrative stocks-like intrusions.

C_2 (Alvikite): Usually shows marked flow banding and medium to fine Grain, form: conesheets and dykes.

C_3 (Ferro-carbonatite): Contains Fe-bearing carbonate minerals, some rare earth and radioactive minerals, form: conesheets and dykes.

C_4 (Late stage Alvikites): Usually barren.

5. Economic Significance

At first considered to be petrological oddities, the carbonatites are now becoming important economic targets. Evidence of constitution, eruptive phenomenon and fenitization carries with it the likelihood of the concentration of economically valuable minerals.

Only intrusive carbonatites (not lava flow or pyroclastics) are associated with mineralisations of economic consideration. These occur either as primary igneous bodies, replacement deposits or residual weathering accumulations from either igneous or replacement protores. Pyrochlore, apatite and REE- bearing minerals are typically the most sought after mineral constituents. However, a wide variety of other minerals, including magnetite, fluorite, calcite, bornite, chalcopyrite and vermiculite, occur in economic concentrations in a few carbonate complex. Carbonatites are heterogeneous with regard to their mineralogy and show considerable diversity in terms of their major, minor and trace element compositions.The SrO and BaO contents of average carbonatite are 0.26 wt.% and 0.40 wt.%, respectively, their concentration is > 1 wt.% in many carbonatites. Other chemical characteristics are their high P_2O_5 (typically ≤ 2 wt.%), F (< wt.1% but quite often > 2 wt.%), Cl (0.10- 0.20 wt.% but over 2 wt.% in natrocarbonatite), SO_3 usually >1 wt.%, but in natrocarbonatite > 2 wt.%), CO_2 content (28 32 wt.%). Carbonatites are enriched in Nb (< 0.2 wt.%), Zr (0.10 wt.%), La (0.05 0.10 wt.%) and Ce (0.15 wt.%).They are characterized by the high concentration of Na_2O, K_2O, BaO, SrO, P_2O_5, CO_2, F, Cl and SO_3. The average Na_2O content of carbonatite is 0.4 wt.% but in natrocarbonatite it is > 7 wt.% and in extreme cases it may be upto 30 wt.%. In addition to the abundant carbonate minerals, they contain the following among their minor components and accessories: phosphate (apatite), Fe (magnetite), manganese oxide and small but highly significant quantities of niobium (pyrochlore), uranium, thorium, REE (monazite, basnasite), copper, flourite, barite, and strontionite. It is a useful source of lime in the area devoid of good limestone. There are at least five major stages of economic mineralisation that can be recognized during the evolution of carbonatites (Heinrich, 1966 and Le Bas, 1987). In the Indian context, many complexes show distinct and at times overlapping stages of mineralisation.

(a) Apatite-magnetite mineralisation: This is an early stage of mineralisation and well developed in plutonic complexes of Sevattur and Sung valley (Krishnamurthy, 1985).

(b) Pyrochlore mineralisation: middle stage

(c) REE mineralisation: It is reported from late stage ankeritic or sideritic phases occurring at Amba Dungar and Samalpatti. REE mineralisation in the Indian carbonatites belongs to a very late stage low temperature hydrothermal or pegmatitic phase.

(d) Flourite and barite mineralisation: Late stage silica rich barite- fluorite carbonatites of Amba Dungar.

(e) Uranium and thorium mineralisation: This appears to be poorly developed in most of the Indian complexes.

Geological structures, governing carbonatite emplacement of carbonatites are peculiar and directly related to stress created by the magma. Intrusive forms and cooling history (fractional crystallization) controls the primary igneous deposits. Tectonics and the local structure influence the forms of the metasomatic mineralisation. The depth of weathering and drainage pattern decide the nature of residual pyrochlore and apatite whereas vermiculite deposits are residual alluvial and eluvial placers. Carbonatites

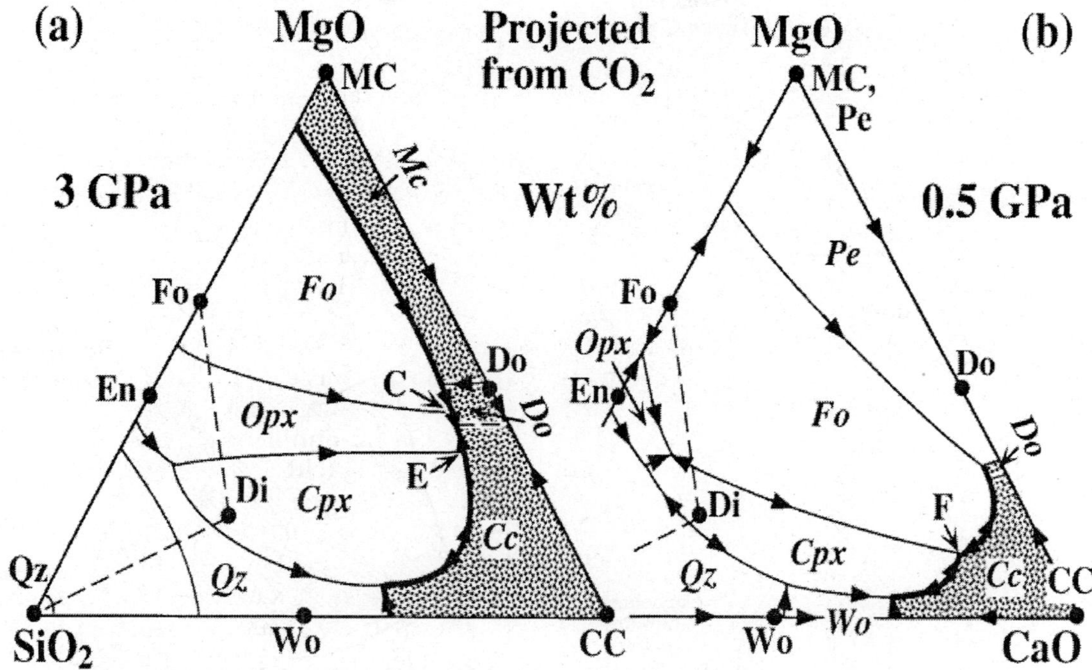

Fig. 1 Projection of the vapor-saturated liquidus surface of $CaO-MgO-SiO_2-CO_2$ (Wyllie and Huang, 1975, 1976; Wyllie and Lee, 1998). **(a)** the system at 3GPa showing in stippled area the carbonate liquidus (solid solutions of calcite Cc, dolomite Do, and magnesite Mc), silicate liquidus (forsterite Fo, orthopyroxene Opx, clinopyroxene Cpx, quartz Qz), and the boundary (bold line) between them. The figure shows that at higher pressure, CO_2 removes the thermal barriers and the crystallizing liquids may follow paths down to the silicate-carbonate field boundary and coprecipitate dolomitic carbonates. The carbonate liquidus region is a forbidden zone for derivatives of silicate parental magmas; hence silicate-CO_2 liquids cannot cross the silicate-carbonate field boundary by fractionation. **(b)** Haplobasaltic liquids forms minerals plus vapor at crustal pressures because a thermal barrier prevents them from reaching the silicate-carbonate field boundary and coprecipitating carbonates. It can be seen that at crustal pressures, the liquidus field for periclase expands at the expense of magnesite. The liquidus field for calcite ends at a field boundary for coprecipitation of wollastonite at low pressures, and quartz at high pressures (Huang et al., 1980). Subsolidus magnesite and dolomite break down successively with decreasing pressure (Irving and Wyllie, 1975), which is responsible for the reduction of the length of this field boundary (cited in Lee and Wyllie, 1998b)

weather with ease, and are commonly are associated with topographic lows. Presence of fenite is one of the important hallmarks of a typical carbonatite complex. Not all carbonatites cause fenitization and there are unfenitized silicate xenoliths in some carbonatites. A fenitization haloe of alkali metasomatised country rock commonly surrounds the carbonatite intrusions. Alteration mineralogy depends largely on the composition of the host rock. Typical minerals are sodic amphibole, wollastonite, nepheline, mesoperthite, antiperthite, aegirine augite, pale brown biotite, phlogopite, and albite. Most fenites are zones of desilicification with addition of Fe^{3+}, Na and K. Nepheline syenites are typically related to some corundum deposits and sodalite, REE and placer deposits of zircon. Wollastonite and fluorite may also be present. Kimberlite and lamprolites may also be present along with some tectonic features.

EXPERIMENTAL STUDIES ON CARBONATITES

In recent years, a large number of experimental studies have been undertaken to understand the role of various components in the petrogenesis of carbonatite. Figure 1. shows a projection of the vapour-saturated liquidus surface of $CaO-MgO-SiO_2-CO_2$ (Wyllie and Huang, 1975, 1976; Wyllie and Lee, 1998), partly estimated on the basis of experimental data at mantle and crustal pressures (Lee and Wyllie, 1998b). The miscibility gap between silicate and carbonate melts (Freestone and Hamilton,1980; Kjarsgaard and Hamilton, 1988, 1989a) are shown in Figures 2,3 and 4 , taken from Lee and Wyllie (1998b). Lee and Wyllie (1998b) further modified the temperature and composition prism for

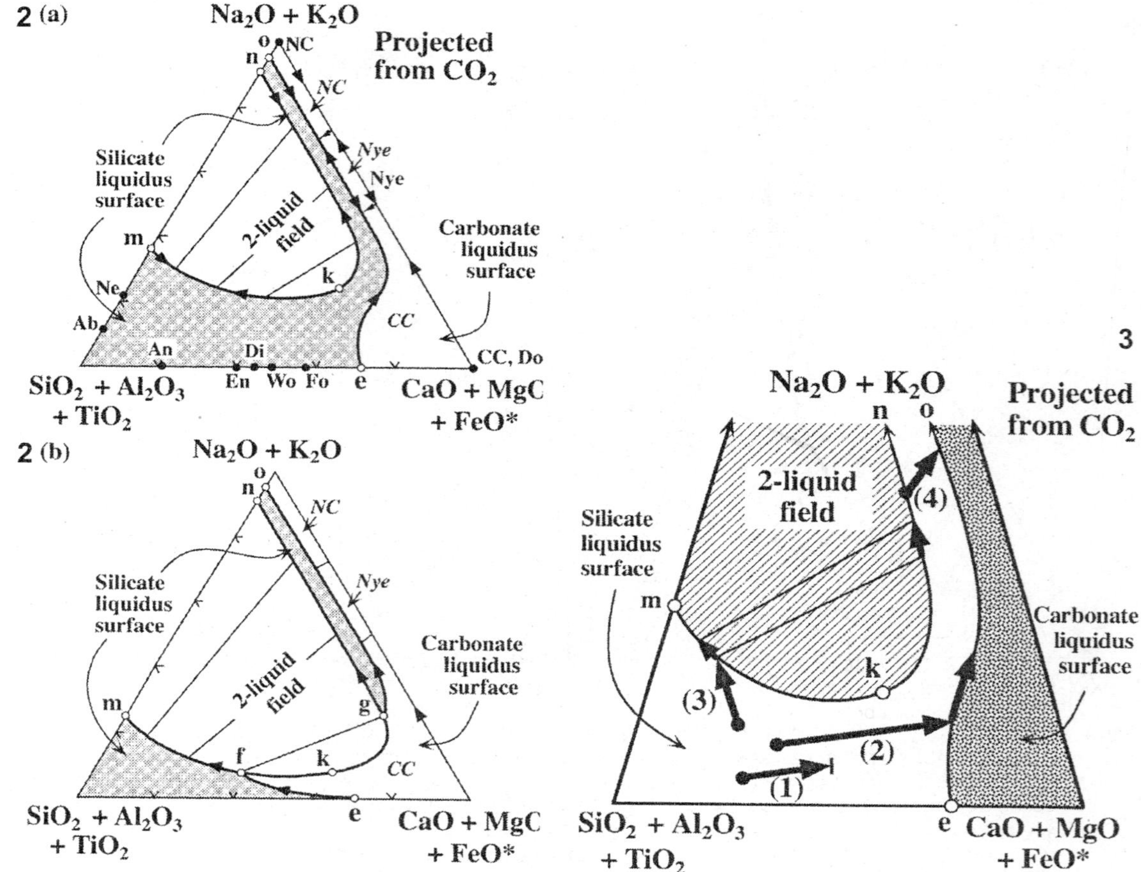

Fig. 2 and 3 Hamilton projection to represent miscibility gap between silicate and carbonate liquids (Freestone and Hamilton, 1980; Kjarsgaard and Hamilton, 1988, 1989a). In Fig. 2, pseudoternary phase, distorted for the purpose of illustration, showing silicate-carbonate field boundary e-o (e-f/g-o) and the miscibility gap field boundary m-k-n (m-f-k-g-n) (Lee and Wyllie, 1996, 1997a, 1997b). In Fig. 2(a) the miscibility gap (two-liquid field with tie-lines) is completely surrounded by the shaded silicate liquidus surface and separated from the carbonate liquidus surface. In Fig. 2(b), the two field boundaries intersect at f and g, dividing the silicate liquidus surface into two parts. **Fig. 3.** Showing lower portion of the pseudoternary diagram (as in Fig. 2 (a)), distorted and vertically exaggerated (by two) to show the shaded miscibility gap, the stippled carbonate liquidus surface, and various crystallization paths, arrows ((1) to (4)), of parental silicate-CO_2 liquids on the silicate liquidus surface. (Both Figs. 2 and 3 are cited in Lee and Wyllie, 1998b).

the Hamilton projection, proposed by Lee and Wyllie (1994) as shown in Figure 5. The system CaO-CO_2-H_2O, suggested by Wyllie and Tuttle (1960) influenced by the effect of H_2O on the silicate-carbonate field boundary is shown in Fig. 6A and the vapour- saturated liquidus surface projected onto the join Ca_2SiO_4-$CaCO_3$-$Ca(OH)_2$ (Wyllie and Haas, 1965) is shown in Figure 6 (b) (Lee and Wyllie, 1998b).

Experimental constraints governing the formation of the three major types of the carbonatites, will be discussed in the following sections.

1. Dolomitic Carbonatites

Experimental studies suggest that dolomitic carbonatites cannot be produced as immiscible liquids from silicate parental melts (Lee and Wyllie, 1998a). Carbonate minerals occur in mantle rocks as e.g. dolomite $(CaMgCO_3)_2$ and magnesite $(MgCO_3)$. The experimental studies of Wallace and Green (1988) support mantle metasomatism by carbonate melts which may occur between 21-31 kbar and 930-1080°C. Canil and Scarfe (1990) subsequently suggested that the temperatures and pressures are dependent upon the CO_2 content. These conditions are traversed by normal oceanic or young continental lithosphere geotherms. A major source of carbonate-

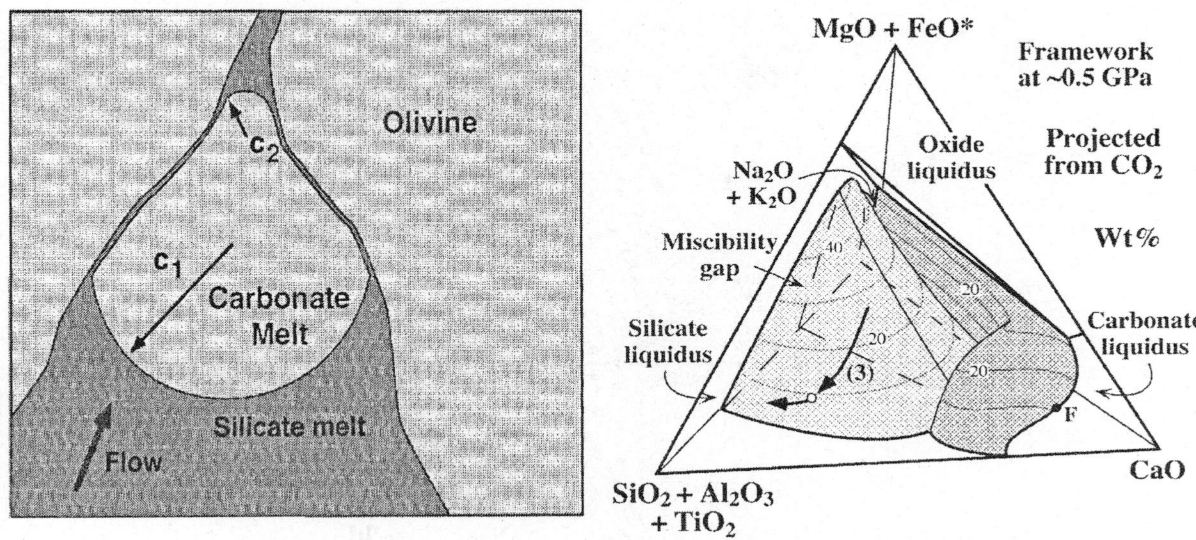

Fig. 4 Generalized pseudoquaternary phase diagram at 0·5 GPa (projected from CO_2), showing dark- and light-shaded liquidus surfaces for the silicate-carbonate field boundary and liquid miscibility gap, with contours for wt % (MgO + FeO*). Bailey, (1994) demonstrating carbonate-silicate immiscible melts is inhibited by the increase in mean curvature C_1 to C_2 (carbonate-silicate melt interface), but interfacial tension will counteract. A stable mineral is in equilibrium(compositionally) with both conjugate silicate-carbonate melts in a stable system. In dynamic systems where minerals are growing fast relative to diffusive homogenization of the melt compositions, a growing mineral could deplete the silicate melt in contrast with that composed of those elements preferentially partitioned into the carbonate melt (LREE and LILE; Jones et al., 1995, cited in Lee and Wyllie, 1998b).

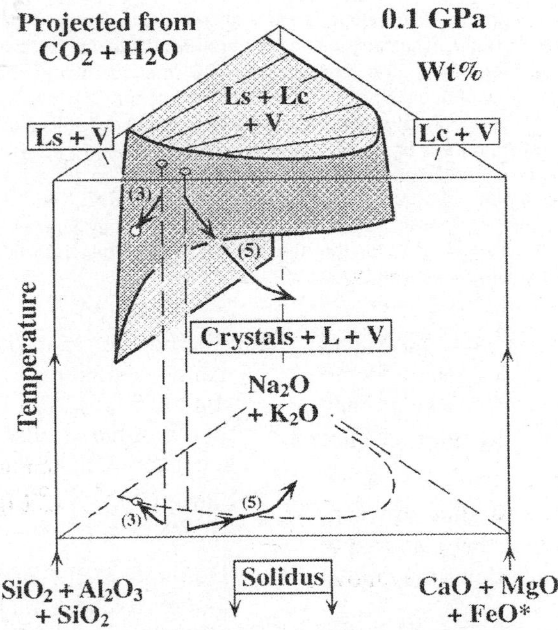

Fig.5 Temperature-composition prism for the Hamilton projection (after Lee and Wyllie, 1994 and modified by Lee and Wyllie, 1998b), showing the miscibility gap (shaded volume with two-liquid tie-lines), liquid paths (3) and (5), and their projections on the base of the prism. Path (3) intersecting the miscibility gap (at open circle) corresponds to that in Figs. 3 and 4, and path (5) for magmas with H_2O passes below the gap, and changes direction when it reaches the hydrous silicate-carbonate field boundary.

rich melts in the convecting mantle may be subducted carbonate (Barker, 1996). Carbonate rich magmas may be generated at depths >70 km by very small degree partial melting of carbonated (dolomitemagnesite) peridotite.

Experimental studies of e.g. Wallace and Green,

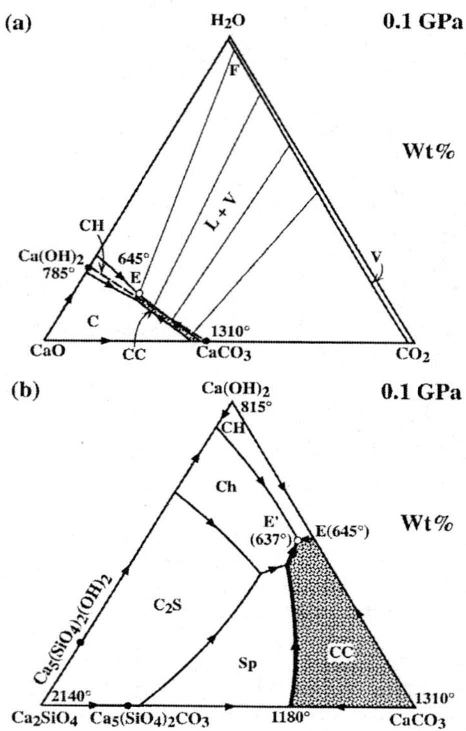

Fig.6 (a) The effect of H_2O on silicate-carbonate field boundaries is shown in the system (CaO-CO_2-H_2O, Wyllie and Tuttle, 1960) demonstrating the effect of H_2O on calcite melting temperatures. With marked decrease in liquidus temperature from 1310°C to 645°C by addition of H_2O, the composition of CC (melted calcite) changes towards portlandite CH (portlandite). The coexisting vapour phase changes progressively from pure CO_2 to almost pure H_2O (F) at the eutectic temperature, with partition of H_2O strongly favouring the vapour. This result provides a simple hydrous liquid capable of precipitating only calcite through hundreds of degrees, with the potential to yield a pure söviite cumulate. The final liquid at E would also precipitate a non characteristic carbonatite mineral, portlandite. (Cited in Lee and Wyllie, 1998b). Fig. 6 (b) Vapor-saturated liquidus surface projected onto the join Ca_2SiO_4-$CaCO_3$-$Ca(OH)_2$ (Whyllie and Haas, 1965), showing the shaded calcite liquidus surface, the bold silicate-calcite field boundary, and the eutectic liquid at E' (open circle). Haplocarbonatite magmas cannot enter the shaded calcite liquidus, a forbidden zone. C_2S, Ca_2SiO_4; Ch, calciochondrodite; Sp, spurrite. (Cited in Lee and Wyllie, 1998b).

(1988) and Dalton and Wood (1993) have shown that carbonatitic melts, like silicate melts, can undergo 'thermal death' before they reach crustal levels. They may also undergo 'chemical death' because of their highly reactive nature.

i) Phase-equilibria experimental studies by e.g. Dalton and Wood (1993) suggest that carbonatite melts react with lithospheric mantle as they move upwards from ~85 km depth.

ii) The solidus inflection or 'ledge' at 21 kbar is controlled by the decarbonation reaction :

$$CaMg(CO_3)_2 + 4\,MgSiO_3 = CaMgSi_2O_6 + 2Mg_2SiO_4 + 2CO_2$$

(Dolomite melt + orthopyroxene = clinopyroxene olivine fluid)

This carbonate metasomatism converts depleted mantle (harzburgite) to wehrlite (ol+cpx) at <21 kbars (70 km). At higher pressures of 25-30 kbar, the dolomite + enstatite assemblage is unstable and reacts with magnesite.

$$CaMg(CO_3)_2 + 2\,MgSiO_3 = CaMgSi_2O_6 + 2MgCO_3$$

(Dolomite melt + orthopyroxene = clinopyroxene + magnesite)

iii) Between 21 and 31 kbars in the garnetspinel transition zone, melts derived from metasomatised mantle change from carbonatite to olivine melilitite, as amphibole breaks down or melts and the degree of melting increases.

At low pressures <21 kbar instability of carbonate in the presence of pyroxene produces a high-temperature solidus at the amphibole melting reaction. The stability of amphibole and phlogopite is influenced by the CO_2:H_2O ratio. It seems highly plausible that melting of amphibole or phlogopite-bearing lherzolite at <21 kbar may generate lamprophyres and explain the occurrence of lamprophyres and carbonatites in the field. If melt segregation and aggregation occur within the carbonatite melt window between 65-95 km, this low viscosity melt may move rapidly along dykes and veins to the Earth's surface yielding primary dolomitic carbonatite magma. Ascent rates have been estimated at 20-65 m/s.

2. Calcic Carbonatites

They may result from :

(a) Melting of secondary calcitic carbonatite in recycled oceanic crust, i.e. carbonated eclogite in an upwelling mantle plume (Hoernle et al., 2002);

(b) Low pressure <8 kbar liquid immiscibility: According to Wyllie and Lee (1998) calciocarbonatites are not generated directly from the mantle. They are immiscible liquids separated from liquids corresponding to many different natural silicate magmas. At low pressures of 1-5 kbar a range of carbonate liquids with respect to Ca and Na appears to separate from a residuum to form a range of carbonated fractionating silicate magmas such as melanephelinite/melilitite-phonolite.Carbonate/silicate liquid immiscibility varies as a function of pressure and magma composition especially with respect to Mg/Ca and Al/Si ratios. The compositions of crystallizing phases may be used to infer immiscibility pressures and temperatures.

(c) Crystal fractionation: If crystallization in the silicate nephelinitic melt is controlled by nepheline, phlogopite and clinopyroxene, then the liquid composition will be driven towards the calcite/silicate phase co-tectic. In this case, calcite will precipitate and calciocarbonatites can be generated by fractionation without immiscibility.

3. Natrocarbonatites

The natrocarbonatite lavas from Oldoino Lengai are chemically different from other known volcanic and plutonic carbonatites because they contain large amounts of alkale up to 33 wt.%, Na_2O of 5 wt.% and K_2O of 14 wt.% as well as low CaO of 14 wt.% (Dawson, 1966). This has led to a lengthy debate whether the parental magma was sodic or calcic carbonatite. The principle mineral in the porphyritic lavas is nyererite (Na, K) Ca $(CO_3)_2$ along with other alkali carbonates and chlorides. The natrocarbonatites appear to have been formed by immiscibility, following extensive crystallization. In the Ca-Na carbonate system there is a thermal barrier between calcite and nyererite, which precludes a parent-daughter relationship. However, it is removed by addition of F which may explain the occurance of nyererite, e.g., at Oldoinyo Lenghai where halogen occurs in the lavas (Dawson et al., 1994).

FIELD AND PETROLOGICAL CONSTRAINTS

Carbonatite rocks occur on all continents. They are widespread globally, particularly, in Africa and Canada. There are many localities of such rocks in Europe, Asia and America. In India, large massifs of carbonatitic rocks, forming mappable outcrops, are found at (1) the Chhota Udaipur district, Gujarat including the Amba Dungar ring complex, the Siriwasan brecciated sill, and the dykes and plugs of the Panwad Kawant area, (2) Newania in Rajasthan and (3) Sevatthur and Samalpatli in Tamil Nadu. Beside these occurance, there are other massifs of alkalic and carbonatitic rocks which occur as lenses, dykes and veins in the country rocks as at Salem and Vellore of Tamil Nadu, Sung Valley (Assam), Mundwara and Sarnu-Dandali near Barmer (Rajasthan) and isolated patches in the Eastern Ghats.

A brief resume of different carbonatite occurances together with their various field and petrological constraint is given below

1. The Mountain Pass Carbonatite, California, USA

The Mountain Pass carbonatite occurs within the Rocky mountain area, as a 1400 Ma old Precambrian intrusion into Precambrian gneiss. The deposits lie in a belt about 10 km long and 2.5 km wide. One of its deposits, the Sulphide Queen carbonatite body, carries a significant concentration of REEs (Olson and Pray, 1954; Evans, 1993).

The metamorphic country rocks have been intruded by potash-rich igneous rocks and REE-bearing carbonate rocks genetically and spatially related to granites, syenites, and shonkinites. Most of the 200 veins that have been mapped are less than 2 m thick. Howeve, Sulphide Queen carbonatite, a 200 m wide and about 730 m long massif is the largest ore body of REE on Earth. Carbonate minerals make up about 60% of the veins and constitute the large carbonate body, which consists chiefly of calcite, dolomite, ankerite, and siderite. Other constituents are barite, bastnasite, parasite and quartz. The REE content of much of the ore body ranges from 5 to 15 wt.%.

2. Palabora Igneous Complex, Transvaal, South Africa

Palabora is one of the most economical, low-grade Cumine in the world. No other mine can match Palabora for the variety of its resources and potentially extractable metals (Th, Al, K and REE; Russell et al., 1954; Verwoerd, 1966; Stanton, 1972).

This complex is located in the Archean Kaapvaal craton of the north- eastern Transvaal. It resulted from alkaline intrusive activity successive stages of emplacement of pyroxenite, syenite and ultramafic pegmatoids.

The first intrusion is that of apatite-rich phlogopite pyroxenite, now preserved as a 6 x 2.5 km kidney-shaped outcrop. Ultramafic pegmatoids were then developed at three centres within the pyroxenite pipe. In the central one, phoskorite a magnetite-olivine-apatite rock and banded carbonatite were emplaced to form the Loolekop carbonatite- phoskorite pipe, that is about 1.4 x 0.8 km in area.

The fracturing of this pipe led to the intrusion of the dyke-like body of carbonatite and development of a stockwork of carbonatite veinlets. The zone along which the main body of carbonatite was emplaced suffered repeated fracturing, and mineralizing fluids migrated along fractures where they deposited copper sulphides. Chalcopyrite with minor cubanite is the principal copper mineral in the core of the carbonatite. Vallerite occurs in both phoskorite and carbonatite. Magnetite with up to 4 wt. %. SiO_2 makes up 25-50% of the phoskorite. Apatite is present in economic amounts in phoskorite. Baddeleyite is recovered from both phoskorite and carbonatite. According to Eriksson (1989), the sulphide mineralization may be an intrinsic part of carbonatite magmatism. Aldous (1986) suggested that Curich carbonatites at Palabora could have arisen from pyroxenitic melts by liquid immiscibility.

3. Alkaline Igneous Complex, Kola Peninsula, Russia

Inward dipping, layered intrusions of various alkaline rocks of the Kola Peninsula are an excellent example of alkaline igneous complexes in the northern Karelia alkaline province. Associated carbonatite, show a spatial relationship to hot-spot activity accompanied by doming and fracturing. The upper Palaeozoic alkaline igneous and carbonatite complex hosts a number of extremely large ore bodies, of which the most important are Khibina, Kovdor, and Sokli. Khibina is a ring complex of about 40 km radius (Notholt, 1979). At Kovdor there is a late-stage development of apatite-fosterite rocks and magnetite ores. Baddeleyite and vermiculite are also recovered (Rimskaya-Korsakova, 1964). Across the border of Finland is the Sokli Complex with an aerial extent of 18 km^2 forming the world's largest carbonatite body (Notholt, 1979; Vartiainen and Parma, 1979). Pyrochlore with U-Ta and Th enrichment occurs in the carbonatites and apatite-magnetite mineralization in metaphoskorites. Seismic evidence suggests that the intrusion tapers downwards from a diameter of 6 km at the surface to about 1 km at 5 km. This suggests that carbonatites like kimberlite become dyke like bodies within deep fracture zones tapping levels in the mantle from which these alkaline rocks and their REE and associated elements have probably been derived.

4.Natrocarbonatites at Oldoinyo Lengai, Africa

The magma that erupted at Oldoinyo Lengai is very unusual for two main reasons. Firstly, it is an extrusive carbonatite that is an uncommon variety. Besides, there are other sites where extrusive carbonatites occur, i.e., at Kerimasi, Mosonik and Shombole near by eastern rift valley and Kaiserstuhl in the Rhine graben (Germany). Secondly the natrocarbonatite contains high proportion of alkalies, dominated by sodium carbonate and is known as natrocarbonatite.

Natrocarbonatites are made up largely of two minerals, nyerecite and gregoryite. Both minerals are carbonates with significant amounts of Na and K. They are anhydrous and react quickly with fluids. As a result the colour of the rock changes from black to white and texture changes from hard to soft and crumbly.

Bell and Keller (1995) suggested that lava from Oldoinyo Lengai probably represents a unique case of volcanic activity. It reflects an extreme composition in a protracted range of melts associated with nephelinite carbonatite centres.

Natrocarbonatite is fluid at much lower temperature (540°C to 953°C) than silicate lava. Another unusual feature is its very low viscosity. Pinkerton's measurements have shown it to be lowest of any terrestrial melts. It has been suggested that the Venusian landforms have been formed by "unusually fluid, low temperature lavas with extraordinary composition, such as molten mixture of carbonates, sulfates and other salts".

5. Amba Dungar, Gujarat, India

This occurance is located 37 km south of Chhota Udaipur in the Vadodara district of Gujarat. The alkaline carbonatite complex (21° 58'-22° 3' N and 74° 74° 8'E) occurs within the Deccan Trap region of Malwa. The area has been studied by several investigators (Sukeshwala and Avasia, 1972; Sukeshwala and Udas, 1963). The Amba Dungar carbonatite intruded into the Deccan Trap basalts. Isotope dating indicates an age of 37.5 ± 2.5 Ma for the Amba Dungar Complex, Sukeshwala and Udas (op. cit.).

The chief associated igneous rocks are phonolite, tinguaite and plagio- fayaites. The Amba Dungar carbonatite are sovitic in nature. Next to carbonates of Ca, Mg and Fe compositions; apatite and magnetite are the most common minerals. Workable deposits of fluorite are also associated with the carbonatites.

In Amba Dungar, potash fenitization is predominant. According to Sukeshwala and Udas (op. cit.), perhaps both the immiscible fraction of magma, silicate and carbonate, were responsible for the fenitization of the country rocks. The carbonatite occurs along the Narmada fault that conceals all formations underlying the thick flow of basalts. These fault zones provide easy passage for the migration of the alkaline melt to form dykes. The lithounits forming minor dykelets/plugs in the entire region include lenticular outcrops of nephelinites and phononephilinites, besides tinguaites between the regions of Khadla-Kharni and Padwani. The compact phonolite types form conical pulgs in several hills around Kadipani, Chikhli and Khadla. These exposures are related to concentric fractures of cone sheet type. Brecciated/agglomeratic types with fragments of basalt, sandstone, phyllite and gneiss, set in a nephelinite matrix, occur at ground level.

Texturally, there are two main types: (1) Porphyritic volcanic rocks that include felsic phonolite/phononephelinite and mafic melanephelinite and (2) Non-porphyritic plutonic rock including syenite and nepheline syenite. From the analysis of major elements the following points have emerged:

1. The alkaline rocks are sodic in nature i.e. Na wt.% > K wt.%.
2. All analyses contain appreciable amount of Co_2 with presence of carbonate minerals.
3. They are low in MgO, total Fe and TiO_2.
4. They have a higher content of P_2O_5.
5. Only a few samples have acmite in the norm, besides being fairly under saturated.
6. Trace element data are similar to those of alkaline rocks associated with carbonatite complexes found in the African Rift Valley and the Oldoinyo Lengai complex of Tanzania.

The Amba Dungar alkaline rocks are associated with carbonated nephelinite. The rocks appear have formed by a process of liquid immiscibility, followed by differentiation that probably has given rise to the carbonatitic magma. It is evident from $^{87}Sr/^{86}Sr$ ratios that the parent magma from which the carbonatite and silicate rocks have been derived might be either kimberlitic or alkali-basaltic in composition, Sukeshwala and Udas (op. cit).

The process of liquid immiscibility separates the silicate and carbonate magmas. The carbonate fraction produces a variety of carbonatites whereas the silicate fraction derived by fractionation and differentiation coupled with metasomatic processes has given rise to a wide range of alkaline rocks viz., tinaugite, phonolite/phonohephelinite, melanepheli-nite and syenite/nepheline syenite in addition to fenites.

6. Newania, Rajasthan, India

A carbonatite pluton occurs at Newania (24° 38' 22" N, and 74° 08' 44" E) ~35 km east of Udaipur city, between the villages Rundera and Tikawas. It is an elliptical body of ~4 Km^2 area trending NW-SE (Chattopadhyay, et al., 1992).

The SE contact of the carbonatite body is marked by fenitization and the fenitized zone is marked by:

1. Occurrence of stringers and veinlets of carbonatite in the host granite.
2. Broadening of the carbonatite veins is followed by neo crystallization of mafic alkaline constituents.
3. Brecciation of granitic host.
4. Depletion of quartz, recrystallization of K-feldspar and the presence of alkali amphiboles.
5. Existence of syenite veins with K-feldspar and

alkali amphiboles.

6. Presence of amphibole-free syenite and finally occurrence of carbonate- rich syenite.

Two types of carbonatite occur at Newania:

(i) Pyroclastic carbonatite that occurs as circular patches of about 1.5 m diameter, composed of calcite and goethite, embedded in a groundmass of K-feldspar, plagioclase and alkali pyroxene, and

(ii) Massive carbonatite of two types:

a) Rauhaugite and beforsite

b) Rodbergite a medium to fine grained rock comprising carbonate and hematite.

Major, trace and rare elements suggest that melt differentiation followed the sequence rauhaugite-beforesite- rodbergite to soviet. There is a significant increase in the REE with LREE enrichment from magnesio to sovitic variants. The decrease in Eu from early to late fractionates is a distinctive character of fractional crystallization.

7. Murud Janjira, Maharashtra, India

The Murud-Janjira area (18° 18' 06" N and 72° 58' 02" E) is located in the Deccan Volcanic Province. It is situated about 65 km to the south of Mumbai and a few hundred metres east of the Panvel flexure. The area is covered by flows of tholeiitic basalt. The basaltic flows are intruded by nepheline syenite dykes and stock like intrusion of an ijolitic body. Veinlets of carbonatites are also associated with the nepheline syenite and ijolite (Sethna and Christopher, 1991).

The ijolite body is elliptical in shape, 1 km long and 100 to 200 m wide. The nepheline syenite dyke is situated close to the northern portion of the ijolite (Sethna and Christopher, 1991). The ijolite rocks shows holocrystalline, porphyritic with hypidiomorphic texture. They show megacrysts of aegirine- augite in the groundmass containing slender needles of aegirine-augite and nepheline. Olivine is also present as phenocrysts. Primary calcite occurs as tiny blebs, giving an impression of its formation by liquid immiscibility of a carbonate liquid in a silicate melt (Le Bas, 1987). A carbonatite variety comprises ankeritic, calcite and dolomite as a constituent minerals of the groundmass. In the southern part of the ijolite outcrop, there are veinlets usually less than 1 cm thick consisting primarily of calcite and augite.

The Murud ijolite is chemically similar to the average ijolite reported by Hyndmen, (1985) except that these rocks contain low amounts of TiO_2, Al_2O_3 and Na_2O and high amounts of total FeO, MgO and CaO. The sample has less silica (45 wt.%) with total alkali concentrations varying from 5 to 7 wt.%, CaO 10 to 14 wt.%, MgO ~ 10 wt.%, Al_2O_3 ~ 11 wt.% and TiO_2 of~1 wt%).

PETROGENESIS

(1) Magma : Source and Composition

Gittins (1986) postulated that primary carbonatite magmas are formed by partial melting of carbonate bearing peridotites at pressure greater than 27 Kb (~ 80 Km) from mantle regions enriched in CO_2, Nb, REE, Sr, P and F and with high heat flow under continents. He once concluded that calciocarbonatites were primary magmas, but Harmer and Gittins (1997) on the basis of field occurrences of dolomitic carbonatites, argued that they were precipitated from a dolomitic parent magma. Wyllie and Lee (1998) concluded from the experimental data that primary carbonatite magmas from the mantle must be dominated by calcic dolomite, but if this magma is protected from lherzolite by metasomatic wehrlite, it can react progressively between 80-70 km to reach a calciocarbonatite composition. (Dalton and Wood, 1993). Many carbonatites that occur in alkaline igneous complexes, especially those where the volume of carbonatite is small, are derived from a silicate parental magma, yet it would not rule out the possibility of the existence of primary carbonatite magmas.

(2) Asthenosphere Verses Lithosphere Status

It is known that nephelinite and ijolitic magmas have an independent primary status and evolve separately. Thus, carbonatite magmas also have primary status.

The relationship of plume activity and carbonatite magmatism is strongly supported by Nd, Pb and Sr isotope data on the Indian carbonatites. The isotope data from both carbonatites and alkaline rocks from western India, which are coeval with or slightly younger than the Deccan volcanism are attributed to mantle sources, of which one is lithosphere and the other associated with Mid Ocean Ridge Basalt sources and the Reunion mantle plume (Basu et al., 1993; Roy, 2003). The 134 Ma Sung valley carbonatites in the Shillong Plateau are similar in age as the breakup of Gondwana and the enriched isotope signatures associated with these carbonatites were correlated by Veena et al. (1998) with igneous and metasomatic activities related to the Kerguelen-Head mantle plume.

It has became apparent from isotopic studies (cf. Powell et al., 1996; Bell and Blenkinsop, 1987a, 1987b) that the primitive magma is of mantle origin. The isotope evidence available at present thus suggests that the parental magmas of carbonatites are produced from interacting asthenosphere- lithosphere. Most carbonatites still preserve their mantle Nd and Sr isotope signatures and many volcanic silicate rocks associated with carbonatites show little evidence for magma contamination unlike many plutonic rocks.

(3) Parental Magmas and their Products

Using several phase diagrams for silicate-carbonate-H_2O systems, Wyllie (1996) illustrated the three kinds of magmatic processes, which may be experienced by a CO_2-bearing silicate parental melt:

(1) The silicate melt may be separated from the carbonate-rich melt by a thermal barrier and solidify at a terminal liquidus point with precipitation of silicate minerals and CO_2 vapour.

(2) The silicate melt may follow a path of fractional crystallization for the enrichment in carbonate components, eventually coprecipitating a carbonate when it reaches the silicate-carbonate liquidus field boundary.

(3) The silicate melt may reach a silicate-carbonate liquid miscibility gap producing an immiscible carbonate-rich liquid.

Most carbonatite magmas are derived from carbonated silicate parents, which could thus follow crystallization paths to (i) a terminal point where they precipitate silicates and evolve CO_2 without carbonatites (e.g. basaltic), (ii) a field boundary where silicates and carbonates are coprecipitated (e.g. melilititic), or (iii) a miscibility gap where a carbonate-rich magma is exsolved (e.g. nephelinitic). Carbonatite magmas do not precipitate carbonate-rich rocks until they reach the silicate-carbonate field boundary, whether this is by fractional crystallization as in Fig. 1 a, path (2) in Fig. 3 , and varied paths in Fig. 6 b; or via a miscibility gap as in paths (3) in Figs. 3, 4 and 5. The silicate-carbonate field boundary appears to limit the composition of carbonatite magmas to <~85 wt. % $CaCO_3$.

(4) Raise and Metasomatism of Carbonatite Melt

Carbonatite melts produced by very low degrees of partial melting in the mantle can form interconnected melts of fractions <0.05 wt.% in carbonated olivine rich mantle with grain size of the order of mm because of low dihedral melting angles and low viscosities (Minarik and Watson, 1995). Ascent rates from fluid flow have been calculated at 20-65m/s. (Geage et al.,1995) however, may or may not reach surface. Yaxley et al. (1996) discussed the role of carbonate metasomatism for a variety of mantle xenoliths from the Newer volcanics, Southeastern Australia and concluded that this event was associated with the loss (from the metasomatized volume) of a fugative, Si-, Al-, alkali and large-ion-lithophile element (LILE) enriched silica melt.

(5) Silica Carbonatite Rock Association

Most of the carbonatites are associated with alkali silicate rocks that are usually of nephelinite or melititic affinity. Thus, it is difficult to divorce the origin of one from that of the other. A number of distinct carbonatite- silicate rock associations occur that include melilitite-sovite, nephelinite-sovite, pyroxenite-sovite and olivine-rich ultramafic dolomitic carbonate (Hornig-Kjarsgaard, 1998). Carbonatites too are spatially related to Group-L kimberlite (Premier, South Africa) and ultramafic lamprophyres (Alno,Sweden, Kandaiaskha, Russia).These diverse associations argue against any simple model for the generation of these magmas.

The timing of the emplacement of carbonatites relative to that of their associated silicate rocks, is in almost all cases, later than those of most silicate rocks.

Carbonate melt has a higher melt-solid interfacial energy than does the coexisting silicate melt. The silicate melt therefore selectively wets the grain-edge channels between solid phases, excluding the carbonate melt to the centre of melt pockets, away from grain edges. This prevents the carbonate melt from migrating independently of the silicate melt and the carbonate melt is unable to separate from the silicate melt in a solid-dominated assemblage. The carbonate melt can migrate effectively only after the silicate melt has solidified, or by separating from the silicate melt within liquid-dominated reservoirs (sills, dikes, or chambers), unrestricted by solid interfaces. This relative wetting behavior may help explain the close association of carbonate and silicate magmas in alkali complexes, and their relative timing of emplacement. (Minarik, 1998)

A schematic diagram (Fig. 4) has been prepared for the two immiscible melts flowing in a grain-edge channel after Bailey, (1994). Flow of the carbonate melt is inhibited by the increase in the mean curvature of the carbonate-silicate melt interface (radius c_1 to c_2) as the carbonate melt enters the channel. This interfacial tension will counteract the flow. The carbonate melt, in turn, restricts the movement of the silicate melt by plugging the grain-edge channel. In a static system, a stable mineral is in equilibrium compositionally with both conjugate silicate and carbonate melts. In dynamic systems, where minerals are growing fast relative to diffusive homogenization of the melt compositions, a growing mineral could deplete the silicate melt in contact with it especially, of those elements that preferentially partitioned into the carbonate melt (i.e.) light REE and LILE (cf. Jones et al., 1995). Such mineral zoning might record the onset of immiscibility in an evolving magmatic system.

(6) Evolution of Carbonatite Magma

Present-day status of understanding favours three known ways in which a carbonatite magma might have originated. (a) As the initial product of melting of mantle containing a small proportion of carbonate minerals. (b) As the final product of differentiation of a silicates melt containing dissolved carbonate; or (c) By liquid immiscibility from an initially homogeneous carbonate bearing silicate melt.

Several experimental studies concluded that for many magmatic systems, when carbonate-rich liquids begin to precipitate calcite, they tend to cluster near 70-80 wt. % $CaCO_3$, 10-20 wt.% Na_2CO_3, and 5-10 wt.% silicates. The residual liquids then follow the silicate-carbonate boundary (the e-o and g-o equivalent of Figs. 2 and 3), toward enrichment in alkalies (Lee and Wyllie, 1996, 1997b). Carbonate might be a primary magma produced by partial melting of carbonate-bearing mantle under appropriate conditions. Increasing pressure raises the temperature at which carbonate minerals can exist and the temperature at which dolomite decarbonates by the reactions En+Dol Di +Fo + CO_2 intersects the solids in the system CaO-MgO-SiO_2 CO_2 at the pressure where dolomite can be present in peridotite up to the melting temperature. Partial melt in this system would be carbonate rich with a low silicate content. The carbonatite liquid would only be the initial product of partial melting because with the increasing temperature it would become more silicate rich as more of the silicate constituents of source rock melt. Mckenzie (1985) suggested that carbonate could be extracted from their source rock even if the melt fraction was as low as 0.1%, in contrast to silicate melts which would remain within their source rocks at this degree of melting. Most carbonatites contain silicate phases indicating that sufficient silica must be present in the parental liquid to precipitate such silicate phases. New experimental results (Petibon et al., 1997) from 100-200 MPa melting of a silicate bearing natrocarbonatite recently erupted from Oldoinyo Lengai were used to model fractional crystallization of nepheline + sodic pyx + melanite, leading to generate silicate free natrocarbonatites.

Carbonate-rich magmas become carbonatite magmas (precipitating carbonates) only when they reach the silicate-carbonate field boundary. This may occur via fractional crystallization, as in path (2) in Fig. 3, or via liquid immiscibility through one of the two routes. Under some conditions, a carbonate-rich liquid may be exsolved from a silicate parent through a temperature interval, while coexisting with both silicate minerals and carbonates, corresponding to point "g" in Fig. 2 b, before becoming separated from the silicate parental magma and following a path along "g-o" in Fig. 2 b. The second route is represented by paths (3) and (4) in Fig. 3, where the immiscible liquids coexist only with silicate minerals (e.g. along k-n or g-n in Fig. 2). Only after the carbonate-rich liquid is separated from the silicate liquid it differentiates down the liquidus surface (across k-n-e-o, or g-n-o in Fig. 2) to reach the lower-temperature silicate-carbonate field boundary. Veksler et al., (1998) examined the mineral melt inclusions from silicate and carbonatite samples collected from Gardner Complex and supported an immiscible origin for carbonatite. They found that major and trace element data for a variety of silicate dykes which range from olivine-melilite nephelinites to nephelinites fron Kandalaksha, Russia are considered to have originated via closed system fractional crystallization. They attributed the limited volume of associated carbonatite to an origin by late-stage liquid immiscibility.

Carbonate liquids coexist with olivine and two pyroxenes between the solidus for carbonated lherzolite at 1250°C and 1450°C (system CaO-MgO-SiO_2-CO_2 at 3 GPa.). The Ca/(Ca + Mg) ratio of these melts is 0·64 and the main effects of rising temperature are increasing SiO_2 content from <4·3 to 7·5 wt.% and a decreasing CO_2 content. Between 1475°C and ~1525°C, the SiO_2 content of the liquid increases

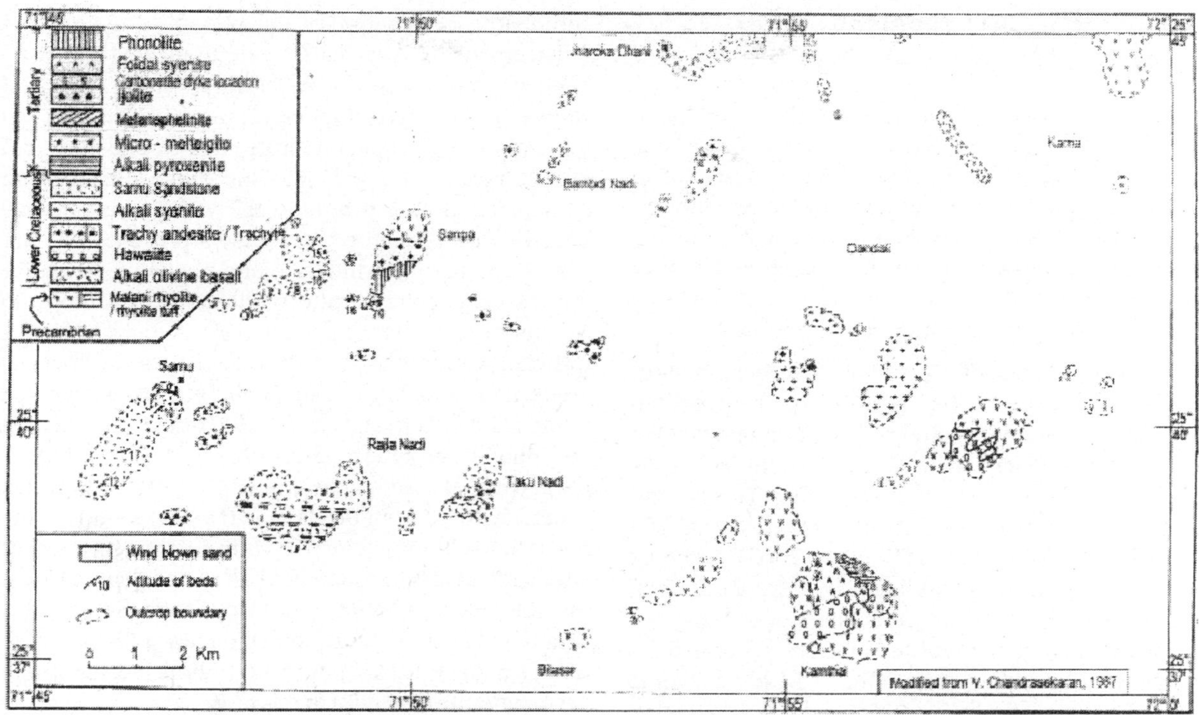

Fig. 7 Geological map of the Sarnu - Dandali area, Barmer district, after (Chandrasekaran, 1987), modified by Bhushan and Chandrasekaran (2002).

dramatically from 10 to 30 wt% and, thereafter, the CO_2 content decreases rapidly as the CO_2-absent invariant point (at >1700°C) is approached. The progression from carbonate to silicate liquids is, therefore, abrupt and the field of transitional compositions (10-30 wt.% SiO_2) is restricted to very narrow temperature intervals at pressures greater than the solidus ledge. All liquids appear to be miscible. The solidus "ledge" between 2·5 and 3 GPa acts as a filter for both carbonatites and transitional melt compositions. Carbonatites, that have a wide stability field at 3 GPa may rise through the mantle if they are isolated from lherzolite by wallrock alteration/reaction and production of wehrlite. Transitional carbonate-silicate melts must also, however, react with the mantle at low pressures. This fact, combined with the small range of physical conditions over which they are generated and their higher than carbonatite viscosity, means that they rarely reach crustal levels. Low-CO_2 silicate melts, in contrast, are not required to react extensively en route to the surface and are abandoned. The binary nature of some carbonatite complexes may be controlled by the compositions of primary mantle melts produced at pressures greater than the solidus ledge.

Crystal fractionation offers yet another way of generating carbonatitic liquids. The sequence of rocks forming a perfect concentric zonation, as at the Saingoroskii ring complex, Kola Peninsula, Russia, is consistent with fractional crystallization (Korobeinikov et al., 1998). As a part of the study on Gardiner primary melt inclusions, Veksler et al., (1998) examined silicate and carbonate samples from Kovdor (Russia) and proposed that the carbonatites at Kovdor were formed by fractional crystallization.

Degassing can also play an important role in magma differentiation by suppressing the generation of carbonatitc fluids. Dawson et al., (1993) argued on the basis of detailed petrological studies on silicates of volcanic rocks from Oldoinyo Lengai that one important factor in controlling magma differentiation is the loss of CO_2. They have suggested that CO_2 degassing inhibits exsolution of carbonatitic liquids from parental carbonated wollestonite- nephelinite melts which concomitantly concentrates alkalis, Ba, Sr and the halogens into more melt composition that generates combeiite nephelinites.

Thus integrated petrological studies support all the three mechanisms outlined above for the generation of carbonatite melts.

(7) Sarnu-Dandali Carbonatite : The Study Example

The Sarnu-Dandali carbonatite-nephelinite complex, covering about 550 sq. km (latitude 25°30'N and longitude 75°45'E) is situated about 40 km east of Barmer city(Fig. 7), at the eastern margin of Barmer-Sanchor Graben. Malani Igneous Suite, the basement of the area is a Neoproterozoic, large igneous province of bimodal but chiefly felsic volcanic rocks, covering parts of the Northwestern India and Pakistan.

The upper Cretaceous-Tertiary alkaline igneous activity appears to be related to rejuvenated taphrogenetic activity along the dominant feeder structures associated with a continental rift zone (CRZ) to produce a carbonatite complex. This second igneous event is believed to be the manifestation of a mantle plume (Basu et al., 1993; Roy, 2003). After the geological succession established by Chandersekaran (1997), there are two alkaline suites, overlying the basement of Malani rhyolites and are separated by sediments. The lower suite, the "Mildly alkaline suite", consists of alkali olivine basalt, hawaiite, trachyandesite, trachyte and gabbro. The sediments in between the two igneous units, consists of Sarnu sandstone and conglomerates. The upper suite termed as the "Alkaline suite", consists of alkali pyroxenite, pulaskite (melanephelinite), ijolite, shonkinite, foyaite, juvite and carbonatites. It is superseded by intrusion of nepheline syenite and subsequently by phonolite dykes. In the present study, it has been observed that yet another phase the of carbonate formation has occurred which has penetrated even the phonolite dykes that were considered as latest activity (Roy and Jakhar, 2002).

Detailed petrological studies of the alkaline unit including carbonatite have been undertaken. At the village Dandali the nepheline syenite is greatly exposed in the form of hillock and demonstrating at least two sets of joints intersecting at right angles (Fig. 8A). One of the sets, phonolite intrude into the nepheline syenite (Fig. 8A). In the thin section, the nepheline syenite shows coarse to medium grained orthoclase and albite with diopside (Fig. 8B). The rocks must have been metasomatised later, as alteration are visible along twin planes of feldspars (Fig. 8B). Better exposures of the carbonatite have been found near the village Bamdi Nadi. Two rock specimen (Fig.8C and E) and their photomicrographs (Fig. 8D and F) show pure calcite along with wollestonite and a little magnetite. A band of carbonatite penetrated in the phonolite dyke show replacement of silicates by calcite at contacts (Fig. 8G), while phonolite in photomicrograph (Fig. 8H) shows almost half orthoclase and albite among the minerals along with nephelinite, acmite and diopside. Differentiation of the alkaline unit silicates are demonstrated at and around the village Sevrano-ki-Dhani. Tephribasanite, here shows presence of pink garnet apart from orthoclase and nepheline as major minerals and albite, acmite, wollestonite and diopside as minor minerals (Fig. 9A and B). A tephrite from a nearby location shows ~40% feldspars (orthoclase, albite and anorthite) with nepheline and diopside (almost 20% each) and ca. 20% Fe oxide; although in the photomicrograph (Fig. 9C and D) a higher proportion of diopside is seen. A juvite specimen coated with carbonatite (Fig. 9E) shows under the microscope ~60% feldspar with notable absence of albite and coarse grained nepheline and acmite (Figs. 9F). Yet another juvite specimen from Sevrano-ki-Dhani (Fig. 9G) shows orthoclase (40%), albite (25%), nepheline (13%) and 32% diopside, anorthite and magnetite (Fig. 9H) in this area.

Geochemical studies in this area (Chandrasekaran et al. 1990; Shastry and Kumar, 1996; Sayyed,2007) bring out that:

(i) Foidal syenites have higher SiO_2, Al_2O_3, Na_2O, K_2O and lower CaO, MgO, TiO_2 and total Fe concentrations as compare to ijolites;

(ii) The "Alkaline Suite" contains low Ni, Cu, U, Sc concentrations; moderate Cr, V, Co, Zn, Ga, Pb, Th, Y and high Rb, Sr, Ba, Li, Zr and Nb abundances;

(iii) Nephelinite and phonolite are depleted in HFSE like Nb, Ta, Hf (except Ti) whereas LILE Ba, Sr are enriched;

(iv) Lithophile elements like Ba, Sr show progressive increase with differentiation in early and middle differentiates;

(v) Rb shows a gradual decrease with progress of differentiation;

(vi) Fractional crystallization is further indicated by the decrease in Fe/Zn, Mg/Li and Ca/Sr ratios; and

(vii) Carbonatites are economically enriched in the REE, Sr, and Ba, respectively with 6, 6 and 1.2 wt.%, with mineralogical expression of strontian calcite, carbocernaite, calcian strontianite and britholite (Ce).

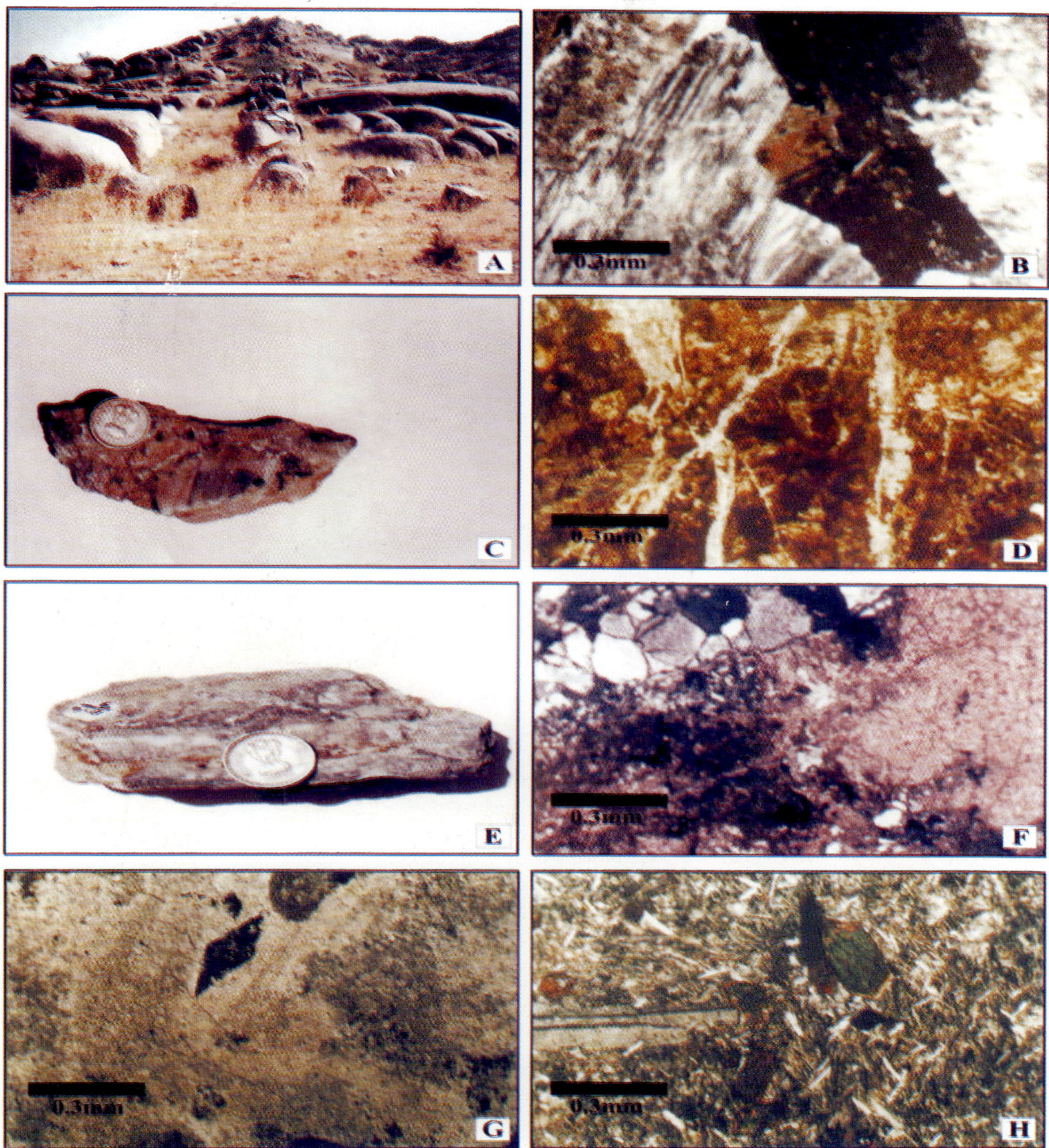

Fig.8 **A**. Field view of a nepheline syenite showing sets of joints (location Dhandali village). A phonolite dyke (dark grey) is penetrating the syenite following fillings of one of the sets of joints. **B**. Thin section photograph of nepheline syenite (Dhandali village) showing coarse to medium grained orthoclase and albite while diopside (at the centre) is medium to fine grained. Alteration effects are visible, specially along twin planes. **C**. Specimen of sovite (cabonatite) collected at Bamdi Nadi village, consisting of coarse grained dog-tooth calcite. **D**. Thin section photograph of sovite from Bamdi Nadi village illustrating calcite with wollestonite and rusty iron oxide intruded by calcite veins. **E**. Specimen of layered type carbonatite from Bamdi Nadi showing coarse grained well crystalline calcite. **F**. Thin section photograph of carbonatite showing calcite accompanied by wollestonite. Isotropic grains are magnetite. **G**. Thin section photograph (transmittent light) of an intermediate ca. 2cm thick band between phonolite and carbonatite bands (each ca.3cm thick) showing tephriphonolite characteristics with addition of fine-grained calcite of carbonatite. The sample is from Dhandali village. Almost 10% iron minerals, 25% nepheline next to feldspars and diopside can be seen. **H**. Thin section photograph of foyalite (peralkaline phonolite) from Sevrano ki Dhani. Orthoclase and albite comprise almost 50% of the total mineral assemblage followed by 30% nepheline, 15% acmite and 7% diopside. Except for few phenocrysts the texture is very fine grained.

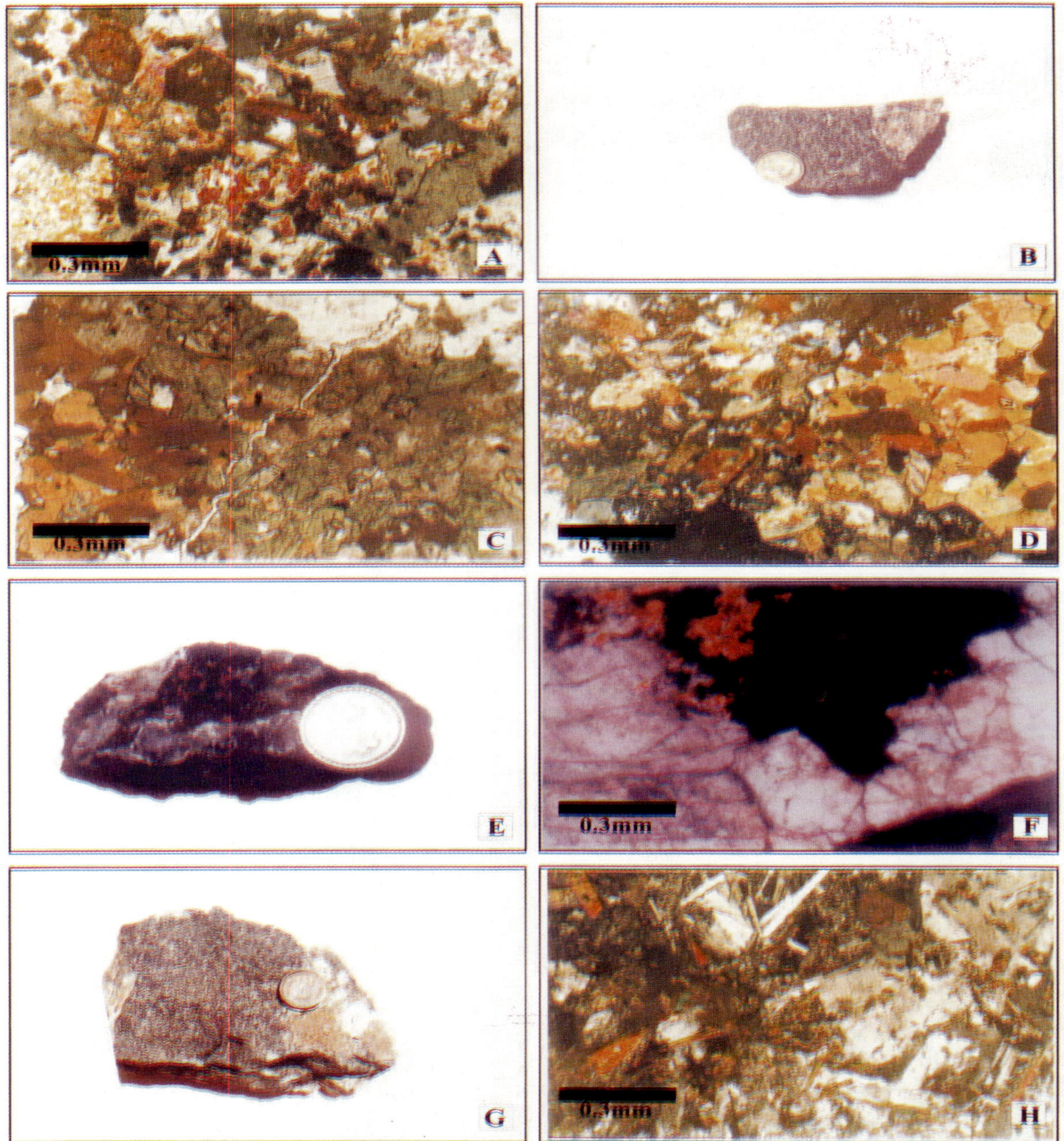

Fig.9 Thin section photograph of phonolite rock of Sevrano ki Dhani. Orthoclase and nepheline are the main among mineral crops. Albite, acmite, diopside, wollestonite are the other minerals present. Some garnets (euhedral) are visible. B. Specimen of tephrite (or tephribasanite) collected near Sevrano ki Dhani. Pink granular minerals are segregated garnets from the main mass which is consisting of feldspars (orthoclase, albite, anorthite) and feldspathoid (nepheline) with diopside and iron minerals. C. Thin section photograph (intermittent light) of tephrite from Sevrano ki Dhani. The rock is having feldspar almost 40% feldspars (orthoclase, albite and anorthite) with nepheline and diopside (20% each) and ca. 20% Fe-oxide. The view is of segregated diopside crystals. D. Thin section photograph of tephrite from Sevrano ki Dhani, same (fig. 2C) under crossed nicols. E. Specimen of Juvite (coarse grained tephriphonolite) rock from Sevrano ki Dhani, coated with carbonatite. F. Thin section photograph of juvite (coarse grained tephriphonolite) rock from Sevrano ki Dhani. Coarse nepheline is present alongwith greenish yellow acmite. The feldspars constituting almost 60% of the mineral crops but albite is absent. G. Specimen of juvite (phonolite) collected from Sevrano ki Dhani. H. Thin section photograph of Juvite (Phonolite) from Sevrano ki Dhani. Almost all minerals are visible including 40% orthoclase, 25% albite , 13% nepheline, and 32% diopside, anorthite and magnetite. Grain size is varying greatly with some radial growth of crystals.

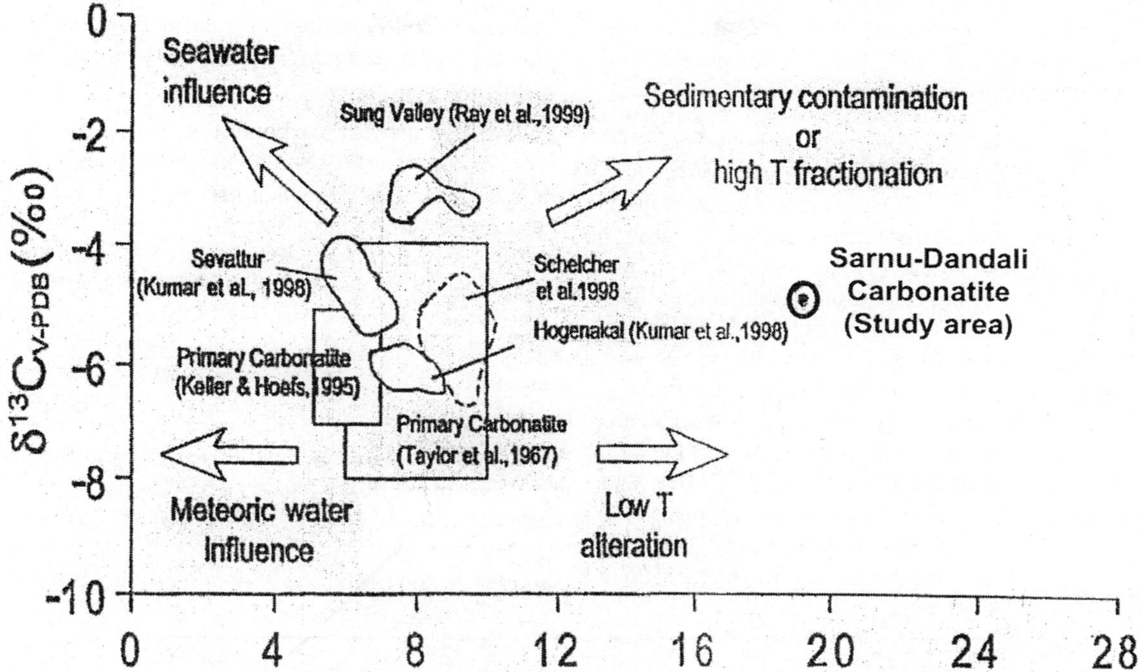

Fig.10 Stable C- and O-isotopic compositions of the Sarnu-Dandali carbonatites (relative to PDB and SMOW, respectively) plotted in the conventional ^{18}O- ^{13}C diagram. The rectangular boxes represent the fields for unaltered primary carbonatites of Taylor et al. (1967) and Keller and Hoefs (1995). The arrows indicates possible effects of main processes on isotopic composition (details in Denis, 1989). The Hogenakal samples plots as a close cluster within the field of primary carbonatites while the Neoproterozoic carbonatites show a spread, mainly due to wide variation in O-isotopic composition. The fields of Sevattur carbonatite and Hogenakal calcite (Kumar et al., 1998) and Neoproterozoic carbonatites (Schleicher et al., 1998) have also been shown. Note uniform isotopic compositions of the carbonatite rock (present study) as both points fall together. Wide O-isotopic composition showing low temperature alteration to high temperature fractionation.

Numerous C- and O- isotopic studies indicate that in contrast to the wide variability in C- isotopic composition of meteorites and mantle material such as diamonds, basalts and hightemperature volcanic gases, the range shown by carbonatites is rather narrow, representing the average composition of a large volume of the mantle source. The carbonatite melts are capable of concentrating volatiles about 1000 times their volume (Deines, 1989). The carbonatites are, however, widely variable in ^{18}O because the O- isotopic compositions are likely to be affected by postmagmatic modifications during various crustal processes. Unlike oxygen, the carbon in a parent carbonatite magma is believed to have completely been extracted from its mantle source (Deines, 1989) and therefore, the C- isotopic variations can be reliably used to model the mantle source. Moreover, the C- isotopic compositions, generally unaffected by a number of crustal processes, can be utilized to model the mantle source even in lowgrade metamorphosed and hydrothermally altered samples (Deines, 1989). However, the carbon budget of the deep mantle reservoir itself can be significantly affected by the ^{13}C- enriched CO_2 from recycled carbonates from a subducting oceanic crust. Carbonatites originating from such magmas are likely to preserve the ^{13}C- enriched character. Experimental studies have shown that carbonates from a subducting

Table-1 Carbon and oxygen Isotope Data (in %o) of carbonatites of the Sarnu-Dandali carbonatite-nephelinite complex, Rajasthan

Sample	Probe	δ C¹³ PDB	Std. Dev.	δ O¹⁸ SMOW	Std. Dev.
1.	34 MA	– 5.00	0.14	+ 19.06	0.22
2.	34 MA	– 5.06	0.12	+ 18.92	0.24

oceanic crust can penetrate to great depths in the mantle (Biellman et. al., 1993) even in the presence of silicate melts (Yaxley and Green, 1994).

The $^{13}C_{VPDB}$ in primary igneous carbonatites range from 4‰ and 8‰, and $^{18}O_{VSMOW}$ between + 6‰ and + 10‰ (Taylor et al., 1967). Based on the data from fresh and unaltered natrocarbonatites from Oldoinyo Lengai, Keller and Hoefs (1995) proposed a much narrower range of C and O isotopic compositions (7‰ to 5‰ and + 5.5‰ to + 7‰, respectively), which can be considered to represent undegassed and uncontaminated (by recycled carbon) mantle compositions. However, a number of carbonatites are known to be more enriched (^{13}C> 4‰, see Deines, 1989, Plyusnin, 1980; Dunworth and Bell, 2001). The isotopic differences within the 'mantle compositions' can be ascribed to mantle heterogeneity (Deines, 1989). 'Primary', as defined by Taylor et al (1967) denotes the carbonatites unmodified by secondary (crustal) processes and does not imply an unmodified partial mantle melt.

Two samples of carbonatites from near village Bamdi Nadi of Sarnu-Dandali area have been analyzed for carbon and oxygen isotopes (C^{13} and O^{18}) using mass spectroscopy at Munich University. The result obtained are shown in Table 1. Fig. 10 shows a stable C- and O- isotopic compositions of the Sarnu-Dandali carbonatites (relative to v-PDB and v- SMOW, respectively) plotted in the conventional $\delta^{18}O$- $\delta^{13}C$ diagram.

In the study area, both the carbonatite samples are showing almost similar values for C and O isotopes (-5.00‰ to -5.06‰ and + 19.06‰ to + 18.92‰ respectively). It shows uniformity in the isotopic composition owing to the fact that the number of the samples is too little.The C-isotopes greatly favour

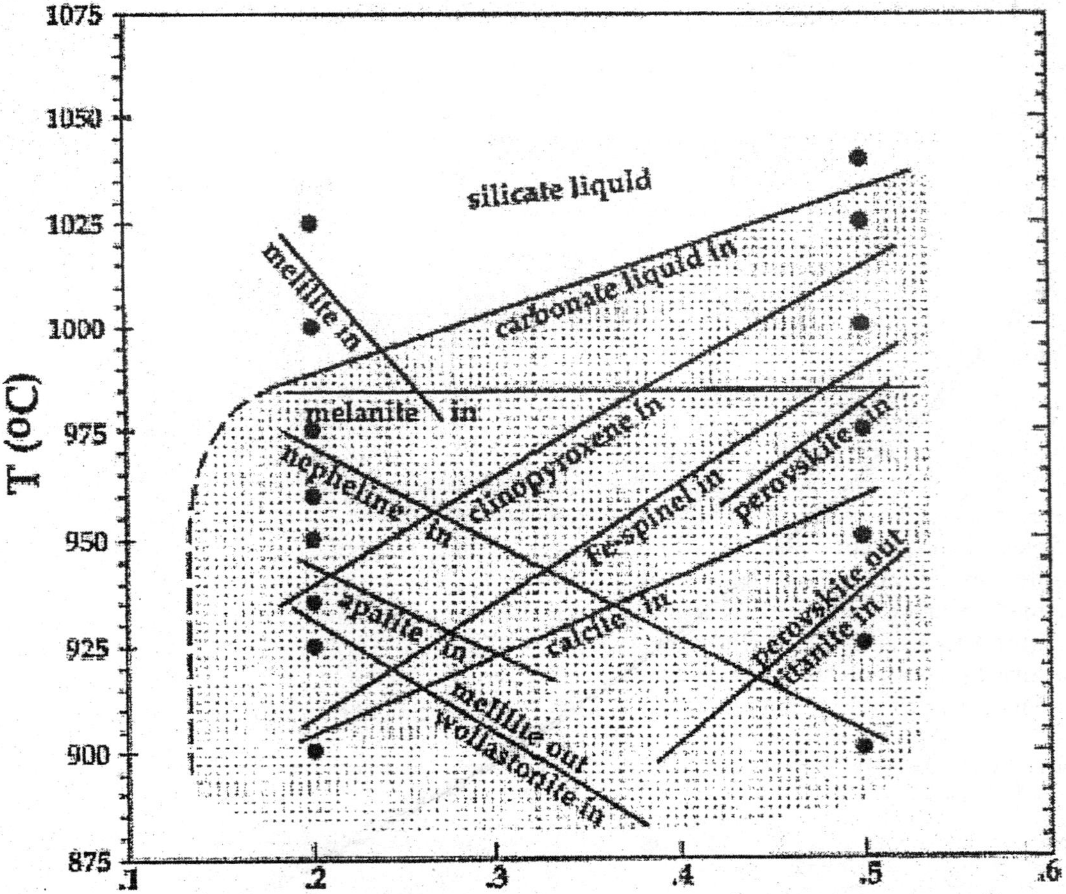

Fig.11 Kjarsgaard, 1998 showed that at 0.2 Gpa, carbonated silicate liquid exists at 1025° C. Melilite precipitates at 1000°C. With further temperature decrease (975°C), melanite and melilite co-precipitate and an immiscible carbonate liquid is exsolved, all runs from 975 to 990°C contain two liquids. Nepheline co-precipitates with melilite and melanite at 960 to 950 °C, and clinopyroxene and apatite join these phases at 935°C. Melilite out and wollastonite in reactions (coupled) occur within the narrow temperature 935 925°C.

mantle sources. These are also been treated as more reliable, being unaffected by the later processes. O-isotopes showing enrichment. A marginal enrichment in ^{18}O is more likely an inherited mantle source characteristics and not a post magmatic alteration phenomenon. In the present study these values are more than marginal data, hence this explanation is unacceptable. Exceptionally, Onnakarai carbonatite (Pandit el al., 2002) show still more enrichment with the values (^{18}O = 24.0 ‰ and 25.4‰). It is therefore, proposed that fractional crystallization has played an important role although the source of carbonatites is mantle.

__Petrogenesis of the carbonatite comprises geological activities of several events in the region. The Malani Rhyolite (~745Ma) has been reactivated, holding a Continental Rift Zone (CRZ) type tectonic set up during the Cretaceous to form the mildly alkaline complex. After a period of quiescence represented by sedimentation, the alkaline suite of economic interest formed. The sequence of formation of various units of the alkaline suite can be represented by a series, ijolite-melanephelinite-phonolite-carbonatite. These events were followed by intrusions of nepheline syenite. Later the whole area was intruded by phonolites and still later by carbonatite dykes and dykelets.

The alkaline series of the Sarnu-Dandali Carbonatite-Nephelinite Complex, demonstrate its formation and evolution through fractional crystallization of an alkaline melt (ijolite or nephelinite type). The conclusion is drawn on the basis of petrology and petrochemistry. The associated "Mildly alkaline suite" possibly has been generated by differential degree of melting of a common mantle source.

Carbonatites, however, represent an immiscible liquid, split from the differentiation of the host phonolite or melanephelinite, from which a metasomatized liquid must have been separated and has produced enough evidences of fenitizations even on phonolites which was till now treated (Chandrasekaran 1987,Bhushan and Chandrasekaran 2002, Roy and Jakhar 2002) as the last phase of the activity in the region. On the basis of isotopic study (Fig. 10) a mantle origin is favoured for carbonatite and the last phase of intrusion due to liquid immiscibility from ijolite/nephelenite/ phonolite magma.

Shrivastava (1989) has found pressure of 30kb and 125 km depth for the formation of alkaline rocks of Sarnu-Dandali area on the basis of the nepheline-forsterite-quartz diagram. It is the depth for the phlogophite-bearing mantle with a very high CO_2/H_2O ratio. These data favour the formation of kimberlite in the region. Utilizing mineral assemblages determined in the present study as clinopyroxene-apatite-wollestonite and simulating field data with extensive experimental data (chiefly, Kjarsgaard, 1998; Fig. 11), it is has suggested that the carbonatites of the area have been formed within a narrow range of temperature from 925 °C to 935 °C.

CONCLUSIONS

There is now a general consensus among geoscientists that the parental magmas of carbonatites are mantle derived, and must be CO_2 enriched (`carbonated'). Most carbonatite magmas are derived from carbonated silicate parents, which could follow one of the three kinds of crystallization paths: (a) to a liquidus-solidus terminal point where they precipitate silicates and CO_2 vapor, (b) to a silicate-carbonate field boundary where silicates and carbonates are co-precipitated, or (c) to a miscibility gap field boundary where a carbonate-rich magma is exsolved, without the co-precipitation of calcite except along the limiting field boundary where the two major liquidus surfaces intersect.

The pseudoquaternary system CaO-(MgO +FeO (total Fe as Fe^{2+})-($Na_2O + K_2O$)-($SiO_2 + Al_2O_3 + TiO_2$) with CO_2 provides a framework for evaluating possible magmatic processes. Experimentally determined phase boundaries in the system indicate that immiscible carbonate-rich liquids separated from liquidus corresponding to many natural silicate magmas tend to be concentrated near calciocarbonatite compositions, with the maximum $CaCO_3$ being ~75-80 wt%, and $(Na, K)_2CO_3$ contents near 15 wt%. The three key features, the liquidus surface enclosing the volume of silicate-carbonate liquid immiscibility, the silicate-carbonate liquidus surface for the co-precipitation of silicates and carbonates, and the curve of intersection between these two surfaces, define conditions for the coexistence of two immiscible liquids, silicates, and calcite. The relationship between the surfaces, which varies as a function of pressure and magma composition (especially Mg/Ca, Ca/Na, Al/Si ratios), controls the nature of carbonate-rich magmas, derived from parent silicate magmas.

Rising carbonate-rich magmas retaining an equilibrium with the mantle will react, crystallize and release CO_2 vapor at depths of ~70 km, with the metasomatic transformation of lherzolite to wehrlite. At shallower depths, wehrlite (but no other peridotite) can coexist with carbonatite magma relatively enriched in Ca/Mg. Carbonate-rich magmas may be generated at depths greater than ~70 km by partial melting of carbonated (dolomite-magnesite) peridotite. The near-solidus magmas lie on the silicate-carbonate liquidus field boundary, with compositions dominated by the liquidus of the system $CaCO_3$-$MgCO_3$, and with minor variations in Ca/Mg and alkali contents, reflecting the peridotite mineralogy. Liquid compositions are dolomitic, with Ca/(Ca + Mg) ~0·5-0·7 from 2 GPa to at least 6 GPa (depth 200 km). Even when the subsolidus carbonate changes from dolomite to magnesite with increasing pressure, the liquid composition remains dolomitic. Carbonatite magmas in the mantle are the first (lowest-temperature) batch of a continuum of small-volume partial melts including melilitites and kimberlites.

The Sarnu-Dandali carbonatite-nephelinite complex is possibly an ideal example, demonstrating the field, petrological and geochemical evidence well in accordance with the existing experimental data. Especially, data on the O- and C- isotopes favour that the carbonatites have been derived from a mantle source. The evolution of carbonatite, through differential mechanism by fractional crystallization of an alkaline melt, following a differential series of ijolite-melanephelinite-phonolite- carbonatite is well documented. Carbonatites represent an immiscible liquid, split from a host melanephelinite or phonolite magma from which a metasomatized liquid has been separated to produce fenitization. The associated, mildly alkaline suite, possibly, may has been produced by different degrees of melting of a common mantle source.

ACKNOWLEDGEMENTS

We gratefully acknowledge Prof. K.L. Rai and Dr. K. Ramesh Kumar, the hon'ble reviewers, for their peer and promt review of the paper. We also thank one anonymous reviewer for his helpful comments and suggestions given on the first draft of the manuscript.

References

Bailey, R. C. ;1994: 'Fluid trapping in mid-crustal reservoirs by H_2O-CO_2mixtures.' Nat., Vol. 371, pp. 238-240.

Barker, D. S.; 1996a: 'Consequences of recycled carbon in carbonatites.' Can. Min., Vol. 34, pp. 373-387.

Barker, D. S.; 1996b: 'Carbonatite volcanism.' in 'Undersaturated Alkaline Rocks: Mineralogy, Petrogenesis, and Economic Potential.' Editor R. H. Mitchell, Min. Asso. Can., Short Course, Vol. 24, pp. 45-61.

Bell, K. and Blenkinsop, J.; 1987a: 'Archean depleted mantle: evidence from Nd and Sr initial isotopic ratios of carbonatites.' Geochem. Cosmochem. Acta,Vol. 51, pp. 291-298.

Bell, K. and Blenkinsop, J. ;1987b: 'Nd and Sr isotopic compositions of East African carbonatites: implications for mantle heterogeneities.' Geol., Vol. 15, pp. 99-102.

Bell, K.; 1989: 'Carbonatites: Genesis and Evolution.' Unwin Hyman, London.

Bell, K. and Keller, J.; 1995: 'Carbonatite volcanism: Oldoinyo Lengai and the Petrogenesis of Natrocarbonatites.' in 'IAVCEI Proc. in Volcanogeny 4, Berlin.' Editor K. Bell and J. Keller, Springer-Verlag.

Bell, K., Kjarsgaard, B. A. and Simonetti, A.; 1998: 'Carbonatite into the twenty- first century.' Jour. Petrol., Vol. 39, pp. 1839-1845.

Brogger, W.C.; 1921: 'Die Eruptiuegecteine des krishaniagebictes IV Das fengebiet in Telomarken. Norwegea, Norsks Vidins, Skrift'. Mat. Natures, K1 No. 9, 408p.

Canil, D. and Scarfe, C. M.; 1990; 'Phase relations in peridotite + CO_2 systems to 12 GPa: implications for the origin of kimberlite and carbonate stability in the Earth's upper mantle.' Jour. Geophy. Res., Vol. 95, pp.15805-15816.

Chattopadhyay, S., Chattopadhyay, B. and Bapna, V. S.; 1992: 'Geochemistry of Newania carbonatite pluton, Rajasthan, India.' Ind. Min., Vol. 46, No. 1, pp. 35-46.

Dalton, J. A. and Wood, B. J.; 1993: 'The compositions of primary carbonate melts and their evolution through wallrock reaction in the mantle.' Ear. and Plan. Sci. Lett.,Vol. 119, pp.511-525.

Dawson, J. B.; 1966: 'Oldoinyo Lengai-an active volcano with sodium carbonatite lava flows.' in 'Carbonatites.' Editor O. F. Tuttle and J. Gittins, John Wiley, New York, pp. 155-168.

Dawson, J. B.; 1966: 'The Kimberlite- Carbonatite relationship.' Min. Soc. Ind., IMA, Editor, P.R.J. Naidu, pp. 1-4.

Dawson, J. B., Pinkerton, H., Pyle, D. M. and Nyamweru, C.; 1994: 'June 1993 eruption of Oldoinyo Lengai, Tanzania: exceptionally viscous and large carbonatite lava flows and evidence for coexisting silicate and carbonate magmas.' Geol., Vol. 22, pp.799-802.

Deines, P.; 1989: 'Stable isotope variations in carbonatites.' in 'Carbonatites: Genesis and Evolution.' Editor K. Bell, Unwin Hyman, London, pp. 301-359.

Dorfman, A. M., Hess, K.-U. and Dingwell, D. B.; 1996: 'Centrifuge-assisted falling sphere viscometry.' Euro. Jour. Min., Vol.8, pp. 507-514.

Dunwarth, E.A .and Bell, K.; 2001: 'The Turiy massif, Kola Peninsula Russia : isotopic and Geochemical evidence for multi source evolution.' Jour. Petrol., Vol. 42, pp. 377 405.

Erikson, S.C.; 1989: 'Palabora a saga of magmatism and

miscibility.' in 'Carbonatites : Genesis and evolution.' Editor K. Bell. Unwin Hyman, London, pp. 227254.

Evan, A.M.; 1993 : 'Ore Geology and Industrial Minerals : An Introduction'. Pub. Blackwell Sci., Ed. III, 390 p.

Genge, M. J., David Price, G. and Jones, A. P.; 1995: 'Molecular dynamics simulations of $CaCO_3$ melts to mantle temperatures and pressures.' Ear. and Plan. Sci. Lett., Vol. 131, pp. 225-238.

Gittins, J.; 1986: 'Genesis and evolution of carbonatite magmas(abstract).' Joint Annual meeting of Geological Association of Canada-Mineralogical Association of Canada-Canada Geophysical Union, Caltron University, Ottawa, 73p.

Gittins, J.; 1989: 'The origin and evolution of carbonatite magmas.' in 'Carbonatites: Genesis and Evolution.' Editor K. Bell, Unwin Hyman, London, pp. 580-600.

Harmer, R. E. and Gittins, J.; 1997: 'The origin of dolomitic carbonatites: field and experimental constraints.' Jour. Afr. Ear.Sci., Vol. 25, pp. 5-28.

Harmer, R. E. and Gittins, J.; 1997: 'Dolomitic carbonatite parental magmas.' Geol. Asso. Can., Min. Asso. Can. Ann. Meet. Ottawa, Geol. Sur. Can., Abstract, Vol., A, 64p.

Heinrich, E. W.; 1966: 'The Geology of Carbonatites.' Chicago, IL: Rand McNally.

Hoernle, K. A. and Tilton, G. R.; 1991: 'Sr-Nd-Pb isotope data for Fuerteventura (Canary Islands) basal complex and subaerial volcanics: applications to magma genesis and evolution.' Schw. Min. Petrol. Mitt., Vol. 71, pp.3-18.

Hoernle, K., Tilton, G., Le Bas, M.J., Duggen, S. and GarbeSchonberg, D.; 2002: 'Geochemistry of oceanic carbonatites compared with continental carbonatites : mantle recycling of oceanic crustal carbonate.' Contri. Min. Petrol., Vol. 142, pp. 520542.

Hornig, I. and Kjarsgaard; 1998: 'Rare earth elements in sovitic carbonatites and their mineral phases.' Jour. Petrol., Vol. 39, No. 11 and 12,pp.21052121. http://www.educeth.ch/stromboli/perm/lengai/

Hunter, R. H. and McKenzie,D.; 1989: 'The equilibrium geometry of carbonate melts in rocks of mantle composition.' Ear. and Plan. Sci. Lett., Vol. 92, pp.347-356.

Jones, A. P. and Wyllie, P. J.; 1985: 'Paragenetic trends of oxide minerals in carbonate-rich kimberlites with new analyses from the Benfontein sill, South Africa.' Jour. Petro., Vol. 26, pp.210-222.

Jones, J. H., Walker, D., Pickett, D. A., Murrell, M. T. and Beattie, P.; 1995: 'Experimental investigations of the partitioning of Nb, Mo, Ba, Ce, Pb, Ra, Th, Pa, and U between immiscible carbonate and silicate liquids.' Geochem. Cosmochem. Acta, Vol. 59, pp.1307-1320.

Keller, J. and Hoefs, J.; 1995: 'Stable isotopic characteristics of recent Carbonatites.' in 'Carbonatite volcanism.' Bell, K. and Keller, J. (Eds.), IAVCEI Proc. in Volcanogeny, Vol. 4, pp. 70- 86.

Kjarsgaard, B.A.; 1998 : 'Phase relation of carbonated high CaO nephelinite at 0.2 and 0.5 GPa.' Jour. Petrol., Vol. 39, No. 11 and 12, pp. 20612075.

Korobeinikov, A.N., Miltroferon, F.P., Gehor, S., Laajoki, K., Pavlov, V.P.and Mamcalov, V.P.; 1998: 'Geology and copper sulphide mineralisation of Salmagorskii Ring Igneous Complex, Kola Penninsula NW Russia.' Jour. Petrol., Vol. 39, No. 11 and 12, pp. 20332041.

Krishnamurthy, P.; 1985: 'Petrology of carbonatites and associated rocks of Sung valley, Jaintia Hills, district Meghalaya, India.' Jour. Geol. Soc. Ind., Vol. 26, pp. 361376.

Kumar, A., Nirmal Charan, S. Gopalan, K. and Macdougall, J.D.; 1998: 'A longlived enriched mantle source for two Proterozoic carbonatite complexes from Tamil Nadu, Southern India.' Geochem. Cosmochem. Acta., Vol. 62, pp. 515 523.

Le Bas, M. J.; 1987: 'Nephelinites and carbonatites.' in 'Alkaline Igneous Rocks.' Editor J. H. Fitton and B. G. J. Upton, Geol. Soc., London, Spec. Pub., Vol. 30, pp.85-94.

Lee, W.-J. and Wyllie, P. J.; 1996: 'Liquid immiscibility in the join $NaAlSi_3O_8$-$CaCO_3$ to 2·5 GPa and the origin of calciocarbonatite magmas.' Jour. Petrol., Vol. 37, pp.1125-1152.

Lee, W. J. and Wyllie, P. J. ;1997: 'Liquid immiscibility in the join $NaAlSiO_4$-$NaAlSi_3O_8$-$CaCO_3$ at 1·0 GPa : implications for crustal carbonatites.' Jour. Petrol., Vol. 38, pp.1113-1135.

Lee, W. J. and Wyllie, P. J.; 1998a: 'Petrogenesis of carbonatite magmas from mantle to crust, constrained by the system CaO-(MgO+ FeO*)-(Na_2O + K_2O)-(SiO_2 + Al_2O_3 + TiO_2)-CO_2.' Jour. Petrol., Vol. 39, pp.495-517.

Lee, W. and Wyllie, P.J.; 1998b: 'Process of crustal carbonatite formation by liquid immiscibility and differentiation, elucidated by model systems'. Jour. Petrol., Vol. 39, No. 11 and 12, pp. 20052013.

McKenzie, D.; 1985: 'The extraction of magma from the crust and mantle.' Ear. and Plan. Sci. Lett., Vol. 74, pp.81-91.

Minarik, W. G. and Watson, E. B.; 1995: 'Interconnectivity of carbonate melt at low melt fraction.' Ear. and Plan. Sci. Lett., Vol. 133, pp.423-427.

Notholt, A.J.G.; 1979: 'The economic geology and development of igneous phosphate deposits in Europe and the USSR.' Eco. Geol., Vol. 74, pp. 339-350.

Olson, J.C. and Pray, L.C.; 1954: 'The Mountain Pass rareearth deposits' in 'Geology of Southern California.' Chap. VIII., Min. Dep. and Min. Ind. Bull., Vol. 129, pp. 2329, Division of mines, State of California.

Pandit, M.K., Sial, A.N., Sukumaran, G.B., Pimentel, M.M., Ramasamy, A.K. and Ferreira, V.P.; 2002: 'Depleted and enriched mantle sources for Paleo and Neoproterozoic carbonatites of southern India : Sr, Na, CO isotopic and geochemical constraints.' Chem. Geol., Vol. 189, pp. 69 89.

Petibon, C., Jenner, G. A., Jackson, S. E. and Kjarsgaard, B. A.; 1997: 'Petrogenesis of Oldoinyo Lengai carbonatites: constraints from trace element partition coefficients.' in 'Geological Association of Canada, Ottawa '97. Geo.

Assoc. Can.-Min. Asso. Can.,Abstracts, A-117, Ottawa.

Phadke, A. V. and Jhingran, A. G.; 1968: 'On the carbonatites at Newania, Udaipur district, Rajasthan.' Jour. Geol. Soc. Ind., Vol. 9, pp. 165-169.

Powell, J. L., Hurley, P. M. and Fairburn, H. W.; 1966: 'The strontium isotopic composition and origin of carbonatites.' in 'Carbonatites.' Editor O. F.Tuttle and J. Gittins, John Wiley, New York, pp. 365-378.

Plyusnin, G.S., Somoylov, V.S. and Golshev, S.I.; 1980: 'The ^{13}C, ^{13}O isotope pair method and temperature facies of carbonatites.' Dokl. Akad. Nauk. USSR, Ser. Geol., Vol. 254, pp. 1241-1245.

RimskayaKorsakova, O.M.; 1964: 'Genesis of the Kovdor iron ore deposit (Kola Peninsula).' Int. Geol. Rev., Vol. 6, pp. 1735-1746.

Russell, H.D., Hiemstra, S.A. and Groeneveld, D.; 1954: 'The mineralogy and petrology of the carbonatite at Loolekop, eastern Transvaal.' Trans. Geol. Soc. S. Afr., Vol. 57, pp. 197-208.

Schleicher, H., Kramm, U., Pernicka, E., Schidlowski, M., Schmidt, F., Subramania, V., Todt, W. and Viladkar, S.; 1998: 'Enriched subcontinental upper mantle beneath southern India : evidence from Pb, Nd, Sr and CO isotopic studies on Tamil Nadu Carbonatites.' Jour. Petrol., Vol. 39, pp. 1765-1785.

Sethna, S.F. and Christopher, P.R.D'sa; 1991: 'Occurance of ijolite with veinlets of carbonatite in the Deccan Trap at Murud Janjira , Maharashtra.' Jour. Geol. Surv. Ind., Vol.37, No. 3, pp. 870- 890.

Stanton, R.L.; 1972: 'Ore petrology.' McGraw Hill, 713p.

Sukheswala, R. N. and Udas, G. R.; 1963; 'Note on the carbonatite of Amba Dongar (Gujrat State and its economic potentialities.' Sci. Cult., Vol. 29, pp. 563- 568.

Sukheswala, R.N. and Avasia, R.K.; 1972 : 'Carbonatitealkaline complex of PanwadKaimut, Gujarat.' Bull., Vol. 35, pp. 564-578.

Taylor, H.P., Frechen, J. and Degens, E.T.; 1967: 'Oxygen carbon isotope studies of carbonatites from the Laachersee district west Germany and Alno district, Sweden.' Geochem. Cosmochem. Acta., Vol. 31, pp. 407 -430.

Vartiainen, H and Parma, H.; 1979: 'Geological characteristics of the Sokli carbonatite complex, Finland.' Eco. Geol., Vol. 74, pp. 1296- 1306.

Verwoerd, W. J.; 1966: 'South African carbonatites and their probable mode of origin.' Ann. Univ. Stellenbosch, Vol. 41 A, No. Z.

Veena, K., Panday, P.K., Krishnamurthy, P.S. and Gupta, J.N.; 1998: 'Pb, Sr, Na isotopic systematics of Sung Valley Meghalaya, NW India: implementation for contemporary plume related mantle source characteristics.' Jour. Petrol., Vol. 39, pp. 1975-1984.

Veksler, I.V., Nielsen, T.F.D. and Sokolov, S.V.; 1998: 'Mineralogy of crystallized melt inclusions from Gardiner and Kovdar Ultramafic Alkaline complexes : implications for carbonatite genesis.' Jour. Petrol., Vol. 39, No. 11 and 12, pp. 2015-2031.

Veksler, I.V., Petibon, C., Jenner, G.A., Darfrmca, A.M. and Deywell, D.B.; 1988: 'Trace element partibony in immissible silicate carbonate liquid system : an initial experiemental study using a centrifuge autoclave.' Jour. Petrol., Vol. 39, pp. 2095-2104.

Wallace, M. E. and Green, D. H.; 1988: 'An experimental determination of primary carbonatite magma composition.' Nat., Vol. 335, pp. 343-346.

Watson, E. B., Brenan, J. M. and Baker, D. R.; 1990: 'Distribution of fluids in the continental lithospheric mantle.' in 'Continental Mantle.' Editor M. A. Menzies, Clarendon Press, Oxford, pp. 111-125.

Wyllie, P. J. and Huang, W. L.; 1975: 'Peridotite, kimberlite and carbonatite explained in the system $CaO-MgO-SiO_2-CO_2$.' Geol., Vol. 3, pp. 621-624.

Wyllie, P. J. and Huang, W. L.; 1976: 'Carbonation and melting reactions in the system $CaO-MgO-SiO_2-CO_2$ at mantle pressures with geophysical and petrological applications.' Contrib. Min. and Petrol., Vol. 54, pp.79-107.

Wyllie, P. J.; 1978: 'Mantle fluid compositions buffered in peridotite-CO_2-H_2O by carbonates, amphibole, and phlogopite.' Jour. Geol., Vol. 86, pp. 687-713.

Wyllie, P. J.; 1987: 'Transfer of subcratonic carbon into kimberlites and rare earth carbonatites.' in 'Magmatic Processes: Physicochemical Principles.' Editor B. O.Mysen, Geochem. Soc., Spec. Pub., Vol. 1, pp.107-119.

Wyllie, P. J. and Lee, W. J.; 1998: 'Model system controls on conditions for formation of magnesiocarbonatite and calciocarbonatite magmas from the mantle.' Jour. Petro., Vol. 39, pp.1885-1893.

Yaxley, G. M. and Green, D. H.; 1996: 'Experimental reconstruction of sodic dolomitic carbonatite melts from metasomatised lithosphere.' Contri to Min. and Petrol., Vol. 124, pp. 359-369.

Yaxley, G.M., Green, D.H.; 1994: 'Experimental demonstration of refractory carbonate bearing eclogite and siliceous melt in the subduction regime'. Ear. and Plan. Sci. Lett., Vol. 128, pp. 313-325.

Hon'ble Reviewers

1. Prof. K.L. Rai
   Former Head
   Deptt. of App. Geol. ISM,
   45, Chatrapati Shivaji Nagar
   Kolar Road
   BHOPAL 462 016

2. Dr. Ramesh Kumar
   Scientist 'G'
   Atomic Mineral Directorate
   Sanganer
   JAIPUR 302 030

Editorial Assistance

Ms. Anju Mathur

Economic Mineralization
K.L. Shrivastava (Editor)
2009. pp.205-209
Scientific Publishers (India), Jodhpur

16

Section-II

Occurrence of Zinc rich A-type granites in the Trans-Aravalli Anorogenic Ring Complexes, Northwestern India

G. VALLINAYAGAM

Department of Geology, Kurukshetra University, Kurukshetra 136119
Email : gvallinayagam@rediffmail.com

Abstract

Granites of A-type affiliations are exposed in the anorogenic ring complexes of Tosham (Haryana), Jhunjhunu, Siwana, Kundal and Jalor (Rajasthan) in the Trans-Aravalli Block of India and belong to the Neoproterozic Malani Igneous Suite (MIS). These granites are subvolcanic in nature, occur in the form of ring complexes, hypersolvus-subsolvus, peralkaline-peraluminous, anorogenic and High Heat Producing (HHP) type. These granites have shown high content of silica, total alkalis and iron, HFS, LIL elements and low in calcium, magnesium, strontium, barium, europium, chromium, cobalt, nickal and vanadium. Present paper describes the enrichment of Zn and the high zinc granites among them are located in Siwana Ring Complex (SRC). Zinc enrichment in these rocks may be attributed to the post magmatic fluid interaction (sodic-metasomatism) process and high concentration of zinc in the source magma. Detailed works are recommended to assess their economic potentiality and the nature of ore.

Key words : Trans-Aravalli Block, Malani Igneous Suite, Anorogenic, A-type Granites, HHP, Zinc Rich Granites, Sodic Metasomatism.

INTRODUCTION

Zawar, Dariba-Rajpura and Rampura-Agucha Lead-Zinc deposits associated with Proterozoic and Archean rocks of Aravalli Ranges are well known in India. The ongoing geological and geochemical investigations by the author (Vallinayagam, 1988; 1998; 1999; 2001; 2003; 2004a; 2004b) in the Trans-Aravalli Block have noted the zinc enrichment in the acid plutonic variants of Neoproterozoic age. Hence the present write up has been taken with the twin objectives of i) to report their occurrence ii) to characterize them in respect of petrology and geochemistry.

GEOLOGY OF THE TRANS-ARAVALLI COMPLEXES

In the Tran-Aravalli region, MIS is one of the predominant magmatic event during the Neoproterozoic (Pareek 1981; Bhushan and Chandrasekaran, 2002; Kochhar, 2004; Singh et al. 2006). The representative rocks of MIS are exposed at Tosham (Haryana), Jhunjhunu, Siwana, Kundal and Jalor (Rajasthan) and also at Kirana near Lahore and Nagar Parkar in Sindh province of Pakistan (Fig.1).

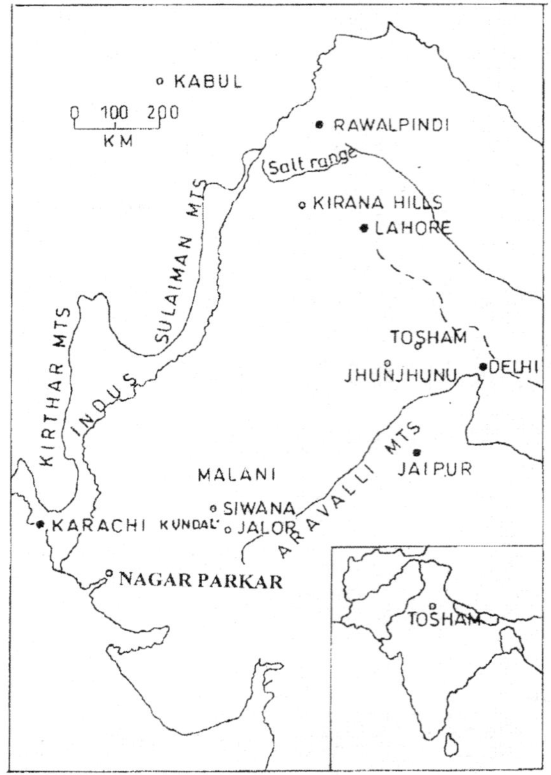

Fig. 1 Location Map of the representative occurrence of Malani Igneous Suite of Rocks.

The rocks are mainly of acid volcano-plutonics with less amount of basic rocks. They occur in the form of ring complexes and dykes. They are characterized as hypesolvus-subsolvus, A-type, anorogenic, HHP, formed by low degree partial melting of crustal rocks, due to hot-spot tectonics related lineaments and bearing rare metals and rare earths (RMRE) (Bhushan and Chandrasekran, 2002).

ZINC RICH GRANITES OF SIWANA

Petrology

Zinc enrichment in the granites of MIS are observed and among them the granites occur in Siwana Ring Complex (SRC) show highest zinc content (Fig 2 and Table 1). The granites shown sharp intrusive relationship with rhyolite/basalt also occur in the form of microgranitic dykes (5070 meter in length; few cm to 2 meter width and trending NE-SW directions) cutting them (Fig.2). They are medium to coarse grained, pink/grey colour and consist mainly of quartz (modal mineral composition - in volume weight: orthoclase (24%), perthite (39%), arfvedsonite (10%), riebeckite (5%), aegirine (4%). The accessory minerals are zircon, apatite, ilmenite, titanite, monazite, xenotime, kataphorite, magnetite amorphous sphene and hematite. They have shown well developed granophyric texture & various hydrothermal alterations characters viz albitization, kaolinization and sericitization. Perthite and K-feldspar occur as phenocrysts in the groundmass of ferromagnesian minerals. Albite occur only as slebs in the perthite component. Quartz occurs as large, anhydral crystal as well as in the groundmass and shows wavy extinction. The alkali amphiboles and alkali pyroxenes exhibit their typical optomineralogical characters viz. arfvedsonite - prismatic crystals, dark green colour with paleochorism (X - dark bluish green Y green Z- yellowish green; X>Y>Z) and extinction angle (XC 12 - 15°); riebeckite - needle shaped, blue colour; aegirine - short prismatic, bright green crystals; aenigmatite- minute crystals of reddish brown colour; kataphorite - bluish green large crystals. The quartzo-feldspathic groundmass and phenocryst of ferromagnesian minerals host large amount of accessory minerals viz. hematite (7%) zircon (5%), ilmenite (4%). The hematites are red colour and occur as fine anhydral crystals. Zircons can be recognized by their prismatic form, absence of cleavage and high relief. Ilmenites are silver grey color, anhyderal form and altered to leucoxene which are seen as white internal reflection. Monazites occur as light yellow small crystals.

Geochemistry

Geochemically they are high in SiO_2, total alkalis ($Na_2O + K_2O$) and iron (ferrous and ferric oxides), Rb, Ga, Nb, Ta, Zr, Hf, U, Th, total REE and low in CaO, MgO, Sr, Ba, Eu, Cr, Co, Ni. These granites (Zn in ppm - Tosham: 106; Jhunjhunu: 89; Jalor: 161 Siwana: 408) are enriched in zinc (Table 1). The granites from SRC show highest content of zinc. In SRC, zinc enriched granites are exposed in the hills of Mokalsar, Gugrot, Piplun and Chappan ka Pahar (Fig.2). They are medium grained, pink colour, presence of cloudy appearance of alkali feldspar, iron oxides and fluorite. They are traversed by numerous pegmatitic and quartz veins and are intruded the rhyolites. The perusal of the Table 1 indicates that among SRC the highest content of zinc bearing rocks are microgranitic and rhyolitic dyke rocks exposed in Chappan ka Pahar hill areas of SRC.

DISCUSSION AND CONCLUSIONS

MIS rocks are characterized as subvolcanic, A-type, anorogenic, HHP and bearing RMRE minerals. The

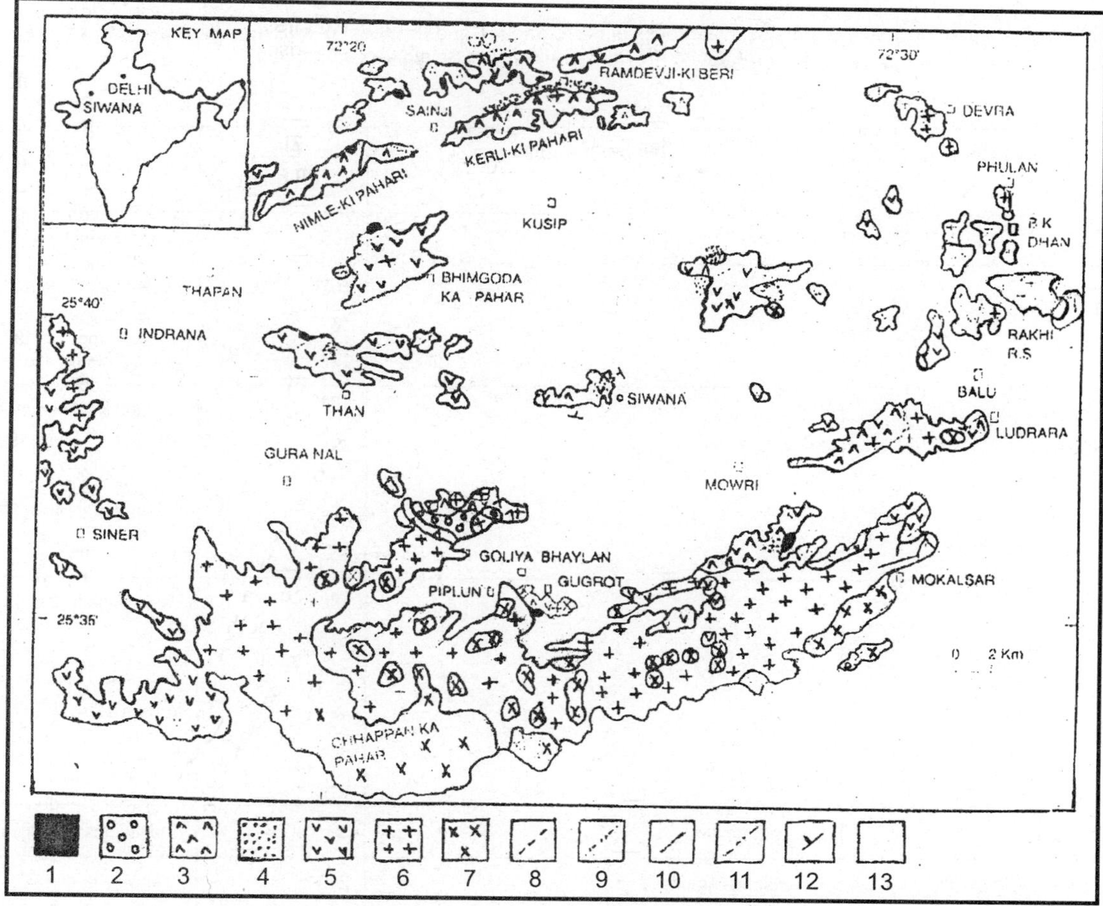

Fig. 2 Geological Map of Siwana Ring Complex, W.Rajasthan. Index: 1- Basalt, 2- Gabbro, 3- Trachyte, 4- Ash Bed, 5-Rhyolite, 6-Pink Granite, 7-Grey Granite, 8- Dacite Dyke, 9-Trachyandesite Dyke, 10-Rhyolite Dyke, 11-Microgranite Dyke, 12-Dip, 13-Alluvium and Blown Sand.

enrichment of zinc in them are highest as compared to similar tectono-magamtic setting HHP granitic rocks of Nigerian Younger Granitic Province and Arabian Shield (Table 1). Also the table shows zinc content of the world average low calcium granites for comparison. The zinc enrichment in this rocks may be due to post magmatic fluid interaction processes. These processes are recognized and described in details by Kinnaird et al (1985), Jackson et al (1985), Kochhar (1992) and Ongunleye et al (2005). Petrological and geochemical data (Table 1) show that each of the hydrothermal alteration processes characterized by a change in specific mineralogical assemblage (s) by addition and depletion of specific element (s). Here the increase in presence of albite, iron oxides, other accessory phases and decrease in alkali feldspar, alumina phases which are characterized by sodic-metasomatism. This sodic metasomatism responsible for the substantial increase of zinc including Na, Fe, F, Ga, Nb, Zr, Hf, REE, HPU (High Heat Producing Units - Kochhar, 1989) and decrease in K, Ca, Mg, Sr, Eu, Cr, Ni oxides. Sodic metasomatism increases the HPU values in term of assist the prolonged circulation of hydrothermal fluids for the ore formation. The increase of alkalinity, Fe and F at the post magmatic stage causes the sodic metasomatism and this responsible for the ore fluid to enrich in zinc incuding the high field strength elements viz. Nb, Ta, Ga, Zr, Hf, U, Th, REE. The large size and high charge of these elements do not allow them to enter into other rock forming elements/ minerals. Thus the change in fluid chemistry and water rock ratio during post magmatic stages perhaps transported and redistributed the ore fluids in the rock.

Zinc enriched granites in Siwana are medium grained, pink colour, turbid & zoned alkali feldspar (i.e whitening around the rim of pink K-feldspar

Table-1 Zinc abundance in the granites, dyke rocks of microgranite+ and rhyolites* of MIS and also from world average low Ca- granites, Nigerian Younger Granite Province and Arabian Shield for comparison.

Location	Sample number	Zinc (in ppm)	References
Tosham	TG2	106	Eby and Kochhar (1990)
Jhunjhunu	MD7*	267	Kochhar and Sharma (2003)
	NP34	38	
	IP1	89	Sharma (1994)
Jalore (Peraluminus granite)	KO1	106	Eby and Kochhar (1990)
	KO2	161	
	KO6	147	
	J117	60	Kochhar and Dhar (1993)
	J19	50	
Siwana (Peraluminus granite)	G1	408	Vallinayagam (1998)
	G46	315	
	C12	368	
	M1a+	759	Vallinayagam (1999)
	P1*	499	
	C3*	816	
World Average	low Ca- granites	39	Turkian and Wedepohl (1961)
Nigerian Younger Granite Province (similar to Siwana granite)	Arfvedsonite granite	249	Ogunleye et al (2005)
	Aegirine arfvedsonite granite	355	
	Albite arfvedsonite granite	1160	
Arabian Shield (similar to Siwana granite)	Jabol microgranite	664	Jackson and Drysdall (1984)

albitization), abundance of iron oxides, F encrustation, quartz vein, pegmatitic and aplitic knots & vugs. These are the field indicators to locate them. Increase in Na, Fe, F, and decrease in K, Ca, Mg are the geochemical tools to assess their potentiality and followed by trace elements geochemistry. These geoexploration tools can be employed to search out here and any other places of similar ore deposit.

ACKNOWLEDGEMENT

The author is grateful to Prof. K.L.Shrivastava for his kind invitation to contribut paper in the edited volume on "Economic Mineralisation", dedicated to Prof. B.S. Paliwal. I also extend my sincere thanks to Kurukshetra University for providing the research facilities and to my son Master V. Kartik Ganesh for his help in drawing the figures.

References

Bhushan, S.K. and Chandrasekaran, V. (2002) Geology and geochemistry of the magamtic rocks of the Malani Igneous Suite and Tertiary Alkaline Province of Western Rajasthan. Mem. Geol. Surv. India, v.126, pp.1-181.

Drysdall, A.R., Jackson, N.J., Ramsay, C.R., Drich, C.J. and Hackett, D. (1984). Rare earth mineralisation related to Precambrian alkali granites in the Arabian Shield. Economic Geology, v.79, pp.1366-1377.

Eby, G.N. and Kochhar, N. (1990). Geochemistry and petrogenesis of the Malani Igneous Suite, North Peninsular India. Jour. Geol. Soc. India, v. 36, pp.109-130.

Jackson, N.J. Drysdall, A.R. and Stoeser, D.B (1985). Alkali granites related Nb-Zr-REE-U-Th mineralisation in the Arabian Shield. In : High Heat Production (HHP) Granites, Hydro-thermal Alternation and Ore genesis, Institute of Mining and Metallurgy, London, pp.479-488.

Kinnaird, J.A., Batchelor, R.A., Whitely, J.E. and Mackenzie, A.B. (1985) Geochemistry, mineralisation and hydrothermal alteration of the Nigerain high Heat Production (HHP) Granites, In; High Heat Production (HHP) Granites, Hydrothermal Alternation and Ore genesis, Institute of Mining and Metallurgy, London. pp.169-195.

Kochhar, N. (1989) High Heat Producing granites of the Malani Igneous Suite, Northern Peninsular India. Indian Minerals, v. 43, pp. 339-346.

Kochhar, N. (1992) Mineralisation associated with A-type Malani magmatism, northwest Peninsular India. In: S.C. Sarkar (Ed.) Metallogeny relate to tectonics of the Proterozoic mobile belts, Oxford & IBH, New Delhi, pp. 209-224.

Kochhar, N (2004) Geological evolution of the Trans-Aravalli Block (TAB) of the NW Indian Shield: Constraints from the Malani Igneous Suite (MIS) and its Seychelles connection during Late Proterozoic. Spl. Pub. Geol. Surv. Ind., no. 84, pp. 247-264.

Kochhar, N. and Dhar, S. (1993) The association of hyperolvus-subsolvus granites: A study of Malani Igneous Suite, India. Jour. Geol. Soc. India, v.42, pp. 449-467.

Kochhar, N. and Sharma, R. (2003) Rare earth and trace element mobility accompanying tourmalisation; evidence from Jhunjhunu, Rajasthan, India. Jour. Applied Geochemistry, v.5, pp. 26-30.

Ogunleye, P.O., Ike, E.C. and Garba, I. (2005) Geochemical characteristics of the niobium-rich arfvedsonite granites, Younger Granites Province of Nigeria. Geochemistry, v. 65, pp. 279-296.

Pareek, H.S. (1981) Petrochemistry and petrogenesis of the Malani Igneous Suite. Geol. Soc. Am. Bull., v. 92, pp. 206-273.

Sharma, R. (1994) High Heat Production (HHP) granites of Jhunjhunu area, Rajasthan, India. Bull. Ind. Geologists' Asso., v.27 (1), pp.55-61.

Singh, A.K., Singh, R.K.B. and Vallinayagam, G. (2006) Anorogenic volcanic rocks of Kundal area of the Malani Igneous Suite, northwestern India: Geochemical and petrogenetic studies. Jour. Asian Earth Sci., v. pp. (Available online at www.sciencedirect. com).

Turekian , K.L. and Wedepohl, K.H. (1961) Distribution of the elements of the some major units of the Earth's Crust. Bull. Geol. Soc. Am., v.72, pp. 175-192.

Vallinayagam, G. (1988) Geology and geochemistry of alkali granites and the associated acid volcanics around Mokalsar, Distt. Barmer, W. Rajasthan, India and their bearing on the rift tectonics. Ph.D. Thesis, Punjab University, Chandigarh, pp. 1-113.

Vallinayagam, G. (1998) Rare metal bearing and barren distinction of alkali granites, Piplun area, Siwana Ring Complex, Rajasthan, India: Petrological and geochemical evidences. In: Paliwal, B.S. (Ed.) The Indian Precambrian, Scientific Publishers, Jodhpur, pp. 492-501.

Vallinayagam, G. (1999) Nb, Zr, Y, REE rich dyke rocks, Piplun area, Siwana Ring Complex, Rajasthan. In: Paliwal, B.S. (Ed.) Proc. Geological evolution of the northwestern India, Scientific Publishers, Jodhpur, pp. 94-102.

Vallinayagam, G. (2001) Geochemistry and petrogenesis of the basic rocks in the Siwana Ring Complex, Barmer District, Rajasthan, India. Indian Mineralogist, v.35, pp.21-133.

Vallinayagam, G. (2003) Basic magmatism of Neoproterozoic Malani Igneous Suite, Western India Craton: Petrological and geochemical modelling. Ind. Jour. Geochemistry, v.18, pp. 1-18.

Vallinayagam, G. (2004a) Peralkaline-peraluminous A-type Rhyolites, Siwana Ring Complex, Northwestern India: Petrogenetic modelling and tectonic implications. Jour. Geol. Soc. India, v. 64, pp. 336-344.

Vallinayagam, G. (2004b) A report on rare metals and rare earths in the Siwana Ring Complex, Rajasthan. Jour. Applied Geochemistry, v. 6, pp. 387-391.

Hon'ble Reviewers

1. Dr. S.K. Bhushan
   Dy. Director General
   Geological Survey of India
   Bandlaguda Complex
   HYDERABAD 500 068 INDIA

2. Dr. K.K. Sharma
   Department of Geology
   Govt. College Sirohi
   SIROHI 307 011 INDIA

Editorial Assistance

Mr. Chandra Prakash Dadhich

Economic Mineralization
K.L. Shrivastava (Editor)
2009. pp.210-216
Scientific Publishers (India), Jodhpur

17

Section-II

Structural Setting of the Nepheline Syenite Pluton, West of Khariar, Eastern Ghats Mobile Belt, Orissa, India

T.K.BISWAL AND HARISH AHUJA

Department of Earth Sciences, Indian Institute of Technology Bombay, Powai, Mumbai-400 076
Email : tkbiswal@iitb.ac.in

Abstract

The Alkaline rocks are associated with various structural settings around the world ranging from subduction zones to continental rifts to uplifted cratonic interiors. The Proterozoic granulitic Eastern Ghats Mobile Belt of India contains nepheline syenite plutons close to its terrane margins with surrounding Cratons. One such pluton, west of Khariar, is emplaced within a fold- thrust belt of the mobile belt that developed on its NW margin. The fold-thrust belt consists of a basal decollement and a stack of nappe sheets structurally resting over the Bastar Craton. The mylonites of the decollement as well as imbricate thrusts are characterized by mylonitic foliation, S-C fabric, mantled porphyroclasts and interclast faults that convincingly show northwesterly vergence of the thrusts. The plutons are marked by magmatic fabric such as magmatic foliation, mineral lineation, magmatic S-C fabric, tiled feldspar grains and magmatic folds which have been superimposed by solid state strain fabrics in a transitional manner. The magmatic and solid state fabrics of the pluton grades into the mylonitic fabric of the host rock, a phenomenon known as coupling that indicates the nepheline syenite magma has been emplaced under a non-coaxial strain. A compressional heating and consequent melting of the lithosphere are proposed for the origin of the alkaline magma that has been brought up through mantle-deep fractures during thrusting.

Keywords : Nepheline Syenite, Khariar, Eastern Ghats Mobile Belt, Synkinematic Emplacement, Shear Heating.

INTRODUCTION

The structural setting of alkaline complexes has long been an issue of debate among scientists world over. Some believe rifting is necessary, particularly of continents, for alkaline rocks to occur (Burke et al; 2003). Others deny that alkaline magmatism is a consequence of rifting or controlled by rifting, they believe that epeirogenic processes were responsible (Le bas, 1987). However, some examples (Khibina and Lovozero alkaline massifs of Kola peninsula, north west of USSR) of alkaline magmatism are not related to a rift environment or to uplift or doming. Rather, they are associated with deep seated fracture zones (Middlemost, 1985). While Briden and Gass

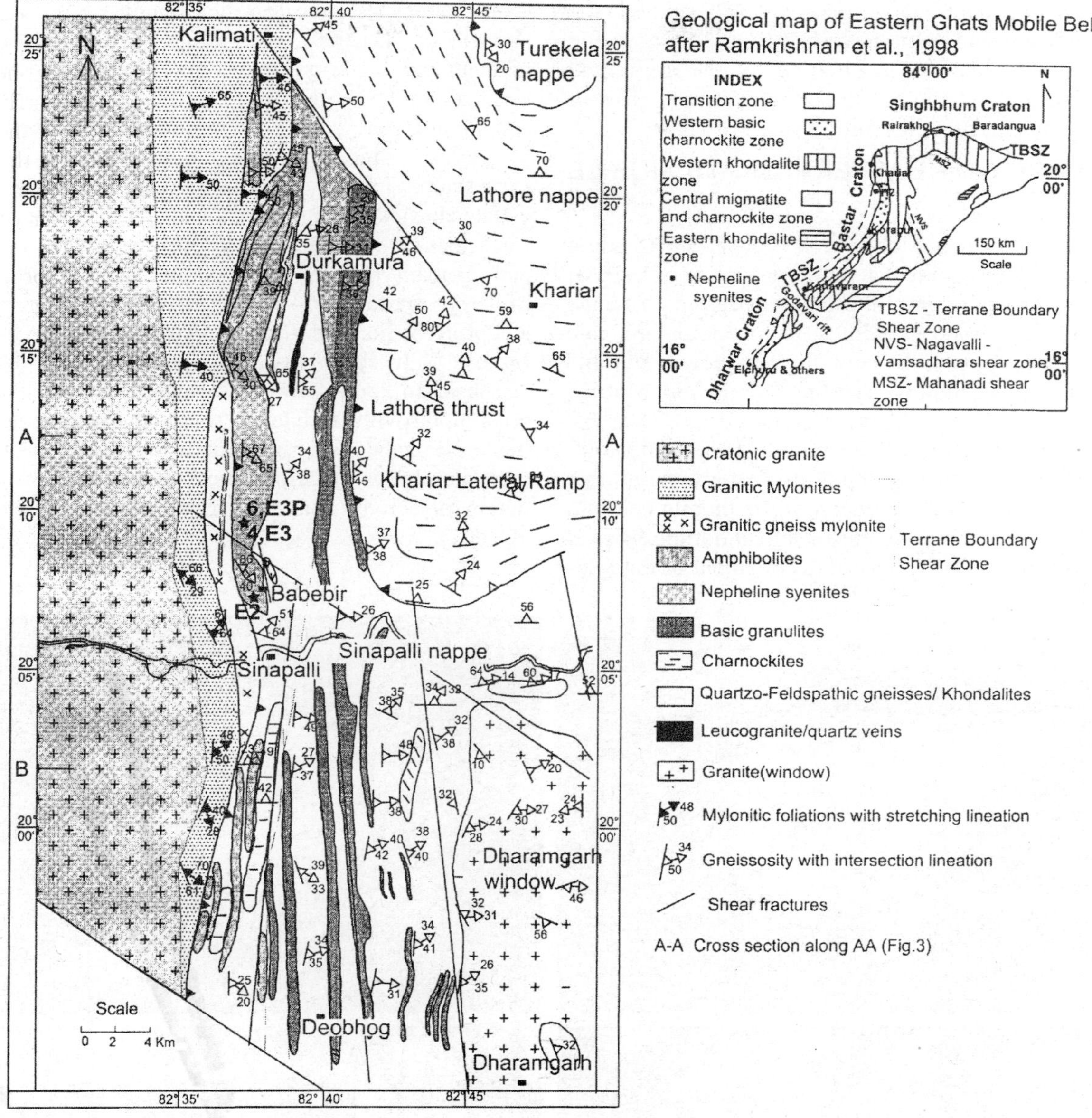

Fig. 1 Geological map study area west of Khariar, Orissa, showing the nepheline syenite pluton.Inset: The generalized geological map of the Eastern Ghats Mobile Belt (modified after Ramakrishnan et al., 1998).

(1974) argued that alkali rocks are associated with the core of a Craton, Burke et al (2003) believed them to occur along the continental margin. Black and Liegeois (1993) suggested the alkali magmatism is related to regional continental lithospheric mantle delamination in a subduction zone while Black et al (1985) suggested they owe their origin to shear heating along pre-existing faults.

The terrane margin between Proterozoic Eastern Ghats Mobile Belt and Archaean Cratons of the Indian Peninsula hosts a series of alkaline plutons that vary in composition from nepheline syenite to syenite to quartz syenite (Fig.1 inset, Madhavan and Khurram Mohammed, 1989; Leelanandam, 1993; Ramakrishnan et al., 1998). The most important fact about these plutons is that they are spatially confined to the margin of the Craton and Mobile Belt. The Eastern Ghats Mobile Belt is considered to be a binding block between India, Antarctica and Australia in the East Gondwanaland assembly. Therefore, the association of alkaline rocks along the margin of the Eastern Ghats Mobile Belt reflects a feature of the

Gondwana continental margin. We studied the nepheline syenite pluton west of Khariar, Orissa and propose a compressional setting for the origin and emplacement of the alkaline magma on the basis of meso and microscopic fabric.

GEOLOGY OF EASTERN GHATS MOBILE BELT

The Eastern Ghats Mobile Belt is a NE-SW trending Proterozoic granulite belt on the east coast of India. It is dominated by granulite facies rocks that include khondalites, basic granulites, charnockites and calc granulites (Fig.1). Structurally it shows multiple generations of folding, shearing (Chetty and Murthy, 1994) and metamorphism (Dasgupta,1995). The most important structure of the mobile belt is the fold-thrust-belt structure on the northwestern margin that has resulted in the protrusion of the mobile belt into the Bastar Craton producing a salient outline (Biswal and Sinha, 2003). The basal decollement of the fold-thrust belt is called the Terrane Boundary Shear Zone (Biswal et al, 2000) and occurs as northwesterly vergent ductile thrust at the interface between the Craton and the Mobile Belt. Overlying the decollement are three granulitic nappes namely Sinapalli, Lathore and Turekela nappes in order of succession. Further, platform deposits occur in isolated basins in the foreland around the mobile belt. These basins are named as the Khariar, Chhattisgarh and Ampani basins, which show an evolutionary history from the Mesoproterozoic to the Neoproterozoic. The granulites of the Eastern Ghats Mobile Belt shows a wide range of ages from Archaean to Pan-African that suggests the Archaean crust along with Archaean-Proterozoic sediments of the terrain have been remobilized repeatedly through multiple phases of deformation, metamorphism and magmatic activity (Sarkar and Paul, 1998; Rickers et al, 2001).

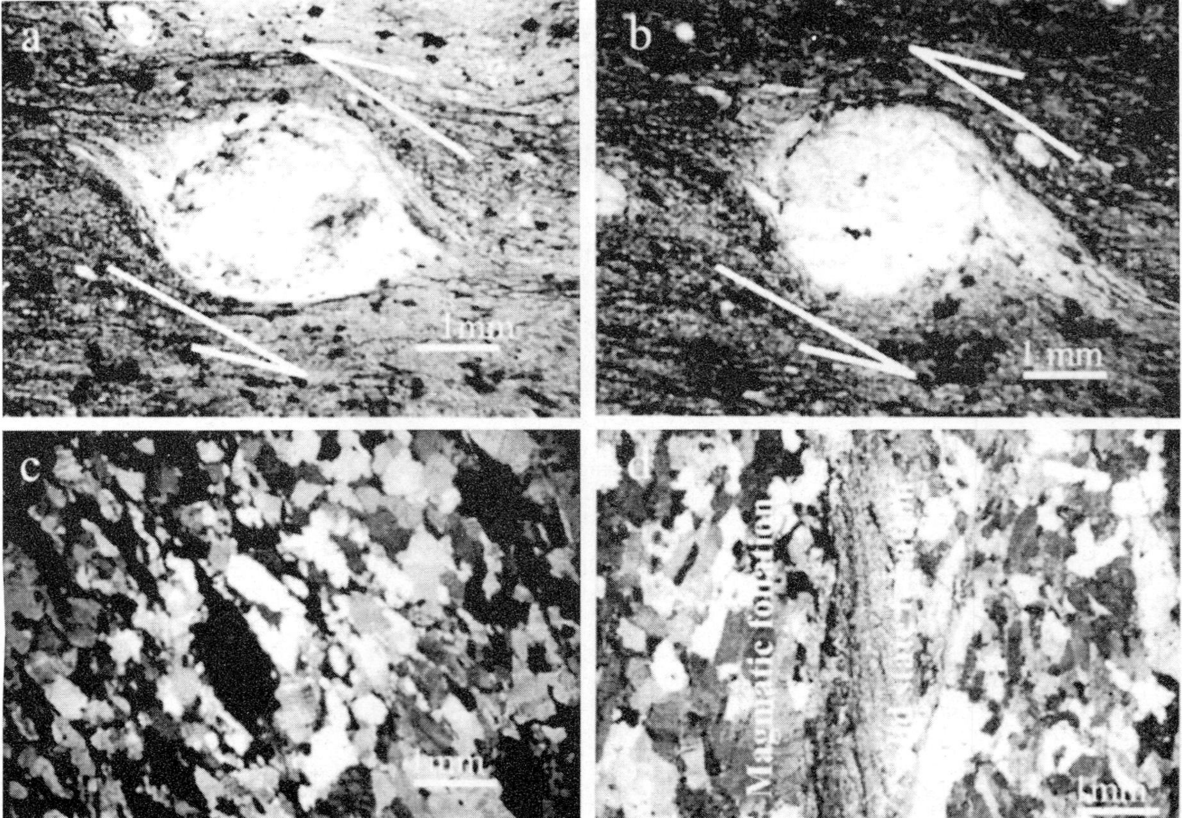

Fig. 2 Photomicrographs and field photographs.
a.b. Sigma and delta type porphyroclasts from mylonites of the Terrane Boundary Shear Zone, showing NW vergence of shear,
c. Euhedral feldspar grains with albite rim in nepheline syenite.
d. Solid state S-C fabric in nepheline syenite.

GEOLOGY OF THE STUDY AREA

Basal decollement

The Terrane Boundary Shear Zone marks the basal decollement of the fold-thrust belt. The shear zone is prominently developed on late tectonic granites and granite gneisses of the Bastar Craton in form of a brittle-ductile thrust (Fig.1). It has a NE to NS trend and shows a kink structure near Boran where it passes over a northerly dipping Khariar lateral ramp. The mylonitic foliations are southeasterly dipping but show a variation in dip amount on either side of the ramp. They are steeper in the north and gentler in the south which has been attributed to the listric character of the decollement at depth. The stretching lineations are mostly down dip. The S-C fabric, porphyroclasts and other shear kinematics suggest northwesterly vergence of the thrust (Fig. 2a, b).

Sinapalli nappe

The Sinapalli nappe is well exposed in the south (Fig.1) consisting of alternate quartofeldspathic gneiss, basic granulites, amphibolites and charnockite units which are folded into an early phase of coaxial folding along NE-SW axis and late fold along NW-SE axis. The folded sequence is dissected by imbricate splay thrusts originating from the sole thrust contemporary with second stage folding. The folds have turned into sheaths at many places. On a larger scale, some of these thrust appear as prominent lineaments running parallel to the westem flank of the basic granulite hills. These lineaments are considered to be deep seated fractures on the Craton, which has been reactivated and cuts through the overlying nappes during thrusting. The nepheline syenite pluton and kimberlite pipes occur along these fractures. The Sinapalli nappe shows a window of basement rocks (Dharamgarh window) to the east around Dharamgarh.

The quartzo-feldspathic gneisses are layered rock showing alternate hornblende-biotite-garnet- sphene-opaque rich layers and plagioclase- orthoclase-perthite-quartz rich layers. The felsic mineral are equigranular, medium to coarse grained. The basic granulites are coarsely layered with hornblende-pyroxene-opaque-garnet layers alternating with plagioclase rich layers. The hornblendes are pleochroic in brown to yellow. Pyroxenes are light green colour. Opx is pleochroic light brown to light green. Garnet is deep brown colour. An EPMA study suggests a PT for granulite facies metamorphism to be around 7-8 kb, 800- 900C. Basic granulites have been retrograded into amphibolites during thrusting.

Lathore nappe

The Sinapalli nappe peters out to the north of the Khariar Lateral ramp, as a result the Lathore nappe comes directly in contact with the Terrane Boundary Shear Zone (Fig.1). The Lathore nappe consists of charnockitic gneiss, basic granulites, khondalites and subordinate amounts of calc granulites. The quartz and feldspar grains in the charnockitic gneiss are extremely elongated parallel to gneissosity and cut by orthogonal fractures. Myrmekitic intergrowth of quartz and plagioclase is common; at places perthitic intergrowth is also observed in orthoclase. Hypersthene and garnet form chains parallel to the layering. Multiple generations of folds are present in the rock. Small to large scale sheath folds have been developed at many places due to axial planar inhomogeneous flow. Shape measurement of the elongate quartz grains indicates a flattening strain affected the rock. The Lathore nappe close to the Terrane Boundary Shear Zone displays realignment of the folds parallel to the shear direction. Furthermore, the granulites have been retrograded into amphibolites. The amphiboles have grown parallel to the XY plane of the strain ellipsoid in the thrust. This suggests thrusting post dated folding and granulite metamorphism.

Turekela nappe

The Turekela nappe represents the top most nappe. It occurs as a klippe surrounded by rock of the Lathore nappe. The root zone of the klippe lies further east. The klippe is dominated by khondalites followed by calc granulites. The khondalites consist of sillimanite, graphite, feldspar, garnet and quartz. Leptynite veins derived from the melting of this assemblage occur concordantly within it. The Turekela nappe exhibits multiple stages of folding consisting of early coaxial folding along an NNE-SSW axis, followed by NW-SE trending folds. Thin sections study of the khondalites show the preferred orientation of sillimanite, garnets and quartz parallel to gneissosity. Quartz grains are distinctly lensoidal and shape analysis of these grains shows a constrictional type of strain in the Turekela klippe, in marked of contrast to the Lathore nappe. The Turekela klippe has been traversed by many close spaced NW-SE trending brittle dextral faults, which cut across the Lathore nappes.

Nepheline syenite pluton

The nepheline syenite pluton west of Khariar occurs as three NS trending en echelon bodies extending from Kalimati in the north to Babebir in the south (Fig.1). The northern body lies within the granite mylonites (Fig.1) while the southern one is flanked by granite mylonite in the west and quartzo-feldspathic gneisses of the Sinapalli nappe in the east. The middle body occurs within the amphibolites of Sinapalli nappe.

The nepheline syenites are leucocratic coarse grained rocks and tend to become pegmatoidal at places. They are marked by distinct compositional layers and largely down - dip mineral lineations. The compositional layers are defined by thin, dark hornblende- or biotite- rich layers alternating with coarse white or pinkish feldspar- and nepheline- rich layers. In many instances the compositional layers are folded into symmetrical to asymmetrical, isoclinal to open, steeply plunging to reclined folds with Class 2 geometry. The hinge zone of the fold exhibits an axial planar arrangement of biotite, hornblende and feldspar grains. Statistical analysis of magmatic foliation and mineral lineation of the pluton shows parallelism with mylonitic foliation and stretching lineation of the host mylonites (Biswal et al., 2004).

Under the microscope the rocks, from the central part of the pluton, predominantly show a hypidiomorphic granular texture with well-developed crystal faces on most of the minerals (Fig.2c). Occasionally, a porphyritic texture is observed due to the presence of small microcline and other minerals around microcline phenocrysts and poikilitic texture is seen with hornblende phenocrysts enclosing smaller grains of feldspars. Perthite structure is very common defined by flame-type albite lamellae inside K-feldspar. In addition, albite rims occur around perthite grains that show optical continuity with the internal albitic lamella. The pyroxene, hornblende and ilmenite grains show a well developed magmatic corona texture indicating lack of equilibrium during their crystallization from the melt . An SEM study showed that the corona is made up of nepheline and Fe- rich amphiboles. Generally, the alkali feldspar and hornblende grains have a preferred orientation in two directions producing a magmatic S-C fabric. However, in some of the sections, they show an arcuate geometry and are tiled in an imbricate fashion suggesting shear flow. Undulose extinction is generally rare in the minerals; feldspars show straight twin lamellae and hornblende and biotite flakes lack any kind of kink in their cleavage. These considerations suggest the minerals are magmatic in origin and have not undergone dynamic recrystallization. The compositional layers in the rock represent magmatic foliations. However, nepheline and feldspar grains with arcuate grain margin and zigzag borders are common. They lack any kind of solid state deformation. This particular shape is attributed to melt enhanced grain boundary migration and brittle fracturing respectively.

The magmatic fabric has been superimposed by a solid state deformation fabric at various scales. At the margin of the pluton, discrete shear fractures (mylonitic C- fabric) are developed in the rock parallel to the magmatic foliation (Fig.2d). Feldspar grains are broken and dynamically recrystallized along these fractures. Biotites are grown along a solid state S fabric defining an S-C fabric in the rock. Biotite and hornblendes are kinked.

The host rock fabric is parallel with the magmatic fabric, a phenomenon known as coupling (Patterson et al., 1998), suggesting that intrusion of pluton was synkinematic with thrusting. Further, the magmatic fabric in the pluton grades into the mylontic fabric in the host mylonite particulary in the northernmost body. Hence, the magma was transported as crystal melt mush under the influence of non-coaxial strain. Due to rotation under high strain (ã~ 5.5) the phenocrysts were aligned parallel to the shear direction producing a magmatic foliation (magmatic C-fabric; Ramsay and Lisle, 2000). The new crystals that developed during transportation of the melt, crystallized parallel to strain axis of the non-coaxial strain producing S-fabric, together giving rise to magmatic S-C fabric in the rock. At places where the crystal to melt ratio is high, feldspar and nepheline grains show melt induced arcuate grain margin and brittle fracturing without any undulose extinction in them. Towards the margin of the pluton which is solidified earlier then the core, solid state strain fabric like dynamically recrystallized feldspar and nepheline grains, S-C fabric and kinks in biotite and hornblende have been superimposed parallel to the magmatic foliation. This is called a transitional fabric.

The crystallization age of zircon (U-Pb) from these plutons has been determined to be Ca 0.5 Ga (Unpublished work of the author). Hence the emplacement of magma and thrusting of Eastern Ghats Mobile Belt belong to that age. Thus a Pan-African juxtaposition of the Eastern Ghats Mobile Belt with Cratons of Peninsular India has been suggested.

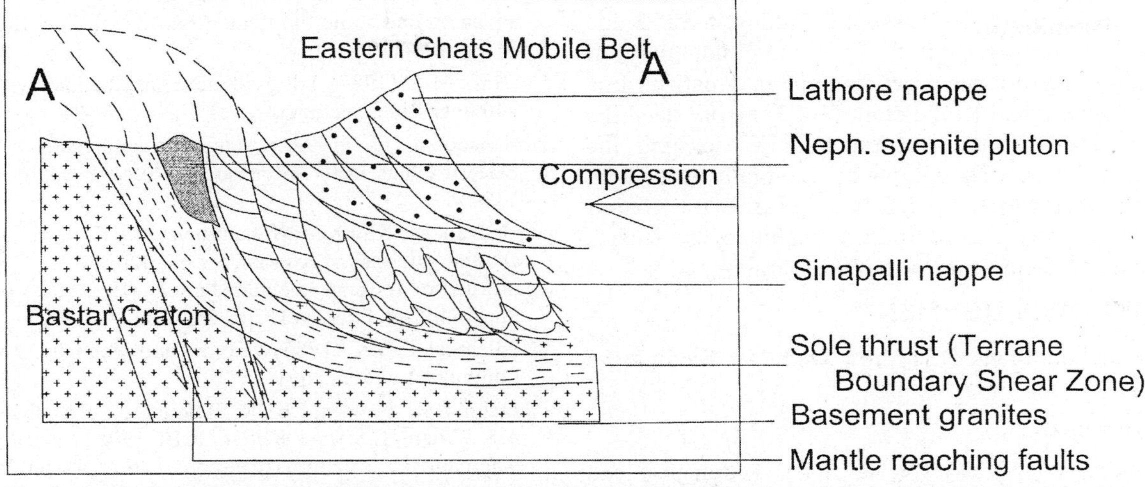

Fig. 3 Cross sections across the fold-thrust belt, showing decollement and leading imbricate thrust structure and mantle reaching fractures. Compression is from the east that reactivated the faults.

STRUCTURAL SETTING : A DISCUSSION

While discussing the structural setting of the Khariar pluton, the following points are to be considered. Secondly, all the alkaline complexes along the terrane margin are to be viewed together.

i) The pluton is confined to the terrane margin of the Eastern Ghats Mobile Belt.

ii) The pluton is localized along shear zones/thrusts of the fold-thrust belt.

iii) The emplacement is synkinematic with thrusting.

iv) Some of the shear zones have large scale manifestation in the form of lineaments. The geomorphology of the area is controlled by shear fractures in terms of shaping the ridges into cuesta. Kimberlites occur in some of these shear fractures. Hence, the shear fractures are assumed to be deep seated in origin and reach the base of the lithosphere.

In view of the above, a shear heating model has been suggested for the origin and emplacement of the pluton, in line with that of African alkaline rocks (Black et al., 1985). This is illustrated in Fig.3 where Eastern Ghats Mobile Belt is shown to have thrusted over the Bastar Craton The interface between the Craton and the Eastern Ghats Mobile Belt acted as the basal decollement for the fold-thrust belt. The granulites of the mobile belt have split into nappes and thrust over the Craton. The Craton to the west of the Eastern Ghats Mobile Belt is dissected by deep seated faults. The foreland platform deposits are hosted in the basins formed due to block faulting along these faults.

The formation of fold-thrust belt on the western margin of the Eastern Ghats Mobile Belt caused extreme compression of the adjoining Cratonic foreland. This led to the reactivation of the deep seated faults, which underlay the fold-thrust belt or close to it. The compression was horizontal and westerly directed. The faults that were steeply inclined towards east were preferentially reactivated rather than westerly dipping ones because of the above stress system. The fractures, which were initially normal faults, were reactivated as thrusts. The compression was transmitted to the base of lithosphere through this reactivation and the lithosphere was heated up. This led to melting to produce alkali magma. The alkali magma was transported upward through these faults and emplaced along the network of thrust planes within the fold-thrust belt. The rise in thermal activity on the Eastern Ghats Mobile Belt front may be attributed to lithospheric mantle delamination beneath Peninsular Craton similar to Tibet plateau (Black and Liegeois, 1993).

The other alkali plutons of the Eastern Ghats are also associated with large scale crustal fractures (Fig.1). The Rairakhol and Baradangua plutons are emplaced along the Mahanadi rift and a group of NE-SW trending lineaments. Kunavaram complex is also thought to be emplaced at the junction of Godavari rift and Sileru Shear Zone. Koraput- (Bhattacharya and Kar, 2005) and Elcheru alkaline complexes are emplaced at the junction of the Sileru Shear Zone with cross cutting fracture. Hence shear heating may be attributed to the origin of all these alkaline rocks.

In view of the importance attached to the

occurrence of nepheline syenite plutons along terrane margins, it is logical to define the boundary of juxtaposing terranes based on such magmatism. In a different context Burke et al, (2003) emphasized the importance of alkali magmatism in defining the ancient sutures. Though we do not totally agree with them with regard to the deformed nature of the Khariar alkali plutons, it is definitely useful to use such a feature for identifying the terrane margins.

ACKNOWLEDGEMENT

The authors thank DST, New Delhi for funding the project.

References

Bhattacharya, S. and Kar R. 2005. Petrological and geochemical constraints on the evolution of the alkaline complex of Koraput in the Eastern Ghats Granulite Belt, India. Gondwana Res., v.8,No.4, pp.596-602.

Biswal, T.K., Jena, S.K., Datta, S., Das, R. and Khan, K. 2000. Deformation of the Terrane Boundary Shear Zone (Lakhna shear zone) between the Eastern Ghats Mobile Belt and the Bastar Craton in the Balangir and Kalahandi districts of Orissa; J. Geol. Soc. Ind. 55,pp, 367-380.

Biswal, T. K. and Sinha, S. 2003. Deformation history of the NW salient of the Eastern Ghats Mobile Belt, India. Jour. As. Earth Sc., v. 22, pp. 157-169.

Biswal,T.K., Ahuja, H. and Sahu H.S. 2004. Emplacement kinematics of nepheline syenites from the Terrane Boundary Shear Zone of the Eastern Ghats Mobile Belt, West of Khariar, NW Orissa, Evidence from meso and microstructures. Ind.Acad.Sci (EPS) 113, 4, pp. 785-793.

Black, R. and Liegeois, J.P. 1993. Cratons, Mobile Belts, Alkaline rocks and continental lithospheric Mantle: the Pan-African testimony. J. Geol. Soc. Lond., 150, pp.89-98

Black, R., Lameyre, J. and Bonin, B. (1985) The structural setting of alkaline complexes. J. Afr. Earth Sci., 3 (1/2), pp. 5-16.

Briden, J.C. and Gass, I.G. (1974) Plate movement and continental magmatism. Nature, v. 248, pp. 650-653.

Burke, K., Ashwal, L.D. and Webb, S.J. (2003) New way to map old sutures using deformed alkaline rocks and carbonatites. J.Geol. Soc. Am., 31(5), pp.391-394.

Chetty, T. R. K. and Murthy, T. S. N. (1994). Collisional structurals in Late Precambrian Eastern Ghat Mobile Belt: Mesoscopic to satellite scale structural observation. Terra Nova, v. 6, pp. 72-81.

Dasgupta, S. (1995). Pressure-Temperature Evolutionary history of the Eastern Ghats Granulite Province: recent advances and some thoughts. Mem. Geol. Soc. Ind., v. 34, pp. 101-110.

Le Bas, M.J. (1987) Ultra-alkaline magmatism with or without rifting. Tectonophysics, 143, pp.75-84

Leelanandam, C. (1993). Alkaline magmatism in the Eastern Ghats Belt- A critique, J. Geol. Soc. Ind., 42, 435448.

Madhavan, V., Khurram Mohammed, Z.A.K. (1989). The alkaline gneisses of Khariar, Kalahandi District, Orissa, in Alkaline rock. In Leelanandam, C. (Ed), Mem. Geol. Soc. India Spl. Publ., v. 15, pp. 265-289.

Middlemost, E.A.K. (1985) Magmas and Magmatic Rocks. Longman, London. 266 p.

Paterson, S.R., Fowler, T.K.JR, Schmidt, K.L., Yoshinobu, A.S., Yuan, E.S. and Miller, R.B. 1998. Interpreting magmatic fabric patterns in plutons; Lithos, 44, pp. 5382.

Ramakrishnan, M., Nanda, J.K. and Augustine, P.F. 1998. Geological evolution of the Proterozoic Eastern Ghats Mobile Belt. Geol. Surv. Ind. Spl. Pub. 44, pp.1-21.

Ramsay, J. G. and Lisle, R. (2000). The techniques of modern structural geology, v. 3. Application of continuum Mechanics in Structural Geology. Academic press. London. 856 p.

Rickers, K., Mezger, K. and Raith, M. M. (2001). Evolution of the Continental Crust in the Proterozoic Eastern Ghats Belt, India and new constraints for Rodinia reconstruction: implications from SmNd, RbSr and PbPb isotopes. Precamb. Res., v.112, pp.183-210.

Sarkar, A., Nanda, J.K., Sarkar, G., Pati, U.C., Bishui, P.K. and Gupta, S.N. (1994). Rb-Sr geochronology of the Khariar alkaline complex Orissa sector Eastern Ghats Belt. Proc. Eastern Ghats Mobile Belt, Visakhapatnam , Geol. Surv. Ind., pp.74-75.

Sarkar, A. and Paul, D. K. (1998). Geochronology of the Eastern Ghats Precambrian Mobile Belt- A Review. Geol. Surv. Ind. Spl. Publ., v. 44, pp. 51-86.

Hon'ble Reviewers

1. Prof. Hari B. Srivastava
   Department of Geology
   Banaras Hindu University
   VARANASI 221 005 INDIA

2. Prof. Mahshar Raza
   Chairman
   Department of Geology
   Aligarh Muslim University
   ALIGARH 202 002

Editorial Assistance

Dr. (Mrs.) Yasmin Sayyed

Economic Mineralization
K.L. Shrivastava (Editor)
2009. pp. 217-223
Scientific Publishers (India), Jodhpur

18

Section-II

Stanniferous Pegmatites on the Eastern Flank of the Bailadila Iron Range, Dantewara District, Chhattisgarh

R. K. TRIVEDI AND L. K. SONI

Department of Applied Geology, Dr. Harisingh Gour, Vishwavidyalaya, Sagar 470 003
Email : rkt_sagar@hotmail.com

Abstract

The occurrence of stanniferous pegmatite association with the Bhansi Granites and amphibolites in relation to geological setup, mineralogical variation and geochemical characterisation have been discussed in this paper. A few new stanniferous pegmatites have been marked on the eastern flank of Bailadila Iron Range on the western extremity of the Baster-Koraput Pegmatite Belt.

Key Words : Stanniferous pegmatites, Metabasites, Bengpal group, Bhansi granite, Baster

INTRODUCTION

Many tin occurrences have been known from different parts of India but the Baster region (incorporating part of Chattisgarh and Orissa states) contains the only stanniferous zone where tin is being recovered at present economically. In Chhattisgarh state A 23 Km wide strip containing tin bearing pegmatite have been mapped along EW and further ESE across the state boundary between Orissa and Chhattisgarh. It extends up to Koraput (Babu, 1994). This strip is named as Baster-Koraput Pegmatite Belt (BKPB) and contains primary as well as secondary tinbearing mineral (Ramesh Babu, 1993).

At the western margin of this strip, pegmatite bearing block of Bacheli-Nerli, (longitude E 81°12' - 81°19' and Latitude N18°38' - 18°47') running parallel to eastern flank of Bailadila Iron Ridge, have been studied. Also, some of the new pegmatite occurrences in this area have been reported in the present paper.

Geological Setup :

The study area belongs to the Bhandara Craton (or Baster Craton of K.L. Rai, this volume) in the central part of India, adjacent to Eastern Ghat Mobile Belt (EGMB) in the south. This craton containing granites and gneisses with engulfed and overlying supracrustal rocks of the Dongargarh, Sakoli, Sausar and Sukma, Bengpal, Bailadila within eight basins including Chhattisgar, Baster etc. of the central India. (Fig. 1)

Iyangar and Banerjee (1971) made an attempt to trace the metamorphic history of Bengpal and Bailadila groups along with the study of intrusions of basic dykes in this area. Metavolcanics also occurs along with intrusive post Bailadila granites and pegmatites. Metapelitic rocks of Bacheli with a prominent BIF are the characteristic litho-units of this area. All the above rock types are properly grouped by

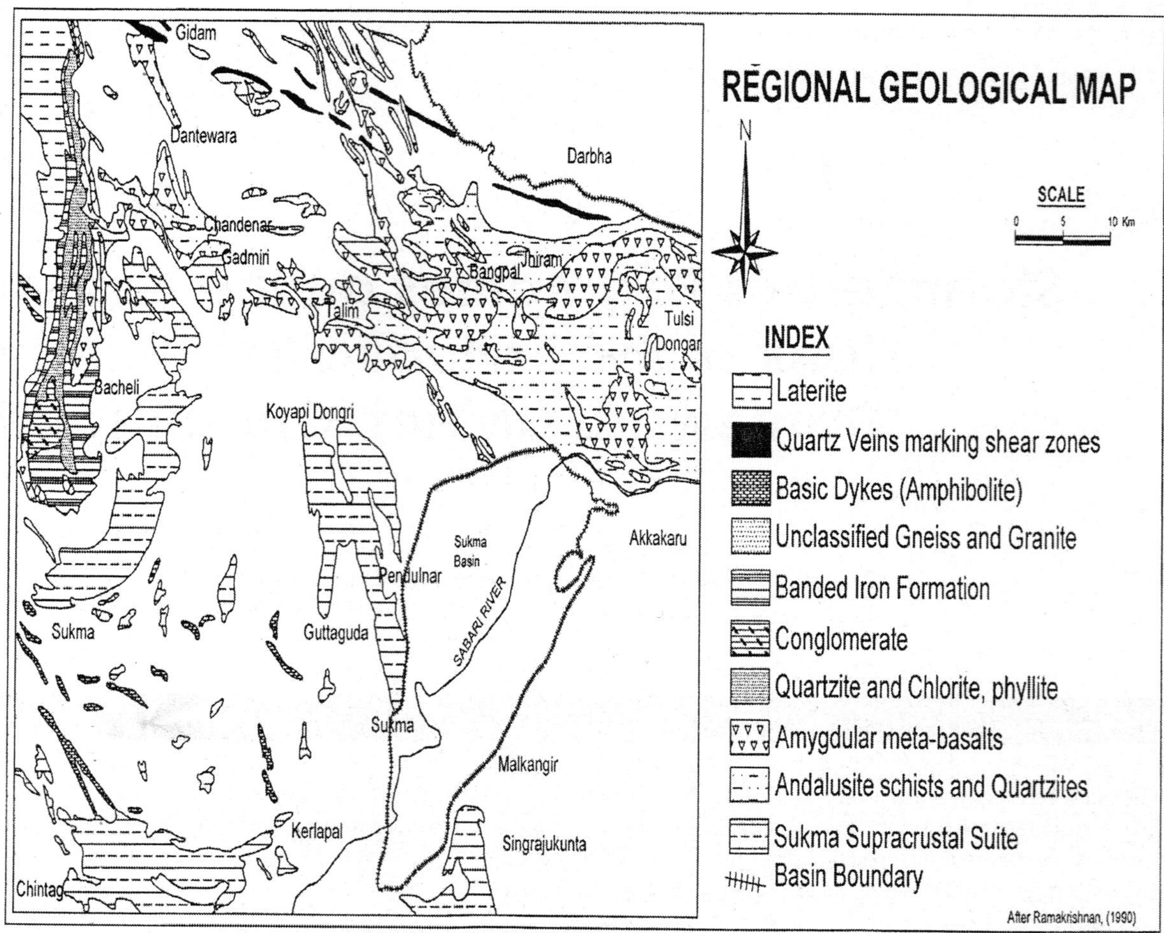

Fig. 1 Regional Geological Map (after Ramakrishnan, 1990)

Table-**1** Generalised Straitigraphic Succession of Central Indian Craton (After Rogger and Naquvi, 1986)

(W of Chattisgarh basin)		(E, SE and S of Chattisgarh basin)	
Nagpur, Bhandara Chindwara	Sausar (950 MY) Dongargarh (2200 MY) Sakoli (2300 MY)	Jaypur–Bastar	Bailadila group Bengpal group Sukma group
Basement Gneissic Complex			

Khan and Bhattacharya (1993) chronologically with the main emphasis on Bailadila formation. The Bailadila Group with its complete sequence is unconformably exposed over to Bengpal meta-sediments, which in turn subdivided in to Bhansi, Bacheli and Loa formation. These Bailadila rocks were later intruded by a suite of younger granite, basic dykes and pegmatites.

As per field observation, the Sukma group of Ramakrishna's (1990) three fold classification seems valid only in the western flank of the Bailadila Ridge. The Sukma outcrops form a garland around southeast Bailadila ridge with intervening granites from Aranpur to Dantewara. Their concordance is the result of the Bailadila deformation.

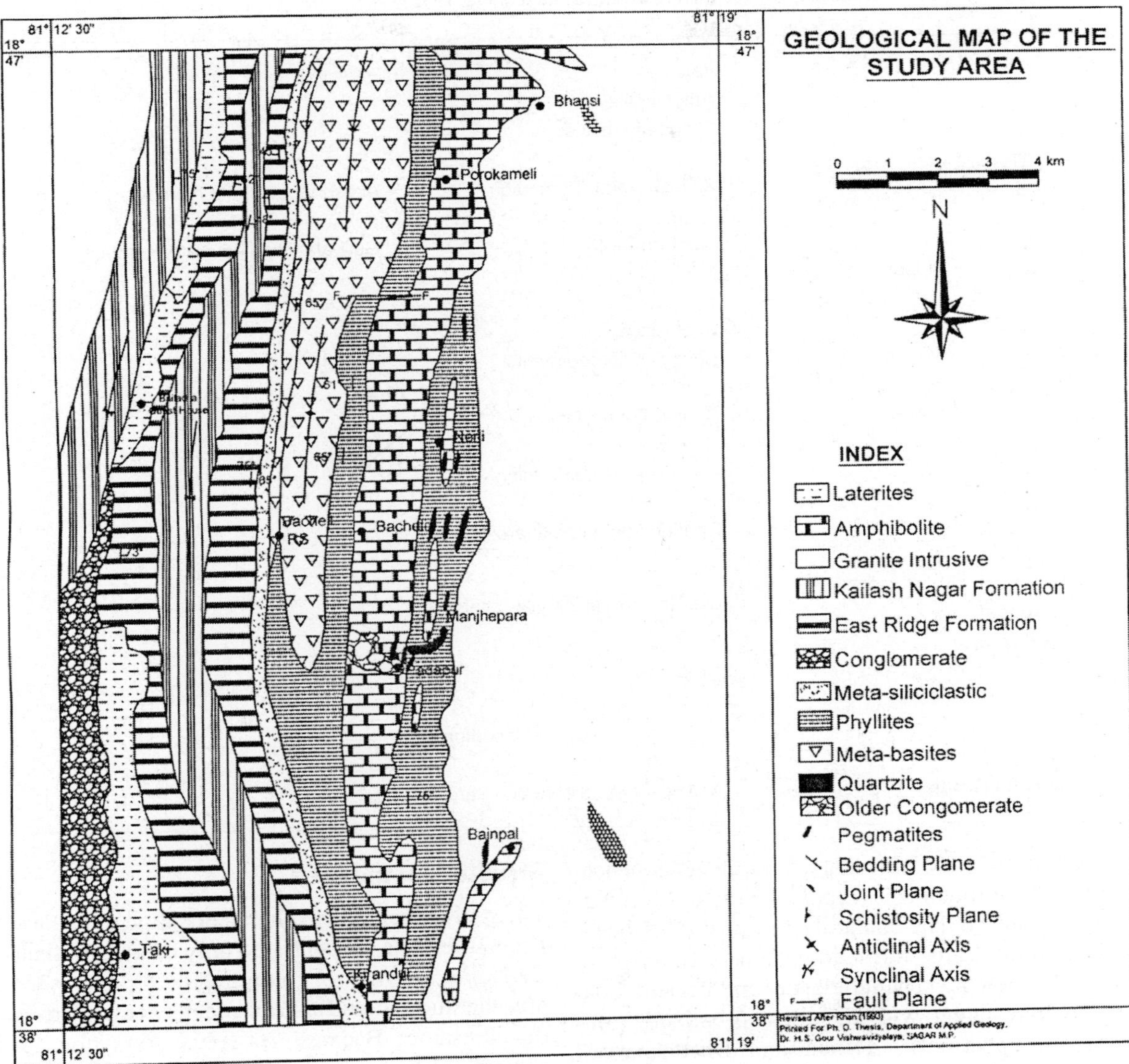

Fig. 2 Geological map of the study area

Bengpal Group

This group is poorly developed sequence of quartzite, andalusite quartzite, chert, chlorite, and sericite-schist and Banded magnetite quartzite exposed in steep valleys. Along the road section from Bhansi towards Dantewara quartz-chlorite- sericite schist are exposed. Authors marked another exposure of quartzite, andalusite-quartzite and crystalline quartzite in a nala section near village Chalkipara. A thin bed of ferruginous conglomerate is exposed just above the nala section south of Bacheli. This conglomerate is also exposed at Padapur and intruded by stanniferous pegmatites. The quartz-chlorite-sericite schist near Bhansi is sandwiched between post Bailadila granites and phyllites. On the other hand quartzite near Bacheli is overlained by conglomerate which in turn is overlained by phyllites. The exposure is near village Beinpal and intruded by pegmatites and quartz veins. (Fig. 2).

Bailadila Group

Bailadila group of rocks form a linear belt trending north south extending from Bailadila in the south to Dalli-Rajhara in the north. They are unconformably overlained by Bengpal group of rocks. In the study area, Bailadila ridge extends from Kirandul to Bhansi.

Table-2 Straitigraphic Succession (Revised after Khan and Bhattacharya, 1993)

Group	Division	Formation	Member
		Laterite Amphibolite Granite Intrusive	
Bailadila Group		Kailash Nagar Formation	Akashnagar Shale member
	Upper	Loa Formation	Gulinala Conglomerate member
	Middle	East Ridge Bacheli Metasediments	
	Lower	Bhansi Formation	
		----------- Unconformity -----------	
		Banded Magnetite quartzite	
Bengpal Group		Quartz Chlorite Sericite Schist	
		Quartzite	
		----------- Unconformity -----------	
Sukma Group		Andalusite–Quartzite and Sericite–Quartzite	

On the eastern flank of this ridge, a complete sequence of Bailadila formation is exposed with intervening younger granites. The Bailadila formation has been subdivided into three formations namely Bhansi the lowest, Bacheli meta silici-clastic middle and Loa conglomerate along with the most promising iron formation, uppermost (Khan and Bhattacharya, 1993).

Further, based on field relationship, the meta-basite and meta-pelites forms the basal part of the Bailadila, at the place of feldspathic quartzite and arkosic quartzite (Bandyopadhyay and Hishiker, 1977). Meta-basites and meta-pelites with intruded amphibolite exposed in eastern part of Bacheli, Nerli, Poro-Kameli, Kameli, and Bhansi. At Beinpal and Manjhepara, Phyllite and quartzite are providing base for meta-basites. These metabasites are intruded by stanniferous pegmatite and quartz vein at Bacheli and Nerli, provided additional support for Khan and Bhattacharya's (1993) model.

Structure of the Area

The study in relation to characterisation of tin bearing pegmatites was concentrated only on eastern flank of Bailadila ridge but for compiling the structures in this area the study should also extended over western flank of the same. Because Bailadila formation forms roughly, N-S trending overturned synclinorium and anticlinorium. A "N" shaped fold is discernable over the Bailadila ridge (Fig. 2). Dip of the limbs is nearly vertical to 60^0 towards E. Minute flexural slip faulting and kinking and dislocation also observed in BHQ. Horizontal displacements of local units (Phyllites and Kailash Nagar Shales) have also evident by crushed matrix near Nerli. Conglomerate of variegated composition marked disconformity with out truncation of beds have observed.

Igneous Formations:

Granite : Bastar craton is characterized by polyphasic acid and basic plutonic activities and some volcanic extrusion as well. The Palim granitic emplacement is responsible for stanniferous

Fig. 3(A) Microcline and Quartz in Pegmatite

Fig. 3(B) Muscovite Quartz and Biotite in Pegmatite (Wavy Extinction in Deformed Muscovite)

Fig. 3(C) Field view of Kaolinized Pegmatite (Length of Photomicrograph is 1.5 mm)

pegmatitic activities (Sarkar et al, 1980). Apart from this, Darbha granite, which is contemporary, but without tin mineralisation. The Dantewara granite, which is western extension of Darbha granite, and also said to be adjacent to Bhansi granite occupied a part of study area (Khan et al, 1993), or post Bailadila granite. The boundary between Dantewara and Bhansi granite can be demarcated on the basis of mineralisation, geochemical relationship and field relationship.

Field relationship reveals that at Chandenar near Dantewara granite is intruded arround metabasite in the south while andalusite-quartzite of Bengpal group in the north. The Bengpal metabasic rocks truncating angularly to different units of Sukma suits (Fig. 1).

Minerologically the granites are non-foliated, medium grained, biotite- hornblande and two-mica granites (Fig. 3 A). They are also characterized by presence of microcline (the chief potasic feldspar) with micro-perthitic intergrowth (Fig. 3 B). The oligoclase variety of plagioclase convincingly suggests low temperature processes.

Geochemically, both Darbha and Palim granites are peraluminous, highly siliceous and low in calcium. Palim granite is characterized by high concentration of Sn, Nb with comparison to Darbha granite (Table 3). Hence it is regarded that Palim pluton is the source of stanniferous fluid and gave rise pegmatitic emplacement with in granite as well as in metabasite (Bacheli-Nerli), adjacent to Bhansi granite, which is westward extension of Palim pluton.

Pegmatites

Pegmatites of Bastar Koraput Pegmatite Belt are highly complex in nature, show variegated mineralogical, structural, mode of emplacement, shape and size, zoning and texture. However, the study area has very little variation in both mineralized and non-mineralized pegmatites. With in the limits of study area, the authors have marked about thirteen pegmatitic exposures out of which eleven are mineralized and two non-mineralized (Fig.2). The non-mineralized pegmatites are found in granitic country rock. These pegmatites are about 10m in width and about 50 m in length, N-S trending and asymmetrically zoned with quartz-alkali feldspar pophyrites with biotite occurs in outer zone, tourmaline and albite in intermediate zone and smoky quartz in core. There is no signature of cassiterite tantalite and columbite found in them.

On the other hand, all the mineralized pegmatites are emplaced in and along amphibolite and dolerite dykes of Bengpal group. Massive pegmatites exposures have seen only at Bacheli and Nerli where local tribal and some authorized private company are quarrying for albitized material, which is used in brick making. These pegmatites are 1 to 5 mts thick and about 30 to 50 mts long, pinching and swelling is common in each pegmatite. In pegmatites, asymmetrical and intercalated mineral growth made it difficult to define the zonation. However complex

Table-3 Compositional variation in Different Granites from Bastar Craton Major oxides (in %) and trace elements (in PPM) After *Singh, 1998, **Babu, 1994, ***Bubu 1994

Major Oxide	Bhansi Granite	Kondagaon Granite*	Palim Granite**	Darbha Granite***
SiO_2	73.5	67.54	71.59	70.26
Al_2O_3	12.26	14.68	14.00	13.59
FeO	--	--	1.71	2.67
Fe_2O_3	2.86	3.05	0.91	1.13
TiO_2	0.14	0.41	0.30	0.55
MnO	0.017	0.04	0.03	0.04
CaO	0.70	1.83	1.92	3.41
Na_2O	2.47	2.1	3.09	2.41
K_2O	5.93	9.17	4.58	3.64
P_2O_5	0.093	0.09	0.11	0.10
H_2O	0.12	0.38	0.69	0.80
Sn	63	<10	60.01	19.07
Nb	63	40	26.70	22.69
Be	35	--	14.14	3
Zr	183	4.5	117.94	114.15
Ti	--	--	749.11	1636.5
Rb	266	380	152.77	214.56
Sr	18	179	43.88	79.32
Ba	545	1010	444.44	612.42
B	<30			

zoning with occurrence of smoky quartz mega-crysts in the core are very common. Only Kaolinized pegmatites are exposed with in lateritic cover on the surface (Fig. 3C).

References

Babu T. M. (1994) Tin in India. Geol. Soc. Ind. Publication.

Bandyopadhyay D. and Hishiker A. K. (1977). Stratigraphic sequence in the southern part of Bailadila range Distt. Bastar M.P. Jour. Geol. Soc. India vol. 18: pp 240-245.

Iyangar S. Y. P. and Banerjee P. K. (1971). The iron ore Bengpal Dharwar group. Rec. of Geol. Surv. India vol. 101 (2) pp 43-59.

Khan M. W. Y. and Bhattacharya T. K. (1993). A reappraisal of the stratigraphy of Bailadila Group. Jour. Geol. Soc. Ind. 42 (6) pp, 549-562.

Rogger J. J. W. and Naqvi S.M. (1986). Precambrian Geology of India. Oxford monographs on Geology and Geophysics, No. 6 Oxford University Press.

Ramakrishnan, M. (1990) Crustal development in Southern Bastar, Central Indian Craton. Geol. Surv. Ind. Spl. Pub. No. 28 pp 44-66.

Ramesh Babu P. V. (1993). Tin and rare metal pegmatites of Bastar Koraput pegmatite belt, Madhya Pradesh and Orissa, India: Characterization and classification. Jour. Geol. Soc. Ind. Vol. 42 pp 180-190.

Sarkar G., Paul D. K., De Laeter J. R., McNaughton N. J. and Mishra V. P. (1990). A geochemical and Pb, Sr isotope study of the evolution of granite-gneisses from the Bastar Craton, Central India. Jour. Geol. Soc. Ind. Vol 35 pp 480-496.

Singh Y. (1998) Early Proterozoic rare metal and tin pegmatites near Kawadgaon, Bastar, MP: an example of vertical pegmatite zonation. Jour. Geol. Soc. Ind. Vol 51: pp 175-182.

Hon'ble Reviewers

1. Prof. P. K. Verma
   Chairman
   Department of Geology
   Delhi University
   DELHI 110 007 INDIA
2. Anonymous
3. Dr. Arun Kumar
   Chairman
   Department of Earth Science
   Manipur University
   IMPHAL 795 003

Editorial Assistance

Dr. Jaimini Swadia

Economic Mineralization
K.L. Shrivastava (Editor)
2009. pp. 224-227
Scientific Publishers (India), Jodhpur

19

Section-II

A Note on Significance of High Rubidium in Lithium Mica From Govindpal Area of Bastar District, Chhattisgarh

POOJA SHRIVASTAVA[1] AND V.K. KHANNA[2]

[1]Department of Applied Geology, Barkatullah University, Bhopal 462 001
[2]232, Napier Town, Jabalpur 482 001

Abstract

A lithium mica with highest ever recorded Rb_2O content (6.98%) was reported from Govindpal area of Bastar district in 1977 by the second author. The occurrence and significance of such high content of rubidium in the mica have been discussed.

Key words: Pegmatite, Lithium Mica, Rubidium, Baster.

INTRODUCTION

The tin bearing raremetal pegmatites of Bastar district are the host of many minerals such as clevalandite, quartz, muscovite, lepidolite, zinnwaldite, spodumene, amblygonite, fluorite, beryl, cassiterite, magnetite, triplite and columbitetantalite (Lamba and Khanna, 1981; Trivedi and Soni this volume) Khanna (1981) reported the occurrence of two unusual mica minerals from Chiruwada area of Bastar district. These include a blue mica (muscovite rich in Mn and Li) and a kidney shaped mica (Alzinnwaldite). A lithim mica with highest ever content of Rb_2O (6.98%) was also reported by Khanna from Govindpal area (lat. 18°42'N, long. 81°54'40°E) (Fig. 1), in 1977. Simulta- neously, Volvoshir (1977) reported Rb rich lepidolites from Russia but the maximum percentage reported by him is 5.16% Rb_2O (Table 1).

The rubidium rich lithium mica is confined to pegmatite offshoots where it is associated with mostly quartz and small grains of cassiterite. It is not found in main pegmatite which contains only normal lepidolite. The pegmatites of Govindpal area have been emplaced in the metabasic sills intrusive into the Bengpals(Fig.1). The Palim granite in the neighbourhood which is a nonfoliated biotite granite appears to be responsible for the pegmatitic activity and the subsequent pneumatolytic deposition of lepidolite (Ramaswamy et al. 1976).

The lithium mica is light grey in colour. It occurs in tabular radiating crystals. The results obtained from calculation of structural formula of the analysis on the basis of 24 (O, OH, F) suggest it to be a lithium mica with low Si (6.658 atoms) and Li (0.5785 atoms). The lithium mica also contains 0.11% cesium.

Rubidium and cesium are two largest alkali metals with extreme values of electronegativity, ionization and ionic potentials. They tend to be dispersed in potassium minerals rather than concentrated in

Table-1 Chemical Analysis of Rubidium rich lithium micas

	1	2	3	4
SiO_2	45.82			
Al_2O_3	29.56			
Fe_2O_3	1.60			
FeO	1.15			
MnO	1.86			
MgO	0.11			
F	2.55			
Na_2O	1.55	0.31	0.18	0.31
K_2O	5.10	7.64	7.98	7.78
Li_2O	1.06	4.78	3.37	4.74
Rb_2O	6.98	5.00	5.06	5.16
Cs_2O	0.11	1.48	1.55	1.51
CaO	0.50			
H_2O^-	0.17			
H_2O^+	1.90			

1. Govindpal area of Bastar District (Khanna, 1977)

2 to 4. Russia (A. V. Volvoshir, 1977; Partial Analysis)

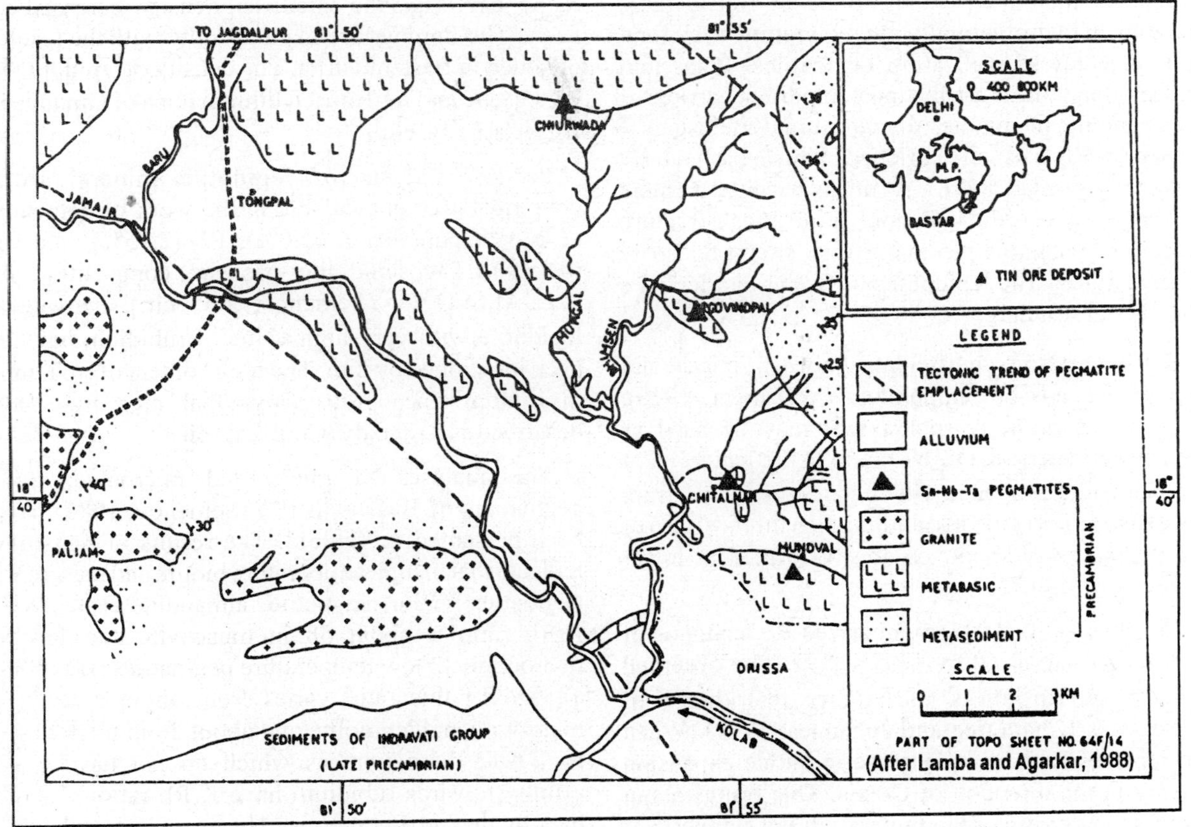

Fig. 1 Occurrence of some Sn-Nb-Ta bearing Precambrian pegmatites of Bastar District, Chhattisgarh (after, Lamba and Agarkar, 1988).

Table-2 Alkalies content and K/Rb ratio in Micas and Microcline of Bastar district

Mineral	Association	K	Li	Rb	Cs	K/Rb	Locality
			MICAS				
Muscovite	Magnetite	7.02	0.06	0.12	0.005	58.50	Tongpal
Muscovite	Amblygonite	6.30	0.19	2.26	0.18	2.79	Chitalnar
Lepidolite	Amblygonite	6.00	2.20	2.15	0.15	2.79	Govindpal
Muscovite (Blue)	Amblygonite	7.97	0.29	1.52	0.10	5.24	Chiruvada
Muscovite (Green)	Amblygonite	7.97	0.21	1.60	0.08	4.98	Chiruvada
			FELDSPARS				
Microcline	Magnetite	8.10	–	0.07	–	115.71	Tongpal
Microcline	Lepidolite	9.50	–	0.37	–	25.67	Govindpal

species of their own. Rb is found in considerable amounts in hydrothermally altered granitic rocks and their aureoles as well as in pegmatites. Potassium feldspars and micas are the most important carriers of Rb in granitic pegmatites. It coexisting K-feldspars + mica pairs Rb is usually enriched in mica, particularly in ferromagnesian and Li, Al micas because it enters the large 12 coordinated potassium mica structure more easily than the most constraint site in feldspars (Cerny, 1982). The lepidolites of Bastar district have $2M_1$ crystal structure.

K-Rb partition coefficients between micas and vapour depends on temperature (Volfringer, 1976). Due to this property such micas may be used as geothermometers. K_2O/Rb_2O ratio in the lepidolite of the pegmatite proper in Govindpal area varies from about 3 to 5. This ratio in rubidium rich lithium mica of the pegmatite offshoot, obviously formed at lower temperature, is 0.73.

Voncken et al. (1987) synthesized Rb analogue of $2M_1$ muscovite at 600°C and 5 kbar. The observed spacing of Rb micas are higher than those of muscovites (Chatterjee and Johannes, 1974). When Rb replaces K there is increase in lattice expansion mainly in the direction of C-axis. This brings about increase in density of the mineral. In the synthesis of Rb analogue of $2M_1$ muscovite, Voncken et al. (1987) have shown that density increased from 2.83g. cm^3 to 3.06g. cm^3. Thus the calculated density increased by 8.1%. The findings are in conformity with the results obtained in Bastar district. The density of lepidolite is 2.79 g, cm^3 and for Rb-rich lithium mica of Govindpal area it is 2.85g. cm^3.

Partial analysis of lepidolite mineral from govindpal area is available in the work of Semenov (1981). It contains K (6.0%), Rb (2.15%), and Cs (0.15%). The lepidolite has the composition of $K_2Li_3Al_5Si_6O_{20}F_4$ (Semenov, op. cit.). However, lepidolite with very high value of rubidium has not been mentioned by him. The high content of rubidium in lithium mica from Govindpal area has been described as an anoaly (Cerny, op. cit.).

The analyses of micas and microcline from pegmatites of Bastar district (Semenov, 1981) have been presented in Table 2. The results of Semenov indicate that in high temperature biotite and muscovite pegmatites (with magnetite, almandite association) K/Rb ratio is about 58 in muscovite and 115 in microcline. In low temperature pegmatites rich in Sn, Ta and Li this ratio varies from about 3 to 5 in muscovite and lepidolite and about 26 in microcline. In Bastar district micas which do not have close affinity towards rubidium have K/Rb ratio of about half to that in microcline. However, micas having close affinity towards rubidium have K/Rb ratio of one fifth to one tenth of that in microcline. This is in

Fig. 1 Geological map of the NathdwaraSarara belt showing distribution of the major lithologic units (based on the geological map of Heron, 1953, Pl. 38, southern sheet and Revised by Roy et al., 1988) Explanations : B Barapal; F Fatehpur; G Gurach; I Iswal; J Jhamarkotra; JR Jharol; K Kapur; KB Kablta; KL Kallashpuri; KT Katar; L Lakhawali; LS Losing; M Madar; NA Nal; ND Nathdwara; NT Nathuwas; R Rama; S Sisarama; SR Sarara; U Umra; Z Zawar

1. Mewar Gneiss (pre-Aravalli basement rocks);
2. Delwara Formation : A. carbonate rocks; B. quartzite/conglomerate; C. chlorite schist and amphibolite (meta-volcanic rocks).
3. Jhamarkotra Formation : A. quartzite, B. dolomite and facies variants, metasilcrete and metaferrecrete, thin bands of stromatolitic phosphorite; C. carbonaceous phyllite; D. argillaceous phyllite.
4. Udaipur Formation : greywacke-phyllite, lithic arenite, local diamictic conglomerate.
5. Debari Formation : A. conglomerate-arkose; B. quartzite with minor bodies of phyllite.
6. Zawar Formation : dolomite, carbonaceous phyllite, quartzite, with local orebodies of galena, sphalerite and silver
7. Macclamagra Formation : quartzite and quartz phyllite.
8. Tidi Formation : slaty phyllite with dolomitic and quartzitic Interbeds.
9. Micaschist, phyllite and thin bands of quartzite (deep-water sequence).
10. Ultramafic intrusions of Kherwar - Dungarpur
11. Delhi Suipergroup

There are several small, elongated, thin, lenticular bands of marble and quartzite associated with schist and gneiss (Gupta et al. 1980; Prasad et al 1997; Sinha Roy 1998).

Talc transformation from Siliceous Dolomite

In the study area, the talc occurs within siliceous dolomites, quartzite. The lenses, irregular blocks and the crystals of dolomite have partially or wholly replaced by talc. Inter-tonguing and gradation of talc

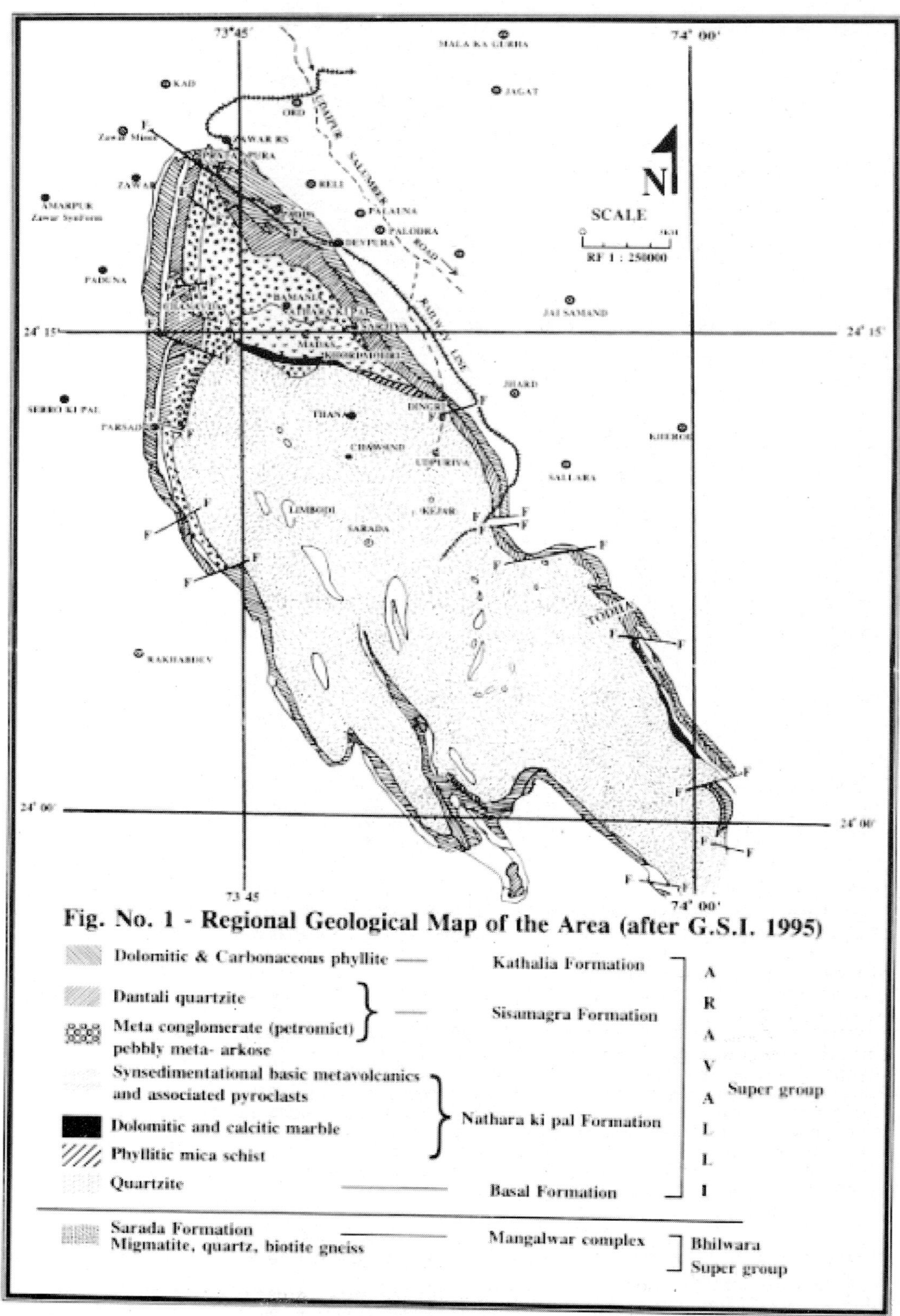

Fig. 2 Regional Geological Map of the Area (after G.S.I. 1995)

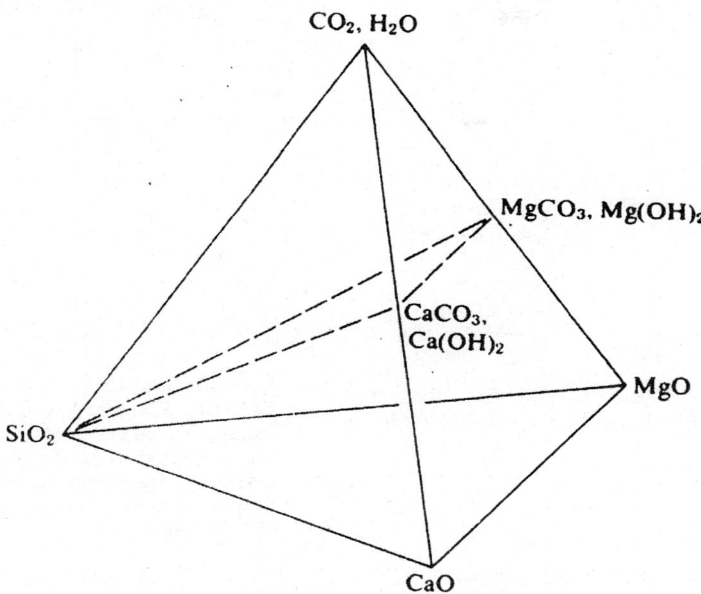

Fig. 3 The compositional tetrahedron MgO-CaO-SiO_2-(H_2O+CO_2). The interior dashed plane includes the carbonate minerals (after Ehlers and Blatt, 1997)

into dolomite and dolomite into talc is also common and perceptible. The dolomite near the talc bodies also contained number of pockets and veins of talc. Talc is also arranged along the dolomite crystal boundaries and the minute fractures and cracks have also developed within the host rock.

The talc in the study area has been formed by the hydrothermal alteration of dolomite. The hydrothermal fluids being of metamorphic origin, appears to have found their way along bedding planes, schistosity planes, crystal boundaries, joints, fractures and foliation planes of dolomite.

It is the hydrothermal solutions played an important role in the talc formation, and has originated during metamorphism. The metamorphic hydrothermal solutions resulted from the metamorphism of the rocks of the area, through the effect of rising pressure and temperature. With increasing grade of metamorphism, the amount of fluid so liberated increased enormously.

The first reaction in a rock consisting of dolomite and quartz (with or without calcite) containing interstitial water leads to the formation of talc and calcite.

$3CaMg(CO_3)_2 + 4SiO_2 + H_2O \rightarrow$
Dolomite Quartz
$Mg_3Si_4O_{10}(OH)_2 + 3CaCO_3 + 3CO_2$
Talc Calcite

For this reaction H_2O is required and CO_2 is liberated, therefore XCO_2 will increase with rising temperature and due to steady increase of CO_2, the state of the isobaric invariant point is reached. It is evident from the experimental data that talc deserves special attention as the first mineral to form during the metamorphism of siliceous dolomite. Calcite is precipitated.

PROGRESSIVE METAMORPHIC REACTIONS IN SILICEOUS DOLOMITE

Plotting the minerals common to this type of metamorphism within à tetrahedron (Fig. 3) is difficult to visualize, so an additional simplification is made; all the mineral compositions within the tetrahedron are projected from the H_2O, CO_2 corner to the triangular base CaO-MgO-SiO_2. This permits both hydrous and carbonates minerals to be plotted together on the same two dimensional triangular representations. Such a plot is shown in Fig. 4, together with some pertinent minerals.

The phase calcite, dolomite and quartz are stable together in the presence of a fluid phase to temperatures up to about 300°-400°C (depending upon fluid composition and pressure). The fluid phase in this situation can be pure CO_2, pure H_2O or any intermediate mixture. This equilibrium assemblage

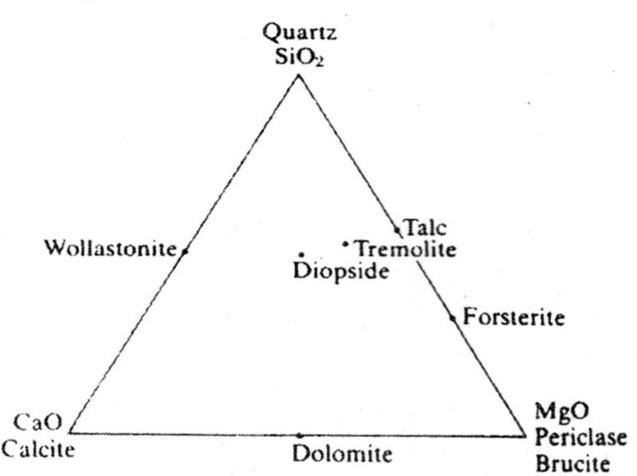

Fig. 4 The compositional triangle MgO-CaO-SiO_2. Phases containing CO_2 and H_2O in addition are plotted within

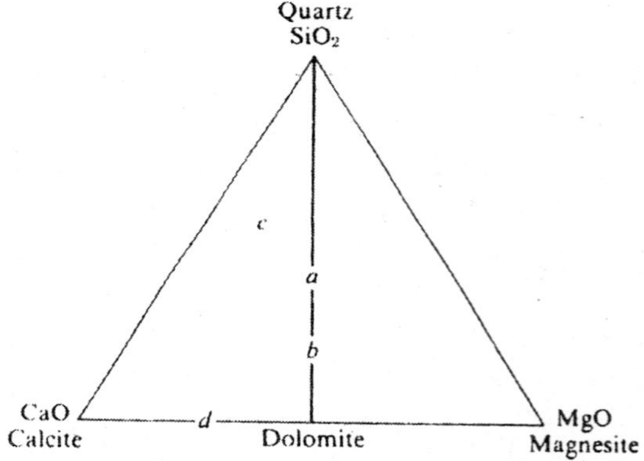

Fig.5 Equilibrium relations within the system CaO-MgO-SiO2-(H_2O+CO_2) at low temperatures. Solid lines indicate compatibility between phases. compositions a and b consist of mixtures of quartz and dolomite; c is a mixture of calcite, dolomite, and quartz, and d is a mixture of calcite and dolomite. (after Ehlers and Blatt, 1997)

can be shown on the compositional triangle CaO-MgO-SiO_2 (with excessCO_2 and H_2O), as in Fig. 5.

Heating of a rock with compositions falling within the compositional triangle CaO-MgO-SiO_2 leads to reaction

$3CaMg(CO_3)_2 + 4SiO_2 + H_2O \rightarrow$

Dolomite Quartz

$Mg_3Si_4O_{10}(OH)_2 + 3CaCO_3 + 3CO_2$

Calcite Talc

The rock composition ***a*** (Fig. 6), consisting originally of quartz, dolomite (in the ratio indicated by the formula above), and fluid phase, reacts to form calcite, talc and a fluid phase. All of the quartz and dolomite are eliminated. If a larger amount of dolomite were present than that of required for the reaction, such as at point ***b***, the same reaction would occur, but dolomite would remain in excess. This can be seen on the diagram, as composition ***b*** fells within the new compatibility triangle joining the phases calcite, talc & dolomite. Similarly the composition ***c***, which consisted before the reaction (Fig. 5) of calcite, quartz and dolomite now it is converted to a mixture of calcite, quartz and talc.

Additional heating (under most circumstances) leads to the formation of tremolite by reaction of talc, calcite and quartz:

$5Mg_3Si_4O_{10}(OH)_2 + 6CaCO_3 + 4SiO_2 \rightarrow$

Talc Calcite Quartz

$3Ca_2Mg_5Si_8O_{22}(OH)_2 + 6CO_2 + 2H_2O$

Tremolite

Tremolite now appears as a stable phase within the compositional triangle (Fig. 7). As tremolite is compatible with quartz, calcite and talc, compatibility lines are drawn from the tremolite composition point to these minerals.

Formation of Diopside and Forsterite with additional rise in temperature, diopside should begin to form roughly at the threshold to amphibohite facies (Fig. 8).

Tremolite + 3 Calcite + 2 Quartz $\rightarrow$

5 Diopside + 3 CO_2 + H_2O

A further rise in temperature causes the formation of forsterite

Talc + 5 Dolomite $\rightarrow$

4 Forsterite + 5 Calcite + 5 CO_2 + H_2O

$Ca_2Mg_5Si_8O_{22}(OH)_2 + 11CaMg(CO_3)_2 \rightarrow$

Tremolite Dolomite

$8Mg_2SiO_4 + 13CaCO_3 + 9CO_2 + H_2O$

Forsterite Calcite

ROLE OF TEMPERATURE, PRESSURE AND FLUID IN THE FORMATION OF TALC

The two reaction discussed above (i.e. A&B) are shown in Fig. 9. Temperature is shown on the vertical axis. The composition of the fluid phase changes

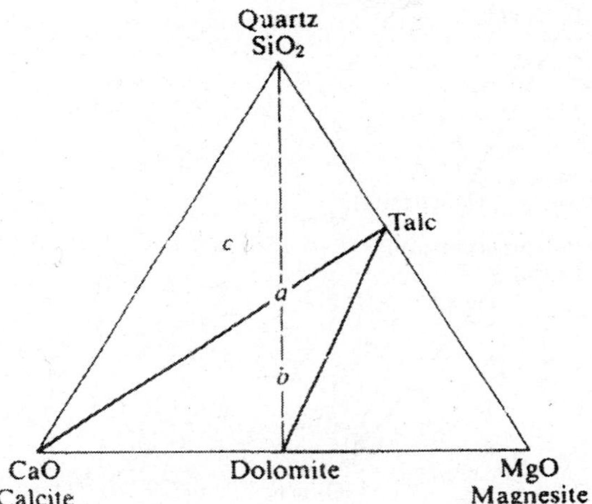

Fig. 6 Equilibrium relations within the system CaO-MgO-SiO2-(H_2O+CO_2) above about 350°-400°C, after the formation of talc. (after Ehlers and Blatt, 1997)

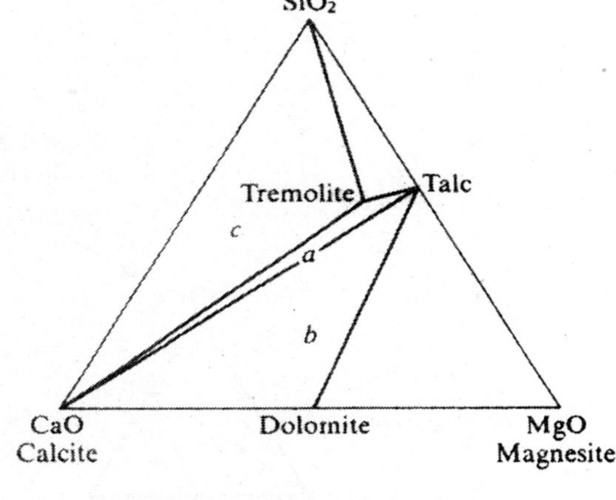

Fig.7 Equilibrium relations within the system CaO-MgO-SiO2-(H_2O+CO_2) after the formation of tremolite. (after Ehlers and Blatt, 1997)

continuously in composition along the horizontal coordinate, being composed of H_2O at the left and CO_2 at the right, with P_{fluid} at all times equal to 1 kbar that is,

$$P_{fluid} = P_{CO2} + P_{H2O} = 1\,\text{kbar}$$

The temperature at which dolomite, quartz and H_2O react to form talc, calcite and CO_2 increases as the mole fraction of CO_2 increases to the right side of the diagram. This is because the assemblage that contains the greatest amount of combined CO_2 (dolomite rather than calcite) is favored as the mole fraction of CO_2 is increased in the fluid phase. Looking at the reaction in the opposite manner, it is seen that the higher-temperature assemblage, that contain the hydrous phase talc, is favored as the mole fraction of H_2O is increased in the fluid phase.

The second reaction, in which tremolite is produced from the combination of talc, calcite and quartz, yields both CO_2 and H_2O as reaction products. In as much as both components of the fluid phase are release during the formation of tremolite, the temperature of the reaction varies to a considerably smaller extent than the previous reaction (which has H_2O on one side of the equation and CO_2 on the other).

It is obvious from the diagram that an attempt to assign *P* and T to a mineral assemblage is very difficult. A rock consisting of talc and calcite, for example, merely assures one of being above the lower reaction curve in temperature. The situation is somewhat improved by a rock composed of assemblage talc-calcite-quartz-tremolite, as this would indicate that the assemblage fell on the upper reaction curve, Still further improvement is possible; note that the two reaction curve intersect at point ***a***. At point ***a***, an isobaric invariant point, the product and reactants of both reaction are found together- a stable assemblage containing dolomite, quartz, talc, calcite and tremolite. If the condition of the diagram is satisfied, namely 1 kbar fluid pressure, the association of these five phases indicates a precise temperature for this assemblage as well as a definite composition of the fluid phase. Unfortunately, not all carbonate rocks are metamorphosed at 1 kbar fluid pressure. As noted earlier, the fluid pressure varies as a function of depth; similarly the temperature of the five-phase assemblage also varies (Fig. 9). If other can estimate the pressure means, such as an approximate estimate of depth of burial during metamorphism, perhaps based on the amount of overburden remove, the temperature can be determined. Or if the temperature can be estimated by other means, the pressure can be obtained from the diagram. The five-phase assemblage can be estimated to have formed between 400°C and 460°C, even if no pressure estimate is possible.

CONCLUSION

Hydrothermal solutions originated during metamorphism, through the effects of rising pressure and temperature, the tectonic activities plays vital role in the formation of talc. The Talc in the study area occurs within dolomitic quartzite in the form of veins,

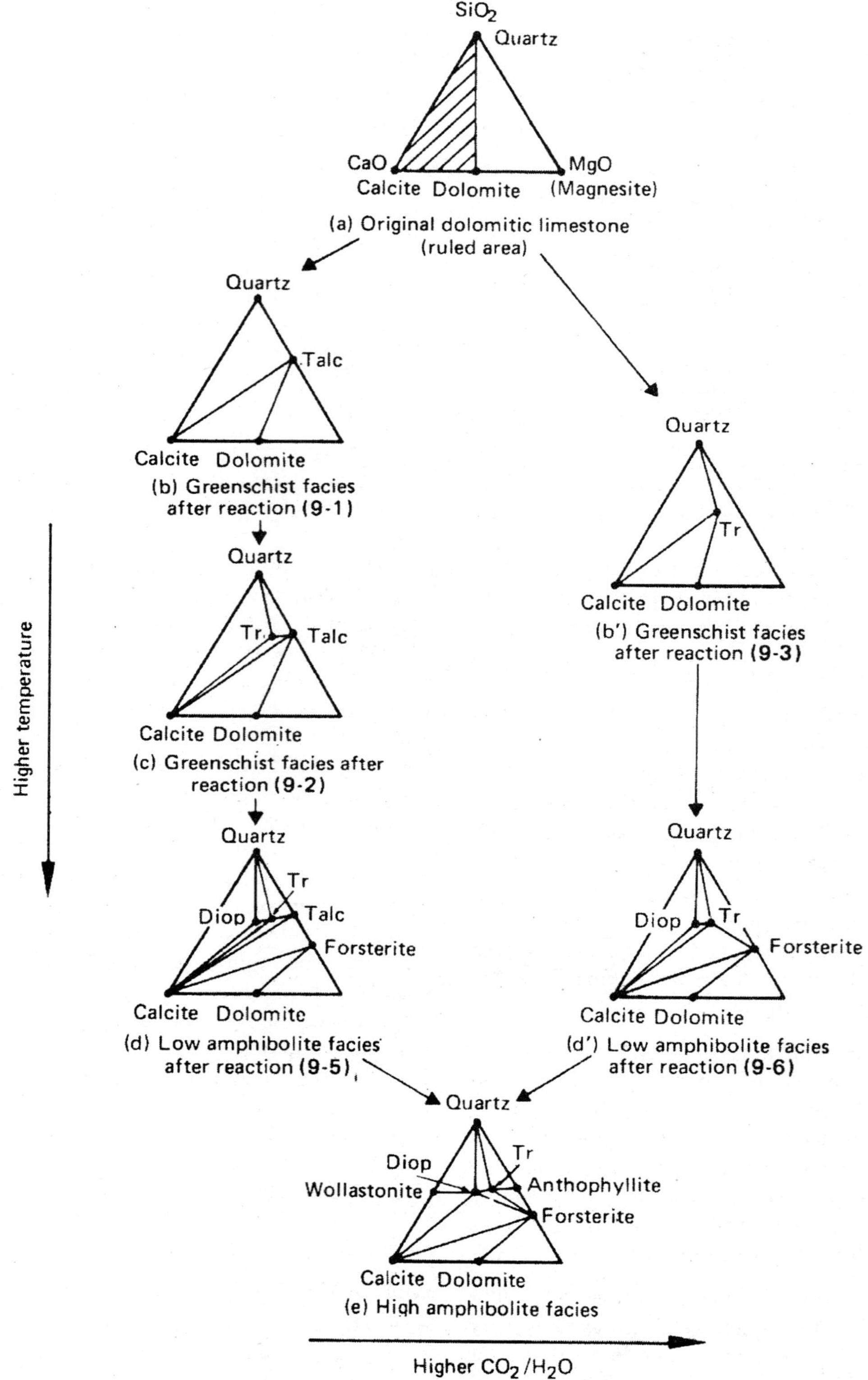

Fig. 8 Progressive changes of paragenetic relations in siliceous dolomite in the presence of mixture of CO_2 and H_2O. *Abbreviation* Diop = diopside, Tr = tremolite. (after, A. Miyashiro, 1973)

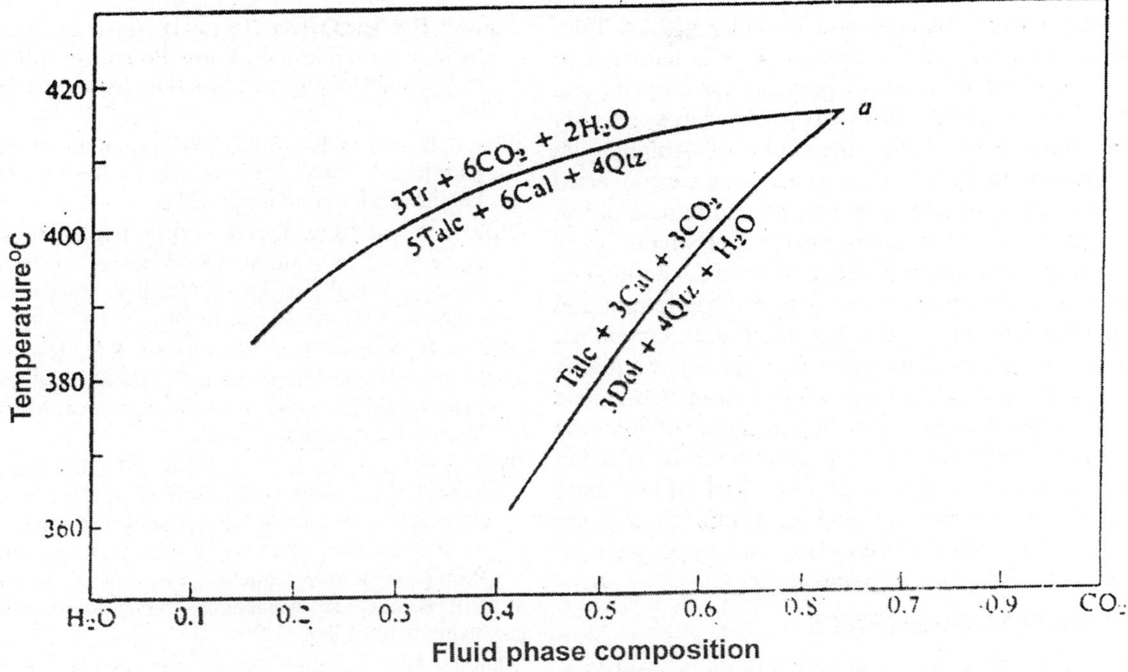

Fig. 9 The talc and tremolite forming reactions at 1 kbar as a function of temperature and CO_2/H_2O. Tr = tremolite, Cal = calcite, Qtz = quartz and Dol = dolomite. (after Ehlers and Blatt, 1997)

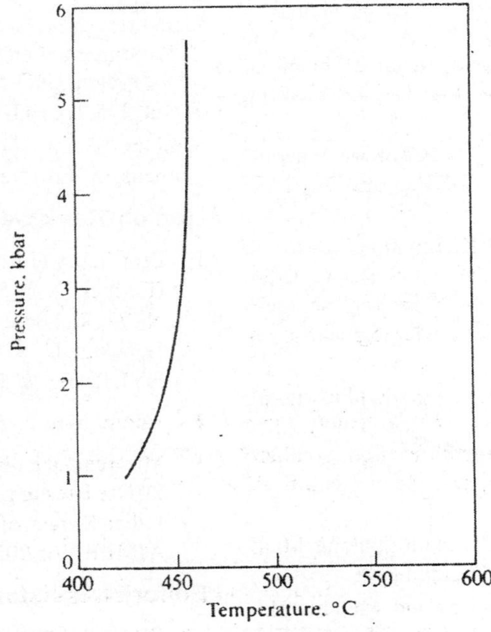

Fig. 10 Change in the stability conditions of the assemblage tremolite-dolomite-calcite-quartz-talc as a function of fluid pressure and temperature. This assemblage is stable only along the curve (after Ehlers and Blatt, 1975)

lenses, pockets, patches and irregular blocks. Talc-Actinolite-Quartz-Calcite assemblage is indicator of green schist facies. Small presence of diopsite and forsterite indicates that the rocks have roughly metamorphosed at the threshold of Amphibolite facies. Dolomite, dolomitic quartzites were converted the formation of talc. The textural and mineralogical observations of metamorphosed carbonate show transformation into talc, tremolite and forsterite, etc. Inter-tonguing and gradation of talc into dolomite, and dolomite into talc is also common and perceptible. The rock composition consisting originally of quartz, dolomite, fluid phase are reacted to form calcite, talc and fluid phase. Wherever a larger amount of dolomite present initially, at such places after the end of reaction the dolomite remains in excess. Appropriate combination of pressure and temperature under the presence of fluid phase leads to the genesis of soapstone (Talc) around Sarada Inlier.

ACKNOWLEDGEMENTS

MKS and NS are highly indebted to there employers (i.e. A.S.D.C.P. Ltd - Golcha Group) for providing necessary facilities during the course of research work. Authors are thankful to Dr. Satish Gupta for extending help in editing figures. The authors are also grateful to reviewers for critically examining the paper and giving constructive suggestions.

References

Bowen, N.L., (1940). Progressive metamorphism of siliceous limestone and dolomite. Jour. Geol.; vol. 48, pp 225-274.

Ehlers, E.G. and Blatt, H. (1997). Petrology Igneous, Sedimentary and Metamorphic. CBS Pub. and Dist. New Delhi, pp. 614-631.

Gupta, S.N., Arora, Y.K., Mathur, R.K., Iqballuddin, Prasad, B., Sahai, T.N. and Sharma, S.B. (1980): Lithostratigraphic map of Aravalli region, Southern Rajasthan and northeastern Gujarat. Geological Survey of India, Hyderabad.

Gupta, S.N., Arora, Y.K., Mathur, R.K., Iqballuddin, Prasad, B., Sahai, T.N. and Sharma, S.B., (1997). The Precambrian Geology of the Aravalli Region, Southern Rajasthan and northeastern Gujarat, Mem. Geol. Surv. India. vol. 123, pp. 23.

Heron, A.M., (1953): Geology of Central Rajputana. Mem. Geological Survey of India, vol. 79, pp. 1-389.

Miyashiro, A. (1973). Metamorphism and Metamorphic Belts, George Allen and Unwin Ltd., London, pp. 270-275.

Paliwal, B.S. and Okada, H. (1993). Aravalli Supergroup of India : an example of the Lower Proterozoic rift tectonics and sedimentation. Jour. Sed. Soc. Japan, vol. 39, pp. 1-30.

Roy, A.B. and Jakhar, S.R. (2002). Geology of Rajasthan (Northwest India) Precambrian to Recent. Scientific Publishers (India) Jodhpur, 421 p.

Roy, A.B. and Paliwal, B.S. (1981). Evolution of Lower Proterozoic epicontinental deposits : stromatolite bearing Aravalli rocks of Udaipur, Rajasthan India. Precamb. Res. vol. 14, pp. 49-74.

Roy, A.B., Paliwal B.S., Shekhawat, S.S., Nagori, D.K., Golani, P.R. and Bejarnia, B.R. (1988). Stratigraphy of the Aravalli Supergroup in the type area. Mem. Geol. Soc. India. vol. 7, pp. 121-131.

Roy, A.B., Sharma, B.L., Paliwal, B.S., Chauhan, N.K., Nagori, D.K., Golani, P.R., Bejarniya, B.R., Bhu, H. and Ali Sabah, M. (1993). Lithostratigraphy and tectonic evolution of the Aravalli Supergroup : A Protogeosynclinal sequence. In : S.M. Casshyap (Ed.) Rift Basins and Aulacogens, Gyanodaya Prakashan, Nainital, India. pp. 73-90.

Shukla, N.K., (2000). Study of Geology, Nature of Mineralisation and Exploitation of Soapstone Deposits at Khori Mohri, District Udaipur Rajasthan. P.G. Dissertation, Deptt. of Geology; MLS Univ.Udaipur.

Sinha- Roy, S., (1988). Proterozoic Wilson Cycles in Rajasthan. In: A. B. Roy (ed), Precambrian of Aravalli Mountain, Rajasthan, India. Mem. Geol. Soc. India, vol. 7, pp. 95-107.

Slaughter, J.D.; Kerrick, M., and Wall, V.J., (1975). Experimental and thermodynamic study of equilibria in the system $CaO-MgO-SiO_2-H_2O-CO2$, Amer. Jour. Sci. vol. 275, pp. 143-152.

Winkler, H.G.F. (1979). Petrogenesis of metamorphic rocks, (5th Ed.) New York: Springer Verlag, 347 p.

Hon'ble Reviewers

1. Prof. T.V.V.G.R.K. Murty
   (Former, Dean, Sagar University)
   N-2/2, Kakatiya Nagar
   Habbsiguda
   HYDRABAD 500 007 (A.P.)
2. Anonymous
3. Mr. Hemant Kumar Jain
   Sr. Ore Dressing Officer
   Indian Bureau of Mines
   AJMER 305 002

Editorial Assistance

Dr. Vijay Gaur

Section III
Economic Mineralization of Sedimentary Affiliation

Preamble

Some sediments, from both sedimentary groups, allochthonous and autochthonous, are sufficiently rich in elements of economic interest to form ore bodies. The section on economic mineralization of sedimentary affiliation includes thirteen papers. It also incorporates chapters with through investigation on the petrogenesis of relevant rock types insofar as it relates to the genesis of affiliated type of mineralization, treating ore genesis as an integral component of a broader petrogenetic process.

The first paper of the section, authored by Dr. T. K. Pandya, Prof. S.L. Solanki and Prof. M.K. Pandya, is entitled 'Myrmekitic textures from the polymetallic sulfide deposit of Dariba- Rajpura, Rajasthan, India. World-class polymetallic sulfide deposit of Dariba-Rajpura, is convincingly sedex type with many depositional and diagenetic features available. The deposit has suffered deformation and metamorphism (P-T, 5kb/525°C) with large varieties of ore minerals. Myrmekitic textures, which are a rare variety of intergrowth texture, even in silicates, have been abundantly observed by the authors in the sulphide ores. It involves significantly interesting genetic consideration as the Dariba-Rajpura deposit is essentially syngenetic sedimentary type. It is proposed that the textures are attributed mainly to metamorphism, still devitrification, collective-crystallisation and diapharesis seem to have played their roles. Hon'ble reviewer Prof. K.L.Rai has greatly appreciated the work and suggested that this significant study is worth correlation with the experimental data.

Prof. G.S.Roonwal has authored the second paper of the section which is entitled 'Beach sand resources and Industry in India'. Large reserves of ilmenite and titaniferous magnetite are known along both eastern and western coasts of India and being mined by the conventional ways since ancient times. The author found nothing great to appreciate until the Government of India initiated the open-door policy recently for exploration and exploitation of these deposits. The paper presents a brief account of beach sand placer resources, recent developments and future prospects of the industry.

The third paper of the section, authored by Prof. K.L. Shrivastava, Dr.Virendra Gaur, Mr.C.P.Dadhich, Prof. D.C. Gupta, Prof. R. P.Tripathi and Dr. Amita Tripathi, is entitled 'Polymetallic polynodule from Indian ocean nodule field'. The polymetallic polynodules have been discussed in the paper covering their uncommon morphology, ore petrography and ^{57}Fe Mössbauer spectroscopy. On the basis of the data obtained, a genetic model has been proposed giving due consideration to the information already available for polymetallic nodules and ferro-manganese crust. It is proposed that the sea mounts near the active volcanoes adjoining oceanic crust are the favourable sites for the polymetallic polynodules to grow. Hence, Indian Ocean Nodular Field has been proposed in the paper, has fertile ground for the polymetallic polynodules to grow. Hon'ble reviewer Prof. G.S. Roonwal has suggested 'first generation' and 'second generation' nodules as new terminology for the host and guest nodules respectively, as proposed in the paper.

"Record of Neo-Pleistocene glaciation, desertification and formation of economic salt lakes-playas in western Rajasthan" is the title of the fourth paper of the section which has been written by Dr.V.P.LauI. The paper has suggested formation of the fresh water lakes in western Rajasthan initially, due to ponding of water originated because of deglaciation during Late-Pleistocene. High rainfall during Holocene and water released during deglaciation produced more groundwater to dissolve evaporates of Marwar supergroup and carried them to make fresh water lakes saline. Before proposing the hypothesis, the paper has discussed record of glaciation, desertification and salt lakes in detail. Hon'ble reviewer Prof. A. B. Roy appreciated freedom of the author to express his views before raising the question which I quote "where are those saltdomes" ? The questions remained unanswered as there are many unanswered questions with all the alternative hypotheses.

Dr. S.C.Mathur, Dr.R.S. Layol, Dr.S.L.Nama and Dr. N.S.Negi have authored the fifth paper of the section entitled "Geological constraints on the bentonite facies of the Banner basin, western Rajasthan, India". Barmer region is known for the occurrence of economically important bentonite at several locations. The paper has critically analysed stratigraphic and sedimentological aspects of bentonite which belongs to Akli formation of the Barmer basin. Well-stratified bentonite sections show evidences of ash fall and alloclastic sedimentation almost simultaneously under restricted marine milieu with high saline evaporitic conditions.

The sixth paper of the section authored by Dr. A.K.Shandilya, Dr. S.K.Gehlot and Dr. N. Shrivastavajs entitled "The clay deposit of Bikaner district, Rajasthan-Geological setting, genesis and quality assessment for industries". The Bikaner region is one of the major industrial clay producing areas of the country with ten individual locations, producing china clay, ball clay, fire clay and fuller's earth. In the first detailed scientific study, the authors have produced an account of geological setting, genesis and quality assessment of these deposits. They have further suggested that most of the individual deposits are suitable for the varieties of industrial purposes and are available in commercially exploitable quantity.

"Basin terminology and classification : a review" is the title of the seventh paper of the section which has been authored by Prof.A.K.Biyani and Dr.Divya Dudeja. The paper has extensively reviewed the terminology and classification of basin and on the basis of geomorphology, tectonics, sedimentology and geographical extent, a new classification has been proposed.

Dr. Meeta Khilnani and Dr.Satvinder Singh are the authors of the eighth paper of the section which is entitled "Depositional environment of Bilara limestone, west of Nagaur, western Rajasthan". The Bilara group, middle division of Marwar supergroup is well-known for the industrial limestone specially around Bilara city, the type area. The authors have identified a new area of the industrial limestone in the west of Nagaur city and performed study on industrial importance and depositional conditions.

The ninth paper of the section authored by Dr. P.K.Jain is entitled "Banded iron mineralization in the Gwalior group of rocks". Tracing the study of area since Hacket's work, the author has suggested that the iron deposits are part of the Morar formation, the upper division of Gwalior group. Four mines of the area have been discussed for their mineral compositions, chemical composition and the mining activity. The paper suggests sedimentary chemical precipitation process for the formation of the deposits some of which are more weathered later, to form the residually concentrated deposits.

The next paper of the section is a small yet important note to report several filamentous and net-like clitellate cocoons of fresh water annelids from two lignite deposits of western Rajasthan. The paper is entitled "A note on the occurrences of the clitellate cocoons from Giral and Barsinghsar lignite deposits, Rajasthan" and authored by Dr. R.P.Tripathi and Prof. B.D.Sharma. The regular net observed is new to the record. On the basis of observations the paper has also discussed paleogeological aspects of the studied areas.

Dr. Pavitra Aravamudhan, Dr.Hema Achyuthan and Dr. Navin Shankar are the authors of the eleventh paper of this section and is entitled "Sedimentary and

Geochemistry of the overbank sediment of the Cooum river, Tamil Nadu". The studies were performed near Chennai to characterize the overbank sediments from the sedimentological point of view also to understand the chemical variation and type of pedogenesis. The authors have suggested that the sedimentation process and textural variations in the lower Cooum river indicate fluctuation in the hydrological regimes and weak pedogenesis with the addition of trace elements due to anthropogenic source and urbanization.

The last two papers are on mineral resources of Rajasthan and Orissa , two very important States of India as far as varieties, quantity and quality of economic minerals are concerned. The earlier one is authored by Dr. Arun Vyas entitled "An overview of the mineral resources of Rajasthan, India". The later one entitled "A profile of the industrial mineral resource potential of Orissa" is authored by Prof. Madhumita Das and Dr. Shreeup Goswami. Both the papers have extensively covered all important deposits; metallic, non-metallic and building stones with a note on quality and genesis for each. Although the States are far apart, there are many similarities emphasized in suggesting the solution of the problems mineral industry is facing regarding exploration and exploitation, in the changing scenario of the growing Indian economy.

Economic Mineralization
K.L. Shrivastava (Editor)
2009. pp.240-247
Scientific Publishers (India), Jodhpur

21

Section-III

Myrmekitic Textures from the Polymetallic Sulphide Deposit of Dariba-Rajpura, Rajasthan, India

T.K. PANDYA[1], S.L. SOLANKI[2] AND M.K. PANDYA[1]

Department of Geology, Mohanlal Sukhadia University, Udaipur 313 001

Department of Metallurgical Engineering, Malviya National Institute of Technology, Jaipur 302 001

Abstract

The polymetallic sedex class sulphide deposit of Dariba-Rajpura belonging to lower Proterozoic age (1799 Ma; Deb, et al., 1989) which lies in Dariba-Bethumni belt of Rajnagar-district in Rajasthan forms part of Bhilwara Supergroup (Gupta et al.1997) or lies in the middle of Bhilwara metasedimentary basin considered equivalent to Aravalli Supergroup (Deb and Pal, 2004) has been studied by a number of workers covering its different aspects.

The deposit is hosted mainly by calc-silicate rocks, garnet, staurolite and kyanite bearing graphite-mica schists, black cherts / cherts and impure marble. The deposit is stratiform and zoned showing number of depositional and diagenetic features (Pandya et al., 1980). The deposit is metamorphosed (P-T, 5 Kb/525°C) and shows several features of metamorphism and deformation (Solanki, 1979, Pandya, 1981 and Pandya et al., 2001). The sulphide ores are dominantly of stratiform type but veins, patches and pods are also seen. A large number of ore minerals, dominated by sphalerite, pyrite, galena, pyrrhotite, chalcopyrite, tetrahedrite-tennantite and arsenopyrite with several sulphides, sulphosalt minerals and native elements are encountered in minor amounts in the deposit. It is in these ore minerals that the authors have encountered the myrmekitic textures between sphalerite, chalcopyrite, galena, tennantite, tetrahedrite pyrrhotite, geocronite, boulangerite, bournonite and other sulphides and sulphosalts.

Intergrowth textures designated as graphic, sub-graphic or micrographic in the literature, in most of the cases are really myrmekitic intergrowths abundantly observed in this deposit. As the Dariba-Rajpura deposit is essentially syngenetic sedimentary in character, the presence of intergrowth textures involve significantly interesting genetic considerations. The textures are attributed mainly to metamorphism, however, devitrification, collective-crystallization and diapthoresis must have also played their roles.

Key words : Myrmekitic Texture, Polymetallic Sulphide Deposit, Sedex Deposit, Dariba-Rajpura.

INTRODUCTION

Polymetallic sulphide deposit of Dariba-Rajpura (24°57'N, 74°08'E) belonging to lower Proterozoic age, located in Rajsamand district of Rajasthan which forms southern most part of the Dariba-Bethumni belt has been studied by a number of workers covering its different aspects. Major contributions are from Raja Rao et al., (1972), Poddar (1974), Solanki (1979), Pandya et al. (1980), Deb and Bhattacharya (1980), Pandya, (1981), Deb and Kumar (1982), Deb and Sarkar (1990), Mishra (2000) and Deb and Pal (2004). This deposit forms part of Archean Bhilwara Supergroup (Gupta et al., 1997). Stratigraphically the Rajpur-Dariba, Pur-Banera, Jahazpur and Sawar Groups lie between Hindoli Group, Mangalwar Complex and Sandmata Complex at the base and rocks of Ranthambhor Group at the top (Gupta et al., Opi. cit.). Based on deformational history Naha et al.. (1987) mentioned that the structural style and sequence of Rajpura-Dariba deposit are identical to those in the type Aravalli rocks. Deb and Pal (2004) considered the Dariba-Rajpura-Bethumni-belt to lie in the middle of Bhilwara metasedimentary basin, and its rocks are taken as equivalent to the Aravalli Supergroup.

The sulphide ore bodies of Dariba-Rajpura deposit are mainly hosted by calc-silicate rocks, graphite-mica schists containing garnet, staurolite and kyanite, impure-marble, carbonaceous cherts (black cherts) at places banded with cherts and quartzites, calcareous-biotite schists and amphibolites. (Fig. 1).

The sedex type deposit of Dariba-Rajpura is stratiform and zoned in character and shows a number of depositional and diagenetic features (Pandya et al. 1980). The deposit is metamorphosed and tectonized showing several features of metamorphism and deformation in ores and sediments (Pandya and Solanki (1975), (Solanki, (1979), Pandya (1981) and Pandya et al., (2001). The sulphide ores occur mainly as stratiform bodies but veins, patches and pods of the ores are also reported (Deb and Pal, 2004). Based on geothermometric estimations Mishra (2000) suggested a P-T condition of 5 Kb/525°C for the deposit. Deb and Pal (2004) mentioned that the peak P-T conditions of regional metamorphism of Rajpura-Dariba ores and host rocks could not have exceeded 5.4 kb/550°C.

The major ore minerals of the deposit are represented by sphalerite, pyrite, galena, pyrrhotite and chalcopyrite with subordinate amounts of tetrahedrite-tennanitite and geocronite. The arsenopyrite is dispersed throughout the ore bodies in minor quantities. Solanki (1979) reported a number of rare minerals viz; boulangerite, bournonite, marcasite, bornite, chalcocite, covellite, graphite, stannite, pyrargyrite, polybasite, neodiginite, enargite, famatinite, cubanite, proustite, Jamesonite, vallerite, violarite, starnbergite, pentlandite, bravoite, owyheeite, gratonite (?), glaucodot, argentite and native silver etc. Subsequent workers (Basu et al. (1981, 1983, 1984); Deb and Bhattacharya (1980) and Mozyava et al. (1992) reported a number of rare minerals viz. owyheeite, argentopyrite, pearcite, electrum, aurostilbite, gudmundite, intermetallic compounds (Ag, Au and Hg), rayite, plumbian tetrahedrite, Mn fahlore, thalcusite and renierite.

It is in some of the above minerals that the authors have encountered for the first time the myrmekitic textures which form the subject matter of the present paper.

MYRMEKITIC TEXTURES IN ORES OF DARIBA-RAJPURA DEPOSIT

The various intergrowth textures encountered in the ores of Dariba-Rajpura deposit are dominantly of myremekitic type. These textures were designated graphic, sub-graphic and micrographic by earlier workers (Bastin, 1950), (Edwards, 1965), Brett and Kullerud (1967) and Craig and Kullerud (1968 and 1969).

Various myrmekitic textures have been reported from the syngenetic sulphide deposits of Yerranderie. N.S.W., Australia; Gladhammar, Sweden; Oravicza, Dognacska and Banat, Rumania; Cornwall, Missouri (USA) and Freiberg in Saxony, Germany. These textures are monotonous and they always show inter-penetrating growth of large grains of two and some times more minerals. The grain boundaries of these minerals are mutually rounded so that in a section the texture resembles finely woven fabric. In many cases characteristic rounded terminations are faithfully preserved to be laminae, in those where there is a gradation into the lamellar growth. In these textures, minerals are interlaced and occur irregularly, showing either similar proportions or one predominant, but largely with uniform orientation and even this may produce interconnected network like skeletal forms (Ramdohr, 1969).

Various mineral assemblages showing myrmekitic intergrowth are listed below:

Galena + chalcopyrite ± pyrrhotite

Galena + tennantite / tetrahedrite

Galena + bournonite

Galena + bournonite ± tennantite

Galena + Quartz

Galena + bournonite ± native arsenic ± enargite and famatinite

Chalcopyrite + tennantite

Sphalerite + galena

Sphalerite + bournonite

Geocronite + native arsenic + polybasite

MYRMEKITIC INTERGROWTH OF GALENA, CHALCOPYRITE AND PYRRHOTITE

Chalcopyrite which is in predominant amount shows myrmekitic intergrowth with galena (Fig. 2), similar to one of Edwards (cf Fig., 145, 1965). Some times galena or chalcopyrite individually or together are intergrown with pyrrhotite (Fig. 3). The associated minerals are sphalerite and tetrahedrite.

MYRMEKITIC INTERGROWTH OF GALENA AND TENNANTITE

Ramdohr (1969) reported for the first time the myrmekitic intergrowth of galena and fahlore. In Dariba-Rajpura deposit the tennantite appears as interwoven grains with galena (Fig. 4). In this intergrowth galena is a predominant mineral and occurs in association of bournonite, enargite-famatinite and native arsenic.

COMPLEX MYRMEKITIC INTERGROWTH OF GALENA, BOURNONITE, ENARGITE-FAMATINTITE AND NATIVE ARSENIC

Galena and bournonite show an excellent coarse myrmekitic texture (Fig. 5) (Ramdohr, Fig. 85 E, 1969). This texture was also mentioned by Odman (1939). The fine grained variety of galena, enargite-famatinite and native arsenic are intergrown with galena and bournonite (Fig. 6). Occasionally the tetrahedrite is also mymekitically intergrown with galena and bournonite (Fig. 7).

MYRMEKITIC INTERGROWTH BETWEEN GALENA AND QUARTZ

In the major phase of quartz, galena is intergrown myrmekitically and shows association of chalcopyrite. sphalerite and tetrahedrite in minor amounts.

MYRMEKITIC INTERGROWTH OF SPHALERITE GALENA, BOURNONITE AND TENNANTITE

In this texture the sphalerite is the dominant mineral and it is intergrown with galena myrmekitically (Fig. 8). Grigorieve (1928), Lindgren (1930) and Ramdohr (1969) have reported similar type of textures from various sulphide deposits. At places, myrmekite intergrowth between tennantite and sphalerite has also been observed. Pyrite and arsenopyrite are other associated minerals in this intergrowth. Occasionally sphalerite and bournonite also show myrmekitic intergrowth.

MYRMEKITIC INTERGROWTH OF CHALCOPYRITE AND TENNANTITE

Myrmekitic intergrowth between chalcopyrite and tennantite (Ramdohr, 1969) is not a very common feature in sulphides of this deposit. Rarely tennantite appears to have intergrown myrmekitically with chalcopyrite.

MYRMEKITIC INTERGROWTH OF GEOCRONITE, NATIVE ARSENIC AND POLYBASITE

The authors for the first time have observed the complex myrmekitic intergrowth of geocronite, native arsenic and poly-basite (Fig. 9) in diopsidite, similar to the texture described by Edward (cf. Fig. 70, 1965). Though geocronite is the dominant mineral, arsenic and polybasite are also present in appreciable amounts. At places, these minerals show an intergrowth with pyrargyrite and argentite.

TEXTURES, WHICH MORE OR LESS APPEAR LIKE MYRMEKITIC INTERGROWTH

(i) Exsolution Texture:

In the literature, ex-solution texture between sphalerite and chalcopyrite has been described exorbitantly, but myrmekitic intergrowth between them has not often been seen. In Dariba-Rajpura deposit the chalcopyrite and sphalerite show the myrmekite like intergrowth.

(ii) Skeletal Intergrowth

Skeletal growth is exhibited by arsenopyrite which lie within the chalcopyrite (Fig. 10), resembling the texture given by Edwards (cf. Fig. 74, 1965). Moreover, arsenopyrite and chalcopyrite also show skeletal intergrowth within tennatite (Fig.11). These minerals occur irregularly in almost dissimilar proportions, but the components show uniform orientations.

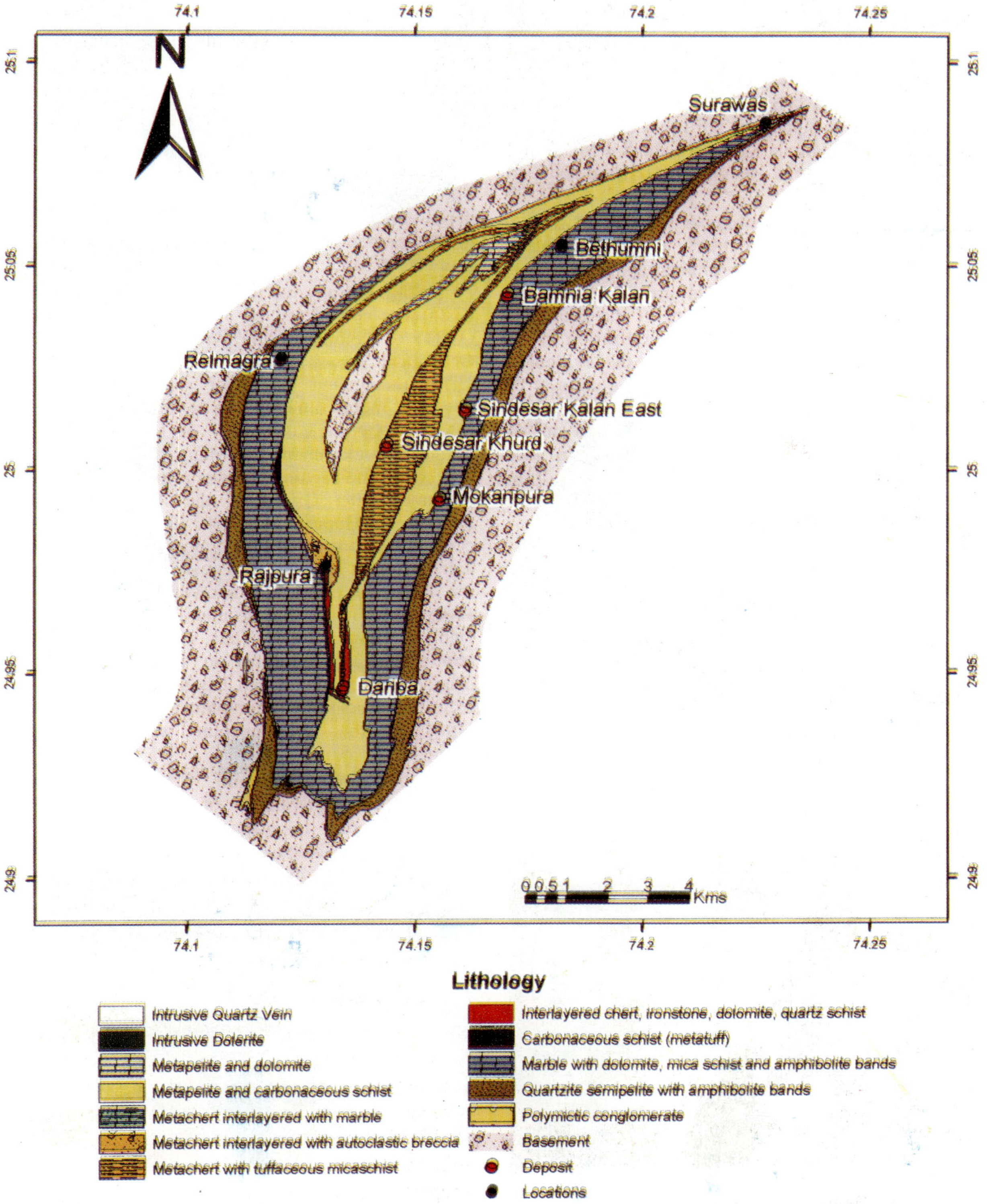

Fig. 1 Regional geological map of Dariba-Rajpura-Bethumni belt (after Gupta el.al, 1995)

Fig. 2 Myrmekitic intergrowth of Galena (grey-etched and chalcoprite (white). Photomicrograph X 190.

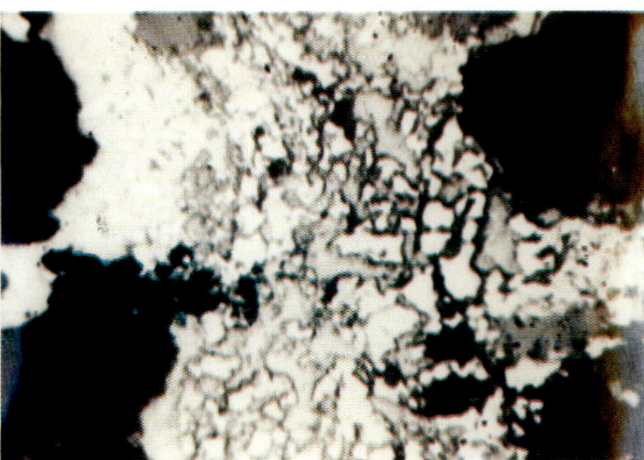

Fig. 3 Pyrrhotite (white) intergrown myrmekitically with chalcopyrite (grey) and a little galena. Photomicrograph X 190

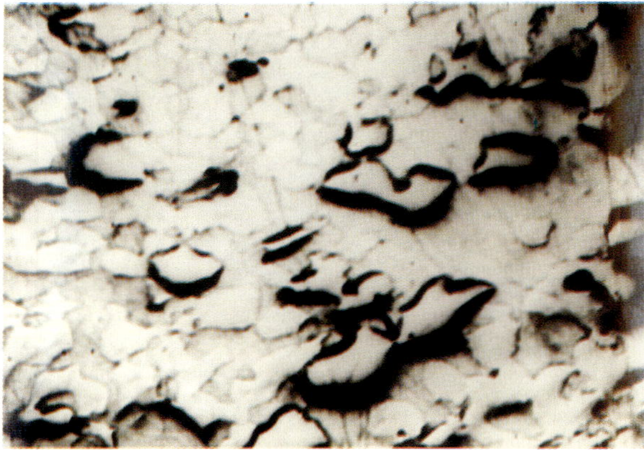

Fig. 4 Coarse myrmekitic intergrowth of Galena (grey etched), tennantite (Greyish- White with high relief), a little native arsenic (clear white) and bournonite (grayish shade Lighter than galena. Photomicrograph X 45.

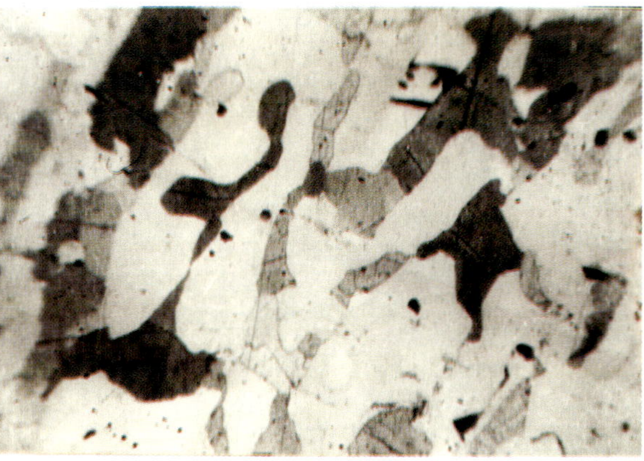

Fig. 5 A very coarse myrmekitic intergrowth of galena (lightgrey and dark grey-etched) and bournonite (grayish white). Photomicrograph X 45.

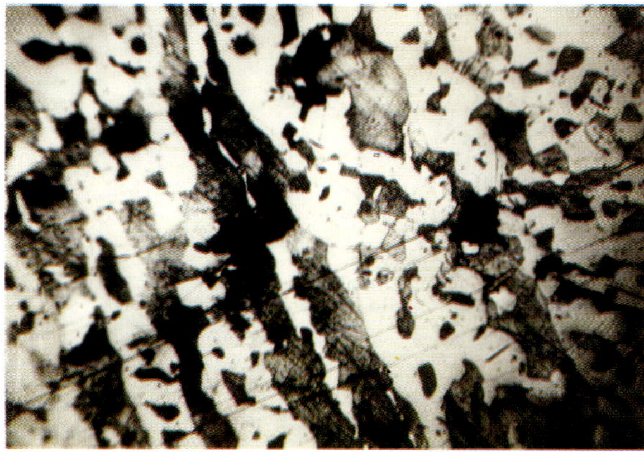

Fig. 6 Galena (grey-etched) shows Myrmekitic intergrowth with native arsenic (clear white in galena), bournonite (white- Shade darker than arsenic) and a little Enargite and famatinite. Photomicrograph X 45.

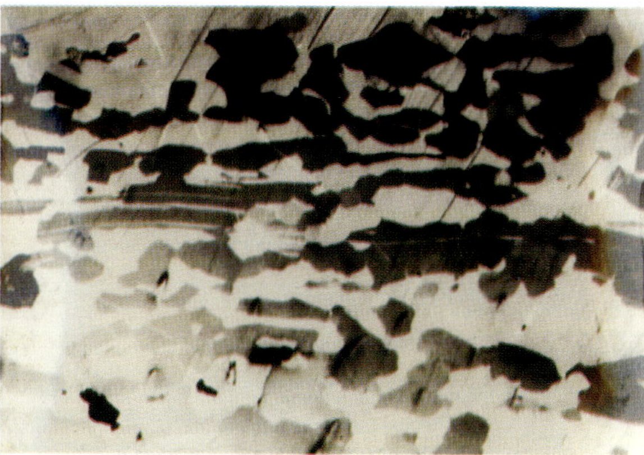

Fig. 7 Bournonite (grey and black) and galena (dull white) grading into lamellar intergrowth, Photomicrograph X 50.

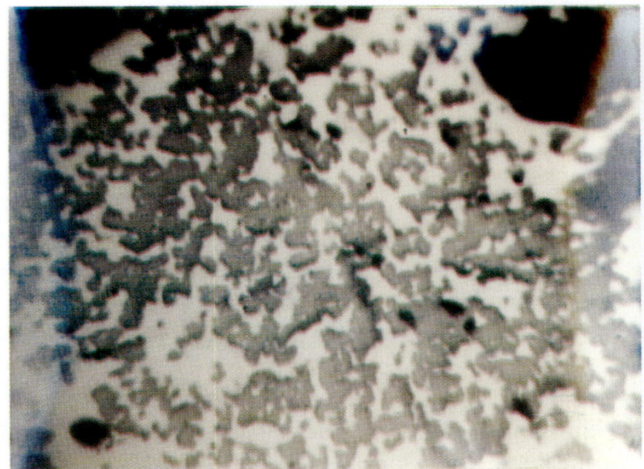

Fig. 8 Myrmekitic intergrowth of galena (white) and sphalerite (grey). Intergrowth of subrounded grains of sphalerite (lower half) in galena are also visible. Photomicrograph X 100.

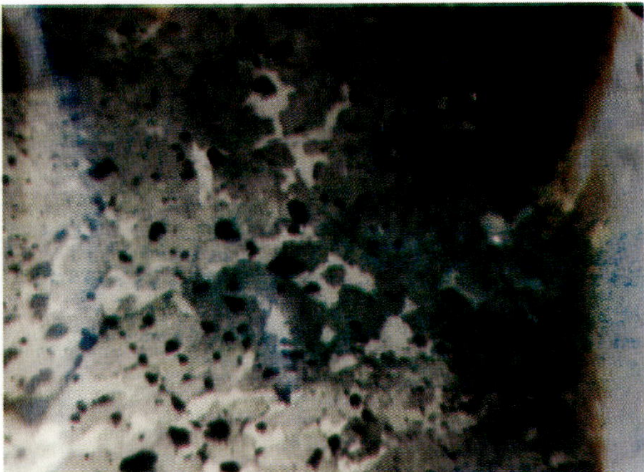

Fig. 9 Complex myrmekitic intergrowth of native arsenic (dirty white), geocronite (light grey), polybasite (dark grey) and gangue (black). Photomicrograph X 190.

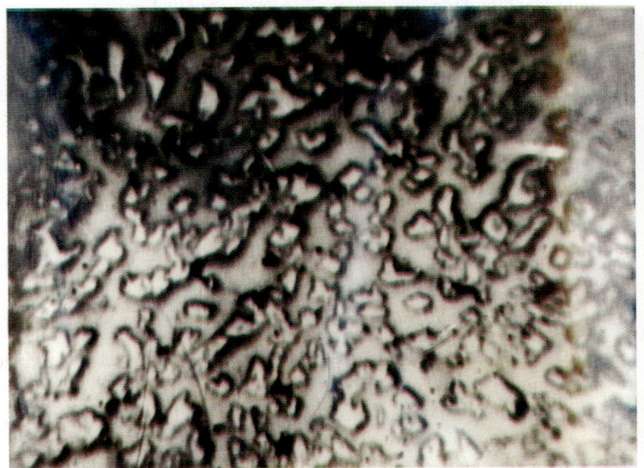

Fig.10 Myrmekite like intergrowth (skeletal) of arsenopyrite in challcopyrite (light grey). Photomicrograph X 190.

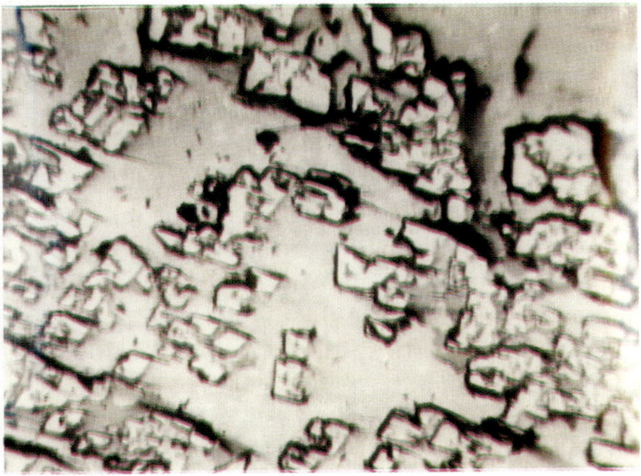

Fig. 11 Skeletal intergrowth of arsenopyrite (white with high relief) and chalcopyrite in tennantite. Photomicrograph X 210.

GENESIS OF MYRMEKITIC TEXTURES

It is difficult to classify precisely the processes which formed myrmekitic textrues in the sulphide ores because the transitions occur in them at many places. Nevertheless, mymekitic textures of the ores in question appear to have been formed mainly by (a) metamorphism-devitrification, (b) collection crystallistation and diapthoresis, must have also played significant roles. Decomposition, replacement, unmixing and interaction of mineral must have taken place due to metamorphism and diapthoresis.

The decomposition can form the reactive product which causes reciprocal attachment on the associated minerals. Hence, transititons to replacement are also found here. Moreover, the decomposition textures can not be sharply separated from ex-solution textures by only appearance therefore other criteria are very necessary for their identification. Unmixing, which produces ex-solution texture in its restricted sense requires host and guest forming from the same original mixed crystals, with atleast one of the two in crystallographic relationship with it and the overall chemistry remains unchanged. This argument or rather supposition strictly is not found to be true. The product can be quite unrelated crystallographically to the original materials, and chemistry can be changed to varying degree by introduction or removal of materials. It is worth mentioning that the myremekitic intergrowth or graphic texture can occur as ex-solution, especially where the lattice structure of the components is very similar.

The myrmekitic intergrowth of sphalerite and galena (Fig. 8) appears to have formed by divitrification of sulphide gel (Grigorive, 1928; Lindgren, 1930 and Ramdohr, 1969).

Myrmekitic intergrowth of galena + chalcopyrite ± pyrrhotite (Fig. 3), galena + pyrrhotite and chalcopyrite + tetrahedrite / tennantite may have resulted due to so called replacement during cooling subsequent to metamorphism and partly due to diapthoresis. Introduction or removal of material during metamorphism, diapthoresis and related decomposition of sulphides and sulphosalts gave rise to myrmekitic intergrowth of galena + quartz, galena + tennantite (Fig. 4) and bournonite + galena (Fig. 5) (Ramdohr, 1969). The phenomena mentioned above can also have the interplay of directed pressure.

Galena-bournonite intergrowth has also formed from geocronite with interchange of materials.

Some minerals are stable at high temperature but react with each other at low temperature to form complex compounds or there may be a supplementary addition of materials to the earlier formed minerals. The myrmekitic intergrowth of sphalerite + bournonite thus formed as a reaction of fahlore and galena (Ramdohr, 1969). Alternatively, bournonite arises as a transformation product with galena and jamesonite, from geocronite. In other cases bournonite appears to be the dissociation product of unstable Pb-Cu-Sb-Sulphides. Quite similarly, geocronite or minerals of fahlore group with much Fe and As is transformed into ordinary fahlore (tetrahedrite and tennantite) + chalcopyrite + arsenopyrite, similar to those of Jabobsbakken sulphide deposit, Sulitjelma, Norway (Ramdohr, 1938).

ACKNOWLEDGEMENT

The authors express their grateful thanks to the officers and staff of Hindustan Zinc Limited Udaipur for providing all kinds of facilities and cooperation during the field work at the Dariba-Rajpura mines. The laboratory investigations were carried out at the Department of Geology, MLS University, Udaipur and the Department of Metallurgical Engineering, Malviya National Institute of Technology, Jaipur. The authors extend their heartly thanks to the Head of the Department of Geology, Udaipur and the Director, MNIT, Jaipur for providing all kinds of laboratory facilities. Late Prof. Dr. G.C. Amstutz, Formar Director, Mineralogisch-Petographisches, Institut, der Universitat, Heidelberg (West Germany) was kind enough to examine some polished sections at Udaipur during his visit for which the authors extend their heartiest gratitude.

References

Bastin, E.S., (1950) Interpretation of ore textures, Geol. Soc. Amer, Mem. 45.

Basu, K; Bortnikov, N.N., Mookherjee, A. and Mozgova, N., Tsepin, A.I., 1981: Rare Minerals, from Rajpura-Dariba, Rajasthan, India, II Intermetallic compound Ag 74,2 Au 16,4 Hg 9.4, Neus Jahrbuch fur Mineralogie and Petrologic Abh. V. 141, p. 217-223.

Basu, K., Bortnikov, N.N., Mookherjee, A., Mozgova, N., Tsepin, A.I., Vyalsov, L.N. 1983: Rare minerals from Rajpura-Dariba, Rajasthan, India IV A New Pb Ag-T,i-Sb Sulfo-Salt, Rayite, Neus Jahrbuch fur Mineralogie and Petrologie: Monatsheft H. 7, p 296-304.

Basu, K., Bortnikov, N.N., Mookherjee, A., Mozgova, N., Sivtsov, A.V., Tsepin, A.I. and Vrubleskoya, Z.V., 1984: Rare minerals from Rajpura-Dariba, Rajasthan, India. The first recorded occurrence of Manganoan fahlore: Neus Jahrbuch fur Mineralogie and Petrologic Abh; V. 149, p. 105-112.

Brett, R. and Kullerud, G. (1967) The Fe-Pb-S system, Econ. Geol., 60, pp 47-68.

Craig, J.R. and Kullerud, G. (1968) Phase relations and minerals assemblages in the copper-lead-sulphur system, Am. Miner, 53, pp. 145-161.

Craig, J.R. and Kullervd, G. (1969) The Cu-Zn-Pb System, Carnegie Inst. Wash. Year Book 67, pp. 177-179.

Deb, M. and Bhattacharya, A.K.; 1980: Geologicalsetting and conditions of metamorphusm of Dariba-Rajpura polymetallic ore deposit-Rajasthan, India in Proceedings of the 5th IAGOD symposium, schewizerbart sche Verlagsbuch handlung, Stutgart, p. 679-697.

Deb, M. and Kumar, R., 1982: The volcano-sedimentary environment of Rajpura-Dariba Polymetallic ore deposit, Udaipur district, Rajasthan in Proceedings of the IGCP Symposium on Metallogeny of the Precambrian. Geological Survey of India, Bangalore, P. 1-17.

Deb, M. and Sarkar, S.C., 1990: Proterozoic tectonic evolution and metallogenesis in the Aravalli-Delhi orogenic complex, north-western India. Precambrian Research, V. 46, p. 115-137.

Deb, M. and Pal. T., 2004: Geology and Genesis of the Base Metal Sulphide Deposits in the Dariba-Rajpura-Bethumni Belt, Rajasthan, India, in the light of Basin Evolution. Sediment-hosted Lead-Zinc sulphide Deposits; attributes and models of some major deposits in India, Australia and Canada. Narosa Publishing House, New Delhi, P. 304-327.

Edwards, A.B. (1965) Textures of the ore minerals Melbourne, Australia, 242 P.

Grigorieve, J. (1928) Structures of mineral intergrowths in ores. Mem. Russ. Min. Soc., 57, pp. 11-56.

Gupta, S.N., Arora, Y.K., Mathur, R.K., Iqbaluddin, B. Prasad, Sahai, T.N. and Sharma, S./. 1995: Geological map of the Precambrians of the Aravalli region, Southern Rajasthan and North-Eastern Gujarat, India, Geological Survey of India, Calcutta, Scale 1:250000.

Gupta, S.N., Arora, Y.K., Mathur, R.K., Iqbaluddin, B. Prased, Sahai, T.N. and Sharma, S.B., 1997: The Precambrian Geology of the Aravalli Region, Southern Rajasthan and North-Eastern, Gujarat, Mem. Geol.Surv. India, Vol. 123, 262p.

Lindgren, W. (1930) Pseudo eutectic structures, Econ., Geol., 30, pp. 331-346.

Mishra, B., 2000: Evolution of the Rajpura-Dariba polymetallic sulphide deposit: Constrains from sulphide-sulphosalt phase equilibria and fluid inclusion studies in Deb, M. ed. Crustal Evolution and Metallogeney in the Northwestern Indian Shield. Narosa Publishsing Home, New Delhi, P. 307-328.

Mozgava, N.N., Borodaev, Yu.S., Nenasheva, S.N., Efimov, A.V., Gandhi, S.M. and Mookherjee, A., 1992. Rare Minerals from Rajpura-Dariba, Rajasthan, India VII: Renierite: Mineralogy-Petrology V. 46, p. 55-65.

Odman, 0. (1939) The gold-copper arsenic ore at Holmtijaran, Skelleftea district, N.Sweden, Geol., Foren., Forth, Stockholm, 61, pp. 91-111.

Pandya, M.K. and Solanki, S.L. (1975) Primary depositional features and metamorphic characteristics of the stratiform multimetal sulphide deposit of Dariba-Rajpura, Udaipur distrrict, Rajasthan, Seminar onRecent Advances in Precambrian geology and mineral deposits with special reference to Rajasthan (Abstract)., Department of Geology, University of Rajasthan, Udaipur.

Pandya, M.K., Solanki, S.L. and Pandya, T.K. (1980) Diagenetic features in sulphides of Dariba-Rajpura deposit, Udaipur district, Rajasthan Bull. Geol. Soc. India, Vol. 21. No. 9, pp. 425-431.

Pandya, M.K., Solanki, S.L. and Pandya, T.K., 2001: Genetic Significance of Depositional, Diagenetic and Metamorphic. Features in Pre-Delhi Sulphide Deposits of Rajasthan. Supplement to Geological Survey of India, Special Publication No. 72, p-6-20.

Pandya, T.K. (1981) Geology of Major lead-zinc deposits of Rajasthan, Unpub. Ph.D. Thesis, Univ. of Raj., Jaipur, 334 p.

Poddar, B.C; 1974: Evolution of sedimentary sulphide shythmites into metamorphic tectonites in the base metal deposits of Rajpura-Dariba, Rajasthan Golden Jubilee Volume, Quarterly Journal of the Geological Mining and Metallurgical Society of India. V. 46, p. 207-221.

Raja Rao, C.S., Poddar, B.C. and Chaterjee, A.., 1972: Dariba-Rajpura-Bethumni belt of Zinc-Lead-Copper mineralization, Udaipur district, Rajasthan. Miscellaneous Publication of the Geological Survey of India, V. 16 Base metal, Pt.2, P. 617-627.

Ramdohr, P (1938) Antimonreiche Paragenesen Von Jakbsbakkon bei Sulitjelma Norsk, Geol. Tskr., 18, pp. 275-289.

Ramdohr, P. (1969) The ore minerals and their intergrowths, Pergaman Press, London, 1174 p.

Solanki, S.L. (1979) Geology of zinc-lead-copper deposit of Dariba-Rajpura, Udaipur district, Rajasthan, Un.Pub. Ph.D. Thesis, Univ. of Raj., Jaipur. 258p.

Hon'ble Reviewers

1. Prof. K.L. Rai
   (Formar Head
   Department of Applied Geology,
   Indian School of Mines)
   45, Chatrapati Shivaji Nagar
   Kolar Road
   BHOPAL 462 016 INDIA
2. Prof. K.K. Singh
   (Formar Vice Chancellor,
   Vikram University)
   702, Technology Apartments
   24 I.P. Extension
   Patparganj
   NEW DELHI 110 001 INDIA

Editorial Assistance

Dr. Sameer Naval

Economic Mineralization
K.L. Shrivastava (Editor)
2009, pp. 248-253
Scientific Publishers (India), Jodhpur

Section-III

Beach Sand Resources and Industry in India

G. S. ROONWAL

(Formaly Chairman, Department of Geology, Delhi University, Delhi 110 007)
Federation of Indian Mineral Industries, 301 Bakshi House,
40-41, Nehru Place, New Delhi 110 019

Abstract

Presence of largescale reserves of ilmenite and titaneferous magnetite all along the continental shelf of India have been proved by various agencies. Since ancient times, small scale mining operation are in practice for these coastal placers. These resources have been in perception of the government of India for some three decades now, but recent open door policy have initiated tremendous exploration and exploitation opportunities to these deposits. A brief account of beach sand placer resources and recent developments and future avenues of the industry is presented.

Key words : Placer deposits, Beach sand resources, Marine resources, Ilmenite, Mineral fuel.

INTRODUCTION

India's resource base in heavy mineral sand (Fig. 1) is sufficiently large for it to become a far more significant force in TiO_2 industry than it is at present. But until recently, the mineral policy of Government of India has prevented necessary private investment. A new opendoor policy suggests that this situation is about to change.

Reviews on the mineral sands of India are available in TZML (2000), special issue on beach and inland heavy mineral sand deposits of India (2001). Qasim and Roonwal (1996), Roonwal and Wilson (1998), Rajamanickam (2001), Mallik (2007), Nambiar et al. (2007) and Nageshwar Rao et al. (2007) in addition to vast information in the various reports of the Geological Survey of India.

Historical Background

The World's first heavy mineral sands production plant was established in India in the year 1911, at Manavalakurichi in Tamil Nadu, near to the southern tip of India, based on large coastal deposits of monazite discovered two years earlier (1909). The first ilmenite shipment was made in 1922, and in the following decade output grew rapidly with the development of Chavara deposits near Kollam (Quilon) in the adjacent southern state of Kerala. According to TZML (2000), India was the world's largest ilmenite exporter during the 1930's and 1940's, operating four plants and exporting 300,000 t in 1940. The industry subsequently went into decline and by 1965 three of the flour plants were closed.

Exports were affected by two factors, namely the unsustanability of the Kollam ilmenite for pigment production through sulphate route, and the passing of

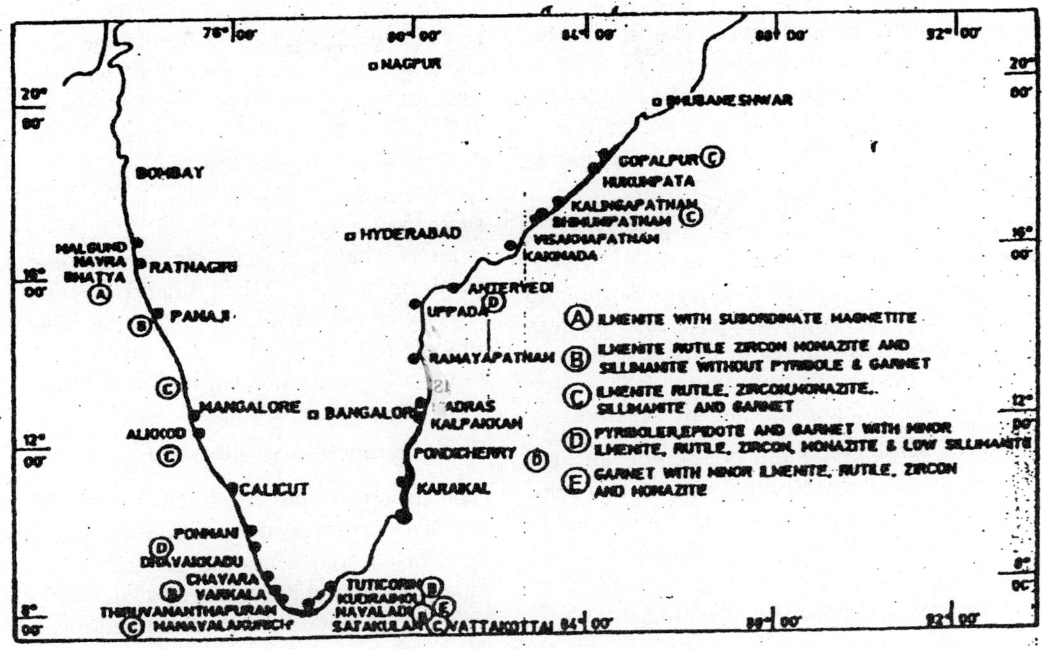

Fig. 1 Location of major heavy mineral sand deposits (after, Ali et al. 2001; in spl. Issue of AMD)

content of ilmenite export to the Atomic Energy Commission (AEC) of India that was established in 1947 after the country's independence. Under the atomic mineral legislation enacted at that time, the AEC took control of all "atomic minerals". These were deemed as those containing in excess of 0.25% monazite and, thus, investment of mineral sand industry was, in effect, private sector stifled.

The current mineral sands industry is dominated by the Indian Rare Earth Ltd. (IREL) that accounts for 95% of output of heavy minerals from sands. IREL was established in 1965 as a Public Sector Undertaking under the Department of Atomic Energy, Government of India and acquired the original operation at Chavara and subsequently the operation at Manavalakurichi. On a global scale, Indian titanium industry is relatively small but highly diverse. There are four mining and processing sites in Tamil Nadu and the north eastern state of Orissa. Collectively, they have an ilmenite production capacity of some 400,000 t/y. During the year 2005, the output was around 395,000 t, with IRE's OSCOM (Orissa Sands Complex) operating Chhatrapur in Orissa contributing 185,000 t. India also possesses four synthetic rutile plants with a combined capacity of some 90,000 t. Production during the year, 2005, was about 50,000 t/y in addition there are four TiO_2 pigment plants, including one in Kolkata (Calcutta) in West Bengal. During the year 2005, they had a combined production of about 45,000 t.

Domestic consumption of ilmenite in the year 1999 was lower than in the previous year as a result of lower off take (90,000 t) by the pigment industry. Demand for ilmenite for synthetic rutile showed little change, at 42,000 t. TZML estimates total exports of ilmenite during the year 2005 at a little over 150,000 t, shipped mainly to Europe, with smaller volumes detained for the US and South Korea. India has yet to capture market in Japan, that at present imports REEs from China (Roonwal, 2000). Of Indian total synthetic rutile output in the year 1999 of 50,000 t, only 15,000 t were exported compared with 30,000 t in the previous year. Japan appears to be the only destination. Domestic usage of synthetic rutile is estimated at 20,000 t.

India's ilmenite resources base, and established by the Atomic Energy Department is reportedly as high on 200 Million Tonne and the ilmenite quality varies with 50% to 60% TiO_2.

New open door policy

In October 1998, a change of The Government of India government policy paved the way for the select entry of Indian and foreign private investors. TZML reports that participation of whollyowned Indian companies is now permitted in all facets of mineral sands mining, separation and value adding projects, and that upto 70% foreign equity participation is possible for valueadding and integrated projects. The policy appears designed to preclude the export of ilmenite, unless the project has a valueadding component.

The change in policy has spurred a number of initiatives, with principal focus on synthetic rutile. TZML further points out that this is not surprising given the constraint on ilmenite exports and the high costs and uncertain availability of electric power in India, especially during the summer months, which makes the production of titanium slag unattractive.

TZML provides details about the number of new developments in the offering but cautious that they should be viewed at best as being under active investigation. This is because the combination of raising adequate finance and securing the necessary government approval has delayed their confirmation.

In Kerala, the state government owns the Kerala Mineral and Metals Ltd. (KMML), and the company already conducts fully integrated operations ranging through mining to synthetic rutile output and pigment production. Production is based on reserves of 60% TiO_2, and exports from the same deposit at Chavara that are mined by IREL. KMML current ilmenite output is just 200,000 t/y, but there are plans to expand capacity to 330,000 t/y, and then raise synthetic rutile production from 200,000 t/y to 130,000 t/y. There are also plans to lift pigment output from 22,000 t/y to 60,000 t/y.

In Orissa, there is an interesting development involving the Austpac Resources NL of Australia and IRE. The Ausrutile joint venture was formed in August 1999 to develop Auspac's ERMS synthetic rutile technology, and the first objective is to build a 10,000 t/y demonstration plant adjacent to IREL's OSCOM mineral sands operation near Chattrapur. OSCOM possesses a 100,000 t/y capacity synthetic rutile plant but it is not currently in production. Recently, Ticor Ltd. announced its decision to participate in the Ausrutile joint venture. Ticor Ltd. is already in a joint venture with Austpac to develop the synthetic rutile technology world class. Ownership of the Ausrutile joint venture will now comprise Austpac and Ticor each with 37%, and IREL holding the remaining 26%.

Elsewhere in Orissa, a privately owned Indian company, Saraf Agencies, is also reported to be considering establishment of a synthetic rutile plant based close to IREL's OSCOM operations. The plant, as envisaged, would be a first in India. A 100,000 t/y capacity operation is contemplated using OSCOM ilmenite as a feedstock, with the grade as low 50% TiO_2. Besides, Saraf has applied to the Orissa Government for the rights to mine an ilmenite deposit lying beyond IREL's lease area, and has signed agreement with MD Mineral Technologies for the latter to conduct a feasibility study for a 500,000 t/y plant operations. MD Mineral Technologies is a subsidiary of EvansDeakin Ltd., the Australian heavy engineering group and is a leading designer and supplier of processing equipment for the mineral sand industry.

Commercial Consideration

Asia has one of the largest reserve of Rare Earth ores. It is also a large producer of mined rare earths, and is the world's major market for rare earths. In 1997, Japan and SE Asia accounted for an estimated 24,500 t (37%) of the 66,000 t of contained rare earth oxides (REO) consumed worldwide with China consuming a further 12,000 t (18%). A decline of these markets could have a serious effect on the supply and demand balance. China supplied more than 60% of the world's rare earth production in 1997. Japan, the largest single consumer of rare earths is forced to rely heavily on China for its rare earth raw material. Japanese imports of rare earth products from China were valued at a record ¥ 14.4 billion in 1997.

Some rare earth producers in China were slow to respond to demand, which mean disruption in supply and demand in 1998-99. Here lies an opportunity for India to get more into the world market. But this is possible only when we make efforts for higher production. At the same time, smaller producers have sometimes expanded too rapidly without evaluating accurately the market's future requirement. As a result, on oversupply has developed for one or more rare earth products. This occurred with yttrium and other "heavy" rare earths in the late 1980s, and with cerium and lanthanum oxide in 1996-97.

Demand for rare earth is now expected to reach 70,000 t/y REO, valued at US$660 million, by 2000. In 1997, S.E. Asia and Japan provided the strongest and most rapidly growing market for key applications such as autocatalysts, permanent magnet, rechargeable batteries, phosphorus and advanced ceramic.

Rare earth prices were relatively static between early 1997 and early 1998, possibly reflecting the stabilizing effects of increased demand for "mischmetal" (mixed metals) for batteries. Mischmetal prices experienced a small increase in their average market price range (from US$6.80, 0.510.5/kg) to (US$6.80, 12/kg) over the same period. The average price ranges from neodymium oxide 96 per cent to 99.99% (US$ 14160/kg) between the first quarter of 1997 and the first quarter of 1998 due to increasing demand for high purity neodymium oxide for neodymium nonboron magnet manufacturers (Table 1).

Table-1 Production of Rare earth minerals (t REO Content)

		1996	1997
Monazite	Brazil	200	400
	India	2700	2700
	Sri Lanka	120	120
	Total	3020	3220
Bastnaesite	China*	55,000	50,000
		20,400	20,000
	Total	75,400	70,000
Xenotime	Malaysia	340	300
Loparite	CIS	6000	6000
	Grand Total	84,800	79,500

* Chinese bastanaesite figure includes production of xenotime, monazite. Based on USGS and Mining J. London, Aug. 1998

Australian Interest

TZML regards the most significant new development in the Indian mineral sands sector to be in Tamil Nadu, where Mineral Deposits Ltd. (MDL) of Australia has outlined proposals to mine substantial deposits in the southeastern part of the state. If this were to go ahead it would constitute a completely new minerals sands province in India and mark the first direct foreign investment. The status of exploration have been given in Table 2 and total Indian reserves are given in Table 3.

The deposits are at Kudiraimozhi, Sattankulant and Navaladi, located 3035 km southwest of the port of Tuticotein, and are reported to contain a total resources of 100 mt of heavy minerals, including 61M of ilmenite. According to TZML, two projects are proposed. At Kudiraimozhi there is an inferred and indicated resource of 370 mt with an average grade of 8.9% heavy minerals. A mining operation producing 270,000 t/y of ilmenite is envisaged which would provide feed stock for a 135,000 t/y capacity synthetic rutile plant. The total project cost is estimated at US $ 150 million.

The second project is based on the Sattankulant-Navaladi deposits, in which there is an inferred and indicated resources of 610 mt an average grade of 11% heavy minerals. The project would also include production of synthetic rutile, plus a 50,000 t/y pigment plant with a total investment of the order of US $ 350 million. TZML says that two Mauritius based companies have been approved by the Government of India as foreign collaborators in the project Ausind Sands Ltd., and Mauritius Titanium Ltd. (MDL). It is understood that MDL proposes to prepare a bankable feasibility study for each project and that appropriate expertise has been recruited to undertake this work.

Further ahead, Tata Iron and Steel Ltd., the largely privately owned Indian industrialist conglomerate has expressed some interest in entering the mineral sands and titanium metal sectors, but no project has been announced. TZML believes that the entry of Tatas, add considerable weight to India's heavy mineral sand industry.

Difficulty to the Industry

Problem facing the industry is illustrated with an example here in Kerala. The Mineral sand mining project, cleared by the state government is now facing opposition because of alleged ecological impact on the coastal belt in the Alappuzha district. This project has been proposed by the Kerala Rare Earth and Mineral Ltd., which is a joint venture of Kochibased Cochin Minerals and Rutile Ltd., Indian Rare Earth's (IREL) and Kerala State Industrial Development Corporation.

According to the project, mining was to take place between Kayamkulam and Thottapally in the Alappuzha district, outside the blocks reserved for IREL and Kerala Minerals and Metals Ltd. It would also avoid the areas demarked for the project proposal by IREL and National Mineral Development Corporation (NMDC).

Table-2 Distribution of placers and exploration status along Indian coasts (after Mallik, 2007)

		PLACERS AVAILABLE	STATUS OF EXPLORATION
	GUJRAT	Mainly Ilmenite and Magnetite	Onshore and Offshore occurrences reported. Still exploration needed.
	MAHARASHTRA	Ilmenite and Magnetite mainly	Onshore and Offshore exploration completed. Appreciable reserves of ilmenite proved.
WEST COAST	GOA	Ilmenite, Magnetite and Zircon	Onshore surficial occurrences reported. Systematic exploration is lacking for offshore. Onshore sampling completed. Exploration being conducted.
	KARNATAKA	Ilmenite, Magnetite, Garnet etc.	Onshore surficial occurrences reported, exploration continued.
	KERALA	Ilmenite, Magnetite, Rutile, Zircon, Monazite, Sillimanite	Detailed on shore, offshore exploration completed. Beach placers being exploited at Chavara.
	TAMIL NADU (WEST COAST)	Ilmenite, Magnetite, Zircon, Rutile, Sillimanite	Onshore and Offshore exploration completed. Placers being exploited at Manavalakurichi.
	TAMIL NADU (EAST)	Ilmenite, Magnetite, Zircon, Garnet, Rutile	Onshore exploration completed. Some deposits of garnet, ilmenite being exploited. Work needed in offshore.
	A.P.	Ilmenite, Magnetite etc.	Onshore and Offshore exploration completed.
	ORISSA	Ilmenite, Monazite, Zircon, Rutile etc.	Onshore and Offshore exploration completed. Exploitation at Chatrapur, Gopalpur being continued.
EAST COAST	W.B.	No systematic data	Exploration needed

Table-3 Total Indian reserves of Beach sand placer resources (after, Mahalik et al. 2006)

Mineral	Resource in million tones
Ilmenite	348
Rutile	18
Zircon	21
Sillimanite	130
Garnet	107
Monazite	8

References

Atomic Minerals Directorate for Exploration and Research, Department of Atomic Energy, Govt. of India, 2001, Expl. Res. At. Miner. Spl. Issue on "Beach and inland heavy mineral sand deposits of India, 159 p.

Mahalik, N.K. 2006, 'Geology and Mineral Resources of Orissa'. SGAT Publication (3rd Edition).

Mallik, T.K. 2007, 'Mineral Resources of the Beaches and Shallow Sea Around Indian Coasts'. Ind. Geol. Cong. Bull. Vol. 1 (2). pp. 80-87.

Nageswara Rao, P.V.; Suryam, R.K. and Ranga Rao, V.; 2007, 'Heavy Mineral Potential and Provenance of Beach Sand Between False Devi point to Suryalanka, southeast Coast of India'. Ind. Geol. Cong. Bull. Vol. 1 (2). pp. 4753.

Nambiar, A.R.; Kumaran, K.; Dinesh, A.C.; Jayaprakash, C. and Venkateswara Rao Ch.; 2007, 'Distribution and Resources of Offshore Placer Minerals in NearShore Area Between Inayam and Midalam, North of Manavalakurichi, southwest Coast, India'. Ind. Geol. Cong. Bull. Vol. 1 (2). pp. 54-58.

Qasim, S.Z. and Roonwal, G.S. (Eds.), 1996, India's exclusive economic zone, Omega Scientific, Pub., New Delhi, 238 p.

Rajamanickam, G.V. (Editor) 2001, Hand Book of Placer Mineral Deposits. New Academic Press, New Delhi 327 p.

Roonwal, G.S. and Wilson, G.C., 1998, India's mineral potential : Present status and future trend, International Geol. Rev. 40 (10), pp. 865-895.

T.Z. Mineral International Pvt. Ltd. (TZML), 2000. Mineral sand report, West Port, 9 p.

Hon'ble Reviewers

1 .Dr. R. Dhana Raju
(Former Add. Director, AMD)
63124, Hastinapuri
Sainikpuri Post Office
SECUNDERABAD 500 094 INDIA

2. Prof. (Mrs.) Madhumita Das
Head
Department of Geology
Utkal University
Vani Vihar

BHUBANESWAR 751 004 ORISSA

Editorial Assistance

Dr. Narendra Singh

Economic Mineralization
K.L. Shrivastava (Editor)
2009. pp.254-263
Scientific Publishers (India), Jodhpur

Section-III

Polymetallic Polynodule from Indian Ocean Nodule Field

K.L. SHRIVASTAVA[1], VIRENDRA GAUR[1] C.P. DADHICH[1] D.C. GUPTA[2], R.P. TRIPATHI[3] AND AMITA TRIPATHI[3]

[1]Department of Geology, J.N.V. University, Jodhpur 342 005
[2]Department of Geology, Barkatullah University, Bhopal 462 026
[3]Department of Physics, J.N.V. University, Jodhpur 342 005
Email : klsgeology@yahoo.co.in

Abstract

A polymetallic polynodule, an uncommon form, provided for the study from unknown location of the Indian Ocean Nodular Field. The paper is aim to study its morphology, ore petrography and ^{57}Fe Mössbauer Spectroscopy to propose its genetic model in the light of information available for polymetallic nodules and ferro-manganese crusts.

Morphologically, the polynodule is almost circular with 5cm diameter, consisting of one major and few small nodules, named here as host and guest nodules, respectively, adhering firmly on the top of 'host' component. Under the ore microscope, both, the 'guests' and the host reveal concentric growth lines and independent nucleus. The colloform bright grey layers of goethite are alternative with the dark grey of manganese oxide minerals almost regularly. The host nodule also shows highly botryoidal pattern. ^{57}Fe Mössbauer spectroscopy performed at 300K, conform the presence of hydrated iron oxide minerals.

Being larger the host nodule is expected to have been buried rapidly leaving little time to scavenge iron then guests, which have more iron minerals. Mn is high in host which might have been supplied from the pore water to the underlying sediments through diagenesis although originally mainly comes from hydrothermal vent. Halmyrolitic process from the decomposition of basalt debris by sea water may have contributed some other metals. The main process of formation of both, host and guest nodule is hydrogenous by slow precipitation. The occurrence of botryoidal pattern is suggesting that there is possible meditation by microbial processes in the genesis.

Sea mounts near the active volcanoes adjoining oceanic crust, are the favourable sites for the polymetallic polynodules to grow. The 'guest' nodules which must have grown at a later date over the 'host' nodule likely to have their parental material supplied by the young volcanoes. This geological set-up producing a fertile ground for the polynodules to grow. Hence, more polynodules are expected in the Indian Ocean Nodule Field.

***Key words**:* Polymetallic Nodule, Polymetallic Polynodule, Indian Ocean Nodular Field, Mössbauer Spectoscopy.

INTRODUCTION

Polymetallic nodules, also called manganese nodules are rock concretions formed of concentric layers of iron and manganese hydroxides around a core. The core may be microscopically small and is sometimes completely transformed into manganese minerals by crystallization (Roonwal, 1986). When visible to the naked eye, it can be small test (shell) of microfossil radiolarian or foraminifer, a phosphatized tooth of shark, basalt debris or even fragment of earlier nodules (Roonwal, 1986). The thickness and regularity of the concentric layers are determined by the successive stages of growth. On some nodules they are discontinuous, with noticeable differences between the two sides. Nodules vary in size from tiny particles visible only under a microscope (called, micronodules) to large pellets more than 20 centimetres across. However most nodules are between 3 and 5 cm in diameter, about the size of potatoes. Their surface is generally smooth, sometimes rough, mammilated (knobby) or otherwise irregular (Roonwal and Friedrich, 1981). The nodules lie on the sea bottom sediment, generally half buried. In that case, the bottom part buried in sediment, is generally rougher than the top. Some nodules are completely covered by sediment. (Pattan and Parthiban, 2006).

Polymetallic nodules were discovered at the end of the 19th century in the Kara Sea, in the Arctic Ocean off Siberia (1868). During the scientific expedition of the H.M.S. Challenger (187276), they were found to occur in most of the oceans of the world.

The Central Indian Ocean Basin (CIOB) hosts the world's second largest and high grade manganese nodule deposit, after the central belt of the Pacific Ocean (Roonwal, 1985). The crust below the manganese nodule field was formed between 60 and 49

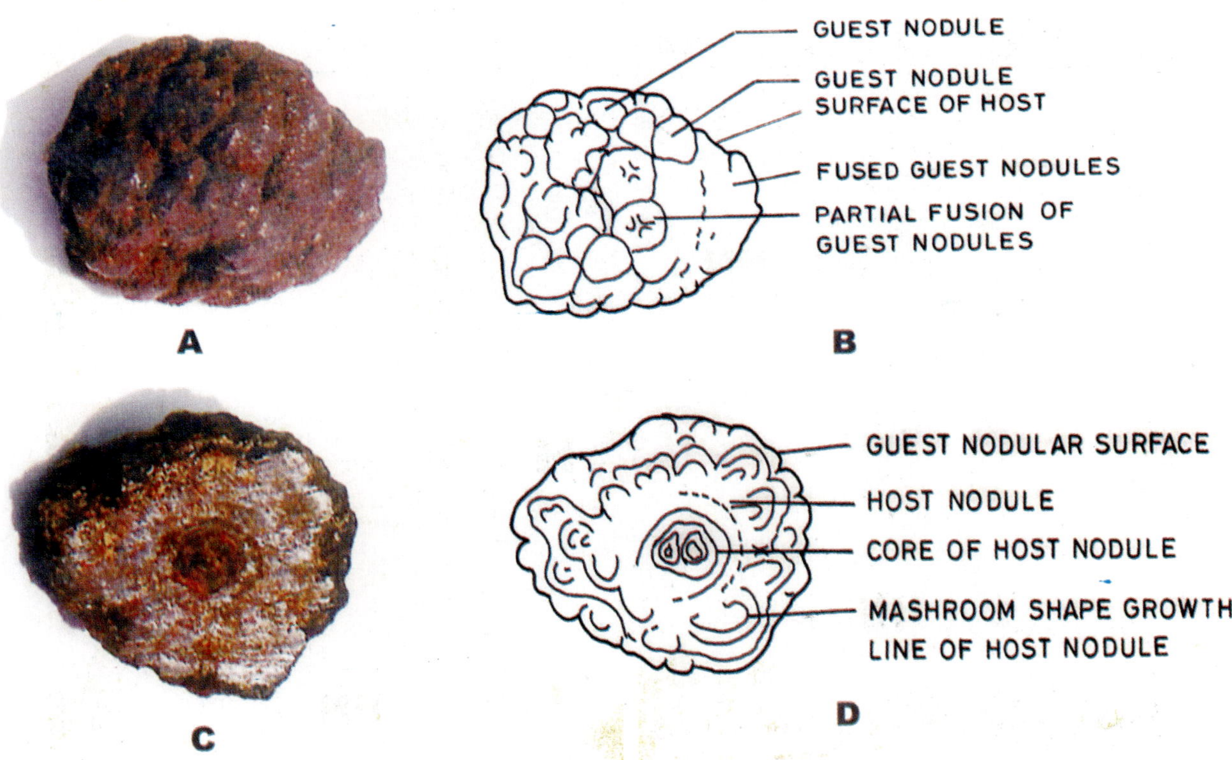

Fig. 1. A and B Top surface of the polymetallic polynodule showing some guest nodules. Few of the guest nodules are showing coalescence. **C and D :** Cross- section of polymetallic ploynodule displaying cores in both the host nodule and the guest nodules.

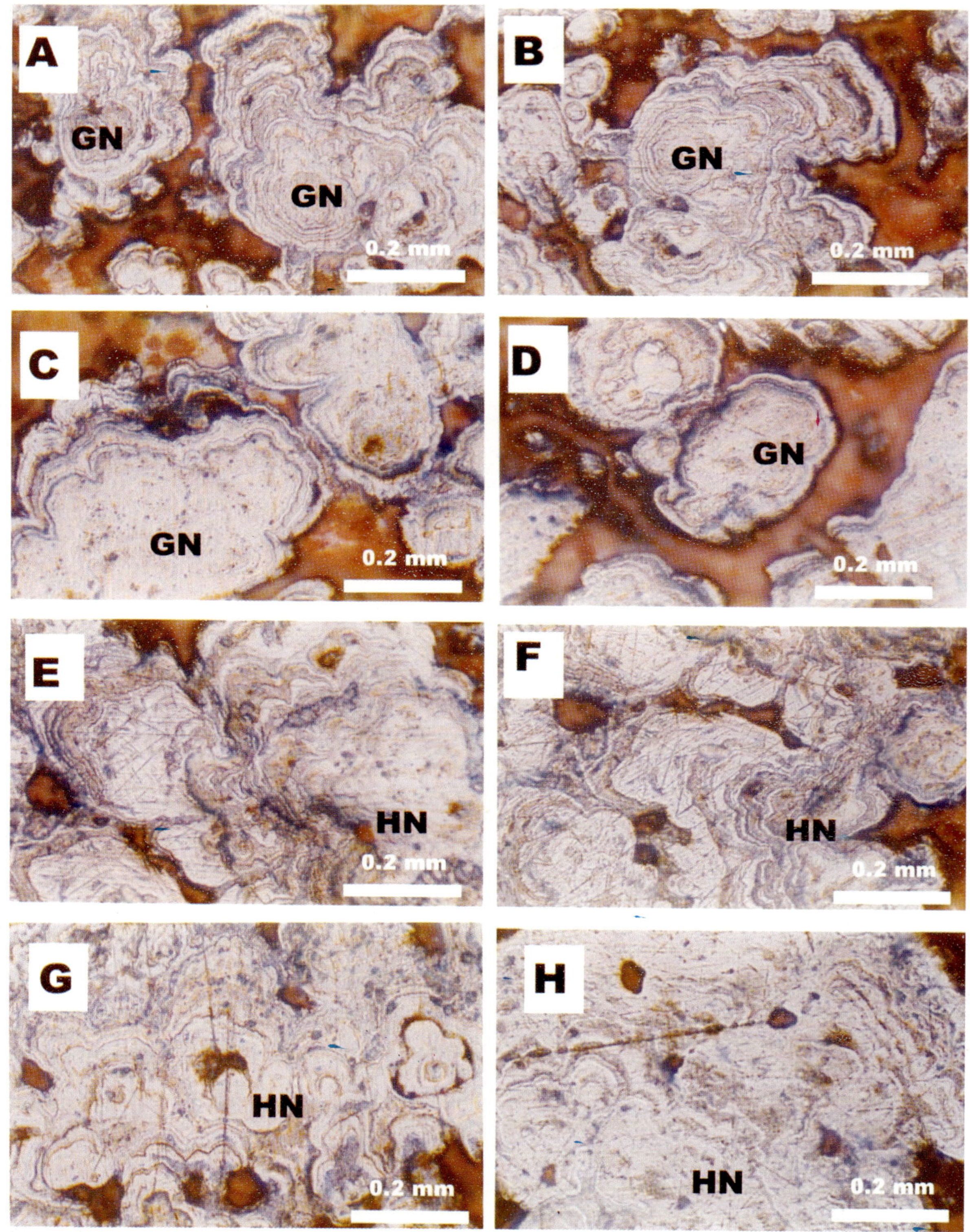

Fig. 2. Polished sections of polymetallic polynodule consisting of host nodule (HN) and guest nodules(GN), under the ore microscope. Displaying concentric growth lines. Bright grey goethite is distinguished from dark grey manganese oxide minerals. Guest nodules (A to D) showing more iron minerals less Mn minerals, and poor botryoidal pattern in comparison with the host nodule (E to H), which is revealing presence of more manganese mineral than iron and highly grown botryoidal pattern.

Ma under fast, medium and slow spreading rates (Naquvi, 2005). More than 0.4 million km^2 of CIOB have been surveyed for the Manganese nodules. The survey also yielded information about the morphotectonic features of the Central Indian Ocean Basin, such as sea mounts, hills, ridges, normal and parallel lineaments. These features have distributed the topographic gradient (1:7000) of the sea floor. Spreading rate variation and formation of new oceanic crust during more than one episodes of India Eurasia collision have influenced the Indian Ocean Nodule Field (IONF) which is divided into four zones from north to south on the basis of nature of the ridge parallel lineations. These zones are related to crustal accretion, rate of crust formation and crustal flexure. The IONF is characterized by both linear and isolated sea mounts (Mukhopadhyay et al., 2002; Naqvi, 2005). We have been provided the polynodule for the study from unknown location of the Indian Ocean Nodule Field. The objective of the present paper is report the polynodule, an uncommon form, from Indian Ocean Nodule Field and study its unusual morphology, Ore-microscopy and Mössbauer spectroscopy to propose the genetic model.

THE POLYNODULE

The polymetallic polynodule is distinctly showing combination of two morphologies namely, the 'host' nodule and the 'guest' nodules (Fig. 1). The main host nodule is oval in shape, a modification of spheroidal shape with 5 cm long axes in the two directions and 6 cm long axis in the third elongated direction. The 'host' nodule is further modified by having one flat surface (bottom), indicating that it must have rested over it. Opposite to the flat bottom, the top surface is most uneven and consist of botryoidal surface texture. Details show that these are many nodules smaller in size, named here as guest nodules; the second component of the polynodule. Each of the 'guest' nodules is again spheroidal (or tending oval) in shape with their radius ranging from 0.5 to 1.0 cm. Each of the 'guest' nodule have its independent identity, although now adhere firmly with the top of the host nodule. There is olive grey clay and dark brown unconsolidated powdery material partially covering surface of both, the host and the guest nodules. These possibly are part of the deep marine sediments. Some of the large guest nodules appear to have formed as a result of the coalescence of two or three smaller ones. (Fig. 1, A and B). Cross section of the polymetallic polynodule is demonstrating a core, consisting of dark brown mineral and many concentric growth lines. Some of the growth lines have unusually grown, forming a mashroom shape in the host nodule. (Fig. 1, C and D). Cross section clearly showing that the guest nodules have their independent identity alongwith their own core. Otherwise, they are genetically unrelated to the host nodule.

ORE-PETROGRAPHY

Polished sections of both the host nodule and the guest nodules were studied under an oremicroscope. The guest nodules (Fig. 2, AD) and the host nodule (Fig. 2, EH) reveal concentric growth lines. Each layer is composed of a dense and sparse sublayer. The thickness of layers varying greatly. There is no clear difference in morphology, mineralogy, composition and growth rates between the guest and host components of the polynodule which makes direct comparison between the two difficult. Nonetheless, following distinctions were observed :

1. Ore microscopy is better demonstrating that the concentric growth lines developed in each of the guest nodules is independent to the host nodule. Each of the guest nodules have their own core. Hence the polynodule is, in fact, an assemblage of some nodules. Occasionally spheroidal spore like structures can be seen in between the spaces of the filaments. No colonial type structures or distinct morphologies leading to identification of microbes were observed. These structures are present only in the host nodule and are totally absent in the guest nodules.

2. Presence of colloform grey or white grey goethite (FeO. OH) is distinct being relatively brighter than dark grey manganese oxides of complex mineralogy (Fig. 2). Among principal manganese oxide minerals, most expected one is 'birnessite'. There could also be 'todorokite' as reported in some nodules. Manganosite if present could only be minor, being unstable.

   Manganosite (MnO-MnO$_{1-15}$) typically occurs as green to black in low temperature hydrothermal deposits (Zhang et al. 1997). It is thermodynamically unstable in marine environment. Its formation may be bacterially mediated in a manner analogous to the formation of magnetite (Fe_3O_4) in magnetotectic bacteria (Stolz, 1992) or of ferrihydrite (FeO. OH) in ferritin micelles of bacteria or fungi (Ghiorse and Ehrlich, 1992) or on bacterial surfaces (Fortin et al., 1998).

3. The composition of both, the guest nodules and the host is heterogenous even within an individual micro layer. Iron rich layers are brighter than dark

Table-1 Mössbauer Parameter of host and guest components of Polymetallic Polynodules

Sample	IS (mms^{-1})	QS (mms^{-1})	LW (mms^{-1})	AREA (%)	χ^2	Assigned mineral
Guest Nodule	0.32 0.29	0.62 0.82	0.46 0.58	80.92 19.08	1.2	Hydrated iron mineral
Guest Nodule	0.29 0.35	0.58 0.97	0.45 0.51	78.94 21.06	1.03	– do –
Host Nodule (Core)	0.25 0.25	0.57 1.5	0.47 0.35	94.36 5.64	1.1	– do –
Host Nodule (General)	0.31 0.46	0.67 0.8	0.49 0.49	93.24 6.76	1.03	– do –

grey Mn rich layers. Visual observation under Ore microscope show that Mn/Fe ratios of these layers are in the range of 312 times, generally. No regular alternation of the Mn and Fe rich layers was observed neither in the guest nor in the host components of the polynodule.

4. Zhang et al., (2002) have consider the high growth rates at host nodule (34 order of magnitude higher than those of the guest nodules) being shallow at the water-sediment-interface. The high growth rate, to be an indirect argument in favour of their biogenic formation from purely in organic processes would not be expected to be significantly high to lead the observed growth rates (Glasby, 2000).

5. After etching the polished section in an ion coater, these layers were seen to consist of many very tiny floccules. Individual detrital particles displaying irregular features are scattered between the floccules. Occasionally, a pseudo-column with a margin surrounded by discrete detrital particles (clastic sediments) was observed. The occurrence of floccules and botryoidal growth patterns in the concretions indicates that the formation of the ferromanganese concretions is not the result of purely physico-chemical precipitation. The suspected micro structures allow suggesting that the host nodule is probably mediated by microbial processes.

Such suspected fossils consist of birnessite with minor amount of manganosite (MnO). The formation of the manganosite may be bacterially mediated. Bacterial processes are also known to play a role in the formation of the metastable manganosite phase. (Zhang, et al., 2000). In the present study, no further distinction could be made among oxides of iron and manganese due to the limitations of the technique used.

MöSSBAUER SPECTROSCOPIC STUDY

Mössbauer spectroscopy is a reliable method to characterize iron bearing minerals, non-destructively. Mössbauer spectroscopy has been used to understand many complex problems of geology. It has been used for mica (Tripathi et al., 1978), coal (Shrivastava et al., 1992), lignite (Tripathi and Shrivastava, 1990; Verma and Tripathi, 2000), meteorites (Shrivastava et al., 1997; Tripathi et al., 2000) and petroleum source rock (Morup et al., 1985; Tripathi et al., 1998; Tripathi et al., this volume).

Four samples; two from different guest nodules, and rest two from the core and remaining (general) parts of the host nodule; have been separated and grounded to fine powder of 53 micron size. For the ^{57}Fe Mössbauer spectroscopic study, the Mössbauer absorbers were prepared by sandwiching these powdered samples in between two paper discs of a 25 mm diameter in a sample holder. All, four spectra were recorded at room temperature (300 K) using constant acceleration spectrometer and ^{57}Co source in Rh matrix. The spectra were analysed by fitting an appropriate number of discrete doublets and sexlets with Lorentzian peaks using a least square fitting programme.

The Mössbauer parameters viz. Quadrupole splitting (QS), Isomer Shift (IS), Area and Chi square were obtained for all the four samples. Isomer shifts were determined with respect to centroid of standard alpha iron foil absorber. The Mössbauer parameters obtained are presented in Table 1 and the spectra in Figs. 3 and 4. The results have been interpreted by making a comparison of parameters obtained with the already available standard data. As the technique is limited to detect the iron minerals only if the spectra are recorded at 300 K, no mineral other then iron minerals

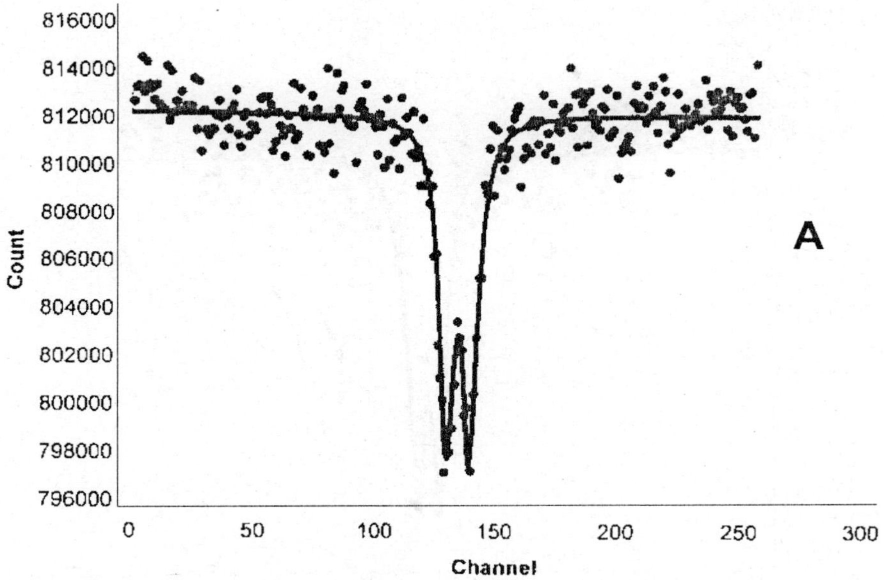

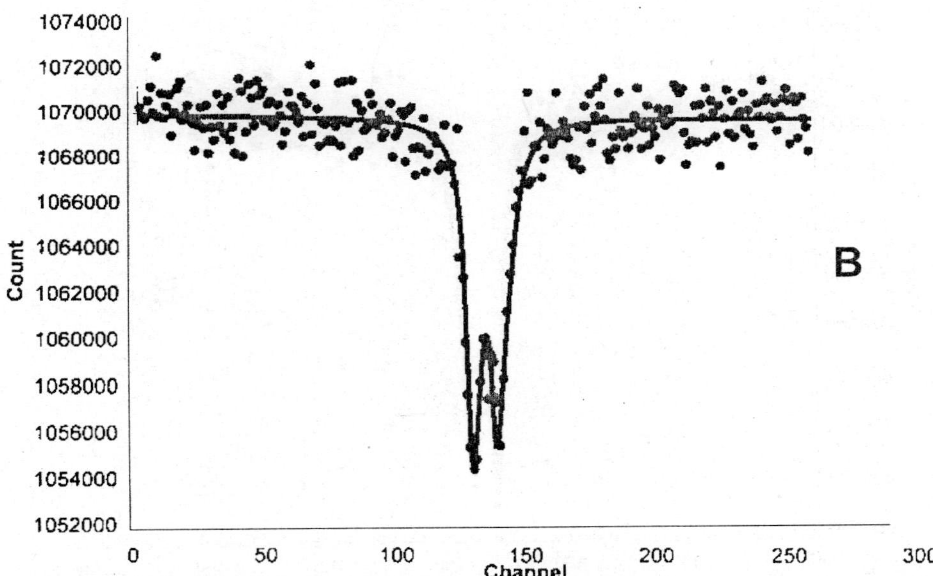

Fig.3.(A and B) : ^{57}Fe Mössbauer Spectra of two separate guest nodules of Polymetallic Polynodule at 300 K.

could be detected in the present study. All the four samples showing presence of similar hydrated iron oxide mineral.

GENESIS OF POLYMETALLIC POLYNODULE

Polymetallic nodules and ferromanganese crusts are the two major marine mineral resources available at sediment ocean water interface. Polynodule is third, uncommon such resource. In the light of genetic models available for polymetallic nodules and ferromanganese crusts, attempt has been made here to suggest a genetic model for the third type, namely polymetallic polynodule.

There are many views which have been proposed for the formation of polymetallic nodules. First theory

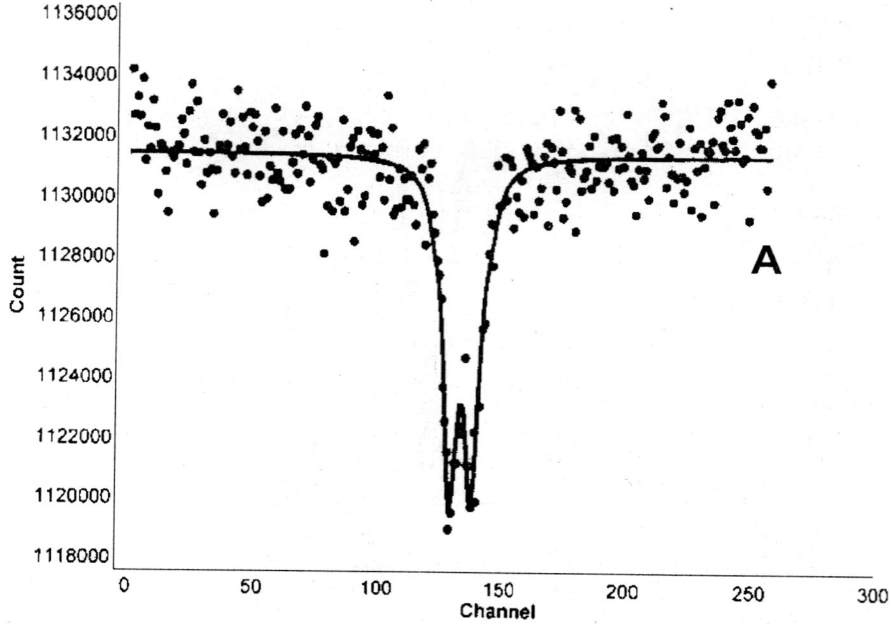

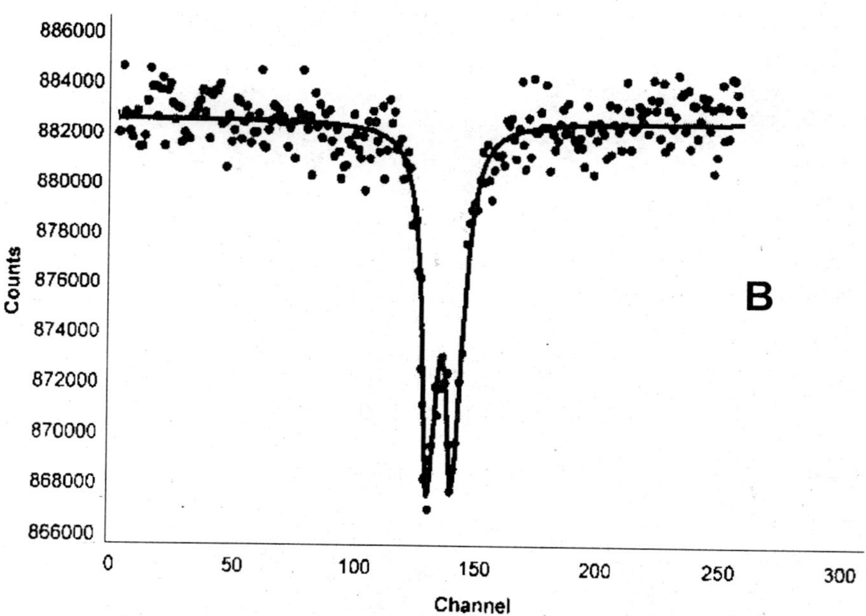

Fig. 4. ^{57}Fe Mössbauer Spectra of core of host nodule at 300 K (A) host nodule in general (B) Polymetallic Polynodule

out of two most accepted ones is a hydrogeneous process in which concretions are formed by slow precipitation of the metallic components from sea water (Barnes and Dymond, 1967). A diagenetic process is the second theory, in which the manganese is remobilized in the sediment column and precipitates at the sediment-water-interface. It is thought, in the first theory, to produce nodules with similar iron and manganese content and a relatively high grade of nickel, copper and cobalt. The nodules formed by the second theory are characteristically rich in manganese but poor in iron, nickel, copper and cobalt.

Apart from these two theories there are atleast three more mechanisms suggested in the literature for the formation of nodules. First one is a hydrothermal process in which the metals derived from the hot springs which are normally associated with the volcanic actively. Halmyrolitic process, in which the metallic components come from the decomposition of basaltic debris by sea water, is the second mechanism. The third mechanism is a biogenic process in which the activity of microorganism catalyzes the precipitation of metal hydroxides.

In fact, most of these processes, may operate concurrently or may follow one another during the formation of a nodule. Whichever of them is operative in particular cases, a number of common factors have been established.

(i) Nodule formation requires a low rate of sedimentation or some process for removing sediment before it accumulates. This enables concretions to grow before they are buried, when they would be cut off from the conditions that allow them to develop.

(ii) Plankton concentrates trace elements such as copper and nickel. The organic matter that falls to the sea bottom when these organisms die is a probable source of the metals incorporated into the nodules.

(iii) Manganese in sea water comes mainly from hydrothermal vents (black smoker), where it is leached out of the under lying basalt as the superheated fluids percolate upwards through the oceanic crust.

(iv) Concretion process is further accelerated by the activity of micro organisms.

Nodule growth is one of the slowest of all geological phenomena in the order of a centimetre over several million years (Barnes and Dymond,1967). The age of Pacific Ocean nodules is therefore between 2 to 3 million years. However, rapid formation of ferromanganese crust has been reported near ships wrecked during the First World War. Such fast growth has implications for the origin of the components of nodule and the way they come together. If nodules are formed slowly, hydrogenous or diagenetic processes might be at work. If they are formed quickly, metal sources other than sea water or sediment seem to be the probable cause. In latter case, hydrothermal or even halmyrolitic processes are operative.

The ferromanganese crusts are found in all ocean basins that contain hard, sediment free rocks. Crusts occur on sea mounts, ridges and plateau where currents have kept the rocks swept clean of sediments for millions of years. In the areas where ferromanganese crust formation occurs in part, as a result of hydrothermal precipitation associated hydrothermal activity tends to dilute the economicaly interesting elements that are sought from crusts.

The sea mounts obstruct the flow of oceanic water masses, thereby creating a wide array of sea mount generated currents of generally enhanced energy relative to flow away from the sea mount. The effects of these currents are the strongest at the outer rim of the summit region of the sea mount, the area where the thickest crusts are found. The sea mount generated currents could be traced for at least several hundred metres above their summits. Other water column features produced by the interaction of sea mounts and currents are density inversions, isotherm displacements, enhanced turbulent mixing and upwelling. The upwelling process moves cold nutrient rich waters to shallower depths. Up welling increases primary productivity, which in turn increases the size and magnitude of the Oxygen Minimum Zone (OMZ) making the sea mounts a potential zone. In anoxic conditions, where sea water has no oxygen, crust would never form. The metals in crust precipitate under cold ambient sea water. Metals in crust are not floating around in sea water as pure metal but are dissolved in sea water as a complex compounds like chlorides or hydroxides.

The occurrence of botryoidal growth pattern in the concretions indicates that the formation of the ferromanganese concretions is not the result of purely physiochemical precipitation (Zang et al., 2002). The microstructures in the concretions as revealed in ore microscope allow suggesting that these concretions are probably mediated by microbial processes. Recent studies describing the bacterial mediation of manganese and iron oxidation have been presented by Ghiorse and Ehrlich (1992), Ehrlich (1996 a, b, 1998), Tebo et al., (1997), Konhauser (1998), Bargar et al. (2000) and Brouwers et al., (2000).

There are clear differences in the morphology, mineralogy, composition and growth rate between shallowmarine ferromanganese concretions and deep sea manganese nodules which makes direct comparison between the two difficult (Glasby, 2000). The high growth rate of the shallow water concretions (34 orders of magnitude higher than those of deepsea nodules) to be an indirect argument in favour of their biogenic formation since their rate of formation from purely inorganic processes would not be expected to be sufficiently high to lead to the observed growth rates (cf. Glasby, 2000).

Following the discussion, regarding genesis of polymetallic nodules and ferromanganese crusts, the genesis of the polynodule can possibly be attempt here giving due consideration to its both the components, guest nodules and the host nodule.

Pattan and Parthiban (2006), has clearly demonstrated that in the CIOB, larger buried nodules occur in upper part and smaller size nodules in the lower part of the sediments in the ocean. The host nodule of the polymetallic polynodule might have been buried rapidly and thus exposed for a short time to scavenge Fe from the sea water or might have been diluted by detrital material. On the other hand, the guest nodules of the polymetallic polynodule appear to have been exposed for a longer time to sea water resulting in higher concentration of Fe. Host nodule have higher Mn content than guest nodules, which might have been supplied from the pore water to the underlying sediments through diagenesis.

Also, the genesis of the polynodule should be viewed in terms of geology of the ocean basin. In Central Indian Ocean Basin (CIOB), the distribution, morphology and growth patterns of the sea mounts are related to the spreading rate varying between 5.5 cm/yr and 2.6 cm/yr, suggesting their formation at the ridge crust. Many of the sea mounts show more than one stage with local intraplate eruptions contributing to the growth of the earlier formed sea mounts. These sea mounts and also the adjoining oceanic crust consist of basalts, spilites, ferrobasalts and pumice. The addition of the younger volcanic rocks at the base of sea mounts, in consistent with the growth of volcanoes, eruption of ferrobasalts and presence of volcanogenic hydrothermal material as young as 10 Ka age have significantly contributed to the polynodule formation. It seems that the guest nodules have grown over the host nodule at a later date, possibly have more contribution from the younger volcanic activity. As IONF is geodynamically and volcanically unstable in the recent past since 8 Ma (Mukhopadhyay et al., 2002; Iyer et al., 1999), it seems a fertile ground for the guest nodules to grow, hence more polynodules are expected only in this region.

ACKNOWLEDGMENTS

Authors want to express their sincere thanks to the hon'ble reviewers for prompt and critical review. One of us, KLS, is thankful to DST for providing special financial assistance to the Department under SAP.

References

Bargar, J.R., Tebo, B.M., Villinski, J.E.; 2000 : In situ characterization of Mn(II) oxidation by spores of the marine Bacillus sp. Strain SG1. Geochem. Cosmochem. Acta. Vol. 64. pp. 2775-2778.

Barnes, S.S., Dymond, J.R. ;1967: Rate of accumulation of ferromanganese nodules. Nature, Vol. 213, pp. 1218-1219.

Brouwers, G.J., Vijgenboom, E., Corstjens, P.L. A.M. de Vrind, J.P.M., de Vrindde Jong, E.W.; 2000 : Bacterial Mn2+ oxidizing systems and multicopper oxidases : an overview of mechanism and functions. Geomicrobiol. Jour. Vol. 17. pp. 124.

Fortin, D., Ferris, F.G., Scott, S.D.; 1998 : Formation of Fesilicates and Feoxides on bacterial surfaces in samples collected near hydrothermal vents on the Southern Explorer Ridge in the northeast Pacific Ocean. Am. Min. Vol. 83, pp. 1399-1408.

Ghiorse, W.C., Ehrlich, H.L.; 1992 : Microbial mineralization of iron and manganese. Catena Suppl. Vol. 21. pp. 75-99.

Glasby, G.P. and Read. A.J.; 1976 : Deep sea manganese nodules in K.H. Wolf (Ed.) Handbook of Strata Bound and Stratification ore deposits. Elsevier. Amsterdam. pp. 295-340.

Glasby, G.P.; 2000 : Manganese : Predominent role of Nodules and crust in Schulz, HD and Zabel M (Eds.); Marine Geochemistry. pp. 335-372 SpringerVerlag. Heidelberg.

Graham, J.W. and Cooper, S.C.; 1959 : Biological origin of manganese rich deposits on the sea floor. Nature Vol. 183. pp. 1050-1051.

Iyer, S.D., Gupta, S.M., Charan, S.N. and Mills, O.P.; 1999 : Volcanogenichydrothermal iron rich materials from the southern part of the Central Indian Ocean Basin. Mar. Geol. Vol. 158. pp. 15-25.

Konhauser, K.O.; 1998 : Diversity of bacterial iron mineralization. Earth Sci. Rev. Vol. 43. pp. 91-121.

Morup, S., Frank, J., Wonterghem, J., Poulsen, R.H. and Larsen, L.; 1985 : Mössbauer spectroscopy study of the chemical state of iron in Danish Mesozoic sediments. Fuel Vol. 64. pp. 528-539.

Mukhopadhyay, R., Iyer, S.D. and Ghosh, A.K.; 2002 : The Indian ocean nodule field. Petrotectonic and ferromanganese deposits. Earth Sci. Rev. Vol. 60 pp. 67-130.

Naqvi, S.M.; 2005 : Geology and evolution of the Indian Plate (from Hadean to Holocean 4 Ga to 4 Ka). Capital Pub. Comp., New Delhi. 450 p.

Pattan, J.N. and Parthiban, G.; 2006 : Composition and origin of buried ferromanganese nodules from Central Indian Ocean basin. Jour. Geol. Soc. Ind. Vol. 68. pp. 605-609.

Roonwal, G.S. ;1985: Mangenese nodules in the the Indian Ocean : Prospects and challenge; in 'The Ocean- realities and prospects.' (Ed. R. C. Sharma). Rajesh . New Delhi pp. 83-90.

Roonwal, G.S. ;1986: The Indian Ocean : Exploitable mineral and petroleum resources. Springer-Verlag. Berlin Heidelberg Germany. 198p.

Roonwal, G.S., Friedrich, G.H. ;1981: Surface characters and growth patternin some manganese nodules from central Pacific. Ind. Jour. Mar. Sci., Vol. 10, pp. 57-60.

Shrivastava, K.L., Tripathi, R.P. and Jangid, M.L.; 1992 : Spontaneous combustion of coal : A Mössbauer approach. Fuel. Vol. 71 No. 4 pp. 377-380.

Shrivastava, K.L., Tripathi, R.P. and Vaya, V.K.; 1997 : A Mössbauer spectroscopic study of the Pipliyal Meteorite. Curr. Sci., Vol. 72 No. 2 pp. 139-141.

Stolz, J.F., 1992 : Magnetotactic bacteria : biomineralization, ecology, sediment magnetism, environmental indicator, Catena Suppl. Vol. 21, pp. 133-145.

Tebo, B.M., Ghiorse, W.C., Van Waosbergen, L.G., Siering, P.L. and Caspi, R.; 1997 : Bacterially mediated mineral formation : insights into Manganese(II) oxidation from molecular genetic and biochemical studies in Banfield J.F. and Nealson K.H. (Eds.) : Geomicrobiology : interactions between Microbes and minerals. Review Mineral. Vol. 35. pp. 225-266.

Tripathi, R.P. and Shrivastava, K.L.; 1990 : Characterization of Merta Road lignite and decompositional study of its pyrite using Mössbauer spectroscopy. Fuel. Vol. 69 pp. 928-930.

Tripathi, R.P., Chandra, U., Chandra, R. and Lokanathan, S.; 1978 : A Mössbauer study of the effect of heating Biotite, Phlogopite and Vermiculite. Jour. Inog. Nucl. Chem. Vol. 40. pp. 1293-1298.

Tripathi, R.P., Sharma, S.K., Patel, K.R., Shrivastava, K.L. and Sahi Ram; 1998 : A Mössbauer approach to hydrocarbon prospecting in Jaisalmer basin in Rajasthan, India. Ind. Jour. Petrol. Geol. Vol. 7 No. 1 pp. 47-58.

Tripathi, R.P., Sharma, S.K., Shrivastava, K.L. and Verma, H.C.; 2000 : Mössbauer spectroscopic studies of the Piplia Kalan (eucrite) and Lohawat (howardite) meteorites. Meteor. Planet. Sci., (USA). Vol. 35 pp. 201-204.

Usui, A. and Ito, T.; 1994 : Fossil manganese deposits buried within DSDP/ODP cores, Legs 1126 Mar. Geol. Vol. 119. pp. 111-136.

Verma, H.C., and Tripathi, R.P.; 2000 : Characterization of iron bearing minerals in Giral lignite from Rajasthan and study of their decomposition during combustion using 57Fe Mössbauer spectroscopy. Fuel. Vol. 79. pp. 599-606.

Von Stackelberg, U.; 1984 : Significance of benthonic organisms for the growth and movement of manganese nodules, Equatorial North Pacific. GeoMar. Lett. Vol. 4. pp. 37-42.

Zhang, F.S., Bian, LZ., Lin, CY., Zhao, XN., Hong, JM., Chen, JL., Shen, HT., and Han, XQ.; 1997 : Biomineralization in pelagic manganese nodules from the East Pacific Ocean. Proc. 30th Int. Geol. Cong. Vol. 9, pp. 169-176.

Zhang, F.S., Lin, C.Y., Bian, L.Z., Glasby, G.P. and Zhamoida, V.A.; 2002 : Possible evidence for the biogenic formation of spheroidal ferromanganese concretions from the eastern Gulf of Finland, the Baltic Sea. Baltica Vol. 15. pp. 23-29.

Hon'ble Reviewers

1. Prof. G.S. Roonwal
   Former Chairman
   (Deptt. of Geol., Delhi Univ.)
   C-520, SFS, Flats
   Sheikh Sarai
   NEW DELHI 110017

2. Prof. V.K. Vaya
   Former Head
   (Deptt. of Geol. JNV Univ.)
   L-7 Income Tax Colony
   Tonk Road
   JAIPUR 302001

Editorial Assistance

Ms. Veena Chaudhary

Economic Mineralization
K.L. Shrivastava (Editor)
2009. pp.264-273
Scientific Publishers (India), Jodhpur

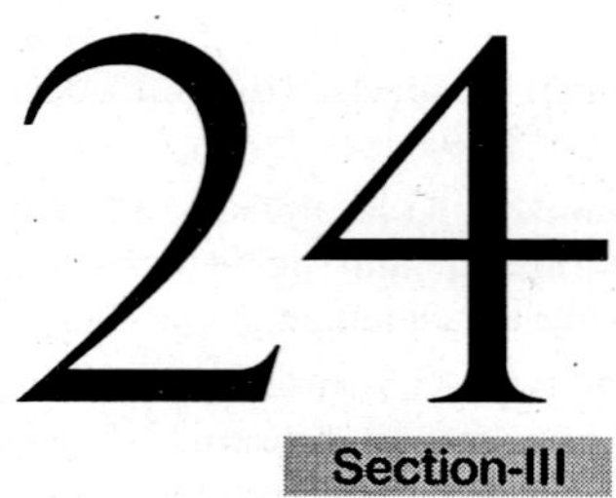

Section-III

Records of NeoPleistocene Glaciation, Desertification and Formation of Economic Salt Lakes-Playas in Western Rajasthan

V.P. LAUL

Retd. Director, Geological Survey of India, 10/805, Malviya Nagar, Jaipur 302017

Abstract

In recent past, glacial boulder and cobble-pebble spreads have been recorded as evidences of Neopleistocene (Late Pleistocene) glaciation in different parts of western Rajasthan. Records of Aeolian activities, pollen studies of lake sediments and paleoclimatic studies suggest that area witnessed large scale Aeolian accumulation during Late Pleistocene glacial period and mid-late Holocene period. Initially fresh water lakes were formed due to ponding of water during deglaciation, particularly in semi-arid region while in arid region playas/ranns developed due to river segmentation. Water released during deglaciation (Late Pleistocene) and high rainfall of early Holocene significantly recharged ground water which dissolved evaporates of Marwar Supergroup (Upper Proterozoic Cambrian) and carried them to fresh water lakes and introduced salinity in them. In Jurassic terrain highly recharged saline ground water reached the surface and become hypersaline playas / ranns during dry periods.

Key words : NeoPleistocene Glaciation, Desertification, Salt Lakes-Playas.

INTRODUCTION

In western Rajasthan, the major aeolian activities during Neopleistocene and Holocene times suggest that old or Paleo-desert was formed during cold climate of last glacial period while younger or Neo-desert was formed during mid-late Holocene period. The fresh water lakes-playas developed due to ponding of water in depressions and some playas / ranns were created due to river segmentation. Lakes in semi-arid region received salinity from evaporates of Nagaur-Ganganagar evaporite basin while in Jurassic terrain highly recharged saline ground water introduced salinity in Khara-Kanod-Kharariwala ranns. The paper is aimed to produce evidences and discuss the formation of economic salt lakes-playas, Neopleistocene glaciation and desertification in western Rajasthan.

RECORDS OF GLACIATION

Western Rajasthan witnessed Neopleistocene glaciation as evidenced by the polymictic, striated, facted boulder, cobble-pebble spreads with erratics in different parts of western Rajasthan having source areas in Marwar sedimentary rocks, Malani Igneous Suite and Delhi metamorphites. They are heterogenous in composition comprising mainly granites, rhyolite, tuff, chert, dolomite, quartzite, slate and phyllite. Their association at places with Quaternary sediments suggests Pleistocene age.

Major boulder and cobble-pebble spreads include Bap boulder spread (Mukhopadhyay and Ghosh, 1976) and Hamira-Bhojasar cobble-pebble spread (Laul and Kumar, 1980 and 1999). The gravelly spreads in pediments occurring extensively near Pokaran, Bap, Phalodi, Kolayat and Chandan in west and Jayal in east are possibly of glacial and glacio-fluvial to fluvial origin. Increasing records of glaciation in lower latitudes and areas of lower altitudes support the new thinking that during the ice ages such areas of lower latitudes and altitudes have also witnessed glaciation. Major spreads of Bap Boulders and Hamira-Bhojasar are discussed here.

Bap Boulder Spread

This spread is observed from Khirwa (27°17'30" : 72°12'15") in the south west to Nokhra (27°38' : 72°39') in the northeast for a distance of 55 km with width 1.5 km to 10 km. This assorted assemblage of spread mainly include pebble, cobble and boulders of Siwana and Jalore granites, Malani rhyolite dolomite (Marwar Supergroup) and Precambrian Delhi Quartzite, slate and phyllite. At places erratics of Delhi quartzite have been observed. The distribution and provenance of Bap boulder spread is shown (Fig. 1). Occurrence of similar boulders and erratic are observed beyond Nokha in the northeast upto Kolayat (27°50'20" : 72°57'30") of Tertiary basin as shown in Fig. 1 (Mukhopadhyay and Ghosh, 1976). They also observed that Bap Boulders are associated with soil, loam and kankar horizons of Quaternary age. The regional field evidences suggest that Bap Boulders may be of post-Eocene and possibly of Pleistocene age (Fig. 1).

Hamira-Bhojasar Pebble Spread

This spread (Fig. 2) is located southwest of Bap Boulder spread and mainly consists of cobbles and

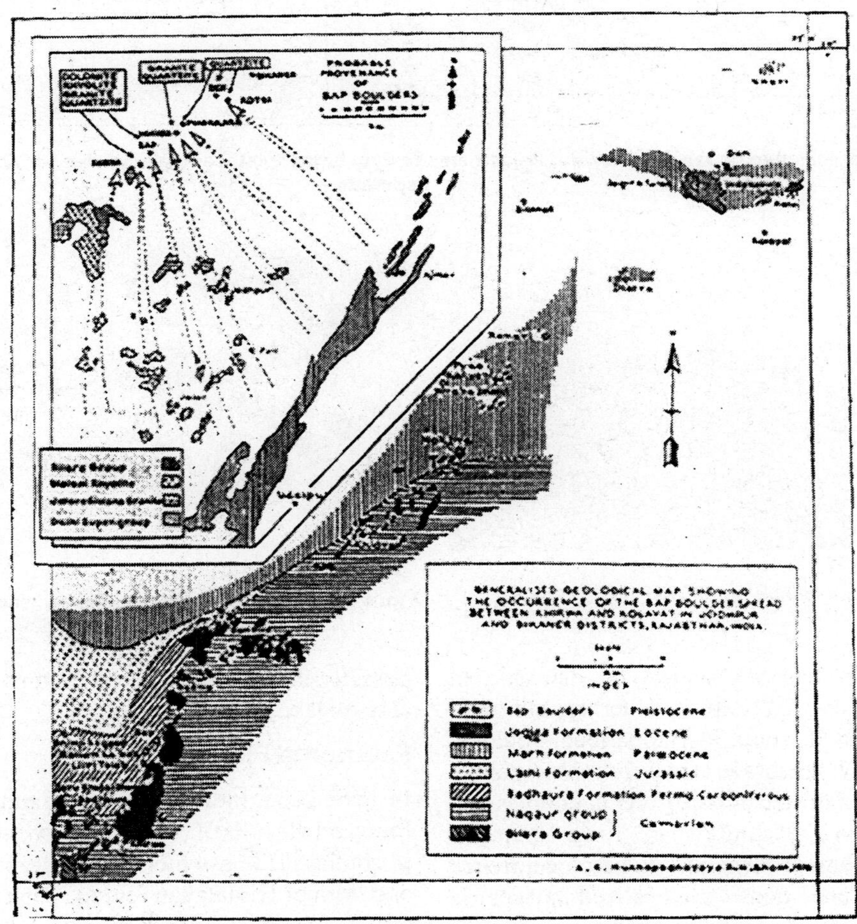

Fig. 1 **Generalized geological map showing the occurrence of Bap Boulder spread between Khirwa and Kolayat (Jodhpur and Bikaner districts) and their probable provenance. (after Mukhopadhyay and Ghosh, 1976)**

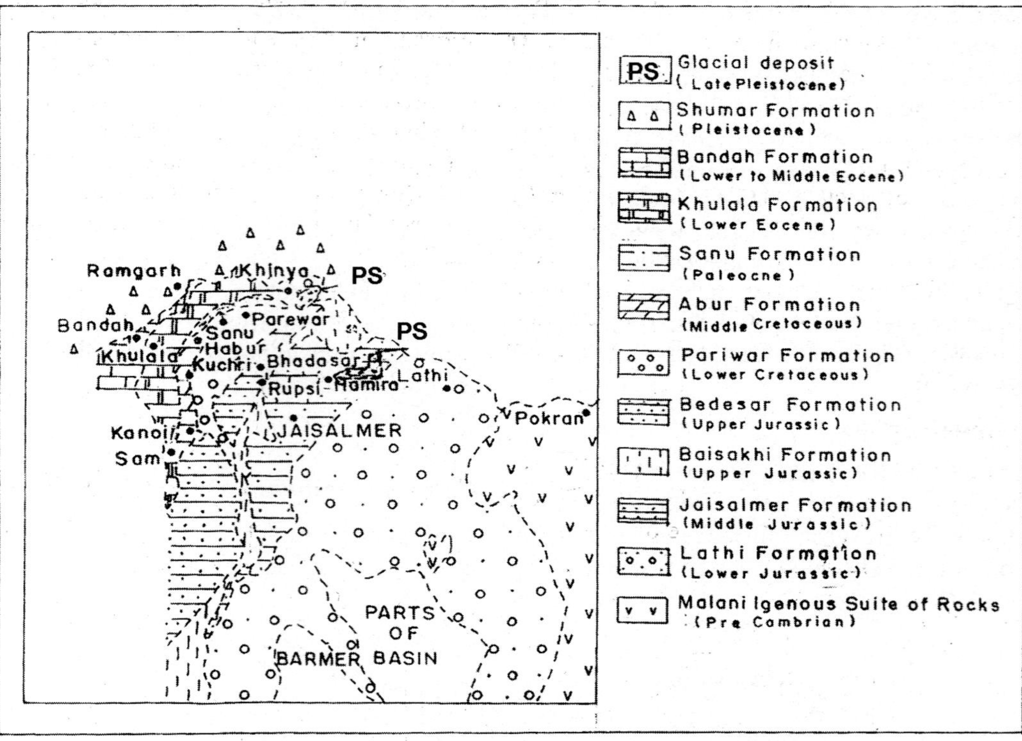

Fig. 2 Geological map of Jaisalmer basin, Rajasthan showing Late Pleistocene glacial deposit cobble pebble spreads

Fig. 3 Cobble-pebble spread at Hamira variation

Fig. 4 A collection from Hamira spread showing size of constituent cobbles-pebbles

pebbles (Fig. 3 and 4) which are similar in composition as that of Bap Boulder spread with addition of silicified wood. In Bap spread, boulders are common while this spread is mainly cobble-pebble spread. Reduction in size of constituents-cobbles and pebbles may be attributed to long distance transportation from provenances. Their occurrence over a Pleistocene beds at localized places is suggestive of possibly Late Pleistocene age.

Major depressions in Western Rajasthan, between Aravalli range and Jaisalmer low plateau region have possibly facilitated the localization of glacial spreads and salt lake and playas.

Fauna and Flora

In peninsular India temperate Fauna and Flora are found on the hills of Nilgiri and other isolated plateaus in southern India which are closely related to fauna and flora of Himalayan region. Such Flora also exists on Mt. Abu in the Aravalli range. Such fauna and flora have been discussed as evidence of ice age (Wadia, 1975).

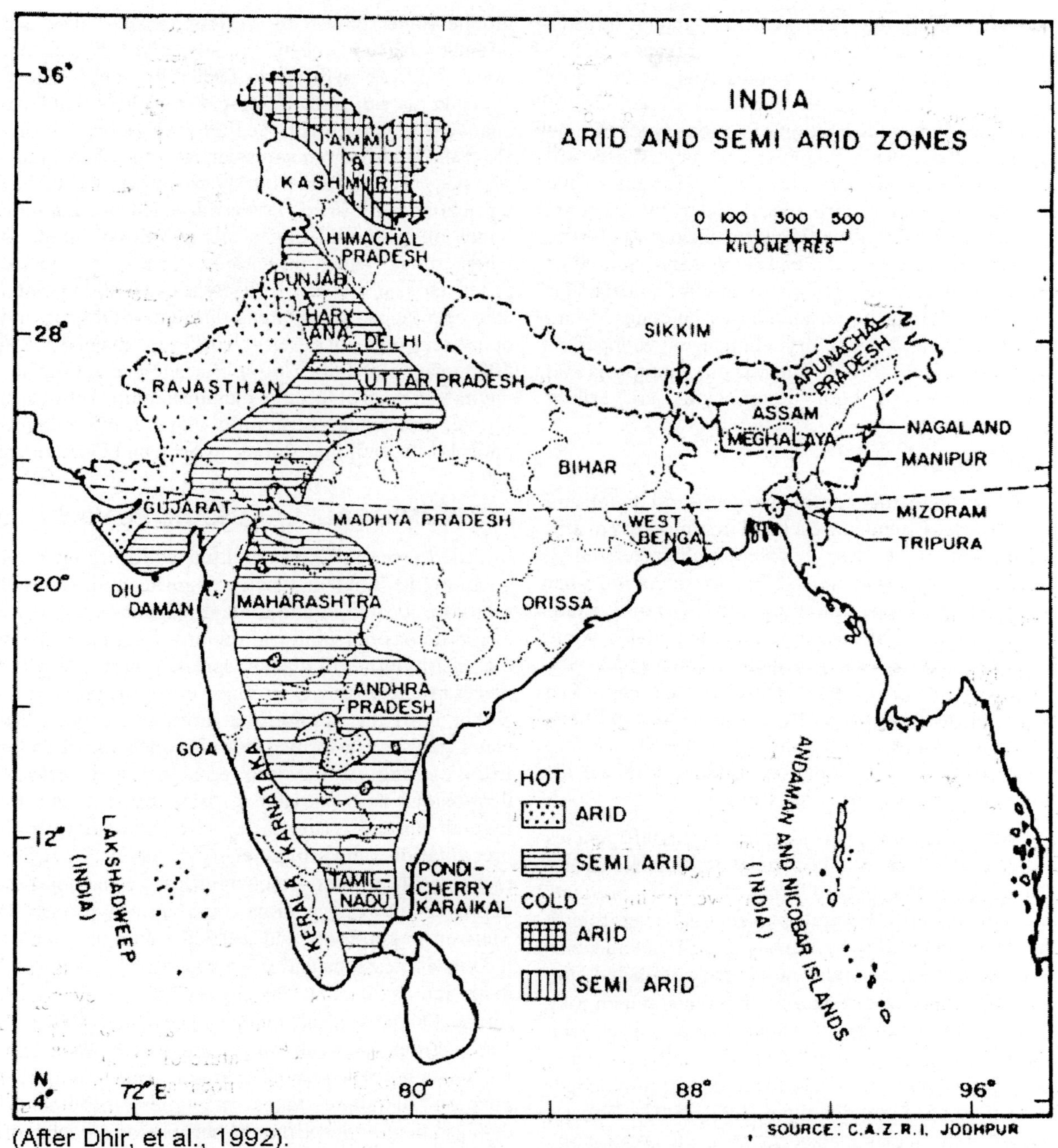

(After Dhir, et al., 1992).

Fig. 5 Arid and semi-arid zones of India (after Dhir, et al. 1992)

The reconstruction of the Late Quaternary plant life and its habitat indicates that vegetation first comprised an open steppe rich in grasses like Artemisia species. Species of Artemisia, Mimosa rubicaulis and Oldenlandia, now growing in area of comparatively high rainfall (>500 mm) appeared to have flourished in the presently drier 250-500 mm rain fall zone (Singh et al., 1974). High frequency of Cyperaceae was recorded in the earliest stage. *Mimosa rubicaulis* a species common in the Punjab plains and the western Himalayas reached its highest frequency during the period. The climate was cooler and wetter than at present (Dhir et al., 1992)..

DESERTIFICATION

The desert in general may be cold desert or hot desert.

In Western Rajasthan, the records of glacial spreads, ages of aeolian sediments, floral evidences, pollen studies and paleoclimatic studies suggest that area may possibly represent two major periods of desertification and formation of various types of sand dunes and intervening depressions, sand sheets and hard pan like desert features. Desertification in the western Rajasthan took place under two distinct climatic periods (i) desertification under cold climate of Neopleistocene period and (ii) hot dry climate of mid to late Holocene period. The desert formed under cold climate may be designated as Paleodesert and that formed under hot dry climate as younger or Neodesert (Laul, 2007). The various landforms and other features of both desert lands are briefly described here.

Paleo-desert

This desert is mainly confined to semi-arid region and possibly also extended west ward in some parts of arid region (Fig. 5). The area at present is characterized by Jayal gravel spread in the east, Bap Boulders and other spreads in the west, salt lakes playas / ranns, sand-dunes (mainly stablised) and inter dunal plains, some gravelly pavements and pediments, buried pediments. The preliminary results of thermolumine-scence (TL) dating of sand sequences suggest that the conditions optimum for the dune building in the Thar were reached only for short duration sometime after the last glacial maximum (Dhir et al., 1992).

During Holocene period, rainfall, fluvial agencies and aeolian activities have formed new landforms and also modified the old land forms, weathering products and their features. Rainfall played significant role in the stabilization of sand dunes while fluvial agencies produced fertile lands and also modified the characteristic glacial features of spreads which may appear like products of glacio-fluvial and fluvial agencies. Aeolian activities have modified the old sand dunes and at places sand dunes of younger generation have covered them (Laul, 2007). Mineral deposits (surface and subsurface) of the area mainly include lignite, gypsum, limestone, potash and salt deposits (Marwar Supergroup), clays, Quaternary salt and gypsite, and building stones.

Neo-desert

Hot desert/Neo-desert formed during mid-Holocene onward is mainly confined to arid region and at places encroached the semi-arid zone (Fig. 5). High rising sand dunes are frequently seen in Tertiary terrains of Rajasthan. The area is characterized by stabilized and mobile sand dunes of various shape, size and orientations, sand sheets, younger generation of dry ranns / playas gypsicrete, ferricrete, rocky desert, Hamada/plateau showing desert varnish, buried river channels covered with aeolian sediments. During Holocene marine transgression at Saurashtra coast, sea possibly invaded low lying areas in Tertiary terrains and left sand. Powerful N-E-ward blowing winds aligned sand dunes along NE-SW direction while other wind directions during the period produced sand dunes of different orientations. Sand also encroached other areas through wind-gaps and open areas. The hot desert (Neo-desert) may be 5000-4000 years old (Laul, 2007). Surface and subsurface natural deposits of the area include natural oil and gas, lignite, limestone, bentonitic clays, fuller's earth, phosphorite, building stones, gypsite and Quaternary salt.

ECONOMIC SALT LAKES AND PLAYAS

In desert areas salt lakes and playas form important geomorphic units and some significant lakes and ranns are located in and around Nagaur-Ganganagar evaporite basin, (semi-arid region). Keeping in view the distribution of glacial spreads, particularly at places in and around, Quaternary evaporite sequences, author is of the opinion that initially several fresh water lakes were formed due to collection of water from melting of ice into erosional and tectonic depressions and also due to ponding of water by glacial spreads (moraine). Subsequently high precipitation and high energy fluvial agencies recharged the ground water significantly and possibly dissolved thick evaporate (salt and gypsum) of Marwar Supergroup and carried them to suitably located depressions and fresh water lakes and turned them into saline/salt lakes which on evaporation formed Quaternary salt and gypsite, (Laul, 2005). This also justifies the fresh water nature of Pushkar lake where such saline water could not reach. High level position of some lakes may be attributed to displacement and upliftment because of neotectonic activity.

Geochemical factors played important role in the development of soda lakes. In Mesozoic-Tertiary terrains, the gypsum bands in Jurassic and Tertiary shales/bentonitic clays became sources for gypsite while brines drawn from highly saline ground water became hypersaline for salt formation during dry periods (Example Mohangarh gypsite and Kharariwala-Kanod-Khara saline ranns of Jaisalmer). The above observations and evidences indicate that

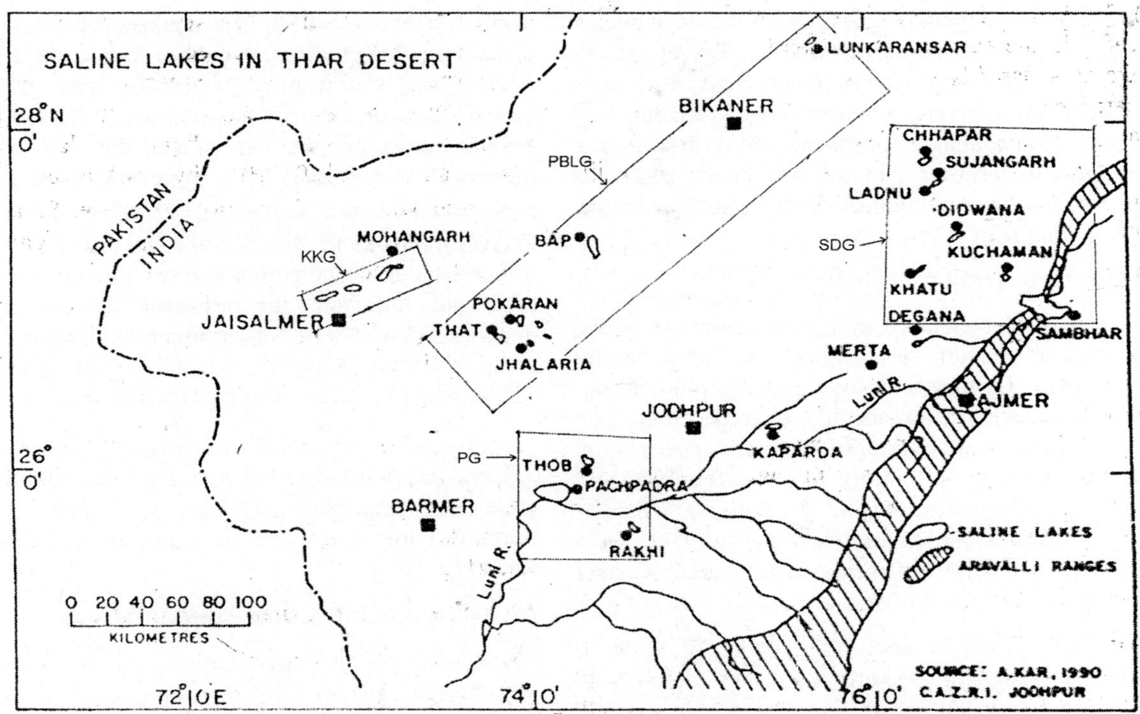

Fig. 6 Groups of saline lakes and playas in Rajasthan desert Sambhar Didwana Group (SDG), Pokaran Bap Lunkaransar Group (PBLG), Khara Kanod Group (KKG) and Pachpadra Group (PG). (after Kar, 1990 and modified by Dhir et al. 1992)

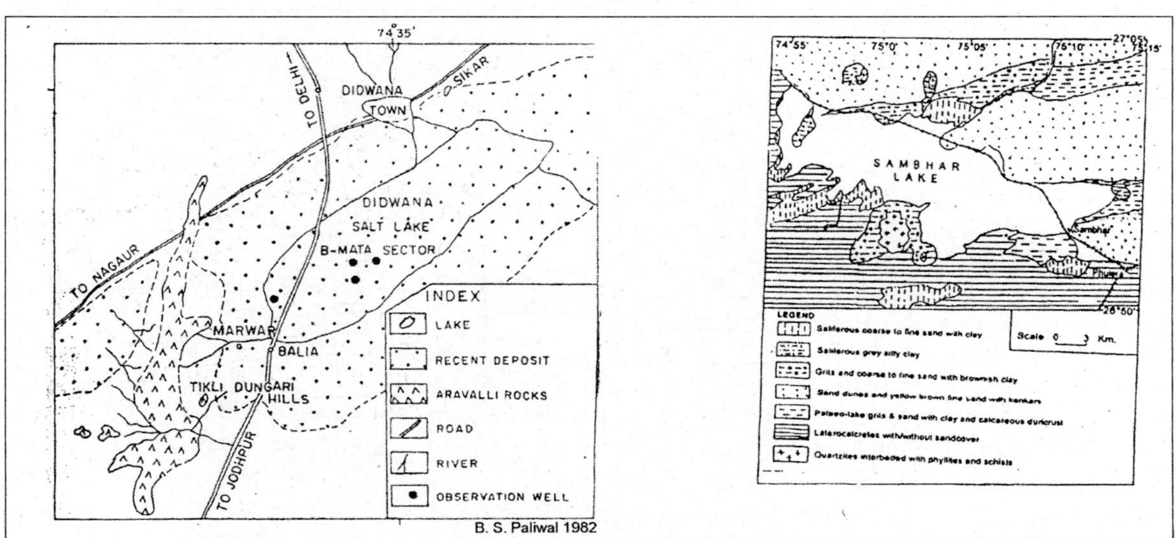

Fig.7 Geological Map of Didwana lake) (after B.S. Paliwal, 1982)

Fig.8 Geological map of Sambhar lake area. (After Yadav et al. 2007)

pre-Quaternary evaporate beds became sources for formation of gypsite and Quaternary salt and salt lakes/playas. With the discovery of huge salt and gypsum deposits (Marwar Supergroup) in recent past, the inland source explains better the formation of widespread deposits of gypsite and salt lakes/playas particularly in eastern part of Rajasthan desert. River segmentation also contributed in the formation of salt lakes/playas - Example Pachpadra Lake, (Laul, 2007).

In the opinion of author the salt lake-splayas/ranns may be broadly classified into groups (Fig. 6) such as

(i) Sambhar - Didwana group: salt lakes - playas located in the eastern part of semi - arid region (ii) Pokaran - Bap / Lunkaransar group : playas / ranns located near or in the vicinity of western limit of Nagaur - Ganganagar evaporite basin and in the vicinity of outcrop area of Marwar Supergroup (iii) Khara - Kanod group : ranns located in Jurassic terrain and (iv) Pachpadra Group.

(i) Sambhar-Didwana Group

These include salt lakes and playas formed in the eastern part of semi-arid region and have mainly developed in Quaternaries over metamorphic terrain. They are located in tectonically disturbed area. The area has good record of rainfall and activities of fluvial agencies. This group mainly includes Sambhar and Didwana Soda lakes, Khatu, Kuchaman, Chappar playas / ranns. Sambhar and Didwana salt lakes have been studied from time to time. The results of some studies are briefly described.

Paliwal (1982) related the source of salts of Didwana lake (Fig. 7) and salt lakes playas of adjoining area to salt domes associated with Aravalli rocks. Precambrian evaporates must have been pushed out at number of places during different phases of tectonic deformation. Based on the study of Late Quaternary sediments, minerals and inferred geochemical history of Didwana lake, Wasson et al. (1984) indicated that at about the last glacial maximum (Ca 18000 BP), hypersaline conditions prevailed with the deposition of halite. Sinha and Raymahashay (2004) studied the evaporite mineralogy and geochemical evolution of Sambhar lake and indicated the presence of non-marine evaporite minerals of Na-carbonate and Na-sulphate such as Trona ($NaHCO_3$ $NaCO_3$ $2H_2O$), Nahcolite ($NaHCO_3$), Mirabilite ($NaSO_4.10H_2O$) etc.

Non-marine evaporite minerals of Na-carbonate and Na-sulphate have also been reported from bittern crust (Sambhar lake) by Yadav et al. (2007). Such minerals suggest non-marine nature of Sambhar salt lake (Fig. 8).

(ii) Pokaran-Bap-Lunkaransar Group

These lakes are developed near or in the vicinity of western margin of Nagaur-Ganganagar evaporite basin and at places in vicinity of sandstone and

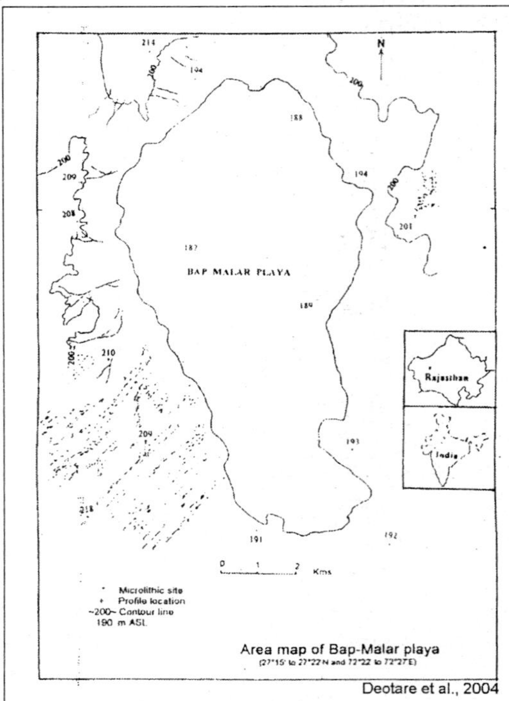

Fig.9 Area map of Bap-Malar playa (after, Deotare, et al. 2004)

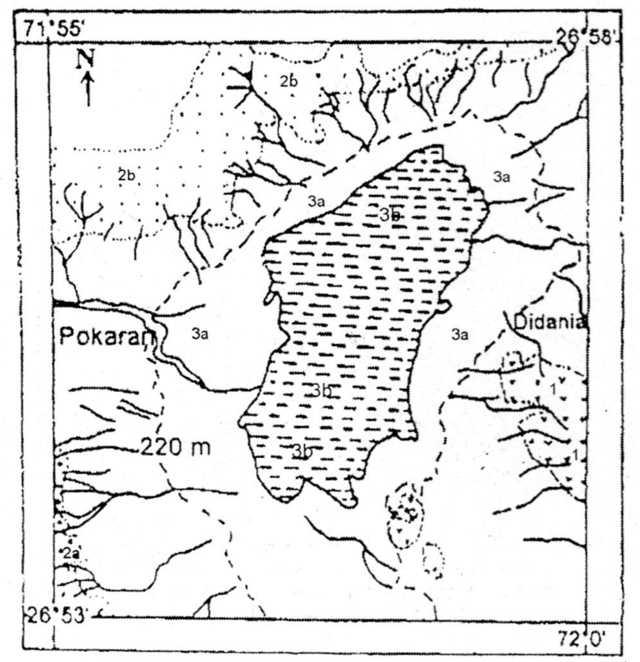

Fig.10 Geology and drainage around Pokaran playa (Rai, 1990) showing 1. Malani volcanics, 2. Marwar Supergroup-2a Conglomerate and 2b Sandstone, 3. Quaternary Formation, 3a Blown sand / fluvial deposits, 3b. Playa deposits.

limestone outcrops of Marwar Supergroup.

Palaeoenvironmental history of Bap Malar playa has been studied by Deotare et al (2004). AMS ^{14}C dates of >15 Ka on pollen from sediments layers indicate that the Bap Malar playa possibly existed even during the LGM. The playa was full of water during the early Holocene (8 Ka-5.5 Ka BP). In the opinion of author, this playa was formed due to ponding of water released by melting of glacial ice during last phase of Late Pleistocene. Bap glacial boulder spread facilitated the collection of water in depression. The early Holocene wet period maintained the water level of playa and highly recharged saline water from evaporites of adjoining Nagaur - Ganganagar basin got way into this playa and this playa became saline (Fig. 9).

Lunkaransar lake (Bikaner district) occurs near Nagaur sandstone (Marwar Supergroup). The brine contains 16% NaCl and 2.6 3% $NaSO_4$. (GSI, 2001). In the opinion of author, geochemical environment facilitated the formation of Na-sulphate.

Pokaran playa has developed over Marwar Supergroup of rocks and is located in vicinity of conglomerate and sandstone (MSG) and Malani volcanics. The streams draining the outcrop areas collect their water in this depression.

Isopach map of Rai (1990) shows a E-W linear depression which was part of Palaeochannel coming from the NNE. The development of saline depression probably took place when the river dried out during arid climate (Fig. 10). In the opinion of author, highly recharged saline ground water of Marwar Supergroup of rocks reached the surface and introduced salinity in Pokaran rann. Drainage system contributed in recharging ground water and increasing salinity.

(iii) Khara-Kanod Group

This group of Khararíwala-Kanod-Khara (Fig. 11) and Mitha ranns from east to west form series of ranns. The southern limits of Khara and Kanod ranns roughly match with contact of Jaisalmer Formation-limestone, Unit-D (Middle Jurassic) with Baisakhi Formation, shale, Unit-A (Upper Jurassic) (Fig. 12). Mitha rann is located in lower and upper Baisakhi shales (unit A and C). The Kharaiwala rann probably received salinity from Kanod rann as these ranns together formed Palaeo-channel during early to mid Holocene period. Erratics of quartzites are seen near Kanod rann. During early-mid Holocene, rainfall recharged saline ground water which reached surface of ranns and introduced salinity. In the opinion of author, during mid-Holocene onward, rann water became hypersaline due to low rainfall and high evaporation and subsequently dried up as saline ranns.

Investigation of Kanod playa was carried out by Deotare et al. (2004). Their AMS ^{14}C dates suggest that playa originated during the Pleistocene-Holocene transistional period. The playa was semi-perrenial to ephemeral between 7.5 Ka and 5.5 Ka and dessicated completely after 5.5 Ka.

(iv) Pachpadra Group

This group of playas and ranns have developed due to segmentation of palaeochannels which may be attributed to reduced flow of streams and fluvial sedimentation, and aeolian accumulation and increasing aridity.

Pachpadra playa is good example of Palaeochannel segmentation. Based on interpretation

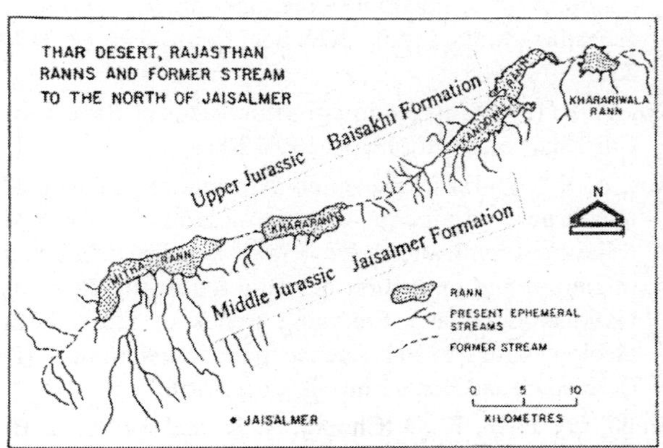

Fig.11 Ranns in Jurassic terrain, Jaisalmer district

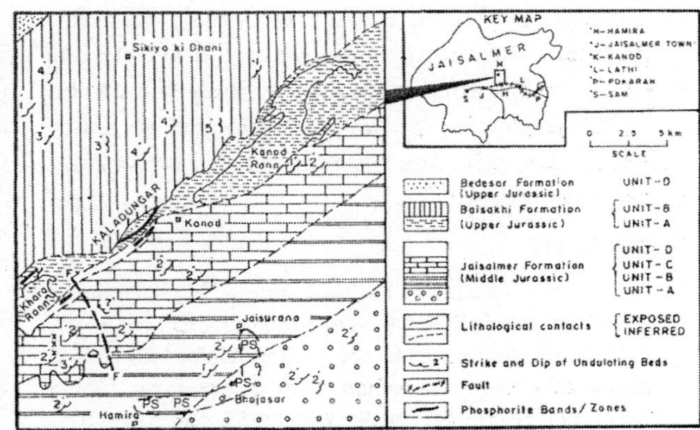

Fig.12 Geological map of Hamira - Bhojasar – Kanod area Jaisalmer district, showing localization of Kanod and part of Khara rann in Baisakhi Shales (Unit A) and cobble – pebble spread (PS) in Hamira – Bhojasar area.

of aerial photographs Ghose (1964) proposed that Pachpadra was confluences of two flourishing streams that flowed as palaeo-Luni in the south. When aridity increased the drastically reduced stream flow could not carry silts load beyond the confluence and trapped silts caused blockage of the river mouth, resulting into shallow saline depression. Thob playa may also fall under this group. Roy (1999, 2007) also discussed the evolution of different salt lakes and saline playas of Rajasthan, and views of various workers.

CONCLUSIONS

Based on above discussion, it may be concluded that western Rajasthan witnessed Neopleistocene (Late Pleistocene) glaciation as indicated by wide spread boulder and cobble pebble spreads. Aeolian activities during cold climate of glacial period resulted in the formation of Paleo-desert. Water released during deglaciation in last phase of Neopleistocene and high rainfall of Holocene significantly recharged the ground water and dissolved the evaporites of Marwar Supergroup and carried the highly saline water to suitably located depressions on either side of Nagaur Ganganagar evaporite basin and introduced salinity in fresh water lakes which were formed due to collection of fresh water in depressions during deglaciation. In the mid-Holocene onward area witnessed hot dry climate with prevalent high speed winds. Aeolian activities produced Neo-desert and also resulted into the formation of younger generation of Lakes-playas.

ACKNOWLEDGEMENTS

Author is thankful to Prof. B. S Paliwal, for fruitful discussion. Author also extends his thanks to Prof. K. L Shrivastava, for kind consideration of paper for publication. Dr. Vivek Laul, DST young Scientist, Department of Earth Sciences, University of Pondicherry, Pondicherry has always helped me to be in touch with the latest developments in Earth Sciences, specially after my retirement. I convey my sincere thanks to him.

References

Deotare, B. C., Kajale, M. D., Rajaguru, S. N., Kusumgar, S., Jull, A.J.T and Donahue J. D. (2004). Palaeoenvironmental history of Bap-Malar and Kanod playas of western Rajasthan, Thar desert. Proc. Indian Acad. Sci. (Earth Planet. Sci), vol. 113, No. 3 pp. 403-425.

Dhir, R. P., Kar, A., Wadhawan, S. K., Rajaguru, S. N., Mishra, V. N., Singhivi, A. K., and Sharma, S. B. (1992). Thar desert in Rajasthan-land, man and environment. Geol. Soc. Ind. pp 1-191.

Geological Survey of India (2001). Geology and mineral resources of Rajasthan, Misc Pub., Geol. Surv. Ind. No. 30 pt 12, pp 1-113.

Ghose, B. (1964). Geomorphological aspects of the formation of salt basins in western Rajasthan. Proc. Symp. Problems of India Arid zone, Govt. of India and UNESCO, CAZRI, Jodhpur pp. 79-83.

Laul, V.P. (2005). Last Himalayan glaciation and neotectonic activity in NW-India vis-a-vis Quaternary events in western Rajasthan. Symposium on sedimentary basins of Himalaya: challenges for future and xxii convention of Indian Assoc of Sedimentologists -IAS 2005 (December 21-23, 2005), Dehradun. Abst Vol.p.75.

Laul, V. P. (2007). Last Pleistocene glaciation and climatic conditions of desertification in western Rajasthan Proc. of National Seminar Emerging trends of Research in Geology with special reference to northwestern India (P. C. Avadich and Harsh Bhu eds.). pp. 79-80.

Laul, V.P. and Kumar, V. (1980). Marine transgressions, regressions, glaciation and desertification events of Jaisalmer basin. 3rd convention, Indian Assoc. of Sedimentologists, Varanasi, Abst. Vol.p.28.

Laul, V.P. and Kumar, V. (1999). Some major Quaternary events in Jaisalmer area, western Rajasthan. Geological evolution of northwestern India (B.S. Paliwal ed.), Scientific Publishers (India) Jodhpur, pp.220-223.

Mukhopadhyay, A.K. and Ghosh, R.N. (1976). The problem of the stratigraphic position of the Bap Boulders, Rajasthan. Ind. Earth Sci. vol. 3 No. 2 pp. 220-228.

Paliwal, B. S. (1982). The source of salt in Rajasthan An investigation of the salt lake in Didwana. Proc. Workshop on the problems of the deserts in India, Misc. Pub. Geol. Surv. Ind. No. 49 pp. 99-104.

Rai, V. (1990). Facies analysis and depositional environment of Pokaran saline rann, district Jaisalmer, Rajasthan, India. Journ. Geol. Soc. Ind. vol. 36, pp 317-322.

Roy, A. B. (1999). Evolution of saline lakes of Rajasthan, Curr. Sci. vol. 88 No. 12, pp. 1995-2000.

Roy, A. B. (2007). Neotectonic deformations causing of Quaternary drainage disorganization in western Rajasthan-implications for origin of saline lakes and migration and extinction of vadic Saraswati. Proc. of National Seminar Emerging trends of research in Geology with special reference to northwestern India (P. C. Avadich and Harsh Bhu eds.), pp. 47-56.

Singh, G., Joshi, R. D. Chopra, S. K. and Singh, A. B. (1974). Late Quaternary history of vegetation and climate of the Rajasthan desert, India, Phil. Trans. Roy. Soc. 267 B pp. 467-501.

Sinha, R. and Raymahashay, B. C. (2004). Evaporite mineralogy and geochemical evolution of the Sambhar salt lake, Rajasthan, India Sedimentary Geology. vol. 166, pp 59-71.

Wadia, D.N. (1975). Geology of India. Tata McGraw-hills Publishing Co. New Delhi, 4th Edition : 508 p.

Wasson, R. J., Smith, G. I., Agarwal, D. P. (1984). Late Quaternary sediments, minerals and inferred geochemical history of Didwana Lake. Peleogeogr. Paleoclim. Paleoeco. vol. 46, pp. 345-372.

Yadav, D. N., Sarin, M. M. and Krishnaswami, S (2007). Hydrogeochemistry of Sambhar Salt Lake, Rajasthan : Implication of recycling of salt and annual salt budget. Journ. Geol. Soc. Ind. V. 69 pp 139-152.

Hon'ble Reviewers

1. Prof. A.B. Roy
   Emeritus Professor
   Department of Geology
   Presidency College
   KOLKATA 700 073 INDIA

2. Prof. Viquar Hussain
   Department of Geology
   University of Karachi
   KARACHI PAKISTAN

Editorial Assistance

. Ms. Tina Prajapat

Economic Mineralization
K.L. Shrivastava (Editor)
2009. pp.274-286
Scientific Publishers (India), Jodhpur

Geological Constraints on the Bentonite Facies of the Barmer Basin, Western Rajasthan, India

S.C. MATHUR[1], R.S. LOYAL[2], S.L. NAMA[1] AND N.S. NEGI[3]

[1] Department of Geology, Jai Narain Vyas University, Jodhpur 342005, India
[2] Centre of Advance Study in Geology, Panjab University, Chandigarh, India
[3] Giral Lignite Project, RSMM, Barmer, Rajasthan-344001, India
Email : sureshsushma@yahoo.com

Abstract

The stratigraphic and sedimentological context of the bentonite lithofacies of the Akli Formation of the Barmer Basin is described in the present paper. The Akli Formation typically represents a clay-dominated lithology, mainly deposited in the central part of the Barmer Basin. The Akli Formation has a rich storehouse of bentonite along with lignite, siliceous earth, phosphorite and sandstones. The bentonite succession of Akli Formation shows a well bedded and stratified sequence contains evidences of ash fall and alloclastic sedimentation almost simultaneously under restricted marine milieu with highly saline and evaporatic conditions.

Key words : Bentonite, Akli Formation, Barmer Basin, Ash fall, Marine environment, Alloclastic sedimentation.

IINTRODUCTION

The economically important Bentonite deposits belonging to the Akli Formation are widely distributed in the Barmer Basin. Out of twenty two localities, the main bentonite occurrences of the Barmer Basin are located at and near Dharvi, Shiv, Harwecha, Hathi Singh ki Dhani, Bishala, Bhadres, Barmer, Sonari, Akli-Thumbali-Giral, Mahabar, Bisu Kalan and Pusad area (Fig.-1). The Bentonite of the area was studied in the past by the several workers on different aspects (Singh, 1952; Sogani, 1952; Siddique and Iqbaludin, 1963; Siddique and Bahal, 1965; Pareek, 1975, 1981 and 1984; Dasgupta, 1974; Verma, 1982: Sisodia and Singh 2000; Roy and Jakhar, 2002; Gour, 2005). An attempt has been made in this paper to present an elaborate account of bentonite of the Barmer Basin, covering basic informations on geology, mode of occurrence, stratigraphic and sedimentological constrains.

GEOLOGICAL FRAMEWORK

The bentonite sequence comprises one of the most extensive deposits in India, falling in the area covered by the survey of India toposheet No. 40 O and 40 N between latitude 25°40' - 26°24' and longitude 71°07'-71°30'. These stratiform deposits form a NNW-SSE

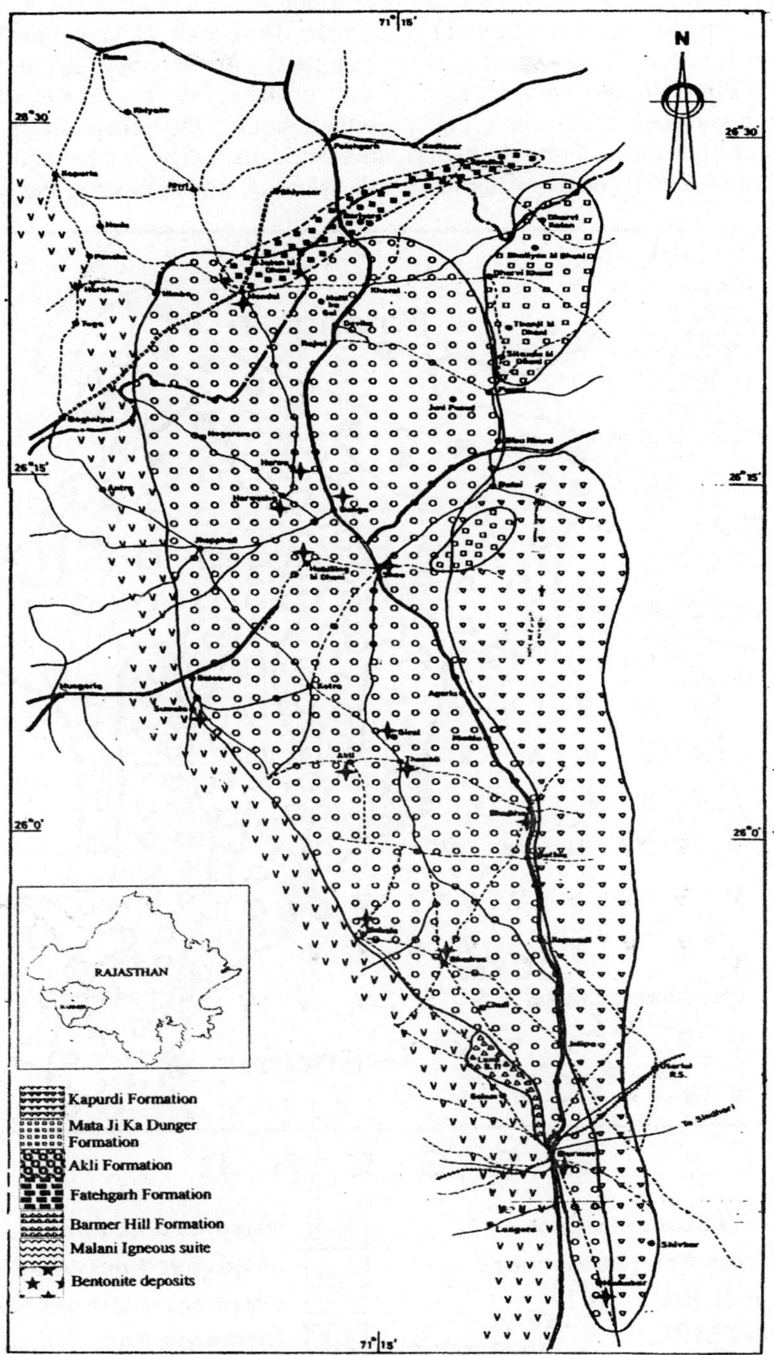

Fig.1 The Geological map of the Barmer Basin showing bentonite occurrences of Akli Formation (Modified after, Dasgupta, 1974, Pareek, 1984 Sisodia and Singh, 2000; Roy and Jakhar,2002; and Mathur et. al.; 2006).

extensive belt of about 8 km. wide, and extending from Bariyara in north to Mahabar in the south for a distance of about 120 km. in the Barmer Basin (Fig. 1).

The Barmer Basin of which bentonites form a distinct entity, itself is a part of Indus Shelf, essentially with Mesozoic Cenozoic sediments cover. In the northern Bikaner and western Jaisalmer Basin, the Barmer Basin principally evolved as a tectonic graben in Early Mesozoic due to rifting of the Indian plate from Gondwana Supercontinent (Dutta, 1983; Mathur and Kumar, 2003). The Barmer Basin is an elongated basin, having roughly north-south trend, which developed partly overstepping in to the Jurassic rocks of the Jaisalmer Basin in the north. It extends for about 100 km. in north south direction and about 50 km. along east west direction as its maximum width (Fig. 2). To the north, Devikot high forms a gravity high structure (Mathur and Kumar, 2003) and Sarnu high in the south east. The horst is in its eastern flank is Barmer Birmania Nagarparkar high. The basin is connected with Cambay Basin through Sanchore Basin in the south. The Fatehgarh Fault marks the northern limit of the Barmer Basin. In the south east, the basin is marked by another fault exposed at Barmer hill, whereas, a fault passing through Sarnu mark its

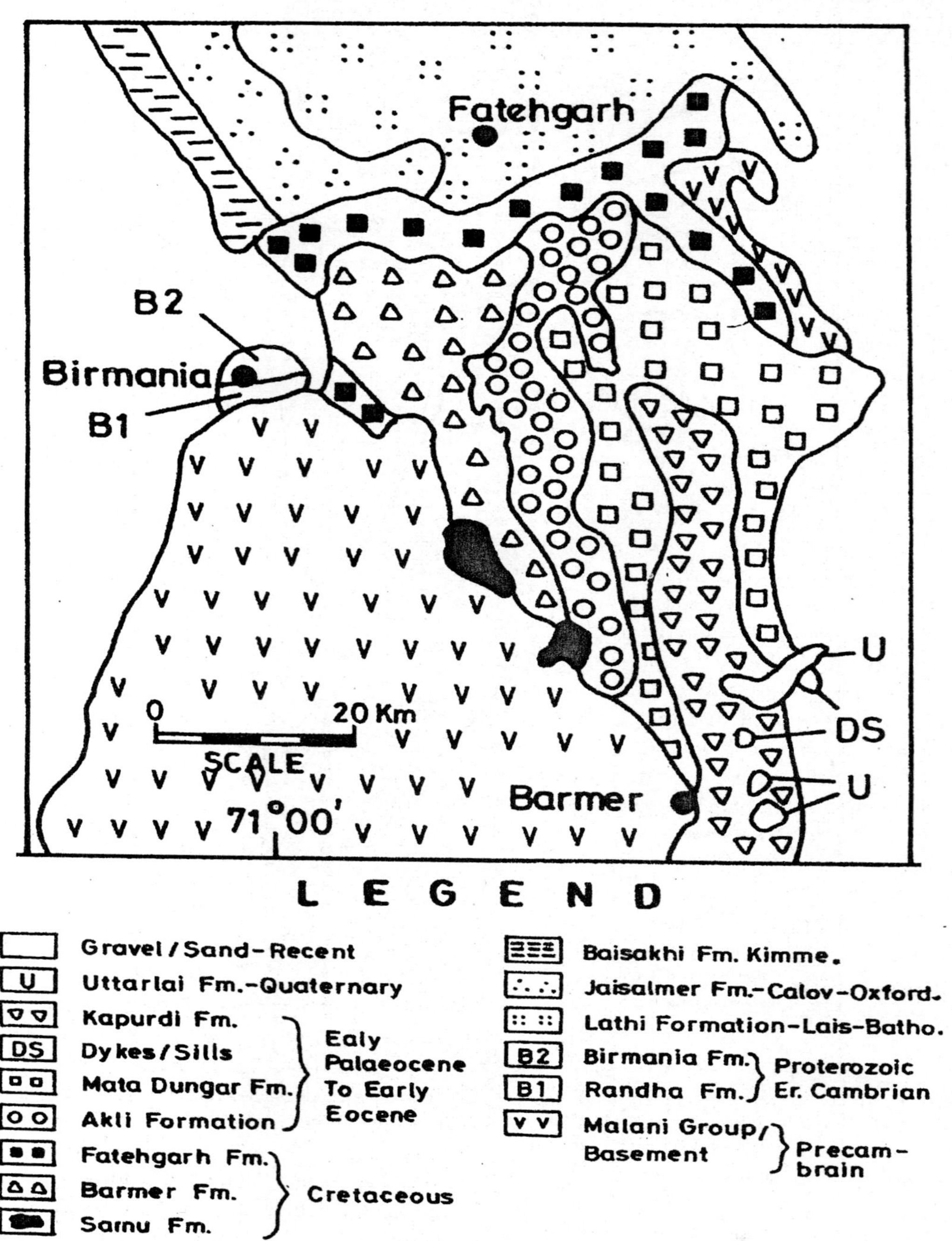

Fig.2 Geology of the Barmer Basin (after, Roy and Jakhar, 2002)

eastern limit. A younger Late Neogene Quarternary fault truncates the original western margin of the basin. (Dasgupta, 1974; Dasgupta and Chandra, 1978; Sisodia and Singh, 2000; Roy and Jakher, 2002; Gour, 2005 and Mathur et. al., 2005a).

The Cretaceous Early Tertiary stratigraphy of the Barmer Basin is not only constrained by geodynamic process (as aforementioned), but also assume interesting palaeontologic and chronologic significance *vis a vis* of economic importance as it yielded petrolium, gas, phosphorite, bentonite, clay, gypsum, lignite, siliceous earth and fuller's earth deposits.

STRATIGRAPHIC FRAMEWORK

The Barmer Basin experienced first sedimentation over the basement of Malani igneous suite of rocks during Early Mesozoic period. In establishing the stratigraphy of the Barmer Basin, a persistent problem has been found in formalizing the strato types of particular formations. The type sections of definite locality and complete section of a formation is not available at one place, nor all boundary contacts observed in the field. In such circumstances, the concept of type locality as pointed out by international subcomission on stratigraphic nomenclature is involved and essentially such type locality were taken into consideration where lower and upper boundaries could be marked (boundary stratotype). Frequent good surface exposures are observable with quite clear boundary surfaces and stratification. Dasgupta (1974) was one of the earliest workers to classify the sediments of Barmer Basin into pre rift, syn rift and post rift sediments owing to the basin being bounded by various faults rifts systems. The pre- rift Neoproterozoic Early Cambrian sediments of the Birmania Basin crops out at southwestern margin (Pareek, 1984; Mathur, 2002, Maheshawari et. al, 2002). The Jurassic sediments of Lathi and Jaisalmer Formation are exposed in the northeast and northwest margin of the Barmer Basin (Pareek, 1984; Mathur, 2002).

Sarnu Formation, Barmer Formation and Fatehgarh Formation of Cretaceous age (Gour, 2005 and Mathur et. al., 2006) represented by the syn-rift sediments in the northern and southern periphery of the basin. The Akli Formation, Mata Ji ka Dunger Formation and Kapuradi Formation represented the post rift sediments and occupying most of the north eastern and central part of the Barmer Basin. (Fig. 1 and 2; Dasgupta, 1974; Pareek, 1984; Sisodia and Singh, 2000; Roy and Jakhar, 2002 and Mathur et al., 2006).

Akli Formation

The Akli Formation is a principle formation, which hosted the bentonite occurrence of the Barmer Basin. The Lithostratigraphy of Akli Formations was previously conducted by many workers (Siddique and Bahal, 1965; Dasgupta, 1978; Pareek, 1981 and 1984; Sisodia and Singh, 2000; Roy and Jakher, 2002; Gour, 2005; Mathur and Kumar, 2003 and Mathur et. al., 2005a). Earlier, Sogani (1952), first include mapable units of bentonite, clay and associated rocks separately as Akli Formation. Siddique and Bahal (1965), during systemic geological work and suggest Akli Thumbli Giral area as type locality of the Akli Formation. The present work suggests various argillaceous, phosphorite, silisiclastic and lignite facies in the Akli Formation. The Akli Formation has underlying unconformable contact with rocks of Malani Igneous Suit, Lathi sandstone, Barmer sandstone and Fatehgarh quartz arenite (plate1a). The overlying contact of Akli Formation with cross-bedded sandstone of the Mata ji ka Dunger Formation is marked by a Palaeosole of red ocher (plate1b). The various facies of the Akli Formation occur at most of the places along the ground level. At some places they occur as continuous to discontinuous small flat ridges and hillocks. Most of the exposures are concealed under desertic sand. There are some nala sections, fault scrap, land incision and quarry section (plate 1c) which provide ample opportunity to study the vertical and lateral relationship among the lithofacies of the study area. It has been found that bentonits display a wide range of colours and representing by white, whitish gray, red and yellow colours (Plate 1d). The entire sequence of Akli Formation is estimated to be about 265 m. thick and the total inferred quarable reserves of bentonite are of about 67 MT. (Siddique and Bahal, 1965).

It has been observed that the stratigraphic section of the Akli Formation is not available at one place. The only best section for geological observation of various units of Akli Formation are available at Giral mine due to quarrying of lignite by open cast method (plate- 1c). The generalized lithostratigraphic section of Giral is presented in Fig. 3. However, based on mode of occurrence, stratigraphic position and lithological correlations, Akli formation can be divided in to two members:

(i) Bariyara-Dharvi Khurd member, consist of siliceous earth with minor bentonite, sandstone and phosphorite facies.

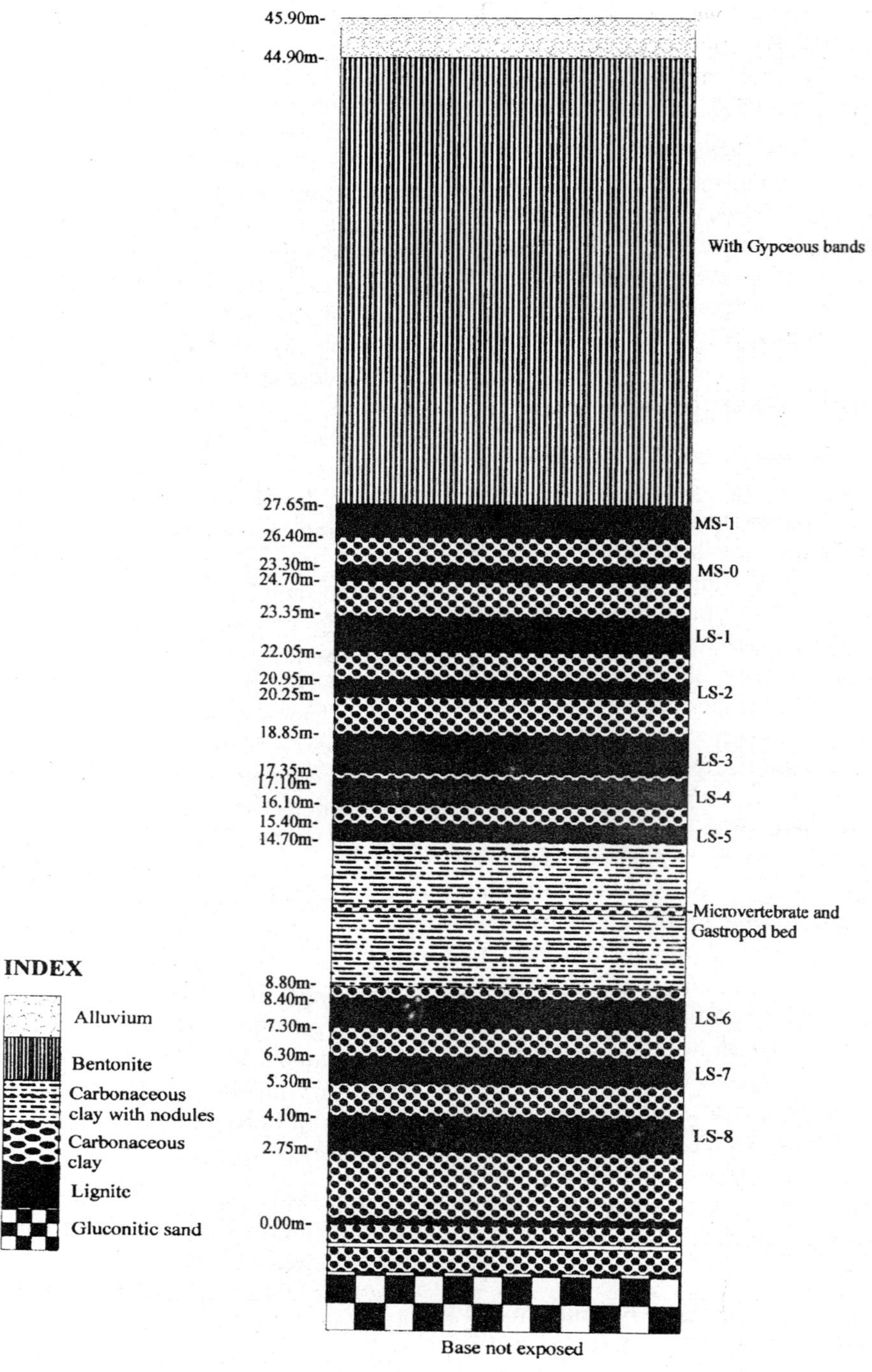

Fig.3 Generalized lithostratigraphic section at Giral

(ii) Akli-Giral-Thumbli member, consist of repetitive sequences of clay, sand, lignite and bentonite facies. (Table -1)

Regarding age of Akli Formation, earlier workers (Sogani, 1952; Siddique, 1963; Dasgupta, 1974 and Pareek, 1984) assigned Tertiary age. Tripathi et al., (1998) first systematically recorded number of palynological assemblage from Giral section and assigned Middle to Late Palaeocene age to lignite

Table-1 Generalized geological succession of the Akli Formation

Mataji ka Dungar Formation (Late Palaeocene)	Lower coursening upward facies	Conglomerate, pebbly sandstone, medium to coarse sandstone.
Akli Formation (Early to middle Palaeocene)	Thumbli – Akli -Giral member	----------------Paleosole -------------- Bentonite with Gypsum Bentonite Repetitive succession of carbonaceous clay, lignite and sand Glauconitic sandstone
	Dharvi - Bariyara member	Phosphatic sandstone Ferruginous sandstone Minor Bentonite Siliceous earth ----------------Unconformity -------------------
Fatehgarh Formation (Cretaceous)	Upper siliciclastic facies	Quartz arenite and sandstone

facies. Subsequently, based on stratigraphic position and lithological correlations, Bhandari (1999) assigned Palaeocene to Early Eocene age to the Akli Formation. Recently, Tabaci and Singh (2000), recorded rich and diversified palaynological assemblage containing ninety-nine species in the Akli Formation and assigned Palaeocene to Eocene age. More recently based on microvertibrate fauna Rana et al. (2005) suggested Late Paleocene to early Eocene age to the Akli Formation. Out of twenty-two occurrences of bentonite of Akli Formation of the Barmer basin, about nine major sections were selected for the present investigation. Lithostratigraphic section of each bentonite bearing locality of the Barmer Basin were prepared (Fig. 4). Systematically, sampling of various lithofacies along and across the strike from each section were carried out.

Geological Framework of Bentonite

The variegated colors like whitish gray, white reddish and yellow colored in the bentonite are quite characteristic, (plate 1d) are widely distributed and covering about 700sq. km in the Barmer Basin. The bentonite beds generally showing N-S strike and having variable dip (10^0-25^0) towards west, NW and SE directions. The bentonite beds varies in thickness from 0.5m to 6.5m and occur as far as up to Khoral in north and Barmer in the south (Fig.1), seems to be a spread of bentonitic sheet uniformly distributed through out the central part of the Barmer Basin. The lithostratigraphic sections of bentonite bearing sequence at various localities are presented in Fig - 4.

Stratigraphically, bentonite-clay beds rest unconformably overlain the Barmer sandstone (Cretaceous) at Sonari, Mahabar, Bishala, Bhadres; Malani Rhyolite at Barmer and Lathi sandstone at rest of the localities (Fig.-4). As such the stratigraphic position of the bentonite suggest that they are post Cretaceous in age. This is further suggested by the occurrence of siliceous earth which has intertonging relationship with bentonite, unconformably overlain the Fatehgarh Formation. The Fatehgarh Formation contain diverse assemblage of microvertibrate and gastropod fossils along with dinosaur, crocodile and turtle bones, phosphatic dung and scute of Late Cretaceous age. (Mathur et. al; 2005 a & b; Mathur et al., 2006 and Gour et al., 2007) The author observed that at all localities, the stratigraphic position of bentonites and associated facies are same with variable thickness only (Fig. 4).

The bentonite deposites describe below except at Dharvi are mainly represent Akli Giral Thumbli member presumably deposited in central part of the basin. These deposits are extending from Harwecha Hathi Singh ki Dhani in north to Mahabar in the south

(Fig. 1). The bentonites of all localities described here indicate homogenous laterally and vertically with less mineralogical variations. The minor changes are regarded as due to temporal variation in the solution chemistry.

(i) Dharvi

About 2 km. south of Dharvi kalan, the bentonite occurrences are located at just north of Mataji ka Dungar. The sharp contact between bentonite of Akli Formation and cross bedded sandstone of Mataji ka Dungar Formation at the base of Mataji ka Dungar hill is marked by a red coloured lateritic palaeosole horizon (Plate 1b). The outcrops of bentonite and siliceous earth are also occur at the base of scattered hills right from Dharvi Kalan in north through Dharvi Khurd to Thanji ki Dhani, Pusad, Bhiyar and Bissu Kalan in the south. The bentonite beds alternate with and grades into Siliceous earth can be seen at these places. Pareek (1984) and Roy and Jakhar, (2002) reported the presence of five bands of Phosphorite facies in nala cutting near Dharvi Khurd (26^0 23' : 71^0 19'). The sequence containing phosphate constitute the lower part of the Akli Formation, which is extending over an area of about 2 sq. km. The thickness of each phosphate band varies from 10cm to 25cm. The P_2O_5 contents in these bands varies between 5 to 20 percent, averaging 10 percent (Balasundram, 1972).

(ii) Harwecha-Sheo-Hathi Singh ki Dhani

Bentonite deposits in these localities are situated about 60 km. north of Barmer at Sheo and at Hathi Singh ki Dhani on Sheo Harsani road (plate 2b). They are covering about 260 sq. km. area in the Barmer Basin. The incisions done for pipeline in the area indicate that bentonite deposits are extended up to south of Bariyara along the national highway No.-15. In these localities, the bentonite uncomformably overlies the Lathi sandstone. The isolated outcrops of Lathi sandstone can be seen about 1.5 km. east of Gunga. The outcrops of petrified wood of Lathi Formation can be seen near Khoryal also.

The lithostratigraphic settings at Hathi Singh ki

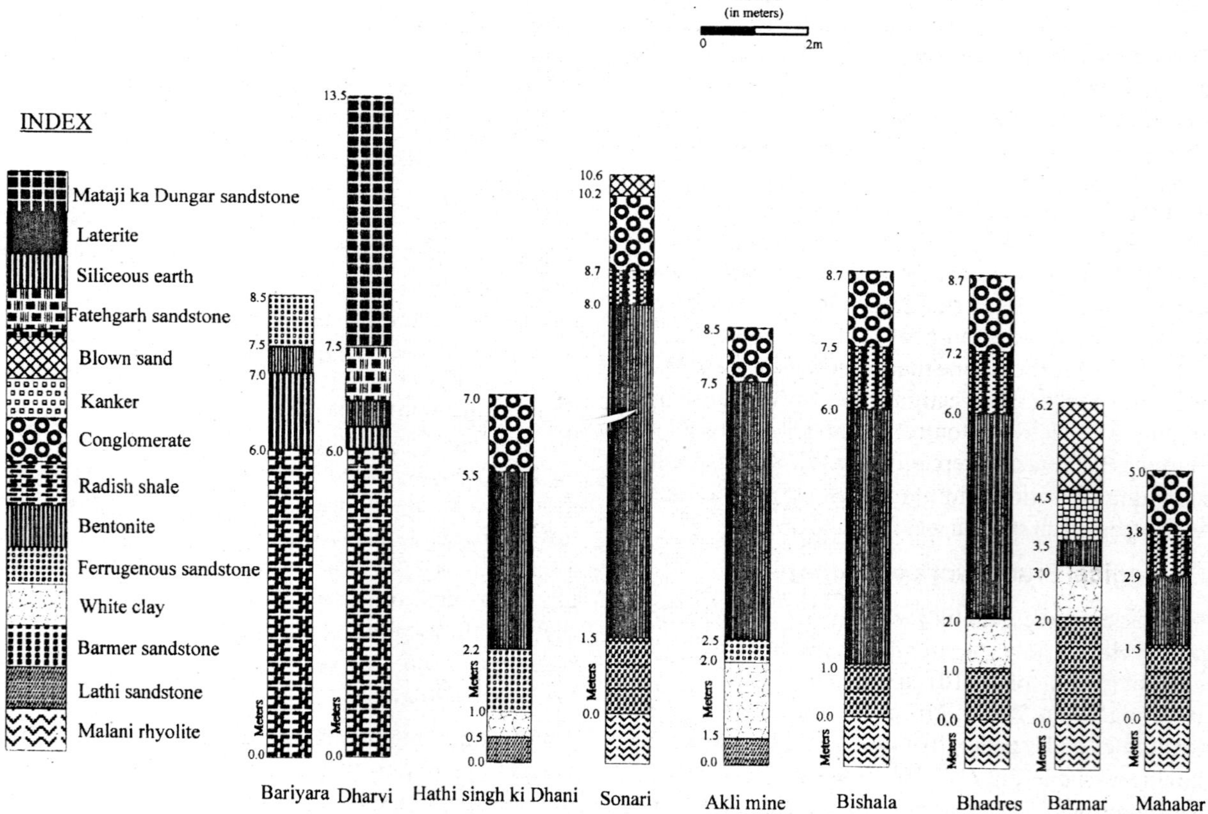

Fig.4 : Lithostratigraphic sections of bentonite bearing Akli Formation in the Barmer Basin

Fig.4 - Plate 1 : (a) Photograph showing sharp contact between quartz arenite of Fatehgarh Formation and Siliceous earth of Bariyara member, Jaton ki Dhani. (b) The contact between bentonite and the cross bedded sandstone of Mataji ka Dungar Formation is marked by a palaeosole of red ochur. Note : Unconformity between Akli Formation and Mataji ka Dungar Formation. (c) Field photograph showing interbedded sequence of sand, lignite and bentonite (Giral section). (d) Field photographs showing different colours in Bentonite.

Dhani section indicate 7m. thick sequence comprises of yellowish brown colored medium grained Lathi sandstone at the base followed by bentonitic clay, ferruginous sandstone, bentonite and conglomerate at the top (Fig.- 4; plate- 2a). The conglomerate horizon seems to be of Quaternary age (Siddique and Bahal, 1965).

The outcrops of bentonite in these areas are very limited and their presence can be seen as lences and pockets in contact with the ferruginous sandstone. Verticle sections for study are available only at Hathi Singh ki Dhani quarry and near Harwecha village. The whitish gray to yellow colored bentonite seems to be extended up to south of Sheo. Bentonite beds in these areas have N-S strike with 15^0 dips towards east. The overlying ferruginous sandstone is yellowish brown in color. It is a medium grained sandstone, mainly composed of quartz detritus with some feldspar. The framework elements are dominantly cemented by iron oxide. A conglomerate bed unconformably overlies the bentonite-clay beds at all these localities. The conglomerate contains pebbles of quartz and ferruginous sandstone seems to be derived from rocks of Lathi Formation. The framework elements of the conglomerate are ranging in size from 2.5mm to about 10cm. and most of them are rounded. The pebble-matrix ratio is 3:2 as such they have mostly clast supported framework. At Hathi Singh ki Dhani and near Sheo where outcrops of bentonite occur at surface, GSI drilling data indicate that up to 243m. Lathi sandstone was not encountered indicating huge thickness of bentonite and clay sequence (Siddique and Bahal, 1965).

(iii) Akli Giral deposits

These deposits are located about 35 km. north of Barmer. The contact between Lathi sandstone and clay bentonite of Akli Formation is not seen. Since Lathi sandstone is exposed in the vicinity of Suwala, it is inferred that they underlies the bentonite in the area. At Akli mine about 8 m. thick section of bentonitic-clay, bentonite, ferruginous sandstone and conglomerate is exposed (Fig. 4).

The best outcrops of bentonite can be seen at Giral mine, where they are associated with 2 to 3 m cycles of bentonitic clay, bituminous claystone, clay and Lignite. At Giral a 60 m. thick section is exposed during opencast mine for lignite excavations (plate 1c). The sequence started with glauconitic sand at the base

Fig.4 - Plate 2 (a) Field photograph showing contact of bentonite (base) and sandstone (top). Hathi Singh ki Dhani. (b) Resudue recovered from bentonite comprises of wide variety of detrital grains. (c) Thin section of bentonite showing detritus spreaded in fine grained ground mass of clay. Minerals. Crossed Nicols. X 5. (d) Typical glass shard recovered from residue of bentonite X 5.

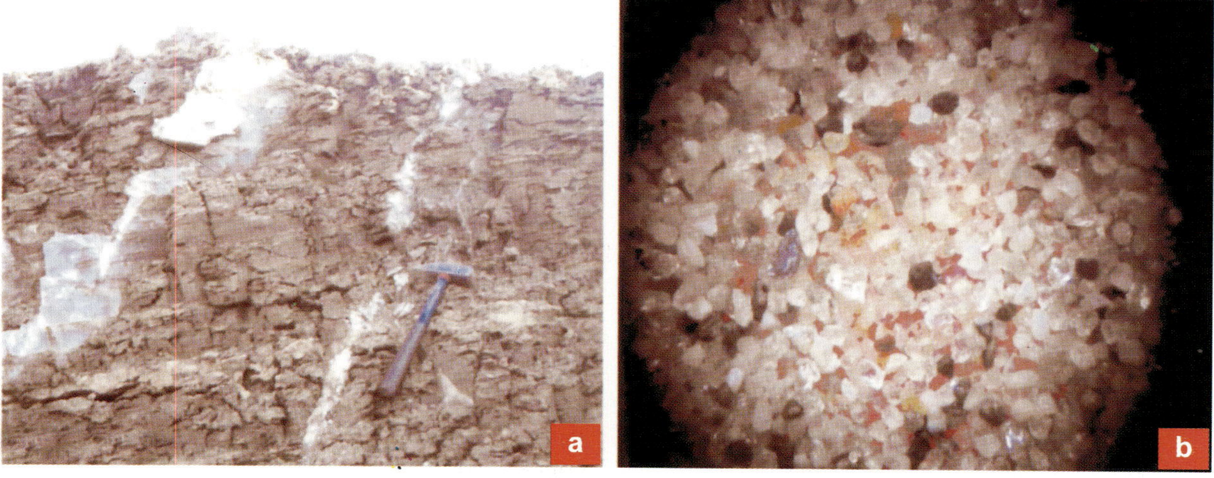

Fig.4 - Plate 3 (a)Field photograph of Giral section showing deposition of gypsum (white) in bentonite. (b)Equigranular quartz grains having subrounded to rounded shape as part of detrital grains recovered from bentonite X 5 (Stereomicroscopic view)

with repetitive sequence of lignite (9 seam) and clay beds. The bentonite of about 30 m. thickness mark the top of the sequence. The typical feature, which is not seen at other localities is the presence of gypsum and salinite in bentonite. Gypsum and Saliniteare occure mainly as veins (up to 50 cm. thick) along and cutting across the bentonite beds. This indicates that at the end of deposition evaporatic condition were presumed leads to formation of gypsum at the top of the sequence (plate 2b).

(iv) Sonari

The bentonite deposits are situated at Sonari about 32 km. north of Barmer. At Sonari and Nimbala, bentonite beds rest on the Barmer sandstone of the Cretaceous age. The contact between both can be seen at east of Nimbala. The Barmer sandstone here is about 1.5 m thick. It is brownish colored medium grained sandstone, mainly composed of quartz, feldspar with rhyolitic lithoclast. The sandstone is dominantly cemented by iron oxide and occasionally with phosphate. A 6m thick persistent horizon of bentonite is unconformable overlain the Barmer sandstone. In general, it is very fine grained in nature but some times it is impure with ferruginous clay. In bentonite beds streaks of gypsum can be frequently observed. The bentonite is overlain by a thin (0.5m) reddish bentonite and 0.75m thick shale horizon which in turn overlain by 1.5m thick compact and thick (3m) conglomerate bed. The conglomerate is composed dominantly of rounded quartz, rhyolite, ferruginous clay stone etc. the adjacent low lying area is marked by the salt pan having salt deposition locally.

(v) Bishala

Bentonite deposite are situated about 1.5 km. north east of Bishala on Barmer Harsani road. A good section is exposed in a quarry (local known as Dodra) near the foothill of the Malani rhyolite. A persistent horizon of bentonite can be traced here for about 1.5 km. The bentonite bed is dipping between 10° - 15° towards west. The most of the bentonite beds are covered by the sand dunes. In quarry section yellowish-white coloured bentonite beds are unconfermably rest over the Barmer sandstone. Yellowish-white coloured bentonite grades upward in to 1.2m horizon of reddish bentonite, which in turn overlain by a persistent horizon of conglomerate.

(vi) Bhadres

The bentonite beds are located at east and south east of Bhadres, about 17 km. from the Barmer on Barmer Harsani road. The exposures of bentonite and associated lithofacies are very poor. The bentonite beds have strike NNW- SSE to NW- SE with 20^0-25^0 dip towards west. The 4-5m thick bentonite horizon occurs from east of Sonari to near Lunu Hills. The exposures are occurring as pockets and lances. A small outcrop of poor exposures of Bhadres section indicate that clay-bentonite sequence of about 5m seems to occur on Barmer sandstone. The contact of both is though not visible but scattered outcrop of Barmer sandstones in the vicinity indicate that they are resting over them. Here also bentonite is grades upward in to reddish clay having some gypseous beds. The overlying conglomerate shows some different composition additionally they contain sandstone fragments and phosphatic nodules along with quartz and rhyolite pebbles.

(vii) Barmer

White colored bentonite is located one km. west of the Barmer Railway station. The bentonite bed is very thin (0.5m) and occurs at the surface in association with white clay. The bentonite beds here directly rest over the Barmer sandstone. The contact between Barmer sandstone and white clay is exposed at low laying area near Barmer-Munabao railway track. Though vertical section is not visible here but field relationship indicate that Akli clay resting over the Barmer sandstone which is followed by 0.5m thick bentonite horizon. The recent kanker bed mark the top of the section. The conglomerate beds are not exposed in this area.

(viii) Mahabar

At Mahabar bentonite are located at and 1.5 km. S-W of Mahabar in the Barmer Dhorimana road. The bentonite bearing formation in this area does not appear to be very thick and exposers are also very limited. Mahabar bentonite are unconfermably overlain the Barmer sandstone. Here bentonite are grades upwards in to highly ferruginous rock with phosphatic nodules. This facies is not persistent and occur in form of pockets, which in turn is overlain by 0.9m thick beds of shale. Conglomerate like in other area mark the top of the sequence. Here younger conglomerate bed is persistent which is a marker horizon to trace the bentonite in the area. The conglomerate bed strike N 30° W S 30° E with 30° - 40° dip towards SW.

Petrographic attribute and genesis of Bentonite

About two hundred samples of bentonite and

associated lithofacies were collected from above described localities for their petrographic studies. The bentonite of Barmer Basin represents wide colored range i.e. white, gray to yellow and light brown color. It is a fine grained, friable, soft and seen commonly in compact nature. It is mainly composed of smectite and kaolinite group as major clay minerals. Clay minerals constitute about 46.5 to 54.5 percent of the rock (Siddique and Bahal, 1965). To observe additional minerals, the samples of bentonite were seen under stereomicroscope. For which samples were place in oven for 12 hours at 90°C and than put in distill water for about 8 hours (c.f. Ruffell et al., 2002). The residue shows presence of various detrital grains and lithoclasts in fare amount (plate 2b).

To confirm these detritus and framework elements, thin sections of bentonite from all localities discussed above were prepared. Petrographic studies indicate that detrital grains are spread randomly in very fine matrix of clay probably rich in smectite and kaolinite (plate 2c). The clay minerals has a micaceous habit, high birefringence and a textured inherited from the volcanic ash. The presences of relict shard structures are observed seem to be characteristic of volcanic source (plate 2d). The framework elements are mainly quartz, feldspar and heavy minerals (tourmaline, zircons, augite diopside, monazite and pyrite etc) along with clasts of ferruginous sandstone, rhyolite, basalts, quartz, arenite etc. The matrix to framework element ratio is 4:1.

The framework grains are ranging in size from 0.02 mm to 2.1 mm indicating silt to coarse sand fraction (plate 2b). The lithoclasts are ranging in size from 1.2 mm to 2.8 mm represent the admixture dominantly of coarse sand to fine pebbles. Among detrital fraction, quartz forms the most dominants admixture (about 80 percent), while feldspar, heavy minerals and lithoclasts form less dominant fraction.

Other minerals include calcite, gypsum, phosphate, muscovite, biotite flacks corresponds to only a very minor components of bulk of rock. The presence of gypsum and salt indicate saline and evaporatic conditions. It further corroborate from the presence of gypsum and salinite associated with upper level of bentonite bed (plate 3a). It indicates that at the end, the basin water was shallower and shallower with increasing salinity and evaporatic conditions.

The silt sized quartz grains shows a characteristic outline with corroded border and few inclusions, seems to be of volcanic origin. Few quartz grains have straight bordered and are equi-dimensional in shape (plate 2c and 3b). They have with numerous inclusion of minute grains seems to be of granitic origin (c.f. Zimmermann and Bahalburg, 2003).

The most of the quartz, feldspar and heavy minerals are subrounded to rounded in nature (plate 3b) while the lithoclasts of rhyolite, granite, sandstone and quartz arenite are angular to subrounded in nature (Plate 3b). The presence of sorted to poorly sorted admixture in very fine-grained clay matrix indicate that these framework grains are allogenic in origin.

The few fine silt sized quartz grains often show resorption embayment also pointing to derivation from volcanic sources and are not of detrital nature. Elongated idiomorphic zircon with gas bubbles inclusion also indicates to be a volcanic origin (Siddique and Bahal; 1965). The preservation of such admixture of volcanic-plutonic and sedimentary source in same strata is indicative of a minor weathering and sorting influence acting as the detritus on its way from source to sink. In turn; this implies; (1) relatively short transport path and thus potentially marked relief. (eg. Nilson and Sylvester, 1965; Zimmermann and Bahalburg, 2003). Further the presence of kaolinite and volcanic quartz associated with smectite in bentonite is an evidence of ash layer deposited in shallow marine conditions (c.f. Teale and Spears, 1986; Spears et al. 1999; Deconinck et al., 2000; Pellenard et al., 2003). This is further evident from the occurrence of monotonous and homogenous bentonite sheet through out the Barmer Basin. The ash fall and deposition of detritus simultaneously in marine conditions is also evident from:

(i) The stratigraphically older bentonite beds are rich in Na-montemorillonite and younger one is rich in Mg-montemorillonitic composition. These geochemical relations suggest that Na-Bentonite when remained under saline water, due to cation exchange, converted in to Mg-Bentonite (Siddique and Bahal, 1965;).

(ii) The association of Bentonite at Giral with interbedded sequence of glauconitic sandstone, black shale and lignite along with wide spread iron minerals associated (Verma and Tripathi et al.; 2000) and various fauna with lignite indicate aqueous marine conditions.

(iii) Further the mixture of rounded quartz, feldspar, lithoclasts of variable composition and heavy minerasl etc. support of the concept of this being neither primary ash fall nor basic rocks completely reworked and is not subjected to in situ alternation from the land.

The ash fall that was preserved as bentonite could have occurred during a period of current activity as detrital quartz grains are common and sandstone beds below Bentonite are cross stratified. The cross beds in Akli sandstone at Hathi singh ki Dhani suggest that provenance of detritus was situated in south and south-east directions, pointing possibly towards Malani and younger sediments, which are wide spread in southern and south-west part of the Barmer Basin and contribute sediments almost contemporaneous with ash fall. (c.f. Ruffell et al., 2002)

(iv) The foraminiferal lining in Giral sections occurred under a low salinity condition and at a comparatively shallower depth perhaps in an arm of an epicontinental sea during Paleocene time (Roy and Jakhar, 2002).

As such the evidences discussed above support that bentonite of the Barmer Basin is outcome of simultaneous deposition of ash fall and detrital material in shallow marine conditions. In order to understands the volcanic source of bentonite, the understanding of geochronology of the Barmer Basin is important. In this context, Mathur et al. (2005a and 2006) established Late Cretaceous age to Fatehgarh Formation over which the Akli formation is deposited. It is to be noted that above Maastrichtian fauna reported by Mathur et al; (2006) from bone bed of the Fatehgarh Formation there is a thick siliciclastic sequence over which siliceous earth of volcanic origin were accumulated in a sort span of time. It has been observed that siliceous earth and bentonite has intertonging and interchangeable relationship. The major fraction of bentonite of the area has volcanic source. The major issue is about the stratigraphic source of the material. In this regard, the geochronology and mineralogical composition of near by volcanic activity of Sarnu-Dandali, Tavidar, Mundwara and Daccen volcanism suggest that these may be the source for the volcanic component of the Barmer bentonite. Further, the existence of siliceous earth bands within bentonite and their graded and mutually merging relationship with each other indicate existence of acidic flows within these volcanic provenances.

ACKNOWLEDGEMENT

The authors express their thanks to Prof. M.S.Sisodia, Head, Department of Geology, for providing facilities to carryout this work. The authors express their gratitude to Prof. D.S. Chouhan, Prof. Ashok Sahani, Prof. K.L. Shrivastava and Dr. S.K.Trivedi for valuable comments. SCM, RSL and SLN acknowledges DST, New Delhi for financial support in form of a research project(SR/S4/ES-58).

References

Bhandari, A., 1999. Phanerozoic stratigraphy of Western Rajasthan basin: A review. Proceedings of the Seminar on Geology of Rajasthan Status and Perspective (A.B. Roy Felicitation Volume). P.Kataria (ed.) Geology Dept, M.L.S.U. Udaipur 126-174.

Dasgupta and Mahesh Chandra, 1978. Tectonic elements of the western Rajasthan self and their stratigraphy. Proc. Golden Jubilee. Sym. Geo. Min. Met. Soc. Ind. Calcutta. 50: 1-16.

Dasgupta, S.K., 1974. Stratigraphy of western Rajasthan shelf: Proc. IV Indian Colloq., Micropal. Strat., Dehradun, India, 219-233.

Datta, A.K., 1983. Geological evolution of hydrocarbon prospects of Rajasthan basin. In: Bhandari, L.L (ed.), Petroleum Asia Jour. 1: 92-100.

Deconinck, J.F., Amedro, F., Robaszynski, F., Pellerard, P. and Recourt, P., 2000. Influences detritiques et volcaniques sur la mineralogy de la fraction argileuse des formations crayeuses traversees par le forage de Poigny (Project craie 700). Resultats preliminaries. Bull. Inf. Geol. Bassin, Paris, 37, 107-111.

Gour, S.D., 2005. Geological Investigation of Phosphorite and associated Lithofacies of Sandha Bhiyar area of Barmer District, Western Rajasthan, India. Unpublished Ph.D. thesis, J.N. Vyas Univ, Jodhpur, 179.

Maheshwari, A., Sial, A. N. and Mathur, S. C. 2002. Carbon isotopic Fluctuations through the Neoproterozoic Lower Cambrian Birmania Basin, Rajasthan, India:. Int. Jour of Carbonates and Evaporates, 17, 1, 53-59.

Mathur S. K., Mathur, S. C., and Loyal R. S. (2006). First Report of microvertebrate assemblage from Cretaceous Fatehgarh Formation. Geological Society of India. vol. 67, No. 4, 759 769.

Mathur S.C., Gour, S.D., Loyal, R.S., Tripathi., A., Tripathi, R.P. and Ajay Gupta, 2005 b. Occurrence of magnetic spherules in the Maastriction bone bed of the Fatehgarh Formation, Barmer Basin, India. Current Science. vol. 89, No.7, 1259 1268.

Mathur, S. C. and Sudhir Kumar. 2003. Sedimentation in Barmer Basin. Jour. Geol. Soc. India. 61, 3, 368-369.

Mathur, S.C., S.D. Gaur, R.S. Loyal, A. Tripathi and M.S. Sisodia. 2005a. Spherules in the Fatehgarh Formation (Cretaceous) of Barmer Basin, India. Gondwana Research. 8, 4. 579-584.

Mathur, S.K., 2002. Geological investigation of phosphorite and associated lithofacies of Fatehgarh area, western Rajasthan, India. Unpub. Ph.D. thesis, J.N. Vyas

Univ., Jodhpur. 147.

Nama, S. L., 2007: Sedimentological and paleontological investigation of rocks of the Akli formation of Barmer Basin, Western Rajasthan, India. Unpublished Ph.D. thesis, J. N. Vyas University. 185p.

Nilson, T. H. and Sylvester, A.G. 1995. Strike-Slip basin. In: Tectonics of sedimentary Basins (Eds. C.J. Busby and R.V. Ingersoll), pp. 425-457. Blackwell Science, Oxford.

Pareek, H.S., (1981). Basin configuration and sedimentary stratigraphy of western Rajasthan; J. Geol. Soc. India, vol. 22, pp. 517-527.

Pareek, H.S., 1984. The Quaternary geology and mineral resources of North-western Rajasthan. Memmo. Geo. Sur. India 115: 1-99.

Pellenard, P., Deconinck, J.F., Huff, W.D., Thierry, J., Marchand, D., Fortwengler, D. and Troviller, A., 2003. Characterization and correlation of Upper Jurrassic (Oxfordian) bentonite deposits in the Paris Basin and the Subalpine Basin. Sedimentology, 50, 1035-1060.

Rana, R.S., Kumar K., Singh H. and Rose K.D., 2005. Lower vertebrates from the Late Palaeocene Earliest Eocene Akli Formation, Giral Lignite Mine, Barmer District, western India. Curr. Sci., vol. 89, 9, pp. 1606-1613.

Roy, A.B. and Jakhar, S., 2002. Geology of Rajasthan (North West India) Precambrian to Recent. Scientific Publishers, (India), Jodhpur. 421.

Ruffel, A.H., Hesselbo, S.P., Wach, G.D., Simpson M.I. and Wray D.S. 2002. Fuller's earth (bentonite) in the Lower Cretaceous (Upper Aptian) of Shanklin (Isle of Wight, southern England). Proceedings of the Geologist's Association, **vol.** 113, pp. 281-290.

Siddiqui, H.N. and Bahl, D.P., 1965. Geology of the bentonite deposits of Barmer Distt. Rajasthan: Mem, Geol Surv. India, 96.

Siddiqui, H.N. and Iqbaluddin, 1963. Occurrence of Palaeocene and Eocene Beds in Barmer district, Rajasthan. Curr. Sci., vol. 32, p. 575.

Sisodia, M.S. and Singh, U.K., 2000. Depositional environment and hydrocarbon prospect of Barmer Basin NAFTA. Zagreb, Croatia. 51(9): 309-326.

Sogani, P. C., 1952. Bentonite and other clays in Barmer district, Rajasthan. GSI report (Unpublished).

Spears D. A., Kanaris-Sotiriou, R., Riely, N. and Krause, P. (1999) Namurian bentonites in the Pennine Basin, UK-origin and magmatic affinities. Sedimentology, **vol.** 46, pp. 385-401.

Tabaei, M. and Singh, R. Y., Palynozonation of Lower Tertiary sediments of western Rajasthan, India, with reference to Akli lignite. 44th Annual Meeting, Palaeontol. Assoc., 2000, p. 27.

Teale, C.T. and Spears, D.A. 1986. The mineralogy and origin of some Silurian bentonites, Walesh Borderland, UK, Sedimentology, **vol.** 33, pp. 757-765.

Tripathi, S.K.M., Singh U.K. and Sisodia M.S., 1998. Palynological investigation and environmental interpretation of Late Paleocene sediments from Barmer Basin, Western Rajasthan, India XII Indian colloq Micropal Strat Goa, 148 (Abstract).

Verma, H.C., and Tripathi, R. P., 2000. Cheracterization of Iron bearing minerals in Giral lignite from Rajasthan and study of their decomposition during combustion using ^{57}Fe Mössbauer spectroscopy. Fuel, **vol.** 79, pp. 599-606.

Zimmermann, U. and Bahlburg, H. 2003. Provenance analysis and tectonic setting of the Ordovician clastic deposits in the southern Puna Basin, NW Argentina. Sedimentology, **vol.** 50, pp. 1079-1104.

Hon'ble Reviewers

1. Dr. Narendra Bhandari
Retd. Sr. Scientist PRL
Basic Sci. Res. Institute
AHMEDABAD 380 009 INDIA

2. Prof. Tarmo Kiipli
Institute of Geology
Tallinn University of Technology
19086 TALLINN
ESTONIA

Editorial Assistance

Mrs. Mamta Chauhan

Economic Mineralization
K.L. Shrivastava (Editor)
2009. pp. 287-294
Scientific Publishers (India), Jodhpur

Section-III

The Clay Deposits of Bikaner District, Rajasthan Geological Setting, Genesis and Quality Assessment for Industries

A.K. SHANDILYA[1], S. K. GAHLOT[1] AND N. SRIVASTAVE[2]
Department of Geology, Dungar College, Bikaner-334003
[2]A.S.D.C.P. Ltd (Golcha Group); Devpura, via Zawar Mines; Udaipur-313905
Email : arunshandilyabikaner@yahoo.com

Abstract

Clays are vital earth materials, which ares in human use since centuries. Its resources are found in various states including Rajasthan. Bikaner district supplies more than 80% production of Rajasthan state. The district has important varieties of clays like china clay, ball clay, fuller's earth and fire clay. The clay deposits of Bikaner district are located mainly near Kotri, Chandi, Marh, Gurha, Deh, Indo-Ka-Bala, Suraj-Ki-Dhani, Nal, Palana, and Barsingsar etc. Geologically these deposits are present in Early Tertiary formations. The better quality of china clay and ball clay are present in Marh Formation whereas the fire clay occur in Palana Formation and fuller's earth found in Jogira Formation. The chemical analyses of clays and the reserves of different places of district shows that at most of the places the quality and quantity is good; and can be utilized by a varieties of industries for diverse purposes.

Keywords : Industrial Clay Deposits, Bikaner Clay Deposits, Tertiary, Refractory.

INTRODUCTION

The term clays is given to the earthy substances (an aggregate of minerals, i.e. a rock) consisting mainly of hydrous aluminium silicate minerals colloidal in nature, consisting of specks of rock fragments which generally become plastic when wet, and become hard like stone under fire. These properties give clays their usefulness since they can be moulded into almost any form that they retain after firing. Wide range of industrial application is associated with the clays especially in paper, plastic abrasives, refractory, rubber textile, cement, pigment, oil and chemical industries etc.

The basic characteristics for any material to be called as clay are, that it must be a natural material with plastic properties, having essential composition of particles of very fine size (Grain Size generally to be less than 2 micron) and should be composed of crystalline fragments of hydrous alumino-silicate (or to some extent hydrous magnesium silicates) minerals.

The important clay minerals are kaolinite (Al_4 Si_4 O_{10} $(OH)_8$), halloysite (Al_4 Si_4 O_6 $(OH)_{12}$), montmorillonite (Al_4 $[Si_6$ Al_2 $O_{20}]$ $(OH)_4$. nH_2O) and illite ($K_{1-1.5}$ Al_4 ($Si_{7-6.5}$ Al $_{1-1.5}$ O_{20}) $(OH)_4$), Other clay minerals found in small quantities are dickite, nacrite, anauxite, saponite, beidelite, allophone, palangorite (attapulgite) etc. The non-clay minerals present in clay are micas, quartz, tourmaline, feldspars, chlorite, siderite, limonite, hematite, etc.

Extensive deposits of clays are found mainly in the states of Rajasthan, Bengal, Orissa, Kerala, Gujarat, Andhra Pradesh, Bihar, Haryana, and Madhya Pradesh etc. The clays produced in Bikaner district alone contribute about 80 % of total clay production of Rajasthan.

The widespread availability, ease of extraction and adaptability for many uses of clays has resulted into wide-range of its uses in human civilization such as:

1. In domestic life, clay is used in making of porcelain pottery, earthen wares, cooking wares, vases, ornamental tiles, bricks, linoleum, wall papers, scourging soaps and polishing bricks etc.
2. In house building, it is used in making of building bricks, vitrified bricks, conduit tiles/pipes, chimney pots, foundation blocks etc.
3. In industry, it is used in making of conduits, sockets, switches, insulators, spark plugs, refractory wares, filling, cloth & paper making, paving bricks, rail-road, ballast, Portland cement and innumerable minor purposes.

CLAY DEPOSITS IN GLOBAL PERSPECTIVE

On the basis of physical and refractory properties, the important varieties of the clays are Kaolin or China clay, Ball clay, Bentonite, Fuller's Earth, Fire clay, lithomarge, etc. The varieties found in the Bikaner district are China clay, Ball clay, Fuller's Earth and Fire clay; and are described below:

Kaolin

Kaolin or china clay contains kaolinite, a hydrous alumino silicate ($Al_2O_3SiO_2.2H_2O$), with a 1:1 silica alumina layer, as the major mineral associated with other minerals like diekite, halloysite, nacrite and anauxite. Impurities like Fe and Ti heavy minerals and quartz if present, significantly affects its industrial utilization. The versatility of this mineral is due to its unique properties like chemical inertness, white color, good hiding power (opacity), softness and non abrasive nature, fine particle size and shape, plasticity, hydrophilic nature and low cost. The important property of china clay is that, it has good plasticity and does not swell in water, stand high temperature (upto 1300^0 C) and is non-conductor of heat and electricity. Its specific gravity varies between 2.55 to 2.65. The major user industries of kaolin are paper, ceramics, rubber, polymers, plastics, textiles, paints, cosmetics and pharmaceutical etc. It is also used in insecticides, white cement manufacturing.

Primary kaolinite bearing deposits are formed through the alteration or kaolinization of feldspar rich rocks such as Granites, Gneisses by weathering or hydrothermal processes. Mostly intense weathering in hot, humid climates forms kaolinite by hydration of anhydrous aluminum silicates accompanied by removal of alkalies. Once formed, kaolinite may remain at or close to the source rock as residual or primary deposit or it may enter the erosion cycle and deposited at some distance forming sedimentary deposits. Iron bearing mineral like biotite gives up their iron during weathering, the resulting oxides may color the clay and restrict commercial value.

USA is the chief producer of Kaolin, followed by Russia and Brazil. The world receives estimated kaolin at 12385 million tones. India possesses fairly large insitu reserves of Kaolin, in the order of 2353 million tones, distributed mainly in Kerala, West Bengal, Rajasthan, Orissa, Karnataka and Jharkhand (IBM, 2002). Kerala is the leading producer, followed by West Bengal. Rajasthan ranks third in total production of kaolin of the country.

The Rajasthan state possesses a total of 336.6 million tones reserves of kaolin.

Proved Reserves -	40.9 million tonnes
Probable Reserves -	33.13 million tonnes
Possible Reserves -	262.5 million tonnes

The total production of Kaolin from the country in 2001-2002 was 807,559 tonnes, out of which Rajasthan contributed 167,530 tonnes.

Ball Clay

The name derived from an early custom in the UK of extracting clay as cutting into $1F^3$ blocks that eventually became rounded to form Balls. Ball clay is highly plastic variety of kaolin, having binding power, tensile strength and shrinkage ability. Ball clay is fine-grained mixture of 70% disordered kaolinite including illite (primary mineral is kaolinite). Micaceous minerals and quartz are the dominant minerals present as impurities along with minor amounts of tourmaline, feldspar, chlorite, montmorillonite, smectite, siderite, limonite, hematite, and 2-3%, carbonaceous material like lignite. The organic matter may darken the clay colour to light brown black etc. Presence of organic content together with the fine-grained nature of Kaolinite and the presence of other clay minerals makes the ball clay high plastic and superior in green strength than kaolinite.

Most of the ball clay deposits are sedimentary. Their association with carbonaceous matter indicates a swampy environment of deposition. Ball clay is most important material in the Ceramic Industry because of its: (i). High Plasticity, (ii). Good Dry Strength and (iii). Long Vitrification Range. The ball clay, when fired, because of presence of carbonaceous material, burns off and produces white or near white products. Ball clay are also used in refractories as anti-caking agent and commonly used in sanitary wares and tiles, animal feed, as fillers in rubber and Plastic, etc.

The main utility of ball clay lies in the plasticity, high bonding qualities and tensile strength. It is therefore, used in blend with non-plastic to semi-plastic clays for obtaining requisite plasticity. It is added in various proportions for the preparation of body composition of various sanitary wares, bathtub tiles, porcelain, etc.

India has about 88.4 million tones of reserves of Ball clay. Andhra Pradesh has the highest reserves but the Rajasthan State is the leading producer of the Ball clay in the country. Rajasthan has about 24.4 million tonnes of reserves, under following categories:

Proved	-	5.52 million tonnes
Probable	-	5.62 million tonnes
Possible	-	13.3 million tonnes

Approximately 91% of the total country's production comes from Rajasthan. During 2001-2002 the total production from the state was 525140 tonnes.

Fuller's Earth

Fuller's earth is also known as 'bleaching clay' due to its intrinsic bleaching properties, therefore, has great commercial value like Bentonite. The clay mainly composed of montmorillonite mineral. Indian reserves of Fuller's earth are reported to the tune of 256.7 million tonnes. Rajasthan state has India's largest reserves approximately 190 million tonnes.

Fire Clay

A variety of clay, mainly used for refectory purposes is known as Fire clay. The essential quality of Fire clay is that it has high fusion point and good plasticity. Fire clay consists mainly of mineral kaolinite with minor amounts of other clay minerals, quartz, iron, titania and alkali as impurities.

India has fairly large reserves of Fireclay approximately 706 million tonnes. Orissa has the largest reserves (177.3 million tonnes), followed by Tamil Nadu (112.8 million tonnes) and Madhya Pradesh (111.58 million tonnes). The state of Rajasthan has nearly 55 million tonnes of reserves out of which 9.9 million tonnes are Proved, 5.0 million

Table-1 Regional Geology of Bikaner District (Revised after Ghosh 1983)

Age	Formation
Pleisto cene to Recent	Quaternary formations
--------------- Unconformity -----------------	
Lower to Middle Eocene	Jogira Formation (calcareous facies)
--------------- Unconformity -----------------	
Upper Palaeocene	Marh Formation (Arenaceous facies)
--------------- Gradational facies ------------	
Lower Palaeocene	Palana Formation (Carbonaceous facies)
--------------- Unconformity -----------------	
Permo -Carboniferous	Bap - Badhaura Formations
--------------- Unconformity -----------------	
Neoproterozoic Eocambrian	Marwar Supergroup

tonnes are Possible and rests are Probable reserves. During 2001-2002 the total country's production of Fire clay was 367625 tonnes out of which Rajasthan contributed 130685 tonnes.

GEOLOGY OF BIKANER AREA

The rock units found in Bikaner district belong to Lower Tertiary, and correspond to upper Palaeocene and Eocene age of the stratigraphic sequence identified by Ghosh (1983). This assemblage generally does not exposed due very thick Quaternary sand and alluvium. Geology of the area is mainly studied on the basis of subsurface studies and outcrops found at places. The basement rocks of Malani Supergroup i.e. Malani rocks and/or Precambrian metamorphics, the Marwar Supergroup and Bap-Badhaura Formation (Pareek, 1981, 1984) are not cropped out in nearby areas; hence contact is not observable in the field but studied in the subsurface studies. However rocks of the Marwar Supergroup are exposed at Dhulmera. The carbonaceous facies the Palana Formation is the earliest Tertiary formation in the area, consists mainly of carbonaceous clays/shale and lignite. The Marh Formation unconformably overlies the Palana Formation. This arenaceous facies of upper Palaeocene age comprises of clay beds, gritty/lateritic sandstone and siltstone. Overlying the Marh Formation with an angular unconformity is the Jogira Formation consisting mainly of fuller's earth and foraminiferal limestone they belongs to calcareous facies of Eocene age. The beds of ball clay and sandstone are devoid of fossils except few leaf impressions in clay beds of Nal and Kolayat areas. Various types of Quaternary formations such as beds of sand and alluvium, boulder spread, and dunal sand etc. overlies unconformably the Lower Tertiaries of the region. The lithological succession of rocks exposed in the area is presented in Table 1.

CLAY DEPOSITS AROUND BIKANER

The clay deposits in Bikaner district are located near Kotri, Chandi, Marh, Gurha, Deh, Indo-Ka-Bala, Suraj-Ki-Dhani, Nal, Palana, Barsingsar etc. the estimated reserves of clays of Bikaner district are more than 55 million tones near working mine sites. There are many more reserves are present in other parts of the district. The clay deposits of all these places in the district are of sedimentary nature and may be classified as secondary clay. The clays available below and above lignite seam are classified as fire clay, e.g. at Barsingsar, Gurha and Palana, where the alumina content in the clay is above 30% lower green strength as compared to plastic ball clays, and a higher temperature of vitrification. The chemical analysis of clays of different areas of Bikaner is given in Table No. 2.

Kotri

A good extensive deposit of ball clay occurs in an area of about one sq km in the form of two seams with an average thickness of 3 meters (varies between 1 to 5 m) near village Kotri (near Kolayat) in Bikaner district. The estimated reserves are about 15 million tons. The clay slakes in water quickly giving a very good plastic mass. It is white in colour having few pinkish specks. It shows an Atterberg No. 22, water of plasticity - 31.2%, dry linear shrinkage - 6%, modulus of rupture (green) - 45.7 Kg/sq cm, and Vitrification at 1305^0 C - colour dull white major amount of vitrification T.L.S. 16%, at 1400^{0}C - colour grayish, completely vitrified T.L.S. 17%. The primary mineral is Kaolinite and impurities are fine silica, hematite and rutile. The ball clay can be used in the manufacture of sanitary ware, cream earthenware, low and high-tension electrical porcelain bodies, cold face mica insulation bricks and potteries etc.

Chandi

A large deposit of clay located one km north of village Chandi (near Kolayat), where two clay beds are associated with siltstone and ferruginous sandstone and exposed for about 500 m strike length. The clay beds are 1 to 8 m thick. The estimated reserves are about 6 million tons. The Chandi clay is white, highly silty and highly plastic.. The clay shows an Atterberg No. 22, water of plasticity -32%, Dry linear shrinkage - 6%, modulus of rupture (green) - 44.8 Kg/sq cm, and vitrification at 1305^0 C colour - creamish white fairly T.L.S. 17% vitrified, at 1400^0 C colour grey completely vitrified T.L.S. 18%. The primary mineral is kaolinite and impurities are fine silica, hematite and rutile. This white ball clay is used in making low-tension electrical insulator, cream earthenware, ceramic tiles and sanitary ware etc.

Marh

A deposit of clay is located near village Marh about 5 km from Kolayat Railway Station. This deposit covers an area of about 4 sq km with an average thickness of about 3 m. The estimated reserves are about 6 million tons. The clay is white. The clay posses very good slaking characteristics plasticity, having very fine particle size. It shows following physical properties: Atterberg No. - 23, water of plasticity - 33%, dry linear shrinkage 7%, modulus of rupture (green) - 68.3

Kg/sq cm and vitrification at 1250°C dull white colour with major amount of vitrification T.L.S. - 16%, at 1305° C colour grayish completely vitrified T.L.S. 19%. The primary mineral is Kaolinite and impurities have fine silica, hematite, and minor quantity of lime. This ball clay is used for the manufacture of cream earthenware, low-tension electrical insulator vitrified dinnerwares, sanitary ware and tiles etc.

Gurha

The village Gurha is located about 10 kms North West of Kolayat where a large clay deposit is located. The deposit is quite extensive with an average thickness of about 3 m. The estimated reserve is about 6 million tons. The clay shows soapy hand feel, dull white colour and has very good plasticity. It slakes in water readily. The clay shows an Atterberg No. - 17, water of plasticity - 29.2%, Dry linear shrinkage - 6% and vitrification at 1250°C colour- creamish fairly vitrified T.L.S. - 15%, at 1305°C colour gray completely vitrified T.L.S. 16%. The primary mineral is Kaolinite and impurities are fine silica, iron, rutile and carbonates. The carbonate impurities could be removed by levitation. This low to medium grade ball clay is used in making sewer pipes, common art potteries and in moulding purposes.

Deh

A large deposit of clay having estimated reserve of about 3 million tons is located at village Deh, near Kolayat in Bikaner District. The clay is white in colour with pinkish specks unevenly distributed on the lumps. The clay slakes completely within few minutes. It has very good plasticity with a soapy hand feel. It shows an Atterberg No. 20, water of plasticity 32%, dry linear shrinkage - 6%, and vitrification at 1250°C colour - dull white with major amount of vitrification T.L.S. 15%, at 1305°C colour grey completely vitrification T.L.S. 16%. The primary mineral of this clay is Kaolinite and impurities are fine silica, iron-oxide, , titanium oxide and minor amount of lime. This clay is used in manufacture of S.W. pipes, cold face insulation bricks, as filler in soap or detergent. This may also find suitable for rubber industries.

Indo-Ka-Bala

A clay deposit exists at village Indo-Ka-Bala near Kolayat in Bikaner District. There are four clay beds with an average thickness of 2.5 -3 meters. The estimated reserves in this deposit are about 3 million tons. This plastic clay occurs in association with ferruginous sandstone. The slaking characteristics are moderate having the following physical properties: good plasticity, Atterberg No. - 21.0, water of plasticity - 31.2%, dry linear shrinkage 6%, and vitrification at 1250° C colour gray with fairly vitrified T.L.S. - 14%, at 1305° C - colour grey with complete vitrification T.L.S. 16%. The primary mineral is Kaolinite and impurities are fine silica, iron oxide, rutile and carbonates. The carbonate impurities could be removed by levitation. This low-grade ball clay finds its use in making sewer pipes, cold face mica, insulation bricks, and common art potteries

Table-2 Chemical analysis of clays of different areas of Bikaner District

	KOTRI	CHANDI	MARH	PALANA	BARSINGSAR	GURHA	NAL	SURAJ KI DHANI	INDO KA BALA	
SiO_2	56.90	45.15	47.71	59.24	56.98	64.32	49.18	57.74	47.71	53.84
Al_2O_3	24.17	37.06	35.40	25.4	28.16	20.23	32.90	28.03	35.40	29.27
Fe_2O_3	2.50	1.80	1.57	1.35	2.40	2.24	1.74	2.10	1.57	1.75
CaO	0.56	0.11	0.15	0.13	0.56	0.56	1.20	0.56	0 .56	1.12
MgO	0. 90	0.31	0.28	0.30	0.40	0.20	0.60	0.20	0.20	0.40
Na_2O	2.90	0.10	0.51	0.07	0.40	0.67	0.20	0.20	-	0.20
K_2O	-	0.96	0.76	0.90	0.84	-	0.20	0.20	-	0.85
TiO_2	0.63	2.56	2.00	0.34	0.80	2.33	0.30	2.56	1.20	1.24
LOI	6.00	12.33	12.78	12.28	9.20	8.82	10.65	10.68	10.68	10.50

including in moulding purposes. Calcined clay can be used as porcelain material.

Suraj-Ki-Dhani

The clay deposit present near Suraj ki Dhani with a strike length of about 3 km. The colour of this clay is creamish white and it is associated with ferruginous patches. The clay beds are 1 to 2.5 meters in thickness. The reserves of clay are about 5 million tons. The clay of Suraj ki Dhani is a low grade plastic clay; having following physical properties: colour- creamish white, plasticity by hand feel-good, Atterberg No. 20, Water of plasticity - 35.20%, Dry linear shrinkage - 5%, Vitrification at 1305^0 C colour- dull white complete T.L.S. - 17%, and vitrification at 1400^0 C colour-grayish, bloating observed T.L.S. 17%. The primary mineral in this clay is Kaolinite and impurities include fine silica, hematite and rutile. This low-grade ball clay is used in making sewer pipes, cold face mica insulation bricks, low vitrification earthenware bodies, and common art potteries etc.

Nal

The Nal is a village located about 12 km west of Bikaner city. The deposit exists 2 km further west of this village. There are two clay beds found to occur associated with siltstone and ferruginous sandstone, and exposed for about 500 m strike length. The clay beds are 1.50 and 2.00 m thick being white highly silty and highly plastic. The reserves are about 6 million tons. The Nal clay shows following physical properties: colour dull white, plasticity by hand feel good, Atterberg No. 20, water of plasticity - 31%, dry linear shrinkage - 3% and vitrification at 1305^0 C colour is dull white, complete T.L.S. - 18.5% and vitrification at 1400^0 C colour-dull white complete T.L.S. 18.5%. The primary mineral is Kaolonite and impurities are fine quartz, silt, iron and rutile. This plastic ball clay is used in making earthenware bodies, low-tension insulators, in cold face mica, insulation bricks, in a batch of sanitary ware body etc.

Palana

The Palana of district Bikaner is well known for its lignite deposits. The fire clay is also found below and above lignite as 0.60 to 0.90 meter thick seams at varying depths from 39 to 76 meters. The estimated reserves at Palana are about three million tons. The fire clay can be extracted as a byproduct by after scrapping the over burden. It will also expose the lignite seam underneath. It is blakish grey in colour having good plasticity and slaking characteristics. It shows an Atterberg No. - 31.0, water of plasticity - 29%, % Dry linear shrinkage - 4%, and vitrification at 1305^{0}C colour brownish white fairly vitrified, at 1400^{0}C - colour brownish white complete vitrification. The primary mineral is Kaolinite and having impurities of fine silica, iron oxide and hydroxide. This fire clay is suitable for manufacture of fireclay bricks. It is also suitable for manufacture of saggars and kiln furniture. It can be used for manufacture of SW pipes insulation bricks *etc.*

Barsinghsar

The Barsinghsar area is located about 5 km of Palana in Bikaner district. The fire clay occurs here with an average thickness of 0.5 mts. to 0.8 mts. at the various depths of 30 to 120 meters below and above lignite. The reserves are about 2million tones. The plasticity of clay is good. It shows dull white colour, atterberg no - 23, water of plasticity 38%, dry linear shrinkage 6%, modulus of rupture (green) 53.2 Kg/Sq cm, and vitrification at 1305^0 C colour dull white with major amount of vitrification T.L.S 17%, and at 1400^{0}C - colour creamish completely vitrified T.L.S. 20%. The primary mineral is Kaolinite and impurities are fine silica, iron oxide and minor amounts of carbonates. The fire clay can be used after washing in the manufacture of low-tension electrical porcelain, whiteware tiles, sanitaryware and refractories.

GENESIS

Clays are most widespread and earliest mineral substance utilized by human being. It carries the records of various civilizations of the world in a great variety of forms. Clays are the member of hydrous aluminium silicate group, which results from alteration of weathering of rocks chiefly of Igneous Origin. The weathered material either remain insitu, where it is formed and so gives rise to residual clays, or is transported by various agencies - water, glaciers, wind and deposits as beds in the sea or lakes.

Clay minerals are generated by weathering of aluminous silicates, but they may also be formed in the environment of sedimentation (early diagenetic) or in the post-burial environment (late diagenetic). Some may be formed by precipitation from ionic solution, others by the crystallization of a gel.

In general, clay minerals seem to form by alteration of some pre-existing minerals. This is especially true for those formed in the soil profile. Kaoline, montmorillonite and illite all appear to arise from the weathering of many kinds of rocks under various conditions. Any aluminium silicate parent material can yield kaolin by weathering provided that leaching removes K, Na, Ca, Mg and ferrous iron and that hydrogen is added. Kaolin is readily formed from granites, whereas gabbro tends to yield minerals of the montmorillonite.

The Litho-units in the area show bedded pattern of sedimentary deposition and lie above one another, although formations along with clay bed shows swelling and pinching behavior reflects undulating basin characteristics. The clay beds are consistently present in dip as well as in strike direction. No major structural disturbance noticed in the area except inclined strata and minor local faults (observed in open mines in the district).

The deposition took place in the Nagaur-Ganganagar Basin, which evolved initially in Neoproterozoic period to deposit Marwar Supergroup of rocks. Later there were at least three more distinct phases of evolution of this basin. In the last phase during Lower Tertiary a thick pile of rocks deposited in a part of this basin. At first the Palana Formation followed by Marh Formation and in the last the Jogira Formation. The Lower Palaeocene Palana Formation and Upper Palaeocene Marh Formation has gradational contact hence shows continuity of deposition whereas the Marh Formation and Lower to Middle Eocene Jogira Formation show an angular unconformity is a clear indication of a period of non-deposition. The Palana Formation has carbonaceous matter in abundance in the form of lignite and carbonaceous clay/shale shows a near shore deposition. The later depositions of Marh and Jogira took place in slightly far away from shore in the sea. Most of the clays of the district are found in the Marh Formation (some are available in Palana Formation also). All the clays found in the district are essentially transported clays (not residual) as the clays are formed with the degradation of feldspar bearing rocks (igneous rocks) that are not found in the nearby areas. The parent material must be from the rocks of Malani Igneous Suite, which are quite far away from this area.

CONCLUSION

The clays are colloidal in nature, consists of very fine-grained hydrous aluminium silicate minerals and specks of rock fragments. It becomes generally plastic when wet, and become hard under fire. Due these properties the clays can be moulded into any form, which they retain even after firing. The clays have a very wide range of industrial application especially in paper, plastic abrasives, refractory, rubber, textile, cement, pigment, chemical and oil industries etc. The important clay minerals are kaolinite, halloysite, montmorillonite, and illite. Other clay minerals found in small quantities are dickite, nacrite, anauxite, halloysite, saponite, beidelite, allophone, palangorite (attapulgite) etc. The non-clay minerals present in clay are micas, quartz, tourmaline, feldspars, chlorite, siderite, limonite, hematite, etc. Good deposits of clays are found in in various states in which Rajasthan is most important. The Bikaner district supplies about 80 % of total clay production of Rajasthan because its quality is better than other places. The important varieties of clays present in the district are china clay, ball clay, Fuller's earth and fire clay. These clays occur in Palana, Marh and Jogira Formations of Paleocene and Eocene age. Since the clay deposits are found at shallow depth hence exploited by open cast methods, which are economical as well as safe comparatively. The clay deposits of the district are sedimentary in nature and occur in bedded form and are located near Kotri, Chandi, Marh, Gurha, Deh, Indo-Ka-Bala, Suraj-Ki-Dhani, Nal, Palana, Barsingsar etc. the estimated reserves of the district are more than 55 million tones. Further explorations may reveal many new and abundant deposits in the district. The chemical characteristics of the clay of the district make them suitable for a great number of industries.

References

Anon. (1988) Bikaner Clays. Rajasthan Mineral Bulletin, Publ. D.M.G.R., v. 12 (Jan.- June1988), pp.1-15.

Anon. (1999) Clays in Rajasthan. Rajasthan Mineral Bulletin, Pub. D.M.G.R., v. 19-20 (Oct. 1998-March 1999), pp. 11-17.

Anon. (1999) Clays in Rajasthan-II Fire Clay. Rajasthan Mineral Bulletin, Pub. D.M.G.R., v. 20, (April-June 1999) pp. 11-14.

Anon. (2001) Geology and Mineral Resources of Rajasthan. G.S.I. Misc. Pub. No. 30, Part 12., 113p.

Ghosh, R.N. (1983) Tertiary Clay Deposits of Kolayat and adjacent areas in District Bikaner District, Rajasthan. Ind. Min.37 (4), pp56- 69.

IBM (2002): Mineral Year Book. Indian Bureau of Mines Publication.

Jensen, M.L. and Bateman, A.M. (1981) Economic Mineral Deposits. John Wiley and Sons Singapore, 593p.

Khandelwal, D.K., Sharma, S., Shandilya, A.K. and Khandelwal,.K. (2004) Ceramics and Ceramic Minerals. Proc. Int. Semi. on Industrial Prospects of Clay and Ceramic Minerals, Bikaner. 15-17 Feb. 2004. pp. 3- 13.

Krishnaswamy. S. (1979) India's Mineral Resources. Oxford & IBH Publication.

Kumar, K., Rana, R.S. and Paliwal, B.S. (2005). Osteoglosid and Lepisosteid Fish remains from the Paleocene Palana Formation, Rajasthan, India. Paleontology, V. 48, pt. 6, pp. 1187-1209.

Kumar, V. (1999) Evolution and Geological set up of the Nagaur - Ganganagar Basin, NW Rajasthan. In: B.S. Paliwal (Ed.) Geological Evolution of NW India. Scientific Publisher (India) Jodhpur, pp. 34-60.

Paliwal, B.S. (1998). Fish skull from Palana Formation at Hadla-Bhatiyan, District Bikaner, Rajasthan. Curr. Sci. v. 76 no. 12, pp. 1536-1538.

Pareek, H.S. (1981). Basin configuration and sedimentary stratigraphy of Western Rajasthan. Jour. Geol. Soc. India. v. 22, pp. 517-527.

Pareek, H.S. (1984). Pre-Quaternary geology and mineral resources of north-western Rajasthan. Mem. Geol. Surv. India. v. 115, pp. 1-99.

Pettijohn F.J. (2000) Sedimentary Rocks, Third Edi. CBS Pub. and Dist. New Delhi, pp 268-269.

Read, H.H. (1976) Rutley's Elements of Mineralogy. Thomas Murby and Co. 560p.

Roy A.B. and Jakhar, S.R. (2002) Geology of Rajasthan. Sci. Pub. Jodhpur, 421 p.

Shandilya, A.K., Shrivastave, N., and Gahlot, S.K. (2005) Geology Mining and Environment around Ball Clay Mine Near Kolayat, Bikaner, Rajasthan. In Agarwal, V. and Shekhawat, M.S. (Ed.) Proc. Nat. Semi. On Environmental Planning and Management in Mining and Mineral Industries, March 11-12.2005, Udaipur. pp. 73-82.

Sharma, M.M., Khandelwal, D.K., and Gahlot, V.P. (2004) Prospects of Ceramic Minerals in Rajasthan. Proc. Int. Semi. on Industrial Prospects of Clay and Ceramic Minerals, Bikaner. 15-17 Feb. 2004. pp. 61-66.

Sinha Roy, S., Malhotra, G. and Mohanty, M. (1998) Geology of Rajasthan. Geol. Soc. India, 278 p.

Hon'ble Reviewers

1. Prof. B.L. Sharma
   Emeritus Professor
   Department of Geology
   51, Saraswati Marg
   MLS University
   UDAIPUR 313 001

2. Dr. M.P. Shrivastava
   Vice President Mining
   Zoom Developers (P) Ltd.
   7, Factory Road, 3rd Floor
   Safdarjang Enclave, Ring Road
   NEW DELHI 110 029

Editorial Assistance

Dr. Virendra Gaur

Economic Mineralization
K.L. Shrivastava (Editor)
2009. pp.295-317
Scientific Publishers (India), Jodhpur

Section-III

Basin Terminology and Classifications : A Review

A. K. BIYANI AND DIVYA DUDEJA
Department of Geology, DBS College, Dehradun-248001 (U.A.)

Abstract

A basin is a common, distinctive, expansive geomorphic term that has been used in many branches of earth sciences. Large number of terms specifies the different types of basins. Numerous classifications of basins, particularly in sedimentology and petroleum geology, are given in the context of plate setting which are essentially based on tectonic origin of basin forming depressions; ignoring the continental situations of variable dimensions where depositional records address the paleoclimate, paleoseismology, environmental analysis and operations of continental exogenic processes investigations. In fact every global segment, irrespective of size and location, is a part of one basin or other. Various terms for basins and geosynclines critically analyzing available tectonic based classifications of basins and pleas for inclusion of small depositional basins in basin classification are discussed. Based on the geomorphology, tectonics, sedimentology and their size a classification is proposed. Valley, lakes and geosynclines are other forms of basins having specific characters and distinctions between lake and basins and between geosynclines and basins are highlighted.

Key words: Basin, Lakes, Geosynclines, Classifications of Basin, Terminology of Basin.

INTRODUCTION

Basin (origin Latin bacinus) is a geomorphic term implying to a present (as well former) natural depression. Basin is widely used in many branches of geology such as structural geology and tectonics, sedimentary petrology, stratigraphy, geohydrology, engineering geology, petroleum geology, economic geology, paleontology. Other variations of basins like lakes, valley and geosyncline are other very popular and frequently used terms in the geological writings. Lake is defined as a mass of still water situated in a depression in the ground without direct communication with the sea. Geologically speaking, most of the lakes are temporary bodies of water, which are usually formed rapidly and filled up quickly and sometimes leaving only scant evidences of their existence in the geological records. Valley (origin Latin vallis) in larger sense, is used for the low area between the water divides restricted to negative land forms of varying size and shape occupied by either perennial or intermittent streams. A class of basin designated as geosyncline, not preferred in current geological writings, is a specific tectonic term used for severely endogenetic turmoiled site of sedimentation that in later stage gives rise to mountain. Later on geosyncline had become an umbrella term and in many of the geosyncline's varieties endogenetic turmoilling was not a precondition. Perhaps the term basin is so common in geology that few bothered to define this term precisely. The paper reviews the terminology, classifications and their origin of above units. Some of the definitions of the basin are:

(i) Sedimentological definitions

Friedman and Sanders (1978) - A basin is one in which sediments can accumulate, are accumulating or have accumulated. They have identified three kinds of sedimentary basins viz. morphologic, bathymetric and tectonic.

Chapman (1987) - Sedimentary basins are areas in which sediments accumulated to a significantly greater thickness than the sediments of the same age in neighboring areas. The sediments accumulate by virtue of subsidence.

Ingersoll (1988) - A sedimentary basin is an area in which sediments have accumulated during a particular time period at a significantly greater rate and to a significantly greater thickness than surrounding areas.

Einsele (2000) - Sedimentary basins are, in a broad sense, all those areas in which sediments can accumulate to considerable thickness and be preserved for long geological time periods.

(ii) Structural definitions

Billings (1978) - A basin is a synclinal depression that has an indistinct trend.

Jaroszewski (1984) - The term basin is used in many geological senses (e.g. sedimentary basin) and even in structural term it is differently conceived by different workers. For this reason in case of doubt it is advisable to add a qualifying adjective e.g. synclinal basin.

(iii) In Petroleum Geology

Laverson (1977) - Areas of the world where substantial amount of unmetamorphosed sediments occur. Some sedimentary basins are depositional and other structural. They are regionally closed synclines.

Biswas (1992) a sedimentary basin is structurally controlled morphotectonic depression capable of receiving and preserving sediments and having a geological history.

(iv) Geohydrological definition

Hamblin (1989) Basin is a depression into which the surrounding area drains.

Constraints in the Definition of Basin

The above definitions basin point out that basin has different meaning in different but fairly precise meanings in various branches of geology. Laverson (1977) rightly pointed out that sedimentary basins are not as simple as the name might suggest. In geomorphology it is a depression in which surrounding areas drain. In structural geology it is a broad region behaving like a syncline having indistinct trend. In hydrology it is used as a specific term for rainwater transfer or for exploitation of ground water in a broad homogeneous geologic setting. In petroleum geology, basins are deeply buried which were on the surface once. In sedimentology, it is a depression in which accumulation of sediments take place. If sedimentation or water/snow collection is considered as criteria for basin then entire earth can be considered as a basin because everywhere such activities are taking place whether it is Mount Everest or Antarctica where snow is depositing, considerably thick desert loess deposits of China and Mongolia on relatively flat terrain or sediments also deposited on the river beds in fact no place on the earth is immune from wind deposition. River beds deposits have greatly increased our understanding of geomorphic, sediment evolution and deposition and hydrolological processes Study of above cited instances may have great utility in their own regime but they are geologically temporary or thicknesses is less or have no possibility of compaction/ induration in a geologic sense, hence they are not part of traditional basin definition. Tectonic evolution of a basin and subsequent deposition of sediments are two different unrelated aspects

These definitions are also silent on the size of the basins. However, Einsele (2000) opines that basins should have at least 100 km length and several tens of km width and ideally they must have circular to elongated shape but most of the basins are irregular in shape. Ingersoll (1988) opines that sedimentary basins of one type or other today cover about 70% of the earth's surface and contain sediment thickness that ranges from about a kilometer to several tens of kilometers. Some basins are geologically young while others have existed for hundreds of millions years. Sometimes very narrow and thin occurrence of sediments may have disproportionate significance as in the case of marine Paleozoics present in the Gondwana Supergroup of India. Soft sediments exposed in scarps along the river valley length have tremendous utility in the paleoseismology. Upliftment of basinal sediments or even surface sediments e.g. loess or sands in deserts or sediments present of the sides of valleys practically cannot escape the deformation that may be subhorizontal tilting of beds to mild folding or faulting as outer surface of the earth has to face the slow or rapid tectonic movements at one time or other. Basin as a form should be recognized irrespective of their size and very short geological histories, of few to some ten thousands of years, as many small basins, like minor/ trace elements in geochemistry, are proving very useful in inferring local paleoclimatic and tectonic conditions as well as their sensitivity to geomorphologic changes though they may not be of use in petroleum and mineral explorations. In hilly region geomorphology, the first order basins are generally less than 1 sq km in size that form the basis of further analysis and plays valuable role in microzonation of mass wasting (landslides) affected area and in present approach of balanced development small drainage basins of few sq km area called watersheds are used as a unit of homogenous area demarcation. Thus the significance of small basins cannot be overlooked despite wide variations in their size, as general terms like faults or glaciers used irrespective of their size, the same may be applicable to basins also. Additionally for sediment accumulation, a depression is not always a pre-requisite; it can accumulate even on a flat surface in case inability of depositional agencies to transport the load to final destination. This may be because of loss of gradient and volume of water, decrease in wind velocity, melting of glaciers, presence of obstacle, excess loading etc. Such areas like alluvial plain without recognizable depression though may be a part drainage basin, and shape also presently gaining attention and may be designated as *incorporeal basin.*

The above discussion may be the reasons for avoiding comprehensive definition of basin in the geological literature. The workers have not constrained shape and size of the basin, thickness and time period of deposition of the sediments in basin, quantitatively as it is not possible because basins exhibit all sorts of variations.

General Characters of Basins

1. Basins have branch specific meanings and applications.
2. It is a present or former depression on surface or on ocean floor or even it may be surface with fluctuating boundaries.
3. The shape and size varies in wide limit and the areal coverage varies in several magnitudes of kilometers and can be studied on different/any chosen scales.
4. Basins have distinctive sediment fill.
5. The depression may have vastly different time span.
6. Basins have history of isostatic, thermal or tectonic subsidence.
7. They may be tectonic or non tectonic in origin; tectonics play significant role in morphogenesis of large basins.
8. It may have mono- or multicyclic history.
9. A basin study may or may not concern with tectonics, particularly when changes occurring in short duration are investigated. Tectonics is not a matter of concern in study like study of Late Holocene changes in climate and sedimentation or snow regime.
10. Basins may be evolved by surfacial or subsurfacial or by combinations of these processes.
11. Basins may have continuously changing characteristics due to surfacial or subsurfacial processes influencing them.
12. Basin evolution is a very slow process in which thermal regime plays an important role.
13. It may be subdivided into several components those varies from location to location such as continents, transitional zone and oceanic. The components may be nature of basin margin, basin slope, bottom configuration etc.
14. The early history is vastly different from the later history.
15. Sediment deposition in basin is a very slow process and the cumulative thickness of sediments in giant basins may be of several km but rarely exceed ten kilometers.
16. Magmatism, metamorphism and mineral genesis may be associated with the basins of long history.
17. It may have asymmetric cross profiles and is usually deeper towards center.
18. Periphery may be blunt or sharp.
19. It is temporary feature on the earth but their relicts are capable of weathering all challenges thrown by the nature.

20. It is a place of collection/ deposition/ formation of water, sediments or petroleum or mineral deposits.
21. Present day size of former basins, in general, progressively decreases with age.
22. Many smaller basins would have vanished from the surface due to various actions of nature.

A succinct definition based on above characters would be

A basin is a present or former partially or fully closed depressive geomorphic or structural trap, or suitable surface configuration on the earth of variable shape or size having characteristic geological identity for water regulation or for purposeful sediment accumulation.

Mechanisms of Major Basins Formation

The three major mechanisms for basin formations are identified. They are isostasy, crustal loading and deep tectonic effects. **Isostatic** mechanism implies basin formation by change in crustal or lithospheric thickness. Crustal thinning is caused by extensional stretching in normal fault setting, erosion of uplifting part of the earth and subsurface withdrawal of magma resulting into a depression on the surface; lithospheric thickening caused by cooling of lithosphere due to cessation stretching or heating and crustal densification i.e. density increase due to changing pressure/ temperature conditions and/ or emplacement of higher density melts into crust of lesser density. **Loading** or addition of mass in large portion of the earth is caused by transport of thrust sheets, formation of volcanic beds, increase in load of sediments and subcrustal loading by underthrusting of dense lithosphere. **Deep tectonic** causes include asthenospheric flow caused by descent of delamination of subducted lithosphere, mantle convention, plumes and meteoritic impact.

Minor basins can be formed by erosional and depositional, tectonic, volcanic and meteoritic and human impacts.

Horizons of Basin Study

Investigations of basins require collaboration between the different disciplines of the earth sciences viz. geomorphology, structural geology and tectonics, stratigraphy, Intracratonic, paleontology, hydrology, remote sensing, geophysics and geochemistry. These branches play an important part for the complete analysis of a basin. Basin analysis is a major field of activity in petroleum geology where it has to establish the physical chronostratigraphic framework of the basin by interpreting stratigraphic sequences, system tracts and Para sequences and/ or simple sequences on outcrops in fields and subsequently by the well logs, seismic data and age data with high resolution biostratigraphy; construct geological history by working out total subsidence and tectonic subsidence, complete tectonostratigraphic analysis including major transgressive regressive phases and their relation to tectonic events; identify the causes of tectonic unconformities and magmatism; construct subsidence curves, relate changes in rates of tectonic subsidence curves to plate- tectonic events, map tectono stratigraphic units and determine style and orientation of structures with tectono stratigraphic units (ONGC, 2007).

Terminology for Basins

Basins occur on the terrestrial or continental surface, ocean floor, incidentally ocean itself is a giant basin and in between true oceanic and continental area i.e. transitional basins extends from ocean to continent or vice versa. Different terms have been proposed according to geomorphologic and geological nature of the basins.

Terrestrial Basins

1. Morphologic basin is a superficial closed depression on land, of any origin.

2. Rift related basin the term rift is used for regional sized normal fault and several types of basins developed in the rift environment. The East African Rift Valley, Red Sea, Gulf of Suez, Upper Rhine Graben etc. are some of the examples of large sized rifts. If these events occur in platform areas then resulting basin is described as

3. Platform basin. Thermal contraction following doming results in platform basins e.g. Arabo- Nubian Shield. During magmatic doming or upliftment a narrow, V-shaped tectonic depression, extending into a craton from an orogen, which subsides deeply and fills with sediment representing the failed arms of triple junction called **Aulacogen** (Friedman and Sanders, 1978). Recently formed aulacogens occur at both ends of the Red Sea. The Gulf of Suez is one of them that contain 4 km thick succession of Neogene sediments. **Protooceanic gulf-** The deep inlet almost surrounded by land, present prior to the formation of the ocean e.g. Red Sea has the floor of basalt and evidences point out that new oceanic crust is being termed along its median zone. **Continental or**

Interior fracture basins are a generalized term for basins produced by faulting. Einsele (2000) included graben, rift valleys and aulacogen in this category. **Chinese type basin** is resulting from block faulting in the hinterland of continental-continental collision. Such type of basins is present in Tibetan region. **Pericratonic basin** forms during continental rifting by subsidence of Atlantic type cratonic margin or trailing edge of drifting continent e.g. East coast basins of India as Krishna-Godawari. **Marginal aulacogen** is a type of pericratonic basin where basin axis is oblique or orthogonal to the rifted continent margin. In India, Camay basin is one of the examples. Kutch and Narmada basins are other Indian examples.

4. Intracratonic rift basin occurs on the continental crust and preferably forms along reactivated palaeosuture caused by divergence and tension within the continent. Gondwana graben basins such as Pranhita-Godavari basin and Mahanadi basin are the typical Indian examples of Intracratonic rift basin.

The sequence of events in development of continental rift basins:

1. Development of a broad, shallow depression prior to doming or vulcanism.
2. Injection of diapiric asthenosphere caused by adiabatic decompression into the lithosphere setting the stage for partial melting and the onset of magmatism.
3. Formation of domal structure by the buoyant isostatic upliftment heated lithosphere.
4. Crustal attenuation and thinning of the lithosphere.
5. And lastly volcanism and magmatic dyke injection and faulting. This phases retards with time. The initial stage of graben development is marked by deeply penetrating faults forming pathways to the mantle and giving rise to volcanism.

Rift valleys develop and may be associated with voluminous extrusive and intrusive felsic magmatism. These fault-bounded basins are relatively narrow and there is rapid subsidence during early rifting. According to Artyushkov and Baer (1989), there are two main modes of stretching in the development of rift basins. One is S-type stretching, in which the fault blocks are not or only slightly tilted toward the downthrown side of a boundary fault. Another is A-type or antithetic stretching, in which the blocks are markedly tilted and separated by listric faults. Only A-type processes can attain substantial subsidence due to significant crustal thinning. The multi directional rift systems prograde towards each other to interfere and form megarift basins.

5. Foreland basin- In front of a growing fold-thrust belt depressing and flexuring the continental crust under the load of the overthrust mountain belt generate this type of basin. These basins are also called **peripheral basins**. The length of these asymmetric basins tends to increase with time. But influx of clastic material from the rising mountain range often keeps pace with or exceeds subsidence and thus causes basin filling e.g. Siwalik foreland basin. North Alpine foreland, Swiss intraarc basin etc. are some other examples.

6. Intracontinental basins- The other names for this category are **Cratonic interior basin** or **Intracontinental Sag basin** or **Intracratonic basin** or **Interior sag basin-** These are large, slowly subsiding, long lived basins. They are known from all continents and may accumulate thick sediments (up to 10 km), undeformed or little deformed sedimentary successions deposited over long periods of time of the order of 200- 800 Ma. Many basins have a saucer shape and range in size mostly from 2×10^5 to 10^6 km^2. In case of large interior sag basins; major fault systems or a central rift zone may be absent that form the boundaries of the depositional area. Slow subsidence occurs predominantly in response to crustal thinning or to a slightly higher density of the underlying crust in comparison to neighboring areas. In addition, slow thermal decay after a heating event and sedimentary loading can promote and maintain farther subsidence for a longtime or long-term subsidence of Intracratonic basins might be related to a decrease of the mantle heat flow (abnormal cooling) above a cold spot. Basins of this type such as northeastern Asia, around the Siberian platform, and the southern Urals are of Late Proterozoic to Middle Paleozoic (Leighton and Kolata, 1990). Vindhyan basin is the Indian example of Interior sag basin.

7. Cratonic multicycle basin- Cratons are the region comprised of Precambrian/ Archaean mobile belts and platforms, which have not undergone any severe deformational movement since the beginning of Paleozoic. A basin having several cycles of sedimentations belongs to this category. Aravalli region of Rajasthan can be put in this category where at least five cycles of sedimentation have been recognized.

8. Intramontane basin- These narrow troughs or basins evolve by extension within or behind thrust belts on top of the overriding plate parallel to strike orogenic regions. They mark the late stage of mountain building processes e.g. orogenic collapse and continental escape. Some basins developed on top of the over thrust belts.

9. Wedge back basins or piggyback basins- Indian examples are Doon Valley of Uttarakhand and Kashmir Valley. Another variety is **Transverse strike basin,** which is developed perpendicular to regional strike of the area. Examples of Intracratonic basin are reported from the Andes of Ecuador, Peru and Bolivia.

10. Rim basin- In divergent margin setup basin developed on the periphery of rift valley is called rim basin (Miall, 1984). Godawari basin may be included in this category.

11. Evaporite basin- Occurs in continental arid to semi-arid zones and loses more water by evaporation then it receives by precipitation and inflow of river water. Example-Sambhar Lake of Rajasthan in India.

12. Hydrologically closed basin- These basins are both hydrologically and sedimentologically closed. The entered water cannot escape from the basin. The water level is sensitive to climate change and as a result the basin areas vary with time. The later phenomenon is particularly pronounced in cases where the basins are shallow and have gently sloping wide margins. Luni river basin of Rajasthan-Gujarat in India.

13. Lateral Extrusion Basin- Continental or tectonic extrusion as a result of plate collision has been observed within and adjacent to several large orogenic belts. The most spectacular example is the extrusion of Indo-china that is caused by the indentation of the Indian plate along the length of the Himalaya into Asia. A part of the shortening of the colliding continental crust was transformed into lateral extrusion of neighboring crustal fragments. This process led to several long extended strike slip zones, for example 1000 km long Red River strike-slip zone along the northern boundary of Indo-china showing sinistral displacement of 700 km.

14. Swamps- In humid climate and on very gently sloping alluvial plains, the ground water reach the surface and swamps are formed. Tarai region, now dried up due to human efforts, along Siwalik Himalaya was one of the best fresh water swamps in the world. In Ghana Bird Sanctuary of Rajasthan, swampy area is present. Kutch area of Gujarat in India and of Pakistan also becomes swampy in post rainy season.

15. Impact basins- Powerful impacts of giant meteorites instantly produce several meters to several hundred meters deep depression both on the land as well as in the sea/oceans and such depressions may be called impact basins. 100 km diameter Popigai structure of the Anabar Shield, North Siberia is formed by the impact of meteorite about 35.7 million year ago. 40 km diameter Mjolnir Crater of the Central Barents Sea formed by impact of meteorites about 142 million year ago in an epicontinental basin with 300-500m palaeo water depth. Lonar lake of Maharastra has been formed by impact of meteorite about 50-60000 years ago.

(ii) Oceanic Basins

Oceans in themselves constitute primary basins on the globe. Geologically oceans characterized by thin (5-6 km thick) oceanic i.e. basaltic crust (abyssal plain). Global tectonic engine is propelled by oceanic elements. In between ocean and continent a transitional crust, of intermediate crustal thickness, comprising of shelf, slope and rise regions exists. Oceans are water filled giant depressions on the earth that have highly irregular bottom surface with varieties of secondary depressions. Oceanic or continental margins have been divided into active (convergent) and passive (divergent) categories of vastly different tectonic characters.

1. Subduction/Island arc related basin- This group of basins is dominated by convergent plate motions and orogenic deformation. Basins related to development of subduction complexes occur along island arcs (tectonically generated curved chain of island with active andesitic volcanic emission) or active continental margins that include deep-sea trenches, forearc basins, backarc basins and intraarc basins. **Deep-sea trench** are >6000 m deep located at the junction of descending oceanic and stationary continental plates; few tens to few hundred km away from the continent and many of them represent the deepest elongate basins examples are Mariana trench, Sunda trench, Northern Japan trench, Peru trench, Himalayan Trench etc. Due to sediment fill some morphologically resemble continental rise. Trenches commonly do not subside like many other basin types and tend to maintain their depth, which is controlled mainly by the subduction mechanism and also by the geometry and volume of the accretionary wedge on their landward side. **Forearc basins** are marine depositional basins located between magmatic front of the arc and slope break for trench. They vary in size and length that have strike length up to 700 km and have many km thick chiefly turbidites sediments with sources in the adjacent arc system develop between deep-sea trenches and magmatic arcs, either on trapped oceanic crust or on the inner part of an accretionary wedge (Condie, 1987). Forearc basins on modern continental margins are the South American forearc basins and Boso Peninsula while Great valley of California and Xigaze forearc basin, Tibet are ancient forearc basins on land. Indian examples are Andaman forearc basin, Kohima and Tripura-Mizoram forearc basins. **Accretionary basin** is a type of forearc basin. They evolve at the top of the accretionary wedge and their sediments lie unconformably on older, deformed strata. Constructed basin is another type of forearc basin developing on both the arc massif (inner side of the basin) and deformed strata of the accretionary wedge (outer side of the basin). Its sediments lie unconformably on the basement of the arc and accreted sediments of the subduction complex. **Intra-arc basin** situated within the volcanic island constituting the arc and typically containing sediments rich in volcanic debris (Friedman and Sanders, 1978). Sedimentary rocks of Karakoram area of India belongs to this category of **Backarc/Interarc basins** are located between continental margin and hinterland of island arc and are divided into active and inactive categories. Active backarc basin occurs over descending slab behind arc systems. These basins are characterised by high heat flow, relatively thin lithosphere and in many instances the size of the basins is enlarged by an active ocean ridge. Inactive backarc are greater in age, have thicker pelagic sediment and lack evidences of recent seafloor spreading compared to active backarc basin (Condie, 2000). Backarc basins form on pre-existing or newly accreted oceanic crust and they tend to be closed by subduction with limited lifetime. (Einsele, 2000). In southwestern Pacific, Tasman Sea basin adjacent to Australian continent, New Hebrides basin, in the northeastern Indian Ocean Andaman Sea are modern backarc basins. Black sea and Caspian Sea are remnants of backarc basins. In Indus and Karakoram sutures of Himalayan region have sedimentary rock of back arc environment. **Retroarc basins** are formed by late-stage compressional deformation and overthrusting behind the continental-margin arc systems i.e. small marginal ocean basins develop by trapping of the parts of the former back arc basin by retreat of the former active arc. They are filled largely with clastic terrigenous sediments derived from nearby magmatic arc and the more distal intraarc sediment sources and magmatic arc (Condie, 1987, Einsele, 2000). The sub Andean zones in South America on the eastern flanks of the Andes display the Retroarc basin. **Infrasuture embayment basin** develops in the suture zone during continental collision and suturing (Miall, 1984). **Superposed basin** develops on a suture zone. Post collision collapse of orogenic infrastructure followed by subsidence lead to the development of superposed basin within the collision complex. Karewa basin is a typical Indian example. (Biswas, 1993) **Pannonian type basin** originates from post- orogenic divergence between two fold-thrust zones. They are associated with an A-subduction zone and are floored by thinning continental or transitional crust. They have thinned continental crust and have more complex structural and sedimentary evolution. The Aegean Sea behind the Hellenic arc in the eastern Mediterranean may be regarded as a modern example. **Remnant Basin-** They are narrowing ocean basins shortly before closure. They collect sediments from the neighbouring magmatic arc and form on the approaching passive continental margin. They are also called **relic basins.** Examples are fore deep of the Apennines, Italy and eastern part of the Bengal Basin in India.

2. Oceanic strike-slip basin-In oceanic strike-slip regime transform fault and transcurrent fault basins are present**. In transform fault** basin the motion relative to offset along ocean ridge axes is opposite to that predicted by transcurrent motion (Wilson, 1965). Faults show apparent lateral displacements of many hundreds of kilometers and anatomizing network of faults characterizes these basins. Cambay basin of Gujarat is an example of this type. **Transcurrent fault** are very large size strike-slip faults, which are almost vertical and do not terminate against ridge or trench as is the case of transform fault at times basins are developed in them (Condie, 1989). Dangarjong fault and Arun Fault of Nepal are transcurrent in nature and sedimentation has taken place around these faults.

3. Atlantic type basin- The basin, which has very gentle slope and reach through continental shelf to abyssal plain, can be called as Atlantic type basin. Indian examples are Bay of Bengal and Arabian Sea. Morphologic basin on the sea floor is a **bathymetric basin**. **Oceanic abyssal basins** are present in abyssal plain that is more than 2000 m deep in ocean.

4. Oceanic fracture basin A generalized term used for that ocean floor basin which is result of large scale fracturing.

5. Coastal Basin refers to those basins that are result of differential subsidence and/or uplift of the landward side of the stretched continental crust. They are present on Atlantic type continental margin.

6. Barred basin/Seepage basins are the lagoons, which are separated from the open sea by a permeable sea bar or ridge. Evaporative draw down of water level in the closed lagoon then causes sea water to flow underground into the lagoon and denser brine to flow back into the ocean. Due to the restricted inflow and reflux this system creates a transitional situation between a hydrologically closed and an open sea basin. Such basins occur along the eastern and western coasts of India.

7. Oceanic sag basin occupies the area between mid-oceanic ridges to continental rise i.e. outer edge of the transitional crust along a passive continental margin. They commonly accumulate deep sea fan or basin plain sediments.

8. Tilted block Basin during crustal collision, some foreland and retro arc basins are broken up into smaller blocks and subsided blocks form basinal depressions, which are called tilted block basins.

(iii) Transitional Basins

1. Narrow oceanic gulf- Gulf is a deep inlet of the sea almost surrounded by land with a narrow mouth. e.g. Gulf of Kutch in Gujarat, India

2. Immature Ocean Basin/Hypersaline/ Hyposaline Basins are shallow or deep-water basins of limited extent situated adjacent to sea and connected with the oceans through a gateway or strait. The opening may be deep and wide or characterized by a narrow sill protecting the adjacent basin from the direct influence of the open sea. Depending on the dimension of this opening and the climate, adjacent basins may have a more or less normal salinity, or they tend to become brackish (hyposaline) or more saline (hypersaline). Some adjacent seas represent immature ocean basins, which may evolve into wider ocean basins in later stage. The subsequent uplift and erosion, providing land exposures or drill holes in such sequences can reveal the early climatic and oceanic history of such a basin development. Gulf of Mannar in India is a hypersaline basin.

3. Arid Basins are characterized by excess evaporation over the sum of precipitation and river runoff. In some cases the seawater flows as a warm surface current into the arid basin and tends to become increasingly saline due to continued high evaporation e.g. Red Sea, Mediterranean Sea, Persian Gulf and Sambhar Lake and lakes of Ladakh region in India.

4. Humid Basin- In Humid basins, precipitation and river discharge into the basin are commonly greater than water loss by evaporation; examples are Baltic Sea and Black Sea. A narrow, shallow entrance, allowing the denser seawater to enter the basin as subsurface inflow, while the less saline water from the basin flows as surface current out into the sea, characterizes them.

5. Extra continental marginal basin- This term is used by for a group of basins named as closed basin, open basin and fore deep basin and the name implies that these basins are formed near the margin of continent, they may or may not have connection with ocean or sea

6. Rifted continental marginal basin- These basins result from the breakup of Pangea and the stretching and thinning of a 500 to 1000 km wide zone of continental crust during the Mesozoic prior to continental breakup and present sea floor spreading. Basins of this type are found below the continental shelves of passive continental margins. Some of these basins are exposed on land where they commonly display simple structures and relatively short time periods of basin filling. (Einsele, 2000). Western coast of India fulfills this condition.

(iv) Terrestrial and Oceanic Basins

1. Physiographic basin is a depression in the surface of land or seafloor that may or may not be in filled with sediments. **Asymmetric rift basins** are developed in the half-graben settings and basin fill thickness is more on footwall side than hanging wall side. **Pure and simple shear basins** occur in rift environment and are produced by pure shear or without rotation of fractured block and simple shear; some blocks are rotated in non-coaxial deformation condition, mechanisms. Pure shear basins arte formed rapidly formed in old, thermally equilibrated crust and showing uniform but limited extension while simple shear basins are formed in the geologic young and hot crust and show non uniform large extension.

2. Wrench basin/ Strike-slip Basin- Transform motions may be associated either with a tensional component (transtensional) or with a compressional component (transpressional). Transtensional fault systems locally cause crustal thinning and therefore create narrow elongate pull-apart basins. **Pull-apart basin** if they evolve on continental crust continuing transform motion may lead to crustal separation perpendicular to the transform faults and initiate accretion of new oceanic crust in limited spreading centers. Until this development occurs, the rate of subsidence is usually high. The increasing sediment load drives later subsidence mainly. Continental pull-aparts commonly collect alluvial fans, fluvial and lacustrine deposits. These basin fills are mostly asymmetric in both longitudinal and transverse crosssections. Their life span is commonly short and sedimentation rates their thickness and structural complexities are often greater then rift basins. Gulf of Mannar is a pull apart basin.

3. Transpressional systems generate wrench basins of limited size and endurance. Gulf of California, Caymen trough in the Caribbean Sea, San Andreas Fault system in California are some examples of pull-apart basins.

4. Pre-, syn- and post depositional basins are related with sediment deposition. Rapid tectonic movements predate significant sediment accumulation forms a pre-depositional basin or morphological basin, which is later, filled by post-tectonic sediments. In syn-depositional basins sediment accumulation is affected by syn-depositional tectonic movements. In post-depositional basins the deposition of sediments largely predates tectonic movements forming a distinct basin structure.

5. Terrane related Basin- They are situated between micro-continents consisting at least in part of continental crust (Nur and Ben Avraham, 1983) and larger continental blocks. The substratum of these basins is usually oceanic crust. They may be bordered by a subduction zone and thus be associated with basins related to either to subduction or collision

6. Starved basin- Sediment starved basins evolve far away from efficient, mostly high relief terrigenous sediment sources. Their upward fining and thinning vertical facies successions indicate deepening of basins, and their slowly deposited sediments tend to become increasingly rich in biogenic material and sometimes also in organic matter. Examples include central parts of large oceanic basins and smaller basins in the vicinity of mid-oceanic ridges; isolated submarine highs and platforms; epicontinental seas surrounded by lowlands; arid adjacent seas with drainage areas of low sediment yield and closed lakes receiving little terrigenous sediment. e.g. East Sardinia Continental margin and Submarine plateaus of the southwest Pacific.

7. Polyphase basins have experienced a polyphase history i.e. two or several phases of structural evolution and basin filling. Tarim Basin of western China, Salin Basin of Myanmar and Magdalena basin of Colombia are examples of this type. They are subdivided into **superposed basins** which develop on a suture zone or formed by subsidence after collapse of orogenic terrain in post collisional period; **successor basins** consisting of changing basin type along a long persisting tectonic weakness and

8. Resurgent basins in which original basin style is repeated and maintained along order tectonic boundaries and trends (Klien, 1987). **Polyhistory basin-** Over long period of time a sedimentary basin may evolve from one basin type into another and thus exhibit a complex tectonic and depositional history.

(v) General Terms for Basins

1. **Tectonic basin** is a basin containing a thick body of sedimentary and/or volcanic strata, which could be deposited where crustal movement antedated or accompanied accumulation of the strata.
2. **Balanced basins are** those in which sediment supply and vertical build up approximately compensates for subsidence. No long-term drastic change in the proportion of terrigenous and biogenic sediment components occur in these basin.
3. **Extensional Basins** are rift basins, growing ocean basins and pull-apart basins. For all of them their size increases with time leading to the general trend of decreasing rates of terrigenous sedimentation.
4. **Successor basin** characterized by changing basin type along a long persisting tectonic weakness
5. **Closed basins** are those, which have no supply of sediments now.
6. **Open basin** are small and display negligible subsidence of their basin floor in relation to the surrounding land areas compared with the high sedimentation rates is open basin.
7. **Morphologic basin** may be of tectonic or nontectonic origin. Nontectonic morphologic basins include features formed by differential erosion, by collapse above dissolved subsurface evaporates and by many other processes. A morphologic basin of tectonic origin forms where crustal movements antedate significant sediment

accumulation. Basins without physical form and shape may be called **Incorporeal basin. Basin inversion** denotes to the trend of upliftment of an originally subsiding extensional basin. Basin in later stage is subjected to compression or transpression type of deformation for example High Atlas rift in Morocco.

Classifications of Basin

Basins have been classified according to nature of branch dealing with the basins. Except sedimentary petrology, in all other branches classification scheme is simple. In geomorphology drainage basins have classified by categorizing them in different orders. In most popular scheme of classification of Strahler (1964) the smallest trunk channel or stream is described as first order basin and where two first order streams join each other, the unified stream becomes a second order stream; like wise stream order jumps finally major rivers like Ganga, Amazon, and Nile get the highest order, usually seventh or eighth order. The basins are also identified according to dominant operative processes in the region like cirque in glaciated region or playa in the arid regions or sinkholes and their varieties in karst region. A good attention was paid in case of sea and ocean where varieties of depressed geomorphic features have been identified on the floor. In geohydrology basins are identified on the basis of aquifer characteristics or hydrological properties of the rocks. Basin classifications in Intracratonic and petroleum geology are dealt with utmost seriousness and all recent classification schemes are based on large size basin forming plate tectonic processes. These schemes also involve lithospheric substratum (continental, oceanic transitional), location on plate (interplate or

Table 1 : Basin classification by Dickinson (1974)	Table 2 : Basin classification by Klemme (1975)
Oceanic Basin	Cratonic interior basin
Rifted Continental margins	Cratonic multicycle basin
Pre-rift arch	Cratonic rift basin
Rift Valley	Extracontinental margin basin
Proto-oceanic gulf	Closed basin
Narrow oceanic gulf	Foredeep basin
Narrow ocean	Open basin
Aulacogen	Pull-apart basin
Arc Trench system	Intermontane basin
Trench	Transverse strike basin
Subduction zone	Delta basin
Forearc basin (arc-trench gap)	Successor basin
Intraarc basin	
Backarc basin (Interarc basin)	
Retroarc basin	
Suture Basin	
Foreland basin	
Intracontinental Basin	

Table-3 : Basin classification by Bally and Snelson (1980)

1-Basins located on rigid lithosphere, not associated with formation of megasutures.
- 1.1-Related to formation of oceanic crust
  - 1.1.1- Rifts
  - 1.1.2- Oceanic transform fault associated basin
  - 1.1.3- Oceanic abyssal plains
  - 1.1.4- Atlantic-type passive margins with straddle Continental and oceanic crust
    - 1.1.4.1-Overlying earlier rift systems
    - 1.1.4.2-Overlying earlier transform systems
    - 1.1.4.3-Overlying backarc basins of
- 1.2-Located on pre-Mesozoic continental lithosphere
  - 1.2.1- Cratonic Basins
    - 1.2.1.1-located on earlier rifted grabens
    - 1.2.1.2-located on former backarc basins of 3.2.1type

2-Peninsular basins on rigid lithosphere, associated with formation of compressional megasuture
- 2.1-deep sea trench or moat on oceanic crust adjacent to B-subduction margin
- 2.2-Foredeep and underlying platform sediments or moat on continent crust adjacent to A- subduction margin
  - 2.2.1-Ramp with buried grabens, but with little block faulting
  - 2.2.2-Dominated by block faulting
- 2.3-Chinese type basin associated with distal block faulting related to compressional or megasuture and without associated A- subduction margin

3-Episutural basins located and mostly contained in compressional megasuture
- 3.1- Associated with B-subduction zone
  - 3.1.1- Forearc basins
  - 3.1.2- Circum Pacific backarc basins
    - 3.1.2.1-Backarc basins floored by oceanic crust and associated with B-subduction
    - 3.1.2.2- Backarc basins floored by continental or intermediate crust associated with B-subduction
- 3.2- Backarc basins associated with continental collision and on concave side of A-subduction arc
  - 3.2.1- On continental crust or Pannonian type basins
  - 3.2.2- On transitional and oceanic crust or W. Mediterranean type
- 3.3- Basins related to episutural megashear systems
  - 3.3.1- Great basin type basins
  - 3.3.2- California type basins

intraplate) and the nature of plate margins (divergent, convergent, transform) for classification. However, many depositional environments have been ignored.

Large numbers of classifications have been proposed for basins, few important classifications are given in Table 1 to Table 8

Proposed Classifications of Basin

Classifications based on tectonics, though precisely explain the origin of large sized depressions, as tectonic processes do not operate on small scale. Mere presence of depression will perhaps not serve a meaningful purpose unless it has the presence of sediments and sediment deposition is work of

Table- 4 : Basin classification by Kingston et. al. (1983)
Interior sag basin
Interior fracture basin
Marginal Sag basin
Wrench basin
Trench associated basin
Oceanic trench basin
Oceanic fracture basin
Oceanic sag basin
Polyhistory basin

Table- 5 : Basin classification by Miall (1984)
Divergent margin basins
Rift basins
Rifted arch basin
Rim basins
Oceanic margin basins
Red Sea type basins
Atlantic type basins
Aulacogens
Convergent margin basins
Trenches and subduction basins
Forearc basins
Interarc & Backarc basins
Retroarc (Foreland) basins
Transform and Transcurrent fault basins
Basin setting
Plate boundary transform fault basin
Divergent margin transform fault basin
Convergent margin transcurrent fault basin
Suture zone transcurrent fault basin
Basin type
Basins in braided fault systems
Fault terminations basins
Pull apart basins in en echelon fault system
Basins developed during continental collision and suturing
Foreland basins
Peripheral basins
Infra-suture embayment basins
Associated transcurrent fault basins
Cratonic basins

exogenic processes. Thus these two are complementary processes and one cannot be excluded from the other. Hence a twin fact based basin classification would be a required one. Einsele (2000) also stated that' On the surface of our present day globe, on land and below the sea, hundreds of the depositional areas are known which meet the definition of the sedimentary basins ...' (pp.12) and 'sedimentary basins originate from various endogenic processes and can be classified on the basis of plate tectonic models... however, the depositional environments and sedimentary facies of the basins are to a large extent controlled by exogenic processes... an additional basin classification is needed which is based on the various depositional environments..' (pp. 15). This statement may not be true in many cases as every area of the earth falls in the definition of basins of one kind or other and many present depositional places have nothing to do with endogenic processes such as present day deltaic deposits or channel

Table 6 : Basin classification by Klein (1987)
Plate interior
Interior intraarc basin
Cratonic margin basin
Passive margin
Rift basin
Aulacogen
Flexure basin
Active margin
Trench basin
Trench slope basin
Forearc basin
Intraarc basin
Backarc basin
Retro arc basin
Transform margin
Pull- apart basin
Transform basin
Collision margin
Foreland basin
Superposed basin
Margin independent
Polyhistory basin
Successor basin
Resurgent basin

Table 7 : Basin classification by Mohriak, Hobbs and Dewey (1990)	
Intracratonic sag basin	C no major fault control
Rift basin	A basins associated with active deep faults
	B Basins associated with major faults that do not control subsidence
	D Basins associated with low angle detachment
	E Basins associated with crustal thinning and Moho uplift
	F Basins of almost unstretched crust carrying a thin veneer of sediments
	A, D and E
Divergent margin basins	
Oceanic basins	Not foreseen or C
Flanking ocean ridges	A, D and E
Convergent margin basins	

deposition, as such basins are of exogenic origin. The attempted classifications address to these aspects and size of basins also. Three primary settings viz. terrestrial (characterized by continental crust), transitional (transitional crust) and oceanic (oceanic crust) of basins have been recognized; these primary settings of the basins are categorized or subcategorized using geomorphic and tectonic aspects in their formation further divided and subdivided as follows:

1. Terrestrial basins

A. Exogenetic

i) **fluvial** meander belt deposits (chute cut off, neck cut off, and oxbow lake), back swamp deposits, braided deposits deltaic plain deposits

ii) **glacial**- cirque, niviation cirque, glacial troughs, hanging valleys, fjords, piedmont lakes, out wash plain, kettle, fosse

Table-8 Tectonic basin classification (after Kingston et. al., 1983; Mitchell and Reading, 1986 adopted by Einsele, 2000)

Basin Category	Special Basin type/Synonyms	Underlying crust	Style of tectonics	Basin Characters
Continental/ Interior Sag basin	Epicontinental basins/Intracratonic basins	Continental	Divergence	Large areas, slow subsidence
Continental/ Interior fracture basin	Graben structures, rift valleys& rift zones, Aulacogens	Continental	Divergence	Relatively narrow basins, fault bound rapid subsidence during early rifting
Basins on passive continental margins, margin sag basin	Tensional rifted basins, tension sheared basins, sunkmargin basins	Transitional	Divergence+ shear	Asymmetric basins partly outbuilding of sediment, medium to low subsidence during later stages
Oceanic sag basins	Nascent ocean basin	Oceanic	Divergence	Large, asymmetric, slow subsidence
Basin related to subduction	Deep sea trenches	Oceanic	Divergence	Partly asymmetric, greatly varying depth of subsidence
	Forearc basins Backarc basins Interarc basins	Transitional, Oceanic		
Basin related to collision	Remnant basins	Oceanic	Convergence	Activated subsidence due to rapid sedimentary loading
	Foreland basins (peripheral), ntracra (intermontane), broken foreland basins	Continental	Crustal flexuring local convergence or Transform motion	Asymmetric basins, trend to increasing subsidence, uplift & subsidence
	Terrane related basins	Oceanic		Similar to backarc basin
Strike slip/ Wrench basin	Pull apart basins (transtensional) and transpressional basin	Continental/oceanic	Transform motion divergence or convergence	Relatively small, elongated, rapid subsidence

iii) eoline- various sand, silt and clay depositional sites

iv) lakes

v) karst- caves, tunnels, solution valley, dry bed, blind valley, uvala, sink hole, doline, solution pan, karst lake

vi) marine-transgressive and regressive

vii) others- anthropogenic, bomb and mine and meteoritic crater

B. Endogenetic

i) volcanic craters, calderas and volcanic-tectonic depressions

ii) faulted- rift or graben, pull-apart

iii) folded- synclinal

iv) complex continental extrusion, epicontinental, foreland, transverse, longitudinal, Intra-arc type

2. Transitional

bay, adjacent sea, seepage, barred, lagoon estuarine,

drowned valley, submarine canyon, fan valley, delta front trough, slope gully deep-sea channels, meteoric craters.

3. Oceanic

a. plate boundary zone

i) **divergent**- fore arc (accretionary, residual, constructed and intramassif), back arc, Retroarc, remnant, pericontinental or marginal, composite, deep-sea trenches

ii) **convergent**- graben,

iii) **transform fault**- pull-apart

b. within plate pull-apart

c. geomorphic or non tectonic sea trough, abyssal plain, drowned valley, basin (Thornbury, 1969)

d. oceanic meteorites

Basins have extreme variations in their size ranging from few tens of square meters (e.g. desert and in glacial regions) to millions of square kilometers (e.g. Pacific Ocean). To avoid expansion of classification, it is proposed to categorize the basins in limited orders where area difference between lower and higher limits is of 100 times, with exception of I and II orders, thus seven orders are proposed (Table 9).

Giant basins cover the entire ocean like Indian Ocean or Pacific Ocean or continental portion of this scale, however, the largest Amazon basin of South America covering an area of 6.1 million sq km even do not fall in this category as its area is less than 10 M sq km. In case of dealing with a small basin in relation to a bigger basin then an adjective can be added like V b basin that means dealing with Tiny or II order basin in context of a Large or V order basin.

Indian Sedimentary Basins

Biswas et al. (1993) recognized (Fig. 1) thirty eight Indian sedimentary basins in south of the Main Boundary Thrust of Himalaya. However, in the Himalaya several sedimentary basins are present to the north of MBT and these basins cannot be ignored from Indian list, if included, then number reaches to forty-five. These basins are Main Lesser Himalayan Basin, probably extension of Vindhyan basin and smaller Kakara and Subathu basins; Higher Himalayan basin, which is later, metamorphosed in varying grades; Tethyan Himalayan Polyhistory basin and Indus and Shyok Suture Zones basin. In addition to these basins in entire Himalayan territory as well as in peninsular India several small size basins were also present having geological evidences for example Kashmir Basin, Doon Basin, Bagh bed Basin, Lameta basin etc. The Higher Himalayan basin poses a peculiar situation because of its low to medium grade polymetamorphic sedimentary fill. These Higher Himalayan rocks are mainly metamorphosed during Himalayan orogeny but they have been debated for pre Himalayan metamorphism dating back even into Precambrian. If Higher Himalayan metamorphic rocks are included in basin classification then it is logical to include peninsular metamorphic rocks also in the given table. However, peninsular basins having metamorphosed sediments are not included in the table because identifications and number of basins are not clear due to polyphase deformations and metamorphisms. The modified classification of Biswas et al. (1993) have been given in Table 10.

Difference Between Lakes and Basins

Lake is a common used term in geology. They are the

Table -9 Proposed Orders of Basins

Order	Sub order	Name	Area in square kilometers
I	a	Micro	<1
II	b	Tiny	1 - 10 (10^1)
III	c	Small	10 - 1000 (10^1-10^3)
IV	d	Medium	1000 - 100000 (10^3-10^5)
V	e	Large	100000- 10000000 (10^5-10^7)
VI	f	Giant	>10000000 (10^7)

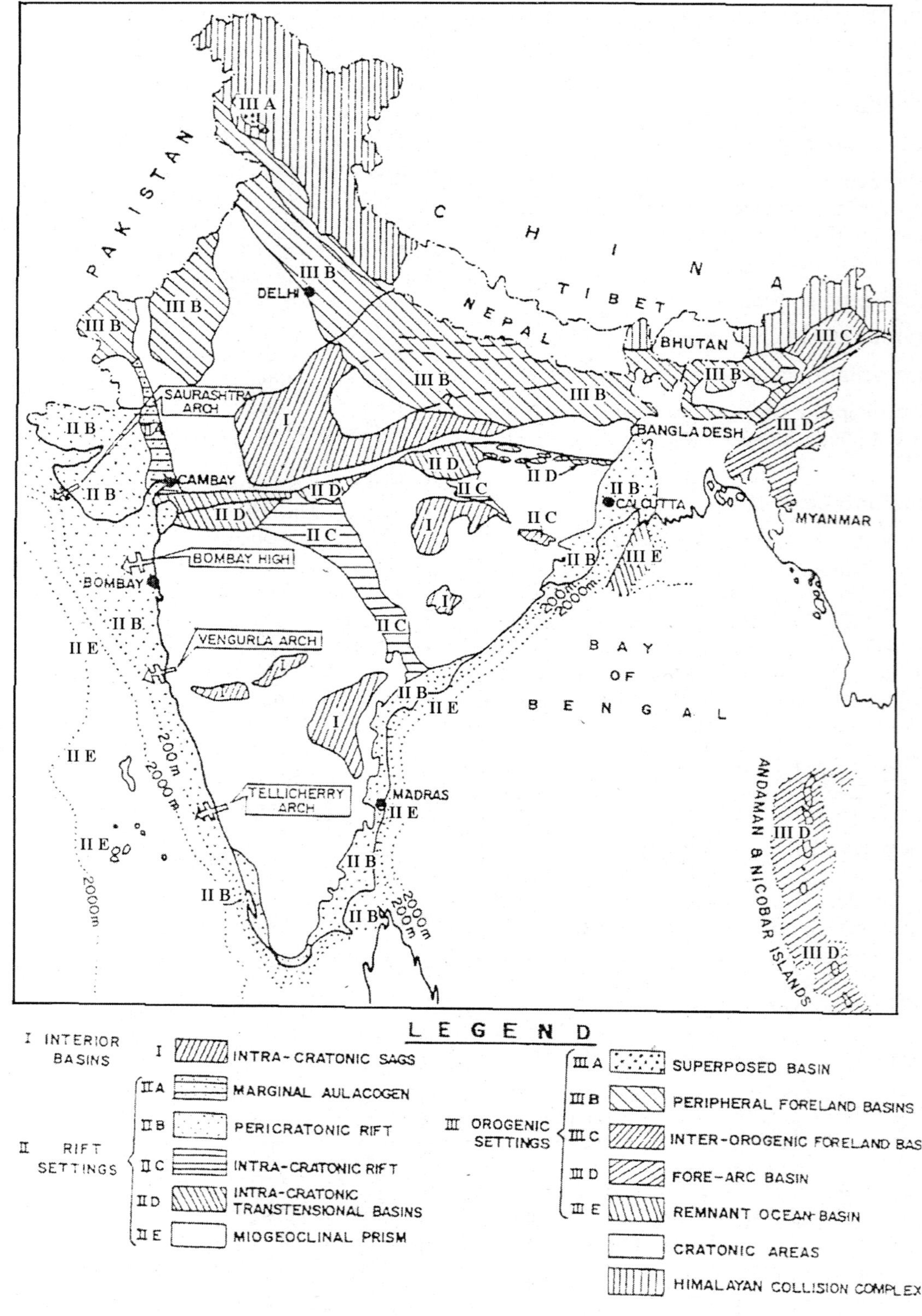

Fig. 1 Sedimentary basins of India and their tectonic classification (after Biswas et al., 1993)

Table-10 Classification of Indian Sedimentary Basins (After Biswas et al., 1993)

Category	Type	Basins
I. Interior basins	Intra-Cratonic Sags	1. Vindhyan basin 2. Chattisgarh-Bastar basin 3. Cuddapah basin 4. Kaladgi basin 5. Bhima basin 6. Main Lesser Himalayan basin
II. Rifted Settings	II_A Marginal Aulacogen	1. Cambay basin
	II_B Pericratonic Rift	1. Kutch basin 2. Saurashtra basin 3. Surat basin 4. Bombay Ratnagiri basin 5. Konkan basin 6. Kerela basin 7. Western Bengal basin 8. Mahanadi basin 9. Krishna-Godavari basin 10. Palar basin 11. Cauvery basin 12. Mannar basin 13. Tethyan basin
	II_C Intra-Cratonic Rift	1. Pranhita Godavari basin 2. Mahanadi Graben 3. Purnea basin
	II_D Intra Cratonic Rift	1. Satpura basin 2. Son basin 3. S. Rewa basin 4. Damodar basin 5. Narmada basin
	II_E Miogeoclinal Prism	1. Deep water basins of west coast 2. Deep water basins of east coast

Table-10 (Contd.)

	III$_A$ Superposed basin	1. Karewa basin
		2. Kakara basin
		3. Subathu basin
	III$_B$ Peripheral Foreland basin	1. Rajasthan basin
III Orogenic Settings		2. Punjab basin
		3. Ganga basin
		4. Lower Assam basin
	III$_C$ Inter-Orogenic Foreland basin	1. Upper Assam basin
	III$_D$ Forearc basin	
		1. Kohima Forearc basin
		2. Tripura- Mizoram Forearc basin
		3. Andaman basin
	III$_E$ Remnant Ocean basin	
		1. Eastern Bengal basin
	III$_F$ Suture Zone basin	
		1. Shyok Suture basin
	III$_G$ Undifferentiated basin	2. Indus Suture basin
		1. Higher Himalayan basin
		2. Peninsular basins containing metamorphosed sediments

non-marine water bodies on the land and are a variety of basins. Geological evidences of palaeo lakes are also reported from the very ancient terrains. The difference between lakes and basins is qualitative rather than sharp as some lakes are like lake Baikal, Caspian sea and other similarly large sized lakes grade into basins. Lakes are divided into open (with water outlet) and closed (without water outlet) and they are of tectonic and nontectonic origins. The important difference between lakes and basins are have been given in Table 11.

Geosynclines

Definitions :

1. Geosyncline is an extensional term for the basin and was proposed for explanation of origin of orogenic regions (mountain) characterized by severe deformation and huge thickness of sediments in comparison to their surroundings. The simple idea of Hall and Dana, in Auboin, 1965, was so elaborated by the later workers to the extent that term lost its relevance particularly in the first half of last centaury. The late attempt of Auboin (1965) to precise the meaning of geosyncline was masked in the exciting development of plate tectonics in that period.

2. Pettijohn (1984) Geosynclines are the mobile belts, which exhibit a characteristic sedimentary, tectonic and magmatic features. He considered most basins as tectonic and essentially is a study of the geotectonic elements of the earth and their behavior.

3. Muratov (1977) Geosynclines are the strips and areas on the earth's crust that were distinguished from the neighbouring ones by a stronger down warping/wrapping throughout the geologically vast stretch of time, resulting in much thicker pile of

Table-11 The Difference between Lakes and Basins

Lakes	Basins
Contain non-marine waters	Marine water dominates in the basins
Highly variable size but generally small	Size is generally large
Subsidence only in large size tectonic lakes	Subsidence is common
Small lakes have short life ranging between 10 ka-100 kilo years (ky)	Very long life in terms of million years
High rate of sedimentation1-10 m/ky	Low rate of sedimentation
Lakes are sedimentological denudation-accumulation system	-
Marked lithological variations and pronounced cyclicity	Limited variation in lithology and cyclicity
Detrital clastic sediment (siliciclastic and reworked carbonate) dominates along with autochthonous biogenic and bioinduced carbonate, autochthonous biogenic silica and organic rich sediment	Sediment characteristics depend upon the environment of deposition
Extremely variable salinity ranging between 1 g/ l to300g/l	Salinity varies in narrow range or constant in open sea
pH of water ranging between 4-10	Mostly mildly alkaline
Short duration climatic change recorded accurately in the sediments	Less sensitive to climatic change
Limited faunal assemblage and less common is bioturbation	Wide variation in faunal content, bioturbation is common
Little or no effect of tides and current on accumulated sediments	Considerable effects of tides and current on loose sediments
Repeated interval of desiccation and soil formation	Very rare
Sodium carbonate presents in evaporite lakes	Sodium carbonate is totally absent

sediments.

4. Weller (1960) rightly pointed out that the word geosyncline has been used for an area, defined as a structure or identified by the strata present in a region. It has been applied to a great variety of synclinal situations. He also suggested that non-committal term basin is available for synclinal situations that do not warrant recognition as geosynclines.

Geosyncline from cross road to confirmed path

Weller (1960) succinctly summarized the distortions in meaning of geosyncline by stating that to 'some geologists it obviously has meant a little more than a large syncline generally filled with sediments. The shape, kind of sediment filling and the possible subsequent deformation of a geosyncline have been considered unimportant by some persons. The variety of structural and stratigraphic concepts identified by this name has been most confusing. From time to time attempts have been made to clarify the situation by classifying so called geosynclines and new name has been given to designate different types. Distinction generally has been based on shape, location, structural relations, kind of sediment and the source of sediments'. Auboin (1965) sorted out the then prevailing confusing situation by constraining the essential prerequisites for designation of a geosyncline. These constraints are that geosynclines should be located between two cratons not within the cratons, it should give evidence of mobility i.e. rock deformation, sediment should be metamorphosed and intrude with igneous rock and it may occur in pair (eugeosyncline and miogeosyncline).

The concept of geosyncline was successfully shadowed by then emerging concept of plate tectonics. The plate tectonics initially restricted to an explanation of global level tectonic phenomenon occurring after break up of Gondwanaland land or Pangea but enthusiasm associated with plate tectonics stretched the plate operation even to Archaean time and many geoscientists have discarded the idea of geosynclinal driven orogenic activity even in ancient time. Rogers and Santosh (2003) documented three super assemblies of continents namely the Columbia (~ 1800- 1500 Ma), Rodinia (~ 1100-800Ma), Gondwana (~ 600-500Ma) with Lauretia (~ 250Ma), Pangea (~ 250 Ma), and 8 partial assemblies of continents viz. Ur (~ 3000 Ma), Kenorland (~ 2500 Ma), Arctica (~ 2500 Ma), Atlantica (~ 2000 Ma), Nena (~ 1800 Ma), Mawson continent (~ 1700 Ma), Paleopangaea (~ 1100- 1000 Ma), in the earth history of past 3000 million years. If extended further back then at least four more partial or super assemblies are possible. The calculated ratio of stagnation to mobility of continents astonishingly reaches to 1:1, which is mechanically and thermally untenable. This can also be described by the term continental mobility factor, which is nearly 1.0. However, the extension of plate type tectonic model for pre-Paleozoic tectonic set up has been doubted (Sankaran, 2006). Had the rate of plate processes in the past was similar or one magnitude less than entire Precambrian and older terrain should have been consumed in stated three super assemblies. The southern tip of India, 3000 km away from Himalayan subduction zone, would finally

Table-12 Difference between Geosynclines and Basins

Geosynclines	Basins
* They are mainly long and linear in shape.	* They are mainly circular and elliptical in shape.
* Thickness is comparatively greater than that of basin (> 10 km).	* Less thickness (generally < 5 km but rarely 10 km)
* Dominant compressive deformation	* Tensional deformation dominates over mild compressive deformation
* Severe folding is commonly seen.	* Mild folding is common.
* Presence of thrust pack.	* Generally thrusts are absent or rare.
* Episodic metamorphism occurs.	* Mild metamorphism occurs
* Long period of deformation.	* Short phase of deformation
* Well-defined sequences of igneous rocks	* No such sequences
* Igneous rocks are present in huge volumes	* Intrusions are of small sizes and having less volume

Table 13: Classification of Geosynclines

Schuchert (1923)	Stille (1913)	Kay (1951)	Peyve and Sinitzyn (1950)	Mitchell and Read ing
Mesogeosyncline Mono, poly and parageosynclines	Alpine Germino	Orthogeosyncline Miogeosyncline Eugeosyncline Parageosyncline Exogeosyncline Autogeosyncline Zeugogeosyncline Late cycle geosyncline Epieugeosyncline Taphrogeosyncline Paraliageosyncline	Primary Geosyncline Secondary Geosyncline Residual Geosyncline	Red Sea type Aulacogen Island Arc Andean Type Japan Sea Type Atlantic Type Mediterranean Type

vanish in 60 Ma time from today, if Indian plate drifts in north with current velocity of ~ 5cm/yr. A simple calculation like this cast doubt on pre-Phanerozoic plate tectonics. If plate tectonic gets invalidated, then one is left with geosyncline model to explain origin of orogenies. The difference between Geosyncline and Basin is given in Table 12 and classification of the Geosyncline have been given in Table 13.

Terminology for Geosynclines

As stated, an elaborate terminology was proposed for the geosynclines, definitions of proposed terms are as follows:

1. Mesogeosyncline is a complex geosyncline located between two continents having considerable thickness of severely deformed sediments with floor characterized by several geanticlines e.g. Mediterranean.

2. Mono-, poly- and parageosynclines- They are situated along the outer margin of, but lying within nuclear areas, emergent borderlands rising along the ocean border of the adjacent syncline and supplying detrital material to it. Mono represents to the simple geosyncline; poly to a complex geosyncline that is divided by one or more geanticline and parageosyncline is an ill-defined situation on the continental periphery.

3. Alpine type geosynclines built of thrust/fold nappes.

4. Germino type geosyncline is characterized by block faulting.

5. Orthogeosyncline- They are situated between cratons either high (continental) or low craton (oceanic). They give rise to mountain chain and subdivided as miogeosyncline is situated near the craton and show no/little volcanic activity and eugeosyncline that are situated away from craton charecterised by ophiolite volcanism. This type of geosyncline gives rise to orogenic mountain.

6. Parageosynclines are intracratonic geosynclines that do not produce orogenic mountain. They are subdivided into **exogeosyncline** situated on the edge of the craton and receive terrigenous from mountain developed from an Intracratonic zone; **autogeosyncline** located mainly in the interior of the craton and are resulted from subsidence independent of Intracratonic zone with carbonates sediments and has little terrigenous sediments and **zeugogeosyncline** which are situated within the craton and receive the terrigenous sediments from Intracratonic mountain.

7. Late cycle geosyncline They did not fit conveniently into any of the above categories. Classes in this category are **epieugeosyncline** that receive detrital sediments from eugeosyncline mountains but volcanic activity was absent; **taphrogeosyncline is** fault bound and **paraliageosyncline** is situated along the continental margin and merging into the coastal plain. They were limited towards the continents by flexure analogous to those bordering miogeosynclines.

8. Primary Geosynclines are characterized by elongated, narrow asymmetric furrows probably controlled by major features of basements, filled with sediments of various facies and having several km thicknesses with basic magmatic activity marked by ophiolitic intrusions that eventually folded and shows intense dynamic metamorphism that resulting in the alteration of ophiolites to serpentine.

9. Secondary Geosyncline- They peripheral primary geosyncline, filled with thick and terrigenous sediments, feebly folded, characterized by metamorphism confined to central zones of granitic batholiths. They are the sites of intense igneous activity marked by granitoid intrusions.

10. Residual Geosyncline They are vast basin filled with epicontinental marine deposits, showing little or no folding, characterized by rare acid volcanic eruption along fracture and absence of metamorphism.

11. Thalassogeosyncline - They are geosynclinal troughs initiated and developed in the oceanic crust and filled with a specific complex of metamorphosed basic, volcanic, siliceous and greywacke formations.

Attempts have been made to fit the concept of geosyncline in the plate setting. With in plate setup several situations have been identified of possible position of geosyncline. The identified set ups are:

12. Red Sea type- they are formed by upwelling of magma as a result of that tension in the earth crust crates three fractures at 120° angle.

13. Aulacogen- The initial stage of graben development is marked by deeply penetrating faults forming pathways to the mantle and giving rise to volcanism.

14. Island Arc Type- in front of these geosynclines trenches are present, Benioff zone dips towards the continents and characterized by andesitic magmatism.

15. Andean type geosynclines are located on the margin of continents with the oceanic plate dipping beneath it and a trench formed on the descending plate.

16. Japan Sea type- In the back arc region behind island arc, the basement may be oceanic or continent; if oceanic a spreading axis may be present.

17. Atlantic type- Part of the geosyncline has of course very early in its history passed through the graben and Red Sea stage.

18. Mediterranean type- This is defined as originating in a plate margin setting when lateral (strike slip) plate motion dominates.

Auboin's characterization of geosyncline should be accepted without modification or addition of terminology. These characters are their locations either on periphery of continents or between two continents but not in the cratons, mobility manifested by intense folding and thrusting with initial silicic magmatism, presence of ophiolites and syn- and post orogenic acidic and basic magmatism in the central part and within geosynclines furrows and ridges demarcate miogeosynclinal and eugeosynclinal zones. Thus he identified miogeosyncline and eugeosyncline as two subdivisions of geosyncline. Extending operation of plate process even in Archaean era creates several problems like explanation of mechanical and thermal instability of the earth, very high mobility factor, weak geomagnetic evidences and preservation of Archean crust in the continents. Thus a clear time demarcation is required for applicability of plate and geosynclinal related activities. The present plate operation started from the onset of Triassic and all other orogenies prior to present plate breakup can be explained by the geosyncline mechanism.

CONCLUSIONS

Basin, a trap for water or sediments, is a widely used term having a specific sense in different branches of earth sciences. Valleys, lakes and geosynclines are the variants of basins but differ significantly in implied geological meaning. The paper brings out clear distinction between these terms and advocates applicability of the term basin irrespective of size, shape and location as entire global surface acts as traps of one kind or other of vastly different magnitudes. Areas such as alluvial plain or of wind deposition having no distinct physical form have been designated as incorporeal basins. The present overview, with restricted scope, critically deals with various definitions, characters, terminology and recent tectonic based classification of the basins and geosynclines and highlights existing ambiguity. Many geological aspects related with basins have intentionally not been included in this review. Basin is defined in the above larger context and two classifications of basins are proposed based on the geomorphic and tectonic characters and on the aerial size. Development of plate concept in last four decades describes the large-scale movements of continents dating back to 3000 Ma and has completely eclipsed the geosynclinal origin of orogen. Validity of pre-Paleozoic plate operation is questioned on the mobility factor, thermal and mechanical ground and existence of Archean crust and prefers geosynclinal model for the origin of orogens.

ACKNOWLEDGEMENTS

The authors are grateful to Dr. M. N. Joshi, Reader, Department of Geology, DBS College, Dehradun for his constructive comments and suggestions. Thanks are also due to Mr. S. P. Bhatta, Research scholar, for his cooperation at different stages of work. Thanks are also due to two anonymous reviewers for their constructive comments and suggestions and to Prof K.L. Shrivastava for cooperation.

References

Aubouin, J. (1965) Geosynclines; Developments in Geotectonics, Elsevier Amsterdam, V. 1, 335 p.

Bally, A. W. and Snelson, S. (1980). Realms of Subsidence. In. A. D. Miall (ed.) Facts and Principles of World Petroleum Occurrence Can. Soc. Pet. Geol. Mem, V. 6, pp. 9-75

Billings, M. P. (1972). Structural Geology, (3rd edition), Prentice Hall, Inc., Englewood Cliffs, N. J., 606 p.

Biswas, S. K., Bhasin A. L. and Ram, J. (1993) Classification of Indian Sedimentary basin in the framework of Plate Tectonics. Proc. II Sem. On Petroliferous Basins of India, V. I, KDMIPE, ONGC, Dehradun, pp. 1-43.

Chapman, R.E. (1973) Petroleum Geology A Concise Study. Elsevier scientific publishing, 304 p.

Condie, K. C. (1989) Plate Tectonics and Crustal Evolution, Pergamon Press, 476 p.

Dickinson, W. R. (1974) Plate Tectonics and Sedimentation: In. W. R. Dickinson (ed.), Tectonics and Sedimentation. Soc. Econ. Paleontol. Mineral., Spec. Publ., V. 22. pp. 1-27.

Einsele, G (2000) Sedimentary Basins Evolution, Facies and Sediment Budget (II edition), Springer Publishers, pp. 781.

Friedman, G. M. and Sanders, J. E. (1978) Principles of Sedimentology, John Wiley & Sons.

Hamblin, W. K. (1989) The Earth's Dynamic Systems (5th edition), Collier Macmillan Publishers, London, pp. 576

Ingersoll, R. V. (1988) Tectonics of Sedimentary Basins, Bull. Geol. Soc. Amer. V. 100, pp. 797- 806.

Jaroszewski (1984) Fault and Fold Tectonics, Wendy L. Kirk (ed), PWN publishers, pp. 391.

Kay, M. (1951) North American Geosynclines, Geol. Soc. Amer. Mem. V. 48, pp. 143.

Kingston, D.R., Dishroon, C.P. and Williams, P. A. (1983) Global Basin Classification System. Am. Assoc. Petrol. Geol. Bull., V. 67, pp.2175-2193.

Klein, G. (1987) Current Aspects of Basin Analysis, Sed. Geol., V. 50, pp. 95-118.

Klemme, H.D. (1975) Giant oil fields related to their geologic setting: a possible guide to Exploration. Bull. Can. Pet. Geol., V. 23, pp. 30-66.

Laverson, A. I. (1967) Geology of Petroleum (2nd edition), San Francisco, Freeman, 924 p.

Leighton, M. W. and Kolata, D. R. (1990) Selected Interior Cratonic Basins and their place in the scheme of Global Tectonics- a Synthesis. Am. Assoc. Petrol. Geol. Memoir, V. 51, pp.729-797.

Miall, A. D. (1984) Principles of Sedimentary Basin Analysis, Springer New York, 490 p.

Mitchell, A.H.G. and Reading, H.G. (1986) Sedimentation and Tectonics. In Reading (ed.) Sedimentary Environments and Facies. Blackwell Scientific Pub., pp 471- 519.

Muratov, M. V. (1977) The Origin of Continents and Ocean Basins, Mir Publishers, Moscow, 191 p.

Nur, A. and Ben- Avraham, Z. (1983) Displaced terranes and Mountain Building, In. Hsu KJ (ed.) Mountain Building Processes, Academic Press, London, pp. 73-84.

Peyve and Sinitzyn (1950) In Aubouin, J. 1965 Geosynclines; Developments in Geotectonics, Elsevier Amsterdam, V. 1, 335 p.

Oil and Natural Gas Corporation (2007). Basin analysis: power point presentation; 95 slides.

Schuchert, C. (1923), Sites and nature of the North American Geosynclines, Bull. Geol. Soc. Amer., Vol. 34, pp.151- 229.

Stille (1913) In Aubouin, J. 1965, Geosynclines; Developments in Geotectonics, Elsevier Amsterdam, V. 1, 335 p.

Strahler, A.N., 1964, Quantitative geomorphology of drainage basins and channel networks. In: V.T. Chow(ed), Handbook of Applied Hydrology. Mc Graw Hill Book Company, New York, Section 4-11.

Weller, J. M. (1960) Stratigraphic Principles and Practice, Harper & Row Publishers, 725 p.

Wilson, J. T. (1965) Transform Faults, oceanic ridges and magnetic anomalies southwest of Vancouver Island, Science, V. 150, pp. 482-485.

Hon'ble Reviewers

1. Prof. P.S. Saklani
(Former Vice Chancellor, Kumaon University)
Emeritus Professor
Department of Geology
Delhi University
DELHI 110 007

2. Dr. M.N. Balasubrahamanyan
(Former, Dy. Director General, GSI)
Flat 1A Gurukripa
117/2B Srinivasamurthy Avenue
Adayar
CHENNAI 600 020

Editorial Assistance

Chandra Prakash Dadhich

Economic Mineralization
K.L. Shrivastava (Editor)
2009. pp.318-327
Scientific Publishers (India), Jodhpur

Section-III

Depositional Environment and Industrial Importance of Bilara Limestone, West of Nagaur, Western Rajasthan

MEETA KHILNANI[1] AND SATVINDAR SINGH[2]
[1]Director, Hydro-Geosurvey Consultants Pvt. Ltd.,C-103 Shastri Nagar, Jodhpur 342 001
[2]C-136, Hanuman Nagar, Jaipur 302 001
Email : meetavk@hotmail.com

Abstract

High grade limestone deposits belonging to Bilara Group extending from Sojat in Pali district to Mundwa in Nagaur district, through Bilara, Borunda and Gotan are well known, but similar quality of limestone occurring in west of Nagaur are relatively less known. In the study at west of Nagaur the high grade limestone occurs as isolated outcrops within sandy landscape.

These carbonates of Bilara group were laid down under shallow marine conditions during late Proterozoic period in the Bilara part of the Nagaur basin in the south. Almost during the same time, more than 650 m thick Hanseran evaporites with magnesite and dolomite were deposited in the north in the Nagaur Ganganagar sector.

Key words : Bilara Limestone, Industrial Minerals, Depositional Environment.

INTRODUCTION

Rajasthan is well known for its limestones reserves, Western Rajasthan has huge deposits of high grade limestone, mainly chemical grade, steel grade and white cement grade limestones. Bilara limestone extending from Sojat tc Nagaur through Bilara, Gotan and Mundwa has been supplying chemical grade limestone, (over 98% $CaCO_3$), steel grade (less than 1.5% silica) and white cement grade (less than 0.1% Fe_2O_3) in addition as sweetner for portland grey cement and building purposes. Jaisalmer limestone is another source of steel grade limestone meeting the requirement of all major steel plants.

Although Sojat-Nagaur belt is a well known deposit of high grade limestone, similar deposits of Bilara limestone occurring west of Nagaur are less known for its quality, reserves and industrial importance. Depositional environment of these deposits has not been investigated so far.

In this paper attempts has been made to interpret depositional environment of these limestone deposits.

THE STUDY AREA

Nagaur is situated in central Rajasthan has 5.18% area

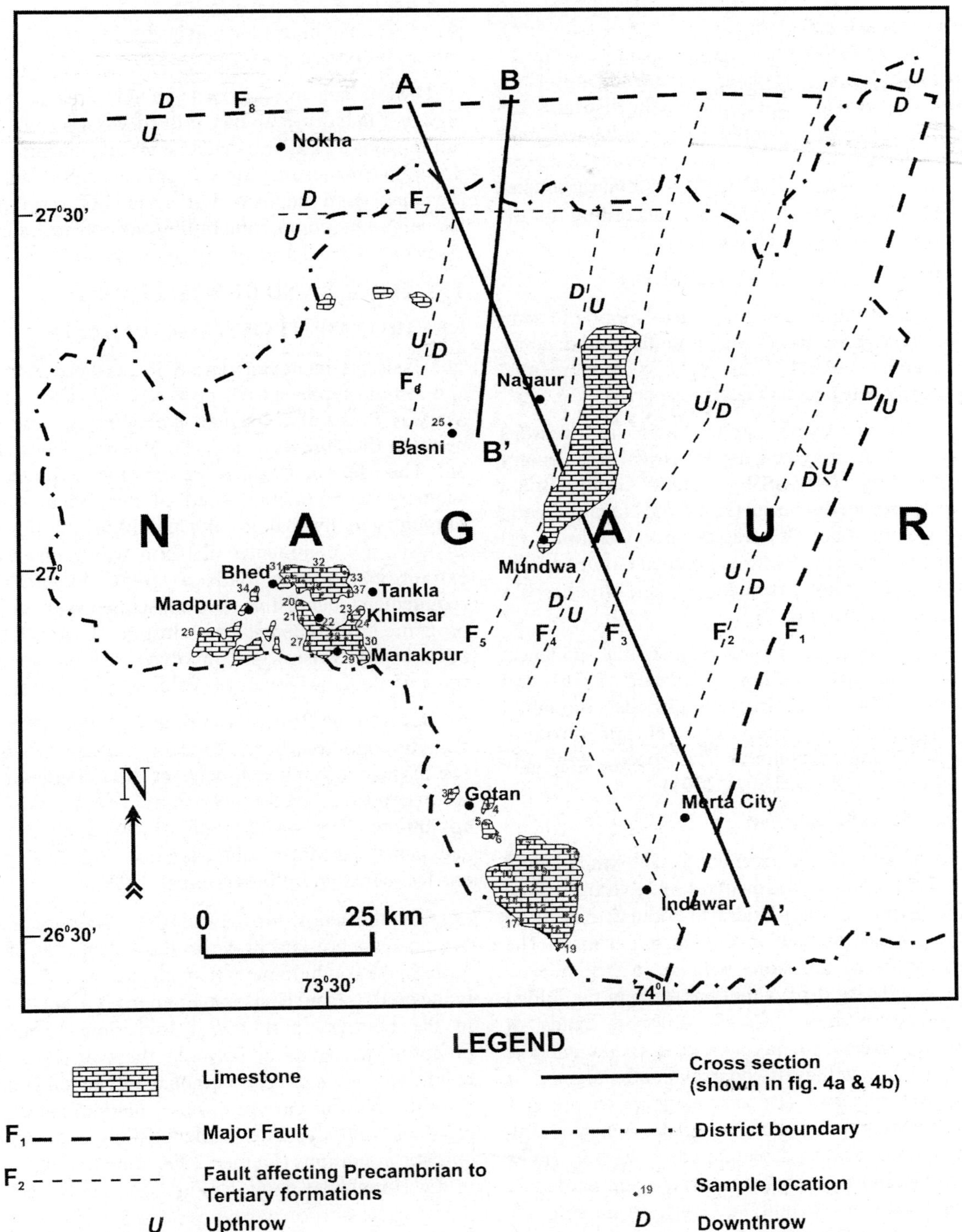

Fig.1. Map showing limestone deposits and geological structures of the western part of Nagaur District (After United Nations, 1976)

of the Rajasthan state (i.e. 17,718 sq. km). Climatically, the mean minimum and maximum temperature of the Nagaur is 23.5°C and 47°C, respectively; and average rainfall in the district is 388 mm. 37 limestone samples were collected form different locations (Fig.1) for their chemical analysis, i.e. loss of ignition, SiO_2, CaO, MgO, total carbonates etc. All samples were analyzed according to the methodology as per Indian standards.

GEOLOGY OF THE STUDY AREA

Major part of the area is covered by mobile to semi stable, NNW-SSE to NE-SW trending sand dunes which overlie calcareous sand pseudo-conglomerate, kankar or gypsite bed of Quaternary period.

The Nagaur-Ganganagar basin in western Rajasthan is an elongated, asymmetrical sedimentary basin trending NNW-SSW. Jodhpur Group, Bilara Group, Hanseran Evaporite Group and Nagaur Group constitute the Late Proterzoic-Early Cambrian (?) sedimentary package (Marwar Super Group) which rests on the Rhyolitic basement (Malani Rhyolite) has been dated 680 Ma (Fig. 2).

Owing to lack of characteristic faunal evidence, except algal stromatolites (Khilnani ,1966) and radiogenic dates from Bilara rocks, its correlation with globally recorded late Neoproterozoic carbonates and evaporates has been enigmatic (Mazumdar and Bhattacharya, 2001).

Jodhpur-Pali-Nagaur belt

Carbonate belt extends over about 160 km in length and is 0.8 to 12 km in width trending NNE from Sojat in Pali district through Bilara in Jodhpur district to Rol-Quazia and beyond in the Nagaur district. The general strike of limestone belt is NNE-SSW with gentle, undulating dips which are steep in the folded zones. The thickness of the different bands of limestones varies from half a meter to six meters. The limestone is extensively quarried and calcined at Gotan and Mundwa (District Nagaur) to produce quicklime which is of chemical grade. Bands of high grade limestone were encountered in the bore holes drilled in Rol-Quazia area but otherwise the carbonate is largely dolomitic and siliceous, unsuitable for cement making (Department of Mines & Geology, 1976).

Nagaur District

The limestone deposits occurring west of Nagaur district are discontinuous in nature and belong to Gotan and Pondlu formations of the Bilara group (Fig. 2). Gotan limestone belt is considered part of the Marwar Supergroup.

NMDC has mining lease in this area covering village Madpura and has proved reserves of 6.36 million tonnes. Earlier NMDC was carrying mining of SMS grade limestone for SAIL plants. Although, this area has good deposits but systematic geological investigations comprising pitting and core drilling has not been carried out so far.

TECTONICS AND DEPOSITIONAL ENVIRONMENT OF NAGAUR BASIN

The Lesser Himalayan terrain in northern Pakistan and Jammu represent the northerly extension of the Marwar Basin of Rajasthan indicating a close link between the Himalaya and the Marwar plains (Fig. 3a). The Marwar Basin represents the northwestern submergent continental shelf of the Indian shield. Through geophysical probing and lithostratigraphical analysis, the continental platform was found to be extraordinary wide with a great thickness of sedimentary succession. As a consequence of Tertiary orogenic movements, the frontal part of this sedimentary prism got involved in the tectonic framework of the Himalaya (Valdiya, K.S., 1999).

During Late Proterozoic time and soon after, the climatic conditions became semi-arid to arid when the rest of the overlying sediments of the Jodhpur Group were deposited under near-shore marine platform conditions. The provenance of these arenaceous sediments was Malani Igneous Suite (Fig. 2) of rocks and Delhi metamorphites (Kumar, 1999).

After the deposition of sands of the Jodhpur group, carbonates were laid down in the Bilara part of the basin in the southern parts (Chauhan et al., 1991), and halite predominant Hanseran Evaporite Group (Fig. 2) in the northern part and a transitional Sulphate predominant facies in between the two. Later the conditions of evaporation gradually changed and the youngest Nagaur Group of rocks, predominantly red argillites, were deposited under shallow water in arid climatic conditions (Kumar, 1999). In the Salt Range in the Himalayan domain (Fig. 3b), nearly 2000 m thick succession of gypsum and salts of the Saline Series (Salt Range Formation) was laid down (Gee, 1989). The widespread occurrence of potash salts in the contemporaneous Marwar and Salt Range successions (Jones, 1970) on the similar Precambrian crystalline floor (Shah, 1977) is very noteworthy. Doubtless, the shallow platform with prevailing evaporatic condition extended from the Jodhpur-

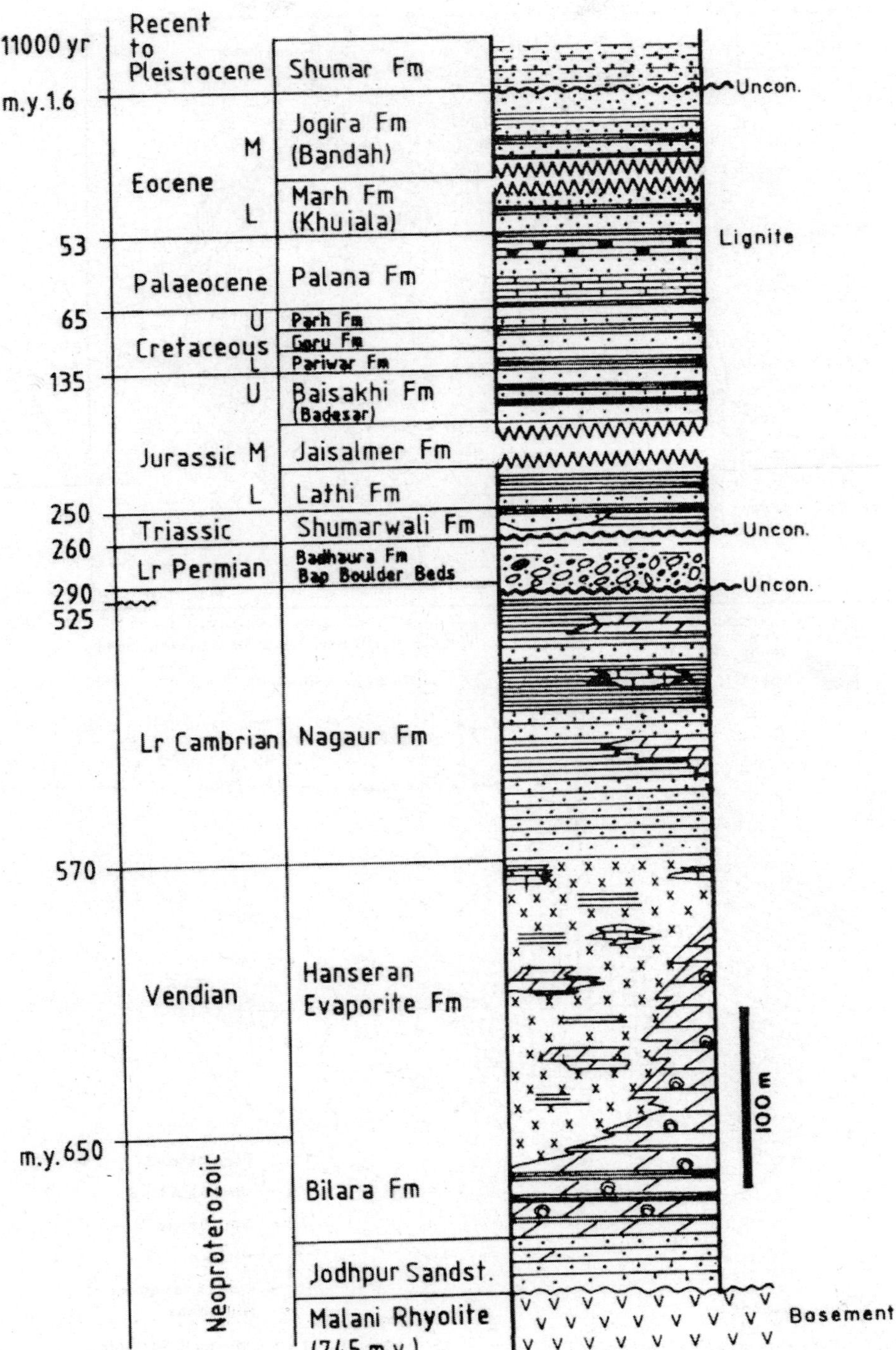

Fig. 2.Integrated lithological column of the sediments of the Marwar basin (Based on Dasgupta et al., 1988; Dasgupta & Bulgauda, 1994, and Mohan and Sangai, 1995)

Nagaur tract in the south through the salt Range to the Abbotabad belt in Hazara (Virdi, 1990) (Figs. 3 b), the Ramban-Sinohu area in Jammu (Jangpangi et al., 1986) and Dharamkot-Ropri-Mandi belt in the Shali domain of Himachal (Srikantia and Sharma, 1972). Deep drilling at Kasganj in the Ganga Basin revealed presence of gypsum layers in the Proterozoic sedimentary rocks forming the floor (Sastri et al., 1977). Presence of algal laminites, stromatolites, ripple lamintes, mudcrack and mudchip

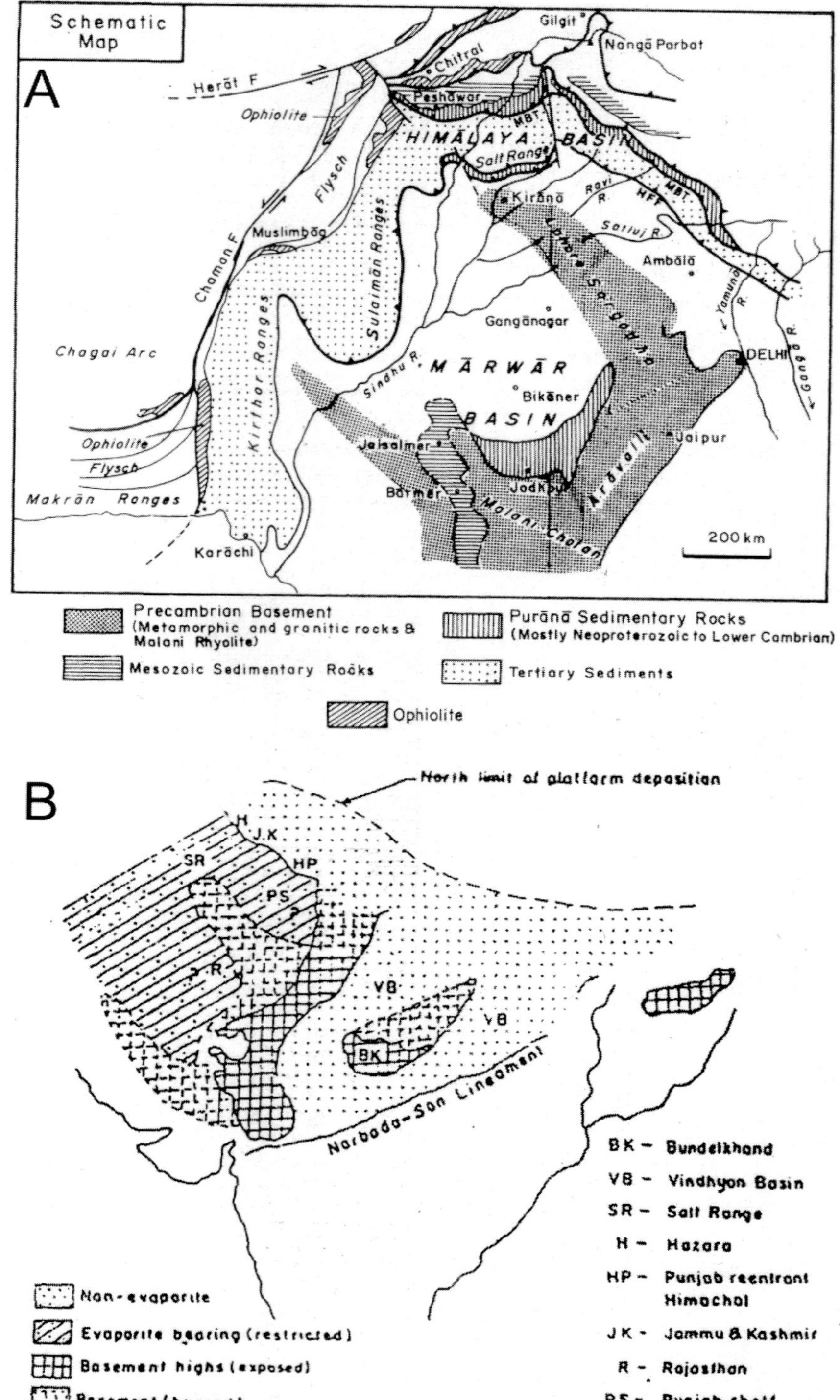

Fig. 3A. Sketchmap of Marwar and adjoining sectors of the Himalaya, indicating close geological connections of the two domains (Based on Gansser, 1964; Valdiya, 1995, and Virdi, 1990)

Fig. 3 B. Palaeogeographical map showing the extent of stratigraphic formations which bear evaporite deposits (Virdi, 1990)

conglomerates indicate shallow marine peritidal depositional environment for the Bilara carbonates (Mazumdar and Bhattacharya, 2001).

Isotope geochemistry of unmetamorphosed Bilara carbonates (when sections are thinly fossiliferous) shows positive oxygen isotopic values (up to 8 $^0/_{00}$) indicating existence of evaporitic condition during deposition. This is further corroborated by the presence of anhydrite and gypsum inclusions in carbonates (Mazumdar and Bhattacharya, 2001).

RESULTS AND CONCLUSIONS

The Nagaur basin came into existence by the rejuvenation of NNE-SSW trending Archaean and Proterozoic lineaments (Fig.1, 4a and 4b). Initial rifting along these lineaments resulted in the interbasinal faults and formation of the horst and graben structure, during post-Delhi orogenic and later Malani igneous activities, between 745±10 Ma to 600±70 Ma (Kumar, 1999). Results of chemical analysis of thirty seven samples are shown in Table 1. All sample shows SiO_2 and MgO content is very less (i.e., 0.78 to 2.46 %) and CaO content is very high (i.e., 52 to 54%). Loss of ignition is less then 42% and total carbonates are more than 95% in all the samples.

The limestone of western Nagaur has been considered to be one of the best grade chemical limes produced in India suitable for sugar refining, chemical industries and manufacture of calcium carbide, precipitated calcium carbonate etc. However, the quality of

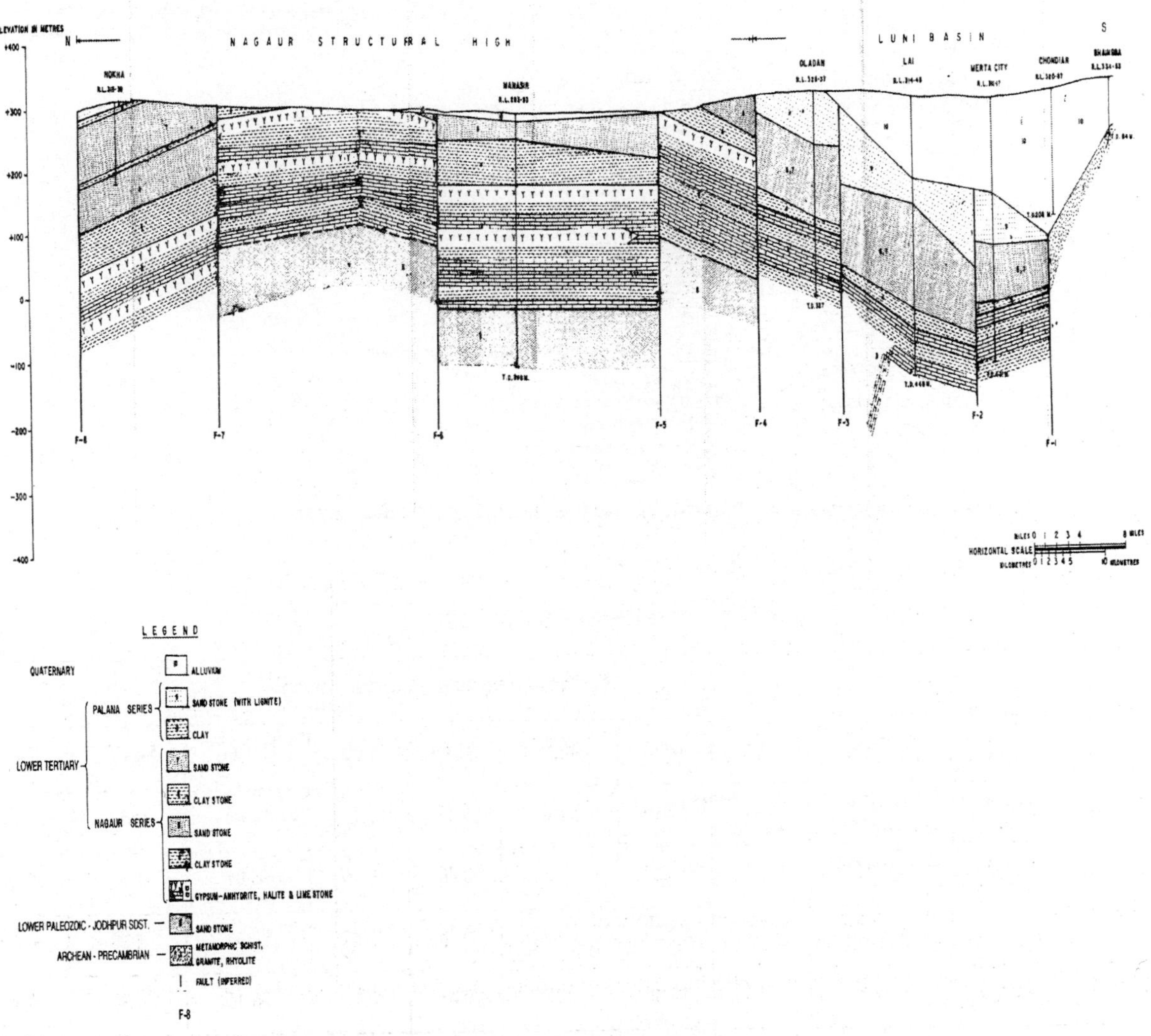

Fig. 4a. Geological cross section along line A-A' of Fig.1 (after United Nations, 1976)

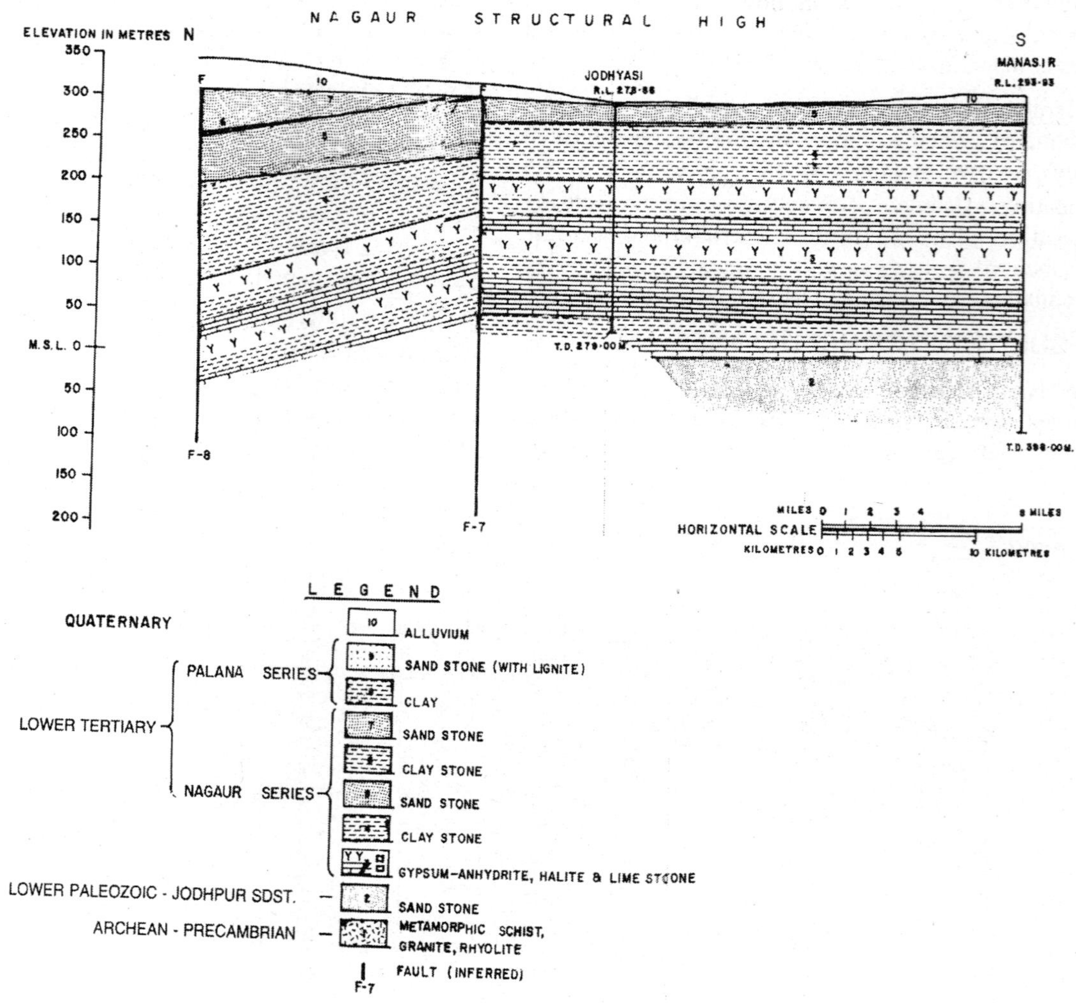

Fig. 4b. Geological cross section along line B-B' of Fig.1 (after United Nations, 1976)

Table-1 Chemical analysis of limestone samples of Nagaur District

S. No.	Location	Chemical Analysis (Parameter Tested)								
		LOI	SiO_2	CaO	MgO	Al_2O_3	Fe_2O_3	$CaCO_3$	$MgCO_3$	T.C.
1.	Jakh-A	42.66	1.44	53.20	0.80	0.39	0.11	94.94	1.68	96.62
2.	Jakh-B	42.38	1.18	53.24	1.12	0.40	0.11	95.01	2.35	97.36
3.	Gotan-A	43.58	0.82	54.05	0.60	0.36	0.096	96.46	1.20	97.66
4.	Gotan-B	42.60	1.03	53.88	0.62	0.47	0.11	96.16	1.30	97.46
5.	Gotan-C	42.78	1.08	52.97	1.12	0.40	0.12	94.55	2.32	96.87

Table-1 (Contd.)

6.	Gotan-D	42.88	1.12	53.38	1.01	0.38	0.11	95.26	2.12	97.38
7.	Gotan-E	42.40	0.96	53.06	1.54	0.38	0.12	94.69	3.23	97.92
8.	Gotan-F	42.66	0.89	53.40	1.15	0.36	0.11	95.30	2.41	97.71
9.	Gotan-G	42.51	1.08	52.77	1.19	0.40	0.12	94.18	2.50	96.68
10.	Gotan-H	42.62	0.89	53.31	1.00	0.38	0.10	95.14	2.10	97.24
11.	Gotan-I	42.90	1.26	52.22	1.34	0.41	0.12	93.19	2.81	96.00
12.	Gotan-J	42.71	0.81	53.45	0.89	0.42	0.13	95.39	1.87	97.26
13.	Gotan-K	42.90	0.81	53.58	1.09	0.41	0.096	95.62	2.29	97.91
14.	Gotan-L	42.66	1.0	53.04	0.89	0.39	0.12	94.66	1.87	96.53
15.	Gotan-M	42.78	1.12	52.63	1.37	0.44	0.13	93.93	2.88	96.81
16.	Gotan-N	42.56	0.79	53.31	1.00	0.39	0.11	95.14	2.10	97.24
17.	Gotan-O	42.80	0.96	53.04	0.89	0.43	0.11	94.66	1.87	96.53
18.	Gotan-P	42.81	1.17	52.50	1.39	0.40	0.12	93.69	2.92	96.61
19.	Gotan-Q	42.56	1.09	52.77	1.19	0.38	0.13	94.18	2.50	96.68
20.	Khinvsar-A	42.80	1.02	53.62	1.09	0.33	0.11	95.69	2.29	97.98
21.	Khinvsar-B	42.66	1.12	53.35	0.98	0.36	0.096	95.21	2.06	97.27
22.	Khinvsar-C	42.94	1.32	52.52	1.50	0.44	0.12	93.73	3.15	96.88
23.	Khinvsar-D	42.80	0.94	53.56	1.03	0.42	0.096	95.59	2.16	97.75
24.	Khinvsar-E	42.64	1.21	53.04	1.12	0.43	0.12	94.66	2.35	97.01
25.	Basni	42.68	1.09	53.22	1.13	0.40	0.11	94.98	2.37	97.35
26.	Madpura	42.82	1.06	52.74	1.06	0.40	0.11	94.12	2.22	96.84
27.	Manakpur-A	42.56	1.12	53.12	0.96	0.42	0.096	94.80	2.01	96.81
28.	Manakpur-B	42.66	0.97	54.07	0.67	0.32	0.088	96.50	1.41	97.91
29.	Manakpur-C	42.58	0.98	53.40	0.76	0.38	0.11	95.30	1.60	96.90
30.	Manakpur-D	42.92	1.28	52.70	1.08	0.46	0.14	94.05	2.27	96.32

Table-1 (Contd.)

31.	Bhed-A	42.64	1.08	52.86	1.12	0.38	0.12	94.32	2.35	96.69
32.	Bhed -B	42.80	1.04	53.41	0.87	0.41	0.1	95.32	1.83	97.15
33.	Bhed-C	42.66	1.29	53.16	0.87	0.42	0.096	94.87	1.83	96.70
34.	Bhed-D	42.95	0.97	53.65	0.96	0.39	.011	95.75	2.02	97.77
35.	Bhed-E	42.79	1.08	53.04	1.40	0.43	0.12	94.66	2.94	97.60
36.	Bhed-F	42.31	2.46	52.43	1.14	0.40	0.11	93.57	2.39	95.96
37.	Tankala	42.81	0.78	53.31	1.37	0.40	0.11	95.14	2.88	98.02

limestone is not uniform and selective mining is to be done for different grades of limestone.

ACKNOWLEDGEMENT

The authors express their sincere thanks to Department of Science and Technology, New Delhi for providing the research grant under SERC Fast Tract Scheme for Young Scientists, and Department of Geology, Jai Narain Vyas University, Jodhpur for providing the necessary institutional support. My sincere thanks to Professor M.S. Sisodia, Head, Department of Geology and Professor K.L. Shrivastava for their constant encouragement. Last but not the least sincere thanks are due to Professor B.S. Paliwal for his guidance and fruitful suggestions in finalizing the paper.

References

Chauhan, D.S., Dubey, J.C., and Ram, B., 1991. Geological analysis of Nagaur basin in the vicinity of Jodhpur City, In: S.K. Tandon et al. (Eds.) Sedimentary Basins of India, Gyanodaya Prakashan, Nainital, pp. 63-73.

Das Gupta, S.P., Kumar, V., Chandra, R., and Jairam, M.S., 1988. A framework of the Nagaur-Ganganagar evaporate basin. Indian minerals, vol. 42 pp. 57-64.

Dasgupta, U. And Bulgauda, S.S., 1994. An overview of the geology and hydrocarbon occurrences in the western part of Bikaner-Nagaur Basin. Ind. Jour. Petroleum Geol., vol. 3 pp. 1-17.

Department of Mines and Geology, 1976. Limestone deposits of Rajasthan. Department of Mines & Geology, Government of Rajasthan, Govt. press, Udaipur, 20 p.

Gansser, A. 1964. The Geology of the Himalayas, Interscience, New York, 289 p.

Gee, E.R., 1989. Overview of geology and structure of Salt Range, with observations on related areas of northern Pakistan, In: L.L. Malinconico and R.J. Lillie (Eds.) Tectonics of Western Himalayas, Geol. Soc. Am. Spel. Paper, 232 pp. 95-110.

Jangpangi, B.S., Kumar, G., Rathore, P.R. and Datta, S., 1986. Geology of the autochthonous folded belt of Kashmir Himalaya. Jour. Palaeont. Soc. India, vol. 31 pp. 39-59.

Jones, C.L., 1970. Potash in halitic evaporate salt, western Pakistan. U.S.G.S. Professional Papers, 700D, pp. 140-145.

Khilnani, B.V., 1966. Algal stromatolitic structures of limestone of Trans-Aravalli Vindhan basin, Western Rajasthan. Proc. 53rd Ind. Sci. Cong. Asso. Chandigarh Pt. III, Abst. Sec. V. p. 190.

Kumar, V., 1999. Evolution and geological set up of the Nagaur-Ganganagar Basin, North Western Rajastha. In: Geological Evolution of Northwestern India, Paliwal, B.S. (Ed.), Scientific publishers, Jodhpur, India, pp. 34-60.

Mazumdar, A., and Bhattacharya, S.K., 2001. Carbon isotope stratrigraphy of Late Neoproterozoic carbonate rocks of Nagaur-Ganganagar basin, Western Rajasthan, India. Eleventh Annual V. M. Goldschmidt Conference.

Mohan, S. and Sangai, P., 1995. ONGC through Thar Desert: an exploration view, In: Proc. First International Petroleum Conference and Exhibition, Vol. I, B.R. Publishing Corp., Delhi, pp. 133-145.

Sastri, V.V., Bhandari, L.L., Raju, A.T.R., and Datta, A.K., 1977. Tectonic framework and subsurface stratigraphy of the Ganga Basin. Jour. Geol. Soc. Ind., vol. 12 pp. 222-233.

Shah, S.M.I., 1977. Stratigraphy of Pakistan. Mem. Geol. Surv. Pakistan, vol. 12 pp. 1-138.

Srikantia, S.V., and Sharma, R.P., 1972. The Precambrian salt deposits of Himachal Pradesh Himalaya: Its occurrence, tectonics and correlation. Himalayan Geology, vol. 2 pp. 222-228.

United Nations, 1976. Ground-water surveys in Rajasthan & Gujarat, India (Technical Report No. DP/UN/IND-71-614/4). United Nations, Newyork.

Valdia, K.S., 1995. Proterozoic sedimentation and Pan-African geodynamic development in the Himalaya. Precamb. Research, vol. 74 pp. 35-55.

Valdia, K.S., 1999. Marwar's connection with the Himalaya. In: Geological Evolution of Northwestern India, Paliwal, B.S. (Ed.), Scientific publishers, Jodhpur, India, pp. 1-18.

Virdi, N.S., 1990. Vendian-Lower Cambrian desiccation along the southern margin of the Proto-Tethys. Jour. Himalayan Geol., vol. 1 pp. 103-114.

Hon'ble Reviewers

1. Prof. I.B. Singh
   (Former, Head of the Department)
   Emeritus Scientist
   Department of Geology
   University of Lucknow
   LUCKNOW 226 007 INDIA

2. Prof. Ravinder Kumar
   Chairman,
   Department of Applied Geology and
   (Centre of Advance Studies)
   Panjab University
   CHANDIGARH 160 014 INDIA

Editorial Assistance

Dr. (Mrs.) Yasmin Sayyed

Economic Mineralization
K.L. Shrivastava (Editor)
2009. pp.328-335
Scientific Publishers (India), Jodhpur

Section-III

Banded Iron Mineralization in the Gwalior Group of Rocks

P.K. JAIN
School of Studies in Earth Science, Jiwaji University, Gwalior 474 011
Email : dpkj_2005@yahoo.co.in

Abstract

'Par' (lower) and 'Morar' (upper) formations. Par is are arenaceous and Morar consists of argillaceous rocks. These are underlained by Bundelkhand Granitic/Gneissic Complex and overlained by Upper Vindhyan Supergroup. The litho-units are identified by the techniques of remote sensing through the interpretation of RSD. Mineralized Zone (BIF, Laterite and Ochre beds) is present observed within 'Morar formation'. BIF (Banded Iron Formation) consists with BHJ and BHC. Residually concentrated Iron rich loose are laterite and thin interbeded ochre are present in three distinct colours, white, red and yellow within the mineralized zone. Small scale mining activity is goingon for exploitation of BIF. It is proposed that the mineralized zone is formed by the sedimentary origin of chemical precipitation, as part of crustal evolution.

Keywords : Banded Hematite Quartzite, Laterite, Gwalior group.

INTRODUCTION

The first account of geology of Gwalior and its vicinity is given by Hacket (1870), who has mentioned the name "Gwalior Series" and classified it into two divisions, lower and upper, namely Par and Morar Series, respectively. Heron (1936) regarded the Gwalior series as an isolated outcrop of unmetamorphosed Aravalli series, which owe their horizontality and absence of metamorphism to their distance from the main axes of folding of the Aravalli range and their protection by the resistant mass of Bundelkhand gneiss over which they rest. In terminology (Kumar, 1998) it is called as "Gwalior Group" and its divisions are named as Par and Morar formations. Par and Morar formations are having average thickness 60m.and 600m., respectively. Small scale open cast mining activity is going on at and around the Gwalior city for the exploitation of the Banded Iron Formations (BIF). Approved mining activities by the Govt. of M.P. at various places are named as Santau Iron Ore Mine, Panihar Iron Ore Mine and Maharajpur Laterite Mine.

The rocks of Gwalior group are exposed at and around the famous Gwalior city. The area under investigation lies between the Cartesian Coordinates 26°00' to 26°05' N and 78°00' to 78°05' E on the Survey of India toposheet No. 54 J/4 of 1:50,000 scale, and same is on remotely sensed data (RSD) IRS 1D, Std., Geocoded, B 234,L-3, P 093/R 053, 12 Feb., 2002.

METHODOLOGY

Field investigations were carried out to spot check of individual lithounit as exposed in outcrops and various mines. Visual interpretation has been carried out from the passive remote sensing data (image) on the basis of different recognition elements for thematic mapping. Geochemical analyses has been taken from the various approved unpublished 'mining plans' by the RCM, Indian Bureau of Mines, Jabalpur (M.P.).

GEOLOGY OF THE AREA

On the basis of field and laboratory study the geological set-up of the area is given in Table 1 (after, Hacket, 1870).

Geological map of Gwalior area is shown in Fig. 1 (after Pateriya and Singh, 1994). The entire lithological succession from Basement to Morar formation is exposed in the type area around Par village, and the name 'Par formation' is given only after the said village. A lithological map has been prepared for the investigated area around Par village through the interpretation of remote sensing data (Fig. 2) and field checks (Fig. 3). Different lithounits as seen in and around the Par village can be easily observed on the map. Broad classes are separated to each other by distinct marker horizons of conglomeratic unconformity. At some places doleritic intrusives are seen in the Upper Morar formation. The chief mineralized zone is also located in the Morar formation (UGF).

MINERALIZATION

Mineralized zone is present within Morar formation. Commercial classification of mineralized zone is as follows :

1. BIF (Banded Iron Formation) is the main mineralized zone which covers a large area. The chief minerals/ores of the zone are BHJ (Banded Hametite Jasper, Fig. 4) and BHC (Banded Hametite Chert, Fig. 5). Low grade iron ore having < 50% (47.44 to 47.88) iron (Srivastava and Gosavi, 1999) is present in them.

2. Laterite deposits are also seen frequently within the BIF. It is the iron rich soil (22.38% of Fe) derived from the BIF.

3. Thin ochre bands are also seen in three colours namely white, red and yellow. Generally these beds occur in small thin bed within the formation (Fig. 6).

Table 1 : GEOLOGICAL SETUP IN AND AROUND THE GWALIOR CITY (after Hacket, 1870)

Group/Super Group	Formation	Lithology
	Recent	Soil & Laterite
Vindhyan Super Group (Upper)	Kaimur Formation	Sandstone
	---------------Unconformity -------------	
GWALIOR GROUP	**Morar Formation (Upper)**	Dolerite dykes & Sills, Slate, Limestone, Siliceous Shale, Clay (ochre), BHJ & BHC, Black Chert, Banded Ferruginous Shale, Banded Chert Shale.
	---------------Unconformity -------------	
	Par formation (Lower)	Sand Stone, Shale and Quartzite.
	---------------Unconformity -------------	
Bundelkhand Granitic/ Gneissic Complex (Basement)		Granite, Gneiss with Quartz Veins and Reefs.

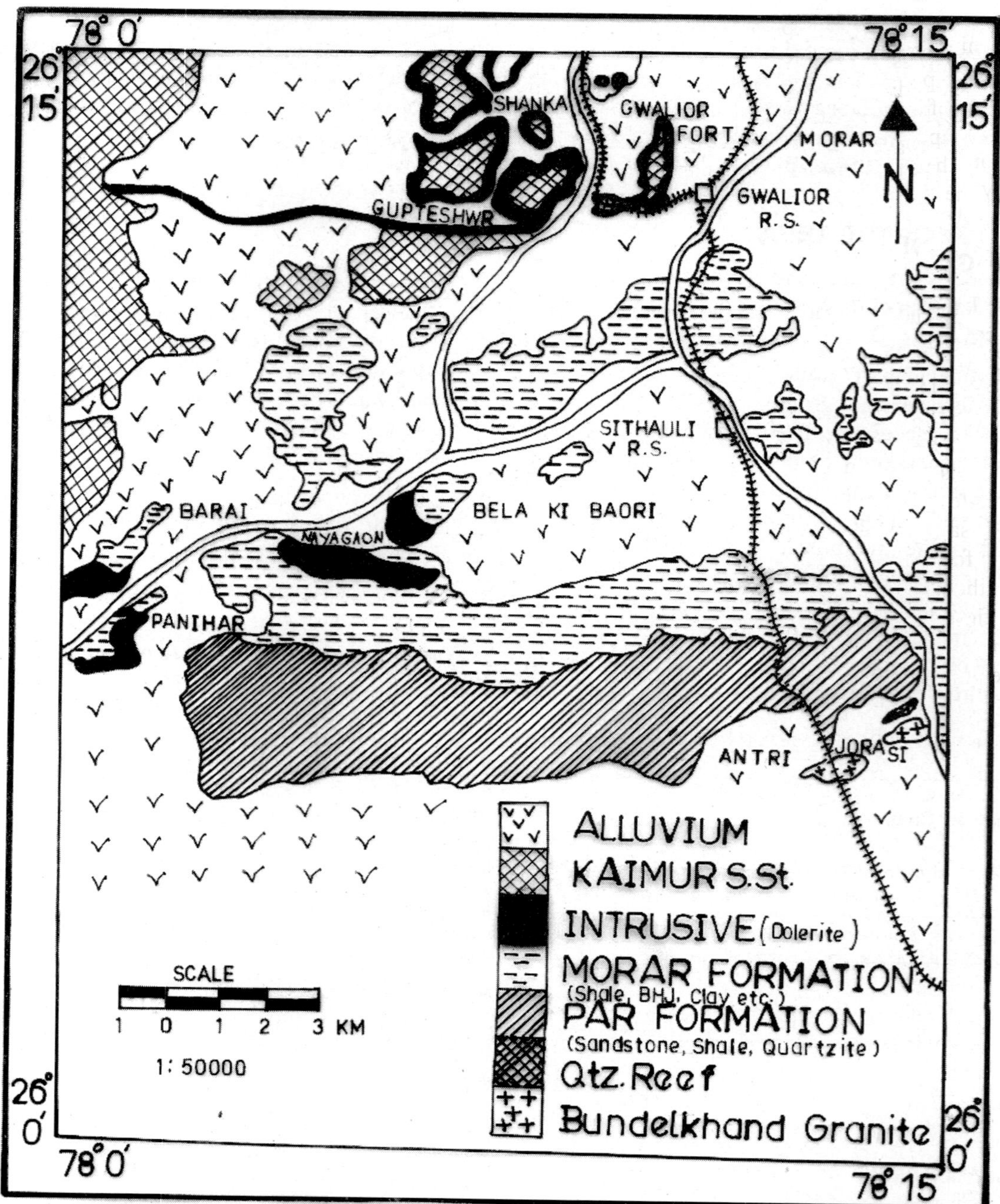

Fig.1 GEOLOGICAL MAP OF GWALIOR AREA (After Pateriya and Singh, 1994)

MINEROLOGY

BIF contents have three sets of following mineral groups:

1. Iron Mineral/Ore : Represented by Hematite, Goethite, Lepidocrosite and Pyrite (Pateriya, 1990). (Fig.7)
2. Silica Minerals : Mainly Jasper and Chert.
3. Clay minerals : Interbeded fine grained clay minerals namely White Ochre, Red Ochre and Yellow Ochre.

GEOCHEMISTRY

Chemical analysis of BIF and laterite for aforesaid

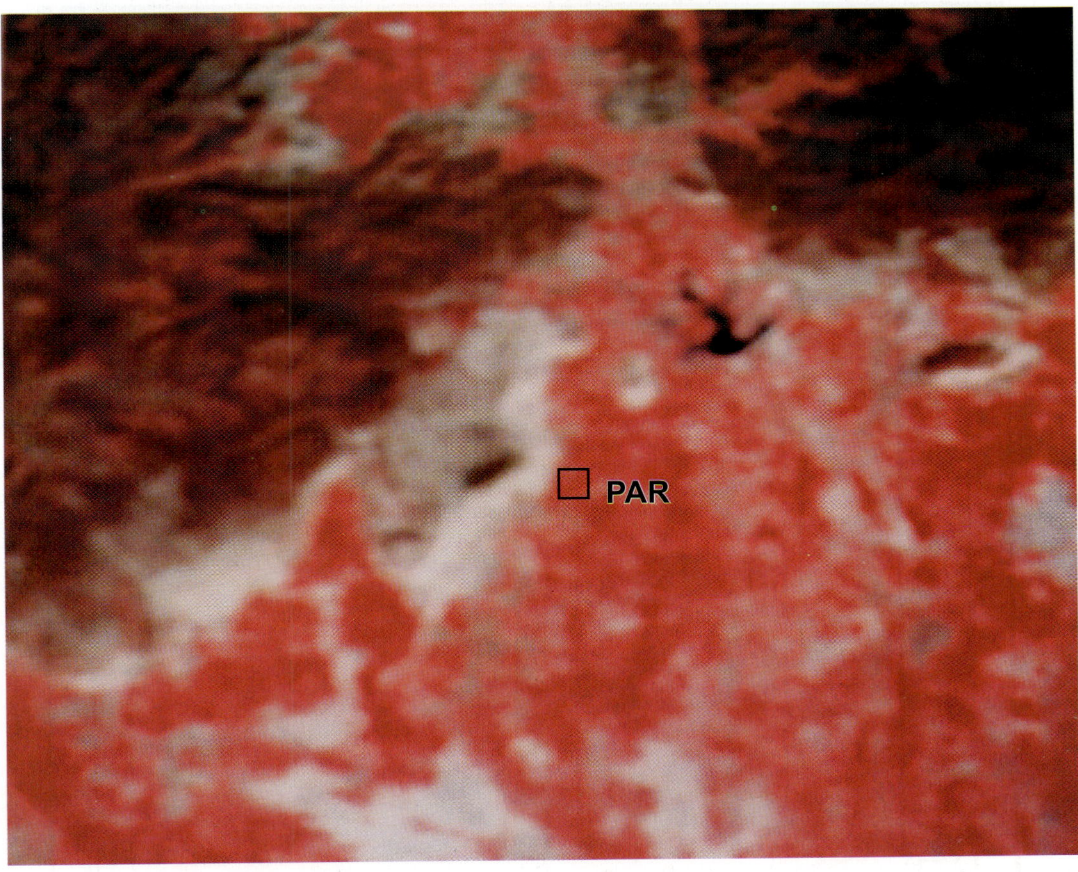

Fig.2 (IRS 1D, Std, Geocoded, B 234, L-3, P 097 / R 053, 12 Feb. 2002)
(Map ID 54 J/4, Scale - 1 : 50,000)

mines are as given in the Table 2.

There is hardly any variation in chemical constituents of BIF in Santau and Panihar mines. On the other hand, the chemical constituents of laterite mine are showing great differences with BIF. It is because of the general properties of laterite as it must have been derived by the process of weathering from the BIF and concentrated on the surface. Weathering effects are relatively high on the directly exposed surface. Laterite is an iron rich soil and always found in loose form. SiO_2 and Al_2O_3 concentration are high in laterite and Fe_2O_3 is very low because of variation in their specific gravity and resistance properties to the atmosphere.

MINING ACTIVITIES

Open cast mining activity is going on in and around the Gwalior city for exploitation of the said minerals in the Morar formation. Name of some approved mining activity at different places approved by the Govt. of India (IBM, Regional Office, Jabalpur, M.P.) and the Govt. of M.P. (Deptt. Of Geology & Mines, Bhopal, M.P.) are as follows :

1. Santau Iron ore mine: It is located in the southern direction from the Gwalior City and falls (Fig. 7) on the Survey of India Toposheet No.54 J/4 in between the latitude: 26^0 05' 52"-26^0 06' 14" N and longitude: 78^0 07' 40"-78^0 07' 58" E (Srivastava and Gosavi, 1999): The lease area is 20.235 hects (Tehsil - Gird, Distt. Gwalior, M.P.).

2. Panihar Iron Ore Mine: It is located in the south-western direction from the City and falls (Fig. 8) on the Survey of India Toposheet No.54 J/4 in between the latitude: 26^0 04' 53"-26^0 05' 23" N and longitude: 78^0 02' 12"-78^0 02' 42" E (Srivastava and Gosavi, 1999): The lease area is 20.235 hects. (Tehsil - Gird, Distt. Gwalior, M.P.).

3. Maharajpur Laterite Mine: It is located in the northern direction from the Gwalior City and falls (Fig. 9) on the Survey of India Toposheet No.54 J/3 in between the latitude: 26^0 16' 35"-26^0 16' 46"

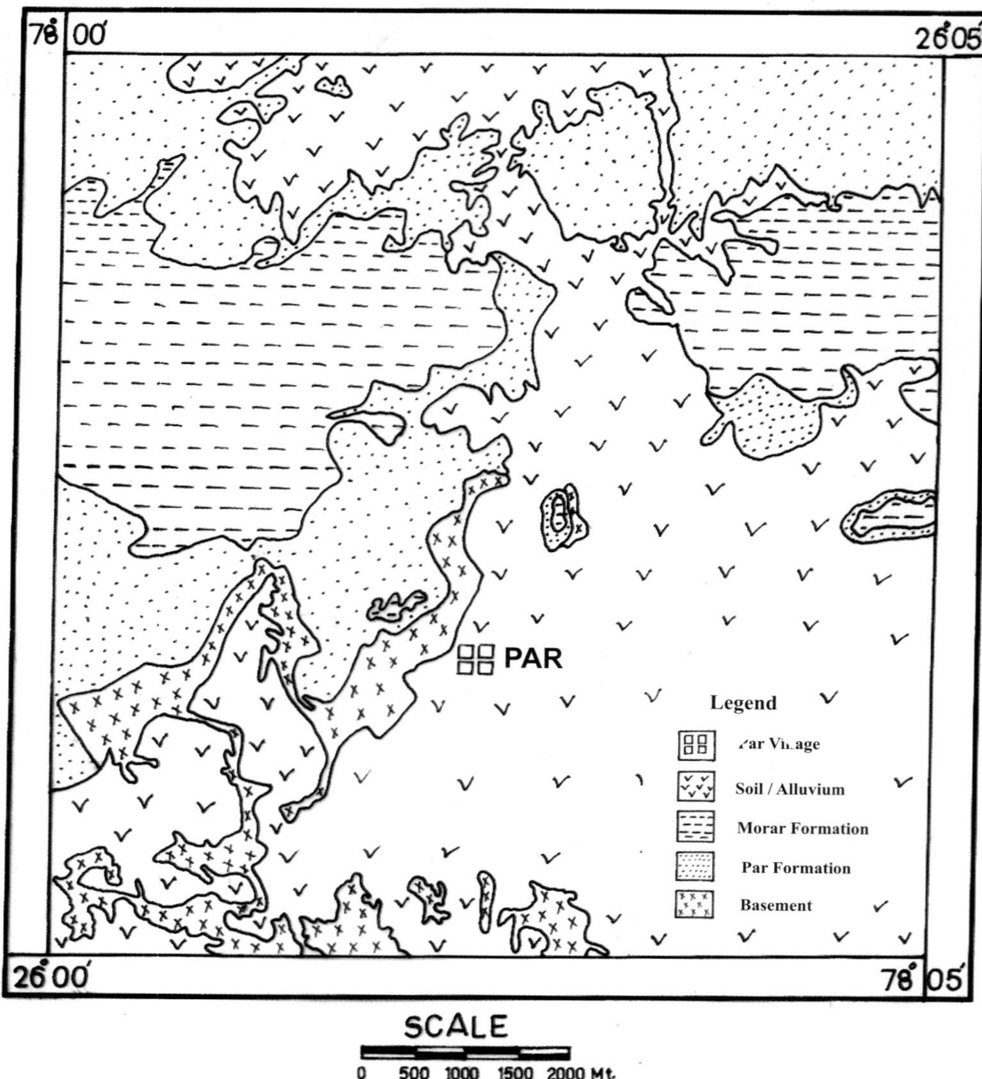

Fig.3 LITHOLOGICAL MAP IN AND AROUND PAR VILLAGE DISTT. GWALIOR (M.P.)
Interpretation from IRS 1D, Std, Geocoded, B 234, L-3, P 097 / R 053, 12 Feb. 2002)
(Map ID 54 J/4, Scale 1:50,000)

N and longitude: 78° 14' 44"-78° 14' 57" E (Jain and Srivastava, 1999): The lease area is 4.840 hects. (TehsilGird, Distt.Gwalior, M.P.).

ECONOMIC USES

The different minerals formed in Morar formation have following uses:

1. Iron ore : Low grade iron ore (BHJ & BHC) having < 50% 'Fe' content, now a days is frequently used in making of 'Sponge Iron" and to meet out the 'Fe' percentage by the different cement industries.
2. Laterite : It's a weathered product of BIF, found as red soil is chiefly used in making of "Iron Oxide" (Red Paint) and commonly used as "Muram" for building roads and buildings.
3. Clay Minerals : Occur as white, red and yellow ochre, generally used in "White wash" of building, sometime used by the different paints & chemicals industries. Fertilizer company uses only white ochre for making of special type of fertilizer.

ORIGIN

Rocks of Gwalior Group consists of arenaceous and argillaceous material. Lithologically Par and Morar formations are comprising by sandstone, shale, quartzite and BHJ, BHC, limestone, ferruginous shale, clay (ochre) etc. respectively. All the said rocks are formed by the sedimentary process (Heron, 1917).

Location : Santau Iron Ore Mine

Fig.4 Field sample of BHJ, Jaspar Band is shown by coin

Location : Santau Iron Ore Mine

Fig.6 Field Photograph is showing interbeded Ochre bands in three colours White, Red and Yellow

Location : Santau Iron Ore Mine

Fig.5 Field sample of BHC, Chert band is shown by coin

Fig.7 Panoramic view of Santau Iron Ore Mine

Layering structures are very common within the formation. Geology and field study of the area suggest sedimentary origin of the mineralized zone (Fig. 10). It is formed during the metallogenesis process of crustal evolution (Dennis, 1967). BIF of the investigated area falls well within the "Indian Macrozonations Metallogenetic Provinces" specially in epiplatformal orogenic belts and stretches (Mishra, 2005).

DISCUSSIONS AND CONCLUSION

The chief mineralized zone (BIF) is formed in the UGF (Upper Gwalior Formation) which might have been derived from the LGU (Lower Gwalior Formation). LGF consists with shale, sandstone and quartzite. BIF comprises chiefly of BHJ and BHC, that is the indication of the main source of material i.e. LGF. The ferruginous shales have been used as iron ore (Krishnan, 2001). Banding nature of BIF it self shows sedimentary origin.

Laterite is the weathered product, derived from the BIF found in loose formation at the top of the UGF. Chemically it is iron rich soil in comparison to other. Major oxides SiO_2 and Al_2O_3 are showing increasing while Fe_2O_3 shows decreasing trend.

Table 2 : Chemical analysis of BIF from various mines of the study area

Mine Name →	Santau Mine	Panihar Mine	Maharajpur Mine
Constituents ↓	Wt. %	Wt. %	Wt. %
SiO_2	12.58	13.40	40.54
Al_2O_3	10.60	09.38	21.45
Fe_2O_3	68.50	67.88	32.03
TiO_2	00.80	00.94	00.36
LOI	07.20	08.10	04.30
CaO	–	–	00.65
MgO	–	–	00.15
Undermined(UD)	00.32	00.30	00.52
Fe	**47.88**	**47.44**	**22.38***

Location : Panihar Iron Ore Mine

Fig. 8 Field Photograph is showing manual mining activity (open cast)

Location : along railway cutting section (Gwalior Jhansi) near SMS Govt. Science College, Gwalior, M.P.

Fig. 10 Field Photograph of BIF, showing layering structure

Location : Maharajpur Laterite Mine

Fig.9 Field Photograph is showing mechanised open cast mining activity

Fine grained thin clay (ochre) beds occur interbeded with slate and shale, showing variation in colour varying from white, red and yellow. They might have been formed from the thin quartz veins/reefs ferruginous shale and feldspar by the process of weathering.

ACKNOWLEDGEMENT

Author is very grateful to Prof. U.C. Singh, Head, School of Studies in Earth Science, Jiwaji University, Gwalior (M.P.) for providing necessary facilities and help during the work. Thanks are also due to Prof. Madhumas Khare, SOS in Earth Science, Jiwaji University, Gwalior for his valuable discussions and suggestions. Author wishes to thank Er. D.K. Shrivastava, Retd. Director, Geology and Mines, Govt. of M.P. for providing chemical analysis and his kind support.

References

Dennis, J.G., (1967). International Tactonics Dictionary, Memoir No. 7, International Geological Congress Publication (American Association of Petroleum Geologists).

Dr. Heron (1936). Mem. GSI Vol. Lxviii. Pt. I (Cited in the textbook "Geology of India", by D.N. Wadia, 118 p. Tata McGraw Hill Pub. Co. Ltd., New Delhi, 4th Ed. 1996).

Hacket C.A., (1870). Geology of Gwalior and Vicinity, Rec. Geol. Sur. Ind. Vol. 69, 34 p.

Heron A.M. (1917), The Geology of northeastern Rajputana and adjacent district. Mem. Geol. Surv. India, 45, pp. 171-188.

Jain P.K. and Shrivastava D.K., (1999). Maharajpur Iron Ore Mine, unpublished "Approved Mining Plan", 1999, by RCM, Indian Bureau of Mines (IBM), Regional office, Jabalpur (M.P.), Lease Area 4.840 hects, village Maharajpur, Tehsil Gird, Distt. Gwalior (M.P.)

Krishnan M.S., (2001). Geology of India and Burma, CBS publishers & distributors, New Delhi (6th Edition).

Mishra R.N. (2005). A. Keynote Lecture on "An Overview of Geological & Tectonic Framework of India with Special Reference to Mineralization": in 4th NNRMS course on 'Application of Remote Sensing & GIS in Mineral Exploration' held at GSI Training Institute, Bandilaguda, Hyderabad (A.P.) 19th Jan 2005 31st March, 2005, pp. 1-20.

Pateriya B. and U.C. Singh, (1994). Geology of the Gwalior area, Asian J. Exp. Sci., Vol. 8, No. 1 & 2, pp. 9-13.

Pateriya B., (1990). Geology of the Gwalior formation with special reference to the origin of Banded Iron Formation, unpublished M.Sc. thesis, Vikram University, Ujjain (M.P.)

Kumar Ravindra (1998). Fundamentals of historical Geology and Stratigraphy of India, p. 89, New Age International Publishers, New Delhi (6th Edition).

Shrivastava D.K. and Gosavi, M., (1999) Panihar Iron Ore Mine, unpublished "Approved Mining Plan", by RCM, Indian Bureau of Mines (IBM), Regional office, Jabalpur (M.P.), Lease Area 20.235 hects, village Panihar, Tehsil Gird, Distt. Gwalior (M.P.)

Shrivastava D.K. and Gosavi, M., (1999) Santau Iron Ore Mine, unpublished "Approved mining plan", 1999, By RCM, Indian Bureau of Mines (IBM), Regional office, Jabalpur (M.P.), lease area 20.345 hects, village Santau, Tehsil Gird, Distt. Gwalior (M.P.)

Hon'ble Reviewers

1. Prof. D.C. Gupta
   Head
   Department of Applied Geology
   Barkatullah University
   BHOPAL 462 026 INDIA
2. Prof. P.K. Verma
   Department of Geology
   Vikram University
   UJJAIN 456 010 INDIA

Editorial Assistance

Dr. Jaimini Swadia

Economic Mineralization
K.L. Shrivastava (Editor)
2009. pp.336-339
Scientific Publishers (India), Jodhpur

A Note on the occurrence of Clitellate Cocoons from Giral and Barsinghsar Lignite deposits, Rajasthan

R.P. TRIPATHI AND B.D. SHARMA[2]
Department of Botany, J.N. Vyas University, Jodhpur 342 005

Abstract

Several filamentous and net like clitellate cocoons of fresh water annelids are reported from the lignite deposits of Rajasthan, India. Cocoons with the irregular nets are comparable with Dictyothylakos pesslerae Horst and the regular nets are new to the record. Palaeoecology of the area is discussed.

Key words : Clitellate annelids, Lignite, Palaeoecology, Western Rajasthan.

INTRODUCTION

A number of fossil clitellate cocoons of the fresh water deposits were described by Manum et al. (1991, 1992) from the Mesozoic period, and assigned to the taxa Burejospermum Krassilove, Dictyothylakos Horst and Pilothylakos Manum et al. Earlier workers, Michael (1936), Horst (1954), Hughes (1955), Dijkstra (1961) and Kovach & Dilcher (1988) considered similar fossils either as plants or of plant origin, e.g., algal nets and palynomorphs.

In the present investigation, fossil cocoons were obtained during the maceration of lignite samples for the purpose of micropalaeontological studies. In addition to the cocoons, the lignite also preserves algal filaments, fungal spores, fruiting bodies and hyphae, bryophytic thalli and capsules, pteridophytic sporangia and spores, gymnospermous cuticles, and a number of fossils of angiosperms, like pollen grains and seed cuticles (Rao & Vimal 1952, Tripathi & Sharma 1996, 1997, 2000, 2000a; Tripathi et al. 1998, 1999).

The lignite samples were collected from Giral (Seam No. RJ9) in Barmer district and Barsinghsar (borewell) in Bikaner district of Western Rajasthan. Location and brief geology of the sites are given in Fig. 1. Maceration of the samples was done by the usual techniques using conc. HNO_3 for oxidation of carbonaceous material, followed by 510% aqueous solution of KOH. Isolated fossils were mounted in glycerine jelly.

The cocoons are fragmentary pieces of different sizes ranging from 170 x 100 μm to 1,120 x 840 μm. These are basically divided into two types, i.e., irregular nets of thick and thin filaments (Fig. 2B) and regular nets of uniform filaments (Fig. 2 A and C). The former are comparable to Dictyothylakos pesslerae Horst (Manum et al. 1991, Fig. 7K, I). The outer filaments (hapsine) are thick and branched while the inner (alytine) are thin and divided. The second type of nets have regularly arranged obovate to subtriangular 56 cavities surrounding a point (Fig. 2A and C) or in other words 56 filaments coming from all directions meet at a point and make a regular pattern in the net. The filaments range 832 μm in thickness and the points of meeting are 3055 μm thick. No separation is

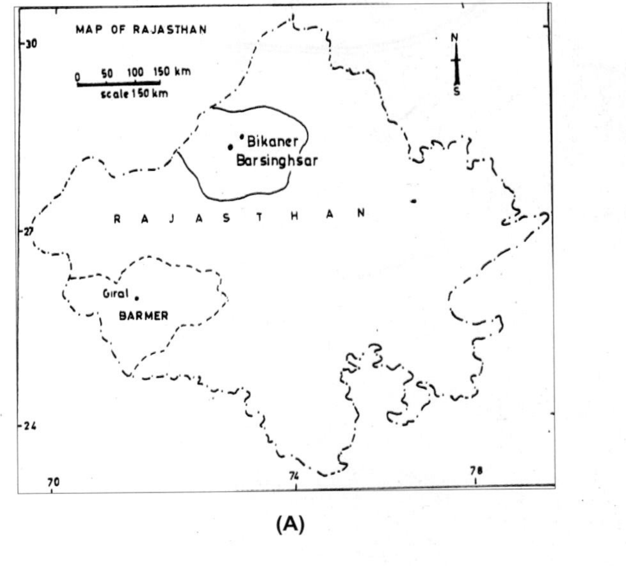

(A)

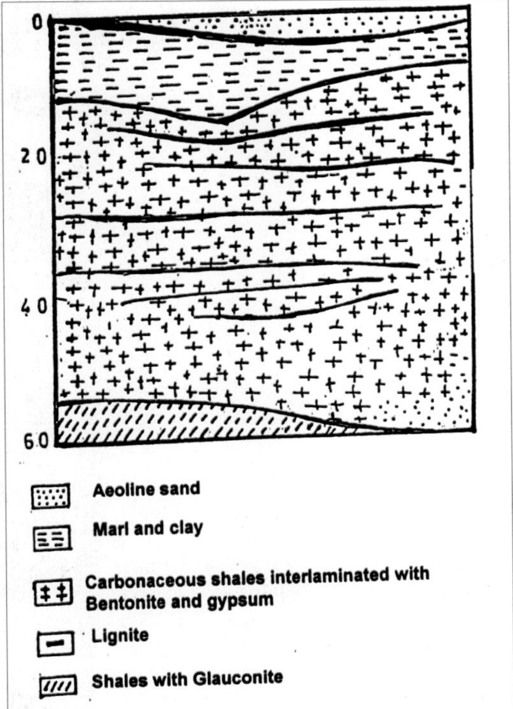

(B)

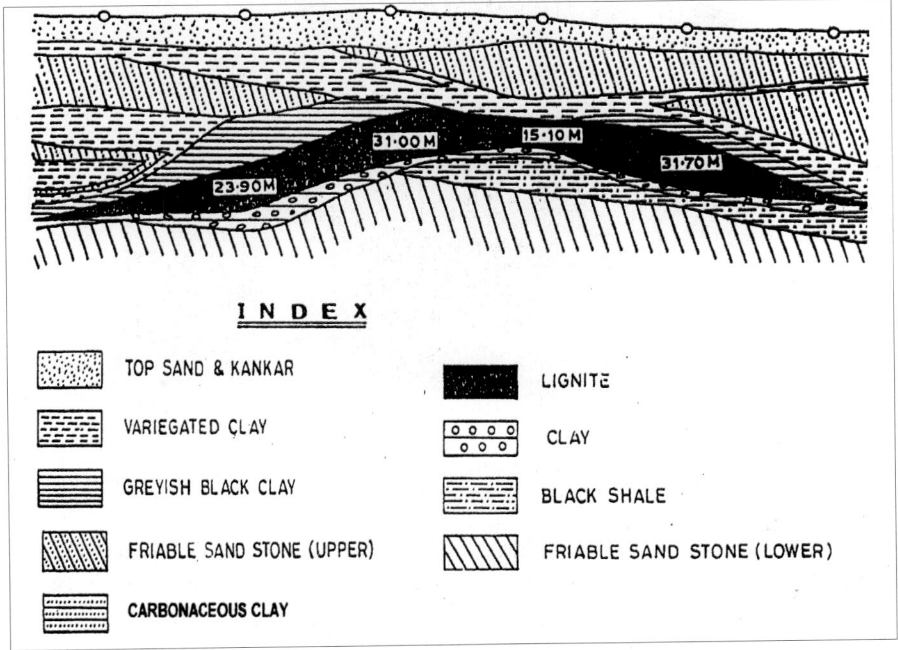

(C)

Fig.1 Location and Geology of Giral and Barsinghsar lignite deposits (after Tripathi et al. 1999) **(A)** Locations **(B)** Geology at Giral **(C)** Geology at Barsinghsar

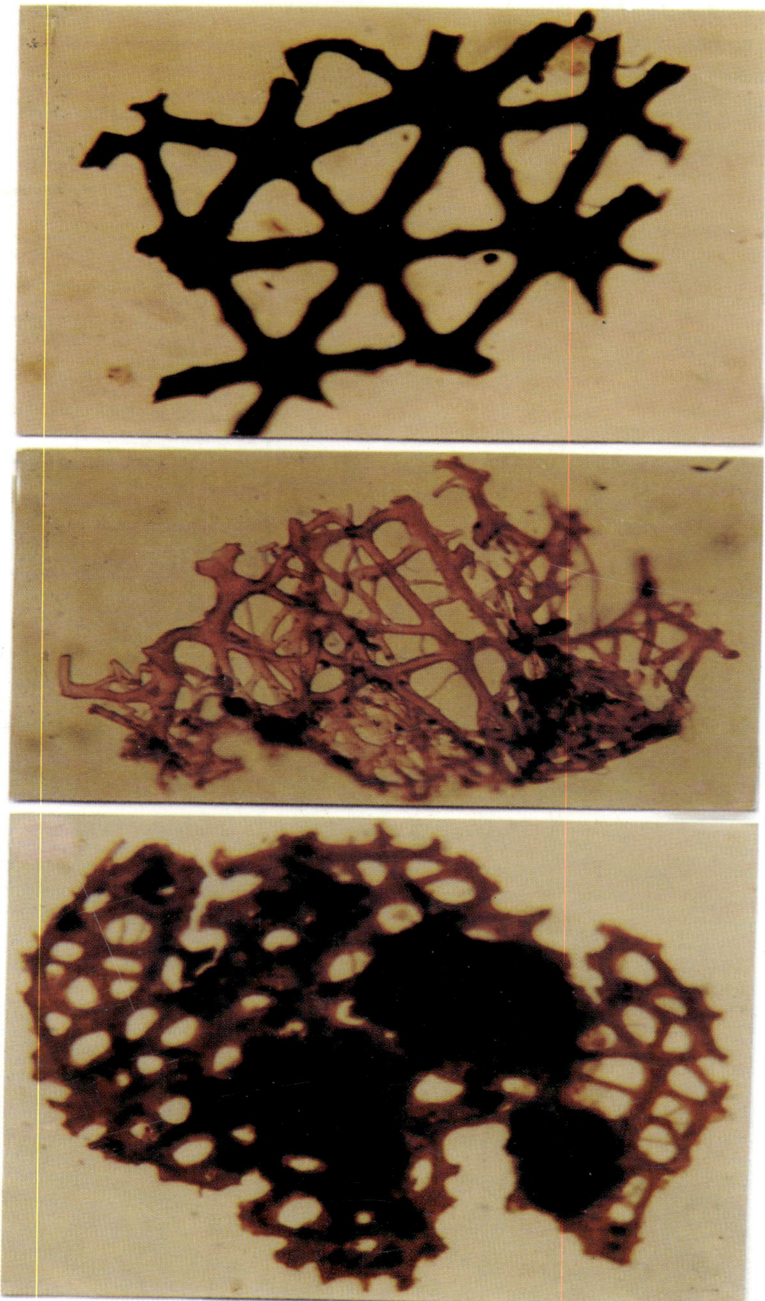

Fig.2 **A.** Regular net with subtriangular cavities surrounding a point, x 240. **B.** Irregular net of thick and thin branched filaments, x 120. **C.** Regular net of filaments surrounding points, x 120.

seen anywhere in the filaments. Similarly, no differentiation of wall could be observed in filaments.

The presence of clitellate cocoons of fresh water annelids/leeches in Giral and Barsinghsar favours nonmarine aquatic palaeoecological conditions during the Eocene in the Thar desert of Rajasthan. This is also in confirmation with the earlier palaeobotanical findings by Harsh & Sharma (1995), Harsh et al. (1992) and Tripathi et al. (1998, 1999).

References

Dijkstra, S.J., 1961. On some megaspores, Charophyta fossils and some other small fossils from the Cretaceous. Palaeobotanist, vol. 8, pp. 818.

Harsh, R. and Sharma, B.D., 1995. Petrified Tertiary woods

from Bikaner (Rajasthan). Indian Jour. Earth Sci., vol. 22, pp. 384-389.

Harsh, R., Sharma, B.D. and Suthar, O.P., 1995. Anatomy of petrified woods of Lacythidaceae and combrataceae from Bikaner (Rajasthan), India. Phytomorphology, vol. 42, pp. 87-102.

Horst, U., 1954. Merkwurdige Gebilde in Kohlen aus dem Wealden. Geologie, vol. 3, pp. 610-616.

Hughes, N.F., 1955. Wealden plant microfossils. Geol. Mag., vol. 93, pp. 201-217.

Kovach, W.L. and Dilcher, D.L. 1988. Megaspores and other dispersed plant remains from the Dakota Formation (Cenomanian) of Kansas, U.S.A. Palynology, vol. 12, pp. 89-119.

Manum, S.B., Bose, M.N. and Sawyer, R.T., 1992. Clitellate cocoons in fresh water deposite since the Triassic. Zool. Scr. Vol. 20.

Manum, S.B., Bose, M.N. and Sawyer, R.T., 1992. Seeds (*Burejospermum* Krassilove) and palynomorphs (*Dictyothylakos* Horst) with netted wall structure reinterpreted : clitellate cocoons. Cour. Forsch. Inst. Senckenberg, vol. 147, pp. 399-404.

Michael, F., 1936. Palaeobotanische und Kohlenpe trographische studien in der Nordwestdeutschen Wealden Formation. Abhandle. Preuss. Geol. Landesanst, vol. 166, pp. 179.

Tripathi, R.P. and Sharma, B.D., 2003. Plant diversity during Eocene in Western Rajasthan. Recent trends in Botanical researches (Prof. D.D. Nautiyal comm.. vol.) D.K. Chauhan (ed.), Dixit Press, Allahabad, pp. 118-128.

Tripathi, R.P. and Sharma, B.D., 1996. Pteridophytic plant microfossils from Eocene lignite deposits of Rajasthan. Indian Fern J., vol. 13, pp. 36-39.

Tripathi, R.P. and Sharma, B.D., 1997. Megaspores from the lignite of Rajasthan. J. Palynol., vol. 33, pp. 181-184.

Tripathi, R.P. and Sharma, B.D., 2000. Seed cuticles from Eocene lignite deposits of Rajasthan, India. Phytomorphology, vol. 50, pp. 180-183.

Tripathi, R.P., Shrivastava, K.L. and Sharma, B.D., 1998. Plant microfossils from the lignite deposits (Eocene) of Barsinghsar in Bikaner district, Rajasthan, India. Palaeobotanist, vol. 47, pp. 110-115.

Tripathi, R.P., Sisodia, M.S., Shrivastava, K.L. and Sharma, B.D., 1999. Micropalaeontological studies of lignite samples from Western Rajasthan. Geological evolution of Northwestern India, B.S. Paliwal (ed.), Scientific Publishers, Jodhpur, pp. 118-128.

Hon'ble Reviewers

1. Prof. David Dilcher
Florida Museum of Natural History
Florida
USA
2. Dr. Rakesh Srivastava
Bhoj University
BHOPAL 462 001 INDIA

Editorial Assistance

Mrs. Anju Mathur

Economic Mineralization
K.L. Shrivastava (Editor)
2009. pp.340-351
Scientific Publishers (India), Jodhpur

31

Section-III

Sedimentology and Geochemistry of the Overbank Sediments of the Cooum River, Tamil Nadu

PAVITRA ARAVAMUDHAN, HEMA ACHYUTHAN AND NAVIN SHANKAR

Department of Geology, Anna University, Chennai 600 025
Email : hachyuthan@yahoo.com

Abstract

Overbank sediments exposed in the lower Cooum River, near Chennai are represented by mud and coarse grained channel facies. Lithosections exposed along the Cooum River between Pudichchathiram and Velappanchavadi were studied to determine sedimentological characteristics and also to understand the chemical variations and type of pedogenesis. The sedimentation process and textural variations in the lower Cooum River indicates fluctuations in the hydrologic regimes and weak pedogenesis. Trace metal concentrations of the overbank surface sediments indicates intense urbanization and anthropogenic source especially around Velappanchavadi.

Key words : Overbank Sediments, Chemical and Textural Characteristics, Pedogenesis and Anthropogenic Activity, Cooum River, Tamil Nadu.

INTRODUCTION

Overbank sediments occurring along rivers and streams are indicative of variable water discharge (Reineck and Singh, 1980; Schaetzl and Anderson, 2005). These are deposited on flood plains and levees during floods, when the discharge exceeds the carrying capacity of the channel During the flooding, velocity of the stream is reduced by overtopping the banks, causing deposition of the suspended fine sediment on the river banks with finer grains depositing away from the channel and are often few centimeters to several decimeters thick. Overbank sediments and in particular flood basin deposits (lowest lying part of a river flood plain) represent the long continued accumulation of fine-grained suspended sediment. This allows for reconstruction of sediment composition backwards into pre-industrial times dates (Schaetzl and Anderson, 2005). The deeper part of the lithosections reflect chemical composition of the sediments deposited before the appearance of any human activity. In this paper sedimentological characteristics and trace metal content of the overbank sediments in the lower reaches of the Cooum River are reported. The purpose of the study was also to identify the intensity of pedogenesis and evaluate the chemical variations in these overbank deposits.

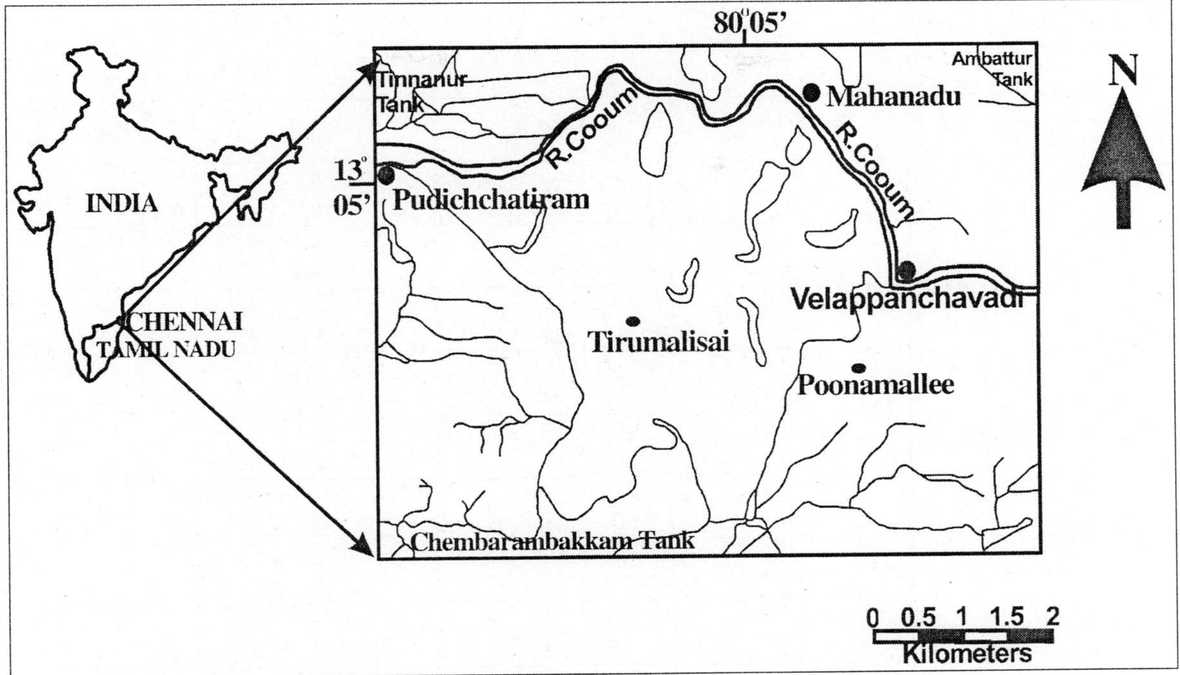

Fig.1 Location map of the study area

THE STUDY AREA

The study area falls within the jurisdiction of Thiruvallur district, enclosed between 13° 00' N 80° E: 13° 05' N 80° 08'E and forms the lower part of the Cooum River (Fig.1). The area comprises unexposed charnockite basement rock overlain by the Upper Gondwana sandstone and shales at many sites. These are covered by the Quaternary alluvium that forms the present day surface. In the upland region the Quaternary alluvium cuts and fills the late Neogene ferricritised surface (Nagalakshmi, 2004). The study area experiences a humid hot climate. The average temperature ranges from 30°C-40°C during the summers and 25°C-30°C during the winters. The district receives an annual average rainfall of 1264 mm with maximum rain during October to December due to NE monsoon.

The Chennai city is intersected by two languid streams, the Cooum and the Adyar. Cooum runs through the heart of the city and enters the sea. Presently, these two rivers are almost stagnant and do not carry water except during rainy seasons. Cooum River arises from Kesavaram Anicut in Kesavaram village built across Koratallaiyar River. The length of the river is about 65 km, of which 18 km, drains the Chennai city.

MATERIALS AND METHODS

Samples at every 5 cm intervals were collected for textural analyses following Ingram (1970). Calcium Carbonate ($CaCO_3$ %) content and Organic matter (OM %) were obtained following Muller (1967) and (Dean, 1974; Bengtsson and Enell, 1986). Major oxides and trace metals were analysed following Shapiro and Brannock (1962). The results are presented in Table 1. Trace metals (Cu, Co, Ni, Pb, Zn and Cr) were determined by using Atomic Absorption Spectrometer GBC 932 plus. Reference sediments standard BCSS^{-1} from NRC, Canada, was analysed repeatedly to determine the accuracy (Table 3). The standard was treated and analyzed for all the elements in a similar manner to those with each batch of sediments studied. The analytical precision value for Cu is 4.21 %, 1.75 % for Co, 2.72 % for Ni, 3.54 % for Pb, 4.46 % for Zn and 4.8 % for Cr. The analytical accuracy is well within 5 % and precision of three replicate measurements for all these metals are within 2 %. Soil samples were collected in boxes, impregnated with resin and 12 soil thin sections covering three lithosections were prepared and analysed following Bullock et al., (1985). Soil features are presented Fig. (4-7).

LITHOSECTIONS

Three short litho sections were studied for their

Table-1 : Textural analysis

	Depth (cms)	Sand %	Silt %	Clay %
Velappanchavadi (A)	10	58	25	17
	20	26	59	15
	30	30	66	4
	40	37	47	16
	50	37	47	16
	60	44	22	34
	70	62	23	15
	80	50	36	14
	90	53	31	16
	100	45	44	11
	110	44	45	11
	120	58	31	11
	130	42	46	12
	140	47	41	12
Mahanadu (B)	10	62	23	15
	20	64	18	14
	30	58	31	11
	40	53	33	14
	50	50	25	25
Pudichchathiram (C)	10	20	69	11
	20	54	34	12
	30	62	25	13
	40	64	18	14
	50	60	25	15

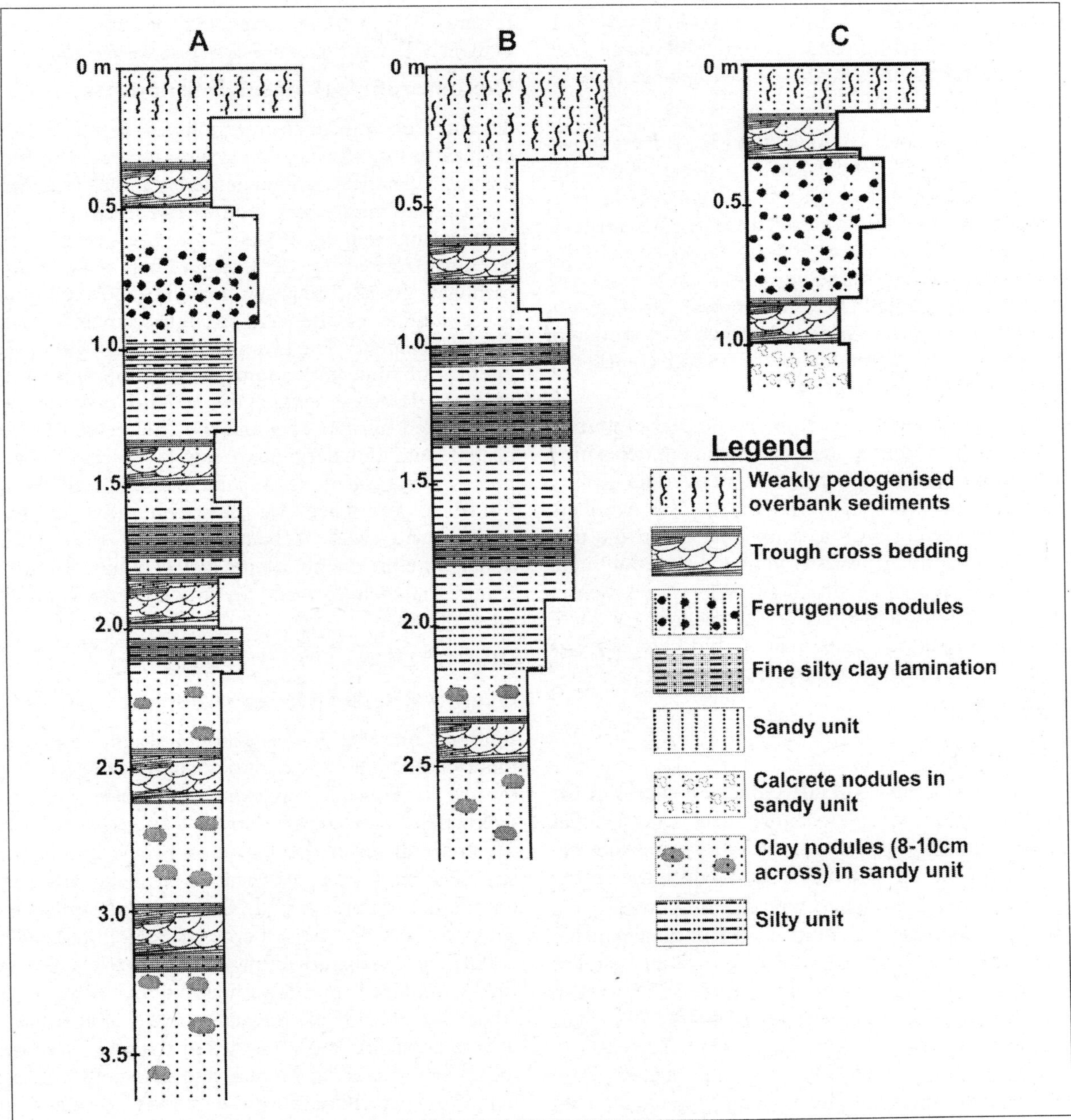

Fig.2 Lithosections studied around Velappanchavadi **(A)**, Mahanadu **(B)** and Pudichchathiram **(C)**

sediemtnary structures, grain size, clasts and sedmientary textures that were exposed along the Cooum river bank. Samples were collected at different depths along the lithosections. The lithosection studied at Velappanchavadi (A) off Thiruvekadu (80° 08' 27.8"E: 13° 03' 45.5" N) is downstream of the Cooum river course. The lithosections at Mahanadu (B) (80° 06' 44.9"E: 13° 04' 46.9"N) and the sediment profile at Pudichchathiram (C), Thiruvallur district (79° 59' 24.0"E: 13° 05' 52.0" N) are located in the upstream of the Cooum river course (Fig. 1 and 2).

The lithosection (~300 cm thick) at Velappanchavadi site (Section A) reveals gravely sand sediments (250-300 cm) at the lower end of the section. This unit is overlain by thick layer (80-250 cm; 170 cm thick) of coarse sand intercalated by clay nodules and this unit is in turn overlain by sand with ferruginous nodules (1-2 cm across). A 50 cm thick unit of dark brown organic carbon rich horizon rests over the lower units with a sharp contact (Fig. 2, Section A).

The lithosection at Mahanadu (Section B) shows calcrete horizon (220 cm-280 cm) at the base of the exposed section. Above this unit a 20 cm thick layer of powdery calcrete unit occurs. This is overlain by thick (50 cm-220 cm) layer of coarse, poorly sorted sand intercalated with clay nodules (4-6 cm across) and ferruginous nodules (1-2 cms across). This unit is subsequently overlain by a dark brown, rich in organic carbon content sity sand unit with a sharp contact (50 cm thick) (Fig. 2, Section B).

The sedimentary profile in Pudichchathiram (Section C) exhibited phases of sediment deposition with sandy gravel (10 cm thick) unit overlain by silty clay unit of nearly 20 cms thick. The overbank sediments resting with a sharp contact on the unit below is dark brownish yellow in color with predominance of fine silt clay poorly sorted, rich in organic carbon content (Fig. 2, Section C). Field characters such as ped formation and clay coatings along the sides of the peds were observed.

RESULTS

Trace metals

Down core variations of silt and clay contents of the three overbank profiles shows that sand content ranges from 26-70 %, the silt content ranges between 13-66 % (Fig 2). $CaCO_3$ % shows decrease in the values down the Velappanchavadi lithosection, while there is an increase in calcium carbonate content in the Pudichchithiram profile and decrease in the value of calcium carbonate content in the third profile as seen in fig. 2. The trace metal content has high concentrations of Cu, Co, Cr, Zn, Ni at Velappanchavadi while in the trace metals from Pudichchathiram and Mahanadu profiles are depleted in Cu and Zn content (3.93 ppm and 1.71 ppm) (Table 2, Fig. 3).

Micromorphology of soils

Micromorphology of soils from all the three lithosections reveal biogeneic features such as root pores and clay coatings around root pores. The clay coatings (2-3 ìm thick) have formed around the detritus grains of quartz and feldspars (Figs. 4-6). Weathering in the plagioclase feldspars and alteration of pyroxene grains are also observed. Iron oxide impregnation within the soil matrix is also a dominant feature (Fig.7). The Overbank sediments reveal evidences of weak clay illuviation and elluviation.

Micromorphology of calcrete nodules

Micromorphology of calcrete nodules reveals features such as dissolution of calcic grains, reprecipitation of calcite in channels, around detrital grains as coastings (micrite and microspar), mottles and sesquioxides of ferruginous impregnations, hair line cracks and fractures related to desiccation and etched non-carbonate grains. The groundmass is microcrystalline to crystalline calcite and strongly cemented with meniscus calcite. The channels and voids within the nodules are filled with microsparite and sparite. Other biologic features within the calcite nodules are represented by root ores and root filaments that are coated with powdery dirty calcite (micrite). Root pores are regular in shape and sometimes elliptical. Voids and pore spaces are filled with micrite or clear microsparite. Many channels within the calcite nodules are lined with calcitic clay coatings. All these features indicate towards the groundwater origin of these calcretes.

DISCUSSION

Overbank sediment deposition

Bedrock in the lower Cooum River basin is represented by the Precambrian charnockite overlain by the Gondwana sandstone and shale and the overbank sediments in the lower Cooum River basin are stratified and are persistent over the bedrock. The stratification in the sediments may have occurred during non turbid period. Coarser textured sandy deposits are stratified and cross bedded in the lower units. Finer overbank sediments are silty to clayey and finely stratified in more horizontally lying beds (Baker et al., 1991). Because of the stratification overbank sediments are progressively younger towards the top of the deposits and an overall increase in the sand percentage at various depths indicates their flood cyclic deposition. High percentage of silt indicates stagnant condition of the water column in the river bank facilitating deposition of suspended particles during those periods.

Overbank sediments shows a close stratigraphic association with the major channel sand body. The lateral extent of the deposition varies from a few 100 m to 6 kms. Overbank facies association has significant proportion of sheet sand with well preserved physical sedimentary structures. Low degree of oxidation and organic matter is preserved.

Table-2 Calcium carbonate, organic matter, major and trace metal concentrations.

	Depth in cm	$CaCO_3$ (%)	OM (%)	SiO_2 (%)	Al_2O_3 (%)	Cu (ppm)	Co (ppm)	Ni (ppm)	Pb (ppm)	Zn (ppm)	Cr (ppm)
Velappanchavadi (A)	10	13.19	3.037	66.0	18.00	57.57	34.58	48.06	3.67	88.1	38.52
	20	12.39	4.456	60.2	17.00	61.5	38.51	60.02	6.66	81.26	35.77
	30	12.45	5.085	32.0	34.18	74.62	46.77	78.31	4.34	68.62	64.42
	40	11.99	5.129	52.6	11.00	19.54	10.12	24.03	3.09	18.91	24.38
	50	11.39	5.978	73.0	11.00	106.61	43.82	68.29	0.68	37.45	50.57
	60	10.79	7.196	78.0	12.00	82.86	39.1	87.36	2.03	35.93	77.14
	70	9.19	5.189	76.0	12.00	72.23	36.75	69.46	1.25	52.27	65.47
	80	7.19	4.078	40.0	55.00	56.8	39.79	59.54	1.35	23.57	81.98
	90	5.99	3.848	74.0	12.00	69.26	28.79	73.74	0	28.04	79.79
	100	4.79	4.218	68.0	14.70	81.23	37.92	80.65	0.39	54.27	110.25
	110	2.79	1.609	42.6	12.41	54.79	25.35	57.01	0.68	50.18	65.56
	120	2.99	1.619	51.4	15.00	50.96	32.91	57.49	1.83	52.18	78.18
	130	2.55	3.768	45.6	16.00	42.63	27.51	47.18	3.47	18.25	62.43
	140	2.78	3.526	65.6	10.90	31.61	13.66	31.62	2.7	13.31	31.5
Mahanadu (B)	10	5.99	5.978	60	10	3.93	0.88	5.74	2.32	5.42	13.09
	20	4.79	5.129	58	7.85	59.68	27.31	56.23	1.74	15.59	34.06
	30	4.15	4.456	64	5.65	48.09	26.33	50.68	0.96	1.71	32.45
	40	4.12	4.62	58	7.45	46.07	27.02	51.66	0.96	6.94	49.15
	50	4.05	4.9	63	5	45.40	25.15	49.03	3.28	20.43	31.41

Table-2 (Contd.)

Pudichchathiram (C)	10	3.599	5.11	62.0	15.45	37.17	15.72	36.87	2.32	13.59	43.46
	20	1.599	3.37	66.2	11.45	32.76	26.04	39.11	1.25	9.6	68.12
	30	1.999	3.53	52.6	7.75	36.97	9.63	30.06	0.96	3.71	31.31
	40	2.398	3.29	55.0	8.35	21.46	12.97	32.3	3.67	11.12	33.68
	50	3.596	4.62	65.6	5.85	30.27	25.25	42.02	4.05	7.7	59.3

Overbank sediment pedogenesis

Micromorphology of overbank sediments revealed biogeneic features such as root pores and clay coatings around root pores. The clay coatings were 2-3 μm thick formed around the detritus grains of quartz and feldspars. Textural features included weathering of plagioclase feldspars (Fig.4) and alteration of pyroxene grains. Iron oxide impregnation (Fig.6) within the soil matrix was also a dominant feature. The overbank sediments reveal evidences of weak illuviation and elluviation which are also signatures of weak pedogenesis.

The channels and voids within the nodules are filled with microsparite and sparite. Other biologic features within the calcite nodules are represented by root ores and root filaments that are coated with powdery dirty calcite (micrite). Many channels within the calcite nodules are lined with calcitic clay coatings. Prominent ferruginous nodules and calcrete nodules have developed. Features such as sesquioxides of ferruginous impregnations, hair line cracks, fractures related to desiccation and etched non-carbonate grains. All these features indicate that these calcretes are formed due to complex action of groundwater and pedogenic processes.

Overbank sediment geochemistry

In the present study, SiO_2 and Al_2O_3% decreases down the profile in all the litho sections (Table 2). The $CaCO_3$ content ranges from 13.19 to 1.59% and OM percent varies between 7.19 to 1.61%). A good correlation between $CaCO_3$ and OM is discerned (Fig. 8) and this is due to the groundwater bicarbonates and the occurrence of organic carbon content in the upper surface layers. The high OM content is due to humus and algal activity of the river waters in the study area.

Most of the surface samples show elevated trace metal concentrations (Fig.3) and low content with chemical breaks at depths of the 0-40 cm and 80-100 cm (Table 2 and Fig.3). Significant variations in heavy metal concentration are observed in the samples at the depth of 80-100 cm Velappanchavadi (Table 2).

The sediments (approximately 60 percent) have elevated chromium (24-110 ppm in Velappanchavadi, 31-59 ppm in Pudichchathiram and in Mahanadu 1-24 ppm of Cr) and (24-87 ppm in Velappanchavadi, 30-42 ppm in Pudichchathiram and 6-56 ppm in Mahanadu of Ni respectively) concentrations that reflect the input of these metals from weathering parent rock probably derived from readily weathering

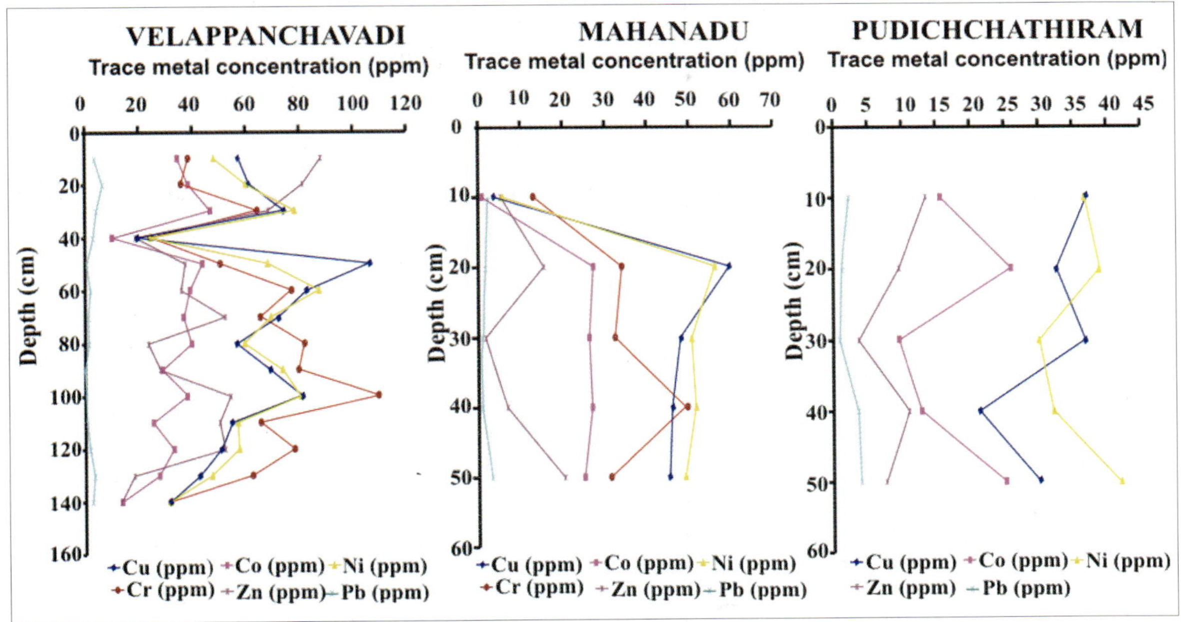

Fig.3 Trace metal data of the Velappanchavadi **(A)**, Mahanadu **(B)** and Pudichchatiram **(C)** overbank sediments.

Fig.4 Plagioclase laths in a coarser matrix (50X)

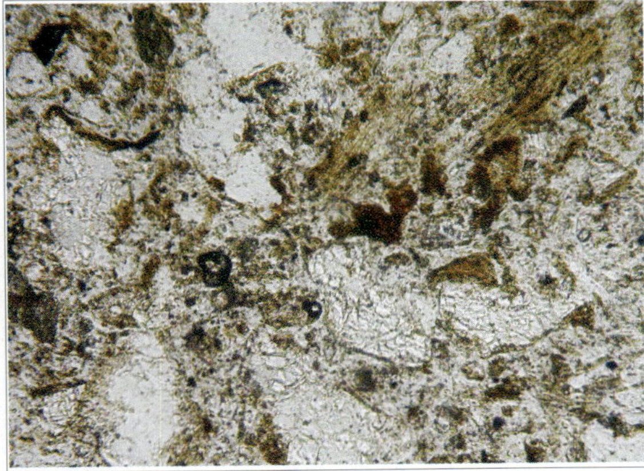

Fig.5 Coatings around the detritus grains quartz grainmatrix. (50X)

Fig.6 Iron oxide impregnation and fine clays forming coating around the grain. (50X)

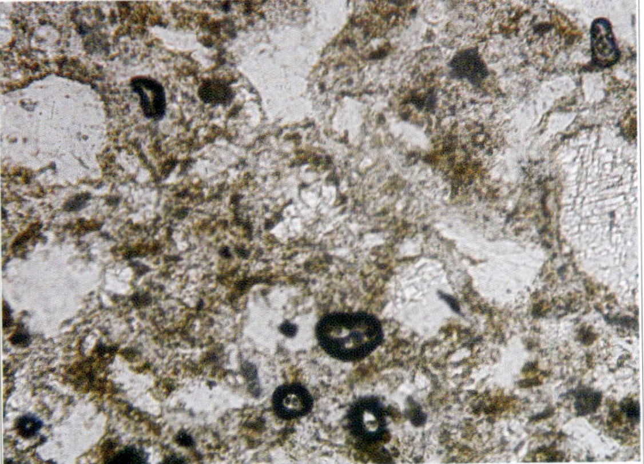

Fig.7 Quartz grains imbedded in fine of (50X)

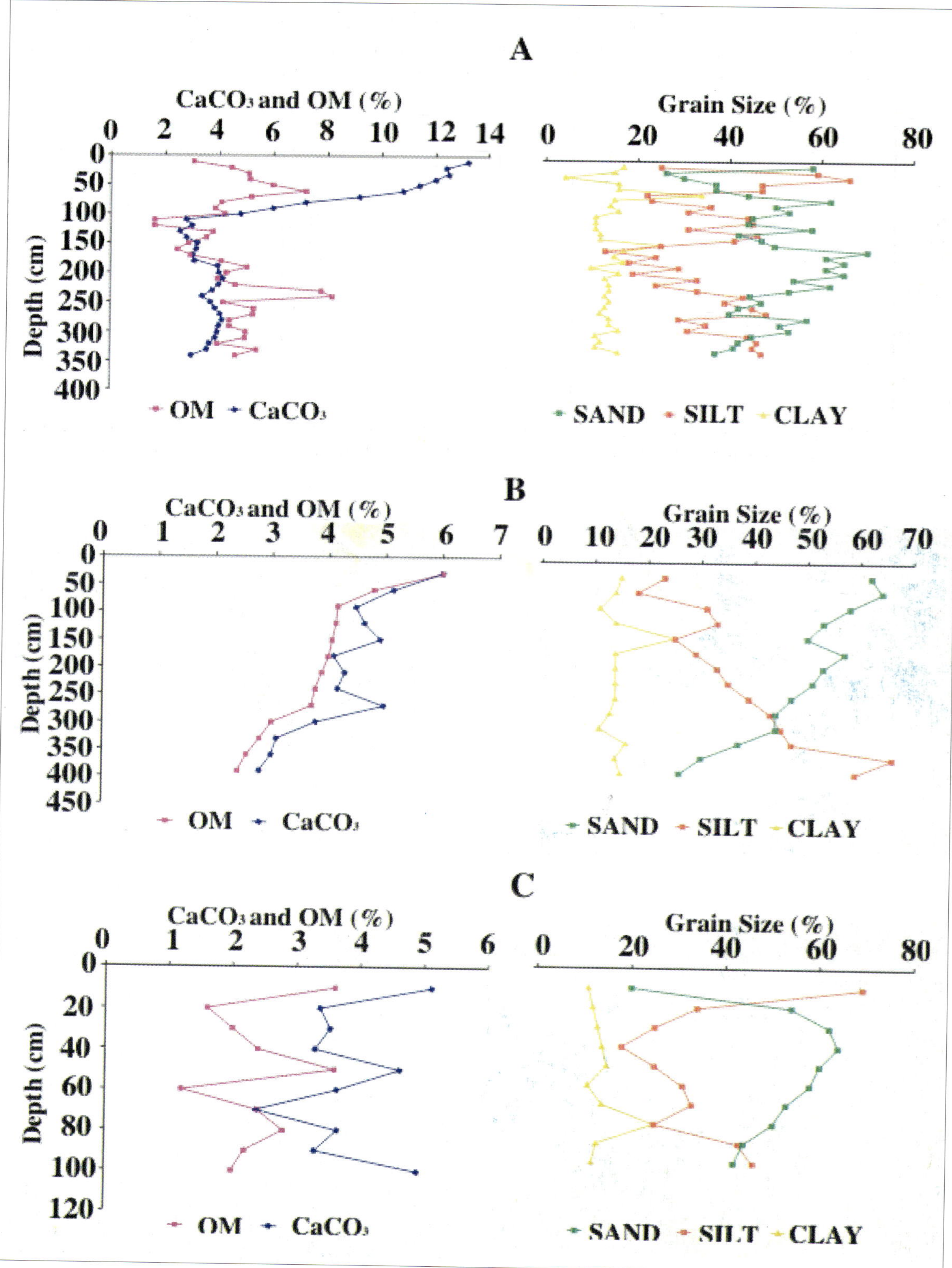

Fig.8 Graphic correlation between $CaCO_3$, OM and Grain size parameters

Table-3 Comparison of BCSS-1 certified values with the present study

Element	Present study (ppm)	BCSS 1 (ppm)	Recovery (%)
Cu	18.2	19	95.79
Co	11.2	11.4	98.25
Ni	53.8	55.3	97.28
Pb	21.9	22.7	96.46
Zn	113.1	119	95.04
Cr	116.7	123	94.88

All elements are in ppm Yoshinaga et al (1996)

sandstones and charnockites and also indicating complexation by soil organic matter (Table 2). Geochemical variations along the vertical sediment profiles sampled based on sedimentological criteria reflect an increase in Zn content before a marked increase in Cu.

There is a strong positive correlation (sand ranges from 20 to 64 %, silt 18-69 % and clay 11 to 17 %) between decreasing grain size and increasing metal concentrations. Fine sediments have a much greater surface area than coarser particles of the same weight (Jones and Bowser, 1978), and a positive correlation between increasing surface area (Horowitz, 1991) and decreasing grain size has been recorded (Horowitz and Elrick, 1987). As a result, fine particles, especially clay minerals, organic matter and iron hydroxides, are able to adsorb large quantities of metals through cation exchange processes (Förstner and Wittmann, 1979; Salomons and Förstner, 1984). It has also been suggested that clays accumulate high concentrations of metals through mechanical processes alone, without the aid of chemical reactions (Jenne, 1976). Several studies, however, have indicated that all size fractions are important contributors of metals (Bradley, 1988; Leenaers et. al, 1988; Macklin and Dowsett, 1989; Moore *et al*, 1989). The results from our study indicate that grain size plays an important role in absorbing trace metals concentrations. A correlation of sand versus trace metals reveals negative correlation except in two samples from the site Velappanchavadi at the depth 40 cm and 100cm. The concentration of heavy metals at the depth 40 cm and 100 cm are low. This may be due to the presence of iron rich particles in the coarse sediments. Trace metals such as Cu, Co, Cr, Zn, Ni concentrations from Velappanchavadi are high as compared to the other two upstream sites (Fig. 3). This is probably due to anthropogenic activity, agriculture and intense urbanization. The trace metals from Pudichchathiram and Mahanadu are lower in Cu, Zn concentrations. All the trace metal concentrations such as Cu, Zn, Co, Cr are high in the surface layers indicating intense urbanization around Velappancahavadi, while the Pb content is low in the lower basin of the Cooum River. Elevated concentrations of Cr and Ni have resulted from the weathering of the bedrock. Stream sediment derived from the erosion of soils from the upland parent material is transported downstream and is deposited in the Cooum river basin. The behavior of metals in fluvial sediments is strongly influenced by a range of physical and chemical processes (Salomons, 1988). A number of physical characteristics, including particle size, surface area, specific gravity, surface charge, and bulk density, influence the behavior of metals within sediment (Horowitz, 1991). Of these, the closely related parameters of grain size and surface area are considered to be the most important. Many metals are closely associated with fine-grained sediments than with coarser material (Gibbs, 1973; Ramamoorthy and Rust, 1978). The data also indicates that overbank sediments are important sinks in the retention of trace metals in the study area through processes including sorption and microbial cycling.

However a better understanding of the environment pollution and steps to be taken for pollution monitoring demands a detailed study covering larger area, several lithosections and rigorous high resolution geochemical data and this will help in better understanding of the environmental shifts of the lower basin of the Cooum River.

CONCLUSIONS

- Observation of the lithosections indicates that sedimentation has taken place in phases and that there has been a change in the hydrologic regime.
- Field characters such as Ped formation of the overbank sediments, clay coatings along the sides of the peds and micromoprhology features such as clay coatings around the detrital grains, root pores in soils and associated calcretes point towards weak pedogenesis process.
- The trace metals from Pudichchathiram and Mahanadu are lower in Cu, Zn concentrations. All the trace metal concentrations such as Cu, Zn, Co, Cr are high in the surface layers indicating intense urbanization around Velappancahavadi, while the Pb content is low in the lower basin of the Cooum River.
- Elevated concentrations of Cr and Ni have resulted from the weathering of the bedrock and anthropogenic sources. Trace metals such as Cu, Co, Cr, Zn, Ni concentrations from Velappanchavadi are high as compared to the other two upstream sites. This is due to anthropogenic activity, agriculture and intense urbanization.

ACKNOWLEDGMENTS

Prof. B.S. Paliwal is known to me since the time I was a Ph.D scholar excavating an archaeological site at Didwana, Rajasthan way back in 1980-1986. I would often meet him to understand the geology of Didwana and he was very helpful. It is a great pleasure to submit a paper on his felicitations. Wishing him and his family all the very best in life. This work was possible due to the funds received from UGC, New Delhi in the form of UGC DRS phase I assistance to the Department of Geology, Anna University. I thank the reviewers for their constructive comments that has helped in the better presentation of the paper. I am thankful to Prof. K.L. Shrivastava for his offer to contribute a paper in this special volume.

References

Baker, R.G., Schwert, D.P. and Bettis, E.A. (1991) MidWisconsin Stratigraphy and paleoenvironments at the St. Charles site in south-central Iowa. Geological Society of America Bulletin, vol. 103, pp. 210-220.

Bengtsson, L and Enell, M. (1986) Chemical Analysis. In: B.E. Berglund (Ed.), Hand book of Holocene palaeocology and paleohydrology, John Wiley and Sons Ltd., Chichester, pp. 423-451.

Bradley, S. B. (1988) Sediment-water interactions; the physical transport of heavy metals in the fluvial system. In: G. Strigel (Ed.), Metals and metalloids in the hydrosphere: impact through mining and industry, and prevention technology, Proceedings of an IHP workshop, Bochum, Federal Republic of Germany, pp. 21-26.

Bullock, P., Federoff, N., Jongerius, A., Stoops, G. and Tursina, T. (1985) Handbook for Soil Thin Section Description, Waine Research Publication, England.

Dean, W.E. JR. (1974) Determination of carbonates and organic matter in calcareous sediment and sedimentary rocks by loss on ignition: Comparison with other methods. Journal Sedimentary Petrology, vol. 44, pp. 242-248.

Förstner, U. and Wittmann, G. T. W. (1979) Metal Pollution in the Aquatic Environment, Springer-Verlag, Berlin, 486 p.

Gibbs, R.J. (1973) Mechanisms of trace metal transport in rivers. Science, vol. 180, pp.71-73.

Gile L.H., Peterson F.F. and Grossman R.B. (1966) Morphological and genetic sequences of carbonate accumulation in desert soils. Soil Science, vol. 101, pp. 347-360.

Horowitz, A.J. (1991) A Primer on Sediment-Trace Element Chemistry. (IInd ed.), Lewis Publishers, Chelsea, Michigan. 136 p.

Horowitz, A.J. and Elrick, K. A. (1987) the relation of stream sediment surface area, grain size, and composition to trace element chemistry. Applied Geochemistry, vol. 2, pp.437-451.

Ingram, R.L. (1970) Procedure in sedimentary petrology. Wiley, New York, pp. 49-67.

Jenne, E. (1976) Trace element sorption by sediments and soils-sites and processes. In: W.Chappell and K. Peterson. (Eds.), Symposium on Molybdenum, vol. 2. Marcel-Dekker, New York, pp.425-553.

Jones, B. and Bowser, C. (1978) The mineralogy and related chemistry of lake sediments. In: A. Lerman (Ed.), Lakes: chemistry, geology, physics, Springer-Verlag, New York, pp. 179-235.

Leenaers, H., Schouten, C. J. and Rang, M. C. (1988) Variability of the metal content of flood deposits. Environmental Geology and Water Science, vol. 11, pp.95-106.

Macklin, M. G. and Dowsett, R. B. (1989) The chemical and physical speciation of trace metals in fine grained Overbank flood sediments in the Tyne Basin, north-east England. Catena, vol. 16, pp.135-151.

Moore, J. N., Brook, E. J. and Johns, C. (1989) Grain size partitioning of metals in contaminated, coarse grained river floodplain sediment, Clark Fork River, Montana, U.S.A. Environmental Geology and Water Science, vol. 14, pp. 107-115.

Muller, C. (1967) Etude de Statistique Lexicale. Larousse, Paris.

Nagalakshmi, T. (2004) Quaternary stratigraphy and depositional environment between the Palar and Koratallaiyar basin. Ph. D. thesis, Anna University, Chennai.

Ramamoorthy, S. and Rust, B. R. (1978) Heavy metal exchange processes in sediment-water systems. Environmental Geology, vol. 2, pp.165-172.

Reineck, H.E. and Singh, I.B. (1980) Depositional sedimentary Environments. IInd Edition, Springer and Verlag, New York.

Salomons, W. (1988) Impact of metals from mining and industry on the hydrosphere In: G. Strigel (Ed.), Metals and metalloids in the hydrosphere: impact through mining and industry, and prevention technology, Proceedings of the IHP workshop, Bochum, Federal Republic of Germany, 21-25th September 1987, UNESCO, Paris, pp. 1-41.

Salomons, W. and Förstner, U. (1984) Metals in the Hydrocycle. Springer-Verlag, Berlin, 349 p.

Schaetzl, R.J. and Anderson, S. (2005) Soils: Genesis and geomorphology. Cambridge University Press, Cambridge.

Shapiro, L. and Brannock, W.W. (1962) Rapid analysis of silicate, carbonate and phosphate rocks. U S. Geological Survey Bulletin, vol. 1144-A, 56 p.

Yoshinga, J., Tanaka, A., Takamatsu, T., Morita, M. and Okamoto, K. (1996) Elements Concentration in NIES candidate marine sediments certified reference materials: reference values for major and trace elements. Analytical Science, vol. 12, pp.993-995.

Hon'ble Reviewers

1. Prof. A.K. Singhvi
   Senior Scientist
   Physical Research Laboratory
   Navrangpura
   AHMEDABAD 380 009 INDIA

2. Prof. L.S. Chamyal
   Department of Geology
   M.S. University of Baroda
   VADODARA 390 002 INDIA

Editorial Assistance

Dr. Vijay Gaur

Economic Mineralization
K.L. Shrivastava (Editor)
2009. pp.352-368
Scientific Publishers (India), Jodhpur

32

Section-III

An Overview of the Mineral Resources of Rajasthan, India

ARUN VYAS
Department of Geology, Govt. Bangur College, Didwana, Rajasthan
Email : avyasgeo@yahoo.com

Abstract

Minerals are the backbone of the industrial civilization and non-renewable resources. The importance to planning and development of these resources are nowadays given utmost priorities through out the world. The various rocks hosting mineralisation represent a long period of geological history (spanning 3500 million years) and have generated mineral resources in the State. Presently there are about a dozen metallic and more than thirty important non-metallic industrial minerals, gem minerals and fuel minerals being worked in the State. Besides, the State is also producer of decorative and building stones like marble, granite, serpentinite, sandstone, limestone, slate, shale, schist and Phyllite. Rajasthan, the biggest State of the country, has an inventory of 69 minerals {46 Major and 23 Minor}under production from different mines. It is the sole producer of emerald, jasper, garnet (gem variety), wollastonite and selenite in the country and is also a leading producer of Lead zinc silver ore, tungsten ore, rock phosphate, asbestos, calcite, soapstone, ball clay and fire clay, feldspar, gypsum, ochre's, barites, marble etc. Field of exploration of base metals, oil and natural gas attracted multinational companies in the state. It is suggested that the information technology should be better utilized for improving effective qualitative and quantitative efficiency in mineral production besides infrastructure facilities and management schemes. The impact of mineral exploitation on environmental degradation is an issue of serious concern therefore the need of the hour is sustainable development of these mineral resources.

Key Words : Mineral deposits, Rajasthan, Lead-Zinc, Marble and Lignite.

INTRODUCTION

Minerals are indispensable not only for the well being of man, but also for his very survival (Chatterjee, 2004). All minerals and mineral products are derived from rocks forming the crust of the Earth (Wadia, 1997). Rocks from Archaean to Recent periods are found in the State of Rajasthan (Paliwal, 1999;Vyas and Paliwal, 2004) except Ordovician, Silurian, Devonian (Palaeozoic), Triassic (Mesozoic), Oligocene, Miocene, and Pliocene (Tertiary). These rocks host various mineral occurrences. The various rocks hosting mineralisation represent a long period of geological history (spanning 3500 million years) and have generated mineral resources in the State. Presently there are about a dozen metallic and more than thirty important non-metallic industrial minerals, gem minerals and fuel minerals being worked in the State. Besides, the State is also producer of decorative and building stones like marble, granite, serpentinite,

Table-1 Mineral Scenario of Rajasthan (after D.M.G. Rajasthan, 2004)

S.No.	Mineral	Distribution	Area of Leases (hect.)	No. of Leases and Production in thousand tones (2003-04)
1.	AMETHYST	Dausa, Pali, Sawai Madhopur, Tonk	NIL	NIL
2.	APATITE	Sikar, Udaipur	NIL	NIL
3.	AQUAMARINE	Bhilwara, Rajsamand, Tonk	NIL	NIL
4.	ASBESTOS	Ajmer, Bhilwara, Dungarpur, Pali, Rajsamand, Sirohi, Udaipur.	2191.69	31 13.459
5.	BALL CLAY	Barmer, Bikaner, Jaisalmer, Nagaur, Pali.	7024.54	74 890.500
6.	BARYTES	Alwar, Bharatpur, Bhilwara, Bundi, Chittaurgarh, J aipur, Karauli, Nagaur, Pali, Rajsamand, Sawai Madhopur, Sikar, Udaipur.	91.58	04 03.581
7.	BENTONITE	Banswara, Baran, Barmer, Jaisalmer, Jhalawar.	210.84	18 75.656
8.	BRICK EARTH	Almost all the districts are having brick earth deposits	581.25	20 9136.186
9.	CADMIUM-SILVER ORE	Ajmer, Bhilwara, Rajsamand, Sirohi, Udaipur.	No separate lease; Recovered as by products from Pb-Zn ores.	No separate lease. **Cadmium**-439 tonnes in 2002 03. **Silver**-29813.00 kilograms in 2003-04
10.	CALCITE	Ajmer, Alwar, Bhilwar a, Jaipur, Jhalawar, Jhunjhunu, Pali, Rajsamand, Sikar, Sirohi, Udaipur.	1136.76	22 124.237
11.	CHALK	Bundi, Jaisalmer, Sirohi	NIL	NIL
12.	CHERT	Alwar, Bharatpur, Bhilwara, Chittaurgarh, Jhalawar, Jodhpur.	5.96	7 01.096
13.	CHINA CLAY	Alwar, Barmer, Bhi lwara, Bikaner, Bundi, Chittaurgarh, Dausa, Jaipur, Jaisalmer, Jodhpur, Jhunjhunu, Karauli, Nagaur, Pali, Sawai Madhopur, Sikar, Udaipur.	4253.41	57 347.513

Table-1 (_Contd.)

14.	COPPER ORE	Ajmer, Alwar, Bharatpur, Bhilwara, Chittaurgarh, Dausa, Dhaulpur, Dungarpur, Jhunjh unu, Pali, Rajsamand, Sikar, Sirohi, Udaipur.	707.55	03 574.28
15.	CORUNDUM	Rajsamand, Tonk.	NIL	NIL
16.	DOLOMITE	Ajmer, Alwar, Banswara, Bhilwara, Chittaurgarh, Dausa, Jaipur, Jaisalmer, Jodhpur, Jhunjhunu, Rajsamand, Sawai Madhopur , Sikar, Udaipur.	765.43	10 195.595
17.	EMERALD	Ajmer, Rajsamand.	230.63	5 No recorded production
18.	FELDSPAR	Ajmer, Alwar, Bhilwara, Jaipur, Pali, Sikar, Tonk, Udaipur.	4955.56	188 201.140
19.	FIRE CLAY	Alwar, Bharatpur, Jhunjhunu, Karauli, Sawai Madhopur.	109.12	04 0.269
20.	FLUORITE	Ajmer, Bhilwara, Dungarpur, Jaipur, Jalore, Jhunjhunu, Pali, Sikar, Sirohi, Udaipur.	1701.07	14 01.786
21.	FULLER`S EARTH	Bikaner, Barmer, Jaisalmer, Nagaur.	1356.79	29 34.100
22.	GARNET	Ajmer, Bhilwara, Jaipur, Pali, Sawai Madhopur, Tonk,	383.49	16 38.000
23.	GOLD	Ajmer, Alwar, Banswara, Bhilwara, Dausa, Dungarpur, Jhunjhunu, Sirohi, Pali, Udaipur.	No separate lease. Recovered as by product from Cu ores.	No separate lease. 114.158 Kilograms in 2002-03.
24.	GRANITE	Ajmer, Alwar, Bansw ara, Barmer, Bhilwara, Bundi, Chittorgarh, Churu, Dausa, Dungarpur, Jaipur, Jaisalmer, Jalore, Jhalawar, Jodhpur, Jhunjhunu, Pali, Rajsamand, Sawai Madhopur, Sikar, Sirohi, Tonk, Udaipur.	655.70	426 210.884
25.	GRAPHITE	Ajmer, Alwar, Banswara, Nagaur, Pali, Sirohi.	382.70	02 No recorded production.
26.	GYPSUM	Barmer, Bikaner, Churu, Hanumangarh, Jaisalmer, Jalore, Jhalawar, Nagaur, Sri Ganganagar, Pali.	15574.48	29 2840.367
27.	IRON ORE	Alwar, Bharatpur, Bundi, Dausa, Jaipur, Jhunjhunu, Karauli, Sawai Madhopur, Sikar, Udaipur.	940.00	16 14.557

Table-1 (_Contd.)

28.	JASPER	Jaisalmer, Jodhpur, Pali.	664.54	15 02.313
29.	KANKER-BAJARI	All along the buried channels in Bikaner district and all along the existing rivers and nala sections throughout the state.	43.33	14 25591.208
30.	KYANITE	Ajmer, Bhilwara, Dungarpur, Udaipur.	05.00	01 0.10 tonnes
31.	LATERITE	Baran, Chittorgarh, Jaisalmer, Jhalawar, Sawai Madhopur.	291.40	01 NIL
32.	LEAD-ZINC ORE	Ajmer, Bhilwara, Dausa, Pali, Rajsamand, Sawai Madhopur, Sirohi, Udaipur.	3058.69	05 3644.163
33.	LIGNITE	Barmer, Bikaner, Jaisalmer, Nagaur.	4518.89	03 658.37
34.	LIME KANKAR	In plains of Churu, Sikar, Jhunjhunu, Nagaur, Jaipur and Bikaner districts and all limestone areas of the state.	NIL	NIL
35.	LIMESTONE (I) Cement, SMS & Chemical grade	Ajmer, Alwar, Banswara, Bhilwara, Bikaner, Bundi, Chittorgarh, Churu, Dhaulpur, Dungarpur, Jaipur, Jaisalmer, Jalore, Jhalawar, Jodhpur, Jhunjhunu, Karauli, Kota, Nagaur, Pali, Rajsamand, Sawai Madhopur, Sikar, Sirohi, Udaipur.	19737.40	33 23225.088
	LIMESTONE (II) Flaggy & Building	Chittorgarh, Jaisalmer, Jhalawar, Kota.	2335.51	207 3756.675
36.	MAGNESITE	Ajmer, Dungarpur, Pali, Rajsamand, Udaipur.	47.37	04 65.000 tonnes
38.	MANGANESE ORE	Banswara	72.27	01 05.902
39.	MARBLE	Ajmer, Alwar, Banswara, Bhilwara, Bundi, Chittorgarh, Churu, Dausa, Dungarpur, Jaipur, Jaisalmer, Jodhpur, Jhunjhunu, Nagaur, Pali, Rajsamand, Sikar, Sirohi, Udaipur.	3075.71	2232 6657.968
40.	MASONRY STONE	All kinds of rocks are being used as masonry stone and available in almost all the districts.	9049.51	2661 48047.515
41.	MICA	Ajmer, Bhilwara, Pali, Rajsamand, Tonk.	1010.45	21 361.000
42.	MILL STONE	Alwar, Bharatpur, Chittorgarh, Karauli, Kota, Tonk.	1076.84	07 05.004

Table-1 (_Contd.)

43.	MURRAM	Available throughout the state.	STP issued over smaller areas.	STP issued 1932.943
44.	OCHERS	Ajmer, Bharatpur, Bhilwara, Chittorgarh, Dausa, Jaipur, Jaisalmer, Jodhpur, Karauli, Nagaur, Pali, Tonk, Udaipur.	2872.35	44 925.734
45.	PATTI -KATLA	Ajmer, Alwar, Bhilwara, Dungarpur, Pali, Sawai-Madhopur Sikar, Udaipur.	370.61	71 203.963
46.	PETROLEUM & NATURAL GAS	Barmer, Jaisalmer, Sri Ganganagar.	275 sq.kms.	03 138928.63 MMS CubM
47.	PYROPHYLLITE	Alwar, Bhilwara, Jaipur, Jhunjhunu, Rajsamand, Udaipur.	249.70	08 14.488
48.	QUARTZ	Ajmer, Alwar, Bharatpur, Bhilwara, Dungarpur, Jaipur, Jhunjhunu, Pali, Rajsamand, Sawai Madhopur, Sikar, Tonk, Udaipur.	7172.29	178 367.382
49.	QUARTZITE (Refractory grade)	Ajmer, Bharatpur, Chittorgarh, Nagaur, Sawai Madhopur, Sikar.	06.00	06 05.595 in 2002-03
50.	RHYOLITE	Barmer, Churu, Jaisalmer, Jalore, Jodhpur, Pali, Sirohi.	125.00	94 821.144
51.	ROCK PHOSPHATE	Alwar, Banswara, Jaisalmer, Jaipur, Udaipur.	3003.61	07 1704.818
52.	SALTPETRE	Bharatpur, Bhilwara, Hanumangarh, Sri Ganganagar	Blanket leases	03 0.365
53.	SAND STONE	Baran, Bharatpur, Bhilwara, Bikaner, Bundi, Chittorgarh, Churu, Dhaulpur, Jaisalmer, Jhalawar, Jodhpur, Karauli, Kota, Nagaur, Pali, Tonk.	270713.44	1199 7337.608
54.	SELENITE	Barmer, Bikaner, Jaisalmer.	1062.45	04 17.371
55.	SERPENTINITE	Dungarpur, Udaipur.	386.63	362 964.427
56.	SHALE/SCHIST/ PHYLLITE	Ajmer, Chittorgarh, Pali, Sawai Madhopur, Tonk, Udaipur.	1.86	02 0.845
57.	SILICA SAND	Alwar, Baran, Barmer, Bharatpur, Bhilwara, Bikaner, Bundi, Chittorgarh, Dausa, Jaipur, Jaisalmer, Karauli, Kota, Nagaur, Sawai Madhopur, Tonk, Udaipur.	4323.86	79 976.929

Table-1 (_Contd.)

58.	SILICEOUS EARTH	Barmer, Jaisalmer.	474.99	13 01.968
59.	SILLIMANITE	Udaipur.	50.50	01 0.000
60.	SLATE STONE	Ajmer, Alwar, Bharatpur, Bhilwara, Bundi, Dausa, Dungarpur, Tonk.	49.32	32 05.311
61.	SOAPSTONE	Ajmer, Alwar, Banswara, Bharatpur, Bhilwara, Dausa, Dungarpur, Jaipur, Jhunjhunu, Karauli, Pali, Rajsamand, Sawai Madhopur, Tonk, Udaipur.	20857.89	303 615.520
62.	STONES FOR CHIPS & POWDER	Alwar, Banswara, Jaipur, Jaisal mer, Jodhpur, Nagaur, Pali, Rajsamand, Sirohi.	36.22	47 184.83
63.	SULPHUR ORE / PYRITE	Sikar, Sirohi, Rajsamand, Udaipur.	NIL	1 200.00 in 2002-03
64.	TUNGSTEN	Ajmer, Jaipur, Nagaur, Pali, Sirohi, Tonk.	NIL	NIL No recorded production
65.	URANIUM ORE	Bhilwara, Sikar, Udaipur.	NIL	NIL No recorded production
66.	VERMICULITE	Ajmer, Barmer.	29.52	02 85.000 in 2002-03
67.	WOLLASTONITE	Ajmer, Dungarpur, Jaipur, Pali, Sikar, Sirohi, Udaipur.	324.65	07 149.089

sandstone, limestone, slate, shale, schist and Phyllite etc (Table.1). Mining of these minerals boosts the economy of the State. The State of Rajasthan is endowed with mineral resources yet mineral sector faces few problems, important among them are mineral extraction in forest lands, delay in approval for aerial prospecting operations, small size of leases, development of infrastructure in mineral bearing areas, restriction on grant of leases for important minerals in tribal areas to tribal only, and availability of finance for development of mines and mineral processing (Bohra, 2001).

Utmost priorities through out the world. The first serious description of Minerals appeared to be in the book entitled "On Stones" by a Greek scholar Theophrastus (371-287 B.C.). About at the same time during the early years of the 4th century B.C. Kautilya alias Chanakya wrote "Arthasthastra" in which he described about few metals, namely, Gold, Silver, Copper, Lead, Tin, Iron and Mercury. Mineral resources are natural gift to its inhabitants, of the earth. Man, since Paleolithic time i.e. more than 50,000 years ago has used minerals for carving implements and tools. He used FLINT about half a million years ago for making hand axes.

Mining industry stands out as the 2nd largest employment sector, next to agriculture, in Rajasthan. Therefore, it was the first State of the country to take initiative towards reforms in the mineral sector and announced the New Mineral Policy to ensure rapid exploration of mineral wealth; to promote scientific and mechanized mining with due regard to conservation, safety, and environmental aspects; emphasis on mineral's export and to provide more job opportunities in this sector particularly for people from weaker sections of the society.

Mineral constitutes a precious heritage of the geological past, playing a very crucial role in the destiny of man. These resources are non-renewable, non-replenishable and finite, occurring only in geographically restricted areas; and merciless, exploitation of these resources create ecological imbalance. Therefore, the need of the hour is to conserve these by scientific mining operations, blending different grades of ores, beneficiation of marginal grades, recycling used metals and scraps, developing metallurgy for substitutes and discovering new deposits on the seafloor (Valdiya, 1987). Environment preservations around the mine site to counter the anthropogenic degradation, caused due to mining activities is another important factor, which needs serious attention.

MINERAL WEALTH OF RAJASTHAN

Mineral wealth of Rajasthan State is classified as under -

(A) Metallic Minerals

(a) Invaluable metallic minerals: Gold and silver.

(b) Ferrous minerals: Iron ore, manganese ore and tungsten.

(c) Non- ferrous minerals: Lead, zinc and copper.

(d) Radio active mineral: Uranium.

(e) Other metallic minerals: Magnesium, beryllium and cadmium.

(B) Non- Metallic Minerals

(a) Refractory minerals: Kyanite, sillimanite, andulasite, graphite, dolomite, fire clay, silica sand, asbestos and mica.

(b) Minerals used in Glass and Ceramic Industries: Quartz, feldspar, clay minerals (Kaolin and Ball clay), and wollastonite.

(c) Fertilizer and chemical minerals: Rock phosphate, pyrite, potash minerals, Barite, fluorite, gypsum and halite.

(d) Filler, abrasive and other useful minerals.

Talc, pyrophyllite, bentonite, fuller's earth, siliceous earth, calcite, garnet, corundum, jasper, Ochre and vermiculite.

(e) Gem Minerals: Emerald, aquamarine, garnet, corundum and amethyst.

(f) Fuel Minerals: Lignite and hydrocarbons - Petroleum and Natural Gas.

(g) Decorative and Building Stones: Marble, granite, serpentinite, sandstone, limestone, slate, shale, schist and phyllite.

(A) Metallic Minerals

Rajasthan is the leading producer of Pb-Zn, Ag and W in the country and also has important resources of Cu, U and Au. Out of the 418 Mt of Pb-Zn resources of India, 95%of Zn and 85% of Pb metals are found to occur in this State (Samar, 1996). Metallogeny in Rajasthan restricted only to the Proterozoic.

(a) Invaluable Metallic Minerals

Gold : In Rajasthan, gold is extracted as a by-product and found at Bhukia (Banswara district); Devtalai-Amargarh-Jhikri area (Bhilwara district); Danva (Sirohi district); RajpuraDariba (Udaipur district); Bajta (Ajmer district); and Khetri (Jhunjhunun district) areas. In Bhukia-Anandpuri, Sadri, and Undawala areas in Banswara district gold is localized along ductile shear zones (DSZ) at the basement-cover contact between the Banded Gneissic Complex (BGC) and the Aravalli supra crustals. In Bhilwara district gold is localized along DSZ at the basement- cover contact between the BGC (Mangalwar Complex) and the Jahazpur belt, near Devtalai and along the faulted contact zone between the BIF unit of the Jahazpur sequence and the Hindoli Group in the Amargarh-Jhikri area. Minor amounts of gold occur in the volcanic, massive, sulfide-hosted (VMS) deposit at Danva in Sirohi district. Gold also occur in sediment- hosted massive sulfides (SHMS) Pb-Zn-Cu deposits at Rajpura-Dariba in Udaipur district. In Ajmer district, at Bajta, gold is found in quartz veins with chalcopyrite-pyrhotite- pyrite association while in the Jhunjhunun district it is found in minor amounts in sediment hosted stratiform khetri copper deposit (Grover and Verma, 1993; Sinha Roy, 1996; Jain, 1987; Raghunandan et al, 1981; Bhattacharjee et al, 1991; Raja Rao and Mathur, 1975; Deb, 1982, 86; GSI, 1994; Vyas and Paliwal, 2004).

Silver : Silver is found in minor amount at Zawar, Rajpura-Dariba (Udaipur district); and Agucha (Bhilwara district) Pb-Zn deposits. Sediment-hosted Precambrian silver deposit is also reported from Bharak (Bhilwara district). Silver is also found in Khetri copper deposit and in volcanogenic, massive, sulfide deposit of the Ambaji Sendra belt at Chitar, Basantgarh, Pipela, Ajari, and Danva areas. (Raghunandan et al, 1981; Bhattacharjee et al, 1991;

1993 Mukherjee et al, 1992; Mukherjee and Bhattacharjee, 1997; GSI, 1994, Vyas and Paliwal, 2004).

(b) Ferrous Minerals

Iron ore: Iron ore deposits occur in the Jaipur, Udaipur, Jhunjhunu, Sikar, Alwar, Bhilwara, Karauli, and Dausa districts. In Jaipur district deposits are at Chomu-Morija-Banol and Nimla areas within the hematite quartzite associated with rocks of the Delhi Supergroup. It also occurs in same geological setup in Lalsot tehsil of Dausa district. Other occurrences are at Tonda and Jamalpura (Jhunjhunu district); Neem ka Thana, Rampur (Sikar district); Rajgarh (Alwar district); Bigod to Jahazpur (Bhilwara district) areas. Reserves of 11Mt averaging 48.52% Fe has been proved in Natharia-ka-Pal in Sarada tehsil of Udaipur district. Its production in the State for year 1997-98 was of 17688 tonnes (Vyas and Paliwal, 2004).

Manganese ore : Manganese ore deposits occur at Kala Khunta and Tripura Sundri, Kevadia, Tambsra, Sagwa, Sivania and Ghatiya in Banswara district. (D.M.G., Rajasthan, 2004).

Tungsten: Its deposits are at Degana (Nagaur district); Balda, Deva Ka Bera (Sirohi district); Koon (Udaipur district); Amartiya (Dungarpur district); Berantiya Khurd, Nana, Bar, Kararavav, Motiya, Pipliya, Babra (Pali district); Ladera (Jaipur district); Padanga, Roopnagar (Ajmer district) and Siras, Sohela (Tonk district) areas (D.M.G., Rajasthan, 2004).

(c) Non-ferrous Minerals

Lead and Zinc : Lead and zinc deposits are found in Udaipur, Rajsamand, Banswara, Bhilwara, Ajmer, Dausa, Pali, Sawai Madhopur and Sirohi districts and are associated with Archaean and Proterozoic rocks. Zawar is famous for its lead-zinc deposits. Situated 40 km. south of Udaipur, these deposits believed to have been worked since 14th century A.D. The mineralisation is found at Mochia Magra, Balaria, Bawa, Baroi and Zawarmala hills. The Zawar leadzinc belt extends for a distance of about 20 km. from Hameta / Magra in the north to Parsad in the south. Significant lead-zinc prospects in Pur-Banera-Bhinder belt, from north to south, are at Devpura, Dedwas, Samodi, Tiranga, Sindersar Kalan and Rajpura Dariba (Rajsamand and Bhilwara districts). Of these, the Rajpura Dariba deposit is the richest. In Ajmer district, the Sawar-Tikhi-Ganeshpura-Bajata belt, Rampura- Agucha deposits, Ghugra prospect, and Lohakhan prospect areas are of interest. Jahazpur lead-zinc prospect (Bhilwara, Tonk and Bundi districts); Basantgarh copper-zinc prospect, Pipela Copper-zinc prospect, Deri lead- zinc copper prospect (Sirohi district); Punagarh lead-zinc prospect (Pali district); Chauth-ka-Barwara lead prospect (Sawai Madhopur district) are other important deposits (Gupta et al, 1997; D.M.G., Rajasthan, 2004;Vyas and Paliwal, 2004).

Copper : Copper occurrences in the Ajmer, Alwar, Bharatpur, Bhilwara, Chittorgarh, Dausa, Dhaulpur, Dungarpur, Jhunjhunu, Pali, Rajsamand, Sikar, Sirohi and Udaipur districts The Khetri copper belt extends from Singhana to Raghunathpura (Jhunjhunu district) is leading producer in the State. Other copper deposits of economic interest are at Kho-Dariba, Bhagoni, Pratapgarh (Alwar district); Sat-Kui, Akwali, Madan-khudan and kolihan (Jhunjhunu district); Balesar, Tejawala, Ahirwala, Chidleta (Sikar district); Ajari-Basantgarh-GoliaPipela belt and Deri (Sirohi district); Pur Dariba (Bhilwara district); Khankhera (Bharatpur district); Jharka-Tarta and Jagpura (Banswara district); Taragarh, and Chainpura (Ajmer district); Chang-Chitar-Kalabar (Pali district); Padar-ki-Pal and Aamjara (Dungarpur district); and Anjani, Majera, Umra, Khedi-Jagat, Sakroda, and Bhinder (Udaipur district) (D.M.G., Rajasthan, 2004;Vyas and Paliwal, 2004).

(d) Radio-active Minerals

Uranium : Uranium occurrences in the state are at Ajmer, Dungarpur, Jaipur, Pali, Sirohi, Udaipur, Chittorgarh and Sikar districts. The important Uranium occurrences are at Putholi, Salera and other localities in the PuthoeiHamirgarh basin in Chittorgarh district; Umra-Udaisagar-Kalamagra-Haldi ghati area of Udaipur district; and Kolihan and Saladipura to Ghateswar in Sikar district (Singh, 1996,D.M.G., Rajasthan, 2004).

(e) Other Metallic Minerals

Magnesium : Magnesite deposits are found in Ajmer, Dungarpur, Pali, Rajsamand and Udaipur districts. In Ajmer district deposits are at Sarupa, Gapha, Kundal, and Kali Dungari are associated with rocks of the Delhi Supergroup. In Udaipur district, Magnesite occurrence at Parsad area is associated with the rocks of the Aravalli Supergroup. It is also reported from Sendra (Pali district); Dewal and Peeth (Dungarpur district); and Daulatgarh (Bhilwara district) areas. (D.M.G., Rajasthan, 2004;Vyas and Paliwal, 2004).

Beryllium : Occurrences are reported from Ajmer,

Bhilwara, Rajsamand, Sirohi and Tonk districts.

Cadmium : Cadmium and Silver are found associated with zinc at Pb-Zn mines in Ajmer, Bhilwara, Rajsamand, Sirohi and Udaipur districts (D.M.G., Rajasthan, 2004). Its ore is found with Pb-Zn sulphide ores at Zawar mines in Udaipur district. Cadmium is obtained as by-product from zinc concentrates at Debari and Chanderia.

(B) Non-Metallic Minerals

(a) Refractory Minerals

Kyanite-sillimanite and andulasite : These minerals, in minor amount, occur in the Ajmer, Bhilwara, Dungarpur, Rajsamand, Tonk, and Udaipur districts. Kyanitesillimanite deposits at Lakhawali, Thoor-Madar, Debari, Chandesara and Brahamao-ka-Gudha Gaon, in Udaipur district, are of lower grade. In Ajmer district, kyanite deposits are located at Keshrpura, Arjunpura, and Chenpura, Srinagar, Beer, Rajgarh, Lachhipur areas. Deposits of andulasite are found at Sardarpura, Sitapura, and Rampura in Dewli Tehsil of Tonk district.

Graphite : Graphite occurs in Ajmer, Alwar, Banswara, Nagaur, Pali, Sirohi and Udaipur districts. Deposits are associated with rocks of the Aravalli and the Delhi Supergroup. In Banswara district deposits are at Kesarpura, Chanduji-Ka-Gura, Bassi, Khandu, Bhoongara, Ghatol, Kotra, Amarthoon, Saskota. In Ajmer district, these are located at Dumada, Lottiyana, Ajeetgarh, Pushkar and Rajoosi. Occurrences of Graphite are reported at Jhadoli, Rogada (Alwar district), Ganwa, and Parbatsar (Nagaur district), Bar, Khiwandi, Desuri, Chhoti Kalab, Guda Bijaji (Pali district) and Reodar in Sirohi district (D.M.G., Rajasthan, 2004;Vyas and Paliwal, 2004).

Dolomite : Deposits of Dolomite occur in Ajmer, Alwar, Banswara, Bhilwara, Chittorgarh, Dausa, Jaipur, Jaisalmer, Jhunjhunu, Jodhpur, Sikar, Rajsamand, Sawai Madhopur and Udaipur districts. In Ajmer district dolomite deposits are at Gagel, Akhri, Makreda, Kotra-kabra, Sarsari, Ralawata, Sawar, Ghatiyali, Hatoondi, Kayampura, Saradhana, Kesarpura, Roopnagar. In Bhilwara district deposits occur at Jahazpur-Chenpura belt, Koshithal belt, and at Chitoriya-Ladpura-Kheri belt. Dolomite (as marble) is mined in Rajsamand district from Rajsamand, Amet, Sardargarh, Kelwa, Morchana, Morwad, Rabacha, Odan, Godhaghat, Ghata, and Thoriya areas (Aravalli Supergroup). In Banswara, dolomite deposits are found in Tripura Sundari belt and OdaBassi belt. Other deposits are at Kho-Dariba, Rajgarh, Nemal, Narayani-Mata, Sariska (Alwar district); Phalodi, Mathania, Bilara (Jodhpur district); Aandhi-Ramgarh (Jaipur district); Dhariyawad, Salumber (Udaipur district); Bhagwanpura, Bambori, (Chittaurgarh district); Dagota, Neemla (Dausa district); Chacha, Odhaniya (Jaisalmer district); Papurna, Taonda (Jhunjhunu district); Maonda, Mahava (Sikar district); Khawa, Chakeri (Sawai Madhopur district) and Iswal, Selu, Morela, Rajpura (Udaipur district) (D.M.G., Rajasthan, 2004;Vyas and Paliwal, 2004).

Fire Clay : Deposits of fire clay's are found in Alwar, Barmer, Bharatpur, Bikaner, Jhunjhunu, Karauli and Sawai Madhopur districts. Deposits are reported at Gudha, Kotadi, Mah and Kolayat (Bikaner district); Khari Charan, Chandi, Gudha, Bothia (Barmer district); Phalodi (Sawai Madhopur district); Kalgaon, Khora Karmali, Baswagate, Phairoti, Bigota (Alwar district); Kutubpura , Girwari (Jhunjhunu district);and Talchira(Karauli district) areas (D.M.G., Rajasthan, 2004;Vyas and Paliwal, 2004).

Silica Sand : Deposits of Silica Sand occur in Alwar, Baran, Barmer, Bharatpur, Bhilwara, Bikaner, Bundi, Chittorgarh, Dausa, Jaipur, Jaisalmer, Jhunjhunu, Jodhpur, Karauli, Kota, Nagaur, Rajsamand, Sawai Madhopur, Sikar, Tonk, and Udaipur districts. The deposits of silica sand are located at Narayani-Mata, Nathusara, (Alwar district); Badodiya, Tikrada, Phoolsagar (Bundi district); Mati-Ka-Gol, Dewaka, Shiv (Barmer district); Jogiwanpura, Dorwali, Karwan, Aacholi, Nawali, Agwali, Bajoli, Tamilpur, Jaswantnagar, Kalwara, Nanerawas (Bharatpur district); Manota, Gumanpura, Samod, Jumma-Ramgarh (Jaipur district); Kundi, Kanpur, Khimach (Kota district); Churli, Narayanpura, Talwara, Rampura, Toda-Bhim (Swai Madhopur district); Madh, Gudha, Jogida Talab, Kolayat (Bikaner district); Kharkhera, Ram Singh-Ki-Dhani, Sonpura (Sikar district); Nijarwali (Jhunjhunu district); Khajwana (Nagaur district); Lathi (Jaisalmer district); and Dantlai (Udaipur district) (D.M.G., Rajasthan, 2004;Vyas and Paliwal, 2004).

Asbestos : Rajasthan contributes about 80% of the total production in the country. Asbestos occurs in rocks of the Aravalli Supergroup in Rajsamand, Udaipur and Dungarpur districts and also in rocks of the Delhi Supergroup in Ajmer, Pali and Alwar districts. Out of the 56 deposits in the State, 8 are of chrysotile variety and remaining is of amphibole variety (Gupta et al, 1997). In Ajmer district the major deposits are located near Mangliawas, Kesarpura,

Kanvali (Kaolai), Kotra Reserve Forest area, Arjanpura, Kotri and Nai khurd. Other important deposits are at Baroli, Uprod, Amberi, Salar, Kirat, Jhanjhar-ki-Pal, Rakhabdev, Jogi-ka-guda, Kagdar Ki pal, Dhelana, Odwas, Mole la (Udaipur district); Jakol, Kheymaru, Bodamaru, Mathumgra, Thana, Pipard dunger, Dewal (Dungarpur district); Deogarh, Tiki, Bharat Singh-Ka-Guda, Padasali (Rajsamand district); Kanotiya-Ramgarh, Manpura, Kotra, Goria, Aasan, Sendra (Pali district); Amalda, Sukhanand Ka Kheda, Nuwaliya, Reh (Bhilwara district); Boribhuj (Sirohi district); Dehlawars, Palpur and Gangi-ka-thana (Alwar district) (D.M.G., Rajasthan, 2004;Vyas and Paliwal, 2004).

Mica : Mica deposits associated with Precambrian granite and pegmatite occur in Ajmer, Bhilwara, Tonk and Rajsamand districts. Important mica deposits are at Sahada, Gangapur, Potala, Mahendragarh, Bagor, Raipur, Mandel, Bhadu, Aamali, Bhunas, Jamoli, Roopa, Paroli (Bhilwara district); Nasirabad, Mangliawas, Sarwad, Kekari, Kishengarh, Jawaza (Ajmer district); Malpura, Dewali, Dholi, Modhorajpura and Bhojpura (Tonk district); Kuwanthal, Sardargarh, Devgarh, and Bhagatpura (Rajsamand district).

(b) Minerals used in Glass and Ceramic industries

Quartz : Quartz is produced in Ajmer, Bhilwara, Pali, and Tonk districts and also occurs in Rajsamand, Udaipur, Sirohi, Alwar, Bharatpur, Dungarpur, Sawai Madhopur, Sikar, Jaipur, and Jhunjhunu districts. It is associated with Precambrian rocks. Important deposits of quartz are at Jawaza, Mangliawas, Rajgarh, Nasirabad, Kishengarh, Beawar, Sarwad, Kekri (Ajmer district); Dewali, Malpura, Toda Raisingh, Aligarh, (Tonk district); Ladera, Sakoon, Bandra-Sindari, Dudu (Jaipur district); Gawardi, Dariba, Dhaneriya, Devgarh, Kuwanthal, Aamet (Rajsamand district); Mavali, Gogunda, Saira, Sanwad, (Udaipur district); Raipur, Bar-sendra (Pali district); Swaroopganj, Rohida, Batera, Koderla, Sanwara, (Sirohi district); Aasind, Badnor, Kareda, Raipur (Bhilwara district); Khetri (Jhunjhunu district); Neem-ka-thana, Chokadi-Chapoli (Sikar district); and Rajgarh (Alwar district) areas.

Feldspar : Deposits of feldspar occurred at Beawar, Nasirabad, Lohagal, Jawaza, Rajgarh, Kalesara, Masuda, Sarwad, Kekri, Kishengarh, (Ajmer district); Tatarpur, Adhira, Thathra, Hamirpur, Angari, Bansur, Rajgarh (Alwar district); Aasind, Badnor, Raipur, Kareda, (Bhilwara district); Bhim, Aamet, Devgarh, (Rajsamand district); Bar-Sendra, Raipur, Haripur (Pali district); Malpura, Dewali, Todaraisingh (Tonk district); Amarpur, Ladera, Sakoon, (Jaipur district); Neem-Ka-thana, Doodawas, Shivpura, Sohanpura (Sikar district); Malpura,Duni,Tondi, Todaraisingh (Tonk district). Feldspar deposits also occur in Udaipur, Banswara, Dungarpur and Sirohi districts. (D.M.G., Rajasthan, 2004;Vyas and Paliwal, 2004).

Clay Minerals (Kaolin and ball clay) : Clay minerals are mainly confined to Precambrian, Jurassic, and Eocene sediments and occur in Bhilwara, Chittorgarh, Jaipur, Sawai Madhopur, Nagaur, Barmer, Jaisalmer, Sikar, Pali, Bikaner, Udaipur, and Kota districts. Kaolin (China Clay) deposits are located in Umedpura, Mangrop, Menpura, Kotdi, (Bhilwara district); Sawa, Bansi, Bari-Sadari, (Chittorgarh district); Torda, Fathepura, Gol, (Jaipur district); Phalodi, Raisena (Sawai Madhopur district); Mahuva, Purshottampura (Sikar district); Chacha (Jaisalmer district); Gunga, Kapurdi, Jaliya, (Barmer district) areas. Ball Clay deposits are found in Nal, Kotri, Madh, Madhogarh, Suraj-Ki-dhani, Kolayat, Gurha, Bithnok, Palana, Barsingsar (Bikaner district); Bhopa paraker Ki dhani, Bothia, Bisala, Gunga, Nimbasar, Jalela, Nimla (Barmer district); Devikot, Kita, Mandai, Boa, Nedai, Rata (Jaisalmer district); Mundwa, Saradhana, Khajwana, Indawar, Luniyas, Rol Chandawatan, Nimori (Nagaur district); Phalka, Lithariya, Kharadi, Jetpura and Shyamgarh (Pali district) areas (D.M.G., Rajasthan, 2004;Vyas and Paliwal, 2004).

Wollastonite : Wollastonite occurrences are in Sirohi, Sikar, Pali, Jaipur, Udaipur, Ajmer, Dungarpur, and Banswara districts. Deposits are associated with Precambrian rocks and are located in Belka hills region at the border of Sirohi, Udaipur and Pali districts and are also situated in Abu road (Sirohi district); Pisangan, Saradhana, Gola, Alipura, Kalesara, Roopnagar (Ajmer district); Barwasa (Dungarpur district); Lohariya (Banswara district); and Sayara, Bhanpura (Udaipur district) areas.

(C) Fertilizer and Chemical Minerals

Rock Phosphate : Rock phosphate deposits occur in different lithological formations varying in age from Precambrian to Tertiary in the Udaipur, Banswara, Sikar, Jaisalmer, Jaipur, Alwar, and Chittorgarh districts. Important deposits are located at Jhamarkotra, Kanpur, Karbaria-ka-Guda, Dakankotra, Maton, Badgaon, Manoharpura, Sisarama, Neemachmata, Kaya, Undari, Lal Madri

(Udaipur district); Kerpura, Salwari (Sikar district); Birmania, Fatehgarh, Kanod, Baisakhi, Roopsi (Jaisalmer district); Achrol (Jaipur district); Aduka, Advani (Alwar district); Sallopat, Ram-ka-Munna (Banswara district); and Jaoda (Chittorgarh district) areas. (Vyas, 1995; D.M.G., Rajasthan, 2004;Vyas and Paliwal, 2004).

Pyrite : In the absence of Sulphur deposits in the country, the only alternative indigenous source of Sulphur is pyrite and its deposits are at Saladipura (Sikar district) which contain 115 Mt of proved reserves averaging about 20% S (Joshi, 1993).

Potash Minerals : Deposits of potash minerals occur in Nagaur, Bikaner, Churu, and Ganganagar districts. These minerals are mainly used in fertilizer industry but are also used in explosives and chemical industry.

Barytes : The deposits of barytes are located in Alwar, Udaipur, Bharatpur, Rajsamand, Jalor, Jaipur, Karauli, Nagaur, Sawai Madhopur, Ajmer, Sikar, Bhilwara, Bundi, Pali and Chittorgarh districts in rocks belonging to Pre-Cambrian, Vindhyan Supergroup as well as in Malani Suite of igneous rocks. Important deposits are located within Alwar and Rajgarh Belts (Alwar district); Relpataliya, Babrmal, Dhariyawad, Jagat, Sakroda, Kharwa (Udaipur district); Hathori to Ghatari (Bharatpur district); Delwara, Nathdwara, Ghoraghati, Karauli - Nerach-Majera (Rajsamand district); Sulia hill, Tavidar hill, Karda (Jalor district); Umar-Pagara, Kishanganj, Babi (Bundi district); Barliyawas, (Bhilwara district); Tirora-Daulatpura, Pushkar (Ajmer district); Naroda, Khargabinapur, Ranasar, Mau (Sikar district); Anoppura, Gyaspur, Rawatbhata (Chittorgarh district); Rela (Jaipur district); Talchira (Karauli district); Gingoli (Nagaur district); Pipalwara (Sawai Madhopur district) Biranthiya Khurd and Punagarh (Pali district) areas. (D.M.G., Rajasthan, 2004;Vyas and Paliwal, 2004).

Fluorite : Its deposits are associated with different Precambrian rocks and Malani pyroclastics rocks and are located in Dungarpur, Jalor, Sirohi, Alwar, Bhilwara, Banswara, Jaipur, Jhunjhunu, Pali, Sikar, Ajmer and Udaipur districts. Important occurrences of fluorite in Dungarpur district are the Mando-Ki-Pal and Kahila deposits extending from Hathai in the north to Kahila in the south, covering a distance of 15 to 20 km. The northern part contains the deposits of Hathai, Nawagaon, Matatiba and Samotha whereas Bhowria-Ka-Naka and Kahila forms the southern part of the belt. The reserves available in the Mando-Ki-Pal deposits have been estimated at 0.4 X 10^6t with 12% CaF_2 (Gupta et al, 1997). Other deposits are of Karda (Jalor district); Balda, Barabera (Sirohi district); Chokadi-Chapoli (Sikar district); Ledera-Sakoon, Mundoti (Jaipur district); Tilora, Richamaliya (Ajmer district); Asind-Badnor (Bhilwara district); Jhallara, Kalamagra, Mandili, Matasola, Bhimroda (Udaipur district); Kherthal (Alwar district).

Gypsum : About 80% of Indian gypsum deposits are in the State, occur in rocks of Tertiary period, and are considered to have formed by evaporation process (Sinha Roy et al, 1998). These deposits are located in Jaisalmer, Barmer, Bikaner, Jodhpur, Jalor, Jhalawar, Nagaur, Pali, Churu, Hanumangarh and Sri Ganganagar districts. Important among these are at Mohangarh, Lakhesar, Hamirwali Nadi, Phalsund, Chandan, Lakha (Jaisalmer district); Jhamsar, Dhirera, Bharu, Dholesar, Akasar, Dher, Dhandhewala, Lunkaransar (Bikaner district); Kawas, Utarlai, Kundal, Shivkar, Bajwa, Chitar-Ki-Pal, Thob (Barmer district); Bhadwasi, Kherat, Manglod, Unthwaliya (Nagaur district); Biramsar, Mahila, Paloo, Bisrasan, Malkasar, Karnisar, Raghunathpura, Desli (Sri Ganganagar district); Hanumangarh, Suratgarh, Faridsar, Dhardu (Hanumangarh district); Bhalone, Devgadhia, Sethon, Bhohiu (Churu district); Khutani (Pali district) areas. Minor amounts of gypsum occur in Khanpur (Jhalawar district); Kathumaria, Gudiakhera (Bharatpur district) areas (D.M.G., Rajasthan, 2004;Vyas and Paliwal, 2004).

Halite (Salt) : Common salt is prepared from the brines of saline lakes in the state. Halite produced from lakes of Sambhar, Didwana and Kuchaman in Nagaur district, Pachbhadra lake of Barmer district and Nanwa lake of Bikaner district. These salt deposits formed due to prolonged aridity in tectonically-controlled depressions in concerning areas (Roy, 1999). The Sambhar lake, lying in the Nagaur district (partly in Jaipur district) covering an area of 243 km^2 annually produces about 2 lac tonnes of common salt while Didwana lake is an oval-shaped depression, 8.04 km long and 6.4 km wide. The Pachpadra lake is a long oval-shaped; lying in the Barmer district covering an area of 82.74 km^2 produces about 1-lac tonnes of common salt per annum (G.S.I., 2001).

(d) Filler, Abrasive and Other useful Minerals

Talc : Talc, steatite and soapstone deposits of Rajasthan are located in Udaipur Rajsamand, Dungarpur, Bhilwara, Banswara, Karauli, Sawai

Madhopur, Pali, Jaipur and Dausa districts. These deposits are associated with either dolomites or ultramafic rocks of Precambrian age. The Talc-Soapstone deposits of Udaipur Rajsamand Dungarpur districts are lying in six belts namely.

(1) Oda-Devura-Dangri-NatharaKiPal belt.

(2) Ushan-Kzgamdar-Jagat-Rabcha-Oden belt.

(3) Salumber-Band-Sanjela-Lobagarh-Padla-Devla-Jagpura belt

(4) Kavita-Lakhawali-Paraya-Rama belt.

(5) Undithal-Uparlatala-Khakhar-Jharol-Dhandhawali belt.

(6) Rishabhadev-Kagdar-Bicchiwara-Sabla-Sisod-Kalyanpur belt.

50% of State's total production comes from these belts. Other important deposits are at Ghevria, Chainpura, Vagwasa, Chabariya, Jahazpur (Bhilwara district); Dagota, Jharana, Gijgarh (Jaipur-Dausa districts); Duta, Kamalpura, Ranjeli, Garni, Pura, Mash Village (Karauli-Sawai Madhopur districts) areas.

Pyrophyllite : The deposits of Pyrophyllite are located in Alwar, Udaipur, Rajsamand, Jaipur, Bhilwara, and Jhunjhunu districts. The two main belts of pyrophyllite are located at the contact of the Archaean and the Aravalli Supergroup of rocks in Udaipur- Rajsamand districts. These are 40 km long belt of Debari Gudli Negariya- Nathuwas and 10-12 km long belt of Bari Madar Rama belt.

Bentonite : The bentonite deposits are located in Barmer, Jaisalmer, Sawai Madhopur, Jhalawar, Banswara, Baran and Bikaner districts. Extensive, high-quality deposits occur in Barmer district only. The bentonite deposits lay above the Lathi and Barmer Sandstones of Jurassic and Cretaceous age, respectively. It is also associated with the Kapurdi Formation (Eocene) and is found as an alteration product of Malani volcanics (Pandya et al, 1999). Important deposits are at Hathi Singh Ki dhani, Akli, Thumbli, Giral, Gunga, Harvecha, Pusad, Sonari (Barmer district); Khuiala, Baliya, Khuri, Haboor (Jaisalmer district); Mathania (Jhalawar district); and Dargawa Pancholi (Sawai Madhopur district) areas.

Fuller's Earth : Deposits of Fuller's Earth (Multani Miti) are located in Barmer, Jaisalmer, Bikaner, Nagaur and Jodhpur districts, in rocks of Mesozoic and Tertiary age. Important deposits are located at Kapurdi, Bhadka, Alamsariya, Sheo, Rohili, Kurlan (Barmer district); Khimasa, Khuiala, Bandha, Mandha, Pangli, Sam, Ramgarh, Dhale Ki Dungri (Jaisalmer district); Palana, Jogira, Madh, Kesardesar, Barsinghsar (Bikaner district); Igyar, Deshwal, Chapri, Arisar, Kuchera, Khajwana (Nagaur district); Rahil, Dip-Ki-Dhani (Jodhpur district) areas.

Siliceous Earth : Its deposits are found only in Barmer and Jaisalmer districts and are associated with rocks of Cretaceous to Tertiary in age. Important deposits are located at Mati-Ka Gol, Bhorasar, Jato-Ki-Dhani, Bariyada, Nadali, Phuleri-Ki-Nadi, and Sindhiyaon-Ki-Dhani areas of Barmer district; Sajeet, Dharvi Khurd, Mandai, Nimba, Bhinyasar, Devda, Chhewada and Kanpuriya in Jaisalmer district (D.M.G., Rajasthan, 2004; Vyas and Paliwal, 2004).

Calcite : The calcite deposits, associated with rocks of the Precambrian age, occur in Bhilwara, Sirohi, Udaipur, Pali, Sikar, Jhunjhunu, Jhalawar, Rajsamand, Alwar, Jaipur and Ajmer districts. Deposits of calcite are located at Belka, Khila, Moras-Malera, Siwera, Wadhera, Wagdri, Bhula Bhesasing, Siyawa (Sirohi district); Rabcha, Gogunda, Sayera, Padrada, Gaiphal, Kotra, Bikarni (Udaipur district); Maunda, Guhasa, Mavandi, Nyarana, Raipur (Sikar district); Mana, Kapil-Ki-Bhagal (Pali district); Paperna (Jhunjhunun district); Barna-Ki-Chowki, Sakoon (Jaipur district); Barga, Bhudol, Mundoli, Siliberi (Ajmer district); Sarbilas, Delwara, and Ghatol (Banswara district) areas (D.M.G., Rajasthan, 2004; Vyas and Paliwal, 2004).

Garnet : Garnet deposits in the State are located at Kalyanpur, Raghunathpura Gaouri, Saroi, Khera, Kundiara, Kushalpura, Tordi Sagar, Bagri, Rajmahal (Tonk district); Agycha, Kuchalwara, Rupaheli, Banasra, Dadiya, Deodi, Maosha, Gharta, Kodukota (Bhilwara district); Sarwar, Sarsari, Bagsuri (Ajmer district) areas in rocks of the Precambrian age (Gupta et al. 1997; Vijaywargiya, 2000). It's occurrences also reported from Jhag, Tickel (Jaipur district); Deokhedi, Karanpura, Alipura-Rawania, Sendra, Kalalia (Pali district); and Badagaon in Sawai Madhopur district. (D.M.G., Rajasthan, 2004; Vyas and Paliwal, 2004).

Corundum : After Diamond, the hardest mineral found in nature. It is used as abrasive. It's red transparent variety 'Rubi' and blue variety 'Sapphire' are gem varieties. It occurs in minor quantities at Juali and Hamotia (Tonk district) and Kuwanthal and Deogarh areas (Rajsamand district).

Jasper : Jasper, mineral of silica group is used as abrasive, and is found in Jalor, Jaisalmer, Pali and

Jodhpur districts. Jasper occurrences are at Chacha, Odania, Ramdeora, Pokaran (Jaisalmer district); Lavera Kalan, Lavera Khurd, Mandai, Mathania, Rundia, Sopra, Ujaliya, Khetawas, Nandra Kalan, Jaleli, Pithawas, Banar (Jodhpur district); Bhambolai, Khimel and Rohat (Pali district) (D.M.G., Rajasthan, 2004;Vyas and Paliwal, 2004).

Ochre : The ochers are oxides of iron found in rocks of all the geological ages. It is known as red ochre, yellow ochre, green ochre, brown-yellow (Sienna) and amber on the basis of its different colors. Deposits are located in Sawa-Nimbahera, Banesti, Basedi, Borundi, Bojunda, Kanthariya (Chittorgarh district); Iswal, Bansliya, Chhotisar, Kumawaton Ka Gurha, Gasiyar (Udaipur district); Jahazpur, Biletha, Mangrop, Jinkli, Onkarpura, Maheshpura, Hameedpura (Bhilwara district); Mandai, Damodra, Chhatrel, Khia-Bandha, Fatehgarh, Chacha-Udaniya (Jaisalmer district) areas and also occur in Alwar, Bharatpur, Bikaner, Dausa, Jaipur, Jodhpur, Jhunjhunu, Karauli, Nagaur, Pali and Tonk districts. (D.M.G., Rajasthan, 2004;Vyas and Paliwal, 2004).

Vermiculite : It's deposits are found at Faisagar, Bhavata, Borand, Lachipura, Guda, Walesara, Rajosi, Nai Kalan and Rajgarh areas in Ajmer district; Simalia, Kitpaliya (Barmer district) and at Amarthoon and Ghatol areas in Banswara district.

(e) Gem Minerals

Emerald : Emerald (Panna), green/pale green transparent variety of beryl, is a very high-priced precious stone available mostly in Ajmer and Rajsamand districts of Rajasthan. Occurrences are at Rajgarh, Bubani, Kasia, Muhami, Barla, Gudas, Chat, Bithur (Ajmer district); Kaliguman, Kabra, Bagmara, Malpaira, Tekhi, Chat Ka Gurha, Machind, Sunkhar, Antaliya, Tikhi, Nimdi and Gam-Gurha (Rajsamand district) areas (Gupta et at, 1997; D.M.G., Rajasthan, 2004;Vyas and Paliwal, 2004).

Aquamarine : Aquamarine occurrences reported in Bhilwara, Rajsamand and Tonk districts. It is light green, blue gemstone variety of beryl. It occurs in Toda Raisingh, Bagri, Hamirpur, Bahriya, Dahriya, Rampura, and Botunda areas of Tonk district, Jamoli in Bhilwara district and Deogarh in Rajsamand district. (D.M.G., Rajasthan, 2004;Vyas and Paliwal, 2004).

Amethyst : Their occurrences in the State are at Sarsop, Isarda (Sawai Madhopur district); Toda Raisingh, Niwai, Hathun (Tonk district); Kishengarh, Arjunpura, Mangliawas (Ajmer district); Singhana (Jhunjhunun district); Manoharpura (Jaipur district); Chokari Chapoli (Sikar district); Didwana, Girota (Dausa district); Nana (Pali district) areas and are also found in Banswara and Chittorgarh districts with Deccan Traps (Vijaywargiya, 2000 D.M.G., Rajasthan, 2004;Vyas and Paliwal, 2004).

(f) Decorative and Building Stone

Marble : Marble deposits in the state are lying in Ajmer, Alwar, Banswara, Bhilwara, Bundi, Chittorgarh, Churu, Dausa, Dungarpur, Jaipur, Jaisalmer, Jodhpur, Jhunjhunu, Nagaur, Pali, Rajsamand, Sikar, Sirohi, Udaipur districts. Marble, sensu stricto, is a metamorphic carbonate rock and the term used here includes relatively softer rocks, capable of taking good polish and suitable for decorative works. Marble deposits of Jaisalmer are of Jurassic age otherwise all other deposits of the State belong to Precambrian age (Vijaywargiya, 2000). The world famous Marble deposits in the State is at Makrana (Nagaur district) exposed for about 15 km along the strike. From Bidiyad Railway station in the south to Bharon Mata temple in the north, it occurs as five parallel low ridges running NNE-SSW (Paliwal et al, 1997). Other occurrences in the State are at Kumari, Chakdungari (Nagaur district); Aandhi, Jamuwa, Ramgarh, Bheslana, Jhiri (Jaipur district); Pratapgarh (Alwar district); Rampura, Neem-Ka-Thana (Sikar district); Saradhana, Hatoondi, Roopangarh, Sanwar, Kayampura, Roopnagar (Ajmer district); Jahazpur, Kotri, Nandrai (Bhilwara district); Bar-Bherunaka (Pali district); Selwara, Bhadechi, Perwa, Serwa, Bhatana Basantgarh- Ajari, Abu-road-Dhanwan, Dhukhana, Khandra (Sirohi district); Aageriya-Sardargarh, Aatma-Thoria, Kelwa, Jhanjhar-Sapot, Tasol, Kharbad, Kunwariya, Pipapli (Rajsamand district); Babarmal, Daroli Midha-Neem, Jagat, Dewla-Dhariyawad, Kesariyaji-Masaro-Ki-Obari, Dachki (Udaipur district); Tripura, Sundari, Paloda, Oda, Bassi (Banswara district); Mandaldeh (Chittorgarh district); and Doonkar (Churu district)areas (Vijaywargiya, 2000).

Granite : Rajasthan has vast granite deposits spread over 23 districts of the State and are of Precambrian age and are generally used to make slabs, pillars and tiles for decorative purposes in building constructions. Important deposits are located in Nabi-Mailawas, Borta, Raniwara, Ratnagiri, Khambi, Tawab, Kalikhan, Bargaon, Jalor (Jalor district); Siwana, Mokalsar, Rakhi, Phoolan, Chautan, Ludrara, Devra, Ramaniya, Piplun, Nalguda, Meli, Mungeriya, Patadi-Nosar, Valu, Redana, Siner, Indrana, Devandi,

Mangta, Suala (Barmer district); Shivganj, Aarasanaji, Revdar, Mandari, Abu, Meerpur, Van, Anapura, Iderla, Sivera, Bala, Sanwara, Siyakara, Sanpur, Jirawal, Dantrai, Magriwada, Nimbaj, Khan-Vasan, Bag, Karjal (Sirohi district); Bar, Sendra, Chitar, Erinpura, Jawariya, Netra, Jakhora, Dani Khimal, Manihari, Sumerpur, Nanabeda, Chatila, Nadol, Narlai, Vinpura, Anna, Panchalwara, Dantiwada, Beda Dari, Jitra, Dharmdhari, Bisalpur, Kharda, Desuri, Rani, Raipur (Pali district); Kanpura, Jiwana, Buharu, Ramaliya, Fatehgarh, Chandma, Morajhari, Dang, Gelpur, Nai Kalan, Dedpura, Bhinai, Beawar, Ajmer, Kishengarh (Ajmer district); Gyangarh, Kidimal, Rampuriya, Daulatgarh, Katar, sabdara, Badakhera, Phakoliya (Bhilwara district); Devgarh, Aamet, Bhim, Malkot, Kakrod, Isharmand, Nhimjhar (Rajsamand district); Ajeetgarh, Neem-Ka-Thana, Mundru, Jetpura, Jitala, Ladi Ka Bas, Dabla, Sirohi Kalan (Sikar district); Khetri (Jhunjhunun district); Dewal, Ganwar, Kadila, Phulmaliya, Hasanpur, Rajpura, Hathgi, Rupaheli (Tonk district) areas and also in Alwar, Banswara, Bundi, Chittaurgarh, Churu, Dausa, Dungarpur, Jaisalmer, Jhunjhunu, Jaipur, Jodhpur, Sawai Madhopur and Udaipur districts (Vijaywargiya, 2000; D.M.G., Rajasthan, 2004;Vyas and Paliwal, 2004).

Serpentinite : Green colored marble used for decorative purposes occur in Rishabdev-Masaron-Ki-Ovari-Dakchi belt in Rajsamand-Udaipur-Dungarpur districts. Deposits occur at Dewal, Manpur, Mathugamra, Rohanwada-Duchki, Sabli, Somnath, Thana, Kochri-Ambada, Bodamli, Hirata, Mal-Surata, Jakol, Khemaru, Lolakpur, Kala Talab, Solaj, Tandi Oberi, Mal- Surata, Navagaon, Khirkhaya in Dungarpur district; Keshriyaji, Kherwara, Masaron Ki Obri, Odwas, Dhelana, Kagdar, Khanmin, Undithal, Saru, Bhauwa, Kakan Ka Gurha, Ramakhet, Akodara, Bhutala, Atamana Kotra, Bansda, Modi, Ugmana Kotra, Chhipala, Sultanji Ka Kherwara, Kalighati, Kankan Ka Gurha, Pipaldara, Bilkhai, Bhadawa in Udaipur district ; Ladnun in Nagaur district and also in Churu district (D.M.G., Rajasthan, 2004;Vyas and Paliwal, 2004).

Sandstone : Rajasthan is the largest Sandstone producing state of Country, it is a sedimentary rock capable of desirable carving and suitable for decorative purpose, is used to make slabs, pillars, and nowadays as tiles in building constructions. Sandstone deposits are located in Baran, Bharatpur, Bhilwara, Bundi, Bikaner, Chittaurgarh, Churu, Dhaulpur, Jaisalmer, Jhalawar, Jodhpur, Karauli, Kota, Nagaur, Pali and Tonk districts. Its deposits in the State belong to Bhander Group (Vindhyan Supergroup), Jodhpur and Nagaur Groups (Marwar Supergroup), and are also of Tertiary age. The Vindhyan sandstone deposits are located at Raverda (Chittorgarh district); Bijoliya (Bhilwara district); Dabi, Budhpura, Lambakho, Khimach, Kanwas, Mandna (Bundi-Kota districts); Karuli, Hindon (Karauli district); Bansi hill, and Saramathura (Bharatpur-Dholpur districts) areas while those of the Marwar Supergroup occur at Fidusar, Soorsagar, Balesar (Jodhpur district); Khatu, Khajwana (Nagaur district); and Dulmedha (Bikaner district) areas. (Vyas and Paliwal, 2003; D.M.G., Rajasthan, 2004;Vyas and Paliwal, 2004).

Limestone : Cement, SMS, and chemical grade limestone deposits in the state are spread over 25 districts. Flaggy and building limestone deposits are lying in Chittaurgarh, Jaisalmer, Jhalawar and Kota districts. The limestone deposits are of Precambrian, Jurassic and Eocene age. It is used as building material and in chemical industries, but extensive deposits of limestone are suitable for cement manufacturing. There are 15 important belts in the State. Ras marble (Delhi Supergroup) belt (Pali district) extends from Ras to Kundal and Tunkara on one flank and up to Asarlai on the other flank for about 80 km Stretch. It hosts good reserves of cement grade lime stone (Paliwal and Vyas, 1999). In Pali district other deposits are at Chandawal, Gudhiya, Nanakararawav and Aamli areas. In Ajmer district limestone deposits are in two belts; eastern belt extending for about 20 km include Shyopura, Lulawa, Shyampura, Kesherpura; where as 140 km long western belt include Roopnagar, Saradhana, Bhagwanpura, Daulatpura, Govindgarh, Pisangan, Kayampura, Rajpura, Jak area. Other deposits of the States are located in Aamli, Kotal, Kundal, Basantgarh, Banas Rohida, Batera Kiwarli, Morthala, Pandoori, Aakarabhatta, Abu road, Siyawa (Sirohi district); Khamera to Undawela, Umrai to Kushapura, Badi-Lalpura, Mokhampura, Kelamela, Oda-Bassi, Amarpura-Prathvipura (Banswara district); Aaspur tehsil (Dungarpur district); Daroli, Mawali, Kaya, Kanpur, Pratapnagar, Gogunda, Kumbhalrgarh and Kotra tehsil (Udaipur district); Hatoon, Jasawantpura (Rajsamand district); Amaragarh to Gelji (Bhilwara district); Dungala (Chittorgarh district); Bheslana to Mohanpura-Jodhpura (Jaipur district); Nawalgarh and Udaiipurwati tehsil (Jhunjhunun district); Rampura, Neem Ka Thana (Sikar district); and Rajgarh tehsil (Alwar district) areas. All these deposits are of Archaean and Proterozoic age.

Limestone deposits of the Vindhyan Supergroup

are at Bhagwanpura, Nimbahera, Chenpura, Manpur, Mewa Hathipura, Binota, Mangrol, Aawal Kheda to Parsoli, Bichore, Katundamod Chittoriya (Chittorgarh district); Rupera to Pathiyal (Bhilwara district); west of Satur to Pipliyantari and west of Bundi to Gandoli (Bundi district); Dolar to Lachmipura and Balwant Ka Tapara to Dumeda (Sawai Madhopur district); Suket, Chechat, Modak, Namana, Ramganj Mandi, Kumbhkot, Nimod, Laxmipura (Kota district); Maholi-Kela devi, Leela Dungari, Dangari (Karauli district); Basaidang to Rajai (Dholpur district); Gagron, Aroliya-Kishenpura (Jhalawar district) areas.

Limestone deposits of the Marwar Supergroup are at Bilara, Ransi, Hariadhana, Panchoti, Borunda (Jodhpur district); Gotan-Maundwa-Rolkajiya belt (Nagaur district); Kalyansar, Aserasar, Sandwa (Churu district); Sarunda - Mandeliya (Bikaner district).

Mesozoic (Jurassic and Cretaceous) and Tertiary (Eocene) limestone deposits are found in Jaisalmer district. Yellow colored Jurassic limestone occurs around Jaisalmer city, Bara bagh, Mool sagar, Amar sagar, Kanoi, Chundi areas. Abur (Cretaceous) limestone occurs in Abur village of Jaisalmer district. The industrial grade limestone of the Tertiary (Eocene) age suitable for cement and chemical industry is restricted to the Khuiala Formation, exposed around Sam, Khuiala, Ramgarh, Hingola, Mohammed Ki Dhani, Sanu, Tulsi Ram Ki Dhani, Khinya, Khivsar and Mandha villages of Jaisalmer district (Bafna and Dhaka, 1999).

Slate-shale-schist and phyllite : These rocks are used in roofing, flooring and decorative works as building material. Shale and slate are mined from Mandan (Alwar district); and Dewali (Tonk district) areas. Slate, schist and phyllite occur in Gogunda, Zawar, Sarada, Delwara (Udaipur district); Jahazpur, Chenpura (Bhilwara district); Bar (Pali district); Bhim-Tatgarh (Rajsamand) district); and in Banswara and Dungarpur districts. All these rocks are of Precambrian age (Vijaywargiya, 2000 D.M.G., Rajasthan, 2004; Vyas and Paliwal, 2004).

(g) Fuel Minerals

Lignite : Lignite deposits occur at Kapurdi, Jalipa, Bothia, Giral, Gunga, Jogeshwar-Tala, Sonari, Sacha-Sauda, Bharka, Bothia-Bhadka, Koslu-Sindhari, Kurla, Hodu, Mahabar, Shivkar (Barmer district); Khuri, Deunga, Chandan (Jaisalmer district); Palana, Barsinghsar, Gurha, Bhoblasar, Bithnok, Gadiyala, Raneri, Mandal Charana, Hadla, Chak Vijaysinghpura, Riri, Badhnu, Sowa (Bikaner district); Merta road (Meera Nagar), Mokala, Kasnau-Igyar, Kuchera, Indawar, Matasukh, Lunsara, Lamba-Jatan-Hansiyas, Kaprion Ki Dhani and Nimbri-Chandawatan (Nagaur district). (D.M.G., Rajasthan, 2004; Vyas and Paliwal, 2004) Total reserves of lignite are at about 1375 Mt (Mehta, 1999). These deposits are of Tertiary age.

Hydrocarbons (Petroleum and Natural Gas) : Exploration for oil and gas in western Rajasthan was started by O.N.G.C. in 1954 and by Oil India Limited in 1983 and since then they have drilled 59 and 23 wells respectively, in Jaisalmer, Bikaner, Barmer, and Sri Ganganagar districts. Gas reserves of 10941 MCM in western Rajasthan are found in Manhera Tibba, Ghotaru, Bhakri Tibba, Kharotal, Bankia, (by O.N.G.C.), Tanot, East Tanot, Ramgarh, Dandewala, Jalalwala, Bagitibba Tibba and Gamnewala (by O.I.L.) gas fields. In addition to natural gas, heavy oil has also been reported in Bagewala, Kabrewala, and Tavriwala blocks of Bikaner- Nagaur Basin (Mehta, 1999). Heavy oil has been discovered from the Bilara and Jodhpur Formations (Marwar Supergroup). Bilara limestone is affirmed to have adequate source potential. (Mukherjee et al, 1995). Oil from Guda Malani, Kosalu (Barmer district); and Nanuwala (Hanumangarh district) areas was discovered in recent years and private sector is increasingly investing in oil and gas exploration in the State (Vyas and Paliwal, 2004; Chauhan et. al. this volume).

(h) Developing Mineral Scenario of the State

After independence, development in mineral sector in State got momentum. In area of mineral concessions, there were only 50 major mineral concessions in the State in 1950-51, which increased, to 1286 in 2000-01. The rise in minor minerals has been from 195(50-51) to 8826 (2000-01). Besides these the growth of quarry licenses (QL) in the State has been from 1200 (50-51) to 23578(2000-01). There has been a tremendous increase in the production of major and minor minerals in the State during last five decades. The total major minerals in the State in 1950-51 were only 6.67 lac tonnes but by 2000-01, it increased to 303.92 lac tonnes. In the case of major minerals, growth in production has been from 8.80 lac tonnes (50-51) to 441.67 lac tonnes (2000-01). In the field of employment and revenue during the year 50-51 about 32250 persons per day were employed in the mining of major and minor minerals in the State. This number increased to about 316220 in 2000-01. The progress in the field of revenue in mineral sector played significant role in the economic and industrial growth of the State. Revenue has recorded an increase from Rs. 48.02 lacs in 1950-51 to Rs. 37000.94 lacs in the year 2000-01. (Vyas and Paliwal, 2004)

ACKNOWLEDGEMENT

I am thankful to Prof. K.L. Shrivastava, Department of Geology, J.N.Vyas University; for giving me opportunity to contribute in the volume, honouring Prof. B.S. Paliwal, an eminent scholar and a best teacher, I have ever seen in my life. Help received from my parents along with Mrs. Anita Vyas, Pallavi, Komal and Abhinav is also acknowledged.

References

Bafna, P.C. and Dhaka, B.S., 1999. Industrial grade limestone deposits of Tertiary period in western Rajasthan. In: B.S. Paliwal (Ed), Geological Evolution of Northwestern India. Scientific Publishers (India) Jodhpur, pp. 210-215.

Bhattacharjee, J., Ramji, Reddy, A.B., Golani, P.R. 1991. Gold mineralisation associated with Cu-Zn massive sulphide in Danva area, Sirohi district, Rajasthan. Indian Minerals, vol. 45(3), pp. 183-188.

Bhattacharjee, J., Fareeduddin and Jain, S.S. 1993. Tectonic setting, Petro chemistry and tungsten metallogeny of Sewariya granite in south Delhi fold Belt, Rajasthan J. Geol., Soc., India, vol. 42, 3-16.

Chatterjee, K. K. 2004. An introduction to mineral Economics. New Age International (P) Limited Publishers. New Delhi. 379 p.

Bohra, N.S. 2001. Amendments and changes in statutes and Policies form development of Mineral sector-Present scenario and future trends. National seminar on strategy for sustainable development of mineral sector, 11-12 August 2001, Organized by Social Policy Research Institute, Jaipur, Rajasthan. pp. 20-31.

Deb. M. 1982. Rare minerals in Rajpura Dariba ores-Some further comments. J. Geol, Soc. India. vol. 23, 251-260.

Deb, M. 1986. Sulfur and carbon isotope composition in the stratiform Zn-Pb-Cu. Sulphide deposits of the Rajpura Dariba belt, Rajasthan, N.W. India: a model of Ore genesis Mineral Deposit, vol. 21, 313-321.

Deb, M. 1999. Metallic mineral deposits of Rajasthan. In P. Kataria (Ed.) Proceedings of the Seminar on Geology of Rajasthan, Status and perspective. (A.B. Roy felicitation Volume) Department of Geology, M.L. Sukhadia University, Udaipur. pp. 213-237.

D.M.G. Rajasthan.2004. Mineral Directory of Rajasthan. Directorate of Mines and Geology, Rajasthan, Udaipur Publication. 193 p.

Grover, A.K. and Verma, R.G. 1993. Gold mineralisation in the Pre Cambrian (Bhukia area) of southeastern Rajasthan-a new discovery. Jour. Geol. Soc. India, 42, 281-288.

G.S.I., 1994. Detailed information on copper-lead-zinc ores in Rajasthan Gujarat (India). P. 1-156 (Rajasthan); pp. 1-13 (Gujarat).

G.S.I., 2001. Geology and Mineral resources of Rajasthan. Geological Survey of India. Misc. Publication No.30 part 12, 2nd-revised edition.72. 113 p.

Gupta, S.N., Arora Y.K., Mathur, R.K., Iqbaluddin, Prasad, B., Sahai, T.N., Sharma, S.B. 1997. The Precambrian geology of the Aravalli region, Southern Rajasthan and Northeastern Gujarat. Mem. Geol. Survey. India, 123, pp. 171-199.

Jain, R.S. 1987. Gold Incidence in Rajasthan, Rec. Geol. Survey, India, vol. 113 (7), 99-103.

Joshi, K.N. 1993. Mineral Resources in Rajasthan. In: T.S. Chauhan (Ed) Natural and Human Resources of Rajasthan. Scientific Publishers (India). Jodhpur, pp. 147-166.

Mehta, D.S. 1999. The Changing scenario of Western Rajasthan through mineral development. In; B.S. Paliwal (Ed), Geological Evolution of Northwestern India, Scientific Publishers (India), Jodhpur. pp. 148-166.

Mukherjee, A.D., Bhattacharya, P.K., and Bhattacharya, H.N. 1992. Petrotectonics of a southern segment of the polymetallic sulphide bearing Proterozoic Delhi mobile belt, India and its implication. In; S.C. Sarkar (Ed) Metallogeny related to tectonic of the Proterozoic mobile belts; A. A. Balkema, Rotterdam, pp. 253-270.

Mukherjee, A.D. and Bhattacharya, H.N. 1997. Strata-bound pyretic copper mineralisation in a fore arc setting in northwestern India-a Proterozoic Besshi type deposit, Trans. Inst. Min. Metall., vol. (Sec B), 106, B45-B52.

Mukherjee, M.K., Bhandari, S.K. and Purkayastha, D. 1995. Hydrocarbon prospects and evidence of presence of Proterozoic Basin in Lunar- Maijlar area, Rajasthan. Proc. Ist Internl. Petrol. Cong. and Exhib. Petrotech 95, New Delhi, 1, 133-145.

Pandya, M.K., Ranawat, P.S. and Pandya, T.K. 1999. Introduction to Industrial Minerals and decorative stones of Rajasthan. In. P. Kataria (Ed.) Proceedings of the seminar on Geology of Rajasthan, Status and perspective (A.B. Roy felicitation volume), Department of Geology. M.L. Sukhadia University, Udaipur. pp. 238-250.

Paliwal, B.S. 1999. Geology of the region to the west of Aravalli Mountain; Some observations. In; B.S. Paliwal (ed). Geological Evolution of Northwestern India. Scientific Publishers (India), Jodhpur., pp. 387-414.

Paliwal, B.S., Pareek, U.S. and Vyas, A. 1997. Structural framework of the Makrana marble deposit and it's bearing on the Precambrian stratigraphy of Rajasthan.

In; Wijayananda, N.P., Cooray, P.G. and Mosley, P. (Eds.), Geology in South Asia-II, Geological Survey and Mines Bureau, Sri Lanka, Professional Paper-7, pp. 123-136.

Paliwal, B.S. and Vyas, A. 1999. Geology of the middle Proterozoic Ras marble deposit: A potential reserve for the cement industry. In; B.S. Paliwal (Ed.) Geological Evolution of Northwestern India. Scientific Publishers (India), Jodhpur. pp. 183-190.

Raghunandan, K.R., Dhruba Rao, B.K., and Singhal, M.L. 1981. Exploration for copper, lead and zinc ores in India. Bull Geol. Survey. India. A., vol. 47, 222 p.

Raja Rao, C.S. and Mathur, R.K. 1975. Rajpura- Dariba lead zinc- copper deposits, Udaipur district, Rajasthan, Jour. Geol. Soc. India. 16. (Suppl.) Abstr. 28-29.

Roy, A.B. 1999. Evolution of Saline Lakes in Rajasthan, Curr, Sci. vol. 67(3) pp. 290-295.

Samar, M.S. 1996. Geological appraisal of lead -zinc mines of Rajasthan. Seminar on "Mineralisation in the western Indian craton". Delhi. Abstr. 40 p.

Singh, R. 1996. Uranium Metallogeny in western Indian craton. In, seminar on "Mineralisation in the western Indian craton" Delhi, Abstr, pp. 47-48.

Sinha Roy, S. 1996. Stratigraphic and tectonic control of gold mineralisation in the Aravalli fold belt, Banswara district, Rajasthan. In Gold resources of India. Geol. Soc. India and N.G.R.I. Seminar, Abstr, pp. 158-160.

Sinha Roy, S., Malhotra, G. and Mohanty. M. 1998. Geology of Rajasthan, Geol. Soc. India. Banglore. 278 p.

Valdiya, K.S. 1987. Environmental Geology Indian context. Tata McGraw Hill Publishing Company Ltd. New Delhi. 583 p.

Vijaywargiya, R.R.P. 2000. Geology and Mineral Resources of Rajasthan. (Hindi). Rajasthan Hindi Granth Acadami, Jaipur. pp. 88-193.

Vyas, Arun. 1995. Structural framework of the lower Proterozoic rocks around Barapal, south of Udaipur city, Rajasthan. Unpublished thesis work. Submitted to M.D.S. University, Ajmer.

Vyas, Arun and Paliwal, B.S. 2003. Jodhpur sandstone: The crowning glory of building marvel. Sandstone spectrum-2003, Souvenir Volume. 2. Published by department of mines and geology, Government of Rajasthan, Udaipur. pp. 15-20.

Vyas, Arun and Paliwal, B.S. 2004. Mineral resources of Rajasthan, India- A developing scenario. Geological Survey of India. Jaipur. Special Publication 72. pp 13-26.

Wadia, Meher D.N. 1994. Minerals of India. Published by National Book Trust, India. 258 p.

Hon'ble Reviewers

1. Dr. Arun Kumar Shandilya
   Department of Geology
   Dungar College
   Bikaner University
   BIKANER 334 001 INDIA
2. Dr. S.C. Mathur
   Department of Geology
   JNV University
   JODHPUR 342 005 INDIA

Editorial Assistance

Dr. Sameer Naval

Economic Mineralization
K.L. Shrivastava (Editor)
2009. pp.369-382
Scientific Publishers (India), Jodhpur

Section-III

A Profile of the Industrial Mineral Resource Potential of Orissa

MADHUMITA DAS[1] AND SHREERUP GOSWAMI[2]

[1]Department of Geology, Utkal University, Vani Vihar, Bhubaneswar-751004
[2]Department of Environmental Sciences, F. M. University, Vyasa Vihar, Balasore-756019
Email : goswamishreerup@yahoo.com

Abstract

Industrial minerals are essential for rapid economic growth. They pervade almost every sector of a modern economy, serving as basic raw materials in many industries where the value added may be several fold or where the mineral itself forms an essential part of an industrial process. The production of industrial minerals is small in India, particularly in Orissa as compared to the production of metallic and energy minerals. Hence, it is necessary to provide the right political and fiscal environment, not only to attract foreign investment for mineral development but also to encourage local entrepreneurs to develop industrial minerals so that they can gradually build up production and capital to supply local markets. There is thus an increasing need for the geological assessment and mineralogical characterization of the industrial mineral resources in the state. Hence a few suggestions have been discussed in this article.

Besides metallic and fuel minerals, Orissa has enormous industrial mineral potential. The article outlines the occurrences and reserve of industrial minerals in Orissa viz., china clay, fireclay, graphite, limestone, dolomite, pyrophyllite, kyanite, sillimanite, andalusite, soapstone, steatite, talc, quartz, quartzite and some dimension and decorative stones such as granite, sandstone, laterite and khondalite etc. The state is well endowed with industrial mineral resources; hence a strong economic growth of the state is expected. Geoscientists cannot ignore this important sector of the mineral industries and the contribution that it can make to economic development of the state.

Key words : Industrial Minerals, Orissa, Dimension and Decorative Stones.

INTRODUCTION

Industrial minerals have always given second place to metallic minerals in developing countries in term of exploration, mining, investment or even in academic interest. But with economic growth of a country, the industrial sector develops. Obviously, the demand for industrial minerals increases.

Industrial minerals are defined as those rocks and minerals of economic value, which are exploited for their physical and/or chemical properties, exclusive of metalliferous minerals, fossil fuels and gemstones (Highley, 1994). Construction minerals, including sand, gravel and crushed rock aggregates are also included in this group. Industrial minerals support a large and diversified extractive industry of worldwide importance, accounting for some 72% of the quantity and 40% of the value of world mineral production, excluding oil and gas (Noetstaller, 1988). With increasing industrialization, the importance of industrial minerals is well recognised. The country in which industrial minerals production exceeds that of the metalliferous minerals, the more recently industrialized the country has become (Bristow, 1987).

The term Industrial Rocks and Minerals covers a wide range of naturally occurring materials with diverse properties. In case of metallic ore deposits, the tenor and grade are vital economic factor, whereas in case of Industrial Rocks and Minerals both physical and chemical characteristics are equally important. However, the accuracy and precision of the physical and chemical properties of the Industrial Rocks and Minerals should match the requirements of the market.

Classification of industrial minerals by major end-use sector is sketched in Table 1. This classification of industrial minerals based on end-uses serves to illustrate the link between industrial minerals and major sector of an economy. Markets for industrial minerals are concentrated in the construction sector, in the production of other primary products such as fertilizers and chemicals, and in the manufacture of non-durable and durable consumer goods. Many industrial minerals are not sold in the open market but are consumed captively in the manufacture of value-added products-for example, the manufacture of structural clay products such as bricks, pipes and tiles, in cement manufacture, in lime, plaster and plasterboard, in the production of heavy inorganic chemicals, such as chlorine and soda ash, and in the production of ready-mixed concrete and coated roadstone. In these the value of the mineral at the quarry or mine forms only a small part of the value of the final product into which they are incorporated (Highley, 1994). However, other products, such as ceramics and glass, consist almost wholly of industrial minerals.

Since every major rock has some use, it is very difficult to classify the Industrial Rocks and Minerals. But some examples are given in the Table 2 for the four major classes of Industrial Rocks and Minerals based on end-use. The table depicts that a number of materials are used in more than one main use category i.e. a single material can be used for more than one category. For example, clays can be used in the construction industry for the manufacture of bricks, as a carrier of fertilizer or pesticides, as a coating or

Table-1 Classification of industrial minerals by major end-use sector (Highley, 1994)

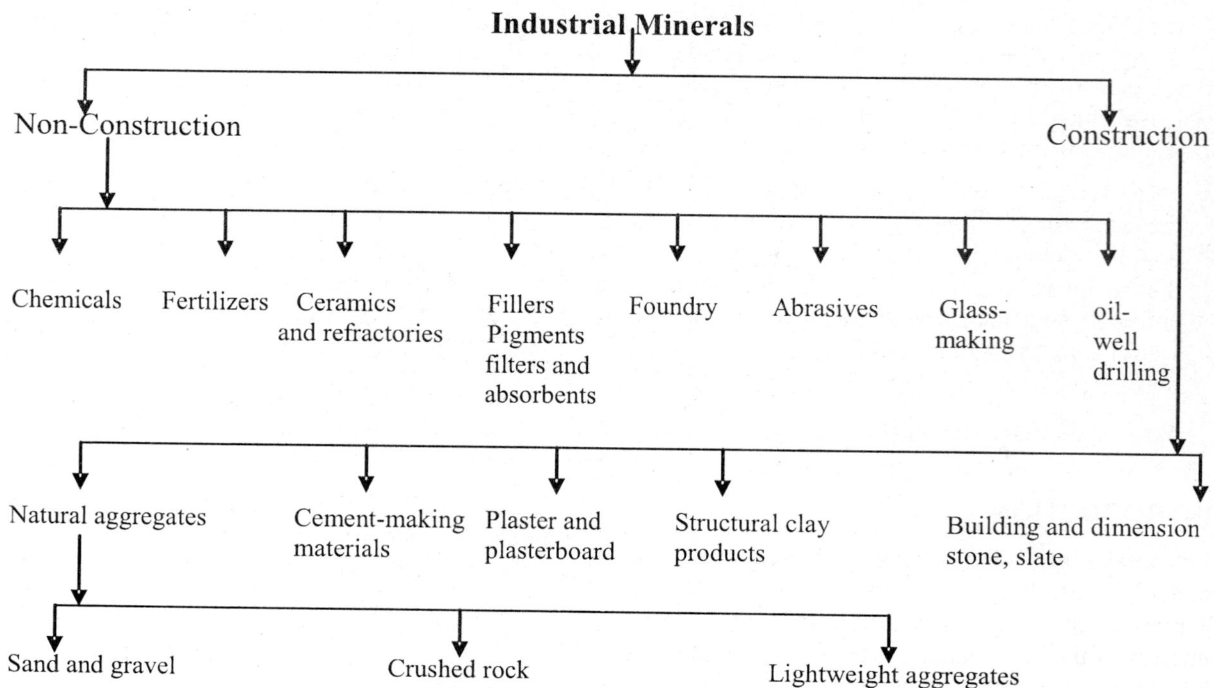

Table-2 An end-use classification of industrial rocks and minerals (Dunham and Scott, 1994)

	Class	Examples
1.	Construction	Sand and gravel, crushed rock aggregate, building stone, clay for bricks, Limestone and clay for cement, Slate, Gypsum and Perlite
2.	Chemicals and fertilizers	Limestone and lime, Dolomite, Halite, Potash salts, Phosphates, Anhydrites, Magnesite Bentonite, Zeolite, Chromite, Fluorite and Borates
3.	Consumer and Industrial products	Kaolin, Plastic clays, pulverized limestone and chalk, Feldspar, Diamond, Celestite, Ti-minerals and Talc
4.	Metal manufacture and shaping	Limestone, Dolomite, Magnesite, Chromite Fireclay, Bauxite, Silica sand, Bentonite, Fluorite, Zircon and Olivine

filling agent of paper or for the manufacture of alumino-silicate refractories. But for each of these uses some specific criteria is required. So, the Industrial Rocks and Minerals must be properly assessed through different tests (eg. Petrographic study, Differential thermal/gravimetric analysis, X-ray diffraction, Scanning electron microscopy, Energy dispersive X-ray and Electron microprobe analysis) before determining the end-use. For some applications they are used after minimum processing, but many uses demand rigid specifications where modification of properties is often necessary. This paper is an attempt to examine the role of Industrial minerals of Orissa state in its economic development.

GEOLOGICAL FRAMEWORK OF ORISSA

Precambrian rocks embrace a vast time span from 3.7 Ga to 0.5 Ga. It occupies about 80% of the land surface of the Orissa State. Precambrians of Orissa are segmented into three geologically significant sectors. They are North Orissa Sector (comprising parts of Singhbhum-North Orissa Craton with Banded Iron Formation, greenstone-granite association), South Orissa Sector (comprising the Eastern Ghats Mobile Belt with high grade granulite facies rocks) and the West Orissa Sector (making the eastern margin of the Bhandara craton with a complex combination of greenstone, granite, granulite facies rocks and Purana sedimentaries) (Naqvi and Rogers, 1987; Mahalik, 1995a; 1995b). Besides metallic minerals, the Precambrians are also storehouse of some industrial minerals/rocks such as graphite, limestone, dolomite, granite, quartzite, khondalite, charnockite and other dimension stones.

The Phanerozoic rocks of Orissa are represented by the Gondwana Supergroup encompassing Upper Palaeozoic to Middle Mesozoic and the Baripada Beds of Lower Tertiary.

The Gondwana sediments in this state occupy an area of about 4, 470 sq. kms and are mainly exposed in two main sedimentary basins, viz. an extensive basin comprising Lower to Upper Gondwana (Permian-Triassic) with coal measures towards West (Talcher and Ib-River coalfields in the districts of Angul, Dhenkanal, Sambalpur, Sundargarh and Jharsuguda) and Athgarh Basin in the East having the rocks of only Upper Gondwana (Early Cretaceous) in the districts of Cuttack and Khurda. In addition to these major basins, there are four more small basins (outliers) in this state. They are 1.Gaisilat, 2. Athmallik, 3. Katringia and 4. Chhatrapur basins / outliers. Besides coal; fire clay and building stones are main industrial minerals /rocks of commercial importance found in these

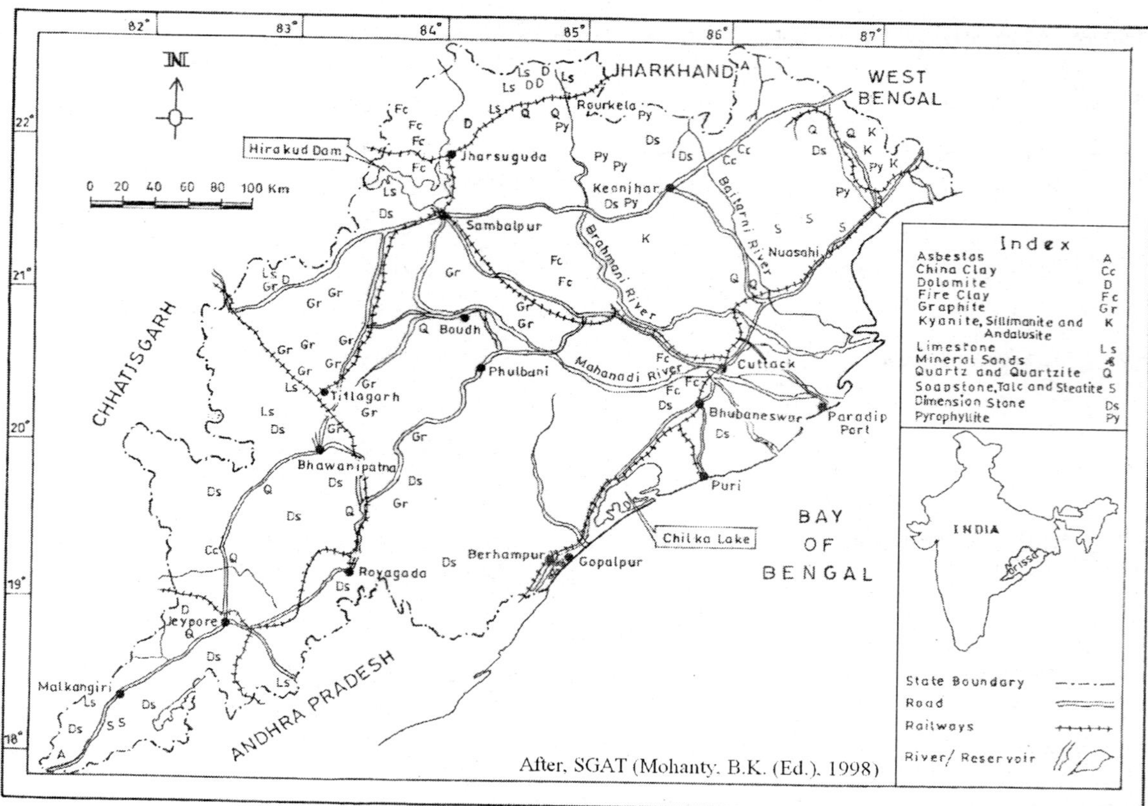

Fig.1 Map showing Industrial mineral deposits of Orissa

sediments (Pandya, 1995; Goswami et al., 2006). The only lithounits of Tertiary age are "Baripada beds". They are exposed within a radius of 8 km around Baripada in Mayurbhanj district. Limestone is the main rock of commercial importance found in these sediments (Pradhan, 1995).

Quaternary Formation and Recent to sub-Recent alluvium are chiefly confined to the coastal tracts of the state from north to south viz., Chandbali, Paradeep, Cuttack, Puri-Konark, Chandipur subbasins. Industrial minerals/ rocks of economic importance like china clay, building blocks of laterites, varieties of sands (filler, foundry, cement), lime shells, soil clay of different types (potter's, oil-mill, brick clay, clay for rural construction and domestic wall wash etc) can be extracted from these sediments.

The stratigraphy of the different groups of rocks of Orissa is shown in Table 3 (Iyengar and Banerjee, 1971; Salop, 1972).

INDUSTRIAL MINERALS

Orissa, a major state of India has enormous mineral potential and is rich in metallic and fuel mineral resources. In addition, Orissa also produces huge industrial minerals. The industrial minerals of Orissa along with their reserve are shown in Table 4. Only about half of these are actively being extracted since much of the manufacturing industry are not in their best condition. The occurrences of industrial minerals (Fig. 1) viz., china clay, fireclay, graphite, limestone, dolomite, pyrophyllite, kyanite, sillimanite, andalusite, soapstone, steatite, talc, quartz, quartzite and some dimension and decorative stones such as granite, sandstone, laterite, khondalite are discussed one by one. In addition to these industrial minerals, a small asbestos deposit is found in the border area of the state to Andhra Pradesh in the Malkangiri District.

(i) China clay

China clay is comprising mostly mineral kaolinite (Al_2O_3, $2SiO_2$, $2H_2O$), where Al_2O_3, SiO_2 and H_2O percentages are 39.5%, 46.56% and 13.94% respectively. It has a specific gravity of 2.6 and a fusion point of 1.8 °C. Kaolin deposits have been formed by decomposition of feldspathic granite. Kaolin occurs in form of pockets or lenses within the

Table-3 Stratigraphic Succession of Orissa

Subdivisions recognised in Orissa.	Age (after Iyengar and Banerjee, 1971)	Age (after Salop, 1972)
Coastal alluvial cover	Recent to Quaternary	
Baripada Beds	Miocene	
Gondwana Supergroup	Upper Palaeozoic to Middle Mesozoic	
Vindhyan Supergroup	Riphean or Vendian (less than 600 m.y.)	Epiproterozoic to Eocambrian
Singhbhum Gangpur Group	Riphean (600-1400 m.y.)	Neoproterozoic
Kolhan Group	Pre-Riphean to Riphean (?) (1500-1600 m.y.)	-
Koira-Simlipal Group	Pre-Riphean (1400-3000 m.y.)	Mesoproterozoic
Eastern Ghats Group	Pre-Riphean (1400-3000 m.y.)	Mesoproterozoic
Iron Ore Group	Pre-Riphean (1400-3000 m.y.)	Palaeo-Proterozoic

main mass of granite. It is observed that the better quality china clay deposits occur near the contact of granite with epidiorite or chlorite schist.

It is an important industrial mineral, because it is soft, non-abrasive and chemically inert over a wide pH range. It has good covering or hiding power when used as a pigment in coated films and has low conductivity of heat and electricity.

China clay has many industrial applications; as filler in paper, textiles, rubber; in the manufacture of potteries, ceramics, sanitary wares, glazed tiles, HT and LT insulations, white cement, insecticides, paints, cosmetics, refractory bricks. China clay is also used in some pharmaceutical industries.

Occurrences

China clay is found to occur in a long belt stretching from southern Singhbhum to Mayurbhanj Granite belt. Badampahar-Joshipur, Karanjia-Ramchandrapur belt with a number of deposits on either side of NH -6 is the most important china clay producing area of the state. The important insitu deposits in Mayurbhanj District are found near Joshipur, Dumuria, Jamba, Kadodiha, Jamkeswar and Thakurmunda. Besides these, there are a few occurrences in Keonjhar, Sambalpur, Koraput, Sundargarh and Dhenkanal districts (Mohanty, 1998).

The reserve estimated by IBM (Year Book, 2004) for Orissa is 278 million tonnes.

Table-4 Industrial minerals of Orissa and their reserve

Sl. No.	Name of the industrial mineral	Reserve of Orissa State as per Mineral year book 2004	% of all India reserve
1.	Limestone	2038.35 million tonne	1.20
2.	Fireclay	175.79 million tonne	25.29
3.	Granite (Dimension stone)	1843.20 million cubic meter	4.89
4.	Graphite	3.65 million tonne	22.98
5.	Kaolin (china clay)	276.11 million tonne	11.98
6.	Sillimanite	19.31 million tonne	30.42
7.	Dunite	4.45 million tonne	2.80
8.	Pyrophyllite	3.79 million tonne	20.81
9.	Quartz, silica sand and quartzite	70.69 million tonne	2.48
10.	Talc and steatite	0.896 million tonne	0.33
11.	Dolomite	840.73 million tonne	11.86

(ii) Fire clay

The term fireclay has been commonly applied to refractory clays, which can withstand of 1500 °C or higher temperature. It is basically kaolinite. On the basis of plasticity, fireclay is classified as plastic, semi-plastic and non-plastic. An important parameter of fireclay is refractoriness expressed in terms of Pyrometric Cone Equivalent (PCE). The only and most important application of fireclay is as refractory. Fireclay-seams in Orissa are mainly confined to the Karharbari, Barakar and Lower Kamthi Formations in Lower Gondwana (Talcher and Ib River coalfields) and Athgarh Formation of Upper Gondwana (Athgarh Basin).

Occurrences

In Orissa fireclay are restricted to three geographical belts.

(A) Talcher Coalfield (Dhenkanal and Angul districts): In Talcher Coalfield, fireclay has been encountered in the Jagannath and South Balanda collieries and near Kaniha and Ghantikhal villages.

(b) Ib River Coalfield (Jharsuguda and Sundargarh districts): The fireclay occurrences of this area have been recorded near Jurabaga, Darlipali, Rampur, Kuropal, Bariapahar, Khinda, Lukopali, Kulda, Siarmal and Belpahar etc.

(c) Athgarh Basin (Cuttack District): Clay of Talabasta area belongs to Athgarh Formation of Upper Gondwana.

The production of fireclay in 2004-2005 is 93872 tonnes (Nanda and Bhol, 2006).

(iii) Graphite

Natural graphite is an allotrope of elemental carbon, which crystalises in the hexagonal system. It is a soft and opaque black mineral with a metallic luster, greasy feel, hardness of 1 to2 in Moh's scale, specific gravity of 2.09 to 2.93, good conductor of heat and electricity and is anisotropic. It has a weak plane of cleavage and has high refractoriness and can withstand temperature up to 3000°C in an inert atmosphere.

Graphite in high grade metamorphic rocks has been variously worked out as a product of (a) granulite facies metamorphism of carbonaceous matter; (b) a direct consequence of granulite facies metamorphism in the presence of CO_2 rich fluid under low O_2 conditions; (c) reduction of oxides of carbon either emanated from magmatic sources or due to silicification of limestones; (d) methanation of carbonate minerals by hydrogen and (e) assimilation of crustal material containing biogenic carbon by magma. In Orissa, neither there is any bedded deposit nor there is any identifiable relic of carbonaceous sediments in the Eastern Ghats country rocks. Thus, no definite sedimentary parentage can be attributed to graphite. However, as the host rocks of graphite are sediments metamorphosed under granulite facies conditions, biogenic origin of some graphite can not be ruled out. Here, metamorphism of carbonaceous material producing graphite is well known. The calc-silicate granulites are probable source of carbon in the Eastern Ghats complex of Orissa. Methanation of carbonate minerals by hydrogen takes place at low hydrogen pressure, under 200-600°C, provided the gases produced (particularly H_2O) leave the system or enter into a solid phase. Subsequent pyrolysis of methane results in the formation of graphite by the reaction, $CH_4 \rightarrow C+2H_2$. Once carbon is formed, it would have a catalytic effect on subsequent methane dissociation (Salotti et al., 1971; Acharya and Das, 1984). Mixing of biogenic and abiogenic source of carbon for the graphite in Eastern Ghats Complex of Orissa are also probable. The present localization of graphite may be due to the remobilization of graphite, after its formation, in solid state through open spaces in rocks to low pressure sites such as foliation planes, fractures, fold hinges boudin necks and joints etc. The textural and micro-structural features show graphite as an infilling of a variety of open spaces in rocks and minerals. A peculiar structure cum lithological feature i.e. the "intrusive breccia" resulting from graphite-mixed breccia near fault zone are common in these belts.

The industrial uses of graphite are as crucibles and foundry facings; lubricants, in brake linings, pencils, electrical and motor brushes, paints, polish, rubber, explosives, batteries. Graphite is used to raise the carbon content and thereby hardening of steel. Natural graphite powder has also been used in powder metallurgy applications.

Occurrences

The major graphite zone of Orissa is coincident with the migmatised quartz-garnet-sillimanite-graphite schists of khondalite suite and its contact with medium-grained granite gneiss. The important graphite belts of the state are -

(a) Sargipalli belt : (21°00'-22°22' : 83°15'-83°40')

Sargipali belt contains more than 150 occurrences, which belongs to the deposits of Sargipali, Dangachancha, Darrhamunda, Sapmuna, Rengali, Mohanilah, Raju- Nagfena etc of Bargarh, Bolangir and Kalahandi districts.

(b) Titilagarh belt: (20°06'-21°24' : 83°00'-83°45')

Titilagarh belt contains more than 120 occurrences in Titilagarh, Boroni, Malisira, Singjharan, Loitora of Bolangir and Kalahandi districts.

(c) Tumudibandh belt: (18°45'-20°00' : 83°00'-83°45')

Tumudibandh belt contains more than 50 occurrences located in Tumudibandh, Lakhajorna, Palur, Raisil, Ambaguda of Phulbani, Rayagada and Gajapati districts.

(d) Dandatopa belt: (20°37'-21°00' : 84°15'-84°45')

Dandatopa belt contains more then 22 occurrences in Dandatopa, Adeswar, Kamalpur, Akharkata and Girida of Dhenkanal and Angul districts (Acharya, 2006).

As a whole, Orissa accounts for 1.68 million tones proved, 1.33 million tones probable and 1.62 million tones possible totaling to 4.63 million tones reserve (Source-IBM year book, 2004). There are 119 graphite mining leases in Orissa varying from less than one hectare to 100 hectares. Of these, only 33 were reported to be in operation in above mentioned four graphite belts of Orissa.

(iv) Pyrophyllite

Pyrophyllite is a minor mineral of secondary origin and is a substitute of talc. It is a hydrous aluminium silicate used mainly in refractories making. The composition of pyrophyllite is Al_2O_3 $4SiO_2$ H_2O (Al_2O_3=28.3%, H_2O =5%). It is an important industrial mineral that finds increasing use in different industries as inert fillers, as raw materials for high grade ceramics, special refractories, electrical insulations, insecticide and decorative tiles in both domestic and export sectors. Pyrophyllite occurs in the form of quartz-pyrophyllite schist associated with granite and quartz reef intruding granite gneiss. It occurs as irregular, lenticular or banded deposits in the area of high grade metamorphism, as foliated masses in crystalline schists and as a gangue mineral in kyanite deposits. It also occurs in slates and tuffs interbedded with metamorphosed volcanic breccias and flows. The origin of pyrophyllite is attributed to metasomatic replacement by hot magmatic solution given off during the hydrothermal phase of granitic intrusion. It may also be formed by hydrothermal alteration of acidic rocks. Pyrophyllite has the highest Si:Al ratio of all known aluminium silicates. Thus for its formation more SiO_2 is required. An acidic environment is necessary for the formation of both pyrophyllite and kaolin; however a relatively high temperature environment facilitates the formation of pyrophyllite (Sahoo, 1991; 2006; Saha and Basak, 1965).

Occurrences

The occurrences of the state are associated with Singhbhum - Bonai granite. The major deposits are Anjor, Balabhadrapur, Nitigotha, Madrangajodi, Roduan, Rampakot and Dalimpur of Keonjhar District, Jashipur, Kankrani, Gorumahisani, Bangiriposi and Manada of Mayurbhanj District, Lahunipada in Sundargarh District and Jamardungi in Sambalpur District.

The total resource of pyrophyllite in Orissa is 3.787 million tonnes (Source-IBM year book, 2004). Production of pyrophyllite in Orissa during 2003-2004 was an all time high amounting to 84046 tonnes which has decreased to only 8253 tonnes during 2005-2006.

(v) Kyanite, Sillimanite and Andalusite

These are aluminium silicate minerals having chemical composition Al_2O_3 SiO_2, produced from the metamorphism of argillaceous sedimentary rocks. These are mostly used for manufacture of high alumina refractories. These are on heating converted to mullite, which remains unchanged on subsequent heating. Hence, these can be used as high performance refractory for a wide range of industrial applications.

Kyanite is characterized by sky blue colour, pearly lustre, bladed form, two sets of hardness (5 along the blades, 7 across the blades). On heating, its volume increases by 18% (as against 2.3% in sillimanite) for which it is not used as refractory mineral in raw state. Kyanite occurs as disseminated grains in schists, gneisses and sometimes in quartzites as a product of regional metamorphism. Very often, hydrothermal activity and contact metamorphism is responsible for segregation of crystalline kyanite. Residual concentration of kyanite bearing rocks and their subsequent separation may result in the formation of massive bouldery deposits containing interlocking, acicular and radiating crystals (Sahoo, 2006). Kyanite is a product of metamorphism under high pressure at great depths. It is resistant to weathering and hence found as beach placer deposits.

Refractory grade kyanite is confined to Panijia in Mayurbhanj District, where it is associated with dumortierite bearing rocks, quartz veins, quartz mica schist and talc-tremolite schist. Since last 50 years, this deposit is being mined and is almost exhausted. Other occurrences of quartz bearing kyanite are confined to Magarmuhan, Torodanali, Jhilli, Golagadia, Sikheswar and Kodabasant in Paliahara and Kamakhyanagar subdivisions of Angul and Dhenkanal districts. The production in Orissa has decreased from 184 tonnes in 2001-2002 to 14 tonnes in 2003-2004.

Sillimanite crystallizes in orthorhombic system. It occurs as fine needle like crystals arranged in parallel or radiating manner. It is transformed into mullite at a temperature between 1550°C-1650°C without much change in its volume. Thus it can be used as a refractory material without calcinations. It occurs in meta-sedimentary rocks formed by both regional and thermal metamorphism. In Orissa, rocks of Eastern Ghats such as khondalite, leptynite, charnockite etc. form the source rocks of which by weathering release many heavy minerals including sillimanite. These heavy minerals are transported into the sea and later concentrated as beach placer deposits along the coastal areas of Ganjam and Puri districts. Orissa has a total reserve of 7.81 million tonnes sillimanite (SourceIBM Year Book, 2004). The production of sillimanite was 13,207 tones in 2005-2006 from IRE's (Indian Rare Earth) sand separation plant near Chatrapur.

Andalusite is of red/pink/deep brown colour having specific gravity 3.15. It is transformed to mullite at a temperature between 1380°C-1400°C without appreciable increase in volume. It is found in contact metamorphic rocks namely andalusite-hornfelses and in regional metamorphic rocks namely schists and as beach placer deposits (Sahoo, 2006). In Orissa no workable deposits of andalusite are reported so far. However, it forms a fraction of the heavy mineral suite with sillimanite in beach placers of Orissa.

(vi) Soap stone, Steatite and Talc

Soap stone, Steatite and Talc are three terms used in trade for one mineral. These are soft hydrous magnesium silicate. The chemical composition of talc is H_2Mg_3 $(SiO_3)_4$ or 3MgO, 4 SiO_2 H_2O containing 63.5% SiO_2, 31.7% MgO and 4.8% H_2O. Soapstone is a term used for a massive form of rock containing about 50% of the mineral talc, whereas the high purity massive and compact cryptocrystalline variety talc is termed as steatite. These are used in the manufacture of ceramics, as filler in rubber, plastics, pigment of plants, coating of paper, insecticides, cosmetics, as roofing material and as refractories. Powdered pure talc is mainly used for manufacture of talcum powder.

Talc occurs as lenses and pockets in metamorphosed dolomite, crystalline schists and gneisses and as large bodies in ultramafic igneous rocks. It is associated with serpentine, chlorite, asbestos, actinolite, tourmaline, magnesite etc. The best grade talc deposits are found in association with the dolomitic limestone and marble.

These minerals are associated with ultramafic igneous rocks of secondary origin by processes of low grade regional metamorphism. Talc and steatite are formed due to hydrothermal alteration and low grade metamorphism of magnesian rocks particularly non-aluminous magnesian silicates. Serpentine, which is the first product of alteration, is transformed into talc by mild hydrothermal metamorphism accompanied by simple process of dynamic metamorphism. However, relatively pure and white form of talc and steatite are derived from sedimentary magnesian rocks (dolomitic limestone, dolomite etc.) by contact metamorphism and hydrothermal alteration usually associated with igneous activity.

Occurrences

These deposits are confined to Mayurbhanj, Sambalpur, Sundargarh, Ganjam, Koraput and Dhenkanal districts. There are ten mining leases in the state. Mayurbhanj District is contributing about 95% of the total production of the state. Important localities in Mayurbhanj District are Tiring, Kendumundi and Kharidamak; in Sundargarh District are Jharbera and Bhaludungri.

Orissa has a total reserve of 0.896 million tonnes (Source IBM Year Book, 2004). Production of soapstone in the state during 2003-2004 was 4393 tonnes which has increased up to 8,883 tonnes in 2005-2006.

(vii) Quartz and Quartzite

The uses of quartz and quartzite depend upon its silica content. Quartz is used for manufacture of silicon alloys, glass, cement and ceramics. Quartz is a piezoelectric material and is used in radio circuit, radars, ultra-sonic devices and chronometers etc. Quartzite is used in the manufacture of silicon refractories, as a flux in iron and steel and ferroalloy industries, abrasive industry and also for manufacture of silicon alloys, glass and sodium silicate. Because of its high refractoriness, it is used for manufacturing acid silica bricks.

Quartz is found as either quartz crystal or as crystalline to cryptocrystalline quartz or as granular form. It occurs in the form of veins and as a constituent of pegmatites. In the Precambrian terrain, quartz veins are common being formed in the last stages of hydrothermal activity. Moreover, river beds, beach deposits and Aeolian deposits contain glass sands in abundant quantities. Weathering and loosening of cementing material in sandstone and mechanical concentration process are also responsible for accumulation of quartz deposits. Quartzite is a monomineralic rock. It occurs as beds interstratified with other metasedimentaries (Sahoo, 2006).

Occurrences

Generally these are found in several districts of the state (Jharsuguda, Sonepur, Bolangir, Kalahandi, Nuapada, Boudh, Baragarh, Mayurbhanj and Keonjhar) in different stratigraphic horizons except in the coastal planes. The reserve of quartz and silica sands is 2.888 million tonnes. Out of which proved reserve is 1.355 million tonnes. In Orissa, 99 mining leases for quartz and quartzite were granted. Out of which only 53 are in operation. These leases are mostly in Kalahandi, Sundargarh, Jharsuguda, Mayurbhanj, Sambalpur, Keonjhar, Boudh, Jajpur, Koraput, Angul, Sonepur, Baragarh and Malkangiri districts. The reserve of quartzite is 5.748 million tonnes (Source IBM Year Book, 2004).

DIMENSION AND DECORATIVE STONES

Any natural stone irrespective of composition and origin, which can be cut, sized and shaped to suit the fancy and specification of the builders, designers and architects, are marketed as Dimension stones. Different dimension stones quarried in the state include dolerite, gabbro, charnockite, syenite, granulite, leptynite, anorthosite, gneisses, sandstone, laterite and khondalite.

(i) Granite

The suitability of granites for utilization as dimension stones depends to a large extent on their physical, chemical and mineralogical properties. The dimension stones under the "Granite" category can be broadly divided into two types viz., acidic (composed of quartz, orthoclase, perthite, biotite, muscovite and garnet etc) and basic rocks (composed dominantly of clino-pyroxene and plagioclase).

(Ii) Sandstone

Sandstone has been used in construction, rough slabs and tiles etc. The cementing material of these sedimentary rocks determines the degree of compaction, colour and thus ultimately the quality of these stones as dimension stones. The Vindhyan and Athgarh Formations constitute an excellent storehouse of this category because of their regular bedding, uniform grain size and durability. Athgarh Sandstone, as the name indicates, comprises various types of sandstones namely gritty, feldspathic, clayey and ferruginous etc. The Athgarh Formation consists of quartz arenite, sub-lithic arenite, lithic arenite and lithic wackes and those are characteristically lack feldspars. Athgarh sandstones have been utilized in construction of temples around Bhubaneswar and Puri.

(iii) Laterite

Laterite is a typical derivative of *in situ* alteration from variable original parent rocks under tropical climatic conditions. It is a porous, pitted with a hard limonitic protective crust containing a large quantity of iron in form of red and yellow ochre. In Orissa, these are used for building purpose as substitute of bricks in the districts of Cuttack, Nayagarh, Khurda, Puri and Ganjam. Aluminous laterites constitute the overburden of bauxite deposits occurring at high level plateaus (Mohapatra and Sarangi, 2006).

(iv) Khondalite

Khondalites are a typical constituent of the Eastern Ghats. It is quartz feldspathic garnetiferous sillimanite schist/gneiss. These have used for construction of temples and are presently being used in the form of tiles for exterior decoration and also for sculptural purposes.

Occurrences

In Orissa, the Precambrian which constitute about two-third of the landmass of the state host a variety of dimension stones like granite, khondalite etc. In particular, the following geological terrains are important.

Table-5 District wise occurrences of Dimension stones in Orissa (Sarangi, 1998)

District (s)	Dimension stones
Ganjam and Gajapati	Granite gneiss, pyroxene granulite, leptynite and augen gneiss, khondalite
Koraput, Malkangiri Nowrangpur and Rayagada	Granite gneiss, pyroxene granulite, dolerite, charnockite, Khondalite and nepheline syenite
Kalahandi	Granite gneiss and anorthosite
Phulbani and Boudh	Granite, porphyries and gneiss
Bolangir	Granite gneiss, augen gneiss and anorthosite
Sambalpur	Granite, dolerite, basic granulite and nepheline syenite
Dhenkanal and Angul	Granite, dolerite and basic rocks
Keonjhar and Mayurbhanj	Khondalite, Granite, Granite gneiss, dolerite, pyroxene granulite and gabbro
Cuttack, Jajpur	Laterite, Charnockite and sandstone
Balasore	Gabbroic rocks
Puri, Nayagarh and Khurda	Laterite, Sandstone, granite gneiss, anorthosite and charnockite
Baragad	Sandstone

A. ChhotnagpurSinghbhum-Bonai: Granites and gneisses of many types, dolerite, anorthosite and gabbro etc.

B. Eastern Ghats: Charnockite, khondalite, leptynite, anorthosite, pyroxene granulite, nepheline and syenite etc. (Sarangi, 1998).

District wise occurrences of different dimension stones are presented in the Table 5.

(v) Limestone and dolomite

Limestone is a calcareous sedimentary rock composed of mineral calcite ($CaCO_3$), which on calcination yields lime (CaO) for a wide range of industrial uses. In its purest form, limestone should contain 56% CaO and 44% CO_2. However, dolomite is a double carbonate of calcium and magnesium containing 30.4% CaO, 21.7% MgO and 47.9% CO_2 in its purest form. Both limestone and dolomite also contain silica, alumina, iron oxides, alkalis, phosphorous, sulphur etc.

Occurrences

Orissa has vast resources of limestone occurring in three distinct geological settings, namely Gangpur Group, Vindhyan Supergroup and Eastern Ghats Group, whereas dolomite is mostly confined to Gangpur Group and Vindhyan Supergroup. These deposits are confined to Sundargarh, Bargad, Koraput, Malkangiri, Nowrangpur, Nuapada and Bolangir districts. Reserves and grades of major deposits are presented in the Table 6 (Rath and Rath, 2006).

HIGH MAGNESIA ROCKS

Dunite, peridotite, serpentinite, pyroxenite and saxonite etc. are common high magnesia rocks. The advantages of these high magnesia rocks over dolomite and limestone in steel making process are (i) smooth operation of blast furnace due to low alkali content, (ii) less consumption of coke, (iii) no additional use of silica, (iv) better slag fluidity and (v) overall 4 to 5% increase in the blast furnace productivity. These high magnesia rocks are mostly found in association with rocks of Iron Ore Group. Dunite, peridotite, serpentinite, saxonite and pyroxenite associated with chromite deposits of Sukinda area of Jajpur district and Boula-Nausahi area of Keonjhar district are high magnesia rocks having the requisite chemical specifications to be used as flux in steel making process. Other areas of occurrence of high magnesia rocks in the Singhbhum craton are (a) Managovindpur of Jajpur District, (b) Bhuban-Asurbandh-Maulabhanj of Dhenkanal District, (c) Bhalukasoni near Nilgiri of Balasore District, (d) Notopahar, Thakurmunda, Amjori, Badampahar of Mayurbhanj District, (e) Rajabasa, Jharbeda, Bonai, Darjing of Sundargarh District, (f) Kaliahata of Keonjhar District and (g) Thelkobadi of Kuchinda subdivision of Sambalpur District. In the Bastar craton these rock are found at Ramgiri of Koraput District and Pagarpani of Bargarh District. Similarly in the Eastern Ghats granulite terrain, high magnesia rocks occur at Kiakata of Boudh District, Ghusramal near Rairakhol in Sambalpur District and Mukundapur of Rayagada District. The production of these rocks has been steadily increasing from 1131 tonnes in 1999-2000 to 2, 09,224 tonnes in 2004-2005 (Sahoo, 2006).

CONCLUSION

For countries with high growth rates in both population and per-capita incomes, such as in India, large increase in their industrial mineral capacities will be necessary to maintain economic growth. Industrial minerals play an important role in the development of the economy of the country and therefore warrant attention of geoscientists, administrators and policy makers.

Traditionally in the case of Orissa emphasis has been placed on the metalliferous and energy minerals, principally as export commodities; but it is now increasingly recognised that industrial minerals can also make a significant contribution to state economic development and are essential for economic growth. Increasing industrialization, improvements in infrastructure, housing and general living standards are creating a demand for a wide range of industrial minerals. Unlike metals and energy minerals, they are mostly not exported but used locally. They therefore constitute local value added products, which enhance local development. Industrial minerals are the single largest input required in the construction of shelter, a basic need for more than millions of people who currently lack adequate housing and of infrastructure and facilities in a poor state like Orissa. This should be realized and accordingly steps should be taken to explore industrial minerals.

Geological Survey of India (GSI), Directorate of Mining and Geology (DMG) and Orissa Mining Corporation (OMC), Universities and RRL should gather information on the industrial mineral potential of our state. Regional Universities must carry out technical studies on the geology, mining and

Table-6 Reserves and grades of major deposits of Limestone and Dolomite in Orissa (Rath and Rath, 2006).

		Reserve	Grade (%)		
		(Million tonnes)	CaO	MgO	Ti
1	Sundargarh District				
a.	Biramitrapur-Raiboga	400 (Limestone) 324 (Dolomite)	46.5 29.0	3.5 20.0	6-10 5-9
b.	Hatibari-Purnapani-Gatitangar-Purnapani	29 (Limestone)	46	3-5	10
c.	Lanjiberna	42 (Limestone)	45-50	1.5-3.4	6-12
d.	Gomardih	30 (Dolomite)	28.3	20.5	6
e.	Khatkubahal-Kiringsera-Kandeimunada-Litibera-Kutra-Teleghana	20 (Limestone) 22 (Dolomite)	45.5	2.5	8-12
f.	Sapai river section	5.4 (Dolomite)			
g.	Lephripara	1.7 (Dolomite)			
h.	Dublabera	3 (Dolomite)			
The alkali content of most of the limestone and dolomite deposits in Birmitrapur belt is high (>3%). Higher alkali adversely affects the linings of the furnace and consequently the productivity.					
2	Umpavalli-Tummiguda-Gondivalsa deposits, Sunki valley, Koraput District	73 (limestone)	45.5-51	0.5-3	2-12
3	Kottameta-Nandivada-Dharampalli-Uskalvagu deposits, Malkangiri District	240 (Limestone) mostly cement and BF grade	46-50	<2	6-14
4	Dumajodi-Binsuli-Parsaguda deposits, Nowrangpur District	14 (Dolomitic limestone)			
5	Nuapada District				
a.	Chandapala	2.26 (Limestone)	47.8	0.5	6-12
b.	Deobahal	1.70 (Dolomite)	48	2.0	6-12
c.	Rohenpadar	3.36 (Limestone)	48	1.8-3.5	6-10
d.	Sagundungri	1.52 (Limestone)	48.5	1.0	8-10

Table-5 (Contd.)

e.	Ghasipara, Chalanpara, San Maheswar, Khariar deposits	(Limestone)	Not estimated		
f.	Bedded deposits of Purana Group of Chhatisgarh Basin (Jogbhata, Kundanjhariya, Gendhulpani, Kharupani, Porapadar, Jholapara, Sindursil)	(Limestone)	All these occurrences are of low CaO and high silica.		
6	Bargarh District				
a.	Dungri-Banjipali-Jampali	56 (Limestone)	45-48.8	0.9-1.2	10-13
b.	Putka-Saramsil	2 (Limestone) 22 (Dolomite)	28	19-20	7-9
7	Dhamandanga, Kuliadora, Karlabahali and Hial, Bolangir District	2.0 (Limestone)	42-53	<2	1.5-12

processing of industrial minerals. Laboratory and pilot plant tests should be conducted with the aim of evaluating the benefits and the feasibility of using industrial minerals.

Government of Orissa should establish an information network and systems aimed at obtaining and disseminating the data on the non-metallic mining resources. Government should draw up agreements for technical and financial cooperation with national and foreign institutions for carrying out these activities. There should be a national industrial minerals workshop meeting in each year to gain instant information, enable producers and consumers to meet each other and advise government of problems.

State Government should also establish an external consultant cell to link together the findings of resource inventory and market study and to propose a development strategy and identify investment opportunities.

ACKNOWLEDGEMENT

We are thankful to Prof. A.K. Paul and Prof. P.P. Singh, P.G. Department of Geology, Utkal University, Vani Vihar, Bhubaneswar for critically going through the manuscript.

References

Acharya, B.C. and Dash, B. (1984). Graphite in Eastern Ghats Precambrian migmatites, Orissa, India. Trans. R. Soc. Edinburgh: Earth Sci., v. 75, pp. 391-406.

Acharya, B.C. (2006). Graphite. In: Mahalik NK, Sahoo HK, Hota RN, Mishra, BP, Nanada JK and Panigrahi AB (Eds)- Geology and Mineral Resources of Orissa. Society of Geoscientists and Allied Technologist, Bhubaneswar, pp. 277-292.

Bristow, C. M. (1987). Society's changing requirements for primary raw materials. Industrial Minerals, v. 233, pp. 59-65.

Dunhamn, A.C. and Scott, P.W. (1994). Appropriate techniques for the evaluation of industrial minerals in developing countries. In: Mathers SJ and Notholt AJG (Eds)- Industrial minerals in Developing countries. British Geological Survey and Associations for Geoscientists for International Development. AGID Report Series Geosciences in International Development, v. 18, pp. 31-38.

Goswami, S., Singh, K.J. and Chandra, S. (2006). Palaeobotany of Gondwana basins of Orissa State, India: A bird's eye view. Journal of Asian Earth Sciences, v. 28 (4-6), pp. 218-233.

Highley, D.E. (1994). The role of industrial minerals in the economics of developing countries. In: Mathers SJ and Notholt AJG (Eds)- Industrial minerals in Developing countries. British Geological Survey and Associations for Geoscientists for International Development. AGID Report Series Geosciences in International Development, v. 18, pp. 1-12.

Iyengar S.V.P. and Banerjee, P.K. (1971). The Iron Ore, Bengpal Dharwar Group. Rec Geol. Surv. Ind, v. 101 (2), pp. 43- 59.

Mahalik, N.K., (1995a). Precambrians. In: Mohanty BK (Ed)- Geology and Mineral Resources of Orissa. Society of Geoscientists and Allied Technologist, Bhubaneswar, pp. 12-41.

Mahalik, N.K. (1995b). Structure and Tectonics. In: Mohanty BK (Ed)- Geology and Mineral Resources of Orissa. Society of Geoscientists and Allied Technologist, Bhubaneswar, pp. 8-11.

Mohanty, B.K. (1998). China Clay. In, Mohanty BK (Ed)- Geology and Mineral Resources of Orissa. Society of Geoscientists and Allied Technologist, Bhubaneswar, pp. 148-154.

Mohapatra, S.K. and Sarangi, S.K. (2006). Dimension and Decorative Stones. In: Mahalik NK, Sahoo HK, Hota RN, Mishra, BP, Nanada JK and Panigrahi AB (Eds)- Geology and Mineral Resources of Orissa. Society of Geoscientists and Allied Technologist, Bhubaneswar, pp. 227-240.

Nanda, A.K. and Bhol, B.B. (2006). Fireclay. In: Mahalik NK, Sahoo HK, Hota RN, Mishra, BP, Nanada JK and Panigrahi AB (Eds)- Geology and Mineral Resources of Orissa. Society of Geoscientists and Allied Technologist, Bhubaneswar, pp. 241-245.

Naqvi, S.M. and Rogers, J.W. (1987). Precambrian Geology of India. Oxford Monographs on Geology and Geophysics. Oxford University Press Inc., pp. 233.

Noetstaller, R. (1988). Industrial minerals: a technical review. Technical Paper No. 76. Washington, DC: The World Bank, pp.117.

Pandya, K.L. (1995). Gondwanas. In: Mohanty BK (Ed)- Geology and Mineral Resources of Orissa. Society of Geoscientists and Allied Technologist, Bhubaneswar, pp. 42-49.

Pradhan, P.K. (1995). Tertiary and Quaternary. In: Mohanty BK (Ed)- Geology and Mineral Resources of Orissa. Society of Geoscientists and Allied Technologist, Bhubaneswar, pp. 50-59.

Rath, B.D. and Rath, R.N. (2006). Limestone and Dolomite. In: Mahalik NK, Sahoo HK, Hota RN, Mishra, BP, Nanada JK and Panigrahi AB (Eds)- Geology and Mineral Resources of Orissa. Society of Geoscientists and Allied Technologist, Bhubaneswar, pp. 318-333.

Saha, A.K. and Basak, A (1965). On the occurrence of pyrophyllite associated with the Singhbhum Granite around Keonjhargarh, Orissa. Jour. Ind. Geosci. Assoc., v. 5, pp. 9-24.

Sahoo, H.K (1991). Geochemical and genetic aspects of pyrophyllite deposits of Keonjhargarh, Orissa. Sambalpur Univ. Jour. Sc. Tech., vol. X, pp.51-57.

Sahoo, H.K (2006). Other Minerals. In: Mahalik NK, Sahoo HK, Hota RN, Mishra, BP, Nanada JK and Panigrahi AB (Eds)-Geology and Mineral Resources of Orissa. Society of Geoscientists and Allied Technologist, Bhubaneswar, pp. 404-414.

Salop, K.J. (1972). A unified stratigraphic scale of the Precambrian. Proc. Int. Geol. Congr. 24 Sess., Montreal 1972. Precambrian Geology, pp. 253-259.

Salotti, C.A. Heinrich, E. W. and Giardine, A.A. (1971). Abiotic carbon and the formation of graphite deposits. Econ. Geol., v.66, pp.929-932.

Sarangi, S.K. (1998). Dimension and Decorative Stones. In: Mohanty BK (Ed)- Geology and Mineral Resources of Orissa. Society of Geoscientists and Allied Technologist, Bhubaneswar, pp. 204-213.

Hon'ble Reviewers

1. Prof. T.K. Biswal
   Department of Earth Sciences
   Indian Institute of Technology Bombay
   Powai
   MUMBAI 400 076 INDIA
2. Dr. Pankaj K. Srivastava
   Economic Geology Research Group
   Department of Geology
   University of Jammu
   JAMMU 180 006 INDIA

Editorial Assistance

Dr. Narendra Singh

Section IV
Prospecting and Exploration

Preamble

Prospecting and exploration of economic mineralization is a vast intra-disciplinary field as there is no single geotechnique that is a panacea for mineral exploration. Reconnaissance to small-scale geological mapping, identification of rock units, structural and stratigraphic relationship constituting primary exploration lead to recommendation as to which tool should be used. The exploration geologist should be knowledgeable about applications, limitations and ramifications of each exploration technique starting from remote sensing, geophysics, geochemistry and laboratory investigations. Therefore, specialization is indispensable.

The first paper of the section is Information theoretic applications in mining is authored by Prof. B.K. Sahu. In fact, Information theory is an independent interdisciplinary subject dealing with random variables (r.v.) and draws from communication (encoding), computer (kolmogorov complexity), gambling (wealth information), statistics (estimation, hypothesis testing), probability (limit and large deviation risk). Entropy (uncertainty of r.v.), relative entropy (kullback-leibler, K-L, distance of two pdfs) and mutual information (doubling rate of wealth) are very useful measures of I.T. The paper has given the most useful information theoretic measures and results that can be adopted to mining industry, omitting more technical measures related to coding, computer complexity and software algorithm which are already available in computer and software systems.

Prof. P.O. Alexander and Dr. D.R. Patel have presented the second paper of the section entitled "Application of Termitaria in Pb-Zn prospecting - a case study from Chauraiya, Dahoh district, Madhya Pradesh". It is determined that the highest content of Pb and Zn in termite mounds overlying buried mineralization is 560 and 389 ppm, respectively; a very favourable contrast of 14 and 6 times from those located in the non mineralized zone. The author emphasis that systematic application of Termitaria studies in mineral prospecting can be very useful, particularly in the termite-infested area.

An excellent review on biological methods of prospecting has been presented by Dr. S.K. Trivedi and Mr. K.K. Solanki with title of their paper "Present status and future trends of biological methods of prospecting for economic mineral deposits". The paper has discussed both the botanical prospecting methods, first, the geobotanical prospecting, to use plant involving visual observation, and second, the biogeochemical prospecting to analyze various parts of plants chemically, for prospecting. Limitations and advantages of both the methods have been discussed by citing some Indian case histories. Geozoological prospecting includes use of mammals (specially, dogs), birds, fishes, insects (specially termite and bees) and micro-organisms (specially, bacteria) for mineral prospecting. A progressive future has been predicted by the authors for the biological methods of prospecting.

The next paper entitled 'An integrated approach for kimberlite exploration in Raipur dist, Chhattisgarh' has been presented by Dr. M.S. Khatediya and Dr. Pramod K. Verma. The paper presents an interdisciplinary approach for kimberlite exploration in the area using remote sensing and field checks apart from establishing lithological, tectonic, geochronological, geophysical and geochemical guides to develop a kimberlite exploration geographical information system (KEGIS). The study has identified and demarcated possible potential target areas.

The rock mass classification which is a backbone for the empirical designing in various engineering projects is not so successfully applicable for excavability, borability and chiefly for the mining of the economic mineral deposits. Dr. Rajiv Badal, while realizing the fact, has proposed a new classification in his paper entitled "Rock mass classification system to rock excavation/cutting/boring application". The proposed classification serves the purpose in its versatility. The paper also outlines earlier rock mass classification systems serving as a guide to arrive at the new classification.

The paper "^{57}Fe Mossbauer spectroscopic study of petroleum prospecting test well rocks from the three basins of India" has been authored by Prof. R.P.Tripathi, Dr. Abhijit Kulshreshtha, Dr. Amita Tripathi and Prof. K.L.Shrivastava. The research group is engaged for the last two decades in search of a Mossbauer log for petroleum exploration. In the present paper ^{57}Fe Mossbauer spectroscopy has been used to characterize the sediment for iron minerals present in petroleum prospecting wells from Jaisalmer, Eastern Krishna Godavari and Cambay basins. The Mossbauer spectrum and parameter obtained are of use to obtain information about palaeoredox environments of the host sediments which will ultimately lead to interpretion of the hydrocarbon prospects,

Ms Mamta Chauhan, Dr Jaimini Swadia, Ms Tina Prajapat and Prof. K.L.Shrivastava have authored the next paper entitled "Petroleum Geology and Hydrocarbon Prospects in the Barmer-Sanchor basin, Northwest India". The study area is a very important potential area for the future and has been discovered very recently.

The first part of the paper provides general geology including tectonics, sedimentation, lithostratigraphy and petroleum geology of the basin. The second provides details of all the 10 fields and their individual prospects. The importance of the region can be appreciated as the Cairn Energy has estimated more than 250 billion bbls of total oil in the region.

In the last paper of the section, Dr. S.K.Haldar has traced the record of exploration and development of mineral deposits, specially base metals, for the first fifty years in India and found it not very attractive. However, he has suggested the application of more modern techniques with better determination for the future exploration to handle more difficult cases, in the growing Indian economy and challenging globalization. He feels the need to inspire and guide the young Indian exploration geologists by senior professionals from industry and academia so as to emerge victorious in this new environment of increasing discovery risks that requires extraordinary creativity, innovation and technical skill coupled with commercial discipline. The incentives, he has suggested, are world economic demand, driving continuously high trends in metal prices, technological necessity and the primeval joy of unveiling the opening chapters of creation.

Economic Mineralization
K.L. Shrivastava (Editor)
2009. pp.385-389
Scientific Publishers (India), Jodhpur

Information Theoretic Applications in Mining

B.K.SAHU
Professor Emeritus, Department of Earth Sciences, I.I.T. Bombay, Powai, MUMBAI 400076
Email : bksahu@iitb.ac.in

Abstract

Mining industry involves exploration, development, production, processing and marketing of mineral resources where the results of information theory (IT) and computations are directly applicable. Information theory is an independent interdisciplinary subject dealing with random variables (r.v.) and draws from communication (encoding), computer (Kolmogorov complexity), gambling (wealth formation), statistics (estimation, hypothesis testing), probability (limit, large deviation, risk) etc. Besides the classical results from communication (Shannon, 1948), IT includes extraction of signal and of relevant information from it as well as actions taken thereof. Most useful measures are Entropy (uncertainty of r.v.), Relative Entropy (Kullback-Leibler, K-L, distance) and Mutual Information (doubling rate of wealth).

Main applications of IT in mining may be listed as follows (i) data compression, channel capacity, (ii)optimization of computational times and program length (Kolmogorov complexity), (iii) doubling rate of optimal investments, (iv) parameter estimation, hypothesis test, risk analysis. Signals optimally extracted from noisy data and optimal decisions (actions) then taken using integrated neuro-fuzzy (Soft) computing/array processing. Optimal systems for information processing are either systems with distinct outputs from a single input or several separate systems yielding distinct output from each one with a single input.

Key words : Integrated neurofuzzy. Mining data variables, Information theory.

INTRODUCTION

Mining data comprise quantitative and qualitative random variables which can be processed by digital computers using suitable soft-computing techniques (see, Sahu 2003a, 2006, 2007). Entropy, H(X) or H(p), relative entropy, D(p||q), and mutual information I(X,Y) are the basic tools required for analyses of random variables in information theory (IT). Signal is considered to be a slowly varying trend in data which are corrupted by noise that is rapidly varying and orthogonal to the signal. Entropy H(X) assesses the uncertainty in the signal (r.v.) which may or may not contain relevant information (Sinanovic and Johnson, 2006). Rate distortion theory (Berger, 1971) requires joint probability distribution of input and output that is characterized by Kullback-Leibler distance measure. This distance is useful optimal classification (or estimation) of statistical parameters (Kullback,1959).Here, we give the most useful information theoretic measures and results (Cover and Thomas, 1999) that can be adopted to mining, omitting more technical measures related to coding, computer complexity, and software algorithm, etc. which are already available in the computer and software systems.

IT MEASURES RELATED TO STATISTICS (STATIC)

Basic concepts of probability and statistics may be obtained from the book (Sahu, 2005). Entropy measures the uncertainty in the r.v.(X,Y,Z) and defined for a discrete r.v. X as:

$$H(X)=H(p)= - \sum p(x) \log p(x), \text{OR } H(X)= -E \log p(X), \quad (1)$$

where the logarithm is to the base 2 (or bits).

Joint entropy H(X,Y) is defined as

$$H(X,Y) = - \sum p(x,y) \log p(x,y) = - E \log p(X,Y). \quad (2)$$

Using results from conditional probability,we get

$$\text{Log}\, p(X,Y)= \log p(X)+ \log p(Y|X) \quad (3)$$

$$\text{OR}, H(X)-H(X|Y) = H(Y)-H(Y|Z) \quad (4)$$

Relative entropy(K-L distance) between p(x) and q(x) is denoted by D(p||q)

$$D(p||q)= \quad p(x) \log (p(x)/q(x)) = E \log (p(x)/q(x)) \quad (5)$$

This is non-negative and becomes zero when p=q, but is asymmetric.

Mutual information is the reduction of uncertainty due to knowledge of another r.v.It is asymmetric and is given by

$$I(X,Y) = p(x,y) \log (p(x,y)/p(x) p(y)) = E \log (p(X,Y) / p(X)p(Y)) \quad (6)$$

$$\text{Hence}, I(X,Y)=H(X) \; H(X|Y)=H(Y)–H(Y|X). \quad (7)$$

$$I(X,X)=H(X)-H(X|X)=H(X). \quad (8)$$

This shows that entropy is self-information. Chain rules for entropy,mutual information and relative entropy are below:

$$H(X(1),\ldots,X(n)) = H(X(i)| X(I-1),\ldots, X(1)) \quad (9)$$

$$I(X(1),\ldots,X(n)) = I(X(i), Y| X(I-1),\ldots,X(1)) \quad (10)$$

$$D(p(x,y) || q(x,y))=D(p(x)||q(x)) + D(p(y|x) || q(y|x)) \quad (11)$$

H(X) is x function whereas D(p||q) and I(X<Y) are concave, where f is convex if we define f as less than equal to sum of $\lambda f(x(1)) + (1-\lambda) f(x(2))$; is between 0 and 1. I(X,Y) is zero if X,Y are independent r.v.`s.Convex (or Concave i.e. negative of Convex)functions are required for any optimization process. For continuous random variables X,Y, we extend these concepts as differential entropy h(f) ,which may become negative, however. Other formulas are similar to discrete cases, where summations are replaced by intergration for the continuous variables,hence omitted.

Data Processing Inequality: If $X \rightarrow Y \rightarrow Z$ forms a Markov Chain(M.C.), then $I(X,Y) \geq I(X,Z)$. This shows that clever manipulation of data cannot improve its information content. For M.C., the relative entropy D(u(n)||u`(n)) always decreases with time.Entropy H(X(n)) increases if the stationary distribution is uniform.H(x(N)|X(1)) increases and H(X(0)|X(n))also increases. These facts are very useful to prove the Second Law of Thermodynamics, that entropy always increase unless the system is reversible.

Sufficient statistics: T(X) is sufficient relative to (f (x)) if $I(\Theta,X)=I(\Theta,T(X))$ (12)

for all distributions on Θ. T(X) is minimal sufficient statistics U(X) if $\Theta \rightarrow T(X) \rightarrow U(X) \rightarrow \geq X$.

Fano Inequality: We wish to estimate r.v. X with a distribution p(x) and r.v. Y related to X by the conditional distribution p(y|x). From Y we calculate a function g(Y)= X hat. We then see that $X \rightarrow Y \rightarrow$ X hat forms a Markov Chain, M.C. with probability of error Pe = Pr(X hat $\neq$ X).Then,

$$H(Pe)+Pe \log (|\chi| -1 \geq H(X|Y). \quad (13)$$

This relation can be weakened to H(Pe) =1 and hence Pe is $\geq$(H(X|Y))-1)/(log |χ|).
Note that Pe= 0 implies H(X|X) = 0. (14)

According to Weak Law of Large Numbers in probability theory/statistics (equivalent to Asymptotic Equipartition Property, AEP, in information theory), we obtain following useful results: 1/n [log{1/p(X(1),...,X(n))}] tends to H. (15)

Thus, for large n, a typical sequence has probability 1 and its elements are equiprobable, the the number of elements are 2^{nHx}. This shows that a typical sequence can be compressed to $2^{nH(X)}$ bits which is used for transmitting messages/pictures.

IT MEASURES RELATED TO STOCHASTIC PROCESSES (DYNAMIC)

For details on time series and stochastic processes, the readers are referred to the books(Sahu,2003b and Ross,2003).In context of mining, the random variables are often dependent on previous state(s) and hence not necessarily independent as is the case of Statistics (see Section II). We can use the concepts of Markov Chain (M.C.) that are stationary(time-invariant and linear and conditionally independent of all previous states except the last or present state).Such Markov chains are characterized by their transition probability matrix (p(i,j)) from state i to state j and their initial state probability. The chain is irreducible if we can reach any state from any other state with positive probability in a finite number of steps. If the probability distribution of states do not change from step n to step (n+1), then the chain is stationary and initial distribution is equal to the stationary distribution. Stationary distribution is easily obtained by powering the transition matrix to large n (M.C. is assumed aperiodic). Entropy rate of a stochastic process is given by

$H(\chi) = n^{\lim} \to\infty$ (H(X(1),...,X(n))/n). (16)

For example, a typewriter is an iid machine and can produce m to the power n sequences of length n, all of which are equally likely. Hence H(X(1)... X(n))= n log (m) and the entropy rate $H(\chi)$=log(m) bits per symbol. Similarly, for a non-iid but independent sequence, we get

H(X)= Sum of H(X(i)) for all i. (17)

If pi is a function of i and not a constant, then H(Xi) oscillates between 0 and 1 and cannot converge. But conditional entropy of the last r.v.given the past is denoted as :

$H`(\chi) = n^{\lim} \to \chi$ H(Xn | Xn-1,...,X1),when exists. (18)

For stationary stochastic processes both H(X) and H`(X) exist and are equal. So, entropy rate is the average ergodic(time parameters= ensemble parameters) processes. For a stationary M.C. the entropy rate (long term behavior) is given by

$H(\chi)$ = H`(χ) = lim H(Xn|Xn-1,...,X1) = lim(Xn|Xn-1) = ...= lim(X2|X1) (19)

Thus if the stationary probability is u and transition matrix is P = H(X2|X1), then

$H(\chi) = -\sum_{i,j} (u_i P_{ij} \log P_{ij})$. (20)

However, we may only have parital information on the process and Y= φf(Xi) may not be a M.C. and its entropy is not well-defined even if entropy of Xi is well- defined. But even then we can compute H(Yn| Yn-1,...,Y1) for each n and find the limit. It is useful to compute the upper and lower bounds converging to its limit from above or below but we can stop computing whenever the difference between these two limits become small(optimal actions based on Markov decision theory).

Information theory can also be used in gambling process(horse race or stock investment systems) in order to build up profits and capital needed for mining industry. In bidding for lease to explore(and mine) a target area, industry loses bid amount if the bid is lower and wins the asset for future capital formation. The net profit on each bid is, say, bi $\geq$ 0 & $\sum b_i$=1. Let Sn be the wealth gained after n such biddings. Then, Sn = $\prod_{i=1}^{n}$S(Xi), (21)

where S(X) = b(X) o(X) is the factor of wealth multiplication on a successful bid.Doubling Rate is W(b,p) = E (log S(X)) = $\sum$(k=1 to m) pk log(bk ok). Then,

Sn = $2^{n W(b,p)}$,using Weak Law of LN.We should maximize W(P) over all choices of b and get

^{Max}b W(b,p) = max ∑Sum(k = i to m) (pk log(bk ok)) (22)

Using Lagrange multiplier (lamda); to maximize wrt bk gives bk = pk for lamda = –1. Thus, the optimal solution **b*** = **p** (proportional gambling scheme) yields W*(p)= ∑(pk log(ok)) H(p). (23)

Mining industry should bid for a block in proportion to the probability of its containing mineral deposit(s) and this probability is estimated by geological assessment of the area. Side information such as reports on previous exploration efforts on the area also influence the bidding process. Increase in doubling rate is given by

ΔW = W*(X|Y) W*(X) = H(X) H(X|Y) = I(X,Y). (24)

A good mining industry must estimate the probability of success pk in K^{th} block very well and hence must compress all data well to get this estimate. We can also use the large deviation theory of statistics and probability, to calculate the risks involved in investing the amount in a mining venture. Currently, wavelet theory with multi resolution analysis(see,Mallat,1999; Sahu,2003b) are used for these purposes under local stationarity and with simultaneous resolution in time-frequency plane.

APPLICATIONS TO THE MINING INDUSTRY

Mining industry has a time-horizon of 20 years or more for a target block and has to go through three stages : (i) Initial(say 3 yrs); (ii)Development(say another 3yrs) and (iii) Production (say 15 yrs).During the initial stage, industry bids for a lease block and if successful goes to explore and mine possible ores therein. During the development stage, industry requires more capital for mining and marketing of ores. In the production stage, it should develop optimal mining and marketing strategies confirming to the existing mining laws. In all these stages, digital processing of various kinds of data and information in order to extract relevant signals and for optimal decisions (actions) are essential. For the success of mining venture, it is apparent that the most efficient, economical, and latest time-saving technologies must be procured. I am, here suggesting a tentative list of computational and information theoretic concepts that could be useful.

(1) Initial Stage

(i) Floating a private or public company with sufficient capital to sustain 3 years and recruitment of technical manpower required at that stage.

(ii) Selection of possible targets among the possible blocks available for bidding and deciding the optimal (maximum) bid amount for each block. This should be based on probability calculations of mineral occurrence using remote-sensing, geology, geophysics and limited drilling of the target and by obtaining all reports on previous mining efforts on the target area. Hence $p_{int} = 1 - (p^c_{rs}\ p^c_{gl}\ p^c_{gp})$.

(iii) Selection of optimal drilling sites to prove quality and 3-D extensions of ore. We should also decide on the optimal mining method to be adopted,; and mineral processing, if any, PROVIDED it is an economically viable prospect(with 20 year horizon).

(2) Development Stage

(i) Recruitment of optimum number of technical personnel for mining, processing and marketing the ores. Contracting labor for various shifts of mine.

(ii) Detailed geological, geophysical, drilling as required for optimal estimation and subsequent mining of the ores. Here soft-computing tools would be most useful (see Sahu, 1999, 2003a,b, 2006).

(iii) Develop optimal transportation, mineral processing and marketing strategies.

(iv) Selection of new targets for bidding and of applying for extensions to the limits of present mine.

(3) Production Stage

(viii) Data updating and taking decisions on optimal mining, mineral processing, and marketing of ores. (see Sahu, 2003a,b).

(ix) Optimal investment of surplus funds in stocks and other financial instruments (wealth can be build by use of doubling rate obtained from Information Theory).

(x) Selection of new targets for bidding using information theoretic results.

(xi) Restoration of the Environment in mining area as required by law.

CONCLUSIONS

(a) Results of information theory integrated with those of digital signal processing are useful for data/image compression or channel capacity (data transmission), optimization of functions using soft-computing methodologies, and for doubling rate of wealth, as well as for hypotheses tests/risk analyses.

(b) Optimal systems for information processing are either (i) several distinct outputs from single input or (ii) several systems acting separately to yield distinct outputs from single inputs. Integrated neuro-fuzzy computing with global optimization by genetic algorithm (soft-computing) seems to be the most suitable for processing the qualitative (and quantitative) mining data available at different times during the lifetime of the mine. The results of this soft-computing would be most useful to take on-line as well as off-line decisions as required at the given time for the mining industry.

References

Berger, T.,1971, Rate Distortion Theory. Prentice-Hall, Englewood Cliff, N.J.

Cover, T.M. and Thomas,J.A.,1999,Elements of Information theory. Wiley, New York Kullback, S., 1959, Information theory and statistics, Wiley, New York.

Mallat, S.,1999, A wavelet tour of Signal Processing (2nd Edn) Academic Press (Elsevier).

Ross, S.M., 2003, Introduction to Probability Models (8th edn) Elsevier (Acad. Pr),735 p.

Sahu, B.K.,1999, Philosophy and Strategies for optimal exploration of mineral deposits. Bull SGAT, Bhubaneshwar, 1, pp. 2-14.

Sahu, B.K., 2003a, Methodologies of reservoir modeling: A comparison. Bull.SGAT, Bhubaneshwar, 4, pp. 1-12.

Sahu, B.K., 2003b, Time Series Modelling in Earth Sciences, Balkema, Netherlands, 283 p.

Sahu, B.K., 2005, Statistical Models in Earth Sciences, BS Publications, Hyderabad, 211 p.

Sahu, B.K., 2006, Groundwater modelling in hard rock terrain. In; N.C. Ghosh and K.D. Sharma (eds). Groundwater Modelling and Management. pp. 368-380. Capital Publ. Co., New Delhi, 594 p.

Sahu, B.K., 2007, Evaluation of Water pollution risks from single and multiple contaminants/ pollutants using statistical methods. In: P.C. Avadich and Bhu, H (eds), Emerging Trends of Research in Geology With Special Ref. To N.W. India. pp. 200-205.

Shannon, C.E., 1948, A mathematical theory of communication. Bell. Sys. Tech. Jour., Vol. 27, pp. 379-423 and 623-656.

Sinanovic, S., and Johnson, D.H., 2006, Towards theory of information processing: Signal Processing (2006), doi.10: 1016, Elsevier Journal on-line.

Hon'ble Reviewers

1. Prof. R. Bhadada
(Professor of Mining Engineering)
(Former, Dean)
D74, Hauz Khas
NEW DELHI 110016 INDIA

2. Prof. A.K. Sharma
(Prof. of Computer Engg.)
Head
Department of Electrical Engineering
MPA Technical University
UDAIPUR 313 001 INDIA

Editorial Assistance

Ms. Tina Prajapat

Economic Mineralization
K.L. Shrivastava (Editor)
2009. pp. 390-395
Scientific Publishers (India), Jodhpur

Application of Termitaria in Pb-Zn Prospecting : A Case Study from Chauraiya, Damoh District, Madhya Pradesh

P.O. ALEXANDER[1] AND D.R. PATEL[2]

[1]Department of Applied Geology, Dr. H.S. Gour Vishwavidyalaya, Sagar 472003
[2]Directorate of Geology and Mining, Chhattisgarh, Raipur, 492 006

Abstract

Multidisciplinary studies at Chauraiya Pb-Zn deposit, Damoh, Madhya Pradesh have shown termiraria studies as an additional prospecting tool with promise. The highest content of Pb and Zn in the termite mounds overlying buried mineralization is 560 and 389 ppm respectively a very favourable contrast of 14 and 6 times from those located outside the mineralized zone. On an average, this figure, however, is lower, 5 and 2 times only. Under the binocular microscope and hand lens (10 x) a few grains of galena can also be seen in the mound material overlying the buried mineralization. It is emphasized that systematic application of termitaria studies in mineral prospecting can be very usefull both for metallic as well as non metallic ores particularly in the termite-infested areas.

Keywords **:** Termitaria, Pb-Zn Prospecting, Chauraiya. Geochemical Exploration, Biogeochemical Exploration.

INTRODUCTION

Herodotus reported in 450 B.C. that certain tribes in India studied termite mounds to detect gold mineralisation. In recent times the use of termiteria has yielded positive results in prospecting for diamonds, gold, and copper deposits. In India termite mound studies from mineralized zones of chromium, vanadium and copper have shown promising results. Inspite of this the use of termitarium in mineral prospecting has been a neglected field. This aims study to bring out the inherent potential of termitaria studies around a small Pb Zn deposit at Chauraiya (N 79°34': E 24°25') in Damoh district of Madhya Pradesh (Fig 1).

The present work is a part of our detailed multidisciplinary investigation on this Pb-Zn mineralisation carried since 1987, in Chert breccia horizon of Palkawan Shale Formation of Semri Group of Vindhyan Supergroup (Fig 2) Except the exploratory pits by the Directorate of Geology and Mining, Madhya Pradesh and some minor mining activity by the villagers, the area has not witnessed any mechanized mining.

TERMITARIUM STUDIES

Twenty three termitaria sites in the area including eight from within the host rock overlying the buried mineralization, three not necessary overlie the buried mineralization and the remaining (Twelve) from outside the host rock were studied and composite

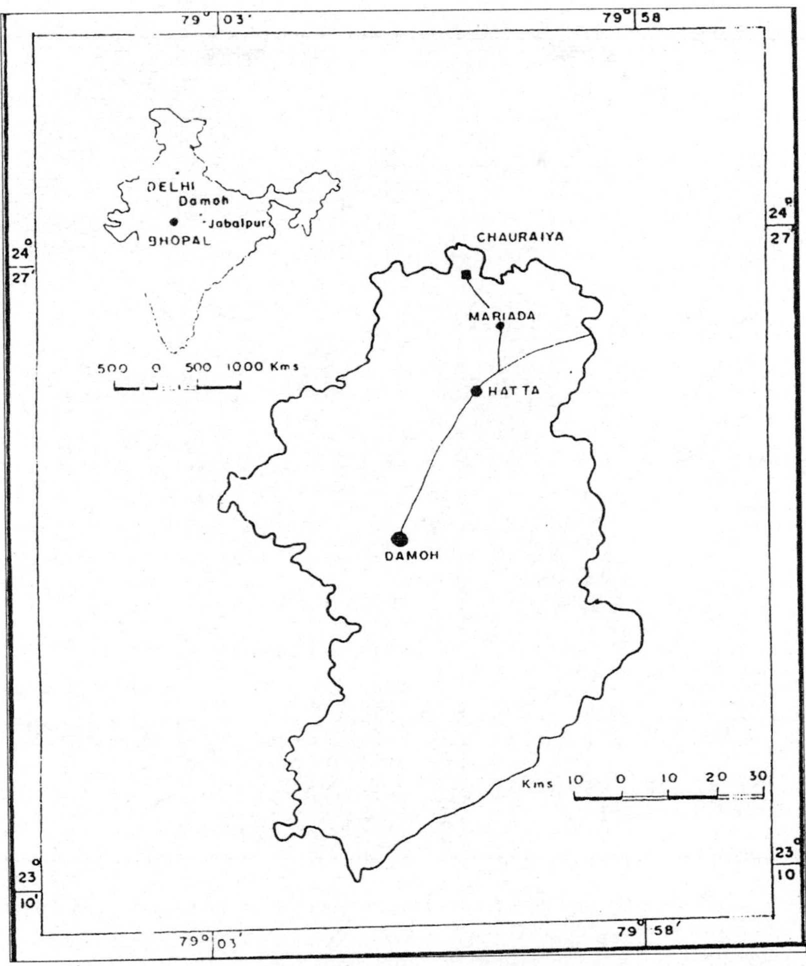

Fig. 1 Location map of Chauraiya area, Damoh district, Madhya Pradesh

sample of about 1000 grams obtained from circumference of each termide mound site. Predetermined regular sampling pattern however, is impossible in termitaria surveys as termite mounds are located where they are located. Geology of the area together with the of termite mounds locations is given in Fig 2. The average height attained by the termite mound is less than a meter while circumference is between two to three meters mostly varying in color varies from the shades of brown and grey.

After drying and removing the vegetal matter, the samples were treated in much the same way as routine geochemical soil samples. Pb, Zn and Cu content were determined by Atomic Absorption Spectrometry at the Geochemical Laboratory of the Directorate of Geology and Mining, Raipur. As also pH of each sample was determined. Details of sample location, lithological control and analytical data Table 1, 2 and 3 for each is tabulated

ANALYTICAL RESULTS

Scrutiny of the analytical data in Tables 1, 2 and 3, shows that for all the three elements (Cu, Pb and Zn) their abundance is decidedly higher in termite mounds that overlie the buried mineralization. (Table 3) Copper content variation, however, is hardly significant because of its overall low values, which is consistent with the mineralogy of the ore body. For the mounds, located outside the host rock (Table 1) Pb varies from 17 ppm to 55 ppm while Zn from 41 to 85 ppm, averaging 38 and 58 ppm respectively. Those located within the host rock and overlying buried mineralization, Pb content varies from 56 to 560 ppm while for Zn from 36 to 389 ppm. The average value for Pb and Zn are 203 and 117 ppm respectively, but ten samples from within the host rock which may or may not be overlying mineralization (Table 2), Pb is

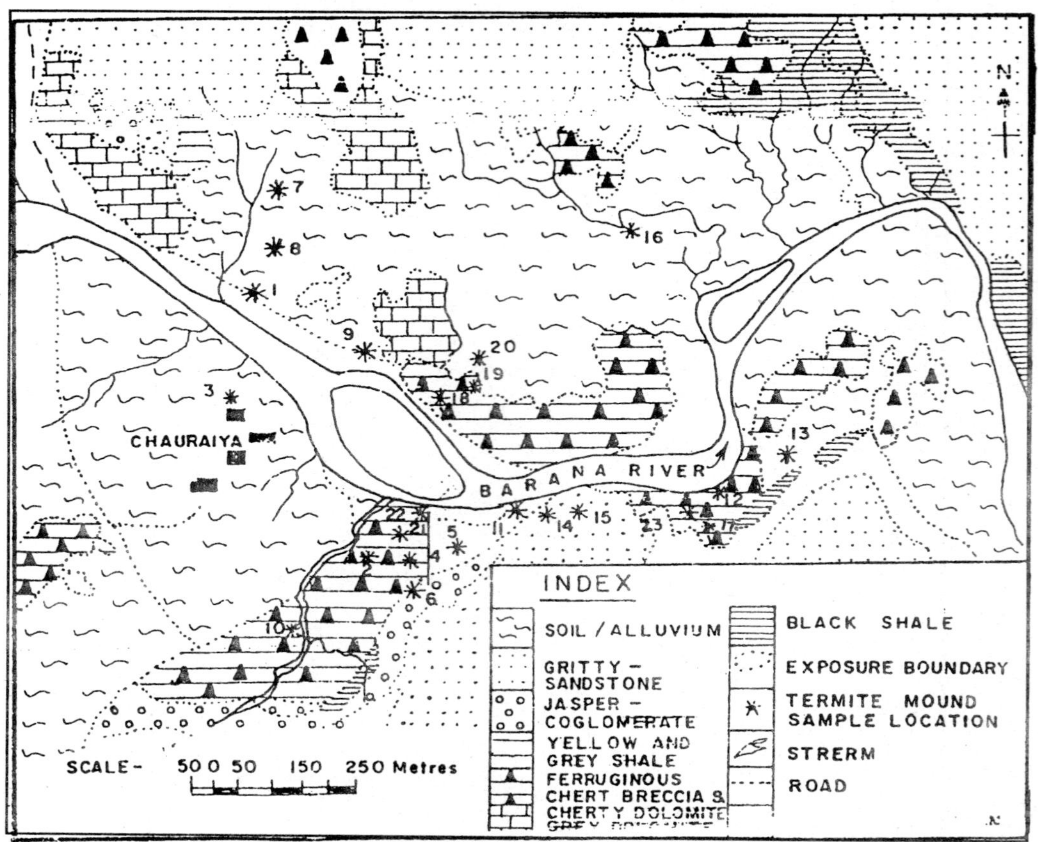

Fig. 2 Geological map of the area showing location of the termite mounds studies and sampled in this work

obviously higher than the samples from the area outside the host rock but Zinc is slightly lower. The average values for Pb and Zn are 57 and 54 ppm respectively. It is reasonably argued that the buried mainly galena mineralization is truly reflected by Pb where is sphalerite, galena does not reflect Zn.

RESULT AND DISCUSSION

The value for Pb in samples 10, 17 and 23 are over 7 times the background of 38 ppm, recorded for the samples outside the host rock area. The maximum content (560 ppm), over 14 times is found in sample 17. It is significant that in this location small galena grains were observed under the binocular microscopic and hand lens (10 x). Lower values for zinc (compared to Pb) from almost every site is consistent with the fact zinc is geochemically more mobile than lead as also for the fact that the Chauraiya deposit is mainly a galena ore body. It is also interesting to note that the pH of most of the samples that are low in Pb and Zn is also low, ranging between 6.71 to 7.92 while termite mounds showing anomalous Pb and Zn have generally higher pH, upto 8.20.

It will be observed that for the termite mounds overlying buried mineralization shown quite a variation in Pb and Zn values. For the former it varies from 56 to 560 ppm and for Zn 36 to 389 ppm. Firstly, it could be due to the nature of the ore body itself, secondly, to some peculiar factor with respect to insect life and activity and thirdly, it can also be a function of depth of the water table upto which limit the termite activity. The chemical study of the mound material does reflect the presence of a buried galena ore body. Since most of the samples are distributed rather erratically, it is not possible to draw an exact section depicting mineralized zone in chert breccia and their relationship to termite mounds along with their Pb/Zn contents. However, a generalized picture is shown in Fig. 3 with Pb/Zn values for a few selected termite mounds which brought out the potential of termitaria studies in locating Pb/Zn ore body serving as a model for similar studies elsewhere.

Geochemical survey of the predominantly residual

Table-1 ANALYSES OF TERMITE MOUND SAMPLE, OUTSIDE THE HOST ROCK CHAURAIYA, DAMOH

S. No	Sample No.	Location of the Sample	Lithological control	pH	Analytical Results ppm		
					Cu	Pb	Zn
1	1	0.5 km North of Chauraiya	Grey dolomite	6.71	26	55	54
2	3	0.1 km NW of Chauraiya	Grey dolomite	7.20	30	17	83
3	6	0.6 km SE of Chauraiya	Jasper conglomerate	7.50	44	35	46
4	7	0.8 km North of Chauraiya	Grey dolomite	7.24	34	27	41
5	8	0.6 km North of Chauraiya	Grey dolomite	7.75	30	46	85
6	9	0.5 km NE of Chauraiya	Grey dolomite	7.92	33	35	37
7	12	0.9 km North of Chauraiya	Grey dolomite	7.80	37	45	47
8	13	1.5 km East of Chauraiya	Grey dolomite	7.90	32	36	53
9	14	0.85 km East of Chauraiya	Gritty sandstone	7.78	25	36	71
10	15	1.0 km East of Chauraiya	Gritty sandstone	7.73	35	46	38
11	16	1.3 km NE of Chauraiya	Grey dolomite	7.77	38	38	57
12	20	0.7 km NE of Chauraiya	Grey dolomite	7.73	34	34	41
			Average, ppm		33.16	38.33	57.75

Table-2 ANALYSES OF TERMITE MOUND SAMPLE, WITHIN THE HOST ROCK BUT NOT NECESSARILY OVERLYING THE BURIED MINERALISATION, CHAURAIYA, DAMOH

S. No.	Sample No.	Location of the Sample	Lithological control	pH	Analytical Results ppm		
					Cu	Pb	Zn
1	11	0.7 km ESE of Chauraiya	Chert breccia	7.75	31	101	71
2	21	0.5 km SE of Chauraiya	Chert breccia	8.14	36	22	51
3	22	0.5 km East of Chauraiya	Chert breccia	7.30	34	47	39
			Average, ppm		33.66	56.66	53.66

soils of the area are also reflects the buried mineralization. A termitaria study yield additional information, and is useful in screening large areas quickly and economically. Since termite mound galleries can extend upto great depth, with a maximum recorded depth of 70m, their use in prospecting can be applicable in areas of transport cover also. This has been observed and studied by the first author in diamond (kimberlite) prospecting in the transported soil cover regions of Kimberley in Western Australia and may be useful for other deposits.

Termite mounds may not be as spectacular a sight in India as they are in the subtropical regions of South America, Africa and Australia (where termite mounds can attain a towering height of 6 metres or even more). However, but they are not common sight in most parts of our country through seldom used as field guide for establishing mineralization.

Table-3 ANALYSES OF TERMITE MOUND SAMPLE, WITHIN THE HOST ROCK AND DEFINITELY OVERLYING THE BURIED MINERALISATION, CHAURAIYA, DAMOH

S. No	Sample No.	Location ofhe Sample	Lithological control	pH	Analytical Results (ppm)		
					Cu	Pb	Zn
1	2	3	4	5	6	7	8
1	2	0.5 km SE of Chauraiya	Chert breccia	6.32	23	62	124
2	4	0.65 km SE of Chauraiya	Chert breccia	6.25	46	114	389
3	5	0.6 km SE of Chauraiya	Chert breccia	7.37	42	80	71
4	10	0.55 km South of Chauraiya	Chert breccia	8.10	39	284	69
5	17	1.4 km East of Chauraiya	Chert breccia	8.20	57	560	106
6	18	0.53 km NE of Chauraiya	Chert breccia	8.04	63	147	97
7	19	0.6 km NE of Chauraiya	Chert breccia	7.72	33	56	36
8	23	1.30 km East of Chauraiya	Chert breccia	8.10	19	319	41
			Average, ppm		40.25	202.75	116.62

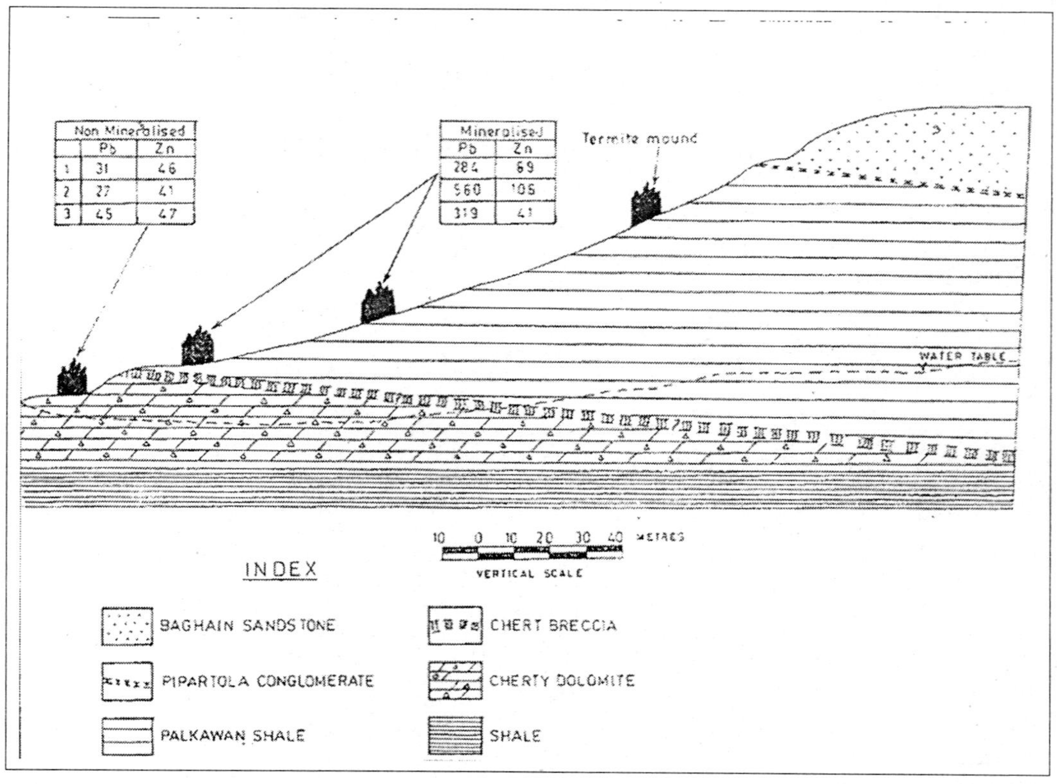

Fig.3 Schematic Representation of Termite Mounds over Lead Zinc Deposit, Chauraiya Valley, District Damoh.

CONCLUSION

This work from a small sized Pb-Zn mineral prospect highlights termitaria studies in establishing buried mineralization. To utilize the full potential of this useful technique further detailed work in mineralized belts of the country will be useful attempt for prospecting both metallic and non-metallic ore deposits.

ACKNOWLEDGEMENTS

We are thankful to Dr. V.K. Jain, Chief Analyst of Directorate of Geology and Mining, Raipur for the analytical facilities provided in this work.

References

Alexander, P.O. (1987) On the need for developing biological methods in mineral exploration, Diamond jubilee Nat. Symp. Devp. India's min. & fuel Res., Indian School of Mines, Dhanbad MRIV, p. 6.

Brooks, R.R. (1993) Geozoology in mineral exploration, Episodes Vol No. 3 pp. 27-32

Bruton, E., "Diamonds" NAG press Ltd., London 532 p.

D'Orey, F.L.C., (1975) contribution of termite mounds to locating hidden copper deposits. Trans Inst. Min, Metal B. 84 No B28, pp. B150 B151.

http://euwivipedia.org/wiki/termites, 2006

http://members.tripod.com/~charles~w/laterite.htnd.2006

www.thefreedictionary.com.2006

Lee, K.E. and Wood K.G. (1971) J Termite and Soils Aca Press New York 257 p.

Prasad, K.E.V. and Saradhi D.V. (1984) Termite mounds in Geochemical prospecting, Curr. Sci., V 53, No 12, pp. 649-651.

Soni, M.K. (1981) On basemetal mineralization in Damoh and Chatarpur districts MP. Misc Pub. 50, GSI, pp. 201-204.

West, W.E. (1965) Some unconventional ideas on prospecting, Chamber Mines Journal Rhodesia V. 7, No 12 pp. 40-42.

West, W.E. (1970) The Bulawayo Symposium Papers no. 2 "Termite Prospecting" Chamber Mines Journal (Rhodesia) V. 12 No. 10 pp. 32-35.

Hon'ble Reviewers

1. Prof. V.K. Verma
(Former Dean, Delhi University)
B-3/1, Janakpuri
NEW DELHI 110 058 INDIA

2. Prof. P.S. Ranawat
Dean, SW
Department of Geology
MLS University
UDAIPUR 313 001 INDIA

Editorial Assistance

Mrs. Mamta Chauhan

Economic Mineralization
K.L. Shrivastava (Editor)
2009. pp. 396-405
Scientific Publishers (India), Jodhpur

36

Section-IV

Present Status and Future Trends of Biological Methods of Prospecting for Economic Mineral Deposits

S.K .TRIVEDI AND K.K. SOLANKI
Department of Geology, J. N. V. University, Jodhpur 342 005

Abstract

An attempt has been made to bring together the element various segments of the diversity of scientific disciplines, to include botany, zoology, geology, analytical chemistry, biogeomorphology, phytochemistry, plant physiology and ecology which are integral components of the biological methods for prospecting of minerals.

Botanical methods incorporate both geobotanical and biogeochemical methods. First, to use plant involving visual observation and second to analyse chemically, various parts of plants for the prospecting. Various limitations and advantages of both methods have been listed apart from discussing various Indian Case histories. Geozoological prospecting includes mammals (specially dogs), birds, fishes, insects (specially termite and bees) and micro organisms (specially bacteria). With the advancement of the technology, it is expected, that the biological prospecting will grow further to become a versatile technique in future.

Keywords : Biological Prospecting. Geobotany, Geozoology, Biogeochemistry, Geomicrobiology.

INTRODUCTION

Mineral wealth is associated with the development of human race since time immemorial. From the Stone Age man used various biological methods using has sense for obtaining a stronger piece of stone to kill animals for his survivals (for his food and protection). In the Iron Age the mineral deposit were localized only by his sense (biological method) and in present era many famous mineral deposits were discovered with the help of plants, animals, birds, ant and even with the excreta of the animals.

There is an unprecedented surge in prospecting for minerals through out the world because of the increasing demands placed on this planet for dwindling resources by expanding economics and burgeoning populations.

Most of the easily discovered minerals were found many years ago. For those hard to detect minerals that remain to be prospected yet tend to be in the inaccessible parts and prospectors needs to work hard. It is well documented that prospectors had always taken help of ants or even humble microbes as field assistants for minerals prospecting.

It has now become a recognizable fact of life that no one technique by itself can give a complete delineation of mineralisation

Biological methods of prospecting for various minerals are the most simple of all methods of prospecting. It is probably true to say that the part decade has withered the coming age of biological methods of prospecting. It no longer seems absurd that scientists should analyze biological materials for various mineral prospecting or should enlist the help of plants, grasses, dogs, ants, or even humble microbes as field assistants for this purpose. (Tyagi, 1981). The variety of biological samples used for mineral exploration is really quite astonishing and ranges from leaves of plants to even the "Final Product", animal faeces. The various sub disciplines of biological methods of prospecting include geobotany; biogeochemistry, geozoology and geomicrobiology are discussed.

Botanical methods of prospecting for mineral incorporate two different techniques, namely geobotanical and biogeochemical. Although it is customary to classify geobotany, involving the visual observation of the vegetation cover and biogeochemistry, invading the analysis of the mineral content of the plants cover and its underlying soil as two separate disciplines, they are in reality closely related since the nature of vegetation and its elemental content are influenced both by the chemical composition of the soil and the physical features of the environments.

Biogeochemical methods of prospecting depend on the chemical analysis of elements vegetation. In such investigations it is assumed that the biological absorption coefficient, that is the concentration of an element in plant material divided by concentrations of the same element in the soil, remains relatively constant. Therefore, anomalies in soil or bedrock will be determinable via the plant material because biogeochemical methods involve chemical analysis of plant material (usually at trace element levels), the technique is much younger than the associated field of geobotany. The successfully utilization of biogeochemical methods of prospecting involve a knowledge of many disciplines, including botany, chemistry, ecology, geology, photochemistry plant physiology, soil science a statistics. It is for this reason, more than any other, that the development of this method has been slower than that of many other prospecting methods.

In the treatment of biogeochemistry, it is not claimed that the method is a universal panacea for the search for minerals, and indeed under some conditions it may not be wise to expect too much from the method.

On the other hand at the first sight the idea of using animals as field assistants might seem to be useless in the extreme. Yet it has been found that such dissimilar animals as dogs and termites can be of use in prospecting for minerals. A survey of the literature shows that not only these but other animals have been used for this purpose cited directly or indirectly.

The most of the world's early discoverable mineral deposits have already been revealed and that those remaining are probably below the sea or in inaccessible part of the globe, perhaps at depth or perhaps covered by ice. Most mineral deposits that are discovered today are smaller than earlier finds, and have become increasingly difficult to uncover. This paper embodied each of the subdivision of biological methods of prospecting for various minerals.

GEOBOTANICAL PROSPECTING

Geobotany is a specialized branch of applied ecology and involves the visual observation of the vegetation cover in to detect mineralization by means of plant distribution, the presence of indicator plants and / or mutational and/ or morphological changes induced by the excess of certain elements in the under lying soil strata and the study of any of the above parameters by aero- visual observation, aerial photography and/ or satellite imagery (Brooks, 1980, 1983).

The Russian geologist Karpinsky, (1841) was the first to recognize that different plant associations exist on varying geological substrates and concluded that greater reliance should be placed on an examination of the whole community rather than any one or a few species within. This classical work led to the development of the science of "Indicator Geobotany". Recent studies in the science of indicator geobotany have been stimulated by satellite imagery and the development of sophisticated computer- assisted techniques which allow for thorough and quick statistical treatment of the raw data in order to evaluate the very subtle variations (Brooks 1983).

Geobotany includes the study of relationship between soil and geology and the overlying vegetation type and its abundance. It has origins in the field of plant geography and includes studies on plant distribution related to their holistic environment (Cain, 1932). Additionally, studies by biologists, geologist, soil scientists, and paleontologist have provided historical perspectives to the geobotanical literature, where geologist and soil scientist have used plant distribution to infer soil and mineral conditions.

In some of the earliest modern ecological literature reported by Venkatesh,(1964) describe the relationship between plant geography and soil from the perspective of plant meaning, morphology, and physiology. Venkatesh, (1964) describe plants whose distribution is limited to particular soil chemistry, like calcareous, saline, siliceous, serpentine, zinc and other. Warming (1892) emphasized the association of characteristic plant species with particular edaphic conditions viz. halophyte, hydrophytes, xerophytes, etc while symptoms of deficiency or excess of specific elements have been described by numerous authors, caution must be used in interpreting visual symptoms of plants because different mineral deficiencies or excesses can cause the same symptoms (chlorosis). Warren et.al. (1966a) provide a good review of the importance of soil chemistry in mineral nutrition and distribution of vegetation.

Most frequently, in North America the term geobotany is used to describe a form of mineral prospecting which relies on characteristics of the vegetation to identify the location and extent of ore bodies (Brooks, 1995). A long list of indicator species have been identified as "hyper accumulative" metals (tolerate and accumulate high concentration of heavy metals) by Brooks (1995).

A few years later Brooks (1995) reported that presence of the mineral indicator plant Minuartia verna that was invariable colonizer of copper lead/zinc mineralization in the Harg region of Germany.

Cannon (1975) reported one of the most successful examples of geobotanical prospecting. She reported the distributions of Astragalus plants as hyper accumulators of selenium (>1% in dry matter). These plants indirectly indicate the presence of uranium also because uranium and selenium are found together in the mineral contrite. This study lead to the discovery of several uranium deposits in the Veteran United States (Brooks, 1995).

Malyuga (1964) has distinguished two types of indicators, namely universal and local. Universal are those, which require a particular metal for their healthy growth and failed to grow in non-mineralized substrates, while local indicators are the species adopted to tolerate mineralized ground but can grow equally elsewhere provided that competition from the other species is not too great.

Early scientific observations dating form the 8th and 9th centuries reported that the morphology and distribution of certain plants were affected by the presence of metals in the soil (Sinha Roy and Paul, 1967). Such visible variations in a plant species are referred to as geobotanical indicators. Many other plants, while not showing any visible variations, are capable of concentrating metals in their tissues and the presence of anomalous metal in the soil or water is often reflected in the metal content of leaves, twigs or other plants organs.

The season full of leaves and needles transfer some of the accumulated metals to the surface soil where they are incorporated in the humus layer. Plants are complex organisms and so is their metabolism. Deep-rooted plants (e.g. the mesquite) are much more effective prospectus for the deeper groundwaters than the shallow rooted flora. Evapo-transpiration has been suggested as a mechanism for movement of metals into the nutrient depth of these plants.

Some species preferentially concentrate metals in specific tissues such as leaves, twigs, barks, or wood. It is therefore very important to establish the most favourable tissues for sampling once a useful species has been identified. This complicity is accentuated by the fact that metal uptake commonly varies with the seasons. In temperate forest regions, accelerated uptake and higher concentration commonly occurs during the spring growth following a dormant winter reason. In hot desert regions, following the exhaustion of available near- surface water during the dry season, deep-rooted plants will tap the deeper, more metal-rich groundwater. (SME mining Engineering hand book 2nd edition volume 1, 1992).

Case Histories

In the view of definite success in the biological prospecting encapsulated already by various abroad, the work has been done in India also.

Venkatesh (1964, 1966) conducted geobotanical surveys in different mineral deposit areas in the Singhbhum copper belt, Bihar. Vernonla cinerea was found to be most common plant over copper lodes and particularly on the ancient mine dumps. In the Agnigundala copper area, Guntur district, Andhra Pradesh. Polycarpea corymbosa was very common near ancient mine dumps. Occasional specimens of Pulicariaglauca and more frequent occurrence of justice gluca were also observed. In the areas of known lead mineralization near Zangamraju-palle, Cuddapah district, Andhra Pradesh. Striga euphrasioides, Justica prostrate, Pulicaria wightiana and Heylendia latebrosa were of common occurrence.

Poddar (1965) struck large population of Impatiens balsimina selectively growing on the mineralized lead-zinc belts of Zawar mines and considered it as probable base metal indicator for the region, but no further geobotanical or biogeochemical studies were made.

Banerjee (1972) and Roy (1974) studied the vegetation overlying the nickel ore deposits of the Sukinda valley, Orissa. Roy (1974) observed that the ore body was characterized by the presence of low shrubby community contrasting very sharply with the surrounding rain forest. Typical components of the shrub associated were Combretum decandrum and Miliusa velutina with scattered dwarf forms of Terminalia tomentosa. This community clearly indicated nickel ore, which was found to lie 19-20 m below the surface.

Gandhi and Aswathanarayana (1975) have reported a possible metal indicator plant, Waltheria indica of sterculiaceae from Mamandur, South India. The distribution of plant is confined to mineralized ground in the area. The study showed that plant is considerably enriched in both copper and zinc.

They have suggested that since the plant is also found to occur on barren areas elsewhere in South India, the mere presence of plant may not be considered as an infallible guide to base-metal mineralization.

Aery (1978) suggested that the usefulness of vegetation in predicting the concentration of the elements in the soil decreases in the order of herbs > shrubs > trees.

Aery (1977, 1978) Aery and Tiagi (1986), Tiagi and Aery (1982) have made extensive geobotanical survey on the known lead-zinc deposits of zawar mines, Udaipur, India. It has been shown that ten plant species namely Impatiens balsamina, Malhania futteyporensis, Triumfetta pentendra, Grewia flavescens, Dyerophytum indicum, Acanthospermum hispidum and Calosio argentia have the highest combined constancy (i.e. as a community) and fidelity values on mineralized ground in the comparison to non mineralized ground. Of these species I.balsamina , M.futteyporensis and T.pentendra have the highest cover value. I.balsamina grows exclusively on mineralized grounds with comparatively much higher density on mineral dumps of the ancient workings. This wild variety of I.balsamina differs from its garden analogue in having small size, low leaf area index, flowers with one whorl of petals, strikingly different seed germination behavior and accumulation of enormously high contents of zinc in its tissues. It can be therefore, regarded as an indicator for the metal zinc at least in the region and thus deserves the rank of a local indicator. (Aery and Tiagi, 1989).

Aery (1990) revealed that certain plants namely Anogeissus latifolia, Cassia auriculata, Cassiatora, Grewia flavescens, Iepdadaenia pyrotechnica, L.muraria, R.mysorensis andTtephrosia purpurea have the highest fidelity and constancy value on metal rich soils in the surrounding areas.

Geobotanical and biogeochemical studies on certain uranium deposits of Rajasthan are being carried out by the Aery (1994), Aery and Jain (1994) under the Department. of Atomic Energy, Govt. of India sponsored research project. Geobotanical studies have been revealed that though no species can be said to be indicator, seven of these deserves the rank of elective pseudometalophyte, metalloresistant or preferential species. These are C. ariculata, C. tora, Diospyros cardifolia, Holoplelea integrifolia, Kriganelia reticulate, L. muraria and W. tinctoria.

Mineral prospectors in North Canada have long believed that plant like "Karipatta" whose leaves (Curry leaves) are used to flavours many south Indian dishes, can be good indicators for the presence of deposits of mineral, especially manganese. Now geologists from Geological survey of India (GSI) have used this traditional knowledge and zoomed in and a workable deposit of manganese has been prospected in North Canada District in Karnataka.

An estimated reserve of 76.975 million tonn, with 25.655 million tonn or just over 30 percent of manganese, was delineated in the district after extensive surveys using "Karipatta" plant as a geobotanical guide. Interestingly, Karipatta not only exhibits versatility in its application in food preparation but also as a geobotanical indicator of minerals. A Karipatta plant with dark colored leaves in likely to indicate manganese ore deposits, while a plant with light green leaves can shows the presence of Iron ore deposits.

The North Karnataka belt in the Karnataka carton, according to geologist, in bestowed with manganese, iron, copper and traces of magnetite and trickle mining activity is already on with the minerals being used for domestic consumption as well as export.

Advantages and disadvantages of Geobotanical Prospecting

A number of advantages of geobotanical method of prospecting can be listed as follows :

1. Once orientation survey has been carried out, further costs are extremely low.
2. Different geological formations, as well as

mineralization within them, can be detected by geobotany.

3. Aerial methods can be applied with consequent saving of time and effort.

4. Indicator plants can sometimes indicate mineralization at depth in cases where there are no surface expressions of this mineralization.

Followings are the disadvantages of geobotanical methods of prospecting :

1. A high degree of skill is required

2. Data may not be universally applicable.

3. Some times the method can be applied only at certain reasons for example when plants are in flower.

4. The method can only be applied where favourable vegetation exists.

5. There is little coordination of these types of work at present.

The Future of Geobotanical Method

It is probable that the most important future developments in geobotany will be in improved satellite imagery combined with more sophisticated methods for evaluations of ground truth. Better resolution remote sensors, combined with more sophisticated statistical and computational procedure should do much to improve the effectiveness of the technique and to remove to a greater extent, the degree of subjective personal judgment, which at present must be applied to the method.

It might at first seem to be difficult to predict what future improvements might be made in the field of study of indicator plants and indicator plant communities, since such operation tend to be visual and do not lend themselves to automation. However, if efficient instruments could be designed for in- situ measurement of elemental concentrations in rocks, soil and mineral, this will be immensely useful in orientation surveys. At present the geobotanist has to rely on chemical analysis of substrate for a thorough study of indicator plants. A further step in this direction has led to the development of portable X-ray fluorescence units (Reeves et.al.1978), which in theory make such measurements a possibility. The problem so far with these units is their relative lack of sensitivity, but if significant improvements in their performance can be made in the future, they may well prove to be useful for geobotanist.

A final factor should not be considered. It time and certainly in the foreseeable future, a very large part of the earth's surface may be surveyed by various geochemical and geophysical prospecting techniques. Geologist will have left until the last what is to them the least familiar of potentials methods. When that time comes, geobotanical methods of prospecting will be the last of the potential techniques to be fully exploited in the search for minerals, which will become progressively more elusive with each passing year.

BIOGEOCHEMICAL PROSPECTING

Biogeochemical methods of prospecting depend on the chemical analysis of elements in vegetation. The assumption is made that an element in the soil and bed rock will be accumulated by the plant in a reproducible manner and that consequently, anomalous amount in the vegetation will indicate anomalies in the substrate. The biogeochemical method has also been called "The ash method" (Tkalich, 1938) and the "soil floristic method" (Malyuga, 1964). In such investigation it is assumed that the biological absorption coefficient that is the concentration of an element in plant material divided by concentration of the same element in the soil remains relatively constant. Therefore, anomalies in the soil or bedrock will be determined via the plant material (Brooks, 1983).

Because biogeochemical methods involve chemical analysis of plant material (usually at trace element level), the technique is much younger than the associated field of geobotany. The history of this technique began with the work of Tkalich (1938) who detected an arsenopyrite deposit after analysis of iron in the overlying vegetation. Warren (1949) is regarded as father of biogeochemical prospecting. During the second worldwar, while digging trenches to reach bedrock, he realized that the tree roots were to reaching the same zone. After series of studies carried out by Warren (1949) and his co-workers for 40 years reported that the soil chemistry was much similar to the composition of plant grown in the soil. In his own words, he was to raise the credibility of biogeochemistry from general disbelief through benevolent skepticism to general acceptance.

Brooks (1965) initiated biogeochemical prospecting in New Zealand and continued the studies for a further 20 years. The studies revealed that chemical analysis of vegetation provided an alternative to soil analysis for delineation of mineralisation anomalies in that country. The studies extended from copper mine Island (Hen and chickens group) all the way down to port Pegasus in the sour of Stewart Island.

Biogeochemical methods have been also widely used for mineral prospecting particularly in boreal forests and semi arid region, but have been fewer applications in tropical areas. (Brent et. al. 1996).

In the former Soviet Union, Kovalevsky based at Ulan Ude in Siberia has been a leading biogeochemist in Russia has been reported main leading center of biogeochemical prospecting in the world. His outstanding to the contribution was the best indication of mineralization below the surface of the soil. He is today the undisputed leader in biogeochemical prospecting (Brooks, 1983). The utilization of birch sap as sampling medium for biogeochemical prospecting for mineral deposit was studied in an orientation surveys (Leo Harju and Stig-Goran Hulden, 1990).

Wareen, et.al (1952) carried out preliminary studies on the biogeochemistry of iron and manganese in the leaves and twigs in order to find out variations, which might indicate mineralization. Iron and manganese variations were studied in different climatic conditions and over different geological formations from the area of Sullivan mine.

It was reported that iron and manganese variations might be most useful in prospecting gold and silver as these belts are accompanied by high contents of iron and manganese but not copper and zinc.

Harbaugh (1950) who was the first to apply the biogeochemical method of prospecting in united states, investigated base metals in soils and vegetation of the tri-state district and found higher average zinc concentrations in plants growing near mineralized zone less than 100 feet in depth. Analysis of 56 samples of twigs of Black jack oak, Quercus marilandica from trees growing above or mineralized zone, showed average zinc concentration, 29 percent greater than 101 similar samples collected in the same areas, but at a distance of 300 feet or more from the mineralized zones. In addition, the samples from above or near mineralized zones showed an average zinc concentration 67 percent greater than 23 similar samples collected in areas presumably derived of mineralization and lying some miles distant. In two-mile square areas, the geographical distribution of zinc concentration in samples of Quercus marilandica twigs showed moderate correlation with known mineralized zones.

Aery and Tiagi (1985) suggest that cadmium can be used as a pathfinder of zinc in biogeochemistry, as higher concentrations in soils as well as in plants are found only on zinc anomalies and it provide a better contrast between the areas of commercial zinc mineralization and the ground background.

Singh (1982) studied the vegetation over Rajpura Dariba zinc belt and observed that several plant species concentrate lead, zinc, copper and cadmium in a reducible and were logically designated a metal accumulators. Some high accumulators are Echinops echinatus, Rostellularia difusa and Vernonia cinerea. Some lead accumulators reported by Singh are Arundo donax, Echinops echinatus and Vernonia cinerea.

Singh (1982, 1990) observed that species over rock phosphate deposits of Udaipur like Lidenbergia muraria, Echinops cenchrus ciliaris and Acanthospermum hispidum accumulate significantly higher concentrations of phosphorus, the range of phosphorus accumulated being 5500-11500 ug/g, 1250-6500ug/g and 12500-20000 ug/g in leaves of certain species such as Auera sangiunolenta, Blainvillea acmelda, Echinops echinatus, Eclipta prostrata, Trichodesma indicum and Zinnia elegans and. Above ground parts of Cenchrus cilliaris gave a good correlation with the phosphorus content of the soil. Like wise the phosphorus content of stems of Echinops echinatus, L.Muraria and T.indicum also correlated fairly well with the soil.

Biogeochemical studies carried out by Tiagi and Singh (1973) and Tiagi and Aery (1986) over the regions of Khetri copper deposits and copper deposits of Madhya Pradesh respectively revealed that Talinum portulacifolium accumulates the highest amount of copper (5253) ppm. The values have been found too higher than accumulated by Becium homblei, the "copper flower" of Zambia (Malaisse et al., 1978). Tephrosia villosa, Aerua tomentosa and Bouchea morubifolia showed a linear uptake of copper.

Biogeochemical techniques have only been used since the late 1930s (Dorn, 1937), since they depend on the existence of speedy and reliable methods of analysis.

Aery and Tiagi (1990) have also discussed the significance of several elemental ratios such ass Co: Ni, Th: U, Cu: Zn, Ca: Mg, and Cd: Zn in biogeochemistry and successful utilization of biogeochemical methods of prospecting with the help of many disciplines, including botany, chemistry, ecology, geology, photochemistry, plant physiology, soil science and statistics.

In the treatment of biogeochemistry, it is not claimed that the method is a universal panacea for search for minerals and indeed under some conditions, it may not be wise to expect too much from the method. However, if used in the right place by suitably trained personnel well versed in its potential and application, it will prove to be a valuable auxiliary or even primary method in the continuity search for minerals throughout the world.

GEOZOOLOGICAL PROSPECTING

The idea of using animals, as field assistants in mineral prospecting might seem to be funny however it has been shown by Brooks et.al. (1995) that, mammals, fish and insect can be used for this purpose. Yet it has been found that such dissimilar animals as dogs and termites can be of use in prospecting of minerals. A survey of the literature shows that not only these but other animals have been used for this purpose either directly or indirectly.

The problem immediately arises as to a suitable nomenclature for what is largely an uncollated new science. To be consistent with botanical methods of exploration, techniques involving direct analysis of animal material (e.g. trout livers) should be termed biogeochemical, where as those involving a visual study of animals or animal disease or behavior should termed as geozoological. There then remain the problems of classification of cases where animals are used directly as field assistants, as in the case of ore detecting dogs. There are further examples of the analysis of termite mounds for inorganic matter brought to the surface by these creatures. Mounds of termite can also be used for mineral exploration as these are made up of inorganic matter dug from the substratum. (see, Alexander and Patel, this volume).

The Science of geozoology, though new in nomenclature has its roots in antiquity when Herodotus (450 B.C) reported that the natives of India studies termite mounds to detect gold mineralization. It is only recently, however, that the field has begun to have a wider application in mineral prospecting. Its potential is obviously not as great as that of botanical methods because of the basic problem of the relative mobility of most (though not all) animals. In only a few cases are animals so restricted in their habitat that they can provide favorable conditions for prospecting for minerals. In geozoology prospecting various animals such as mammals, birds, fish and insects can be used in the search of minerals. In this connection the geochemical maps produced also contribute to an understanding of the structural development of the crust and are of assistance to geological mapping. Prospecting by geochemical methods, from regional to detailed scale, has led to the discovery of many mineral occurrences. (Strenfelt, 1987).

The use of dogs for mineral prospecting was first proposed by Orlov et. al. (1969) and was later also developed in Fennoscandia and Canada. The use of dogs as field assistant in the search for ore deposits seems at first sight to be rather fanciful propositions, however, dogs have been trained to scent out sulfide-rich boulders and ore deposits in our countries. According to Brooks (1972), the pioneering work began in Finland with the investigation of Dr. Aarno Kahma of the geological survey of Finland, who conceived the ideas in 1962. It was based on the assumption that a trained dog, with its very keen sense of smell, should be able to detect a weathered sulfide boulder from a distance of several meters and possibly even below the overburden.

Together with an assistant Kahma selected an Alsatian dog and trained it for two years before putting it to a practical field test in 1964. Preliminary field tests were successful and in 1965, the dog "Lari" was pitted against a human float-boulder prospector on an unprospected field of some 3 sq. km. The field was gridded and the finds marked on a map. The results were very impressive. Lari found 1330 sulfide bearing boulders, some of which were below an overburden of 10-20 cm. The prospector found only 270 surface boulders. In that same summer, Lari also located pyrite and chalcopyrite floats on another prospecting site and as a result of this discovery, a copper ore body of economic significance was later encountered. The geological survey of Finland has a number of prospecting dogs in use and uses these animals in routine prospecting. Kaham's work was not reported until much later (Ekdhal, 1976,Kahma et.al., 1975) but the idea of using dogs quickly spread to Sweden (Nilsson,1971,1973) the soviet Union (Orlov et.al.,1969), and Canada (Brooks,1972) shows the ore dog "Hessu" searching for uranium in Finland.

Marine organism such as seaweeds has also been used in mineral prospecting. These marine algae are able to concentrate levels of trace elements by a factor of at least 10000 compared with the seawater itself. Dunn (1990) has shown that various weeds could be used to detect a zone of copper/gold mineralization on the shores of Texada Island, British, Columbia. They determined 33 elements in the plant material and were able to identify the deposited by anomalous concentrations of arsenic, that is usually associated with other sulfide mineral elements. (Brook, 1998 NZ Bioscience/ Nov. 98). The most promising animal "field assistant" are the humble termites. The Greek historian Herodatus recorded the use of termites for prospecting in 450 BC. He described method of gold prospecting used by Indians (Brooks, 1998).

West (1970) noticed termites burrow down to the water table and in the process bring up a large amount of soil from a depth of 10 meters or more. The termites have been doing the work for the prospector without charge and because their mounds are spread rather uniformly across the terrains, it is possible to carry out a survey on a square grid of termite mounds. As a result of West's work, several gold anomalies have been discovered; the first of these is now appropriately name "termite mine" since the initial work of West (1970). Numerous surveys on termite mounds have been carried out, mainly in Africa and have resulted in some important discoveries (Brooks, 1998).

Birds as indicators

There are numeric references in the pre- 1900 literature to accumulation of noble metals such as gold in birds (Boyle, 1979). Razin and Rozhakov (Razin et.al.,1966) found up to 30mg/g (ppm) gold in five species of Siberian birds. The data are of no practical value in prospecting because of the vast distances covered by the birds during their migratory travels.

Fishes as indicator

Fish may be used more successfully than birds in mineral exploration. (Warren et.al., 1971). Reported the first possibility working on liver of rainbow trout (salmogaridneri) in British Columbia.All the livers with >50 mg/g zinc (wet weight) were form fish living in waters known to associate with economic zinc mineralization of 17 localities where lives showed > 60 Mg/g copper, four were know to contain significant mineralization and another seven merited further investigation on the basis of their known geology.

Bees as indicators

It was first suggested by Makarochkin et.al. (1960) that the composition of honey may indicate the elemental content of the soil in which the nector -producing plants grew originally. They reported concentration for Si, Al, Mg, P, Mn, Fe, Ti, Mo, and Cu in the ash of honey form the lake llmen area of the Soviet Union.

In a later work (Makarochkin, 1974), analysis of honeys from six widely separated districts of the Soviet Union Indicate geographical differences in trace elements of the soils. In a further extension of this work, in a region of the southern Urals (Makarochkin et.al.,1960) it was observed that the molybdenum, copper, and titanium content of honey ash was significantly higher than in the local rocks. The technique at this stage is admittedly somewhat esoteric and will need to be proved by extensive experience and scientific testing until it has widespread acceptance by the scientific community. It is obvious that the analysis of livers can only be effective if the species are confined to small areas only. Shell fish has been reported to be very sensitive indicators of potential mineralization. Heavy metals (Cd, Cu, Fe, Mn and Zn) have also been reported.

GEOMICROBIOLOGY PROSPECTING

In descending the scale of size from trees to dogs to termites, the technique of geomicrobiology uses even smaller biota, the humble microbes. It is well known that microorganism have an important role to play in the processing of minerals. For example the bacterium thiobacillus ferroxidants can convert sulfide to sulfate and can be used to transform insoluble copper sulfide and can be used to transform insoluble copper sulphide to soluble copper sulphate that can be leached with water from a stockpile of excavated ore (Bricaley, 1982). These geomicrobiological techniques are now well established and in the United States alone, some 20% of the annual copper production is derived from microbiological processing of low-grade ores.

A recent development in geomicrobiological prospecting (Prudhin, 1995) involves the discovery that bacteria resistant to heavy metals are also resistant to antibiotics such as penicillin. Prudhin (1995) carried out her studies on "Bacillus cereus" extracted from soil samples taken across mineralized terrain through to background on each side. The methodology is based on shaking a small soil sample with water and, after centrifugation removing small aliquots, for transfer to pour plated containing unamended or penicillin amended Bcerus test media. (Brooks, 1995).

After 14-20 hours, the plates are examined for the appearance of circular opaque spots surrounding each pinpoint bacterial colony. The method is fairly well established in North America and has been tested over a wide range of environments. It has been used for gold exploration, particularly in desert terrains where conventional geochemical methods have limited effectiveness due to leached soil or where mineral deposits are located beneath a thick overburden. (Brooks, 1995).

CONCLUSION

It is although difficult to forecast the future of these studies but biological exploration techniques have had much greater acceptance in recent years. Perhaps this review will lead to generation of multiple new ideas and also new instrumentation to aid the biology to explore the hidden treasure of minerals. There is no reason why the method should not make progress in the near future of prospecting minerals greatly exceeding that which has gone before.

References

Aery, N.C.; 1977 : Studies on geobotany of Zawar mines. Geobios. pp 225-228.

Aery, N.C.; 1978 : Geobotanical studies of regions of zinc ore deposits in the Udaipur region, PhD Thesis university of Udaipur (unpublished).

Aery, N.C.; 1990 : Geobotanical and biogeochemical studies on certain tungsten deposits of Rajasthan, Scientific Report (unpublished) DST New Delhi.

Aery, N.C.; 1994 : Geobotanical and biogeochemical studies on certain uranium deposits of Rajasthan, Scientific Report (unpublished) DAE Govt. of India BARC, Bombay.

Aery, N.C. and Jain, G.S.; 1994 : Geobotenical studies on Uraniferous region of Udaisagar Rajasthan, India. Plant and Soil. Vol.2. pp34 - 41.

Aery, N.C., and Tiagi, YD.; 1985 : Biogeochemical studies on zinc deposit areas in the environs of Udaipur on cadmium -the pathfinder of zinc. Jour. Ind. Bot. Soc. Vol.. 64 pp. 148-155.

Aery, N.C., and Tiagi, Y.D.; 1986 : Bioindicators and accumulators in geobotanical and biogeochemical prospecting of metals, Acta.Biol.Hung Vol.. 37 pp 67-78.

Aery, N.C., and Tiagi, YD.; 1989 : Studi~s on the physiology of seed germination in the cultivar rose flowered double mixed and wild variety of impatients balsamina Linn, Nat. Acad. Sci. Lett. Vol.. 12 pp 108-114.

Aery, N.C., and Tiagi, YD.; 1990 : The signifance of elemental ratios in biogeochemistry, Jour. of Trop. Fore. Vol.. 6 pp 108-114.

Banerjee, P.K.; 1972 : Geology and Geochemistry of the Sukhinda Ultra meffic field: Cuttak District Orissa, Mem. Geol. Surv. India. Vol. 103 pp 11-19.

Boyle, R.W.; 1979 : Geol.Surv.Can.Bull, Vol.. 280, 576 p.

Brent, LA., Mc Innes, C. E., Dunn, E. M., Cameron, L. K.; 1996 : Biogeochemical exploration for gold in tropical forest regions of Papua New Guinea, Jour. of Geochem. Exp., Vol.. 57, (1-3) pp. 227-243.

Brooks, R.R.; 1972 : Geobotany and Biogeochemistry in mineral exploration, Harper and Row, New York. 225p.

Brooks, R.R.; 1980 : In biogeochemistry of nickel, Willey, New York. 314p.

Brooks, R.R.; 1983 : Biological methods of prospecting minerals, Willey and Sons, 323 p.

Brooks, R.R., Dunn, C.E., Hall, G.E.M (eds); 1995 : Biological system in mineral exploration and processing, Ellis Harwood limited, Hemel Hempsted, U.K. 538 p.

Cannon, H.L.; 1957 : Description of Indicator plants and methods of botanical prospecting for uranium deposits on the Colorado Plateaus U.S. Geol. Surv. Bull. 1030- M. pp. 399-598.

Cain, H.; 1932 : Concerning certain phytosociological concepts. Ecol. Monogr. Vol. 2 pp 475-505.

Cannon, H.L.; 1960 a : Botanical prospecting for ore deposits. Science, Vol.. 132, pp. 591-598.

Cannon, H.L.; 1960 b : The development of the botanical method of prospecting for uranium on the Colorado plateau U.S. Geol. Surv. Bull. A 1085, PP. 14-50.

Dom, P.; 1937 : Plants as a guide to mineral deposits (in German), Biologe, Vol.. 6 p. 11.

Gandhi, S.M., and Aswathanarayan, U.; 1975 : A probable base-metal indicator plant from Mamandur, South India. Jour. Geochem. Exp. Vol.. 4 pp. 225-228.

Harbaugh, 1.W.; 1950 : Biogeochemical investigations in the tri-state district. Econ.Geol., Vol.. 45 pp. 548-567.

Kahma, A., Nurmi, A., and Mattson, P.; 1975 : Geol. Surv. Finl. Rep, No. 6.

Karpinsky, A.M.; 1841 : Zhur. Sadovodstva, Nos. 3 and 4.

Kkdhal, E.; 1976 : Geochem. Exp., Vol. 5, 296 p.

Leo Harju, and Stig-Goran Hulden.; 1990 : Birch sap as a tool for biogeochemical prospecting, Jour. Geochem. Explo. Vol.. 37 (3).

Makarochkin, B.A., and Yedinic, D.M.; 1960 : Pchelovodstvo, Vol.. 37, p. 34.

Makarochkin, B.A.; 1974 : apiculture Abs. 25-104.

Malaisse, F., Gregoire, J., Brooks, R.R., Morrison, R.S., and Reeves, R.D.; 1978 : Aeolanthus bioformifolius De Wild: A hyper accumulator of copper from Zaire, Science Vol.. 199 pp 887-888.

Malyuga, D.P.; 1964 : Biogeochemical methods of prospecting, Constultants Bureau, New York, 213 p.

Malyuga, D.P.; 1964 : Biogeochemical methods of prospecting, Consultants Bureau, New York.

Nilsson, 1971 : Geol. For. Stockh. Forh, Vol.. 93, 725 p.

Nilsson, 1973 : Prospecting in areas of glaciated terrain, Institute of mining and metallurgy, London. 108p.

Orlov, A.P., Robonen, V.I., and Kirilenko, G.M.; 1969 : Geological prospecting with ore searching dogs (in Russ.) Nedra press, Moscow.

Poddar, B.C., (1965), A probable plant indicator for zinc mineralization in the zawar Pbzn belt Udaipur district Rajasthan, Curr. Sci. Vol.. 34, pp. 48-49.

Razin, L.V., and Rozkov, L.S.; 1996 : Geochemistry of gold

in the crust of weathering and in the biosphere in the gold deposits of the kuvanakh type, Nauka press, Moscow.

Reeves, R.D., and Brooks, R.R., 1978 : Trace element analysis of geological materials, Wiley, New York. p.

Roy's.; 1974 : Geobotany in the exploration for nickel in the ultramafic of the Sukinda valley Orrisa. Quart. Jour. Geol. Min. Metal l. Soc. India, Vol.. 46, pp. 251-256.

Sinha Roy, S. and Paul, D.K.; 1967 : Geochemical prospecting. Use of "A" hirizone in soil surfaces, Econ. Geol. Vol.. 462, pp. 1102-1103.

Singh, H.B.; 1982 : Geobotanical studies on the vegetation of Rajpura Dariba zinc bearing belt PhD Thesis university of Udaipur, Udaipur (unpublished).

Singh, R.; 1982 : Geobotanical studies on the rock phosphate deposits in the environs of Udaipur PhD Thesis University of Udaipur, Udaipur (unpublished).

Singh, R.; 1990 : Geobotanical and biogeochemical studies on rock phosphate deposits in the environs of Udaipur. In N.C.Aery and B.L.Choudhary (eds) Botanical researches in India. Himanshu Publ., Udaipur, pp. 425-430.

Strenfeld, A.; 1987 : Geochemical mapping and prospecting in Greenland a review of results and experience. Journal of geochemical exploration, Vol. 29, (1-3). pp. 189-205.

Tiagi, YD., and Aery, N.C.; 1981 : Biogeochemical studies in Zinc deposit areas of Zawar mines, Rajasthan, India. Proc. Indan Natl. Sci. Aca. Vol. 47, pp. 867-887.

Tiagi, YD., and Aery, N.C.; 1982 : Geobotanical studies of zinc deposit areas of Zawar mines Udaipur. Vegetatio. Vol. 5, pp. 65-71.

Tiagi, YD., and Aery, N.C.; 1986 : Biogeochemical studies on Khetri copper deposits of Rajasthan India. Jour. Geochem. Exp. Vol. 26, pp. 267-274.

Tiagi, YD., and Singh, B.P.; 1973 : Geobotanical studies ofthe regions of copper ore deposits in Madhya Pradesh, Rajasthan and Bihar, Scientific Report INSA. pp. 30.

Tkalich, S.M.; 1938 : Experience in the investigation of plants as indicators in geological exploration and prospecting (in Russian), vest. Dal'ne Vost.Fil.Akad.Nauk. SSR. Vol. 32, 3 p.

Venkatesh, V.; 1964 : Geobotanical methods of mineral prospecting in India, Ind. Min. Vol.18, p. 101.

Venkatesh, V.; 1966 : Geobotany in mineral exploration, Steel Mines Rev. (India). Vol. 6, pp. 3-5.

Wareen, H.V., Delavault, R.E., Fletcher, K., and Peterson, G.R.; 1971 : Can. Inst. Min. Metall. Spec. Vol. 11, pp. 444.

Waren, H.V. and Delavault, R.E.; 1949 : Further studies in Biogeochemistry. Geol. Soc. Am. Bull. Vol. 60, pp. 531-560.

Warren, H.V., Delavault, R.E., and Irish, R.I.; 1952 : Biogeochemical investigation in the pacific northwest. Geol. Soc. Am. Bull. Vol. 63, pp. 435-484.

Waren, H.V., Delavault, R.E. and Cross, C.H.; 1966 : Geochemistry in mineral exploration II west. Miner, Vol. 39, pp. 36-42.

West, F.W.; 1970 : Chamber of mines Jour. (Zimbabwe), Vol. 12, pp. 32.

Hon'ble Reviewers

1. Prof. D.C. Gupta
Head
Department of Applied Geology
Barkatullah University
BHOPAL 462 010 INDIA

2. Dr. D.K. Sinha
Scientist 'G'
Atomic Mineral Division
Deptt. of Atomic Energy (G.O.I.)
HYDERABAD 500 001 INDIA

Editorial Assistance

Dr. Virendra Gaur

Economic Mineralization
K.L. Shrivastava (Editor)
2009. pp. 406-415
Scientific Publishers (India), Jodhpur

37

Section-IV

An Integrated Approach for Kimberlite Exploration in Raipur District, Chhattishgarh

M. S. KHATEDIYA[1] AND PRAMOD K. VERMA[2]

[1]Department of Geology and Mining, Govt. of Madhya Pradesh, Bhopal 462 001
[2]School of Studies in Geology, Vikram University, Ujjain 456010
Email : drpkverma@rediffinail.com

Abstract

The Kimberlite exploration may involve great risk, when it is successful, returns very high rewards in tenns of diamond recovery. One of the major guideline for Kimberlite exploration is sound. understanding of tectonic control on evolution, emplacement and preservation of kimberlite. The Raipur district of Chhattisgarh state presents an ideal condition for localization of Kimberlites. Four Kimberlite pipes are already reported in the area. The present work is a multidisciplinary approach towards developing strategy for kimberlite exploration in the area. The area has been thoroughly investigated using remote sensing technique supported by field checks. The lithological, tectonic, geochronological, geophysical and. geochemical guides for Kimberlite exploration in the area have been evolved and Kimberlite Exploration Geographic Infonnation System (KEGIS) has been developed. The potential. target areas for Kimberlite exploration have been identified and demarcated.

Key words : Kimberlite, Bastar Craton, Diamond Exploration, Geographic Information System

INTRODUCTION

The science of exploration deals with search for hidden or concealed mineral deposits (vis-a-vis host rock and/or placer deposits). Fonnulation of exploration strategy is an outcome of multidisciplinary approach incorporating geohistorical infonnation, air-photo and satellite data interpretation, geophysical and geological studies, geobotanical and geochemical surveys, exploratory drilling and reserve estimation etc. The primary diamond occurs chiefly in kimberlite, lamproite and related rocks. Accordingly, the exploration for Kimberlite has hidden connotation as - "exploration for diamond". The fundamental requirement for Kimberlite occurrence is a persistently low geothermal gradient with little thermal erosion as the plume events may destroy the diamond potential of post plume Kimberlite magmatism. The formation and ascent of Kimberlite is speculative, however, lithospheric extension is considered to be L'te main cause for trigger of Kimberlite magmatism. As a rule, the Kimberlite is found in clusters, fields and provinces exclusive within thick Archaean cratons and immediately adjacent to Proterozoic mobile belt. The craton covered by relatively undeformed sedimentary rocks is considered to be an especially favorable environment for preservation of Kimberlite diatrems. The exhumation and/or deep erosion of craton leaves

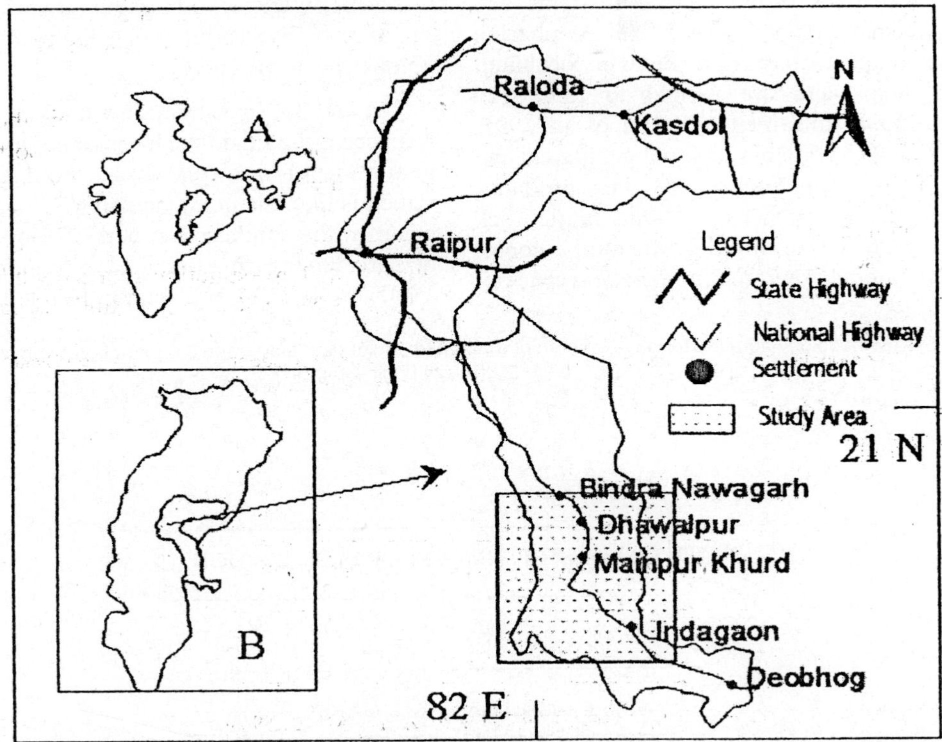

Fig.1 Location map showing area selected for study. Inset : **[A]** position of Chattisgarh State in India and **[B]** position of Raipur district in Chhattisgarh State.

Table-1 SOME IMPORTANT CHARACTERS OF THE KIMBERLITE PIPES OF MAINPUR-DEOBHOG AREA

Kodomali pipe (Root Zone Facies)	Paylikhand, Behradih and Jangra pipes (Diatreme Facies)
Country rock in which "pipe" is intruded represents highly weathered terrain.	The country rock is relatively less eroded.
Elongated irregular shape and in dyke like form trending NE – SW.	The morphology of all the pipes is circular to semicircular in plan.
Pipe rock is rather fre sh than other pipes in the area. The olivine is less serpentinised. Deficient in garnet and chrom-diopside, equigrannular porphyritic texture with rounded to idiomorphic olivine crystals.	Ground mass matrix mostly contain rare mantle derived xenoliths and discrete xenocrysts of garnet, chrom -diopside, picro -ilmenite, chromite etc.
Intrusive character with more kimberlite breccia C > 15% with > 4mm clasts in comparison to few crustal xenolith of granite and dyke.	Presence of rare mantle derived xenoliths and discrete xenocrysts of garnet, chrom-diopside, picro-ilmenite, chromite etc.
Width increasing and decreasing frequently.	Presence of abundant angular to sub rounded country rock xenoliths.
Pyroclastic fragments and textures absent.	The weathered mantle material of pipes is fragmented in nature.

behind the root zones of Kimberlite diatrems and hypabyssal feeder dykes. Often, the Kimberlite occurrences indicate distinct alignment with local fault and dyke systems and are related to frequently reactivated regional lineaments (White et al. 1995; Smirnov 1993). The available regional geological and geophysical maps, remotely sensed data products, indicator mineral occurrences, alkaline mafic rock occurrences, presence of high permeability zones, characteristics of craton and mobile belts, presence of sedimentary cover on craton etc are some of the important observations that help in targeting the Kimberlite occurrences.

The area for investigation within the Raipur district has been outlined on the basis of reported and possible gem mineral occurrences in the region. The gem minerals like, diamond, alexandrite, gamet etc. mainly occur in the southeastern part of Raipur district. The demarcated investigation area is bounded by latitude N 19.45 to N 20.30 and longitude E 82.00 to E 82.30

Fig.2 The methodology adopted for development of Kimberlite Exploration Geographic Information System (KEGIS) based on Fuzzy Logic Operation.

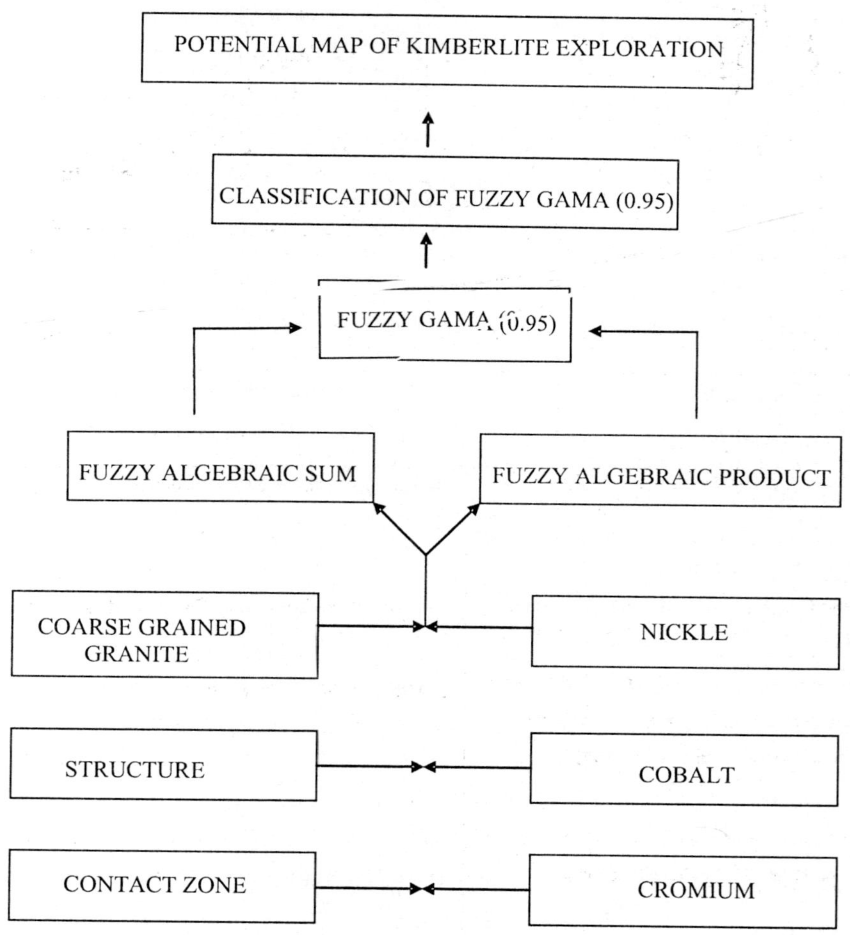

(Fig. 1) and is represented by the Survey of India toposheet nos. 64 L/3,4,7,8 and 65 1/5,9 covering an area about 3600 sq. km. The area is located at about 145 km from Raipur city on Raipur-Deobhog state highway.

The area under investigation, for further reference, has been abbreviated as Mainpur-Deobhog Area (MDA). Tectonically, the MDA is situated close to the boundary of the Bastar Craton (BC) and Eastern Ghat Mobile Belt (EGMB). The boundary is a metamorphic discontinuity and is an east dipping thrust (Chetty and Murthy, 1998). The younger rocks (Proterozoic) of EGMB are thrusted over the older basement rock (Archaean). The EGMB has been envisaged as convergent plate boundary with protracted history of formation and mobility spanning 2600-500 Ma. The metamorphic events in EGMB has been dated at 2500 Ma, 1000 Ma and 500 Ma, out of which the 1000 Ma event is more predowinant (Sarkar and Paul, 1998). The granitic rocks to the west of EGMB are regarded as basement of Archaean age. The overlying sediments of Chattisgarh basin and equivalents follow the western boundary of EGMB and are relatively unmetamorphosed. The Proterozoic platform sediments indicate two cycles of sedimentation, 1100-900 Ma and 700-450 Ma. As indicated by the presence of xenoliths of country rock and its characters, the Kimberlite emplacement is lying between these two protracted periods. The four known kimberlitic occurrences in Mainpur-Deobhog area are along the western shoulder of the Khariar syncline.

EXPLORATION STRATEGY FOR KIMBERLITE

The simplest exploration guide for Kimberlite is the known occurrence of Kimberlite. Once a Kimberlite occurrence is located in any area, the next essential factor for exploration becomes the establishment of regional and local structural control on the emplacement of Kimberlite. Additionally, the Kimberlite occurrences are revealed by observation such as, higher magnetic-gravity values, juxtaposition of cool cratonic mass - mobile belt -intracratonic basin, anomalous enrichment of Nb-Ni-Cr-Ce-Mg-Ba in weathered rock and/or soil horizon, exceptionally rich vegetation in an otherwise vegetation poor country etc. The Kimberlite rocks in Mainpur-Deobhog area are reported from Paylikhand (N20.09.45, E82.20.20), Jangra (N20.I1.00, E82.I9.00), Behradih (N20.13.00, E82.12.00) and Kodomali (N20.I1.35, E82.14.00). All

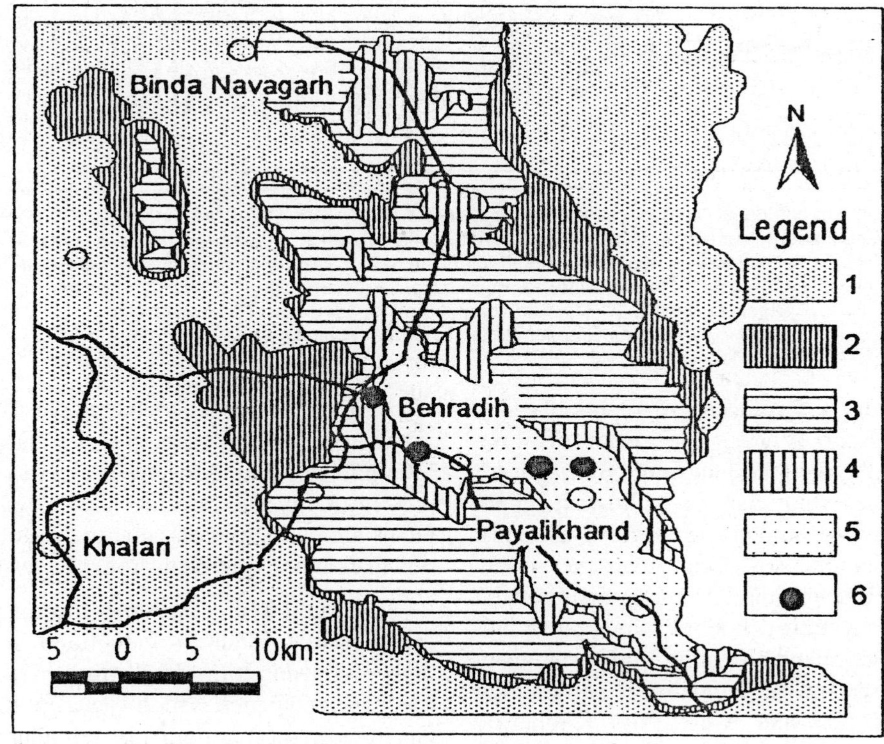

Fig.3 The probability potential map for kimberlite occurrence based on the Kimberlite Exploration Geographic Information System (KEGIS). Legends : 1 <0.64 (very low probability); 2 0.64-0.67; 3 0.68-0.76; 4 0.77-0.80; 5 0.81-0.90 (very high probability); 6 Known kimberlite occurrences.

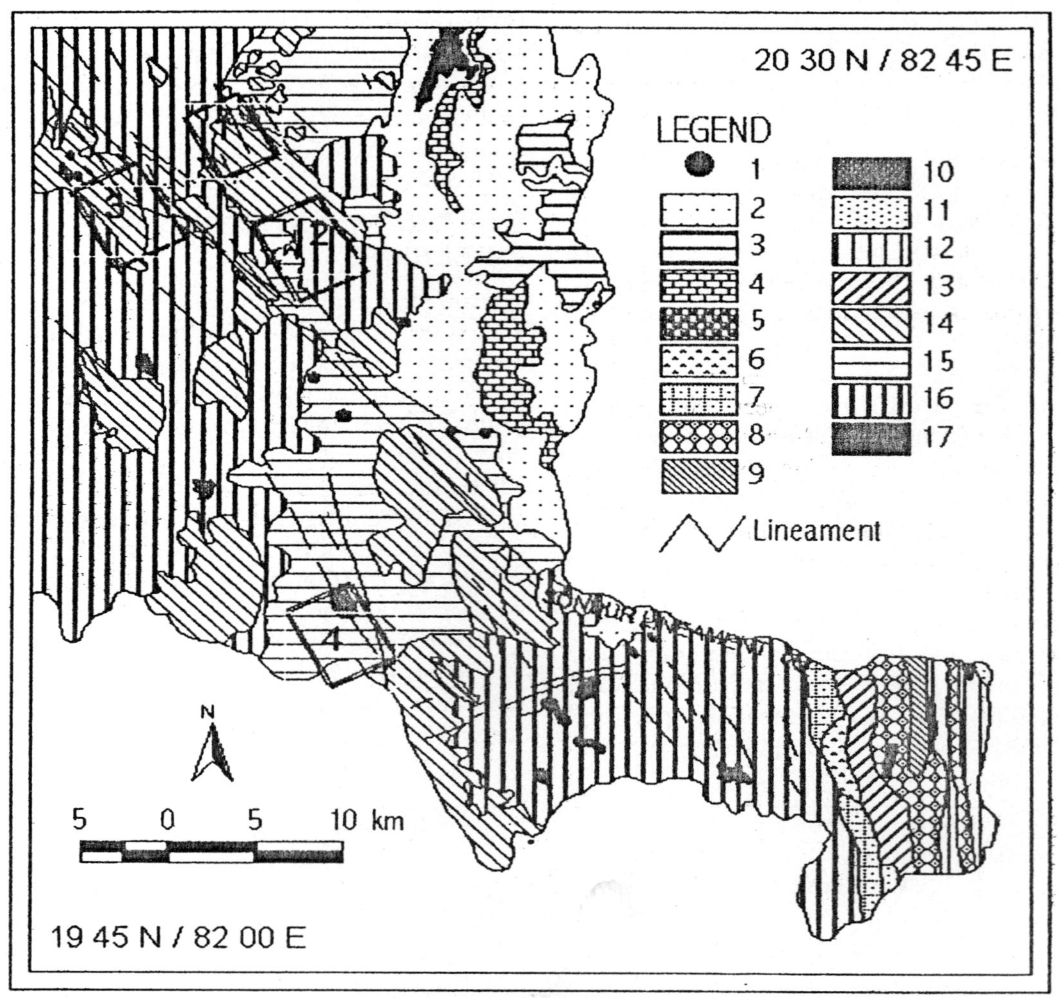

Fig. 4 Geological map of Manipur-Deobhog area, Raipur district, Chhattisgarh State. The superimposed boxes (1,2,3 & 4) represent priority zones identified for Kimberlite targeting. Legend : 1 Known kimberlite pipes, 2 Quartzitic Sandstone, 3 Shale/Siltstone, 4 Limestone/dolomite/chert, 5 Magnetite quartzite, 6 Metabasalt, 7 Augen gneiss, 8 Granulite, 9 Amphibolite, 10 Khondalite, 11 Charnockite, 12 Quartzo-feldspathic gneiss, 13 Mignatite gneiss, 14 Medium grained granite, 15 Coarse grained granite, 16 Granite gneiss, 17 Water bodies.

the four Kimberlite occurrences are reported to be diamondiferous (Babu 1998). The rocks of three localities, namely, Paylikhand, Jangra and Behradih are highly weathered giving rise to yellowish green coloured soil (locally known as yellow ground). Sometimes black colour soil occurs over the yellow ground (known as blue ground). Megascopically in the yellow ground, magnetite, ilmenite, and gamet are identified. The serpentine, phlogopite, and enstatite are visible ~der microscope. Chemically, it shows high values of Ni, Cr, Co and MgO. The Kimberlite occurring at Kodomali is relatively less altered. Megascopically it is hard, dull greyish green in colour and porphyritic in nature. Olivine (altered to serpentine) and garnet are easily observed in this rock. Some important characters of the Kimberlite pipes of Mainpur-Deobhog Area are presented in Table-1. The present Kimberlite occurrences belong to diatreme facies indicating thereby that the crator facies (usually rich in diamond) has been eroded away (Karanth 2000).

According to the available information, Kimberlite is the primary source of diamond in MDA. Due to higher concentration of ferromagnesian minerals in Kimberlite, the rock is highly susceptible to weathering and hence, morphologically circular to semicircular depression with or without water logging and thick vegetation patches are potential exploration target. The pathfinder factors for Kimberlite exploration are

Table-2 EXPLORATION STRATEGY FOR KIMBERLITES

S.No.	Global Exploration Guides for Kimberlite	Exploration Guides for Kimberlite in Mainpur – Deobhog Area
	General	
1.	Archaean craton with abnormally Cool keel of lithosphere.	Area comes under Bastar craton with Abnormally cool keel of lithosphere.
2.	Stable Archaean craton bordered by Orogenic mobile belt.	Eastern Ghat Orogenic Mobile Belt borders Bastar craton.
3.	Contact of proterozoic interacratonic basin with Archaeans.	Archaean rocks of Bastar craton are in contact with the Proterozoic intracratonic Khariar basin.
4.	Zone of high magmatic permeability.	A zone of high magnetic permeability as defined by repeated intrusion of various types of igneous rocks.
	Geochronology	
5.	Thermo-tectonic reactivation in mobile belt is dated around 1500±200 and 800±100 Ma and kimberlite events aro und 1000±100 Ma in central India (Sarkar et al. 1990).	EGMB is thermo-tectonically reactivated in 1500±2000 Ma and 800±100 Ma.
6.	The metamorphic events in EGMB have been dated at 2500 Ma, 1000 Ma and 500 Ma out of which 1000 Ma event is more predominant.	1000 Ma event in EGMB coincides with the kimberlite orogeny.
7.	Thick pile of sedimentary cover rock.	The Khariar Proterozoic platformal sediments indicate two cycle of sedimentation i.e. 1100-900 Ma and 700-450 Ma. The maximum thickness of sedimentary cover rock is about 1000m.
	Tectonics	
8.	Linear grabens / aulacogens with the history of frequent reactivation.	Area is bordered by the Mahanadi, Pranhita and Godaveri linear grabens / aulacogens rifts that have history of frequent reactivation.
9.	Region of crustal tension with mantle reaching deep seated fault.	Lineament study indicates that the area is under crustal tension with mantle reaching deep-seated fault i.e. sondur lineament etc.
10.	Splay of secondary fault and intersection of linear fractu re/fault corridors.	Splay of secondary faults and intersection of linear fractures/fault corridors are common in the area.
11.	An intersection of high permeability zone with transverse structures (Orthogonal fault).	Indicated an intersection of high permeability zone.
12.	Landward continuation of oceanic fracture zones.	Correlation yet to be made.

Table-2 Contd.

13.	A dome or block shaped uplift.	Dome shaped uplifts is displayed in fairly plastic sedimentary cover i.e. Khariar sediments, while block shaped uplifts occu r in more rigid cover rocks or within outcrops of basement on craton (Bastar craton).
14.	Location of radial or intersecting pattern of fracture lineament on a craton.	Lineament map fairly indicated such type of abnormality.
15.	Thin skinned thrust belt.	EGMB mobile belt.
16.	Shoulder part of an intracratonic mobile zone adjacent to a craton.	Shoulder part of a Khariar intracratonic EGMB zone, adjacent to Bastar craton.
17.	**Lithologic** Recognition of lithological anomalies. (Cliffard rules 1966) i.e. **Archon :** favourable places for emplacement of KLP. **Protons :** favourable for lamproites on the margin of craton. **Tecton :** not favourable for tectonic emplacement.	Most of the craton part in the Mainpur –Deobhog Area (MDA) is represented by Archon and Proton.
18.	Platform cover rock on the craton.	Khariar platform cover rock on the Bastar craton.
19.	Known occurrences of Kimberlite Pipes. (First hand discovery of kimberlite).	"4" pipe rocks are already discovered. The kimberlites always occur in cluster.
20.	Alkaline rock province.	MDA is included in alkaline rock province.
21.	**Geophysical** Circular/elliptical ground geophysical anomalies (aeromagnetic).	Geophysical (aeromagnetic) survey is in progress.
22.	High magnetic anomaly.	High magnetic anomaly is reported.
23.	High boundary velocity (v_p) and high gravity anomaly.	Represents high boundary velocity (v_p) and high gravity anomaly.
24.	Region of high heat flow (geothermal gradient 40-45 mw km^2).	Area absolutely comes under low geothermal gradient.
25.	**Geochemical** Dispersion halos of diamond indicator. High concentration of trace element (Ni, Cr, Co), indicator minerals (Phlogopite, Pyrope garnet, Picro -ilmenite, Olivine etc.) and whole rock geochemistry (high value of MgO).	Streams sediments show dispersion halos of diamond indicator.
26.	Finds of diamond within the dispersion halos.	Local diggers recover placer diamonds.

Table-2 Contd.

	Miscellaneous	
27.	Drainage anomaly.	Radial/meander/straight course and other drainage anomaly are noticed.
28.	Morphological depressions.	Morphological depressions are recognised.
29.	Biological / Geobotanical anomaly, (Abnormally luxuriant vegetation in an otherwise vegetation free area).	Such areas are recognised.

Archaean cratons with mantle reaching faults, intracontinental mobile zone with abnormally thick lithosphere and low geothermal gradient etc (Morgan 1995). The Bastar Craton, in general, indicates superimposed Kimberlitic event of possibly widely different age (Lehmann et al. 2006). In Mainpur-Deobhog area, the Kimberlite occurrences are identified through circular to semicircular morphological. features, abnormal drainage pattern and contact zone of basement (granite gneiss)-cover rocks (Khariar rocks). The known four occurrences are apparently controlled by Sondur fault cutting across the area. The identified exploration guides for Kimberlite of MDA in the background of global exploration guide is presented in Table-2.

FUZZY LOGIC IN KIMBERLITE EXPLORATION

A detailed discussion over development of Kimberlite Exploration Geographic Information System (KEGIS) is presented elsewhere (Khatediya and Verma, 2001a) "The kimberlite occurs in clusters" - is the guideline for development of Kimberlite Exploration Geographic Information System (KEGIS) in MDA. The KEGIS permits available data to be manipulated and displayed quickly. The results may be displayed in several formats and can be stored for repetitive application. The study takes into account the topographic, drainage, lithologic, tectonic and geochemical data used for Kimberlite exploration. These data are incorporated into the digital database for further manipulation and development of KEGIS. The digital data used for study are IRS-IC/ID WiFs and LISS-III GEOCODED products. The aerial photographs are used for collection of additional information for data input. The methodology adopted for development of KEGIS is presented in Fig. 2. The existing database for known Kimberlite occurrences in MDA has been prepared. The various controlling factors identified for Kimberlite occurrences in MDA are - close proximity to faults/lineaments, strong affinity to coarse grained granite, close proximity to Precambrian-Archaean contact zone, high concentration of Ni, Crand Co in residual soil. The gravity data is reinterpreted for identification of deep-seated faults. Based on above controlling factors, a probability ranking for Kimberlite occurrences in MDA has been worked out. The probability ranking is based on Fuzzy logic and is directly/inversely proportional to the distance/value in ppm from thecontiol points. Based on the KEGIS potential target areas for Kimberlite exploration have been identified.

TARGET AREAS FOR KIMBERLITE EXPLORATION

The prioritization of smaller potential segments for Kimberlite exploration takes into account the GIS result, permutation and combination of different exploration data and extensive field database. Though, the Kimberlite occurs in clusters, in a given geologic environment the various Kimberlite clusters may be separated in space and time. The concentration of diamond, garnet and other heavy minerals at the suitable placer locations provide indication for intensity of Kimberlite occurrences in any area. The occurrence of diamonds at Sinapali (adjacent to MDA but lying in the state of Orissa and, a trading center for diamonds and other minerals) in placer invariably points towards possible large number of primary sources, located upstream areas falling within MDA. One of the major hindrances in identification of Kimberlite in MDA is its small dimension and level of erosion since emplacement. The Kimberlite pipes, in general, occur with negative topographic expression and are often, picked up in low attitude, closely spaced aeromagnetic data. The aeromagn~tic data is available with GSI, however, fresh data with better resolution is being generated by the company that is undertaking diamond exploration in this area.

The result of KEG IS yields a fuzzy gamma, which is finally classified on the basis of the histogram of fuzzy membership value. The five probability classes

Table-3 Characteristics of identified target areas for Kimberlite exploration.

S.No.	Guide	Prioritised Zones			
		1	2	3	4
1.	Lithologic	Coarse grained granite and younger granite pluton. Coarse grained granite intruded by the acid and basic dyke, indicating the high permeable zone. Dykes are trending NW -SE and NNE-SSW.	Coarse grained granite and granite gneisses is overland by sedimentary formation (Intra-cratonic basin)	Coarse grained pink granite invariable intruded by the dykes. Indicate the high permeable zone (Dyke swarms).	Coarse grained granite with granite pluton. Two huge dykes namely amphibolite and meta gabbros are intruded in the coarse grained granite. In south India many pipes are intruded along the meta gabbro.
2.	Tectonic	Two major faults namely Sondur and Garchiroli are passing through the area, cut across to each other. Granitic rock is highly sheared, mylonitised, fractured, jointed and filling with materials of intrusive. Quartz vein and sheared granite makes a major shear zone (enclose proximity) Clusters of intersection of lineaments/ joi nts/fault. Splay of secondary faults.	Contact zone (litho) between intra-cratonic proterozoic basin and Archean basement. Garchiroli fault passing through the zone. Close proximity with sondur fault. Radial lineament system. Bedding traces making the colla r. Intracratonic two major lineaments namely Garchiroli and BN (N-S trending) apart from these many more intersections are also present.	Two major faults intersected in the region. Granite is highly sheared, fractured, jointed and filled with rocks of the later phase intrusion. Intersections of lineament/faults are moderate to dense. Densities of lineaments are very high. At places mylonitised granites are recorded. Shear zone is running NNW-SSE in the area.	Close promixity with the contact zone with intracratonic basin. Area bounded by two major parallel fault trending NW-SE. Lineament density is low. Proximity with Sondur fault.
3.	Geomorphic	Subdued highly undulatory topography, forming ridge and farrows. Circular and semicircular morphological features are identified.	Subdued topography with some circular to semicircular features are observed.	Circular and semi circular - morphological features are common. Highly dissected and undulatory topography.	Semi-circular, morphological features are identified.
4.	Geochemical	Abnormal value of Ni, Cr, Co. Indicator mineral of pyrope garnet is reported near the village Pandaripani.	Abnormal value of Ni, Cr and Co is observed in soil samples. Indicator minerals namely pyrope garnet, magnetite, diopside, ilmenite, s pinal are identified in the stream sediments.	Analysis of soil samples shows encouraging values of Ni, Cr, Mg and Co. The heavy minerals recovered from the samples are mainly garnet, diopside ilmenite, magnetite, spinel etc.	Soil sample analysis shows satisfactory values of Ni, Cr, Co & MgO. Stream sediment samples shows the pyrope garnet, diopside, ilmenite, magnetite, rutile, apatite etc.
5.	Miscellaneous	Drainage anomaly. Very contrast vegetation cover, some places having very thick, while at some place s it is very sparse vegetation cover.	Anomalous drainage pattern (Radial).	Anomalous drainage pattern, curvature of drainage pattern should be thoroughly studied.	Drainage anomaly.

are identified in which fuzzy membership value less than 0.63 indicate low possibility of Kimberlite occurrence, where as fuzzy membership value in 0.81 to 0.95 range indicates very high possibility of Kimberlite occurrence. The potential zone for Kimberlite exploration based on KEGIS is shown in Fig. 3.

The identified potential zone for Kimberlite exploration through KEGIS has limitations. The technique is primarily centered on known Kimberlite occurrences and is based on averaging of all observed factors. However, based on real world field data and other exploration guides (Table-2), four priority zones for Kimberlite targeting are identified (Fig. 4). The salient exploration guides established for identified zones are presented in Table-3.

CONCLUSIONS

The remote sensing technique has been found to be extremely useful in Kimberlite "exploration program in Raipur district of Chhattisgarh State. As Kimberlites are one of the important primary sources for diamond occurrence, the objective behind exploration exercise was to look for the possibility of diamond occurrence in the area. As the known Kimberlite pipes in the area, in general, indicate." considerable level of erosion, it is emphasized that apart. from search for primary source of diamond, efforts must be made to target potential zones for placer concentration of washed away fraction of diamond (Khatediya and Verma, 2001b). The Raipur district has much potential for future diamond exploration program.

ACKNOWLEDGEMENTS

MSK thanks the DGM, Govt. of M. P. for necessary permission. The present report is an outcome of the doctoral work of MSK.

References

Babu, T.M. (1998) Diamonds in India- Geol. Soc. India, Bangalore, 327 p.

Chetty, T.R.K. and Murthy, S.N. (1998) Regional tectonic framework of the eastern ghat mobile belt: A new Interpretation. Proceeding of workshop on Eastern Ghat Mobile Belt, Special Publication No. 44, Geol. Survey of India. pp. 39-50.

Karanth, R.V. (2000) Gems and Gem industry In India. Memoir - 45, Geol. Soc. India Bangalore. pp 394.

Khatediya, M.S. and Venna, P.K. (2001 a) Development of Kimberlite Exploration Geographic Infonnation System (KEGIS). GISdevelopment.net(http://www.gisdevelopment.net/application/geology/mineral/ geom0004pf.htm)

Khatediya, M.S. and Venna, P.K. (2001b) Strategy for placer diamond exploration in Mainpur-Deobhog area, Raipur district, Chhattisgarh State. Int. conf. on Diamond & Gemstone, Raipur, Chhattisgarh, Nov. 21-26, 2001.

Lehmann, B., Mainkar, D. and Belyatsky, B. (2006) The Tokpal Crater facies Kimberlite System, Chhattisgarh, India: Reconnaissance Petrography and Geochemistry. Jour. Geol. Soc. India, V. 68, pp. 9-18.

Morgan, P. (1995) Diamond exploration from the bottom up: Regional Geophysical signatures of lithosphere conditions favorable for. diamond exploration. In: W. L. Graffin (Editor), Diamond exploration into the 21st Century. J. Geochem. Explor., V. 53, pp. 145-165.

Sarkar, A. and Paul, D.K. (1998) Geomorphology of the Eastern Ghat Precambrian mobile belt - A review. Proceeding of workshop on Eastern Ghat Mobile Belt, Special Publication No. 44 Geological Survey of India. pp. 51-86.

Smirnov, Y.D. (1993) Structural setting of the Kimberlites of the east European craton. Intern Geol. Rev., V. 35, pp. 264-270.

White, S.H., De Boorder, H. and Smith, C.B. (1995) Structural controls on the emplacement of Kimberlites and Lamproites. In: W.L. Griffin (Editor), Diamond exploration intp the 21st Century. J. Geochem. Explor., V. 53, pp. 245-264.

Hon'ble Reviewers

1. Prof. R.P. Tiwari
   Dean
   Faculty of Forestry and Geology
   Mizoram University
   AIZAWAL 796 009
2. Dr. (Mrs.) Beena Tripathi
   Deptt. of Geology
   JNV University
   JODHPUR 342 005

Editorial Assistance

Mr. Chandra Prakash Dadhich

Economic Mineralization
K.L. Shrivastava (Editor)
2009. pp. 416-421
Scientific Publishers (India), Jodhpur

38

Section-IV

Rock Mass Classification System to Rock Excavation/Cutting/Boring Applications

RAJIV BADAL
SMEC India, New Delhi 110 001
Email : dr.badal@smeccig.com

Abstract

In various engineering projects including tunnel, dam, foundation construction etc. the rock mass classification system is a back bone for the empirical designing. Unfortunately, the same is not applicable so successfully to classify the rock mass for excavability, borability and chiefly for the mining of the economic deposits.

A versatile Rock Mass classification system have been proposed for the purpose. A brief outline of earlier rock mass classification systems, now serving as guide to arrive the classification system has also been discussed.

Key words: Rock Mass Classification system, Excavation, Boring.

INTRODUCTION

Rock mass classification system forms the back bone of the empirical design approach, which relates to practical experience gained on previous projects to the conditions anticipated at a proposed site. It has been used for a number of decades as one of the primary methods of empirical design in rock engineering. In fact, on many projects, the classification approach serves as the only systematic and practical basis for the design of complex excavations in rock. They have been applied for a wide variety of projects, ranging from tunnel and cavern design to dam and foundation construction, for rock slope rating, but seldom to classify the ease with which a rock mass may be excavated. To classify rock mass on the basis of excavability or borability has not been successfully applied to any of these projects. When applied to the excavation for borability on a variety of underground projects, none of these classifications come out to satisfy the required level of information, which can provide a clear picture or the confidence to define the rate of progress, i.e., excavability. Hence there is a need for a modified method of classification suitable to cuttability assessment of the many existing classification systems. Those that made important contributions are classification systems proposed by Terzaghi (1946), Deere (1967), Wickham et. al. (1972), Bieniawski (1973) and Barton et. al. (1974). For particular projects, some specialized ground classifications have been developed some of which incorporate seismic velocities (Ikeda and Nisimatsu, 1980; and Mitani. et. al. 1978). Barton (1995 and 1999) modified classification scheme along with compressive strength, using the seismic velocities to predict rock deformational modulus, porosity and Q value for the terrain evaluation. Some of these classification systems which give guidance to arrive upon an excavability / rock cutting classification are explained.

Rock Quality Designation (RQD)

RQD (Deere 1967) has been applied for the selection of steel supports in tunnel design but has limited applications in rock bolt supported openings. RQD does not provide geological structural information for its disregard for the orientation, tightness, infilling of joints etc. The most effective discontinuities for penetration oriented parallel to the excavation face and/or which have average spacing on the order of the disc cutter spacing. This means that indices, such as discontinuities frequency or RQD, evaluated on vertical core may not present information pertinent to penetration rate evaluation. Bilgin et al. (1996) had recommended that the performance prediction of roadheader can be made with the RQD and compressive strength of rocks for instantaneous cutting rate.

Rock Structure Rating (RSR)

The RSR concept (Wickham et al., 1972) represents a significant step in the development of rock, mass classification system for choosing support. The concept gives a quantitative approach to describe the structure of a rock mass and dimensioning the support system for a tunnel. The geological parameters, rock type, joint pattern, discontinuity type, major fault, shear and folds, rock material properties, water inflow and finally weathering or alteration. The construction parameters are size of tunnel, direction of drive and method of excavation. A particular feature of RSR is its recognition of the relevant importance of various factors in the support problem and the subsequent use of a weighting system of each factor.

Rock Mass Rating (RMR)

Also called Geomechanics Classification, the RMR system was developed by Bieniawski (1973). It was evolved out of the many existing systems by concentrating upon the best aspects of each classification to arrive at a value that represents both the properties of rock and the behavior of a rock mass. It is based on six parameters, which can be obtained in the field or alternatively from bore-hole data. The parameters are: Universal compressive strength, state; and orientation of discontinuities. It has been used in a variety of applications including obtaining a class description of rock mass, an average stand-up time (dependent upon unsupported span) and approximate value of cohesion and friction angle of rock mass in question.

NGI Tunnelling Quality Index (Q System)

It was proposed by Barton et al. in 1974. The concept, upon which the Q system is based, depends upon three fundamental requirements: classification of the relevant rock mass quality; choice of the optimum dimensions of the excavation with consideration given to its intended purpose, required factor of safety and also estimation of the appropriate support requirements for the excavation.

There are a few recorded instances of Rock Mass Classification System being used to assess rock that can be excavated. Abdullatif and Cruden (1983) applied both the Q and RMR systems to a number of quarries in England, using excavation methods of digging, ripping and blasting. They felt that the RMR identifies the limit of a particular excavation system better than the Q system.

Excavability Index

A classification based on Q system was developed by Kristen in 1982 for the characterisation of the excavability of soils and rocks. The joint planes allow the ripper to penetrate the rock mass and cut as a release planes from which the individual blocks can be lifted out. The efficiency of this process depends primarily upon the strength of the rock mass, the strength of the joints, the size and shape of the blocks and their relative orientation and related to the direction of ripping / excavation. The expression for the characteristics index takes the form

$$N = M_s (RQD / J_n) J_s (J_r / J_a)$$

where M_s = Strength number of the ground

RQD = Rock Quality Designation

J_n = Joint set number

J_s = Relative ground structure number

J_r = Joint roughness number

J_a = Joint alteration number

The mass strength number Ms, is related to shearing strength of the material and effort to excavate, as if it were homogeneous, intact and dry. Js is relative orientation of individual blocks to the direction of excavation. Other parameters are same as in the Q system. Validity of this ground excavability classification system was found reasonably accurate during actual construction. Bamford (1984) suggested that two geotechnical factors, based on rock mass classifications schemes warrant more study. Group based on RMR, system using the sum of parameters 3 and 4 with Laubscher's adjustment; and G_Q based on Q system using the product of the first two ratios RQD/Jn and Jr/Ja.

The rock mass parameters used in RMR and Q are representative of the site conditions for excavation of rocks. In RMR classification systems the parameters used are more representative to the rock mass properties, which influence more on mechanical rock excavation than the Q system.

Seismic Velocity for Rippability performance prediction

Fracture porosity, fracture index (RQD) and Rock mass quality (Q-system) had typically influenced the propagation of seismic wave velocity (Hatherly, 1983). This makes the firm base to evaluate the ground to characterise their properties including the rock mass classification to the rippable strata. Rippable rock strata are generally characterized by level of stratification, lower strength parameters, high porosity, and higher frequency of planes of weakness, weathering conditions and low seismic velocity. The field seismic velocity is an indication of rippability (Fig. 1), it indicates that weak planes exist for where weathering is found. Igneous rocks typically lack the stratification and cleavages essential for successful ripping, hence, usually rippable, when rock is deeply weathered and highly fractured. Sedimentary rocks contain individual layers varying in thickness, texture, colour, compositions and porosity, which are features of these materials. Whereas metamorphic rocks are varied in degree of stratification and foliation. These generally, have changed mineral composition and / or texture by intense pressure, high temperature, chemical and shearing action. Near surface the rock is generally homogeneous. The ripping characteristics of the rock formations are also very well described as to other parameters of the rocks classification system (Q-System) as given in the Fig. 2. The rock strata exhibiting a seismic velocity of less than 2000m/s are typically rippable, this limits can be increased up to 3000 m/s to a more intense machines. Barton (1995) has developed a relationship with the seismic wave propagation velocity with the ground properties relevant to the tunnel support design (Fig. 2), i.e., Q-system, including porosity and modulus of deformability etc. This can be proved to be useful for understanding the ground conditions which can be further applied to the performance prediction. Once the RMR is obtained for the rock mass corresponding Q-values are estimated from correlations and the performance predicted for TBM's. Similar, seismic velocities, can serve a better scale for the feasibility evaluation for the possible deployment of roadheaders. Fig. 3 shows, that at the closer joints spacing, the roadheader's cutter head is able to rip out block or larger pieces of rock and consumed less energy (Thuro and Plinninger 1999).

Use of Rock Mass Classifications in performance predictions of rock cutting applications

The rock mass classifications were developed and being practiced. It has been observed that all the rock

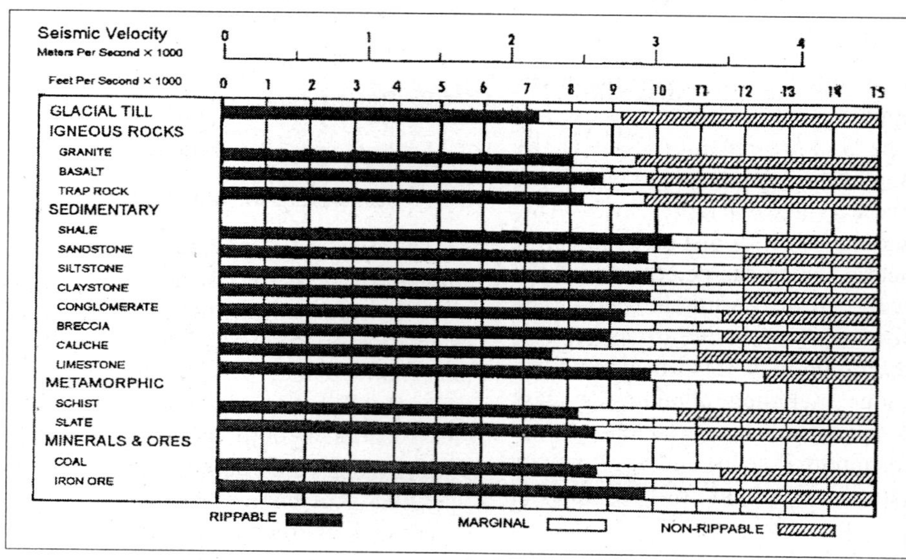

Fig. 1 Rippable rockmass classification system using seismic wave velocity in rock strata (after Caterpillar 1982)

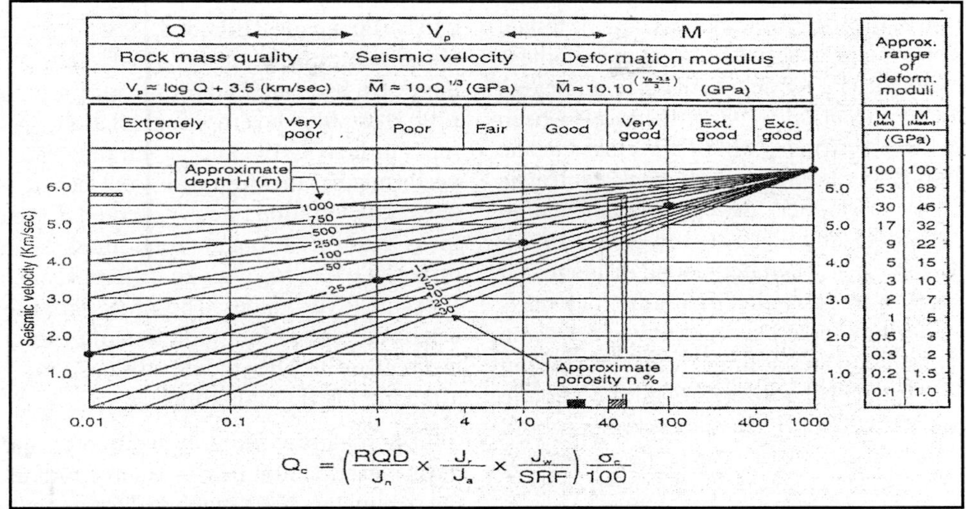

Fig. 2 Integration of Q-value with the in-situ rock properties, seismic velocity and rock mass static deformational modulus (after Barton 1999)

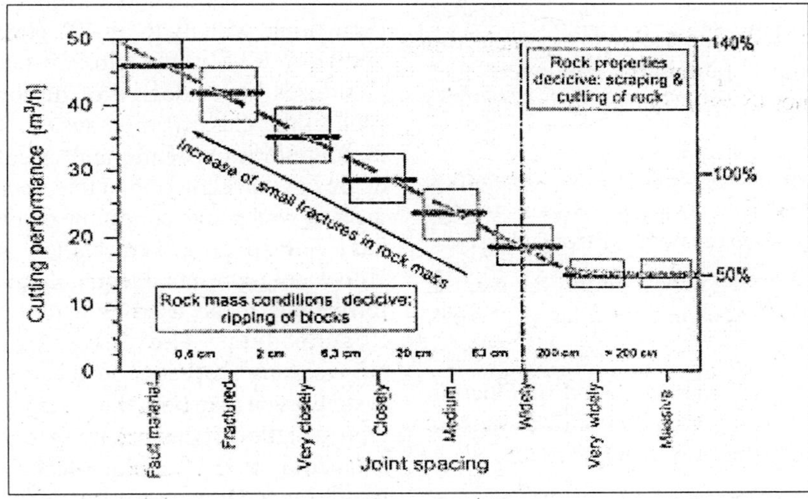

Fig. 3 Cutting performance versus joint spacing in argillaceous slate and quartzite of the Zeulenroda sewage tunnel by Roadheader (after Thuro and Plinninger 1999)

mass classification schemes are based on rock and discontinuities parameters, which are predominantly responsible for the rock mass behavior and accordingly they have been categorised /classified. Rock and rock mass parameters, which are typically responsible for the behavior pattern, because of their intrinsic / inherent geological parameters such as, strength, weathering, texture, porosity, hardness, abrasivity, Joint sets, spacing, orientation, persistence, aperture, roughness, wall alteration, filling materials, stresses, water condition, block size, RQD, etc. The rock and rock mass parameters are used in the various rock-mass classification schemes, based on some set of parameters, some of these parameters have influence on the rock cutting process and effect on the productivity of the mechanical excavation equipments greater than the machine's capability (even some times in many fold of the designed productive capacity). The rock and rock mass parameters, which are affecting the mechanical excavation, the available rock mass classification schemes have been analysed intrinsically and observed that some such classification schemes are primarily designed to assess the stability of underground opening in terms of gravitationally induced movement of discrete rock blocks rather than stress controlled categorising the ground conditions encountered as a means of describing the cuttability of rocks. Around the world, the TBMs tunnelling projects studied on the behavior pattern of the rock formations along the tunnel alignment, it has been observed that some of these schemes have shown a better degree of correlation than the others (Innaurato, et. al., 1988). It has also been observed that these schemes, which are closer to the rock discontinuity's parameters and rocks intrinsic properties, show a higher degree of correlation compared to others (Sapigni, 2002; Innaurato, 1993). Some schemes have also been developed or identified the strength indices or index along with the rock mass classification and they have been employed for some mechanical excavation tunnelling projects (Barton 1999 and 2000). Even, in the absence of any systematic scheme for the TBMs, tunnel excavation prediction and the available schemes are still not adequately explained or popularized to use for the prediction or performance evaluation of TBMs. Tallon (1982) applied the RSR, Q and RMR to the construction of several road tunnels in Northern Spain and found that the various systems yielded the area of application and the some short comings were identified. The following conclusions were drawn for most popularly employed rock classification schemes to rock cutting / boring by TBM as:

-The RSR system has a tendency to concentrate volumes in the medium strength classes to the determent of rocks found in the poor or strong classes.

-The Q system provides the best definition over the whole range of rock volumes in particular very weak rocks.

-The RMR system makes a good distinction between medium and strong rocks but fails to make a good differentiation with the poor strength rocks.

Jalil et al. (1995) employed RMR and Q-System parameters and stated that in the Q-System classification, Q value is more sensitive to variation in the extent of weathering, as reflected by changes in the alteration and roughness parameters. The RMR system is less sensitive to variation in joint surface condition.

Mckelvey et al. (1996) tried RMR, Q and Norwegian Institution of Technology (NTH) fissure classification systems on the two tunnels data from Clermont and Emolweni tunnels found a reasonable trend in relation to NTH fissure classes with RMR and Q system. However, it is desired to carry out further studies to correlate the classification schemes. Zaninetti et al. (1992) adopted RSR to define TBMs performance. In RSR rating, number of discontinuities, joint surface alteration and their orientations were incorporated. An acceptable degree of correlation exists between RSR and TBM net advance. The classification schemes of RMR and Q-System could have also been applied for the purpose. Sapingni, et al. (2002) established empirical relationship between rock mass rating (RMR) and penetration rate that the TBM performance maximum in the RMR 40-70, while slower penetration rate is expected in weak /poor rock-mass and in stronger rock masses. However, the different rocks have given different rate of penetration for same RMR values. The RMR classification scheme can be employed for prediction of penetration rate for a normalized RMR index, the use of basic factors affecting TBM's tunnelling.

The rock discontinuity parameters used in the rock mass classification of Q-system for tunnelling, not all the parameters have much influence on the rate of rock cutting such as SRF, and J_w. But, some of them, such as RQD, J_n, J_a and J_w have established prominently influencing parameters with a higher degree of correlation with ICR or CR (Instant cutting rate or cutting rate). In the similar pattern, RMR-system comprises better degree of discontinuity parameters, used in the classification system such as UCS, RQD, discontinuities spacing, and orientation have a better degree of correlation than the other parameters such as ground water and condition of discontinuities. RQD has a better degree of correlation with the ICR/ CR, but it does not represent adequately to classifying the rock formations to be used as a tool for the selection of Roadheaders. The discontinuities parametric propositions are better in the RMR compared to Q-system, even than both the systems can provide a better representation of the rock mass to be classified up to the extant of ICR/CR to make a decision on the selection of Roadheader. However, is does not replace/ preclude for the intact rock properties relevant to the economic rock cutting process for selection of roadheader.

ACKNOWLEDGEMENTS

Prof. R N Gupta, NIRM, KGF, for his critical suggestions and support during studies.

References

Abdullatif, O. M. and Cruden, D. M. (1983) The Relationship between Rock Mass Quality and Ease of Excavation, Bull. Int. Assoc. Eng. Geology, vol.28, pp.183-187.

Aeberli, U. and Wanner, H. (1978) On the influence of geological conditions at the tunnel boring machine. Bull. Int. Assoc. Eng. Geology, vol.12, pp. 7-14.

Anon, (1982) Caterpillar performance hand book; Caterpillar Tractor Company, Preoria, Illinois.

Bamford, W. E. (1984) tests for assessing Drillability,

cuttability and rippability of rocks are being international standardization. Excavation characteristics Seminar, Melbourne. pp 1-16.

Barton, N., Lien, R. and Lunde, J. (1974) Engineering classification of rock masses for design of tunnel support, Rock Mechanics, vol. 6 No.4, pp. 189-236.

Barton, N. (1999) TBM performance estimation in rock using Q t. Tunnel and Tunnelling. pp 30-34.

Barton, N. (2000) TBM Tunneling in jointed and faulted rocks, Balkema, Rotterdam.

Barton, N., Lien, R. and Lunde, J. (1974) Engineering classification of rock masses for the design of tunnel support, Rock Mechanics, vol. 6, no. 4, pp.189-236.

Bieniawski, Z.T. (1973) Engineering classification of jointed rock masses; Trans. South African Inst. Civ. Engrs. Vol. 15, No. 12, pp. 335-344.

Biligin N, Yazici S and Eskikaya S. (1996). A model to predict the performance of roadheaders and impact hammers in tunnel drivage. Eurock'96, pp. 715-720.

Blindheim, O. T. (1979) Drillability prediction in hard rock tunnelling. Tunnelling'79. UK pp. 284-289.

Deere, D. U. (1967) Technical description of rock core for engineering purposes; Rock Mechanics and Engineering Geology, Vol. 1, No. 1, pp. 17-22.

Gupta, R. N. and Badal, R. (1997) Rock Characterization for the Selection of TBM and Road Header for the Civil and Mining Engineering Projects, Annual Progress Report of the project of Ministry of Water Resources, New Delhi, Vol. I &II. (Unpublished Report).

Hatherly, P.J. (1983) The analysis of shallow reflection of seismograph. Ph. D. thesis, Macquarie university, Australia.

Ikeda K and Nisimatsu Y. (1980). The effect of geological properties on the productivity of tunnel boring machine. Rep. Int. tunnelling Assoc. Working Group Research, 14.

Innaurato N, Mancini R, Stragiotti L, Rondena E and Sampaolo A. (1988). Several year of experience with TBM in the excavation of hydroelectric. Int. Conf. Tunnel and Water, Madris. vol. 1 pp. 1027-1035.

Innaurato, N., Mancini, R., Rondena, E. and Zaninetti, A. (1993) Hydraulic tunnel excavation by TBM in schist the effect variability, as monoyored by the machine control instruments. Tunnelling' 93. pp. 539-546.

Jalil YA, Nelson PP and Laughton C. (1995). TBM performance analysis and rock mass impacts. 2nd Int. Symp. Mine Mechanization and Automation, Lulea. pp. 201-209.

Kristen, H.A.D.(1982) A classification system for excavation in natural material. Siviel Ingenieur in Suid Afrika, pp. 293-308.

Mckelvey JG, Schultz FA, Helin TAB and Blindheim OT. (1996). Geotechnical analysis in South African Tunnels. Tunnel and Tunnelling. Nov. 29-33.

Mitani S, Iwai T and Isahai H. (1978). Relation between condition of rock mass and TBM's feasibility. 6th ISRM Cong. 1:701-704.

Norwegian Institute of Technology (1994) Hard rock tunnel boring. Project report 1-94, Trondheim, Norway.

Norwegian Institute of Technology (1988) Hard Rock Tunnel Boring. Project report 1-88, Trondheim, Norway.

Sapigni, M., Berti, M., Bethaz,, E., Busillo, A., and Cardone, G. (2002) TBM performance estimation using rock mass classifications. Int. J. Rock Mech. Min. Sci. vol. 39, pp. 771-788.

Singh, R.N., Denby, B., Egretli, I. and Pattan, A. G.(1987) Assessment of Ground Rip-ability in Open Cast Mining Operation, Nottingham University, Mining Department Magazine, vol. 38 pp. 21-34.

Tallon EM. (1982). Comparison and application of geomechanics classification schemes in tunnel construction. Tunnelling'82. pp. 241-246.

Terzaghi, K.(1946) Rock defects and loads on tunnel supports; Rock tunneling with steel supports, Ohio, pp.15-99.

Thuro, K. and Plinninger, R. J. (1999) Roadheader excavation performance-geology and geotechnical influence. 9th Int. ISRM cong. On Rock Mechanics.

Wickham, G. E., Tiedmann, H.R. and Skimner, E. H. (1972) Support determination based on geological predictions; 1st RECT, AIME, pp.43-64.

Zaninetti A, Rondena E, Innaurato N and Mancini R. (1992). Hydrolic tunnels: review of the TBM excavation feature on the basis of Italian Electricity Bored experiences. Proc. Toward new world in tunnelling, Acaulco.

Hon'ble Reviewers

1. Prof. Arun Kumar
   Head
   Department of Earth Science
   Manipur University
   Canchipur
   IMPHAL 795 003 MANIPUR

2. Prof. Nikhil Desai
   Head
   Department of Applied Geology
   M.S. University of Baroda
   VADODARA 390 002 GUJRAT

Editorial Assistance

Dr. (Mrs.) Yasmin Sayyed

Economic Mineralization
K.L. Shrivastava (Editor)
2009. pp. 422-432
Scientific Publishers (India), Jodhpur

39

Section-IV

^{57}Fe Mössbauer Spectroscopic Study of Petroleum Prospecting Test Well Rocks from the Three Basins of India

R.P.Tripathi[1], Abhijit Kulshreshtha[2], Amita Tripathi[1] and K.L. Shrivastava[3]

[1]Department of Physics, J.N.V. University, Jodhpur 342005
[2]Jodhpur Engineering College and Research Center, Boranada, Jodhpur 342001
[3]Department of Geology, J.N.V. University, Jodhpur 342005
Email : rpt2002@sify.com

Abstract

Various petroleum prospecting test well rock samples from three petroliferous basins of India, namely Jaisalmer basin, Eastern Krishna Godavari basin and Cambay basin; have been characterised using ^{57}Fe Mössbauer spectroscopy at 300K. All the basins are geographically distant apart and the sediments in which potential source rocks are deposited are lithologically and stratigraphically different form one another. The source rocks, show similarity in the nature of the kerogen in sediments, yet there is noticeable difference in their prospects. In the present study ^{57}Fe Mössbauer spectroscopy have been used to characterised the sediments using the iron minerals present. The results obtained may act as a probe to obtain information about paleo-redox environment in which sediments were deposited, to understand ultimately the hydrocarbon prospects.

Keywords : Mössbauer Spectroscopy, Source Rock, Petroliferous Basin, Petroleum Prospecting.

INTRODUCTION

The exploration of petroleum is a complex process. Every information is vital and important during the course of exploration. To evaluate potential, every new exploration program starts with identification of the extent and kind of potential source rocks present in the sedimentary basin (Tissot and Welte, 1984; Hobson and Tiratsoo, 1981).

Hydrocarbon source rocks may be defined as fine-grained sediments which in their natural setting have generated, are generating, or will generate and release enough hydrocarbons to form a considerable accumulation of oil or gas (Brooks et al. 1987). Whether such an accumulation is commercial largely depends on economic considerations. A potential source rock has the capacity to generate hydrocarbons in substantial quantities, but has not yet reached sufficient organic maturation. Oil shales and coal must have a very high organic matter content to be of economic interest. In contrast, hydrocarbon source rocks may release only very small proportion of oil or gas per unit rock volume to form an important accumulation in reservoir rocks. This is possible if the source rocks represent a large and sufficiently subsided rock mass.

The preservation of organic matter is, among other factors, a function of oxygen content of bottom water, the sedimentation rate, and intensity of benthic life. With water oxygenation and benthic activity decreasing, the zone of methane fermenting bacteria expands upward at the expense of the sulphate reduction zone. As a result, more and also less stable organic matter can be preserved in the sediment.

During burial and under increasing temperature, the organic matter undergoes a series of geochemical reactions leading from "biopolymers" to "geopolymers", often collectively called kerogen (Tissot and Welte 1984; brooks et al. 1987). Besides kerogen, the organic matter of buried sediments still contains a small fraction of organic compounds, which are similar to the compounds originally produced by organism.

Carbonate sediments containing at least 0.3 to 0.6 % organic carbon and shales with 0.5 to 1 % organic carbon may already be regarded as hydrocarbon source rocks. These lower limits of potential source rocks depends, however, on the kerogen type present in these rocks. Low sedimentation rates under well-oxygenated conditions only allow the preservation of the so-called inertinite (usually in very small amounts), whereas anoxic depositional environments enable the preservation of comparatively large quantities of the H-rich liptinite and thus strongly enhance the hydrocarbon potential.

In most of the studies carried out so far the emphasis is usually given on the analysis of the organic part of sediments to correlate it with probability of finding petroleum in a petroliferous basin, in spite of the fact that inorganic part (that also include iron bearing mineral) goes together and can provide valuable information. In present work we have made an attempt to show that study of iron minerals present in sediments can be used as the probe to obtain information about paleoredox environment in which sediment were deposited. We believe that this information when coupled with conventional studies can provide a better picture for hydrocarbon prospecting.

Mössbauer Spectroscopy is a promising means for non-destructive characterisation of iron bearing minerals. At the same time, this method can also characterise the various chemical states and sites in which iron is present. (Tominaga and Minai, 1984). Mössbauer spectroscopy has been used to understand many complex problems of geology. It has been used for mica (Tripathi et al. 1978), coal (Shrivastava et al. 1992), lignite (Tripathi and Shrivastava, 1990; Verma and Tripathi 2000) meteorites (Shrivastava et al. 1997; Tripathi et al. 2000) Polymetallic nodule (Shrivastava et al., this volume) and source rock evaluation for petroleum exploration (Morup et al. 1985; Tripathi et al. 1998; Kulshrestha et al. 2004). Since, limited number of iron bearing minerals are expected to be present in the sediments, Hence, it is even more useful. The technique provide all the information about every iron bearing minerals present in the sample, by least square fitting of the raw spectrum obtained from the original sample using a suitable Mössbauer fitting programme. Hence, in present investigation we have employed this technique for the study of the source rocks sediments.

SOURCE ROCK AND IRON MINERALS

The process of deposition of sediments in a petroliferous basin during its primary stage of deposition are the mixture of large amount of water, dead organic matter, inorganic phases (which also contain iron) and numerous living microorganisms. Such a mixture results from various sedimentary processes and primary components of very different origins so it is out of equilibrium and therefore unstable. System tends to approach equilibrium through different schemes of evolution of organic matter viz. diagenesis, catagenesis and metagenesis, (including, transformations, alterations and reprecipitations) so that sediment becomes consolidated (Fig.1). During this process of stabilization the organic and inorganic interaction occur at different stages of sediments evolution. The diagenesis of inorganic and organic matter in sediments may influence the oxygen ion concentration (Eh) and hydrogen ion concentration (pH) conditions prevailing in the sediments (Fig.2). The degree of diagenesis is related with the generating potential of the source rocks. Hence it is clear that some correlation exist between paleo-redox condition and source rock potential. In perspective of hydrocarbon prospecting the detailed study of paleo-redox conditions is important. Fortunately, the crucial information about reducing or oxidizing conditions in which sediments are deposited can be obtained by study of iron bearing minerals present in sediments, e.g. the presence of pyrite indicates highly reducing condition and hematite/goethite on the other hand show presence of severe oxidizing conditions.

PETROLEUM GEOLOGY OF SAMPLES

In the present work we have carried out study on large number of rock samples collected from different depth interval of the test wells located in three different basins

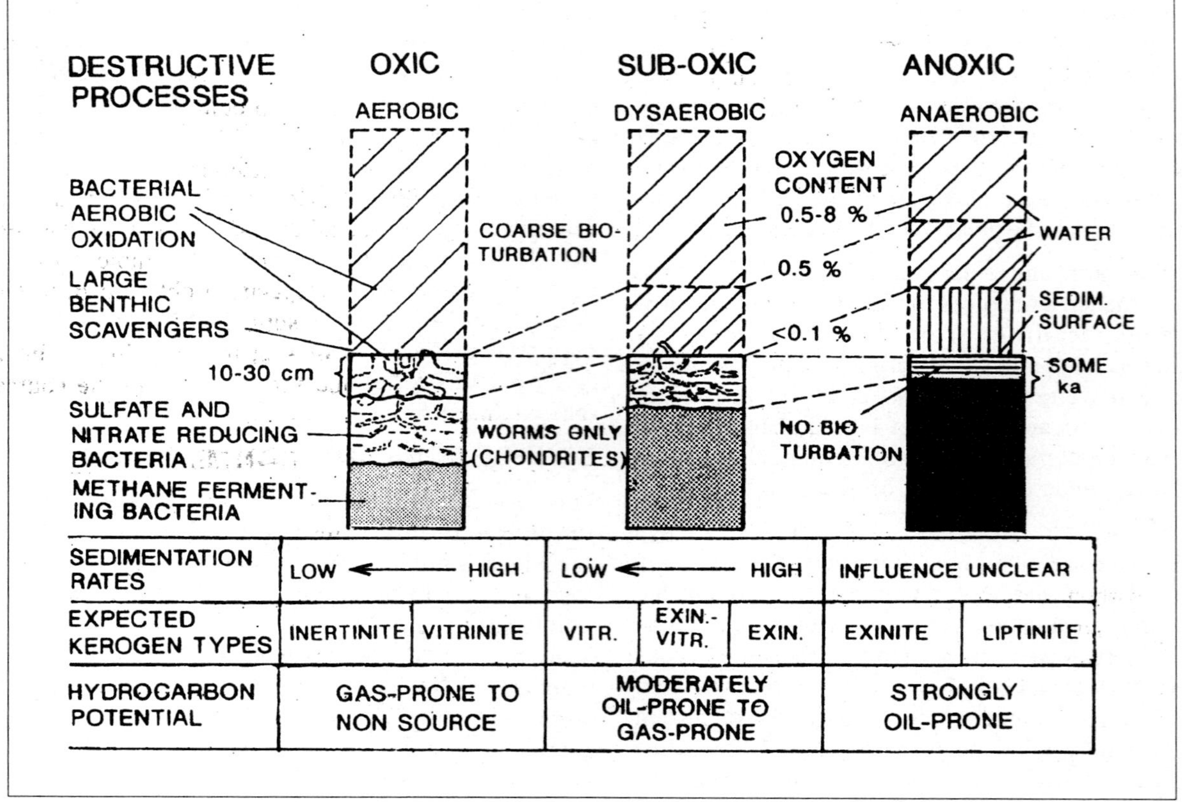

Fig.1 Hydrocarbon potential of petroleum source rocks in relation with kerogen types, benthic environment and sedimentation rate (After Demaison and Moore, 1980; Brooks et. al. 1987)

viz Cambay, Eastern Krishna Godavari (EKG) and Jaisalmer basin of India namely, K-263 of Cambay basin, CTP-1 (Chintalapalli-1) and MDP-1 (Mandapeta-1) of EKG basin and DND-1 (Dandewala), LNR-1 (Lunar-1), and MNW-1 (Manewala-1) of Jaisalmer basin. All these basins are geographically distant apart and the type of sediments in which potential source rocks are deposited are lithologically and stratigraphically different. Litho-Stratigraphic details with depth intervals (m) for wells of three different basins are given in Table-1, 2 and 3. The source rocks studies carried out by exploration agencies (ONGCL or OIL) (Dutta, 1983; Rao et al., 1988a and b; Mitra et al. 1993; and Chandra, 2000). On these rocks, show similarity in the nature of the kerogen in sediments. The studies also indicate that sediments in all these three basins mainly contain Type III kerogen, which is being considered as adequately matured to generate hydrocarbons and favourable conditions for the accumulation of reservoirs. The Cambay and the KG basins, proven the major gas fields. The Jaisalmer basin comes out to be a very poor basin. It is deprive of oil. And the quality of gas is also not good, because of high concentration of CO_2 and N_2; although it possesses a great geochemical similarity with both, KG and Cambay basins.

EXPERIMENTAL DETAILS

For the Mössbauer spectroscopic investigation, Mössbauer absorbers were prepared by sandwiching finely ground sediment powder between two paper discs in a 25mm diameter sample holder. All the spectra were recorded at room temperature (300K) using constant acceleration spectrometer and ^{57}Co source in Rh matrix. The Mössbauer spectra were recorded for sample collected from depths for each well. Spectra were analyzed by fitting an appropriate number of discrete doublets and sextets with Lorentzian peaks using a least square-fitting programme. The solid line in the spectrum represent computer fitted curves and dots represent the experimental points. Each iron bearing mineral show a characteristic doublet on the basis of known assignment of the doublets, we were able to identify different iron bearing minerals present

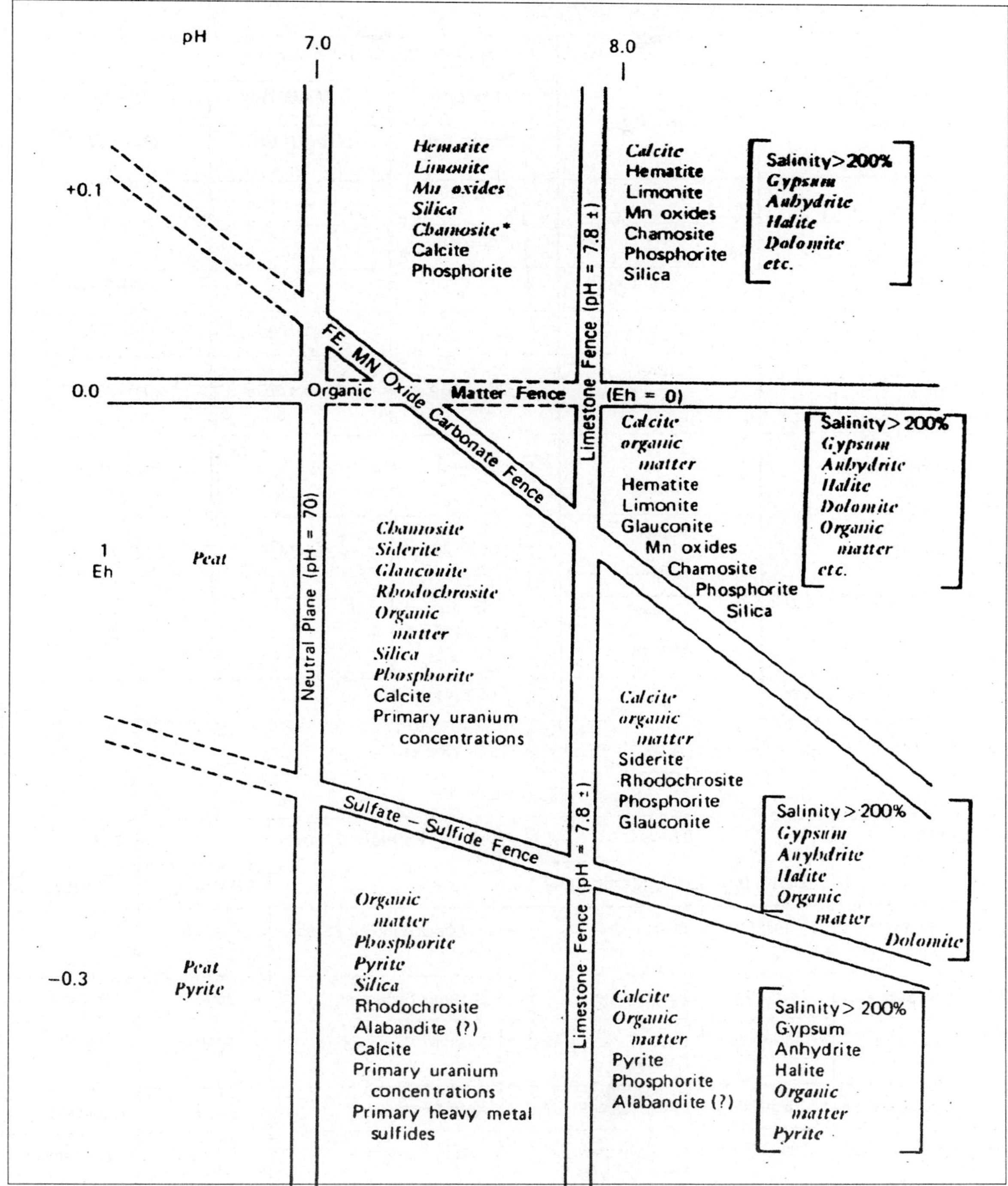

Fig.2 Fence diagrams showing relations of chemical sediments (including iron minerals) to Eh and pH controls for normal seawater conditions (After Krumbein and Garrels, 1952)

in each sedimentary rock samples.

RESULTS AND DISCUSSION

All the spectra obtained have been analysed using its parameters to recognize the iron minerals present in these sediments alongwith their relative amount. Diagrams have been prepared to show the result of relative distribution of various iron minerals as a

Table-1 Litho -Stratigraphy of wells of Jaisalmer Basin

Formation	Stratigraphy	Well of Jaisalmer Basin		
		LNR-1	MNW-1	DND-1
		Depth (m)	Depth (m)	Depth (m)
Khuiala		0-150		
Sanu		150-281		
Parh	Upper Cretaceous			1300-1445
Goru	Upper Cretaceous		1005-1540	1445-2017
Pariwar	Lower Cretaceous	281-710	1540-1779	2017-2509
Baisaki -Badesir	Upper Jurassic	710-1008	1779-2290	2509-3037
Jaisalmer	Middle Jurassic	1008-1722	2290-3450	3037-4363
Lathi	Lower Jurassic	1722-2175		4363-4498
Shuearwah and below		2175-2641		

Table-2 Litho -Stratigraphy of wells of Krishna -Godavari Basin

Wells of KG Basin					
Formation	CTP -1		Formation	MDP -1	
	Depth (m)	Stratigraphy		Depth (m)	Stratigraphy
Rajahmundary Sand Stone	0-430	Post Eocene	Rajahmundary Sand Stone	0-260	Tertiary
Narsapur Sandstone	430-480	Post Eocene	Narsapur Sandstone	260-360	Tertiary
Matsyapuri sandstone	480-1665	Upper Eocene to Post Eocene	Nimmakru Sandstone	360-735	Tertiary
Bhimanapalli Sandstone	1665-2000	Middle Eocene	Razole Formation	735-770	Paleocene
Pasarlapudi Formation	2000-2935	Late Paleocene to Lower Eocene	Tirupati Sandstone	770-1595	Late Cretaceous
Palakollu Shale	2935-3590	Early Paleocene to Late Paleocene	Ragahavpuram Shale	1595-2480	Early Cretaceous
Razole Formation	3590-3940	Early Paleocene	Gollapalli Sandstone	2480-2740	Late Jurassic
Chintalapalli Shale	3940-4500	Late Cretaceous	Manadapeta Sandstone	2740-3315	Permo-Trissic
			Kommugudem Formation	3315-4265	Early Permian

Table-3 Litho -Stratigraphy of well K-263 of Cambay Basin

Formation	Stratigraphy	Depth (m)
Post Kand	Recent to Miocene	665
Post Kand	Mid Miocene to lower Miocene	665-915
Babaguru	Lr. Miocene	915-1157
Tarapur	Oligocene to Upper Eocene	1157-1350
L.Kalol	Middle Eocene	1350-1535
Cambay Shale	Lower Eocene	1535-2595
Olpad	Paleocene	2595-4507

function of depth for all the wells selected in the present study from all the three petroliferous basins.

The relative distribution of the minerals as the function of depth for the representative well K-263 of Kalol basin given in Fig. 3 for well MDP-1and well CTP-1 of KG basin are displayed in Fig. 4 and Fig. 5 respectively; for well DND-1, LNR-1 and MNW-1 of Jaisalmer basin are shown in Figs. 6,7 and 8; respectively.

If one compare the relative distribution of iron minerals in various wells there is a sharp difference in the distribution pattern of wells located in gas prone basins (i.e. KG and Cambay) and wells located in the Jaisalmer basin. The wells of Jaisalmer basin show

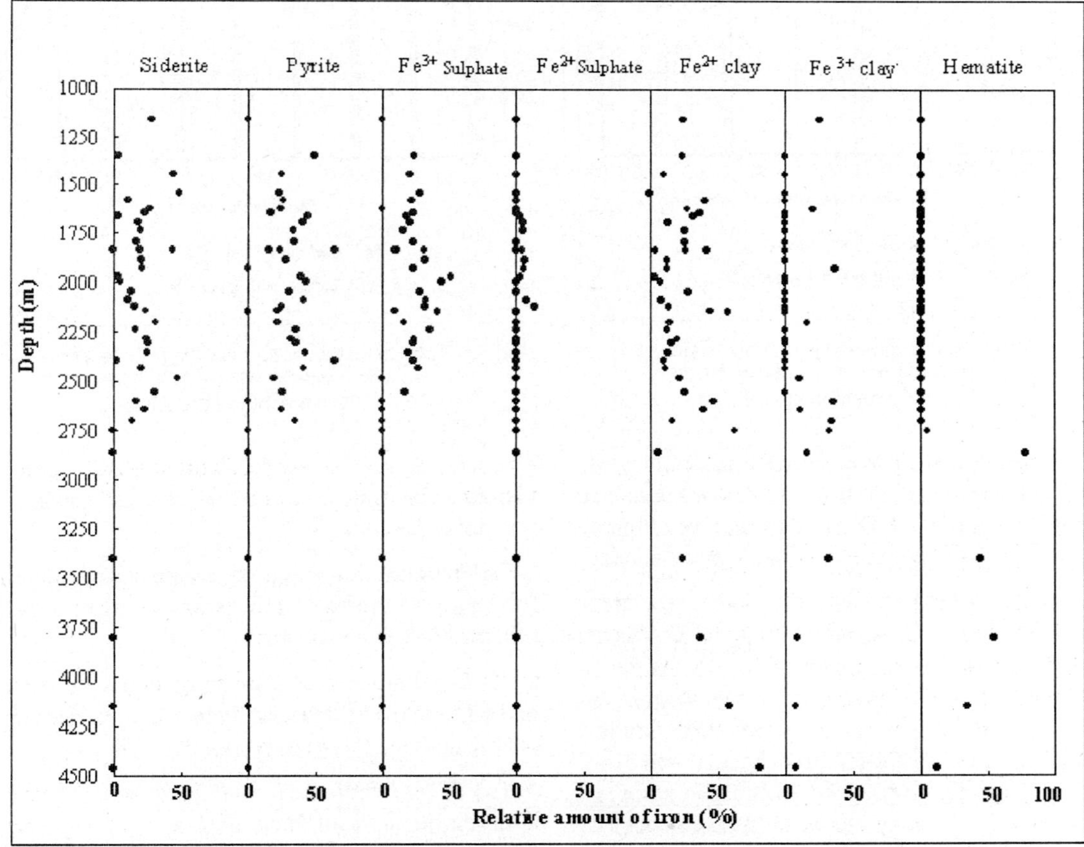

Fig.3 Relative distribution of iron minerals (%) in sediments versus depth in well K-263

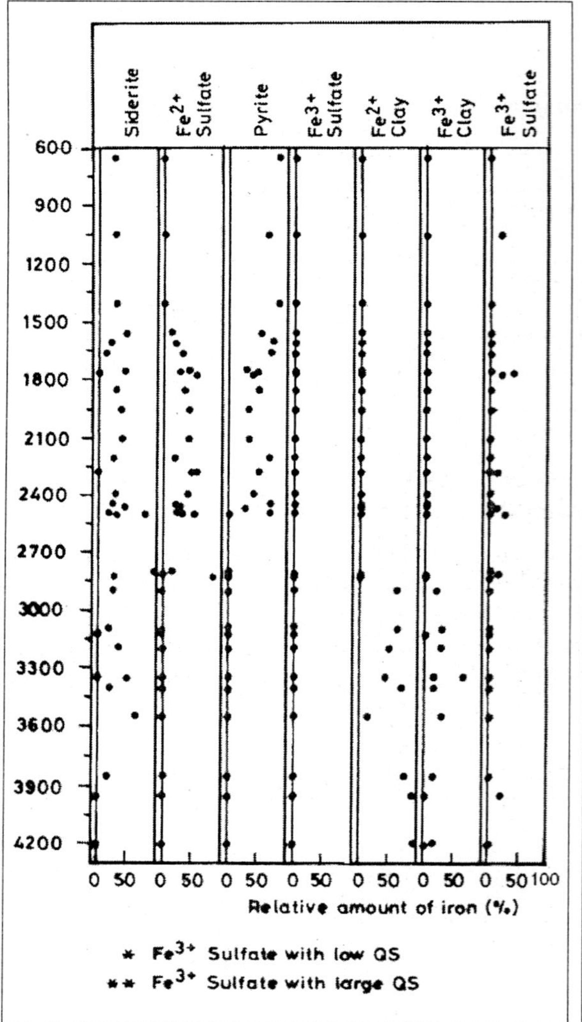

Fig.4 Relative distribution of iron (%) in sediments versus depth in well MDP-1 (After Kulshreshtha et al. 2004)

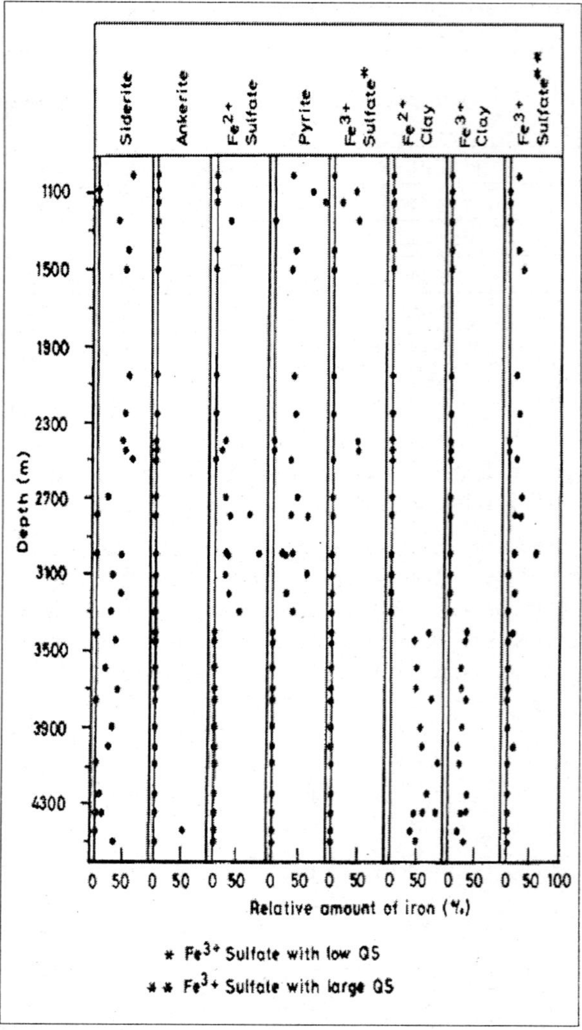

Fig.5 Relative distribution of iron (%) in sediments versus depth in well CTP-1 (After Kulshreshtha et al. 2004)

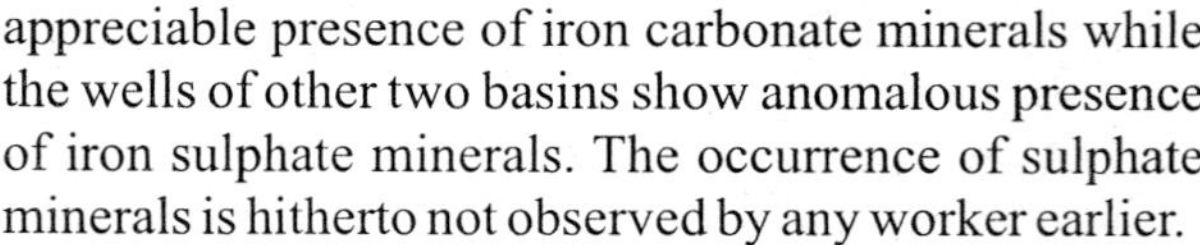

appreciable presence of iron carbonate minerals while the wells of other two basins show anomalous presence of iron sulphate minerals. The occurrence of sulphate minerals is hitherto not observed by any worker earlier.

It is worth pointing that sulphate minerals are hardly found in deep sub surface sediments, as discussed above. Their presence is reported in literature in exposed coal samples, where they are formed by oxidation of pyrite in minute amount. But studies carried on sediment of these three basins, have not exhibited any significant presence of coal in samples. As an example, in Cambay basin, the source rocks of younger and older Cambay shale formation possess shales as the major component of sediments. Mössbauer studies show considerable amount of sulphate minerals even in these shale samples, which was not expected.

By examination of relative distribution pattern, the following important features regarding sulphate minerals have been observed :-

(i) Sulphate minerals are present in wells of Cambay and KG basins and not in wells of Jaisalmer basin i.e. only found in wells of the proven gas fields.

(ii) This indicates that presence of gaseous hydrocarbon and sulphate minerals goes together in the present case.

(iii) From distribution pattern, we can also see that

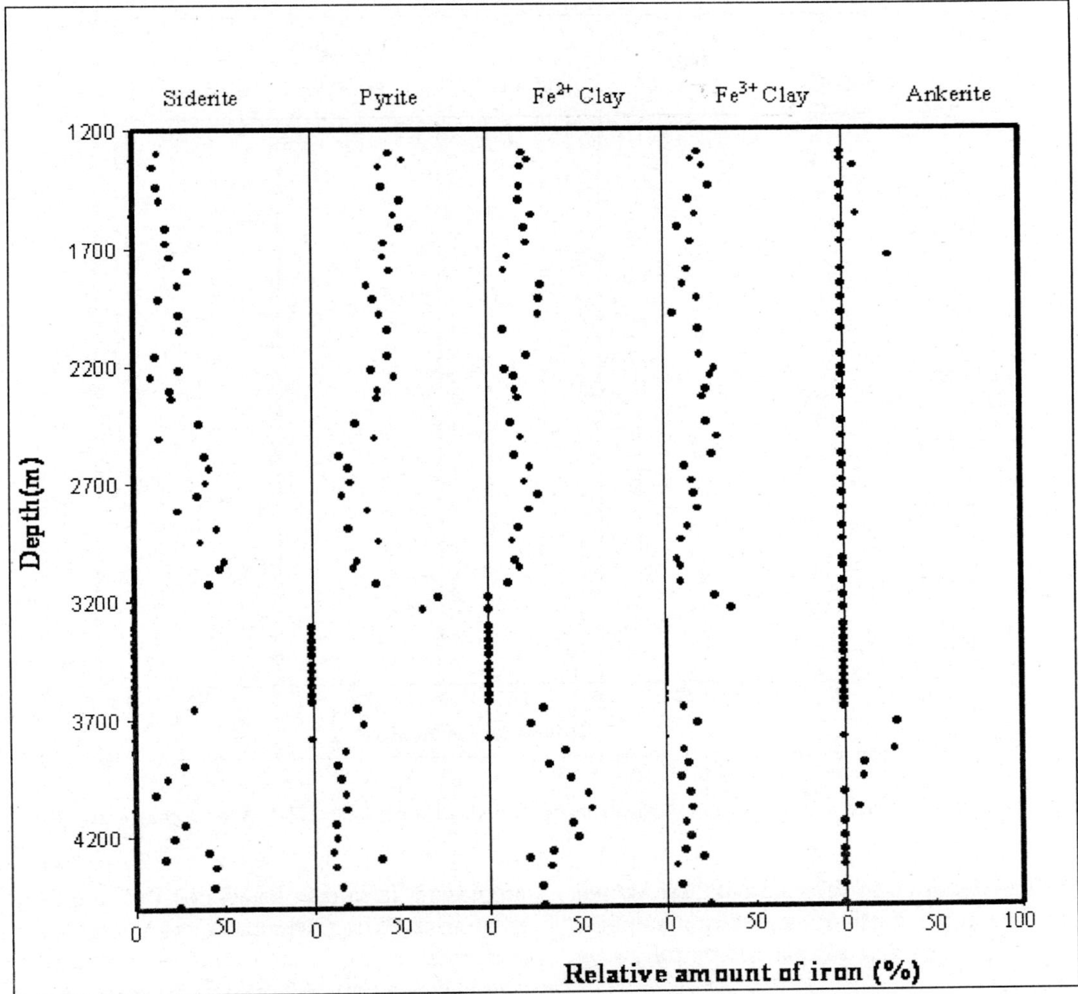

Fig.6 Relative distribution of iron minerals (%) in sediments versus depth in well DND-1 (After Tripathi et al. 1998)

there is some relation between amount of pyrite and sulphates. As sulphate increases quantity of pyrite is reduced.

(iv) The depth interval in which sulphate minerals are found belongs to a marine sections. The marine conditions are always more favourable for deposition of better quality of source rocks.

(v) Iron sulphate minerals are formed below the certain depth in all wells.

It is important to note that wells in which sulphates are discovered, are distant apart from each other and lithologically and stratigraphically diverse but even than invariably in all the wells sulphate minerals are found and that also below a particular depth. This clearly indicates that sulphide iron minerals are not formed during the process of deposition of the sedimentary sequence, looking towards the other geological parameters. If it would have been so, they ought to have been found randomly with depth rather than being confined to a particular depth. Since they are formed below a certain depth it means that temperature might have played a crucial role in creation of sulphate minerals. Below a certain depth source rocks must have reached a temperature, probably this temperature must have also coincided with temperature where transformation of organic matter takes place into gas. So one possible way of there formation can be explained like this:

In the process of transformation, organic matter might have released, O_2 and H_2O, which may have resulted into oxidation of pyrite (Shrivastava et al., 1992). The sulphates are formed on the expense of

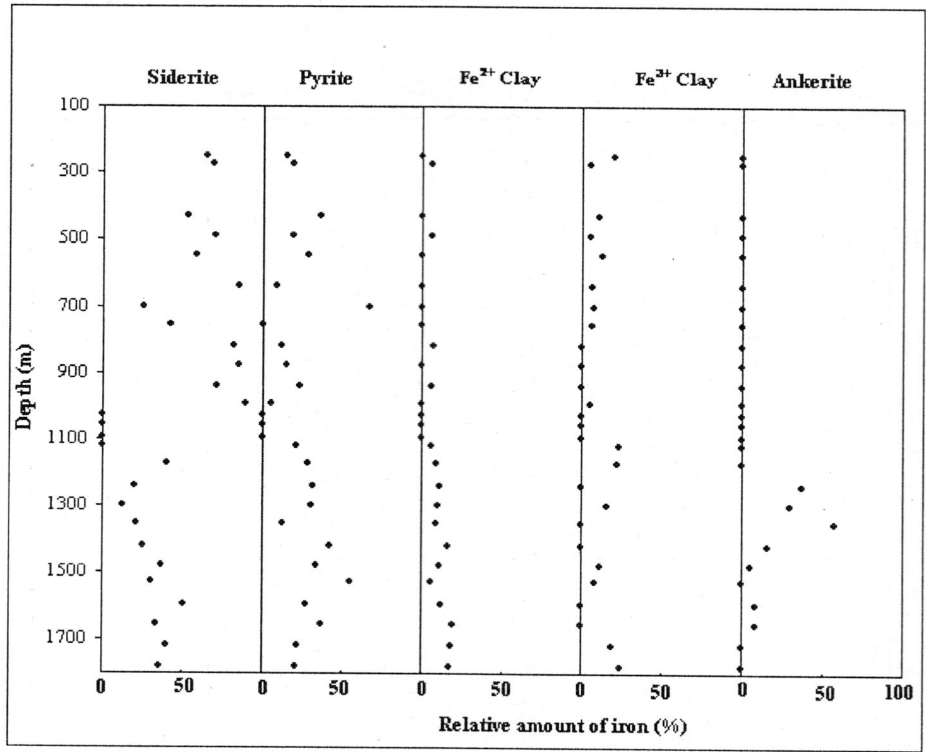

Fig.7 Relative distribution of iron minerals (%) in sediments versus depth in well LNR-1 (After Tripathi et al. 1998)

pyrite is further supported by the fact that there is total absence of pyrite in the distribution pattern if sulphate minerals are present. It can be explained by following chemical reaction:-

$$3\,FeS_2 + 10\;O_2 \rightarrow FeSO_4 + Fe_2(SO_4)_3 + 2SO_2$$

pyrite

It is also possible that there may be one or few intermediate steps in the process.

The predominating presence of sulphate minerals in source rocks in any structure of a basin is indicating presence of gas. May be, because sulphate-bearing minerals are indicative of oxidizing conditions it may indicate unfavourable circumstances for the oil formation. It is interesting to note that in both the basin the quantity of sulphur is significant in gases explored.

It seems possible that the absence of oil and gas in Jaisalmer basin can also be explained by advocating similar arguments. The wells of Jaisalmer basin show dominating presence of iron containing carbonate minerals (siderite). The siderite is known to form in open sea and also indicator of fluctuating redox conditions, its appreciable amount in most of the samples in Jaisalmer basin possibly indicating poor reducing condition, which are unfavourable for oil and gas formation. So in such cases even when the quality of source rocks is good one cannot expect good quality of oil and gas. Recently, a well CT-1 was drilled in Jaisalmer basin it was expected on the basis of conventional studies, that good oil must be present there, but actual exploration could not yield oil. We studied the sample using Mössbauer effect from this well. There is presence of large amount of siderite. These once again confirm our present arguments. The Mössbauer spectra of samples collected from well CT-1 of Jaisalmer basin shows dominating presence of siderite.

CONCLUSION

Iron bearing carbonate (siderite) and sulphate minerals provide crucial information about the paleoredox environment and hence knowledge of their presence/absence together with the quality/quantity of organic matter can provide significant information means even a good organic matter for the petroleum exploration. It seems that even good organic matter in poor paleo-redox condition or vice versa cannot be

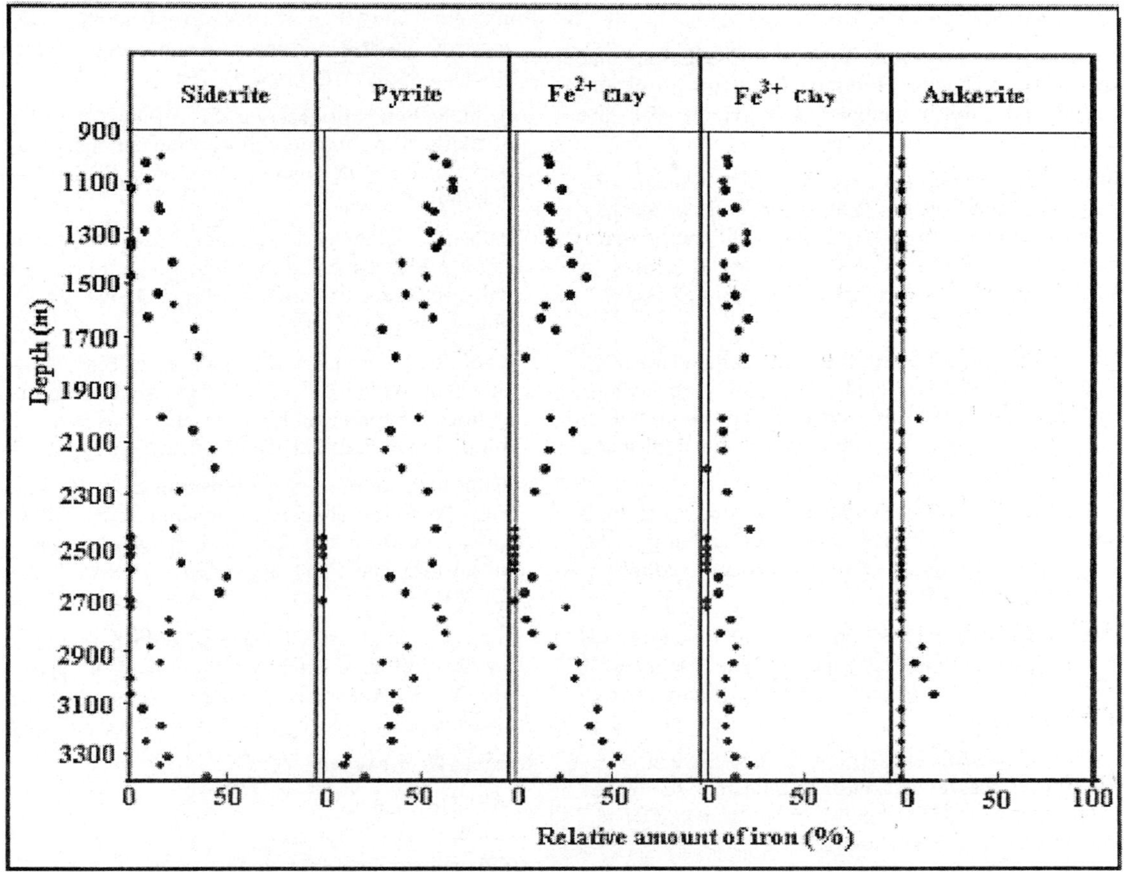

Fig.8 Relative distribution of iron minerals (%) in sediments versus depth in well MNW-1 (After Tripathi et al. 1998)

considered as favourable prospects. Hence, a more reliable picture can be obtained by using the collective studies.

^{57}Fe Mössbauer Spectroscopy studies are of great significance to find the distribution of iron minerals. In addition it may provide crucial information about paleo-redox condition and paleo-temperature of the sediment of petroliferous basin. Mössbauer Spectroscopy coupled with other methods may prove a finger print method for accessing hydrocarbon prospecting in future. To achieve this yet extensive study is needed from several petroliferous basins all over the world. Present work may be useful in this respect.

ACKNOWLEDGEMENTS

We are thankful ONGCL for providing the samples. One of the authors (RPT) is grateful to Department of Science and technology (DST) and University Grant Commission (UGC), for financial assistance to carryout this work. Amita Tripathi is grateful to, Council of Scientific and Industrial Research (CSIR) for providing Research Associateship.

References

Brooks, J., Cornford, C., and Archer, R.; 1987 : The role of hydrocarbon source rocks in petroleum exploration In : Brooks J. and Fleet A.J. (Eds.) Marine Petroleum source rocks. Geol. Soc. Spec. Publ. Vol. 26, pp. 17-26.

Chandra, K.; 2000: Petroleum Systems in the Indian Sedimentary Basins: Stratigraphic and Geological Perspectives, special lecture in 12th Indian Geological Congress.

Datta, A.K., 1983, Geological evolution and hydrocarbon prospects of Rajasthan basin, In: Bhandari, L.L. et. al. (eds), Petroliferous basins of India: Petroleum Asia Journal , Dehra Dun (India), Vol. 6, pp. 93-100.

Demaison, G.J., and Moore, G.T.; 1980 : Anoxic environment and oil source bed genesis. Am. Assoc. Petrol. Geol. Vol. 64, pp. 1179-1209.

Hobson, G.D.; and Tiratsoo, E.N.; 1981: Introduction to Petroleum Geology. Edn. II, Gulf Publishing Company

Houston, Texas, 352 p.

Krumbein, W.C. and Garrels, R.M.; 1952 : Origin and classification of chemical sediments in terms of pH and oxidation reduction potentials. Jour. Geol. Vol. 60, pp. 1-33.

Kulshreshtha, A., Tripathi, A., Agarwal, T.N., Patel, K.R., Sisodia, M.S. and Tripathi, R.P.; (2004): ^{57}Fe Mössbauer spectroscopic study of organic rich sediments (source rocks) from test wells CTP-1 and MDP-1 located in Eastern Krishna Godavari basin, India. Fuel Vol. 83, pp. 1333-1339.

Mitra, P., Mukherjee, M.K., Mathur, B.K., Bhandari, S.K., Qureshi, S.M., and Bahukhandi, G.C. ;1993: Proceedings Second Seminar on Petroliferous Basins of India, Vol. 2, (Eds. Biswas, S.K. et. al.), Indian Petroleum Publishers, Dehradun, India.

Morup, S., Frank, J., Wonterghem, J., Poulsen, R. H., and Larsen, L.; 1985 : Mössbauer spectroscopy study of the chemical state of iron in Danish Mesozoic sediments. Fuel, Vol. 64, pp. 528-539.

Rao, G.B., Narain Lal, and Saxena, R.P.; 1988a: Source rock evolution of sedimentary sequences of exploratory well Mandapeta-1, K.G. Project, ONGC, Rajamundari;. (unpublished report).

Rao, G.B., Phillips, P.C., Chopra, V.S., and Saxena, R.P.; 1988b: Source rock evolution of sedimentary sequences of exploratory well Chintalpalli-1, K.G. Project, ONGC, Rajamundari (unpublished report).

Shrivastava, K.L., Tripathi, R.P. and Jangid, M.L.; 1992 : Spontaneous combustion of coal : A Mossbauer approach. Fuel. Vol. 71, No. 4, pp. 377-380.

Shrivastava, K.L., Tripathi, R.P. and Vaya, V.K.; 1997 : A Mossbauer spectroscopic study of the Pipliyal Meteorite. Curr. Sci. Vol. 72, No. 2, pp. 139-141.

Tissot, B.P.; and Welte, D.H.; 1984: Petroleum Formation and Occurrence. Edn.II, Springer- Verlag, Berlin Heidelberg, New York, 699 p.

Tominaga, T. and Minai, Y.; 1984 : Application of Mössbauer spectroscopy to environmental and geochemical studies. Nucl. Sci. Appl. Vol. 1 pp. 749-791.

Tripathi, R.P. and Shrivastava, K.L.; 1990 : Characterization of Merta Road lignite and decompositional study of its pyrite using Mossbauer spectroscopy. Fuel. Vol. 69, pp. 928-930.

Tripathi, R.P., Chandra, U., Chandra, R., and Lokanathan, S.; 1978: A Mössbauer Study of the effect of heating Biotite, Phlogopils and Vermiculite Jour. Inog. Nucl. Chem. Vol.40, pp. 1293-1298.

Tripathi, R.P., Sharma, S.K., Patel, K.R., Shrivastava, K.L. and Sahi Ram 1998 : 'A Mossbauer approach to hydrocarbon prospecting in Jaisalmer basin in Rajasthan, India'. Ind. Jour. Petrol. Geol. Vol. 7, No. 1, pp. 47-58.

Tripathi, R.P., Sharma, S.K., Shrivastava, K.L. and Verma, H.C.; 2000 : Mossbauer spectroscopic studies of the Piplia Kalan (eucrite) and Lohawat (howardite) meteorites. Meteoritics and Planetary Science (USA). Vol. 35, pp. 201-204.

Verma, H.C., and Tripathi, R.P.; 2000:Characterization of iron bearing minerals in Giril liginite from Rajasthan and study of their decomposition during combustion using ^{57}Fe Mössbauer spectroscopy. Fuel. Vol.79, pp. 599-606.

Hon'ble Reviewers

1. Prof. H.C. Verma
   Department of Physics
   Indian Institute of Technology Kanpur
   KANPUR 208 010 (U.P.)
2. Prof. S.K. Date
   Emiretus Professor
   Department of Physics
   Pune University
   PUNE 411 001 (M.S.)

Editorial Assistance

Dr. Jaimini Swadia

Economic Mineralization
K.L. Shrivastava (Editor)
2009. pp. 433-441
Scientific Publishers (India), Jodhpur

Section-IV

Petroleum Geology and Hydrocarbon Prospects in the Barmer-Sanchor Basin, Northwest India

MAMTA CHAUHAN, JAIMINI SWADIA, TINA PRAJAPAT AND K.L. SHRIVASTAVA
Department of Geology, J.N.V. University, Jodhpur 342 005
Email : mamtageology@yahoo.co.in

Abstract

Recently discovered hydrocarbon prospects in the Barmer-Sanchor Basin of northwest India have rejuvenated hope of Indian economy regarding energy resources. The Barmer-Sanchor Basin is narrow elongated, N-S trending, linear graben, conceived as the northward extension of the Cambay Basin with present disposition grossly modified by Neotectonism of the region. It is essentially a Tertiary Basin, formed over a basement comprising of older rocks which include Late-Mesozoic formations. So far, Thumbi Member, Dharavi Dungar Formation and Barmer hill Formations have been proved economically potential with Guda Shales and Barmer Shales as source rocks, with sandstone and siltstone as main reservoir rocks of both, Structural and Stratigraphic traps.

The first few wells drilled by Cairn Energy and Shell have given good indicators yet there were primary exploration risks, prior to Mangla discovery. So far, Cairn Energy has drilled almost 100 wells and discovered 10 fields. The northern fields comprising of Mangla, Aishwariya, Bhagyam, Shakti, Vijaya and Vandana. The southern fields are Saraswati, Raageshwari Oil, Raageshwari Gas, Guda, GR-F and Kameshwari. The Cairn Energy estimated the total oil to be more than 250 billion bbls in the region with major estimates so far, for Mangla 1067 mm bbls, Bhagyam 355 mm bbls and Aishwariya 219 mm bbls. The production, which can now be started, have been targeted for these three, in the range of 120,000 to 150,000 bopd.

Key words : Barmer-Sanchor Basin, Petroleum Prospects, Petroleum Geology, Tertiary oil.

INTRODUCTION

Economic condition of a country chiefly depends upon its energy resources. In India energy resources and economic growth rate both are increasing simultaneously since independence but after 1980 it took 8% of economic growth rate and 15% of total energy consumption. Traditional fuels are gradually getting replaced by the "Commercial fuels" such as petroleum products and natural gas for the last two decades. At present, India has no good energy resources, hence it is depending mostly, upon other countries like Middle East, U.S.A. and Britain. With introduction of new economic policies, advanced technology and collaboration with multi national companies, new petroleum fields like Barmer-Sanchor, have been discovered.

Global crude oil and gas reserves at the current estimates are likely to last for 45 years and 65 years respectively. In India this is still lower at 19 years and 45 years. The production level have peaked at 33 MMT per year where as the demand of oil has grown from 18.5 MMT per year in 1970 to about 124 MMT per year in 2004 and hence Indian import dependency has increased to a staggering 74%.

India, which imports 80 to 90 million tonns of crude oil annually (cost : over Rs. 85,000 crore Rs. 850 billion) against the total domestic availability of only 3233 million tonnes, desperately needs to discover more petroliferous regions or extention of the older ones.

As per Government of India estimates the country's oil requirement will rise to 3.2 million barrels per day (around 2.2 million barrels now) by 2010. In the process, the country, which is now the seventh largest consumer of oil, will emerge as the fourth largest after the United States, China and Japan.

Cairn Energy Pvt. Ltd., (Scotland) has made almost a dozen discoveries in the Barmer-Sanchor basin. Each discovery is at a different stage of appraisal and evaluation. It is estimated that the total volume of oil found in place is more than two hundred fifty billion barrels. During the year 2004, six discoveries were made, including the three northern fields Mangala, Aishwariya (March) and Bhagyam (August). The Mangala and Aishwariya fields have been fully appraised by six and five appraisal wells respectively and 3D seismic data. Four successful Bhagyam appraisal wells have been drilled between February and April 2005 and two more appraisal wells are planned. A 3D seismic survey over Bhagyam is in progress.

Current life of field reserve estimate of 500 mmbbls is based on secondary recovery only for the three northern fields, Mangala, Bhagyam and Aishwariya. Northern fields peak production target 120,000 to 150,000 bopd. Independent assessment of the three northern fields, assigns 1.64 billion barrels of oil in place. Application to more than double the development area from 1,858 km^2 to 4,743 km^2 includes Bhagyam and Shakti (NC) fields. Total expected production from all fields is about 45 million tones per year and the production from the Barmer-Sanchor Basin is expected to start by the end of 2007.

The aim of the paper is to present the information of petroleum geology, potential and future prospects of this newly discovered petroliferous region. The informations have been gathered from recent literature and internet sites of MNCs available for academic and scientific benefits.

GEOLOGY OF THE REGION

The Barmer Basin opened as a narrow, elongate, roughly NS trending linear graben partly overstepping into the Jurassic rocks of the Jaisalmer Basin in the north. The northern limit of the basin is marked by an arcuate Fatehgarh Fault. In the South, the basin is limited by another fault exposed at the Barmer Hill. Whereas a fault passing through Sarnu marks its eastern limit. The original western margin of the basin is truncated by a younger (late NeogeneQuaternary) fault.

The geological map (Fig. 1) incorporates the lithounits described by Das Gupta (1974) from the Barmer Basin (cf. Singh, 1999). Generalised stratigraphy of the Barmer Basin is given in Table 1.

The Barmer Basin has been conceived as the northward extension of the Cambay Basin connected through the Sanchor Basin, in between. According to Chandra and Choudhary (1969), the Cambay Basin owes its formation to the development of tensional faults along its margins, accompanied by a large scale volcanic activity and trends roughly in a NS direction. The disposition in the basin (particularly the Sanchor Basin) has been grossly modified by the late Neogene-Quaternary neotectonism in the region (Roy and Jakhar, 2002).

The inverted Barmer Basin has a maximum length of about 100 km; while the width is about 50 km. It is essentially a Tertiary basin, formed over a basement comprising a variety of older rocks, which include LateMesozoic formations. The earliest depositories in the basin constituted the Fatehgarh Formation. This Upper Cretaceous-Palaeocene formation is unfossiliferous except in the topmost part where a coquina bed similar to that constituting the top of the continental facies of the Sanu Formation in the Jaisalmer Basin is present. The Barmer Formation which succeeds the Fatehgarh Formation, shows facies variation from continental to marine. This transition is brought about by the 'epeiorogenic' movement related to volcanism during the deposition of the Barmer Formation. The palynological evidence suggests a Paleocene age of the formation (Shah and Kar, 1971). The succeeding Akli Formation consists of volcanogenic bentonite beds in association with clay and thin lignite bed at the base. The arenaceous Mataji Ka Dungar Formation has an interfingering relationship with the Akli Formation in its basal part. The Kapurdi Formation containing fuller'searth is a

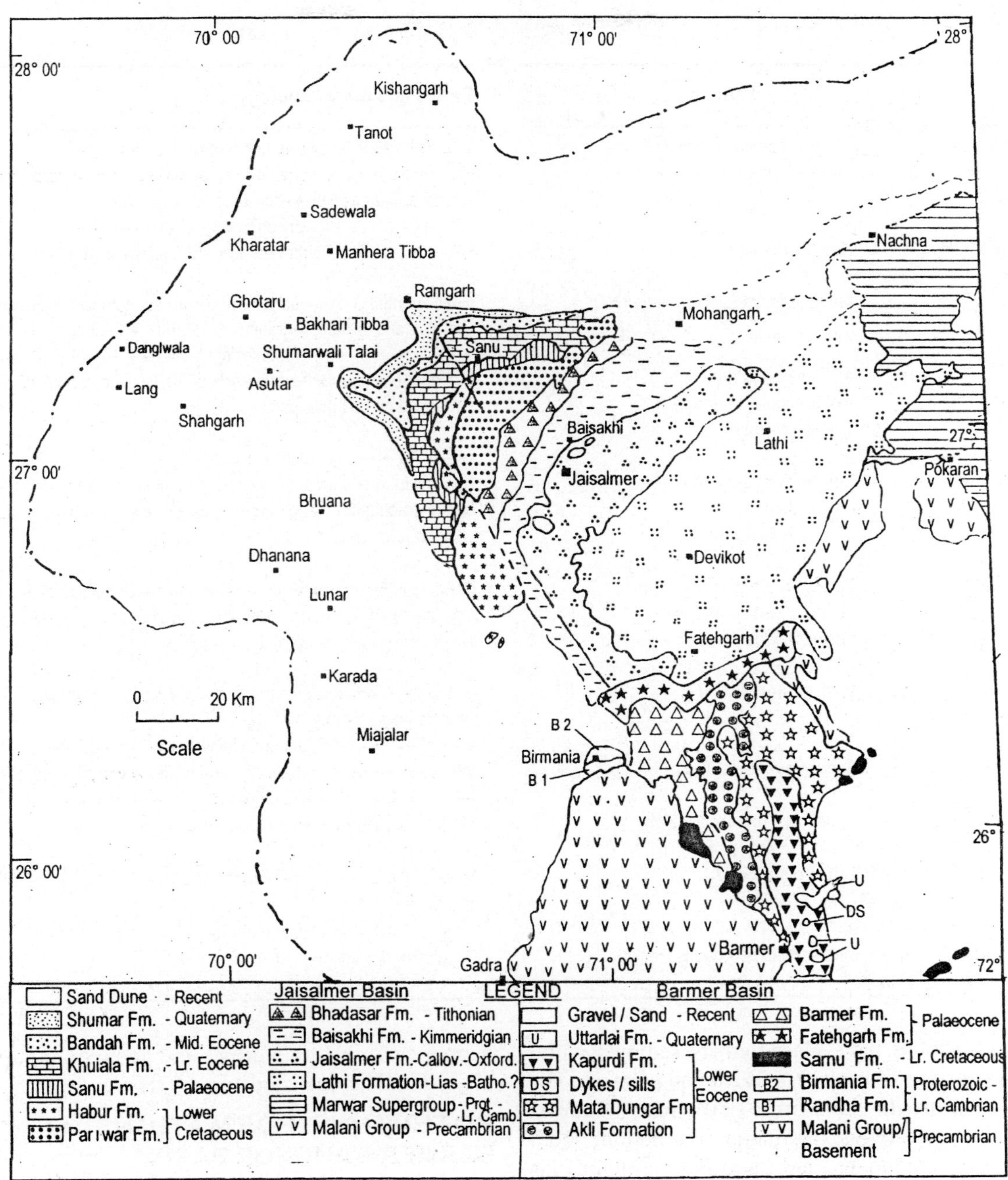

Fig.1 Geological map of western Rajasthan (after Das Gupta, 1974, Singh, 1999, Roy and Jakhar, 2002)

lacustrine deposit succeeded by a thin bed of marine bioclastic limestone of the Ypressian (the earliest Eocene) age (Das Gupta 1974). Thereafter, a total regression of sea took place in the Barmer Basin. The succeeding Uttarlai Formation was deposited after a prolonged hiatus, as a typical continental (evaporite) sequence of the Recent age.

The Late-Cretaceous and Early-Tertiary magmatism and the associated volcanoclastic sediments constitute a unique feature of the Barmer Basin. The Barmer Basin in the west originated due to the break up of Indian craton in the late-Cretaceous and early-Palaeocene time that led to the formation of the Cambay Rift and the constituent basins (Sisodia and Singh 2000). Sisodia

Table-1 Generalised stratigraphy of the Barmer Basin (Das Gupta, 1974)

Age	Formation	Facies and gross lithology
Recent	Dune sands and sediments	Alluvium sands, river alluvium and gravel wash
Sub Recent and (?) older	Uttarlai Formation (3 4m)	Thin gypseous limestone and salt sequence, with unconsolidated sands, kankar and gravel beds
		---------------------------- Unconformity ----------------------------
Lower Eocene	Kapurdi Formation (30)	Lacustrine Fullers Earth deposits (?) interbeded with marine Bioclastic limestone.
	Mataji ka Dungar Formation (180)	Shallow marine orthoquartzite and hard sandstone with pisolites and Ball clay and impure bentonitic clay bands at the base.
		---------------------------- Unconformity ----------------------------
	Akli Formation (280 m) Akli Bentonite Member Thumbli Member	Volcanogenic bentonite sequence at the top and sandstone lignite sequence in the basal part.
		---------------------------- Unconformity ----------------------------
	Barmer Formation (80 m) Madai Member Barmer Hill Member	Shallow marine sandstone with rare plant fossils and orthoquartzite bands grading into conglomerate, sandstone with plant fossils and volcanogenic clays
Paleocene	Fatehgarh Formation (520 m) Fatehgarh scarp Member Vinjori Member	Sandstone layer with mixed bivalve and gastropod casts at the top, Dominantly of ochreous clay bands, variegated sandy siltstone and sandstone sequence with coquina beds.
		---------------------------- Unconformity ----------------------------
	Volcanic Formation (?)	Acid to basic volcanic rocks mainly in form of sills and dykes and local Intrusive porphyrites.
		---------------------------- Unconformity ----------------------------
Cretaceous	Sarnu Formation (80 m)	Indurated, terrestrial sandstone and siltstone with plant fossils
		---------------------------- Unconformity ----------------------------
Callovo Oxfordian	Jaisalmer Formation (15 m)	Marine, fossiliferous, arenaceous limestone
Bathonian to Lias	Lathi Formation (?)	Terrestrial arenaceous sequence with wood fossils and fossiliferous tree trunks
		---------------------------- Unconformity ----------------------------
Precambrian	Malani Suite of Igneous Rocks	Rhyolite and granite

and Singh (2000) have classified its sediments into three categories viz. pre-rift, syn-rift and post-rift. The prerift sediments constitute siliceous Randha Formation, calcareous Birmania Formation; both Lower Palaeozoic in age and Lathi Formation of Lias age. These sediments are deposited on a Neoproterozoic basement of the Malani Igneous Suite. The synrift formations include Barmer Formation and Fatehgarh Formation. Barmer Formation comprises poorly cemented sandstones and intraformational conglomerate and represents alluvial fan environment. Fatehgarh Formation is a mixed sand, mud and phosphorite formation deposited on an intertidal flat environment. It is highly fossiliferrous. The Phosphorite in this formation is correlated with global Cretaceous upwelling. The postrift sediments deposited as thickening and coarseningupward claystone-siltstone-sandstone cycles.

POTENTIAL FORMATIONS OF THE BARMER-SANCHOR BASIN

The stratigraphy of oil bearing formations from the drilled hole data shows the development of Cretaceous to Eocene sequence overlying the fractured Mesozoic sediments and basement with an unconformity. A typical seismic section across the Barmer-Sanchor basin indicates that the basin appears to have been initiated with the extensional rifting, followed by a strike-slip fault reactivation which produced localized inversion. The syn-rift phase of the basin is of Paleocene to Early Eocene age. The post-rift phase is of Eocene to Recent age. The Paleocene is largely fluvial to lacustrine or occasionally marginal marine with biostratigraphic signatures indicative of arid climatic conditions.

The primary pay elements of source, trap availability and a working migration pathway of petroleum were proved by the first few well drilled by Cairn Energy and Shell and other oil prospecting companies in the basin; good reservoir was perceived as the primary exploration risk prior to Mangla discovery. It gives the presence of stacked fluvial sand system.

The well RJE-1 drilled in block RJ/ON/90-1, produced approx 0.2 MM SCMD gas from the interval 1241-1388 m. Besides, additional reserves have been estimated from the deeper (2408-2634 m) tight reservoirs. The tested zones failed to flow hydrocarbons to surface because of low permeability conditions. In addition, the well N-A-1, 8 km SE of Mangla field and 52 km NWW of Saraswati discovery (Cairn Energy) encountered log evaluated oil bearing column of 60-150 m in the Barmer Hill Formation. The section is planned to be further stimulated and for application of treatment for the recovery.

Thumbli Member

The Thumbli Member has good reservoirs of varying quality and occur in the central and southern areas of the block RJ/ON/90-1. The maximum natural flow rate achieved in a Thumbli test has been 1585 bopd of 36 degree API oil in Raageshwari-3. Recent wells such as Raageshwari-6, which tested 550 bopd (billion oil per day), have confirmed the producing potential of this formation which is a target for future exploration and appraisal activity.

Dharvi-Dungar Formation

The Dharvi-Dungar Formation is present in the central basin area and to date has only been targeted by a few wells. The original Rajasthan discovery well, Guda-2 tested 2,000 bopd of 38 degree API oil under natural flow, from the Dharvi-Dungar Formation.

Barmer Hill Formation

The Barmer Hill Formation is a siltstone in the north of the block where discoveries have yet to be evaluated. The relatively poor reservoir characteristics mean that stimulation by freeing will be required. In the central area the Barmer Hill Formation, in Saraswati-1 has produced 62,000 bbls over two long tests totaling 70 days. Development of the southern area fields will require a high density of wells, but with potential for rapid payback of development expenditure.

PETROLEUM GEOLOGY

Only about 2 percent of the organic matter dispersed in finegrained rocks becomes petroleum, and only about 0.5 percent of it ends up in a commercial reservoir accumulation. This emphasizes the inefficiency of the origin, migration and accumulation processes of hydrocarbons. The ratio of dispersed hydrocarbons to reservoir hydrocarbons is about 200 to 1 on a worldwide basis, partly because the volume of potential reservoir rock is small compared to the total sediments of the earth's crust. Within prospective parts of oilforming basins, the ratio generally varies 10 to 100. Petroleum is found from the Precambrian to the Pleistocene periods but is increasingly abundant in the younger sediments for reasons (i) older oil fields are increasingly destroyed over geologic time and (ii) an increase in continental margins and restricted basins occurred when the continents split during the Jurassic period.

In the operational area of Cairn Energy in Barmer-Sanchor Basin oil and gas discoveries have been made in the Tertiary reservoir which is overlying either fractured Mesozoic sediments or the Basement of the Malani Igneous Suite of rocks.

Source Rocks

Based on source rock study, it is inferred that sediments in general comprising mainly of lacusterine and fluvial environments have yielded the oil. Sediments show sinuosity channels and meandering channels of deposition. Low energy environment have also been seen.

They have good presence of oil and gas. Regionally, Guda Shale of Dharvi-Dungar Formation and Barmer Shale of Barmer Hill Formation have been identified as good source rock. Richness increases towards north direction along the Block RJ/ON/90-1.

Reservoir Rocks

A petroleum reservoir is a rock capable of containing gas, oil or water. In Barmer-Sanchor Basin the main reservoir rocks are sandstones and siltstones. These sandstones belong to Thumbli Member of the Akli Formation, Barmer Formation, Barmer Hill Formation and Dharvi-Dungar Formation of Upper Cretaceous to Eocene age.

Raageshwari field is deep with non conventional volcanic reservoir having gas column of 800 m, due to the impermeable nature of the rocks.

At Guda South 05 location, the reservoir rock is

sandstone that belongs to Thumbli Member of the Akli Formation. This sandstone is dark colored, having 18% to 20% porosity. This sandstone is well sorted and medium to coarse grained.

Cap Rocks

The argillaceous sequence of various ages such as shales of Akli Formation and shale of Barmer Formation worked as good cap rocks in the Barmer-Sanchor Basin. These are impermeable rocks which restrict the movement of petroleum and gas from the reservoir rocks. At Guda South 05 location, cap rock is Guda Shale of Barmer Hill Formation.

Traps

In the Barmer-Sanchor Basin the trap is produced by a set of geological conditions which cause oil and gas to be retained in the porous reservoir rocks, or at least allow escaping at a negligible rate. The structure in the Barmer Basin is mainly controlled by the major Fatehgarh Fault and main type of trap in Barmer are (i) Structural traps and (ii) Stratigraphic traps.

The Structural traps are mainly the Fault Traps in this basin, in which we get shales on the reservoir rocks (sandstone) making trap for accumulation of petroleum.

Geological Time

Paleocurrent analysis of the discovered structures have suggested that most of the structures were well defined by the end of Cretaceous to Eocene period.

The Migration of Hydrocarbon

The migration of hydrocarbon is mainly in the Barmer Hill Formation and structures of that time were in favourable position for the migration of oil drops.

PETROLEUM FIELDS OF THE BARMER-SANCHOR REGION

Fig. 2 shows important petroleum fields of the BarmerSanchor region. An independent evaluation conducted by the international firm; DeGolyer & MacNaughton (D&M), Reservoir Engineering Consultant, in April 2005 has assigned a 2P oil in place with a volume estimate of 1.64 billion bbls for the three northern fields comprising Mangala 1,067 mmbbls, Bhagyam 335 mmbbls and 219 mmbbls at Aishwariya. Peak production targets for the three northern fields are in the 120,000 to 150,000 bopd range with Mangala planned to come on stream at the end of 2007. Exploration successes by Cairn Energy in 2004 confirmed the oil and gas potential of the Barmer Basin. While hydrocarbons had previously been found at a number of stratigraphic levels throughout the block, a key to the major success was finding high quality oil bearing reservoirs in the Fatehgarh formation in the north. The high oil flow rates and good reservoir interconnectivity seen on well tests indicated that excellent recovery rates may be expected from these Fatehgarh sands. The Cairn Energy estimated the total oil in place discovered to date on the block to be more than 250 billion bbls. Introduction and petroleum potential of important fields of the region have been discussed in the following sections.

Mangala Field

Mangala was fully appraised by wells between March and July 2004 and 3D seismic survey acquired between May and December in 2004. A range of technical studies to characterize the geological, geophysical and geochemical nature and disposition of the reservoirs and fluids is being undertaken. These studies have been integrated with both subsurface engineering studies in a Front End Engineering and Design (FEED) study that form the basis of the field development plan which was to be submitted to the Government of India in May 2005. The development plan for Mangala Field will target at 35% recovery factor assuming a secondary water flood. Ongoing studies, including water injectivity testing, are aimed at supporting this objective. To combat the high pour point and waxy nature of the crude, injected water is likely to be heated.

Cairn Energy is working closely with the local water authorities to ensure a sustainable water extraction and reinjection process. Extensive deep saline (non drinking water) aquifers have been discovered which can be utilized as a source of water for injection purposes. Extraction will be monitored to ensure minimal impact on the water table.

Tertiary and enhanced oil recovery techniques, such as polymer flooding or the use of surfactants, are under investigation. Such techniques may further raise the recovery factors in addition to the conventional water flood scheme. Analogous fields worldwide, including some in the neighbouring Cambay Basin in Gujrat, are being reviewed to determine the optimum secondary and tertiary recovery methods. An ultimate recovery factor of more than 40% of oil in place is possible.

The current Mangala Field data shows proved and possible (2P) gross reserve estimate as 376 mmbbls, based on secondary recovery by water flood and production up to 2041. Peak production is envisaged to be between 100,000 and 110,000 bopd with field start

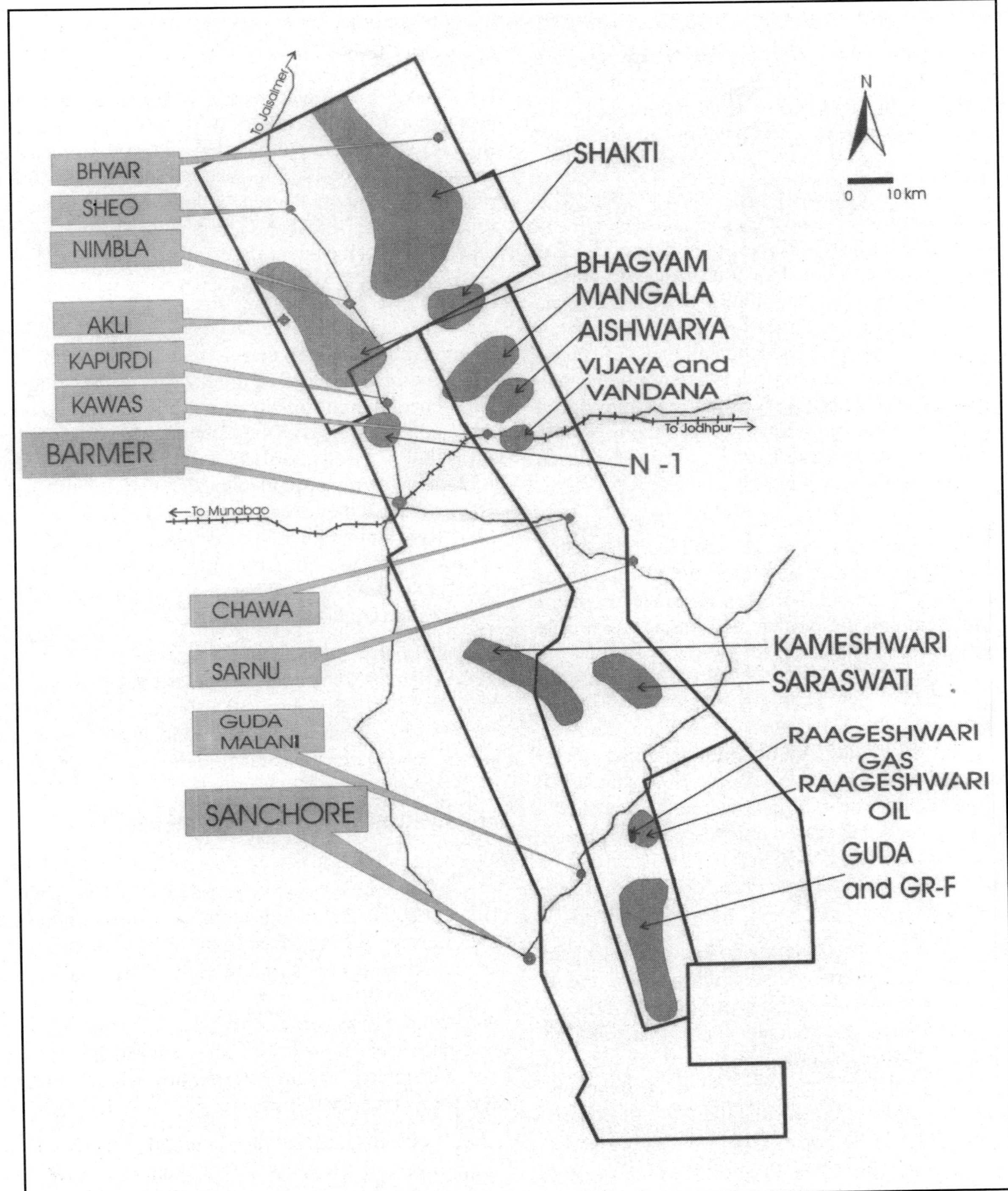

Fig.2 Petroleum Fields of the Barmer-Sanchor Region (Based on Cairn Energy Pvt., data)

up planned for Q4 2007. Prior to the 3D seismic suvery over Mangala Field being completed. An independent analysis of the data by D&M (July 2004) confirmed 2P oil in place volume of 1,005 million barrels. After incorporating the 3D seismic data in April 2005 D&M has revised its estimate to 1,067 million barrels. The D&M 2P current reserve estimate for Mangala is 256 mmbbls based on secondary recovery only assuming the production contract term runs until only 2025.

Aishwariya Field

The Aiswariya Field was fully appraised between April and September 2004. A 3D seismic survey has been acquired over the field and is being interpreted. As a result of the early 3D interpretation, an additional appraisal well was drilled to test the crest of the structure.

The field is slightly deeper than Mangala and the oil quality, whilst similarly waxy and of high pour point, is less variable than at Mangala, with a mean API gravity of 29 degrees. Despite being smaller than the Mangala Field and having a slightly different reservoir configuration, the proposed techniques of dense drilling, secondary recovery via water flooding and enhanced tertiary oil recovery techniques introduced early in field life are equally applicable to the Aishwariya Field. Reserve estimates will be revised once studies are complete.

In April 2005 D&M estimated the 2P oil in place volume for Aishwariya to be 219 mmbbls. The D&M current 2P (Prove and Probable) reserve estimate for Aishwariya is 41 mmbbls based on secondary recovery only and assuming the production contract term runs until only 2025.

Bhagyam Field

The Bhagyam Field has been appraised between January and April 2005 and two more appraisal wells are being planned. A cased hole test in Bhagyam -1-ST established a flow rate of 1,050 bopd of 26 degree API oil from one zone. Two cased hole tests in Bhagyam-4 established a flow rate of 660 bopd of 24.5 degree API oil and 1,700 bopd of 30 degree API oil from two separate zones.

The field is slightly shallower than the Mangala Field and the oil quality, whilst similarly waxy and of high pour point, is more viscous at Mangala. Despite being smaller than the Mangala Field, the proposed techniques of dense drilling, secondary recovery via waterflooding and enhanced tertiary oil recovery techniques introduced early in field life are equally applicable to the Bhagyam Field. Reserve estimates may change once the final appraisal wells have been drilled and studies are complete.

In April 2005 D&M estimates the 2P oil in place volume for Bhagyam to be 355 mmbbls. The D&M 2P reserve estimate for Bhagyam is 85 mmbbls based on primary and secondary recovery only and production until 2025. D&M estimate that an additional 48 mmbbls of reserves could be added with a successful EOR programme.

Shakti Field

The Shakti Field was partially appraised between September and December 2004. The Shakti-2 and Shakti-3 wells were put on pumped test for a four day period in January 2005 to assess the recovery of this relatively viscous 19 degree API oil. Shakti-2 flowed at a sustained rate of 50 bopd and Shakti-3 at a rate of 60 bopd from selected intervals.

Guda and GR-F Field

The Guda discovery well was drilled in 1999 and encountered oil in the Dharvi Dungar Formation. The GR-F-1 discovery well drilled 9 km to the south in 2003 encountered oil in the Thumbli Formation, Subsequent appraisal of these two fields by the Guda3 and GR-F-2 wells drilled early in 2005, when combined with remapping of the 3D seismic survey has established that they are in fact one field at the Thumbli level. Reexamination of the original Guda well has confirmed that the Thumbli reservoir, which was not tested at the time, was oil bearing. The combined field is to be called Guda.

Cairn Energy has currently estimated that the range of oil in place volumes for Guda is between 40 and 400 mmbbls. Further appraisal of the Guda Field, comprising an eight well programme, is being planned to commence later this year to delineate the potential field size.

Saraswati and Raageshwari Fields

The Saraswati Field was discovered in late 2001. It has been covered by 3D seismic and 2 appraisal wells were drilled in 2003. The field is complexly faulted and had oil in Barmer Hill and Fatehgarh reservoirs. In 2004, The Saraswati-4 crestal well successfully established significant oil flows from the fractured basement. Although the reservoir quality in the Barmer Hill and Fatehgarh Formations gave somewhat low initial flow rates, subsequent fracture stimulation has been more successful in increasing the rates.

Oil was discovered in Thumbli reservoirs at Raageshwari in early 2003. The Raageshwari-2 and Raageshwari-3 appraisal wells drilled in 2003-2004 have confirmed the presence of oil and gas down dip of the discovery well in separate fault blocks and with slightly differing pressures and oil water contacts.

The declaration of commercial viability submitted to the Government of India in May 2004 included the Saraswati and Raageshwari Oil Fields. The development plans based on current data have been submitted to the Government of India in May 2005. However a full evaluation programme on these fields is yet to be carried out. Initial fracture stimulation studies have highlighted the potential for productivity increases from the reservoirs in these southern fields.

The southern fields, Saraswati, Raageshwari oil and Guda are all candidates for early production and evacuation of crude by trucking.

Raageshwari Gas Field

The Raageshwari Gas Field and the deep gas exploration potential within the southern area are potential important new resources. Gas has value as an important component of the development plan for the Mangala and northern fields, which will have their own power requirements.

The proven mobile gas column established at Raageshwari is 500 m. However, the low permeability of both the volcanic and Fatehgarh reservoirs has produced mixed results on test. Two Raageshwari appraisal wells are being planned this year to further delineate this gas field. The Raageshwari-4 well is currently being tested across the volcanic and Fatehgarh reservoirs.

The maximum stable rate achieved has been 3.3 mmbbls. A stimulation programme is being planned to be carried out in the second half of 2005 which is expected to improve the productivity.

An exploration review of the Central Basin High prospects on trend with Raageshwari has highlighted potentially significant additional unrisked gas volumes. An exploration well is being planned for this year.

Other Significant Oil Fields

Further evaluation and appraisal of oil discoveries at Kameshwari, Vijaya (NR-1) and Vandana (NR-2) is ongoing. The Vijaya and Vandana discoveries are the first stratigraphic traps to be tested on the block. Both fields have been covered by 3D seismic survey which is currently being evaluated. The combined oil in place estimates for the Vijaya and Vandana discoveries range between 50 and 500 mmbbls.

References

Chandra, P.K. and Chaudhary, LR.; 1969 : Stratigraphy of the Cambay Basin. Bull. Oil and Nat. Gas Comm. Vol. 6, No. 2, pp. 37-50.

Dasgupta, S.K.; 1974 : Stratigraphy of western Rajasthan shelf : Proc. IV Indian Colloq. Micropal. Strati., Dehradun, pp. 219-233.

http://www.cairn energy.plc.uk

http://www.fotojinak.info

http://www.gasandoil.com/goc/news/nts43834.htm

http://www.simmonsco-intl.com/files/02060 3.pdf

Roy, A.B. and Jakhar, S.R.; 2002 : Geology of Rajasthan (Northwest India) Precambrian to Recent. Scientific Pub., Jodhpur, 421 p.

Shah, S.C.D. and Kar, R.K.; 1971 : Polynostratigraphic evolution of the Lowest Eocene sediments of India. Proc. Sem. Paleopalynology and Indian Strat., Calcutta University Publication, Calcutta, pp. 255-264.

Singh, N.P.; 1999 : Relevance of laboratory studies in Geological modeling and field geology. Jaisalmer field guide. IMD, ONGC, Dehradun, pp. 125, Fig. 114, Pts. IV.

Sisodia, M.S. and Singh U.K.; 2000 : Depositional environment and hydrocarbon prospects of the Barmer Basin Rajasthan India. Nafta. Zagreb (Crotia). Vol. 51, No. 9, pp. 309-326.

Hon'ble Reviewers

1. Mr. Alok Tripathi
   Suprtd. Geologist
   Oil and Natural Gas Corporation Ltd.
   ONGC Camp
   JAISALMER 345 001 INDIA
2. Dr. Sudhir Kumar Mathur
   Lead Geologist
   Cairn Energy India
   Vipul Plaza Suncity, Sec. 54
   GURGAON 122 002 INDIA

Editorial Assistance

Ms. Anju Mathur

Economic Mineralization
K.L. Shrivastava (Editor)
2009. pp. 442-450
Scientific Publishers (India), Jodhpur

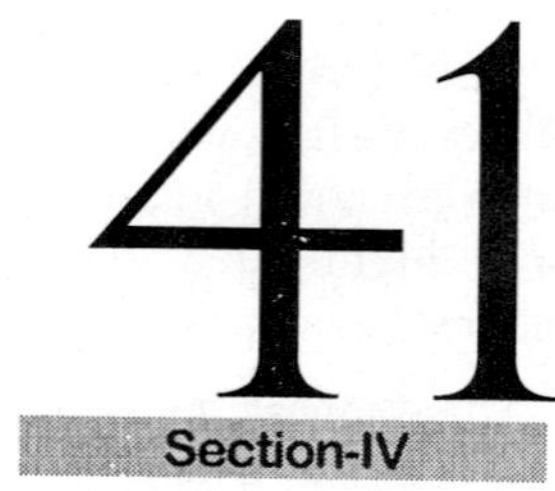

Section-IV

The First Fifty Year's Record and A New Beginning in Mineral Discovery in India with Special Reference to Base Metals

S.K. HALDAR
Emeritus Scientist, Department of Geology, Presidency College, Kolkata 700 001
Email : skhaldar@yahoo.com

Abstract

The paper traces the development of the mineral based industry in India from the forties, and as such deals with more than half a century of our national experience and adventures in this domain. This might seem a lot in mortal years: but as a new-born country, as India Inc, we are still in our first steps bold, and steady steps no doubt, but toddler steps nevertheless. The discovery rate of base metal deposits the world over rose throughout the '60s and '70s. It declined during the '90s and no significant deposits have yet been reported in the twenty-first century. The current Indian base metal deposits are rediscoveries of ancient abandoned mining sites and extensions in the known belts. Most of the ore bodies have cropped out at the surface, either unaltered or exposed as oxidized gossans. Traditional prospecting combined with geology, geophysics and some limited geochemistry gave rise to the discoveries in the recent past.

Our task is to visualize new approaches, and seek deep-seated metal deposits in conducive-geological settings. Forthcoming discoveries will depend on modern exploration technologies applied to an integrated geological model. To emerge victorious in this new environment of increasing discovery risk requires extraordinary creativity, innovation, and technical skill coupled with commercial discipline. The incentives are world economic demand, driving continuously high trends in metal prices, technological necessity, and the primeval joy of unveiling the opening chapters of creation. These are the factors that motivate corporations to pioneer and unearth new deposits, and to upgrade the marginal low-grade deposits for production purposes.

Key words : Mineral Discovery, Basemetal Deposits, Exploration, Reserve Adequacy.

INTRODUCTION

The paper traces the development of the mineral based industry in India from the forties, and as such deals with more than half a century of our national experience and adventures in this domain. This might seem a lot in mortal years: but as a new-born country, as India Inc, we are still in our first steps - bold, and steady steps no doubt, but toddler steps nevertheless. On a geological time scale, mankind is yet in the cradle, and our mineral studies but a glimmering recognition of the baubles presented to us by the stellar sub-system which we call Earth. Having grasped this instant in eternity, let's take a quick look at what we've learnt so far.

Indian mineral explorations, for centuries, had been historically focused mainly on coal, iron-manganese ore, mica and such, a number of interesting events occurred in the 1940's - World War II ended, yours truly was born, and India, impatiently shaking off the crown and shackle, was, at last, free. The emerging Republic of India visualized a new perspective and formulated 5-Year Plans in tune with the infrastructure development of the country as a whole. Over the years, the Government incorporated various Public Sector Undertakings such as National Mineral Development Corporation (NMDC, 1958), Bharat Aluminium Company (BALCO, 1965), Hindustan Zinc Limited (HZL, 1966), Hindustan Copper Limited (HCL, 1967), Uranium Corporation Limited (UCIL, 1967), Bharat Gold Mines Limited (BGML, 1972), Steel Authority of India Limited (SAIL, 1973), Oil and Natural Gas Commission (ONGC, 1956), etc. These Public Sector Undertakings were entrusted with the mission of exploration, development, and production of public consumable and marketable mineral base end product. This triggered extensive new searches for Petroleum, Coal, Bauxite, Gold, Iron, Lead-Zinc-Copper-Uranium and many other minerals. Perhaps for the first time in Indian history since the earliest recordings in Kautilya's Arthashastra (327 BC) was such an extensive search and research undertaken in this field.

A discussion of base metal trending should typically address some fundamental questions. Such as: How are these mineral discoveries made over the past 50 years? What are the significant trends that have emerged from our forays? Where are the prospective geological hosting environments identified for the immediate future? What are some of the rapidly evolving geological concepts? And which are the bleeding edge technologies used to support these new concepts as well as practices?

This paper focuses on the base metal studies in the context of zinc-lead-copper mineral deposits. Rather than sampling a few stray occurrences, the scope of the study covers all the mineral belts as well as the known singular giant-type deposits.

Significance of Investment Framework

In terms of business requirements, the process of mineral discovery is a high risk tolerance, supra-mortal spans of time and deep pockets for a sustained cash flow. A small business unit in this field may often end its brief tenure with a total loss, in case of failure to make an economic return. Indeed, many of the discoveries are not viable at current market prices and prima facie evidence might indicate that investment in these endeavors is a waste. However, one discovery out of 100 or even 1000 attempts may pay back the entire efforts (Haldar, 2007). The task of Government policy is, therefore, to plan timely allocation of funds for exploration and technology research of various mineral types, predicated on long-term demand and supply scenarios. "Character, in the long run, is the decisive factor in the life of an individual and of nations alike". Post-independence, it is this fundamental shift in national character which has empowered the brave new generations.

The Discovery Types

The discovery of an occurrence or a deposit is characterized by a measured quantity and grade, which indicates an estimated amount of contained minerals or metals. The discovery may be immediately beneficial and the exploitation profitable at that point in time, in which case it is classified as a reserve or more specifically ore reserves. If not, if the potential is known but not immediately extractable or advantageous, it is classified as a resource. To quote Ivanhoe (1995), "Explorationists must first find and then ..... engineers converttheoretical resources into producible reserves." Often, the technology exists, but it will toggle between one type and other with the changes in market price. The uneconomic discovery may become economic tomorrow and vice-versa.

"**Greenfield discoveries**" are the findings from a broad based grassroots exploration program well away from known ore bodies or known mineralized belts - in essence, pioneering discoveries in new locales.

"**Brownfield discoveries**" are classified where discovery is made by enhancing the reserve in strike and dip continuity of known ore body or in the vicinity of existing mine. In such cases the economics of development are improved by existing infrastructure. This is an important distinction in the analysis of discovery trends though both types contribute to the rates of depletion (Blain, 2000).

The terms come from the building industry, where undeveloped land is described as Greenfield and previously developed land is described as Brownfield.

Exploration

The geological exploration program can be classified by successive stages, each consisting of activities with a combined specific objective to attain within the schedule and budget constraints, with the outputs of each stage providing inputs to the next successive stage. The stages are herewith presented in the context of any exploratory procedure:

(a) "General Exploration" is the initial to detailed delineation of an identified deposit. The objective is to establish the main geological features of a deposit, giving a reasonable indication of continuity and providing an estimate of size with high precision, shape, structure and grade.

(b) "Reconnaissance" is the "Grassroots" exploration for selection of initial targets. The activities which make up this stage are literature survey, obtaining permission (Reconnaissance Permit RP) from the State Government authority, regional geological mapping, rock chip and regional geochemical sampling, pitting and trenching, acquisition and interpretation of available geophysical data, if any, regional ground magnetic and electromagnetic traverses as required, scout/ reverse circulation/ diamond drilling. The objective is to scan the entire area under lease, and to identify, target and ranking of broad mineralized areas worthy of further investigation towards deposit identification. The original area is thus substantially reduced to fractional units, enabling focus of maximum resources to the target areas in the next stage of exploration. The total area permissible for the RP within a State is limited to 10,000 sq. km. The period for which the license may be granted shall not exceed three years.

(C) "Prospecting" is the next phase of exploration of potential target anomalies after obtaining permission (Prospecting Lease - PL) from State Government. The activities encompasses detailed geological mapping, rock chip and soil sampling, close spaced ground geophysical traverses, diamond core drilling, reserve estimation of Proved/ Probable and Possible category. Other information like rain fall, climate, and availability of infrastructure, logistic facilities including health care, environmental implication, pilot plant beneficiation tests and broad economic studies are also collected. The objective is to identify a suitable deposit that will be the target for further profitable exploration. The activity ends with preparation of a Pre-Feasibility or Feasibility report. The total area for an individual PL application is limited to 25 sq. km and it can be couple of numbers within the RP area of 10,000 sq. km. This is thus a mere part of the original size of the area under reconnaissance permit. The period for which the license may be granted shall be initially for three years and can be extended to a maximum of another two years if the State Government is satisfied that a longer period is required to enable the licensee to complete prospecting operation.

(d) "Detailed Exploration" is conducted before the start of mining phase to outline firm contacts of the ore body, rock quality designation (RQD) for mine stability, planning and preparation of samples for beneficiation test work. The Mining Lease (ML) is obtained from State Government with the clearance of Ministry of Mines (MOM), Ministry of Environment (MOE) and Indian Bureau of Mines (IBM). The works envisaged are close space diamond drilling, trial pit in case of surface mining and sub-surface entry with mine development in case of underground mining. The reserves are categorized as Developed/ Proved/ Probable/ Possible. Diamond drilling is a continuous process through out the entire life of the mine to supplement reserve for the depleted ore. The maximum area permissible under Mining Lease (ML) is 10 sq. km - which could be $1/100^{th}$ of the reconnaissance area. The period for which the license may be granted shall not exceed thirty years: Provided that the minimum period for which any such mining lease may be granted shall not be less than twenty years. A mining lease may be renewed for a period not exceeding twenty years.

In terms of project schedule, we thus see a total span of around 15 to 45 years from beginning to end of these stages. Generations and conditions change in this time, and it is a combination of founding foresight, steady perseverance and agility in adaptation, which, along with a good measure of statistical providence, provide for project success.

Economic Resources

Post independence, the continued efforts by various exploration agencies has resulted in significant augmentation of Indian zinc-lead-silver resources. As a bellwether of this trend, take a look at the Mochia Mines of Zawar Group, Rajasthan, with 175 tons per day production capacity from 0.5 Mt of known reserve in 1947. Since then the resource had risen to 418 Mt by 2000 and further to 430 Mt in 2005 (Fig. 1; Haldar, 2005; Anon, 2006). The Proterozoic sediments of Rajasthan, has been particularly favourate in hosting more than 95% of lead-zinc sulfide deposits known to India.

The economically viable category of reserves stand at 130 Mt having 9.2% Zn and 2.2% Pb, besides 30 Mt of possible resources in producing/ developing mines and deposits under detailed exploration (Bhatnagar et al., 1986).

The total in situ copper reserves and resources have

been estimated at about 10 Mt in 1947 with production of 310,000 tons of ore @ 2 % Cu per annum from a single mine at Mosabani, part of Singhbhum Group of deposits in Jharkhand. The reserve has been increased to 712 Mt (9.4 Mt Cu) in 2005 by continued exploration. The reserve is further categorized as Proved (224 Mt), Probable (222 Mt) and Possible (266 Mt) resources. Geographically, this is shared mostly among the states of Madhya Pradesh - 295 Mt (3.9 Mt Cu), Rajasthan - 195 Mt (2.5 Mt Cu), and Jharkhand - 179 Mt (2.3 Mt Cu). An additional 722 Mt of conditional resources (3.16 Mt Cu) have also been outlined from these states (IBM, 2003).

The reserve status fluctuates with the uneven tides of continued exploration in and around the deposit / ore body, sometimes ebbing with mine depletion, and frequently rising with the new discoveries. What is the reserve which should be reported for a mineral body? Should it continue to reflect the initial discovery? Or

have too low tonnage for making any substantial contribution. Both have already been phased out.

Hindustan Zinc Limited, the largest zinc metal producer in the country with integrated mines and smelters, currently has an annual production capacity of about 400,000 tons per annum. Binani Zinc Limited has another 30,000 tons of metal production capacity that is founded on imported concentrate. These corporations ensure India's significant self-sufficiency in zinc (Fig. 2). Hindustan Zinc Limited has expansion plans on the way for an additional capacity of 258,000 tons by mid 2008. Thus, India is likely to reach near total self-sufficiency and to become a net exporter of zinc by 2008, when additional capacities will have come on stream (FIMI, 2007). It's unlikely that future exploration efforts for reserve build up will give priority to lead, as it is recognized as one of the major environmentally sensitive substances.

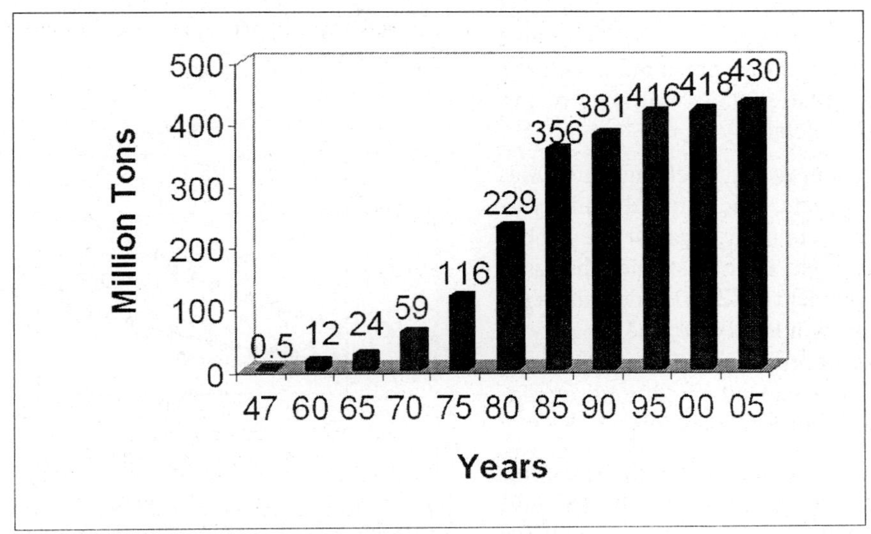

Fig.1 Zinc-lead Reserve builds up

should we assess the current state of the reserve after depletion? Or, most accurately, report the status quo after accounting for both depletion and new finds? The author strongly advocates the last option. Other than the obvious benefits to local explorers and business alike from accurate reporting, this update will qualify the deposit for global comparison and merit-based ranking.

Reserve adequacy

By and large the Indian zinc-lead deposits are comparatively richer in zinc with low lead content. The exclusive lead deposits like Sargipalli and Agnigundala

The Indian copper deposits are either narrow with low grade orebody (Khetri belt), or low angle deep seated body not supporting large scale mechanization in underground mining (Singhbhum belt), or changing over phase from surface to underground method as at Malanjkhand, MP.

Resource requirement

Zinc is the fourth most widely used metal after iron, aluminum and copper in the world. Due to its resistance to non-acidic atmospheric corrosion zinc is instrumental in preventing the corrosion of steel and

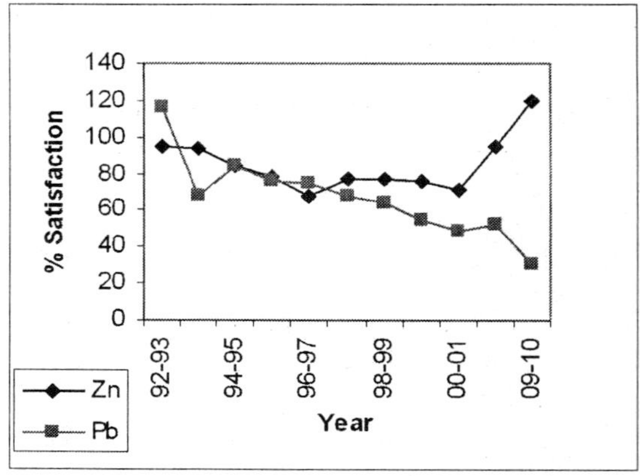

Fig.2 % Demand satisfaction of zinc and lead metal in the country

thus extending the life of buildings, vehicles, ships and steel goods and structures of every kind, ranging from telecom towers to hi-mast lighting. Thus, in a booming economy which encourages construction, the demand for zinc literally galvanizes a nation. Even in India, most of the zinc demand in the country (more than 70%) is in the construction sector. The current output of zinc lead ore being mined is about 4 Mt per annum, with 7% Zn and 2% Pb. This satisfies 95 % satisfies of the national demand for zinc metal.

As per the Multi-Commodity Exchange of India, "Over the next five-six years, zinc demand is likely to grow at 12-15 per cent annually, against the global average of 5 per cent. Even if one assumes that zinc demand grows by 10 per cent till 2010 and at slower 7 per cent thereafter, India would require zinc capacity of 14 lakh TPA by 2020, in order to still be self-reliant. The next round of large capacity additions would, therefore, be warranted from 2008 onwards."

One source of production increase would be the full expansion of the Rampura Agucha Mine, which should be adequate to fulfill National demand and export for the next ten years and more. The reserve/ mine production required up to 2015 at 90% smelting capacity as well as the 100% level of country's satisfaction with zinc is given in Fig. 3. Perhaps excluding the Rampura Agucha cache, the present reserve base may cater to resource requirements only for the next 15-20 years.

Copper metal production has yet another constraint as it is heavily dependent on imported concentrate. This is a good time to pause and remember that mineral resource is a non-renewable asset -- and that the gestation period between discovery and production takes 12 - 15 years. The question arises - What discovery rates would be required, at the current rates of production, if the industry had to replace exhausted reserves with new grassroots discoveries? Regional exploration plans would need to maintain a rolling in-situ reserve inventory of 100 to 190 Mt to maintain self-reliance on long-term basis. If we fail to meet these numbers, we would be left with the alternatives of import of concentrate metal, secondary generation etc. India is fast approaching crisis state, and can be overcome as "The Chinese use two brush strokes to write the word 'crisis.' One brush stroke stands for danger; the other for opportunity."

Discovery trend

It is fitting that the land of philosophers should have the reputation of first extracting zinc metal from philosopher's wool (ZnO). While zinc was probably extracted in India as early as in the 4th to 3rd century BCE, the reduction of zinc-oxide using a condenser and oxygen-devoid container on an industrial scale took place in Zawar, India around the 13th Century AD, after which the technology spread to China and other countries (Anon, 1989). Vasco de Gama's first step in Calicut on the Malabar Coast laid the foundation for future fleets to and from "East India". In 1597, Libavius, a European metallurgist got hold of a sample of zinc extracted from India, and for want of a better name called it Indian / Malabar lead. This is a striking example of India's age-old association with this wonder metal.

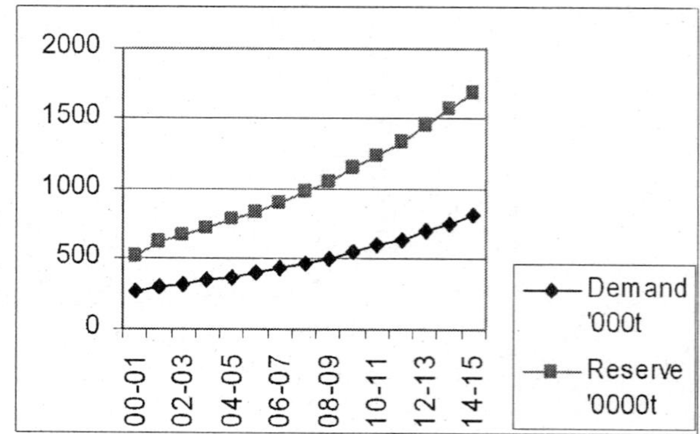

Fig.3 Projected reserves / mine production up to 2015 at 90% smelting capacity

India has always been one of the primary copper metal producers in the world. As per the International Copper Study Group, India accounts for about 3.5 to 4% of the world's copper production and usage. As any school-going student of history knows, the best known artifact from the ancient Indus Valley culture is an approximately four inch high **bronze** figure of a dancing girl. There are ample evidences of ancient mining and smelting remains around all the present zinc-lead-copper mining belts. Carbon dating of wooden implements recovered from the workings has indicated that these operations are 2100 years BP (Gandhi, 2000). Therefore, all the "finds" of Zn-Pb-Ag-Cu deposits are, in a strict sense, "rediscoveries" of abandoned mining belt and may be termed as "Brownfield discovery". Barring a few hidden deposits at a depth of 120 m from the surface at Rajpura- Dariba- Bethumni belt, most of these discoveries were serendipitous. New investigation over the years has enhanced significant economic reserves in and around the known ore body and can thus be termed as "New" discoveries. Thanks to the availability of existing infrastructure, these reserves are of enhanced value.

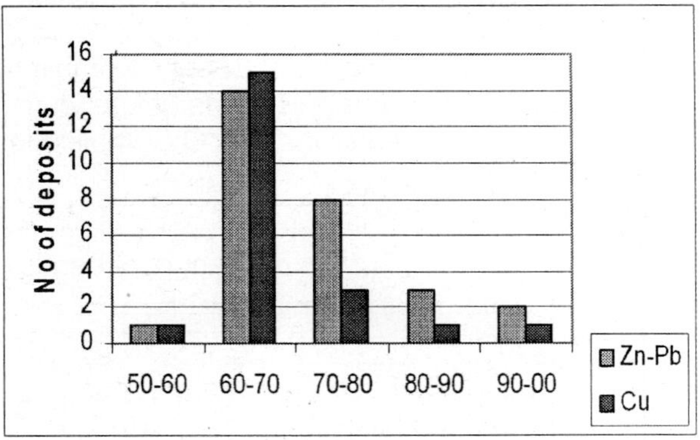

Fig.4 Number of deposits discovered in a decade

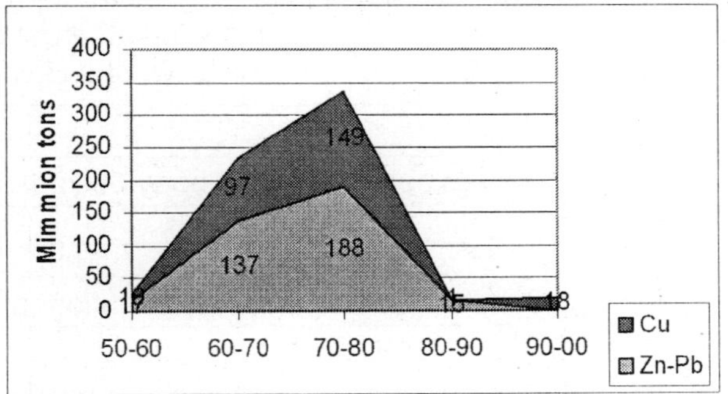

Fig.5 Resource builds up in each decade on discovery

A quick scan of historical discovery trends reveals some interesting results. The overall records of mineral discoveries by number of deposits and associated reserve base created over last 50 years have been depicted in Fig. 4 and Fig. 5 respectively. The point of note is that base metal discovery rates picked during 1960s and 1970s and declined thereafter. Overall, this major reserve enhancement should provide for immediate (short / medium-run) mine development and production.

Most of the Indian deposits are medium sized, the exceptions being the two world-class BIG deposits - Malanjkhand Copper Deposit (145 Mt @ 1.35% Cu), Madhya Pradesh, discovered in 1970 and the often-mentioned Rampura Agucha Zinc-Lead-Silver Deposit (+ 60 Mt @ 13.50% Zn, 2.00% Pb and 50 g/t Ag), Rajasthan, in 1977. These deposits are high in grade, while being low in cost (surface mining), and thus are significant production and profit components of the Indian mining sector. Two other comparatively recent deposits worthy of mention are the Sindesar Khurd (9 Mt @ 6.8% Zn, 2.9% Pb and 155 g/t Ag), adjacent to Rajpura Dariba Mine discovered in 1987, and the Banawas Copper Block (18 Mt @ 1.74% Cu), an extension of Khetri Copper mine that was discovered in the '90s. Both are rich in metal content and amenable to underground mining. As we can see in these 2 cases, "the real voyage of discovery consists not in seeking new landscapes but in having new eyes."

A new beginning

Even as technological breakthroughs enable better and faster method of extracting wealth, the world's non-renewable mineral reserves are steadily depleting at progressively increasing rates of production. The economic growth, in the process, has assumed the softer name of revenue! Greenfield discoveries and expansion of reserves in and around the existing operating blocks must needs to be intensified for a ton-to-two tons substitution of the depleted ore. In India, surface ore bodies have become particularly hard to find any more. Therefore, traditional exploration techniques may no longer yield the required results. We need to take the plunge, and target deep-seated, and hitherto hidden, deposits.

Future exploration will integrate several approaches and techniques in all the phases of discovery. Multidisciplinary approaches will utilize "in-depth" knowledge of regional and deposit geology, studies of structures like lineaments, faults and fractures which control the mineralization in an area, application of rock and soil geochemistry, and interpretation of high-resolution aeromagnetic (Magnetic and Electromagnetic) surveys with deep penetration capacity. These are the new divining rods which will lead us to the hidden metallic bodies.

The exploration projects would still need to be

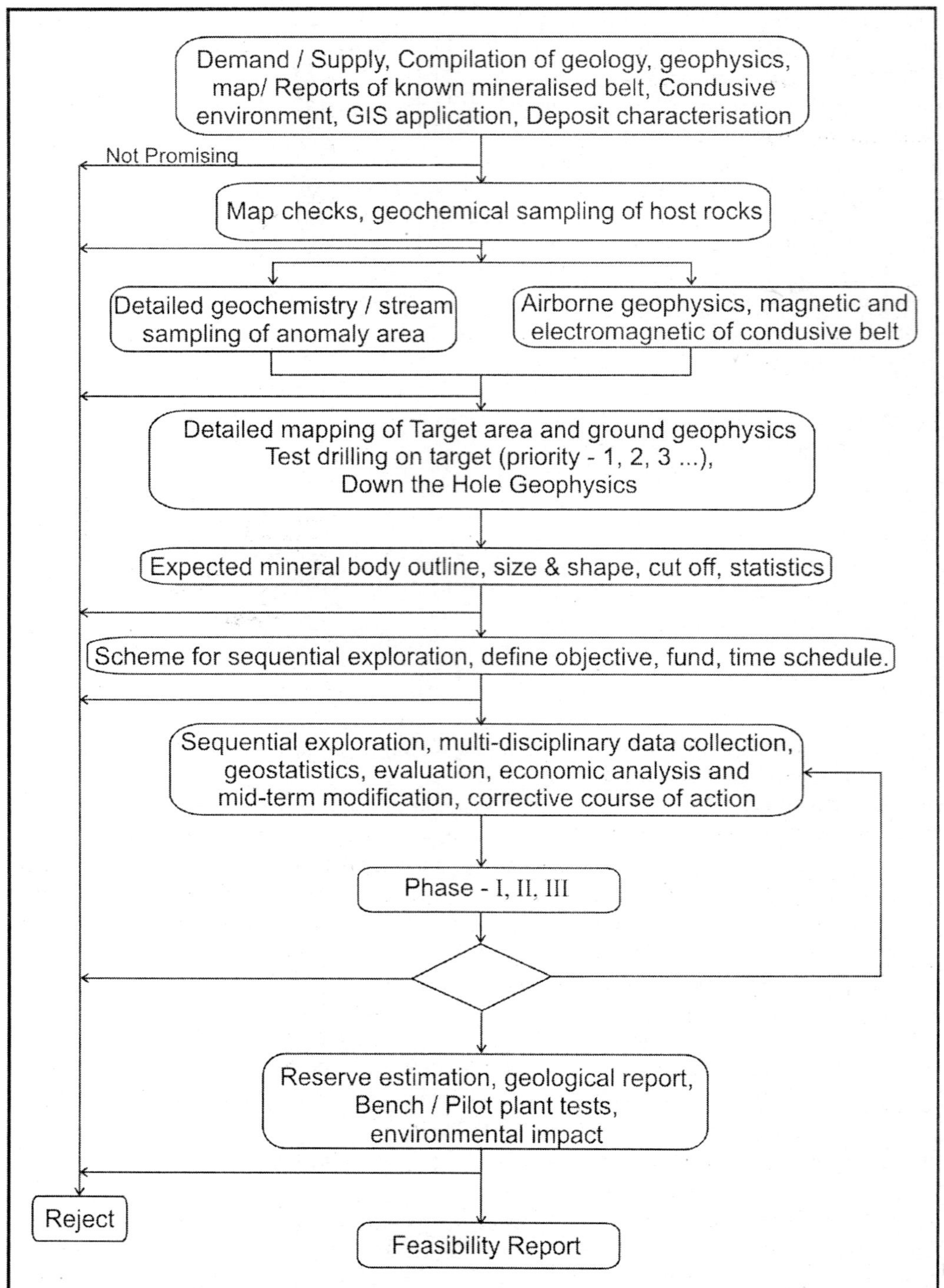

Fig.6 Exploration activity flow diagram - A New Beginning!

sequential and dynamic, with controlled objectives, schedules and budgets for the intermediate phases of activity. The data collected and computed must be interpreted and evaluated at the end of each intermediate phase, with the subsequent phases being rescheduled as and when necessary. In the event that intermediate results do not meet the "Go-No Go" criteria, the project can be shelved till adequate resources are available for the next stage or the market price increased in between. Thus, today's explorer needs to wear the hard-hat of the engineer, the stopwatch of the project manager and the pocket-protector of the statistical analyst. Many deposits such as the Neves Corvo zinc-lead-copper-tin deposits in Portugal and the Cannington zinc-lead-silver deposits in Australia are the results of reinterpretation, and application of new approaches to existing information.

The flow diagram of total exploration philosophy in Fig. 6 builds in checks at the end of each stage, which determine when or whether to move on to the next stage.

CONCLUSIONS

The discovery rate of base metal deposits the world over rose throughout the '60s and '70s. It declined during the '90s and no significant deposits have yet been reported in the twenty-first century. The current Indian base metal deposits are rediscoveries of ancient abandoned mining sites and extensions in the known belts. Most of the ore bodies have cropped out at the surface, either unaltered or exposed as oxidized gossans. Traditional prospecting combined with geology, geophysics and some limited geochemistry gave rise to the discoveries in the recent past.

No surface indications lead to mineral discoveries any longer (Singh, 2004). Chances of finding new deposits at shallow depths are remote in India, as we are an old-world nation. Ancient eyes have already swept their gaze over the timeless tracts and ancient hands have built civilizations from the treasures un-surfaced. Our task is to visualize new approaches, and seek deep-seated metal deposits in conducive-geological settings. Forthcoming discoveries will depend on modern exploration technologies applied to an integrated geological model. Yet future generations will look to the stars. The international race to the moon is already begun again - this time with the stated objectives of mining the moon for resources as varied as rocket propellant, helium-3 and lunar metals (See N. Bhandari, this volume).

The giant basemetal ore bodies (such as Rampura Agucha zinc-lead-copper and Malanjkhand copper) significantly influence the economics of production because these deposits are amenable to relatively low cost surface mining. Experiences derived from these large-scale ventures are useful training for smaller expeditions, and thus contribute to the economic health of the organization and overall expansion program in many ways.

To emerge victorious in this new environment of increasing discovery risk requires extraordinary creativity, innovation, and technical skill coupled with commercial discipline (Blain, 2000). The incentives are world economic demand, driving continuously high trends in metal prices, technological necessity, and the primeval joy of unveiling the opening chapters of creation. These are the factors that motivate corporations to pioneer and unearth new deposits, and to upgrade the marginal low-grade deposits for production purposes. To accept the challenges ahead, the mining companies need fortitude, foresight, fortune, fearlessness - and a fair share of humor.

ACKNOWLEDGEMENT

I am beholden to the Department of Geology, Presidency College, Kolkata, for providing me with all possible logistic support. I would like to express my gratitude to the management and colleagues of M/s Hindustan Copper Limited and Hindustan Zinc Limited, where I had the honour of developing initial mineral exploration concepts. My students, with whose able aid, I applied the concept, supported me through and through, and I would like to thank them. Suratwant has edited the entire document keeping aside his busy schedule at Los Angeles. As always, I acknowledge the Supreme Entity that makes sense of it all, and urges us to seek for ourselves the ultimate Truth. In that pursuit, this treatise is dedicated "To strive, to seek, to find, and not to yield"!

References

Anon; 2006 : 40th Annual Report of Hindustan Zinc Limited 2005-06, pp. 17 18.

Anon; 1989 : Zinc- A Heritage, HINDZINC TECH, v. 1, no. 1, 12 p.

Arthur, Robert Peacocke., 2005 : The Palace of Glory : Gods world and Science, ATF Press, 122 p.

Bhatnagar, S.N., Khamesra, G.S., Haldar, S.K.; 1984 : Lead Zinc resources and development in India; Significance discoveries of geology for mineral industries, Oxford & IBH Publishing Co., India, pt. II, pp. 4268.

Blain, Chris.; 2000 : Fifty-years Trends in Mineral Discovery Commodity and Ore-type Targets, Canadian Institute of Mining, Explor. Mining Geol., v. 9, pp. 1 11.

Boyle, Alan, Wanted: Home-builders for the moon, http://www.msnbc.msn.com/id/16871258/

Formation of Indian Mineral Industry, 2007: India to become next exporter of zinc by mid 2008, News Bulletin, v. XXXX, no. 2, pp. 20 21.

Gandhi, S.M.; 2000, Ancient Mining and Metallurgy in Rajasthan, In: M. Deb (ed), Crustal evolution and Metallogeny in the northwestern Indian Shield (A

Festschrift for Ashok Mukherjee), Narosa, New Delhi, pp. 29-48.

Haldar, S. K.; 2005, Exploration Modeling of Proterozoic Sediment Hosted Zn-P-Ag Mineralization Logical Dynamic Approach with Case Study of Rampura-Agucha Deposit, Rajasthan, Jour. of Economic Geology and Georesource Management, v. 2, No. 1 and 2, pp. 47-70.

Haldar, S. K.; 2007, Exploration Modeling of Base Metal Deposits, Elsevier (in Press), 166 p.

Indian Beuro of Mines, 2003 : Indian Minerals Year Book, pp. 26-1-18, 42-1-16, 52-1-19.

ICSG: http://www.icsg.org/conferences/ICSG11General/Session.htm.

Ivanhoe, L.F.; 1995 : Future world oil supplies: There is a finite limit [World Oil, October 1995]

http://dieoff.org/page85.htm.

Multi-Commodity Exchange of India, http://www.mcxindia.com/products_zinc.htm.

NASA: http://science.nasa.gov/headlines/y2007/02feb_181.htm.

Polish Geological Institute :

http://www.pgi.gov.pl/mineral_resources/glossary.htm

Singh, N.N.; 2004 : Indian Peninsular Precambrian terrain and metal potential : a new approach, Jour. Geol. Soc India, vol. 63, p. 345349.

Wikipedia., http://en.wikipedia.org/wiki/History_of_metallurgy_in_the_Indian_subcontinent#zinc

Hon'ble Reviewers

1. Dr. Arun Kumar Shandilya
   Department of Geology
   Dungar College
   Binaker University
   BIKANER 334 003 INDIA
2. Anonymous
3. Dr. P.K. Jain
   School of Earth Sciences
   Jiwaji University
   GWALIOR 474 011 INDIA

Editorial Assistance

Mrs. Veena Chaudhary

Section V
Mining, Economics and Environment

Preamble

Mining of economic deposits and associated refining and smelting are nowadays facing more severe restrictions to effect economics on environmental grounds that it ever was before. The issue is volatile and emotional but understandable and desirable when rational. Sustainable development is the only answer to make a balance between economic growth and tolerable conditions of survival, without over pressing, either of the two.

Artisanal gold mining which uses mercury amalgamation has serious adverse implication for human health globally. Prof. U. Aswathanarayana in his paper entitled "How to make artisanal gold industry viable and ecofnendly?" has drawn attention to ways and means whereby science and technology can be used to achieve two purposes with regard to the artisanal gold mining industry: (i) better methods of target identification, and (ii) to reduce, if not eliminate, the environmental degradation arising out of the use of the mercury amalgam method. Prof. Aswathanarayana suggests in his paper as much as five highly scientific, yet very practical and effective solutions, to revolutionize the artisanal gold mining.

Dr. D.C. Joshi and Dr. K.D. Sharma in their paper entitled "Characterization of land degraded due to mining and their rehabilitation in Arid region of Rajasthan", have identified that the traditional and indiscriminate mining activities have caused loss of topography, disruption of natural drainage, destruction of natural vegetation and degradation in soil fabric and physical, fertility and microbial properties. Better availability of economic minerals and building stones in environmentally strained, arid regions of Rajasthan has overall degraded the land. The authors have given many scientific and practical suggestions to be adopted for specific case of the rehabilitation, including suitable plantation of tree, shrub and grass by adopting modification of soil profile and proper rain water harvesting.

Low-grade ores and mineral concentrates are now being treated increasingly with bioleaching, considering it as a true economic alternative for extraction of metals. Dr. V. Sheoran has presented an excellent review on bioleaching entitled "Bioleaching of metals from sulphide minerals: a review". The paper emphasizes the contribution made by the microbial community, the bioleaching mechanism and the chemistry involved, factors influencing the process(es) and future directions, is yet to be adopted. The paper practically includes all aspects of the present status of knowledge in this strongly growing microbially-based metal extraction industry, and also evaluates the promising future.

The paper entitled "Environmental impact of the coal mining activity on trace element concentration of natural waters in Pench valley coalfield area, Chhindwara district, Madhya Pradesh" is authored by Prof. D.C. Gupta. The Pench valley coalfield (PVC) consisting of 13 underground and 4 opencast mines, covers an area of almost 240 sq km The author in his exhaustive study of trace elements in its natural water has studied the concentrations of elements like As, Cd, Cu, Co, Fe, Hg, Mn, Ni, Pb, Se, V and Zn. He has found that the concentrations of Hg, Se, Cu, Fe and Pb are above the prescribed limits. It shows that the sustainability of the public living in the region is greatly challenged.

The next paper on impact of mining on biodiversity in India has been presented by an interdisciplinary team of scientists consisting of Dr. D.K. Khandelwal, Dr. Shishir Sharma, Dr. Kalpana Khandelwal and Dr. Rajni Sharma. Before presenting a brief account of the present status of biodiversity in India, the authors have shown their concern for the fast depletion of biodiversity globally, despite the efforts to conserve it. The paper has emphasized environmental degradation in general and land degradation in particular, due to mining activity and suggests a series of preventive measures including legislation.

The mining industry is facing a critical environmental challenge to treat the acid mine drainage (AMD) and polluting other effluents and liquid residues originating in different stages of extractive mining. Active treatment systems which are nowadays in practice, require continual addition of expensive chemicals and produce voluminous sludge as major disadvantage. Constructed wetland systems are the recent technology fast gaining acceptability and are alternative to the conventional way of treating AMD. In fact, a constructed wetland is a shallow basin filled with substrate and planted with vegetation tolerant of saturated condition. The treatment of AMD is obtained by physical, chemical and biological processes which involve sedimentation, settling, filtration, adsorption, precipitation and co-precipitation of insoluble compounds. The paper of Dr. A. S. Sheoran and Dr. V. Sheoran is on passive treatment of acid mine drainage from sulphide mineral mining and processing industry. The paper covers the mechanism of removal of pollutants, design and development, advantage and limitations of the wetland system in detail. Hon'ble reviewer Prof. V.P.Singh has appreciated the paper and added a comment that it covers every minor and important aspect of the wetland system.

Kolayat, one of the important industrial clay mining areas of India, has been studied by Dr. Shishir Sharma, Dr. D. K. Khandelwal, Dr. A. K. Shandilya and Dr. Rajni Sharma for its environmental degradation due to mining. In their paper, they have suggested some effective remedial measures. Hon'ble reviewer and an ecologist of distinction, Prof. D. Mohan, has suggested an unconventional use of abandoned clay mines for pisciculture by storing Indian and exotic carps at least during the rainy season. It will not only generate income but also help in stabilization and reclamation of the degraded land, he has added.

The next paper in the section entitled "A remote sensing study on environmental impact on mining- a case study of Rampura- Agucha mine" has been authored by Dr. N. K. Kalra, Dr. T. S. Sharma, Dr. Ashok Gehlot and Dr. Rajeev Jain. Remote sensing by virtue of synoptic view of a large area, multi-date, multi-resolution capability in different wavelength regions has proved to be of great assistance in providing important information on landscape manifestation including land degradation. The authors have utilized IRS LISS-III and PAN sensors data, to map 15 km radius of Rampura-Agucha mine and to generate statistics of various land use/land cover details and respective changes have been observed.

The last paper of the section entitled "Demand and supply for geoscientists in India for the 21st century is authored by Prof. Mike Katz. The paper predicts that, with the unprecedented growing economy of India and mineral development, there will be more demand for geoscientists with appropriate specialization as trends indicate. On the supply side, the author feels, there are some problems of training facilities and attracting the number and quality of students at postgraduate level. While at graduate level it is expected to occur as alternative periods of shortage and oversupply of geoscientists of cyclic nature. Prof. Katz makes an attempt to assess India's demand and supply of geoscientists in 21st century by utilizing an Australian-based model which correlates the mining in GDP (MGDP) with the professional geoscientific stock and manpower projected increase with the MGDP projected growth rate. The results so obtained seem to suggest reasonable preliminary estimates. Prof. Katz feels that the number of geoscientists necessary to meet the need of the 21st century in India, as estimated by him, are speculative projections, but the increasing technological quality required is unquestioned. Hence, the quality needed will put pressure on demand and supply of geoscientists in 21st century in India. Hon'ble reviewer Prof. S. K. tendon has given his comments "suitable for publication but it is too speculative and data on India has been very weakly presented". Hon'ble reviewer Prof. Aswathanarayan has commented that the paper is of 'high quality' but the Australian model seems to be more appropriate for other countries like Finland than India where circumstances are different. He has further added that the training of Indian personnel has scope to take advantage of India's competence in Information Technology.

Economic Mineralization
K.L. Shrivastava (Editor)
2009. pp. 454-458
Scientific Publishers (India), Jodhpur

42

Section-V

How to Make Artisanal Gold Industry Viable and Eco-Friendly ?

U. ASWATHANARAYANA
(Former Director, Centre of Advance Studies in Geology, Saugar University)
Mahadevan International Centre for Water Resources Management, Hyderabad-500 082
Email : uaswathanarayana@yahoo.com

Abstract

The paper have discussed the ways and means whereby science and technology can be used innovatively in regard to the artisanal gold mining industry as better method of target identification, to make more profitable and less hazardous, and to reduce environmental degradation arising due to the use of the mercury amalgam method. Better understanding of ore genesis and occurrence helps the pattern of gold distribution for the identification of target. Biogeochemical methods and portable XRF analyzer are also in recent practice. Artisanal gold mining using mercury amalgamation has serious adverse implications on human health, and the integrity of the ecosystems.

By adopting recent and innovative ways suggested here, it is possible to improve the efficiency of small scale mining, while concomitantly reducing its harmful consequences. It includes, first the research and development to design improved methods of ore search and ore extraction; second, using mobile units for preconcentration and extraction on site. Newly designed Tanzanian retort, can be used as most effective way to reduce the pollution of airborne mercury produced by the firing of the Au-Hg amalgam. Fourth way, the portable gold plant is a veritable godsend for small scale gold minerals. It is ecofriendly as no Hg is used, can be erected even in remote areas, shifted, reassembled and operated with minimum expertise and is economically viable. As a fifth way, recently designed modular plants can be used which have revolutionized the recovery efficiencies in artisanal mining.

Key words : Artisanal Gold Industry, Mining Environment, Mercury Pollution.

INTRODUCTION

Among the metals, gold has some unique features: there has never been a reduction in the production and demand for gold, and it is 100 % recycled. Considering that the production cost of gold is rarely more than USD 150/oz (1 troy oz = 31.10348 g.), and the market price rarely went below USD 300/ oz, gold mining has always been profitable. Gold mining is currently very profitable, as the gold prices are in the range of USD 620 640 / Oz. The world production of gold is about 2600 t, worth about USD 52 b.

The purpose of the article is to draw attention to ways and means whereby science and technology can be used innovatively to achieve two purposes in regard to the artisanal gold mining industry: (i) better methods of target identification, to make the artisanal mining industry more profitable and less hazardous, and (ii) to reduce, if not eliminate altogether, the environmental degradation arising out of the use of the mercury amalgam method.

The contribution of small-scale or artisanal mining to the overall mining output in the world is estimated to be 10-16%. In general, the percentage contribution with respect to industrial minerals is higher than for metallic minerals. Estimates vary widely about the number of persons involved in artisanal mining (upto 16 million). In some countries, the contribution of artisanal mining is of considerable economic significance. For instance, Peru has about 3,000 small-scale mines which produce 100% of the antimony, 90% gold, 15% tungsten, etc. of the national production (Gocht, 1980, Natural Resources and Development, v.12, pp. 7-18). About 90% of Brazil's gold production is attributable to artisanal mining.

Capital-intensive, mechanized mining is not cost effective for the exploitation of small deposits, even of high grade. Such a deposit is amenable to small-scale mining, particularly where the deposit occurs at or close to the surface, and where the mineral could be won by simple methods, such as hand-picking, panning, sluicing, etc. It therefore follows that artisanal mining may be a cost-effective option in the case of several economic minerals occurring in the soil. For instance, by the nature of its occurrence and its properties, opal can only be mined manually.

Small-scale mining has several advantages: it is labor intensive, can be initiated on any scale, with simple technology, at low capital cost, low consumption of energy, and short lead time, and without expensive imported equipment. It can also promote local industries. This, however, does not mean that small-scale mining is the panacea for the problems of the developing countries. Artisanal mining suffers from the following serious disadvantages: (i) it tends to be haphazard, since in most countries there is no systematic exploration activity to support small-scale mining; (ii) destructive exploitation by the gauging of rich pockets; (iii) low recovery rates; (iv) low labour productivity (about 4% of highly mechanized mines); and (v) non-extraction of valuable byproducts which are therefore irretrievably lost to the country.

In future, vegetative methods of reclamation of mined land may emerge as a significant, economically viable, and employment-generating activity.

TARGET IDENTIFICATION THROUGH AN UNDERSTANDING THE PATTERN OF GOLD DISTRIBUTION

Geochemistry of gold :

The important ore minerals and the processes of concentration to form ore deposits are summarized as follows: Crustal abundance: 4.5 ppb, Concentration coefficient (needed to form a deposit): 2000. The most important commercial mineral of gold is native gold. Though several inter-metallic compounds with Ag (electrum - Au, Ag), Sb, Pt, Te, etc. are known, they have no significant role in Au production. Mantle-derived mafic-ultramafic rocks have the highest gold contents (upto 10 ppb). In the endogenous environments, gold may be mobilized and transported in the form of thiosulphate complexes $[Au(S_2O_3)]^{3-}$, which result in coarsely crystalline gold alloyed with 50-75 % Ag, and chloride complexes ($AuCl_2^{2-}$ and $AuCl_2^-$) which tend to be reprecipitated with concretionary Fe oxides with < 0.5 wt. % of Ag. In the exogenous environments, Au may be transported as Au-humate complexes, and may end up as very fine lateritic gold. Alluvial placers of gold are of considerable economic importance and are the mainstay of artisanal gold mining.

Bacterial techniques of prospecting for gold in soils: *Bacillus cereus* is a common soil bacterium. Gold in soluble form is toxic to microbes, but bacterial spores can survive in harsh environmental conditions better than bacteria in the normal growing stage. The auriferous soils in the northwest Sichuan province of China, are characterized by two to three times more Bacillus cereus spores than normal soils. The analysis of *Bacillus cereus* spores in soils, is much simpler and cheaper than the chemical analysis of soils and rocks for gold. This biogeochemical method is useful not only to artisanal miners in the identification of auriferous soils, but also to mining companies in the search for underlying gold deposits (source: American Society for Microbiology, posted on Internet: May 24, 2002).

In situ methods: Using a portable X-ray fluorescence analyzer, it is possible to make a quick and fairly accurate on-site assay of several ore metals in the material mined or to be mined by a miner. Such an assay can serve two purposes : (i) to make the miner aware of the economic value of the material already mined by him (through a knowledge of what kind of ore metals and in what concentrations occur in the material mined by him), and (ii) to advise him as to what kind of material he should be mining in order to get greater returns.

MERCURY POLLUTION DUE TO ARTISANAL GOLD MINING

Artisanal gold mining using mercury amalgamation has serious adverse implications for human health, and the integrity of the ecosystems. Usually, 6 - 8 kg of mercury is used for amalgamation to extract one kg of gold. Simple squeezing and sublimation recover most of the mercury, but about 1.5 kg of mercury is irretrievably lost to the environment per kg of gold extracted (the Brazilian average is 1.32 kg of mercury). This leads to mercury pollution in soils, sediments, waters, and biota. The bacteria (such as, *Enterobacter aerogenes* and *Escherichia coli*) present in the soils, sediments, human faeces, etc. convert the metallic mercury to the highly toxic form of methyl mercury. Mercury compounds may enter the human body through inhalation, ingestion of food and water, and transfer through the skin. The intake and uptake of mercury is highly species sensitive. The organomercury, particularly methyl mercury, CH_3Hg^+, is more easily absorbed and is far more toxic than elemental mercury (Hg^0) and divalent inorganic mercury (Hg^{2+}). In the food chain, methyl mercury gets concentrated in fish (about 80% of mercury in fish is in the form of the more toxic methyl mercury). The FAO / WHO permissible tolerable level of mercury exposure has been set at 0.3 mg / week, with methyl mercury constituting not more than two-thirds of it (i.e., 0.2 mg).

Artisanal gold mining using mercury amalgamation has increased markedly in developing countries, such as Brazil, Tanzania, Zimbabwe, Philippines, Indonesia, China, Vietnam, etc. involving about 10 million people. According to Akagi (1998), there are 2000 goldfields (or "garimpos") in the Amazon Basin in Brazil, involving about 650,000 gold miners (or "garimpeiros") according to official reports. Since many of the garimpeiros are illegal miners, the actual number may be 1 to 1.2 million. The amount of mercury released into the environment annually is of the order of 130 t about 45 % of it is released into the river systems, and 55 % into the atmosphere. The total amount of mercury released to the environment since 1980s is a mind-boggling figure of 2000- 3000 t. There are two types of garimpos: the lowland garimpo where the garimpeiro digs a large hole in the ground to excavate the auriferous alluvium, and the raft garimpo where a motorized suction pump collects the god-bearing sediment from the river bed.

There are two main pathways of mercury to man in the gold mining areas, gold miners and gold shop workers are exposed to metallic mercury through inhalation. In the secondary environments, the metallic mercury would get methylated, and enters the food chain (say, fish) as more poisonous methyl mercury. The fish-eating populations who may have nothing to do artisanal gold mining, are hence at risk.

Akagi and Naganuma (2000) reported the mercury concentrations in hair, blood and urine of persons in Jacareacanga and Vila Novo Sitio, which are fishing villages, and Alta Floresta, which is the main trading centre in the Tapajos River basin in Brazil.

In the Tapajos River, the mercury levels in fish ranged from 0.08 to 3.82 ppm (the permissible level in Brazil is 0.5 ppm). In the case of fish-eaters, mercury present in hair and blood is almost wholly in the form of methyl mercury. There is clear evidence that the inhabitants of fishing villages have been exposed to abnormally high methyl mercury levels due to the consumption of fish. On the other hand, the ingestion of metallic mercury by gold miners and gold shop workers shows up in urine. Levels of mercury in hair samples were more for males than females.

Since hair grows fast at the rate of about one cm. per month, the longitudinal analysis of mercury along a strand of hair provides timing of the exposure of individuals to methyl mercury. Such a study of the long hair of the women showed that they were exposed to methyl mercury during the fast few years, with an increasing trend with time.

East Africa is another major area concerned with artisanal gold mining (van Stratten, 2000 a, b). Tanzania and Zimbabwe produce about 5 t of gold each from artisanal mining. It has been estimated that 200,000 to 300,000 persons are involved in small-scale gold mining activities in Tanzania, and more than 200,000 persons in Zimbabwe. For every 1 g. of gold recovered, 1.2 to 1.5 g of mercury is lost to the environment (about 70- 80 % of Hg is lost to the atmosphere due to processing, and 20-30 % in tailings, soils, stream sediments, etc.). Approximately 3-4 t of Hg is released to the atmosphere in the Lake Victoria goldfields, and > 3 t in Zimbabwe. Mercury in soils and sediments is in the form of metallic mercury. The dispersal of Hg in soils and sediments is limited (laterally < 260 m, and vertically < 20 cm). Urine analyses show that about 36 % of the gold miners working with amalgam exceeded the WHO limit (of 50 Φ g/g creatinine). Concentrations in fish were low, and > 90 % of the hair samples from the fish-eating populations showed <2 Φ g/g T-Hg. Highest Hg concentrations were found in the fish caught in the rivers draining the gold mining sites, and southern shores of Lake Victoria. Among the fish, lungfish species (*Protopterus aethiopicus*) have the highest concentrations, and tilapia (*Oreochromis niloticus*) the least. The amount of mercury pollution in Tanzania is estimated to be of the order of one mg/ m^2 annually (Aswathanarayana, 1995, p. 177).

The critical organs affected by mercury intoxication are the lungs, kidneys and the brain. The effects of mercury on the respiratory tract are coughing, bronchial inflammation, chest pain, and in severe cases, respiratory arrest. Methyl mercury causes the disintegration of cells within the brain, and consequently affects the sensory, visual, auditory and coordination control functions of the brain. This leads to loss of coordination in walking, slurred speech, loss of hearing, blindness, coma, etc. (Fergusson, 1990, p. 542).

INNOVATIVE TECHNOLOGIES SUGGESTED

It is possible to improve the efficiency of small-scale mining, while concomitantly reducing its deleterious consequences, by the adoption of the following innovative approaches :

(i) Research and development to design improved methods of ore search and ore extraction relevant to small-scale mining. Placer gold is a case in point. An artisanal miner can extract gold only if it is coarse grained (say, 30 μm) and high grade (say, about 25 g / m^3). He uses the mercury amalgam method of extraction which is highly polluting. New carbon-in-pulp and carbon-in-leach technologies have several advantages: (i) they are capable of extracting fine-grained gold (about 10 μm) and at low concentrations (about 2 g/ t); (ii) they are environmentally benign. These technologies need to be adapted for small-scale operations. In extremely dry areas, pneumatic methods of gold separation have to be developed.

(ii) Using mobile units for preconcentration and extraction on site: Truck-mounted, diesel-powered, self-contained, ore-dressing modules are taken to the site of the artisanal mining and the ore is concentrated / extracted on site. The mobile unit can be owned and operated by a cooperative or a private company. A part (say, one-third) of the output could be collected in kind towards service charge due to the mobile unit and the royalty due to the government. As the recovery rates by the mobile unit are at least 2 - 3 times higher than by manual methods, the artisanal miner is still left with considerably more saleable material than he would have been able to recover on his own.

(iii) Through the use of the mercury amalgam method, artisanal gold mining industry in Tanzania has severely contaminated the waters and soils. About 78 % of the water samples analyzed contained concentrations of mercury higher than the permissible level of 1g/l. Mercury levels in the mine tailings range from 1.31 to 18.7 g/g. The Institute of Production Innovation of the University of Dar es Salaam, Tanzania, has developed a rugged, easy to handle, highly portable and locally manufactured, and inexpensive (eq. USD 50) retort which has efficiency of 99.6 to 100 %. The use of such a retort is the most effective way to reduce the pollution of airborne mercury produced by the firing of the Au Hg amalgam (Mpendazoe, 1995).

(iv) The "Portable" gold plant developed by Libenberg, Rundle and Storey of San Martin mining company (*Mining Magazine*, July, 97, pp. 8-10) is a veritable godsend for small-scale gold miners. The salient points of the plant are as follows: San Martin 's claims encompasses two dumps around Bonda, Kenya, with 250,000 t of material, grade: 1-3g/t. Carbon-in-pulp/carbon-in-leach technique; Capacity of the plant: 10,000 t / month; production cost: USD 150 / oz. The total steel requirement (12 t) for tanks, baffles, agitator mountings, and the pumps and piping, were brought from South Africa in one container and erected on site. Dump material is reclaimed by high-pressure water. In the first leach tank, lime (5 kg / t) and sodium cyanide (0.5 kg/ t) are added. The residence time in each of the absorption tanks is approximately 1.3 hr at a throughput of 10,000 t / month, and carbon concentration of 15 - 20 g / l. The eluate is heated in a diesel-fired cast iron burner. Gold is recovered onto steel wool cathodes. Security is ensured by having the recovery cell protected by a 220 V inner cage, and 15,000 V outer cage (powered by solar cells). Doré is 80% pure. Total power consumption: 145 kw (diesel generators). Water is pumped from the Yala

River (3 km from the plant).

The plant has been in operation for more than a year and has the following advantages: environmentally-benign as no mercury is used; can be erected even in remote areas, and shifted and reassembled without much problem; can be operated with minimal expatriate assistance; economically viable. This technology can be used in two kinds of situations : (i) as a private enterprise, for treating dumps, where they exist, and (ii) as a cooperative, by setting up the plant at a central place where a number of artisanal miners (50 to 100) operate.

(v) The modular plants designed by M/S Bateman Project Holdings Limited, Boksburg, South Africa, have revolutionized the recovery efficiencies in artisanal mining of a number of minerals (such as, modular process plants for prospecting, sampling and small scale production of diamonds). The heart of the equipment is the Dense Media Separator (DMS) plant, which uses water-ferrosilicon mixture as the flotation medium. The density of the medium is automatically controlled. The smallest unit has a capacity of 1t / h of diamondiferous gravel in the size range of 1 mm to 8 mm. The modular unit which has become very popular is 5 t/h DMS unit, which weighs only 4.5 t, and can fit into a standard 12 mm container, with all the spares. It is custom-designed depending upon the kind of feed, assembled and tested. It is then dismantled and shipped to site. It can be erected on site in a day or two. The DMS modules are skid mounted and has built-in spillage pumps, and no concrete foundations are needed.

References

Akagi, H.; 1998 : Studies on mercury pollution in the Amazon, Brazil. Global Environmental Research, Vol. **2**, No. 2 pp. 193-202.

Akagi, H., and Naganuma, A. (2000) Human exposure to mercury and the accumulation of methyl mercury that is associated with gold mining in the Amazon Basin, Brazil. J. Health Sci., Vol. 46, No. 5 pp. 323 -328.

Aswathanarayana, U.; 1995 : Geoenvironment : An Introduction. A.A. Balkema, Rotterdam.

Fergusson, J.E.; 1990 : The Heavy Elements: Chemistry, Environmental Impact and Health Effects. Pergamon Press, Oxford.

Mpendazoe, F.M.T.; 1995 : Towards sustainable small-scale mining in Tanzania, Proc. First World Mining Congress, New Delhi, India. pp. 251-265.

van Stratten, P.; 2000a : Human exposure to mercury due to small-scale gold mining in northern Tanzania. The Science of Total Environment (Elsevier), Vol. 259, pp. 45-53.

van Stratten, P.; 2000b : Mercury contamination associated with small-scale gold mining in Tanzania and Zimbabwe. The Science of Total Environment (Elsevier), Vol. 259, pp. 105-113.

Hon'ble Reviewers

1. Dr. S. Bhandari
   (Professor of Mining Engineering)
   (Former, Dean)
   Director , Earth Resource Centre,
   Bundoora, VICTORIA 3083
   AUSTRALIA
2. Dr. Vinod K. Sharma
   Dean
   (Professor of Environmental Engineering)
   Indira Gandhi Institute of Development Research
   Goregaon (East)
   MUMBAI 400 065 INDIA

Editorial Assistance

Mrs. Anju Mathur

Economic Mineralization
K.L. Shrivastava (Editor)
2009. pp. 459-471
Scientific Publishers (India), Jodhpur

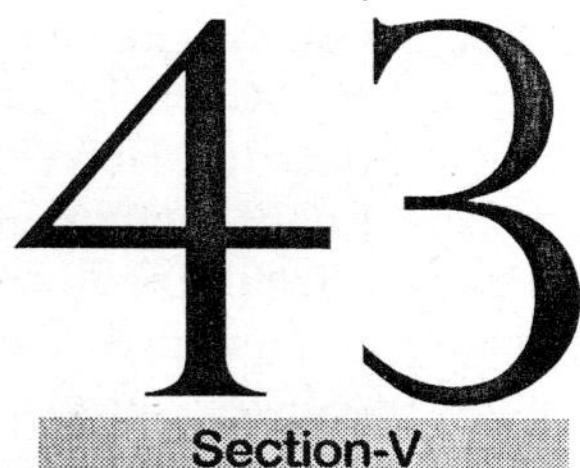

Section-V

Characterization of Land Degraded Due to Mining and their Rehabilitation in Arid Region of Rajasthan

D. C. JOSHI[1] AND K. D. SHARMA[2]

[1]Principal Scientist (Ex), (Central Arid Zone Research Institute), 7A66 Nandan Van, Jodhpur 342 008
[2]Director, National Institute of Hydrology, Jalvigyan Bhawan, Roorkee 247 667
Email : dcjoshi66@yahoo.com

Abstract

The arid region of Rajasthan is gifted with valuable minerals like lime-stone, gypsum, granite, sandstone and lignite. The rich mineral resource is a boon for industrial development and building construction. The traditional and indiscriminate mining activities have caused loss of topography, disruption of natural drainage, destruction of natural vegetation and overall deterioration of surrounding environment. The characteristics of degraded land due to gypsum, limestone, marble, sand stone mining have been presented. The mined sites with heaps of mine muck and ditches appear as vast barren land devoid of natural vegetation. The soil mantle is drastically disturbed resulting in disruption of soil fabric, poor physical, fertility and microbial properties. Growing suitable tree, shrub and grass species by adopting modification of soil profile and proper rainwater harvesting system can revegetate the degraded land. For each of the kind of mined land technologies for rehabilitation have been presented.

Key words : Mining Environment, Land Degradation, Rehabilitation, Arid Region, Gypsum, Saline Soils

INTRODUCTION

The arid region of Rajasthan is endowed with vast deposits of minerals like gypsum, marble, limestone, lignite, sandstone and granite. The rich mineral resource is a boon to desert dwellers, because they easily find alternative employment during the drought period. In arid Rajasthan annual production of gypums and marble is about 95 % and that of limestone is 25% of the national production. Rajasthan occupies the first position in reserves and production of sand stone in our country and contributes 90% of the total production. Lignite is an important source of energy for thermal power generation. In western Rajasthan large deposits of lignite have been located and their mining is at initial stage.

Perhaps after agriculture, mining and related activities provide employment to the largest number of people in the state. Gypsum is used in the manufacture of cement, fertilizer, plaster of Paris, pottery and for reclamation of sodic soils. Limestone is an important ingredient for use as flux in iron ore smelting and for manufacture of cement, mortar, lime etc. Therefore, both these minerals are transported for long distances for processing plants within Rajasthan. Marble and sandstone are used for building purposes and are in great demand in local and distant markets.

The marble blocks are subjected to cutting, grinding and polishing which is generally performed in the vicinity of quarries. However, the sandstone *khandas* and slabs do not require such finishing and directly transported from the mining sites. Marble and sandstone are also used for making various types of artifacts used for interior decoration. With the development activities during last two decades the related activities have enormously increased and provided employment opportunities and prosperity in the region.

Open cast mining is largely adopted for quarrying the rocks and minerals. The indiscriminate mining activities cause loss of topography, destruction of natural vegetation, adverse effects on surface and ground water sources. Dust generated during quarrying causes severe health problems. Proper rehabilitation of such quarried land is required so that it does not cause adverse effects. Characteristics of degraded land due to quarrying these minerals and technologies developed for their rehabilitation have been presented in this paper.

CHARACTERISTICS OF MINED LANDS FOR GYPSUM AND THEIR REHABILITATION METHODS

Gypsum is geologically recent to sub recent in age and occurs as sedimentary deposit in the form of bands intercalated with sand and silt at the depth of 0.5 m to 5 m. Gypsum beds are generally horizontal beds covered by thin veneer of sand and occurs at a relatively lower elevation than the surrounding area. Gypsum is used for making plaster of Paris, pottery, also used in cement plants and fertilizer industry and for reclamation of sodic soils.

Soils having gypsum horizon at certain depth are called gypsiferous soils (Fig. 1). In western Rajasthan these soils occur in 845 km^2 ((Dhir et al., 1997). According to Soil Taxonomy, two great groups viz. Petrogypsids and Haplogypsids of the sub order Gypsids have been recognized (Joshi, 1999 and 2000 a) . Petrogypsids have hard compact substrata at 20-50 cm depth which do no permit arable cropping and these are under silvi pasture. These occur in Barmer, Jaisalmer, Bikaner and Nagaur districts The Haplogypsids have gypsic horizon at 60-100 cm depth and arable cropping is widely practiced. These are wide spread in Sri Ganganagar and Hanumangarh districts. The gypsiferous soils are generally low in major and micronutrients (Sharma et al., 2004 b). Now with the availability of water from IGNP system in these districts irrigation agriculture is being practiced.

Characteristics of gypsum mined area : The gypsum is quarried on common land as well as on private land. On private land, after gypsum mining the fields are back filled and returned to the owner. But on government land mining activities are more vigorous and quarrying extend in large areas for longer time. The mined area appears undulating due to pits and heaps of tailing. Back filling of pits is seldom done. In both the situations the land is severely degraded but nature of degradation is quite different. Growing crops and/or establishing vegetation on such land is very difficult.

Characteristics of gypsum mined backfill: The mining companies acquire land for gypsum quarrying on lease from private land owner. After mining the fields are backfilled and returned to land owners. The backfilled soil materials are drastically disturbed. The solum is generally up side down i. e., finer material filled below and coarse at the surface. Thus the crop growth on such soil is very poor (Fig.2) and yields are low. Comprehensive study was carried out on the characteristics of back filled land at 22 sites in Bikaner, Hanumangarh and Sri Ganganagar districts (Joshi et al., 1996) and the results are presented for some sites.

Soil physical characteristics : The backfilled sites are characterized by uneven surfaces and gypsum nodules strewn randomly. The back fill in different horizons is non coherent, non-aggregated and highly varying in particle size. Soil texture of undisturbed sites is sandy loam with sub-angular blocky structure and soft consistency under dry condition. In comparison the backfill is composed of unsorted, un-aggregated, loose sandy mass. Due to loss of soil fabric, the capillarity for moisture retention is disrupted and large size macro-pores are created which adversely affect soil air-water relationship within the soil profile (Tables 1 and 2).

Physico-chemical characteristics: The physico-chemical characteristics of undisturbed gypsiferous soils and the back fill are presented in Tables 3 and 4 . The pH of backfill (7.7-7.8) and undisturbed soil (7.6-8.1) is in alkaline range, EC values slightly higher for undisturbed soil (0.28-1.11 d S m^{-1}) than backfill (0.51-0.99 dS m^{-1}). The $CaCO_3$ is also less in undisturbed soil (0.9- 4.9 %) than the backfill (1.9-5.2 %). Available water capacity values of back fill was much lower and irregular with depth. Water holding capacity varies from 21 to 31% for different backfills and undisturbed soils.

Fertility characteristics: Available major and micronutrients of undisturbed sites and backfill are presented in Tables 5 and 6. Organic carbon was slightly low in backfill (0.01-0.11%) than undisturbed soil (0.06- 0.21%) but there was no difference in the available P content (tr8.4 kg ha^{-1}). Available micronutrients were slightly low in back fill (Fe 1.39 - 3.94 ppm, Mn 2.2-6.2 ppm, Zn 0.27-3.15 ppm, Cu 0.16-0.45 ppm) than the undisturbed sites (Fe 1.39 - 4.11 ppm, Mn 3.4-9.7 ppm, Zn 0.27-2.94 ppm, Cu 0.16-1.49 ppm). Irregular distribution of organic carbon, major and micronutrients with depth in the backfill can be attributed to heterogeneity of backfill material (Joshi and Sharma, 1997).

Microbial population: The microbial population (Table 7) reveal that in the mine spoils microbe population was very low compared to undisturbed soils. The acid and alkaline phosphatase activity and dehydrogenase activity was also very low indicating disruption of the microbial ecosystem (Rao *et al.*, 1996).

Rehabilitation of gypsum mined land : Rehabilitation of such drastically degraded land can be achieved through an optimum combination of rainwater harvesting, profile modification and plantation of appropriate species. A study was carried out at Kavas (Dist. Barmer) during 1992-97 (Sharma et al., 2001). The surface of disposed gypsum mine spoil land was shaped into terrace and slope provided for rainwater harvesting. For rainwater harvesting,

Table-1 Morphological characteristics of undisturbed sites

Location	Horizon	Depth (cm)	<2 mm fraction (%)		
			Colour	Texture	Structure and consistency
Deshli	A	0 15	Light grey (10 YR 7/2)	sl	Fine weak, subangular blocky, soft
	Cy	15 30	White (10 YR 8/2)		Gypsum powder
Asuwali Talai	A1	0 15	Light brown (10 YR 6/3)	sl	Fine weak, granular, soft
	A2	15 30	Light grey (10 YR 7/2)	sl	Fine weak, subangular blocky, soft
	Cy	>30	White (10 YR 8/2)		Gypsum powder
Mahila D5	A1	0 20	Light brown (10 YR 6/3)	fs	Fine weak, subangular blocky, soft
	A2	20 60	Light grey (10 YR 7/2)	sl	Medium weak subangular blocky, slightly hard
	Cy	>60	White (10 YR 8/2)		Gypsum powder
Basant ki Dhani	A	0 15	Light brown (10 YR 6/3)	fs	Fine weak, subangular blocky, loose
	C1	15 30	Light Yellowish brown (10 YR 5/6)	sl	Medium moderate subangular blocky, slightly hard
	C2	30 50	Light grey (10 YR 7/2)	sl	Fine weak, subangular blocky, soft
	Cy	> 50	White (10 YR 8/2)		Gypsum powder mixed with mega crystal
Purabsar	A	0 25	Light brown (10 YR 6/3)	fs	Fine moderate, subangular blocky, soft
	C1	25 40	Light grey (10 YR 7/2)	sl	Fine weak, subangular blocky, soft
	C2	40 55	Light grey (10 YR 7/2)	sl	Medium weak, subangular blocky, soft
	Cy	>55	White (10 YR 8/2)	-	Gypsum powder mixed with mega crystal

Texture: fs -fine sand, ms- medium sand, ls- loamy sand, sil- silt loam, sl-sandy loam

Table-2 Morphological characteristics of backfill

Location	Horizon	Depth (cm)	>2mm fraction (%)	<2 mm fraction (%)		
				Colour	Texture	Structure and consistency
Deshli	A	0-15	25	Light grey (10 YR 7/2)	fs	Non-coherent mixed with gypsum nodules, non- sticky
	C1	15-45		Light grey (10 YR 7/2)	fs	Non-coherent, non- sticky
	C2	45-75		Light grey (10 YR 7/2)	fs	Non-coherent, non- sticky
Asuwali Talai	A	0-40	17.1	Light grey (10 YR 7/2)	ls	Non-coherent mixed with gypsum nodules, non- sticky
	C1	40-55	6.5	Light brown (10 YR 6/3)	ms	Non-coherent mixed with gypsum nodules, non- sticky
	C2	55-70	15.1	Very pale brown (10 YR 7/3)	ms	Non-coherent mixed with gypsum nodules, non- sticky
Mahila D5	A	0-25	11.9	Light grey (10 YR 7/2)	fs	Non-coherent mixed with gypsum nodules, non- sticky
	C1	25-50		Light grey (10 YR 7/2)	fs	Non-coherent, non- sticky
	C2	50-75		Light grey (10 YR 7/2)	fs	Non-coherent, non- sticky
Basant ki Dhani	A	0-2		Light grey (10 YR 7/2)	sil	Surface crusting, friable
	C1	2-12		Light grey (10 YR 7/2)	sl	Non-coherent, slightly sticky
	C2	12-35		Light brown (10 YR 6/3)	fs	Non-coherent, non- sticky
	C3	35-60		Light brown (10 YR 6/3)	ms	Non-coherent, non- sticky
Purabsar	A	0-30	9.7	Light brown (10 YR 6/3)	fs	Non-coherent, non- sticky
	C1	30-60	11.1	Light brown (10 YR 6/3)	fs	Non-coherent, non- sticky
	C2	60-90		Very pale brown (10 YR 7/3)	fs	Non-coherent, non- sticky

Texture: fs -fine sand, ms- medium sand, ls- loamy sand, sil- silt loam, sl-sandy loam

micro-catchment (Fig. 3) half moon terrace and ridge and furrow structures were created. The augar holes of 15 cm diameter and 1 m depth were dug at 5m x 5m distance and a mixture of top soil, fine sand and FYM in equal proportions were filled. In the augar holes Trees (12 species) and shrubs (10 species) were planted and grasses of 2 species was sown in the open area. During a period of 1992-97, maximum height was gained by the *Tamarix aphylla* (132%) in micro-catchment followed by *Salvadora persica* (102%) in half moon terrace, *Acacia tortilis* (63%) in ridge and furrow and *Pithecellobium dulce* (31%) in the control. Among shrubs *Cercidium floridum* gained maximum height (970%) in half moon terrace, followed by *Prosopis juliflora* (251%) in micro-catchment, *Acacia nubica* (154%) in micro-catchment, *Dichrostachys nutans* (122%) in ridge and furrow and *Colophosphopermum mopane* (19%) in half moon terrace. The maximum increase in plant cover was observed for *Acacia tortilis* (521%) in ridge and furrow, followed by *Tamarix aphylla* (492%) in micro-catchment, *Salvadora persica* (334%) in half moon terrace, and *Pithecellobium dulce* (166%) in control. *Cercidium floridum* (3221%), *Prosopis juliflora* (2217%), *Colophosphopermum mopane* (780%) and *Dichrostachys nutans* ((395%) gained significantly higher plant cover in half moon terrace than did *Acacia nubica* (1370%) in micro-catchment.

Table-3 Physico-chemical characteristics of undisturbed sites

Location	Horizon	Depth (cm)	pH	Salinity (EC dS m)	CaCQ (%)	Moisture equivalent (%)	Water holding capacity (%)
Deshli	A	0-15	7.58	1.08	11.9	5.20	20.9
	Cy	15-30	7.78	1.09	6.6	13.67	32.1
Asuwali Talai	A1	0-15	7.70	0.51	3.4	7.02	22.1
	A2	15-30	7.72	0.55	3.5	6.05	25.1
	Cy	>30	7.70	1.11	3.7	6.50	24.7
Mahila D5	A1	0-20	7.70	0.56	0.9	3.91	21.1
	A2	20-60	7.70	0.28	1.4	5.22	24.0
	Cy	>60	7.71	0.30	1.9	8.20	24.7
Basant ki Dhani	A	0-15	7.85	1.15	7.7	9.8	36.8
	C1	15-30	7.86	1.11	5.6	13.0	31.1
	C2	30 50	8.14	0.32	4.1	4.2	27.1
	Cy	> 50	7.86	1.05	4.9	7.0	27.5
Purabsar	A	0-25	7.78	1.02	0.25	10.6	26.3
	C1	25-40	7.78	1.05	1.79	14.4	24.7
	C2	40-55	7.82	1.04	3.4	6.0	22.7
	Cy	>55	7.78	1.05	2.6	17.7	24.7

After six years, there was slight increase in the soluble salt (EC from 1.3 to 1.5 dSm^{-1}) and build up of organic carbon from 0.09 -0.14 to 0.14-0.38%, available phosphorus from 6.3-8.2 to7.2- 15.0 kg ha^{-1} and available potassium from 158- 208 to 169-484 kg ha^{-1}. Average soil moisture storage was the highest in the half moon terraces (4.6 %) followed by micro catchment (4.4 %) and ridge and furrow (4.2 %) method. The un-mined control recorded the lowest moisture (0.2 %).

CHARACTERISTICS OF MARBLE MINED LAND AND THEIR REHABILITATION

Marble, a elegant looking rock formed by thermally metamorphism of Precambrian age in Rajasthan. Marble is a much sought after rock which has led to significant growth of marble industry. As a result number of marble quarries as well as marble processing units have grown during last 10-15 years. Besides bringing prosperity in the region the marble quarrying and processing wastes have caused environmental degradation.

During the last two decades, marble quarrying and processing units have spread in the Rajsamand areas at very fast rate which has made tremendous effect on the socio-economic status of the society. The spread of the area affected due to marble mining and the spread of slurry in Rajsamand area was studied by using IRS 1 C FCC satellite images and ground checked (Joshi, 2000 b). On the satellite imagery slurry covered areas appear bright white, the quarry areas with the spread of mine muck devoid of vegetation cover appear dull white where as on low hills as bluish white and on hill flanks as light pink with bluish white. Indiscriminate quarrying and processing and dumping of marble slurry could be mapped along road (Fig 4), on open grazing land and in river/ stream beds (Fig. 5). The slurry has also spread in agricultural lands resulting in abandoned of agriculture, decline in crop yield, reducing ground water recharge. It was also noted that

Table-4 Physico-chemical characteristics of backfill

Location/	Horizon	Depth (cm)	pH	EC (dS m^{-1})	$CaCO_3$(%)	Moisture equivalent (%)	Water holding capacity (%)
Deshli	A	0-15	7.84	0.99	4.1	6.62	31.1
	C1	15-45	7.80	0.93	5.2	4.58	25.9
	C2	45-75	7.81	1.07	1.9	7.26	27.3
Asuwali Talai	A	0-40	7.70	0.51	2.9	7.02	22.1
	C1	40-55	7.72	0.55	2.9	9.15	21.1
	C2	55-70	7.70	0.85	2.1	10.50	24.0
Mahila D5	A	0-25	7.70	0.56	1.7	3.91	21.1
	C1	25-50	7.70	0.28	2.3	5.22	24.0
	C2	50- 75	7.60	0.90	2.4	7.20	25.0
Basant ki Dhani	A	0-2	7.60	1.09	2.3	15.8	25.8
	C1	2-12	7.67	1.05	2.2	11.3	24.8
	C2	12-35	7.74	0.97	1.9	11.6	25.0
	C3	35-60	7.60	0.85	2.1	12.5	24.5
Purabsar	A	0-30	7.84	1.08	2.1	6.6	21.8
	C1	30-60	7.84	1.07	1.1	5.6	23.5
	C2	60-90	7.82	1.36	2.1	2.9	22.1

the marble dust has covered foliage of most of the trees and shrubs causing severe physiological disorder.

Waste generated during quarrying : These are overburden, side-burden, inter-burden, ungraded and under size material. The waste generated varies from 30 % mechanized to 65% manual mining.

Waste generated during processing: This include dressing, cutting and polishing wastes. The sludge and slurry generated during processing (Sawing of blocks into slabs and smaller pieces) can be estimated at about 10% of the total stone quarried (20-25% of block) and slurry produced during grinding and polishing varies from 5-7%.

Characteristics of marble slurry: The slurry is a suspension of fines generated during processing and polishing of marble rock. The chemical composition is 28-35% CaO, 10-14% MgO, 1-2.5% R2 O3, 15-20 % acid insoluble and 35-40% loss on ignition (Sinha, 2000). The marble slurry collected from different sites in the Kelwa-Amet, Pasund-Tasol, Amet Sardargarh, and Kelwa-Nathdwara area has been analysed for physico-chemical characteristics (Table 8) and particle size distribution (Table 9). The slurry is characterised by high $CaCO_3$ (40-50 %), alkaline pH (8.9-9.7) and moderate salinity (EC 0.27- 3.01 dS m^{-1}). The slurry particles (50-70 %) are of 150 μm size followed 10-22 % particles of 500 μm, and 9-13 % particles of 250 micron size. The water holding capacity ranges from 20-26% . The non-capillary porosity which stores moisture for plant growth is almost non existent. The slurry does not contain organic carbon. nitrogen, phosphorus and other nutrients and is also devoid of microbial population. Thus it is a very poor media for plant growth and will not support any kind of vegetation unless favourable rooting environment is created by mixing with soil (Joshi and Paliwal, 2004).

Hazards due to marble slurry: Although sites have been identified but marble slurry is discharged on open grazing lands and heaped along roads. Slurry being fine particles is blown away with wind and forms thick cover on natural vegetation and hazardous

Table-5 Fertility characteristics of undisturbed sites

Location/	Horizon	Depth (cm)	Organic carbon (%)	Available (kg ha $^{-1}$)		Available (ppm)			
				P	K	Fe	Mn	Zn	Cu
Deshli	A	0- 15	0.17	4.8	225	3.22	9.31	0.77	1.49
	Cy	15-30	0.21	4.2	101	2.50	9.67	1.31	0.84
Asuwali Ta lai	A1	0-15	0.16	4.2	236	2.72	8.80	0.87	0.67
	A2	15-30	0.16	3.4	240	1.50	4.62	0.65	0.67
	Cy	>30	0.05	Tr	80	1.50	7.90	0.54	0.43
Mahila D5	A1	0-20	0.09	7.0	219	2.44	6.73	0.53	0.31
	A2	20-60	0.08	2.5	124	2.05	7.16	0.44	0.48
	Cy	>60	0.06	2.5	80	1.50	5.40	0.34	0.28
Basant ki Dhani	A	0-15	0.09		123.8	4.11	5.82	0.78	0.45
	C1	15-30	0.12		151.9	2.94	7.73	0.90	0.33
	C2	30-50	0.09		247.5	2.16	4.15	2.94	0.42
	Cy	> 50	0.09		247.5	1.50	4.62	0.65	0.67
Purabsar	A	0-25	0.11		112.4	2.33	5.18	0.46	0.28
	C1	25-40	0.13	2.24	78.8	2.33	3.90	1.30	0.16
	C2	40-55	0.104	8.4	169	2.00	6.16	0.27	0.28
	Cy	> 55	0.007	1.1	202	1.39	4.00	0.88	0.16

to the inhabitants and causes respiratory ailments. The slurry is also carried away with rain water which chokes the drainage lines and obstructs supply to surface water bodies and recharge to ground water. The slurry deposited on the agricultural fields reduces porosity and permeability of the top soil. The fine particles reduce fertility of the soil by increasing alkalinity. The hazardous dumping pose a severe threat on the environment and health of people. Marble slurry should be specified as hazardous waste under the existing rules.

Reuse of marble waste : Although so far the proper use of slurry could not be found out, the slurry is being used as a filler for road and embankment. The slurry also has potential for the manufacture of bricks, cement and ceramic tiles however many more uses need to be found out. There is still scope for research on the rehabilitation of areas degraded due to marble quarrying and processing.

CHARACTERISTICS OF LIMESTONE MINED LAND AND THEIR REHABILITATION

Traditional open cast mining of limestone has caused destruction of land resources. There is severe degradation of natural vegetation cover, creation of uneven topography, disruption of natural water resources. The mine muck is composed of different size rock fragments including cobbles (25-40%), gravels (15-30%), stones/boulders (10-50 %) and fines (10-20%). Spread of mine spoil causes loss of soil fabric and fertility. The mine spoil is highly rich in $CaCO_3$ and alkaline in reaction. The ecosystem of lime stone mine spoil is highly degraded and does not allow growth of natural vegetation. It is estimated that nearly 1, 84, 000 ha have been degraded due to lime stone quarring in arid region during last 60 years (Sharma et. al., 1998, 2000).

Rehabilitation of lime stone quarry areas: The study was carried out by Sharma et al. (2000) to

Table-6 Fertility characteristics of backfill

Location/	Horizon	Depth (cm)	Organic carbon (%)	Available (kg ha^{-1})		Available (ppm)			
				P	K	Fe	Mn	Zn	Cu
Deshli	A	0- 15	0.07	Tr	146	1.72	6.20	0.69	0.36
	C1	15-45	0.03	Tr	146	1.22	2.21	1.61	0.31
	C2	45-75	0.04	Tr	146	1.72	2.49	1.50	0.22
Asuwali Talai	A	0-40	0.08	1.1	202	2.27	5.54	0.96	0.39
	C1	40-55	0.07	2.8	247	2.39	1.57	3.15	0.25
	C2	55-70	0.03	1.1	191	2.00	1.64	2.52	0.25
Mahila D5	A	0-25	0.06	6.2	140	2.05	5.01	1.17	0.45
	C1	25-50	0.02	4.5	191	1.83	2.72	0.58	0.31
	C2	50- 75	0.02	1.7	213				
Basant ki Dhani	A	0-2	0.13	4.8	298	3.94	18.78	0.78	1.49
	C1	2-12	0.09	1.7	152	2.44	12.10	0.90	0.65
	C2	12-35	0.01	3.6	146	1.50	0.93	0.72	0.11
Purabsar	A	0-30	0.11	8.4	169	2.00	6.16	0.27	0.28
	C1	30-60	0.06	7.0	169	1.55	3.47	0.74	0.16
	C2	60-90	0.01	1.1	202	1.39	4.00	0.88	0.16

Table-7 Microbial population of undisturbed soil and gypsum mine spoil After Sharma et al (2001)

Microbes	Undisturbed soil Depth (cm)		Gypsum mine spoil Depth (cm)	
	0 45	45 120	0 45	45 120
Fungi (10^6 kg^{-1}soil)	125	19.5	19.30	4.2
Bacteria (10^6 kg^{-1}soil)	200	38.5	63.6	14.0
Actinimycetes (10^6 kg^{-1}soil)	50.5	18.2	37.0	8.3
Nitrifiers (10^6 kg^{-1}soil)	20.5	1.9	5.4	0.5
Acid phosphatase activity (nkat kg^{-1} soil)	9.0	4.0	4.0	2.0
Alkaline phosphatase activity (nkat kg^{-1} soil)	22	13.2	10.1	6.0
Dehydrogenase activity (pkat kg^{-1} soil)	9.8	4.1	5.3	1.7

Fig. 1 Profile of gypsiferous soil

Fig. 2 Poor crop on gypsum mined backfill

Fig. 3 Micro catchment harvesting for tree plantation

Fig. 4 Marble slurry heaped along road side

Fig. 5 Marble slurry spread in stream bed

rehabilitate mine spoil in 20 ha areas at Barna village. The approach was to combine rain water harvesting , use of soil amendment, selection of appropriate germplasm and method of planting technique. The mine spoils were compacted, and designed variously shaped sloping terraces. The rainwater harvesting technique included inward sloping bench terraces, half moon terraces and a tear drop configuration. Pits of 0.60 m diameter and 0.60 m depth were dug at spacing of 5 x 5 m and filled with a mixture of equal proportion of sand, bentonite clay and FYM. Indiginous tree species viz. *Acacia, catechu, Acacia planifrons, Azadirachta indica, Salvadora oleoides, Holopteila integrifolia, Acacia Senegal, Acacia tortilis* and shrubs viz. *Maytenus emarginata, Commifora wightii, Ziziphus nummularia* and *Grevia tenax* and *Cenchrus ciliaris* and *Cymbopogon jwarancusa* grasses were planted. Live fencing of *P. Juliflora* was provided to stop animal encroachment.

Improvement of moisture storage: The monthly average soil moisture during July- September was the highest in the inward sloping bench terraces (6.4%) followed by the half moon mound (4.5%), tear drop configuration (4.3%).

Growth performance of plantation: Between 1993 and 1998 *Acacia planifrons* gained maximum height in inward sloping bench terraces and half moon terraces treatments followed by *Acacia senegal* in inward sloping bench terraces and half moon terraces, *Acacia tortilis* in tear drop configuration and inward sloping bench terraces and *Pithecellobium dulece* in inward sloping bench terraces. Among shrubs *Cercidium floridum* gained maximum height in inward sloping bench terraces and in half moon terraces followed by *Maytenus emerginata, Dichrostachys nutans* and *Prosopis juliflora* in half moon and *Acacia catechu* in tear drop configuration.

Maximum increase in plant cover was observed in *Acacia planifrons* in inward sloping bench terraces followed by *Acacia senegal* in inward sloping bench terraces and half moon, *Pithecellobium dulece* in inward sloping bench terraces and *Acacia tortilis* in half moon and tear drop configuration. *Acacia tortilis* in half moon and tear drop configuration. *Cecidium floridium* in half moon terraces and half moon, *Maytenus emarginata* in half moon terraces, *Acacia Catechu* in tear drop configuration and *Dichrostachys* in half moon. At the end of six years grass cover was most abundant

Improvement in mine soil quality : Over a period of 6 years between 1993 to 1998 slight increase in pH 7.6 to7.9 and three fold increase in EC 0.20-0.58 dS m^{-1} was observed. $CaCO_3$ content decreased from 20 to 5.2%. The water holding capacity and moisture equivalent increased from 21 to 25% and 5 to 7%, respectively. The build up in organic carbon from 0.10 to 0.50% and increase in available P from 4 to 10 to 15 to 17 kg ha^{-1} and increase in available K from 66 to 100 to 93 to 207 kg ha^{-1} was observed. Over the six year period population of fungi, bacteria and actinomycetes were significantly higher than the original mine spoil (Sharma et al., 2000).

CHARACTERISTICS OF SAND STONE MINED LAND AND THEIR REHABILITATION

Rajasthan occupies first rank in reserves and production of sand stone in our country and contributes about 90% of our country's production. Jodhpur leads in sand stone mining with 6,000 operating quarries followed by Bijoliya (District Bhilwara) and Bundi respectively with 1600 and 1000 number of quarries. Land degradation due to sand stone quarrying has major impact on land disfiguration, soil erosion, creation of barren land as well as by the disposal of waste material. Because the quarrying is shallow it spreads in large areas leaving heaps of low grade stone haphazardly. Drilling, blasting and transportation of sand stone causes lot of air pollution affecting the workers with silico-tuberculosis and silicosis.

Impact on natural drainage system : The surface water bodies like baori, tanks, lakes receive water from the natural catcment through feeder streams which originate from the hills. At present large number of quarries are located in the catchment of these water bodies and indiscriminate quarrying chokes natural flow of water. Besides there is accumulation of rainwater in deep quarry pits. Singh *et al.* (1999) have reported that 32 % catchment area of Balsamand reservoir is damaged by quarrying activity. Out of total catchment area of the water reservoir 11.35 % is not available and subsequently it reduced runoff by 10 percent (Menariya and Bhandari, 1990). It is essential to demarcate area for waste disposal. The disposal sites should be provided with plantation. The waste can be used for filling for road

CHARACTERISTICS OF LIGNITE MINED LAND AND THEIR REHABILITATION

Lignite is an important source of energy for thermal power generation. Large deposits of lignite amounting to 1000 million tonnes is found within 2650 km^2 in Barmer, Jaisalmer, Bikaner and Jodhpur districts (Rajasthan Patrika, 2001). The lignite is found in 8 to

Table-8 Physico-chemical characteristics of marble slurry

Location	pH	EC ($dS\ m^{-1}$)	CaCQ (%)	Available water capacity (%)	Water holding capacity (%)	Non capillary porosity (%)
Slurry from factory tanks						
Puthol	9.29	0.64	41.7	19.0	21.3	2.3
Khatamla	9.54	0.27	49.7	20.9	23.2	2.3
Tunkra	9.67	0.81	59.1	18.7	20.5	2.2
Pasund	8.92	3.01	34.4	22.9	22.9	0.0
Slurry from road side heaps						
Siyana	9.63	1.01	55.1	20.7	20.4	0.0
Kiro-ka-bas	8.89	7.58	33.2	21.6	19.9	0.0
Sardargarh	9.27	0.23	52.0	23.7	23.7	0.0
Amet I	9.16	0.17	45.5	21.9	21.1	0.0
Amet II	8.87	1.50	49.7	21.9	23.5	1.6

Table-9 Particle size distribution of marble slurry

Location	Percent of particle size (μm)								
	500	**250**	**180**	**150**	**106**	**75**	**53**	**45**	**<45**
Slurry from factory tanks									
Puthol	10.0	9.1	2.8	73.5	2.5	0.05	0.8	0.25	1.0
Khatamla	17.6	10.5	3.8	61.8	0.1	0.25	2.05	1.02	2.8
Tunkra	22.1	33.9	20.6	22.3	0.1	0.05	0.71	0.05	0.2
Pasund	20.8	16.2	12.9	46.9	0.5	0.50	1.36	0.15	0.4
Slurry from roadside heaps									
Siyana	15.5	10.7	3.2	66.6	0.5	0.25	1.40	0.75	1.1
Kiro-ka-bas	21.7	20.2	19.0	37.1	0.2	0.41	0.82	0.15	0.3
Sardargarh	20.5	13.1	3.8	58.5	0.5	0.25	1.15	0.35	1.8
Amet I	18.6	8.8	2.5	49.5	2.6	4.54	6.96	1.51	4.9
Amet II	17.1	10.6	3.1	44.5	3.6	4.89	7.65	3.62	5.0

Table-10 The physico chemical characteristics of dump and spoils After Kumar et al. (2005)

	Dump		Spoils		Nonmined	
Depth	0-30	30-60	0-30	30-60	0-30	30-60
pH	3.6-4.1	4.1-5.7	7.1-9.2	7.4-9.2	8.0-8.1	8.3-8.4
EC (d S m^{-1})	8.5-10.5	7.8-10.4	2.8-10.7	2.8-10.9	0.5-0.6	0.4-0.6
Organic carbon (mg kg^{-1})	1.2-1.7	3.3-4.5	0.7-0.8	0.6-0.8	1.8-2.2	1.7-2.0
Avaiable P (µg g^{-1})	5.2-10.8	6.3-9.8	1.8-2.0	1.3-2.0	3.3-4.8	3.2-4.5
Avaiable K (µg g^{-1})	100-350	99-410	75-155	68-99	201-228	198-205
DTPA Fe ((µg kg^{-1})	23.5-32.5	44.4-46.5	11.8-17.6	7.6-11.8	5.4-6.4	5.2-5.6
DTPA Mn ((µg kg^{-1})	18.6-29.8	19.8-30.5	11.6-17.8	15.3-17.6	9.6-10.8	9.0-10.3
DTPA Zn ((µg kg^{-1})	6.4-7.9	6.0-8.3	8.9-9.7	8.7-9.7	0.7-0.0	0.7-0.0
DTPA Cu ((µg kg^{-1})	2.6-5.9	3.5-6.6	0.7-1.1	0.6-1.1	0.3-0.4	0.4-0.5

10 seams between 30 to 50 m depth. Each seam is 0.3 to 2 m in thickness and separated by clays and shale. Rajasthan State Mines and Minerals (RSMM) Ltd. started open cast mining of lignite at Giral (District Barmer) and the mining has created two types of surfaces (1) dumps and (2) spoils. The dumps are of 40 m height and were formed by stock piling material excavated from the pits. The spoils were created by filling the previously dug areas with the material obtained as the mining proceeds. Till now both the dumps and spoils remain barren. The physico chemical characteristics of dump and spoils are presented in Table 10.

Rehabilitation of lignite mined land: During strong wind regime in summer the material from dumps and spoils blow away and deposit on agricultural fields causing soil deterioration, vegetation degradation and associated environmental problems. Therefore, these sites need to be rehabilitated. The spoils are back filled areas with the mining tail. Surface of spoil was covered with fresh top soils from adjoining areas to a thickness of 25 to 30 cm and land shaping was done to create basins to harvest rain water. The micro-catchment water harvesting was adopted. Four month old tree sapling were planted in pits 60 cm diameter and 60 cm deep. Trees and shrubs were planted at spacing of 9 m x 5 m. During 1997 and 2000 the *Azadirachta indica* gained maximum height followed by *Salvadora persica, Colophospermum mopane, Tamarix aphylla, Prosopis cineraria* and *Tecomella undulata*. This was possible because MCWH stored more moisture than control plots (Sharma et al. 2004).

CONCLUSION

The mining has brought prosperity in the region and in near it is likely to be intensified with the development in technology and increasing demand for sandstone, limestone, granite, marble and gypsum. Technologies have been developed only for rehabilitation of gypsum and lime stone mined lands. But now there is urgent need to develop technologies for rehabilitation of sand stone, marble, granite and lignite mined land through on site studies.

References

Dhir, R P, Joshi, D. C., Singh, N., Sharma, B. K. and Kolarkar, A. S. (1997) Taxonomy and distribution of arid soils of Rajasthan. Annals of Arid Zone. Vol. 36, pp. 11-20.

Joshi, D. C. (1999) Gypsiferous soils of Rajasthan. In . B S Paliwal (ed) pp. 338-347, Geological Evolution of North-western India, Scientific Publishers, Jodhpur.

Joshi, D. C. (2000 a) Morpho-genetic characterization of gypsiferous soils of arid Rajasthan. Journal Indian

Society of Soil Science. Vol. 48, pp. 134-139.

Joshi, D C. (2000 b) Progress report of the project "Characterization of soils degraded due to marble quarrying and processing" Central Arid Zone Research Institute, Jodhpur.

Joshi, D. C. and Paliwal, M. S. (2004) Environmental Issues of Marble Mining and Processing in Raj Samand area. Paper presented at the seminar "Mining in Aravalli" Feb. 28-29, 2004; Department of Geology, Jai Naraian Vyas University, Jodhpur.

Joshi, D. C. and Sharma, K. D. (1997) Characterization of gypsum mine wastelands and their rehabilitation. 4th International Symposium on Environmental Geochemistry, United States Geological Survey, Vail, Colorado, USA pp. 40-41.

Joshi, D. C., Sharma, K. D. and Kumar, S. (1996) Rehabilitation of gypsum mined lands in IGNP command area. Central Arid Zone Research Institute, Jodhpur.

Kumar, P., Kumar, S., Sharma, K. D., Chaudhary, A. and Gehlot, K. (2005) Lignite mine spoil characterization and approach for its rehabilitation. Arid land Research and Management. Vol. 19, pp. 47-60.

Menariya, N. K. and Bhandari, S. (1990) Assessment of environmental impact due to sandstone mining of Jodhpur. Report, University of Jodhpur, Jodhpur pp.20-41.

Rajasthan Patrika (2001) Possibilities of lignite based power plant. Rajasthan Patrika, September , 05, 2001pp.3, Jodhpur.

Rao, A. V., Tarafdar, J. C., and Sharma B. K. (1996) Characteristics of gypsum mine spoils. Journal of the Indian Society of Soil Science. Vol. 44, pp. 544-546.

Sharma, K. D., Kumar, S. and Sharma, B. K. (1998) Rehabilitation of lime stone mined land in the Indian desert. Best Management Prcctices bulletin -2, Central Arid Zone Research Institute, Jodhpur

Sharma, K. D., Kumar, S. and Gough, L. P. (2000) Rehabilitation of lands mined for lime stone in the Indian desert. Land degradation and Development. Vol. 11, pp. 563-574.

Sharma, K. D., Kumar, S. and Gough, L. P. (2001) Rehabilitation of gypsum mined lands in the Indian desert. Arid Land Research and Management. Vol. 15, pp. 61-76.

Sharma, K. D., Praveen-Kumar, Gough, L. P. and Sanphilipo, J. R. (2004 a) Rehabilitation of lignite mined disturbed areas in Indian Desert. Land degradation and Development. Vol. 15, pp. 163-176.

Sharma, Manoj Kumar, Joshi, D. C. and Sharma, A. K. (2004 b) Physico-chemical and morphogenetic characteristics of Petrogypsids. Annals of Arid Zone. Vol. 43, pp. 21-27.

Singh, V. et al. (1999) Impact of sandstone mining on natural drainage system in Balsamand catchment area, Jodhpur. Project report, University of Jodhpur, Jodhpur, pp. 25-41.

Sinha, A. K. (2000) Marble industry and environmental issues: Scenario in Rajasthan. Indian Stone Mart 2000, special paper pp. 277-293.

Hon'ble Reviewers

1. Prof. A.S. Faroda
   Professor of Soil Science
   (Former Vice Chancellor,
   M.P.A.T. University, Udaipur)
   Krishna Nagar, New Pali Road
   JODHPUR 342 005 INDIA

2. Dr. Hema Achyuthan
   Department of Geology
   Anna University
   CHENNAI 600 001 INDIA

Editorial Assistance

Dr. Vijay Gaur

Economic Mineralization
K.L. Shrivastava (Editor)
2009. pp. 472-484
Scientific Publishers (India), Jodhpur

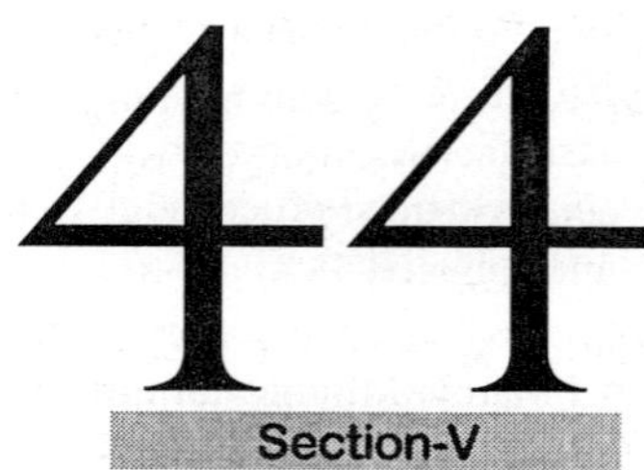

Section-V

Bioleaching of Metals from Sulphide Minerals: A Review

V. SHEORAN
Department of Zoology, Jai Narain Vyas University, Jodhpur 342 005
Email : Vimi_sheoran@yahoo.com

Abstract

Bioleaching is the use of microorganisms to extract metals from sulphide and iron-containing low grade ores and mineral concentrates. Today it occupies an increasingly important place among the available mining technologies and is an actual economical alternative for treating low grade ores and mineral concentrates. This study outlines the current research in strongly growing microbially based metal extraction industry and also evaluates promising developments. The bioleaching of sulphide minerals is reviewed with emphasis on the contribution from the microbial community, bioleaching chemistry of various sulphide minerals, mechanisms, factors influencing and future directions. Microbial dissolution of metals involves direct adhesion of the bacterial cell on the sulphide surface and allowing chemical oxidation of liberated sulphides by ferric iron generated microbially or through oxidation of air. At present the main techniques employed for bioleaching are heap, dump and in-situ leaching. Tank leaching is generally practiced for the treatment of refracting gold ores. Bioleaching has also some potential for detoxification of industrial waste products, sewage sludge and soil contaminated with heavy metals.

Keywords: Bioleaching, Sulphide Minerals, Micro-organisms, Bioleaching Chemistry, Metal Solubilisation.

INTRODUCTION

Applications of biotechnological methods are being increasingly applied for metal recovery from low grade ores and concentrates that cannot be processed economically by conventional methods. The use of microorganism to recover valuable metals/minerals from such ores and mineral concentrates has developed into a successful and expanding area of biotechnology (Sharma and Verma, 1991; Mousavi et al., 2007). It offers both economic and environmental benefits via less costly and more environmental friendly integrated processes. These technologies have been used since prehistoric times and probably the Greeks and Romans extracted copper from mine water more than 2000 years ago (Colmer and Hinkle, 1947). However, it has been known only for about 50 years that bacteria are mainly responsible for the enrichment of metals in water from ore deposits and mines (Bosecker, 1997). In the recent years there has been a strongly growing microbe-assisted minerals biotechnology industry for the recovery of metals (Brierley, 1982; Rawlings and Silver, 1995; Rawlings, 2002).

Microorganisms have geologically been active in

mineral formation via direct action of their enzymes or indirectly through chemical action of their metabolic products. They act as an agent of fractionation/ decomposition of mineral matter in the earth crust (Bandyopadhyay and Banik, 1995). The bioleaching processes are based on the ability of microbes to transform solid compounds and result in soluble and extractable elements which can be recovered, and this occurs in nature where ever suitable conditions are found for the growth of bioleaching microbes. Bioleaching of base metal and sulphides is in use commercially and offers advantages over the conventional techniques.

Most of the base metals that exist as natural resources occur in the form of sulphide deposits. Sulphide minerals have semiconductor properties, galvanic interactions appear when electrical contact among different mineralogical phases occurs. These galvanic effects are important in the aqueous processing of ores and minerals such as flotation and leaching. Galvanic interactions can substantially increase both leaching rate of and metal recovery from different minerals (Cruz et al., 2005; Santos et al., 2006).During dissolution of a mineral assembly of different sulphides, those minerals that have the highest reduction potentials behave as cathodes. The oxidant is reduced on their surface. Simultaneously, electrons are removed from the sulphides having the low reduction potential (Santos et al., 2006). The later behave as anodic sites where usually sulphide is oxidized directly to either elemental sulphur or sulphate. Pyrite is a mineral that is usually directly oxidised to sulphate. (Tripathi and Shrivastava ,1990; Shrivastava et al. 1992; Schippers and Sand, 1999) while the oxidation of pyrrhoitite generates elemental sulphur and microorganisms can oxidise elemental sulphur during bioleaching (Boon, 2001).

The mineral industry is greatly needed to process low grade ores, overburden and waste by economic processing methods such as *insitu*, dumps and heap leaching. According to statistics published in the early 1980s (USGS, 1990) about 18% of copper production in the United States is estimated to derive from in-situ, dump and heap leaching. However, bioleaching of the precious and semiprecious metal sulphides, has yet to be implemented at commercial scale. Since the amount of low grade ores of sulphide minerals is continually increasing, the necessity for application of this bioleaching technology has increased accordingly. Studies on the bioleaching of sulphide minerals have shown the potential of this technique to be applied in the processing of the copper, zinc, uranium, gold and some other precious and semiprecious metals utilizing the heap dump *insitu* or tank bioleaching processes (Brierley, 2000 and Romero et al., 2003) . Metal solubilisation /bioleaching from solid wastes or other solids is achieved through a variety of acidophilic and chemoautolithotropic bacteria. Thus use of microorganisms or its biomass for the recovery of metals from low grade ores have achieved growing attention and been a great success for the mining industry (Shi et al., 2006; Umrania, 2006; Liu et al., 2007).

According to Holmes and Debus (1991) biological processing would need to have >20% advantage over conventional processing to interest the mining industry. Traditionally, the advantages of bacterial leaching technology are compatible with these requirements:

- Moderate capital investment with low operating costs,
- Appropriate recovery of metals from low-grade ores and waste materials,
- Basic equipment and simple operating procedures.

The purpose of this study is to outline the focus of current research in respect to bioleaching that is the role of microorganisms, chemistry and mechanism of various sulphide minerals, factors influencing and to assess its promising developments made in the mining industry.

MINERAL-DEGRADING MICROORGANISMS

Microbiological hydrometallurgy is rapidly evolving since dealing with the extraction of metals like copper, iron, zinc, silver, gold, uranium, selenium and many others from sulphide ores through solubilizing and metal accumulating properties of microorganisms (Hutchins et al., 1986; Clark and Norris, 1996). Various microorganisms possess different trials to bring about metal extraction, metal concentration, metal detoxification and fossil fuel desulphurisation (Hossain et al., 2004). Sulphide ore dumps and heaps, with their varying mineralogical compositions and different climatic environments represent extremely complex microbiological habitats. Yet, only a modest number of iron- and sulphur-oxidising bacteria have been isolated from mineral sulphide ores, characterized physiologically and phylogenetically, and deposited in data banks (Hallberg and Johnson, 2003). These extreme acidophiles that grow optimally at pH<3, must comprise only a small selection of those that have mineral processing capabilities. Various acidophilic, iron and sulphur oxidizing bacteria of relevance in bioleaching have been isolated from the leaching sites. Studies revealed the presence of *Acidithiobacillus ferooxidans, Acidithiobacillus thiooxidans, Leptospirilium ferooxidans, Acidithiobacillus caldus strain, Metallogenium, Acidianus/sulfolobus* species and some other species (Rawlings et al., 1999; Boon, 2001; Johnson and Hallberg, 2003; Lavalle et al., 2005). The bacteria most active in bioleaching is gram negative, non-spore forming rods belong to genus *Thiobacillus* which grow under aerobic conditions. Most thiobacilli are chemolithoautotropic species which use the carbondioxde from the atmosphere as their carbon source for the synthesis of new cell material (Brierley and Brierley, 1973; Brierley, 1978; Bosecker, 1997).

The bacterial leaching is carryout at pH values between 1.5 and 3 at which most metal ions remain in solution state. Therefore, an autotrophic bacterium of acidophilic species *Thiobacillus ferrooxidans* and *T. thiooxidans* has received much attention over the past few decades (Torma, 1971; Torma and Sankaguchi, 1978; Sugio et al., 1985; Nemati and Webb, 1997; Vasquez and Espejo, 1997; Okibe et al., 2003). The bacterium *T. ferroodoxians* was first isolated by Colmer and Hinkle in 1947 from coal AMD (Colmer and Hinkle, 1947). *T. thiooxidans* was isolated in 1922 by Waksman and Joffe (Waksman and Joffe, 1922). The intensive sulphuric acid production leads to a rapid decomposition of rocks so that acid-soluble metal compounds can pass into solution as sulphates.

Leptospirillum ferrooxidans was also reported to bioleach zinc from the marmatite flotation concentrate effectively (Norris et al., 1987; Hansford, and Vargas,2001; Schrenk et al., 1998; Rawlings et al., 1999; Shi and Fang, 2005; Liu et al., 2007; Reyes, 2007).*T. thiooxidans, T. ferrooxidans* and *L. ferrooxidans* are mesophilic bacteria which grow best at temperatures of 25-35 °C. Even some of the bacteria are reported to be thermophilic bacteria which can grow in the range of temperature 50°C. Thiobacillus like bacteria so called Th-bacteria can grow on pyrite, pentlandite and Chalcopyrite at the same temperatures. Other microorganisms like Heterotrophic bacteria and fungi which require organic supplements for growth and energy supply may contribute to metal bioleaching. Leaching of manganese, chromium, from preservative treated wood is promoted by these microorganisms (Ehrlich.1983). Among the bacteria members of the genus Bacillus are most effective in metal solubilisation, with regard to the fungi the genera Aspergillus and Penicillium which are the most important ones (Brierley, and Brierley, 1973; Xu and Ting, 2004; Shi et al., 2006;). The increase in understanding of the role of microorganisms in circulation of mineral in biosphere has made them important biotechnological tools in mining from ores.

CHEMISTRY OF LEACHING OF SULPHIDE MINERALS

Bioleaching of metal sulfides is affected by thiobacteria via the oxidizing action of Fe^{3+} ions dissolving a metal sulfide. During this course, Fe^{2+} and elemental sulfur (S_8) will be generated (Porro et al., 1997; Sand et al., 2001; Tributsch, 2001).

The bacterial action in leaching process is a catalytic one during which the protons, Fe^{3+} ions and other reactants break the chemical bonds and are recycled by the bacterial mechanism. Bacterial oxidation of ferrous has been found to be a function of the ferric/ferrous ratio. By regenerating the ferric iron, the bacteria are able to maintain a high redox potential in bioleaching systems. May et al. (1997) have investigated the reaction between pyrite and ferric iron in sulfuric acid by monitoring the redox potential of an acidic ferric sulfate solution which is related to Nernst equation. Thus, the importance of cyclic regeneration of Fe^{3+} /Fe^{2+} and maintenance of high redox potential was emphasized for better rates of leaching.

The mineral dissolution/leaching reaction is not identical for all metal sulfides. It has been observed that the oxidation of different metal sulfides proceeds via different intermediates (Schnell, 1997; Schippers and Sand 1999). They proposed a thiosulfate mechanism for the oxidation of acid-insoluble metal sulfides, such as pyrite (FeS_2), molybdenite (MoS_2) or tungstenite (WS_2), and a polysulfide mechanism for acid-soluble metal sulfides, such as sphalerite (ZnS), chalcopyrite (Cu_2S) or galena (PbS).

In the thiosulfate mechanism, solubilisation is through ferric iron attack on the acid-insoluble metal sulfides, with thiosulfate being the main intermediate and sulfate the main end-product. Using pyrite as an example, the proposed reactions are shown in equation 1 and 2, (Schippers and Sand, 1999).

$FeS_2 + 6Fe^{3+} + 3H_2O \rightarrow$

$S_2O_4^{2-} + 7Fe^{2+} + 6H^+$ (1)

$S_2O_4^{2-} + 8Fe^{3+} + 5H_2O \rightarrow$

$2SO^{2-}_4 + 8Fe^{2+} + 10\ OH^-$ (2)

In the case of the polysulfide mechanism, solubilization of the acid-soluble metal sulfide is through a combined attack by ferric iron and protons, with elemental sulfur as the main intermediate. This elemental sulfur is relatively stable, but can be oxidized to sulfate by sulfur-oxidizing microbes (Eqs. 3, 4 and 5).

$MS + Fe^{3+} + H^+ \rightarrow$

$M^{2+} + 0.5H_2S_n + Fe^{2+}$ $(n \geq 2)$ (3)

$0.5H_2S_n + Fe^{3+} \rightarrow$

$0.125S_8 + Fe^{2+} + H^+$ (4)

$0.125S_8 + 15O_2 + H_2O \rightarrow$

$SO^{2-}_4 + 2H^+$ (5)

This explains why strictly sulfur-oxidizing bacteria, such as *At. thiooxidans* or *At. caldus*, are able to leach some metal sulfides but not others.

The ferrous iron produced during metal dissolution and bioleaching might also be reoxidized by iron-oxidizing organisms to ferric iron (Eq.6).

$2Fe^{2+} + 0.5O_2 + 2H^+ \rightarrow$

$2Fe^{3+} + SO^{2-}_4 + H_2O$ (6)

The role of the microorganisms in the solubilisation of metal sulfides is, therefore, to provide sulfuric acid (Eq.5) for a proton attack and to keep the iron in the oxidized ferric state (Eq. 6) for an oxidative attack on the mineral.

Copper: Different reactions involved during the dissolution process of copper bearing flotation concentrates are summarized as follows (Natarajan and Iwasaki, 1983; Hansford and Vargas, 2001). In pyrite oxidation, acid forms as a result of the following two reactions (Eqs. 7 and 8):

$FeS_2 + 7/2\,O_2 + H_2O \rightarrow$

$Fe^{2+} + 2H^+ + SO^{2-}_4 + 2e^-$ (7)

$FeS_2 + 14Fe^{3+} + 8H_2O \rightarrow$

$15\,Fe^{2+} + 16H^+ + 2SO^{2-}_4$ (8)

The chemistry of microbial oxidation of sulphides is closely related to that of ferric ion oxidation in acidic solutions (Eq.7). The microorganisms play a catalytic role in oxidising ferrous ion to ferric ion, thus regenerating the oxidant (Eq. 9). They also oxidize sulphur to sulphate, generating acid (Eq. 10).

$2Fe^{2+} + 2H+ + 0.5\ O_2 \rightarrow$

$2Fe^{3+} + H_2O$ (9)

$2S + 3O_2 + 2H_2O \rightarrow$

$2H_2SO_4$ (10)

The costs associated with the maintenance of sulphide bioprocessing microorganisms are minimal because many gain energy from the redox reactions (Eqs. 9 and 10), utilize carbon dioxide from the air as 'C' source and obtain their phosphorus, nitrogen, potassium and micronutrients, etc., from the bioleach environment.

Because the microorganisms are particularly efficient at oxidising ferrous ions to ferric ions, the bioleaching conditions typically exhibit a relatively high redox potential which is less conducive to chalcopyrite dissolution. One consequence of the high solution potential is that ferric ion readily precipitates as a basic sulphate, like jarosite (Eq.11) in an environment containing monovalent alkali cations and sulphate ions.

$3Fe^{3+} + 2SO_4^{2-} + 6H_2O + M^+ \rightarrow$

$MFe_3(SO_4)_2(OH)_6 + 6H^+$ (11)

where $M = K^+ \cdot Na^+$ or NH_4^+

The extraction of selected metals from mineral sulphides such as chalcopyrite can be an oxidative process in which ferric ions are the oxidant and sulphide component of the mineral is oxidized to elemental sulphur (Eq. 12).

$CuFeS_2 + 4Fe^{3+} \rightarrow$

$5Fe^{2+} + Cu^{2+} + 2S^o$ (12)

Chalcopyrite

The reaction is known to be sensitive to redox potential; and interestingly, higher dissolution rates have often been measured at lower potentials (Hiroyoshi et al., 1997; Hiroyoshi et al., 2001). In addition, it has been shown that a parallel, non-oxidative dissolution process that is also sensitive to potential, contributes to the kinetics of chalcopyrite leaching in sulphuric acid solutions (Eq. 13) (Nicol and Lazaro, 2003; Lazaro and Nicol, 2003).

$CuFeS_2 + 4H$ →

$Fe^{2+} + Cu^{2+} + 2H_2S$ (13)

chalcopyrite

In chalcopyrite, both the oxidizable metal moiety and sulphide moiety are simultaneously attacked in the below said reaction Eq. (14):

$4CuFeS_2 + 17O_2 + 10H_2O$ →

chalcopyrite

$4Cu^{2+} + 4Fe(OH)_3 + 8H^+ + 8SO^{2-}_4$ (14)

Zinc: Sphalerite (ZnS) leaches readily in the presence of ferric ion (Eqs. (15, 16 and 17)). Iron occurs as an impurity almost in all natural sphalerite and the rate of dissolution of sphalerite is proportional to the iron content (Crundwell, 1997). Although sphalerite has not typically been subjected to bio-processing, there is strong academic and commercial interest in developing a technology for sphalerite and also for mixed sulphides containing sphalerite (Fowler and Crundwell, 1999; Gomez et al., 1999). A recent study has indicated that sphalerite leaching is diffusion controlled by a surface layer of sulfur (Rodriguez et al., 2003; Deveci et al., 2004).

bacteria

$2FeSO_4 + H_2SO_4 + 1/2\,O_2$ →

$Fe_2(SO_4)_3 + H_2O$ (15)

bacteria

$S^0 + H_2O + 3/2\,O_2$ →

H_2SO_4 (16)

$ZnS + Fe_2(SO_4)_3$ →

Sphalerite

$ZnSO_4 + S^0 + 2FeSO_4$ (17)

Bio-oxidation of complex zinclead sulfides has been widely investigated using samples of sulfide concentrate,pure galena, or sphalerite minerals (Garcia et al., 1995; Chaudhury and Das, 1987; Bang et al., 1995). Among the bacteria involved in bacterial leaching, A. ferrooxidans has been studied the most intensively and has been the most important in commercial operations to lead-zinc sulphides. However, few investigations have examined the application of A. ferrooxidans to oxidize the zinclead sulfide ores (Hus and Harrison, 1995).

Silver: Ferricyanide is a powerful oxidant and can effectively dissolve silver sulfide in cyanide solutions (Eq. 18). The results of general chemistry tests indicate that under the typical cyanide leaching conditions (0.5 g/L NaCN, 0.5 g/L Fe as ferricyanide, pH=11.5, N2, room temperature), the main reaction involved is,

$Ag_2S + 5CN^- + 2Fe(CN)_6^{3-}$ →

$2Fe(CN)_6^{4} + CNS^- + 2Ag(CN)_2^-$ (18)

According to this reaction, at least 6 kg $K_3Fe(CN)_6$ would be needed to extracted 1 kg silver from the pure silver sulphide(Xie and Dreisinger,2007).

Nickel: Nickel concentrates produced from sulphides ores usually contain different proportions of iron sulphides (pyrite and pyrrhotite) and nickel (pentlandite, for instance). As pentlandite usually has high rest potential while pyrrhotite shows lower values, in those ores containing only pyrrhotite and pentlandite, the rest potential indicates that the former dissolves faster than the latter given a certain set of conditions (Lu et al., 2000; Zhu et al.2003). The proposed reactions for pyrrhotite dissolution (Eqs. 19 and 20) (Mason and Rice, 2002) are as follows:

$Fe_{1-x}S + 2H^+$ →

$(1-3x)Fe^{2+} + H_2S$ (19)

$2Fe_{1-x}S + O_2 + 4H^+$ →

$(2-6x)Fe^{2+} + 4xFe^{3+} + 2S^0 + 2H_2O$ (20)

Eqs. (19) and (20) show that as the pyrrhotite dissolves, Fe(II) is being released, providing iron for bacterial growth and further mineral oxidation. Ferrous iron dissolved during pyrrhotite leaching can be further oxidized by *A. ferrooxidans* and *L. ferrooxidans* to maintain an Eh high enough to sustain the leaching process.

The principal nickel-bearing sulfide mineral (pentlandite) gets leached (Eq. 21) as follows:

$[(FeNi)_9S_8] + 175/8O_2 + 13/4H_2SO_4$ →

$9/2\,NiSO_4 + 9/4Fe_2(SO_4)_3 + 13/4H_2O$ (21)

The chemistry involves oxidation of pyrite, chalcopyrite and pentlandite utilizing ferrous iron as a source of energy to support growth and maintenance with an aim to improve the bacterial performance in terms of selective leaching of nickel. The ultimate objective is to enhance the rates of ferrous and sulfur oxidations and correlate its effect on nickel leachability.

MECHANISMS OF BIOLEACHING OF SULPHIDE MINERALS

Bioleaching of metal sulphides is presently based on the activity of the microbes which convert heavily soluble metal sulphides via biochemical oxidation reactions into water soluble metal sulphates (Porro et al., 1997; Sand et al., 2001; Tributsch, 2001; Kodali et al., 2004). In principle metals can be released from sulphide minerals by direct and indirect contact bacterial leaching (Silverman and Ehrlich, 1964 and Silverman, 1967; Tributsch, 2001; Hallberg and Johnson, 2003).

Direct contact bacterial leaching : Direct mechanism requires physical contact between bacterial cell and sulphide mineral surface and oxidation of sulphate takes place via several enzymatically catalyzed steps (Silverman, 1967; Rulkens et al., 1995; Bosecker, 1997).

In this process pyrite is oxidized to iron (III) sulphate according to following reaction (Eqs. 22,23 and 24):

$$\underset{\text{pyrite}}{4FeS_2} + 14O_2 + 4H_2O \xrightarrow{\text{bacteria}} 4FeSO_4 + 4H_2SO_4 \quad (22)$$

$$4FeSO_4 + O_2 + 2H_2SO_4 \xrightarrow{\text{bacteria}} 2Fe_2(SO_4)_3 + 2H_2O \quad (23)$$

The direct bacterial oxidation of pyrite can be summarized as follows

$$\underset{\text{pyrite}}{4FeS_2} + 15O_2 + 2H_2O \xrightarrow{\text{bacteria}} 2Fe_2(SO_4)_3 + 2H_2SO_4 \quad (24)$$

Even non-iron metal sulphides can be oxidized by T.ferrooxidans directly (Torma, 1971; Klauber et al., 2001; Klauber, 2003; Hossain et al., 2004; Parker et al., 2004; Santos et al., 2006).

Therefore direct contact bacterial leaching can be described according to the following reaction (Eq.25

$$MS + 2O_2 \xrightarrow{\text{A. thiooxidans}} MSO_4 \quad (25)$$

MS is the metal sulphide.

Although the mechanism of attachment and the initiation of metal solubilisation are not completely understood, yet it is clear that bacteria prefer specific sites of the mineral surface with crystal imperfection and the metal solubilisation is due to electrochemical interactions (Bennett and Tributsch, 1978; Rodriguez and Tributsch, 1988; Ramachandra Rao and Finch, 1988; Das et al., 1999; Sui et al., 1995).

Indirect bacterial leaching: The bacteria here generate a lixiviant which chemically oxidizes the sulphide minerals. The overall reaction can be written as (Eqs.26 and 27):

$$S_o + H_2O + 1.5O_2 \xrightarrow{\text{A.thiooxidans}} H_2SO_4. \quad (26)$$

$$H_2SO_4 + \text{Material-M} \rightarrow \text{Materials-2H} + MSO_4 \quad (27)$$

Where M was a bivalent metal

In acid solution this lixiviant is Fe (III) which is bacterially generated and metal solubilisation can be described (Eq.28) as:

$$MS + Fe_2(SO_4)_3 \rightarrow MSO_4 + 2FeSO_4 + S^0 \quad (28)$$

In this reaction the bacteria have only a catalytic function because they accelerate the reoxidation of ferrous iron which takes place slowly in the absence of bacteria. The sulphur arising here may be oxidized to sulphuric acid by *T. ferrooxidans,* but oxidation by *T.thiooxidans* which frequently occurs together with *T. ferrooxidans* is much faster (Lacey and Lawson, 1970).

$$2S^0 + 3O_2 + 2H_20 \rightarrow 2H_2SO_4 \quad (29)$$

The indirect bioleaching process is the extraction of uranium from uranium ores (DiSpirito and Tuovinen, 1982)

$$U^{IV}O_2 + Fe_2(SO_4)_3 \xrightarrow{\text{T.ferrooxidans}} U^{VI}O_2SO_4 + 2FeSO_4 \quad (30)$$

Here *T.ferrooxidans* can oxidise U^{IV} to U^{VI} enzymatically and uses some of the energy of this reaction for the assimilation of CO_2.

This approach has been refined or further summarised by Crundwell, (1997) who summed up the entire bioleaching mechanism into three steps by which microorganisms (specially *A.ferrooxidans*) might interact with a sulphide mineral as follows:

1. Bacteria attached to the mineral surface oxidize the sulphide phase by biological means directly, without any requirement for ferric or ferrous ions- Direct contact mechanism.

2. Bacteria attached to the mineral surface oxidize ferrous ions to ferric ions within a biofilm comprised of bacteria and exo-polymeric material and the ferric ions generated within this layer oxidize the sulphide phase- The indirect contact mechanism.

3. Bacteria oxidize ferrous ions to ferric ions in the bulk solution and the ferric ions oxidize the sulphide phase- Indirect mechanism.

In both indirect leaching and indirect contact leaching microorganisms catalyze the oxidation of ferrous ion to ferric ions (Crundwell, 1997). The oxidation of a sulphide mineral by ferric ions may proceed via a thiosulphate intermediate or a polysulphide intermediate depending upon its solubility in acid (Steudel, 1996; Schipper and Sand, 1999; Hansford and Vargas, 2001; Lazaro and Nicol, 2003; Nicol and Lazaro, 2003).

Leaching Techniques: The bioleaching of minerals is a simple and effective technology for the

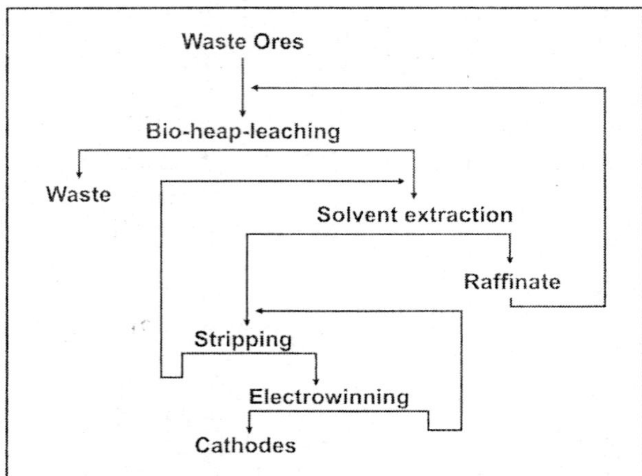

Fig.1 Bio-leaching flow sheet (after Songrong,et al. 2002).

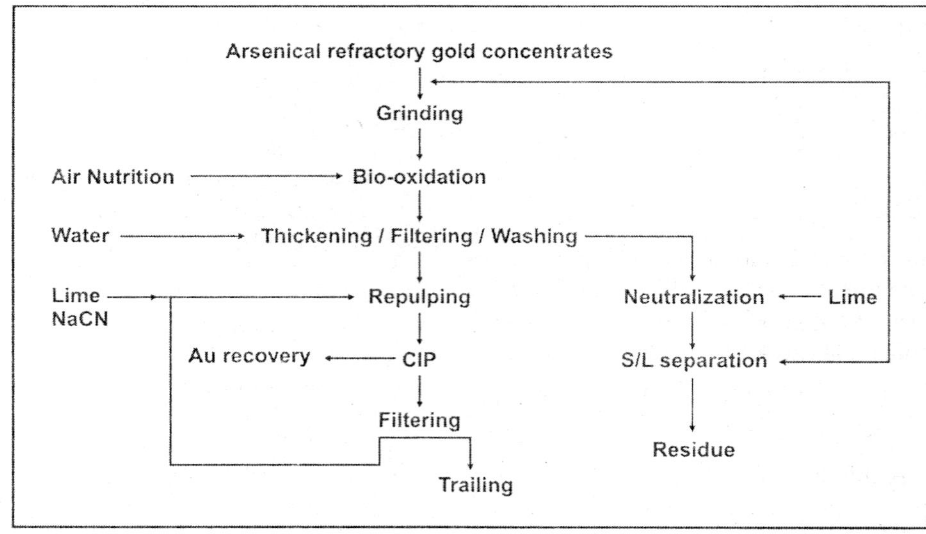

Fig.2 Bio-oxidation flow sheet (after Songrong, et al. 2002).

processing of sulphide ores and is used so far only on a technical scale mainly for recovery of copper and uranium. The effectiveness and economics of microbial leaching processes chiefly depend on the activity of the bacteria and on the chemical and mineralogical composition of the ore. Therefore, processes tested on one ore cannot as such be applied for an other one (Watling, 2006; Bosecker, 1997).

Heap leaching is mainly used for those fine grained ores which cannot be concentrated by flotation. Thin-layer leaching, exemplifies heap leaching, is the foundation of many of the current heap leach operations. Although originally it was developed to process copper oxides, subsequently it was modified to accommodate an increased contribution from sulphide minerals in the ore (Watling, 2006). The key features of thin layer leaching are (i) curing, in which concentrated acid is added to the crushed ore and allow to react with acid soluble and gangue minerals, and (ii) consolidation of fines with larger particles in a rotating drum, which ensures acidification and wetting of the ore prior to stacking. In those cases where organic free raffinate is used for curing/acidification, bacteria may also be distributed through the ore before it is stacked. The ore is stacked in carefully designed heaps so as to create a bed that exhibits good permeability for both solutions and gases. The bed is irrigated from sprinklers or drippers arranged in a grid across the top of the heap (Fig.1).

Dump leaching is the oldest process. The size of the dumps varies considerably and the amount of ore may be in the range of several hundred thousand tons of ore. The top of the dump is sprinkled continuously or flooded temporarily. Depending upon the type of ore the lixiviant may be water, acidified water or acid ferric sulphate solution (Ehrlich, 1980). Dump leaching and insitu leaching can be thought of as sub-sets of heap leaching in which solutions percolate through a bed of ore and metal is recovered from the leachate. Dump leaching is relatively inefficient because large sized particles present small surface areas to lixiviants whilst small particles block solution flows and impede aeration. Haulage vehicles and ageing creat compaction of ore, making it even less permeable (Songrong et al., 2002; Watling, 2006). Modern dump practices that minimise compaction and maximise aeration and solution permeability, can improve metal recoveries (Fig.2). Pre-conditioned dump material may be used with high acid levels. The effectiveness of this strategy depends upon the type of mineralisation and the acid consumption of gangue minerals. There are variable irrigation programs which may be employed when the permeability of the dump decreases with age. The enormous tonnages (and if the cost of removal is already borne by the mine) make it economical to scavenge metal like copper, zinc, uranium from low-grade (0.10.5%) dumps (Schnell, 1997). Studies on dump biodiversity and biodynamics are sparse (e.g. Edwards et al., 1999) but it can be pressumed that bacteria can readily colonise over the exposed mineral sulphide surfaces and engage in iron and sulphur oxidation.

There is only limited evidence that bacteria play a role in the *insitu* processing of undisturbed metal ores (Concha et al., 1991). *Leptospirillum ferrooxidans* is thought to play a role in the leaching of a chalcocite and bornite ore body below an old open-pit mine, relatively near the surface (Concha et al., 1991; Sand et al 1992). However, while solutions injected into deep in situ ore bodies may contain bacteria, it is unlikely that the bacteria will be particularly active at depth under conditions of oxygen limitation. That is not to say that an in situ leach of the future cannot be engineered to meet bacterial physiological requirements (Watling, 2006).

FACTORS INFLUENCING BIOLEACHING

Sulphide dumps and heaps with their varying mineralogical compositions and different climatic environments represent extremely complex microbiological habitats. The effectiveness of leaching depends largely on the efficiency of these microorganisms and the chemical and mineralogical composition of the ore. Thus the limiting factors are necessary for the bacteria to grow.

Nutrients: As the bacteria which play a role in recovery from the sulphide minerals are chemolithoautotropic bacteria and therefore inorganic compounds required for growth. Thus mineral nutrients are obtained from both the environment and the material to be leached. For optimum growth iron and sulfur compound may be supplemented together with ammonium phosphate and magnesium salts (Leduc and Ferroni, 1994; Brierley, 2001; Rawlings, 2002).

Oxygen and Carbon Dioxide: Aerobic environment is a prerequisite for good growth and high activity of the leaching bacteria. Although oxygen may cause some difficulty in case of heap and dump leaching, carbon dioxide is the only carbon source required, yet there is no need for addition of CO_2 (Acevedo et al., 1998; Bosecker,1997).

pH: pH value is decisive for the solubilisation of metals. pH value in the range of 2.0-2.5 are optimum for the bacterial oxidation of ferrous iron and sulphides. At pH value below 2.0, a considerable inhibition of *T. ferrooxidans* will occur but *T. ferrooxidans* may be adapted to even lower pH values by increasing addition of acid (Tuovinen and Kelly, 1973 and 1974; Shen and Jih,2001).

Temperature: The optimum temperature for ferrous iron and sulphide oxidation by T.ferrooxidans is between 28^0-30^0C (Rodriguez et al., 2003; Torma, 1977). Mineral decomposition being chemical process, the rate of chemical reactions approximately doubles with rise of every 10^0C temperature. However the optimal temperature is limited to 30°C, although some of the microbes decompose even at 40^0C or 50^0C.

Mineral substrate: The mineralogical composition of the leaching substrate is of primary importance. The rate of leaching depends on the total surface of the substrate. A decrease in the particle size (means an increase in the total particle surface area) produces higher yields of metal which can be obtained without a change in the total mass of the particles. A particle size of about 42μm is regarded as the optimum size (Chen and Lin, 2001 and 2004).An enlargement of the total mineral surface area can also be obtained by an increase in the pulp density. An increase in the pulp density may result as an increase in metal extraction but the dissolution of certain compounds which have an inhibitory or even toxic effect on the growth of leaching bacteria will also increase.

Heavy Metals: The leaching organisms, especially the thiobacilli, have a high tolerance to heavy metals. Various strains may even tolerate 50g/l Ni, 55g/l Cu or 112g/l Zn. Different strains of the some species may show completely different sensitivities to heavy metals. Often, it is possible to adapt individual strains to higher concentrations of metals by gradually increasing the concentration of the metals (Bosecker, 1997; Wong et al. 2004).

INDUSTRIAL APPLICATIONS AND FUTURE ASPECTS

The bioleaching of minerals has opened up new opportunities for extractive metallurgy and biohydrometallurgy from the last 25 years, is now practiced in the copper and uranium industries, especially for the treatment of low grade ores (Tichy et al., 1998; Fang and Zhou, 2007).

At present, bioleaching is being used commercially only for the recovery of copper, uranium and gold. In the future, however, these processes will become important for zinc, nickel, and cobalt and molybdenum recovery. Investment and operating costs are much lower than for conventional pyrometallurgical and hydrometallurgical processes. The processing plant can be built in the immediate vicinity of the ore deposit, saving transport costs. The procedures are not complicated and are easy to control; extensive technical knowledge is not required. This technology should be of great interest for developing countries (Acevedo et al., 1998).

Besides the metal recovered in the leachate, there is increasing interest in the insoluble metals left in the residues, e.g., lead. Inferior lead sulphide concentrates can be transformed into high-value concentrates by leaching of metals (e.g., zinc, cadmium, copper) which interfere with conventional processes for the recovery of the lead (Torma, 1977). Similar procedures are being investigated for the extraction of the silver and other precious metals which are finely disseminated in iron, arsenic, copper and zinc sulphides. To extract such finely disseminated precious metals the metal sulfides are first removed by microbial leaching and then the precious metals are recovered from the residue(see Aswathanarayana, this volume).

Mineral industrial waste products which often contain high amounts of valuable metals cannot be treated in that way because most metals are present mainly as oxides rather than sulphides. These metal oxides in such residues can be leached by acid produced by T. thiooxidans. Thiobacilli also have some potential for the detoxification of sewage sludge, soil and sediment contaminated with heavy metals and may contribute to diminishing some of our environmental problems (Blais et al., 1993; Bock et al., 1996; Seidel et al., 1998).

CONCLUSIONS

The use of microorganisms to recover metals from low grade ores and mineral concentrates has developed into a successful and expanding area of biotechnology. The current panorama of bioleaching in developing countries is encouraging. Current understanding of the mechanism of metal solubilization is that it is primarily a chemical process, although attachment of microbes to the mineral can enhance the dissolution. Sulphides of metals such as zinc, copper, nickel, cobalt are almost insoluble in water but sulfates of these metals are readily soluble and due to action of microbes metal sulphides are oxidised to sulfate, the metal is leached into solution from where it can be extracted. Bioleaching provides the possibility of recovering metals from low grade deposits which otherwise would have been considered waste. The application hence, greatly depends on the value of metal to be recovered.

The use of bioleaching in metal recovery is now well established. Furthermore, its operational simplicity, minimum environmental impact and low capital and operating cost will make it still more versatile in near future.

References

Acevedo, F.; Gentina J.C.; and Garcia, N.; 1998: CO_2 supply in the biooxidation of an enargitepyrite gold concentrate, Biotech. Lett. Vol. 20, pp. 257-259.

Acevedo, F.; Gentina, J.C.; Algegre, C.; and Arevalo, P.; 1997: Biooxidation of a gold-bearing enargite concentrate, Proc. IBS '97 Biomine '97 (Sydney), Australian Mineral Foundation, Glenside, pp. M3.2.1-M3.2.9.

Bandyopadhyay, N.; and Banik, A. K.; 1995: Optimisation of Physical Factors for Bioleaching of Silica and Iron from Bauxite Ore by a Mutant Strain of Aspergillus niger.' Res. Ind. Vol. 4(3), p 14-17.

Bang, S.S.; Deshpande, S.S.; and Han, K.N.; 1995: The oxidation of galena using Thiobacillus ferrooxidans. Hydrometall. Vol. 37 (2), pp.181-192.

Bennett, J.C.; and Tributsch, H.; 1978: Bacterial leaching patterns on pyrite crystal surfaces. Jur. Bacteriol. Vol. 34, pp. 310-317.

Blais, J.F.; Tyagi,R.D.; and Auclair,J.C.; 1993: Metals removal from sewage sludge by indenous iron oxidizing bacteria. Jour. Environ. Sci. Health. Vol. 28, pp. 443-467.

Bock,M.; Bosecker,K.; and Winterberg, R.; 1996: Bioleaching of soils contaminated with heavy metals. DECHEMA Monographs, VCH, Weinheim. Vol. 133, pp. 639-644.

Boon, M.; 2001: The mechanism of "direct" and "indirect" bacterial oxidation of sulphide minerals. Hydromet. Vol. 62, pp. 67-70.

Bosecker, K.; 1997: Bioleaching: metal solubilization by microorganisms. Microbio. Rev. Vol. 20, pp. 591-604.

Brierley, C.L. ; 1982: Microbiological mining. Sci. Am. Vol.247, pp. 42-51.

Brierley, J.A.; 1978: Thermophilic iron-oxidising bacteria found in copper leaching dumps. Appl. Environ.Microbiol. Vol. 36, pp. 523-525.

Brierley, J.A.; 2000: Expanding role of microbiology in metallurgical processes. Mining Eng. Vol.52 (11), pp. 49-53.

Brierley,J.A.; and Brierley,C.L.; 1973: A chemoautotrophic microorganism isolated from an acid hot spring. Can.Jour. Microbiol. Vol.19, pp.183-188.

Chaudhury, G.R.; and Das, R.P.; 1987: Bacterial leaching complex sulfides of copper, lead and zinc. Inter. Jour. Mine. Proc. Vol. 21 (1), pp. 57-64.

Chen, S.Y.; and Lin, J.G.; 2001a. Effect of substrate concentration on bioleaching of metal-contaminated sediment. Jour. Hazard. Material, Vol. B82, pp. 77-89.

Chen, S.Y.; and Lin, J.G.; 2001b. Effect of substrate concentration on bioleaching of metal-contaminated sediment. Jour. Hazard. Mater. B 82, pp. 77-89.

Chen, S.Y.; and Lin, J.G.; 2004: Bioleaching of heavy metals from contaminated sediment by indigenous sulfur-oxidizing bacteria in an air-lift bioreactor: effects of sulfur concentration. Water Res. Vol.38, pp. 3205-3214.

Clark, D.A.; and Norris, P.R.; 1996: Acidimicrobium ferrooxidans gen.nov., sp. nov.: mixed-culture ferrous iron oxidation with Sulfobacillus species. Microbiol. Vol. 142, pp. 785-790.

Colmer, A.R.; and Hinkle,M.E.;1947: The role of microorganisms in acid mine drainage; a preliminary report. Sci. Vol. 106, pp. 253-256.

Concha, A.; Oyarzun, R.; Lunar, R.; and Sierra R.; 1991: Over a century of bioleaching copper sulphides at Andacollo. Min. Magaz. Vol.165 pp. 325-328.

Crundwell, F.K.; 1997: Physical chemistry of bacterial leaching. In: Rawlings, D.E. (Ed.), Biomining: Theory, Microbes and Industrial Processes. Springer, Berlin, pp. 177-200.

Cruz, R.; Luna-Sanchez, R.M.; Lapidus, G.T.;Gonzalez, I.; and Monroy, M.; 2005: An experimental strategy to determine galvanic interactions affecting the reactivity of sulfide mineral concentrates. Hydromet. Vol.78, pp.198-208.

Das, T.; Ayyappan S.; and Chaudhury, G.R.; 1999: Factors affecting bioleaching kinetics of sulfide ores using acidophilic micro-organisms. BioMet. Vol.12, pp. 110.

Deveci, H.; Akcil, A.; and Alp, I.; 2004: Bioleaching of complex zinc sulphides using mesophilic and thermophilic bacteria: comparative importance of pH and iron. Hydromet. Vol. 73, pp. 293-303.

Di Spirito, A.A.; and Tuovinen, O.H.; 1982: Uraneous ion oxidation and carbon dioxide fixation by Thiobacillus ferrooxidans. Arch. Microbiol. Vol.133, pp. 28-32.

Edwards, K.J.; Goebel, B.M.; Rodgers, T.M.; Schrenk, M.O.; Gihring, T.M.; Cardona, M.M.;. Hu, B ; McGuire, M.M.; Hamers, R.J.; Pace N.R.; and Banfield, J.F.; 1999: Geomicrobiology of pyrite (FeS2) dissolution: case study at Iron Mountain, California, Geomicrobi. Jour.

Vol. 16, pp. 155-179.

Ehrlich, H.L.; 1980: Bacterial leaching of manganese ores. In: Trudinger, P.A., Walter, M.R. and Ralph, B.J., (eds.), Biogeochemistry of Ancient and Modern Environments. Australian Academy of Science, Canberra. pp. 609-614.

Ehrlich, H.L.; 1983: Leaching of chromite ore and sulfide matter with dilute sulfuric acid generated by Thiobacillus thiooxidans from sulfur. Progress in biohydrometallurgy. Cagliari, pp. 941.

Fang, D. ; and Zhou, L.X. ; 2007 : Enhanced Cr bioleaching efficiency from tannery sludge with coinoculation of Acidithiobacillus thiooxidans TS6 and Brettanomyces B65 in an air-lift reactor. Chemo. Vol. 69, pp. 303-310.

Fowler, T.A.; and Crundwell, F.K.; 1999: The role of Thiobacillus ferrooxidans in the bacterial leaching of zinc sulphide. In: Amils, R., Ballester, A. (Eds.), Proceedings International Biohydrometallurgy Symposium Part A. Elsevier, Amsterdam, pp. 273-282.

Garcia Jr., O.; Bigham, J.M.; and Tuovinen, O.H.; 1995: Sphalerite oxidation by Thiobacillus ferrooxidans and Thiobacillus thiooxidans. Canadian Jour. Microbi. Vol.41 (6), pp. 578-584.

Gomez, C.; Blazquez, M.L.; and Ballester, A.; 1999: Bioleaching of a Spanish complex sulphide ore bulk concentrate. Miner. Engi. 12, 93-106.

Hallberg K.B.; and Johnson, D.B.; 2003: Novel acidophiles isolated from moderately acidic mine drainage waters, Hydrometall. Vol. 7, pp. 139-148.

Hansford, G.S.; and Vargas, T.; 2001: Chemical and electrochemical basis of bioleaching process. Hydrometall. Vol. 59, pp.135-145.

Hiroyoshi, N.; Hirota, M.; Hirajima T.; and Tsunekawa, M.;1997: A case of ferrous sulfate addition enhancing chalcopyrite leaching. Hydrometall. Vol. 47, pp. 37-45.

Hiroyoshi, N.; Miki, H.; Hirajima T.; and Tsunekawa, M.; 2001: Enhancement of chalcopyrite leaching by ferrous ions in acidic ferric sulfate solutions, Hydrometall. Vol. 60, pp. 185-197.

Holmes, D.S.; and Debus, K.A.; 1991: Biological opportunities for metal recovery. In: Malik, K.A. Naqvi, S.H.M. and Aleem, M.I.H. (eds), Proc. Intl. Symposium on Biotechnology for Energy (Faisalabad, Pakistan, 1989), NIAB/NIBGE, Faisalabad. pp. 341-358.

Hossain S.M.; Das, M.; Begum,K.M.M.S.; and Anantharaman N.;2004: Bioleaching of zinc sulphide (ZnS) ore using thiobacillus ferrooxidans. IE(I) Jour.chemi.engi.Vol.85, pp.7-11.

Hus, C.H.; Harrison, R.G.; 1995: Bacterial leaching of zinc and copper from mining waste. Hydrometall. Vol. 37 (2), pp.169-179.

Hutchins, S R; Davidson, M. S.; Brierley, J. A.; and Brierley C. L.; 1986: Microorganisms in Reclamation of Metals. Annual Revi. Microbi. Vol. 40, p 311-316.

Johnson, D.B.; and Hallberg, K.B.; 2003:The microbiology of acidic mine waters. Res. Microbiol. Vol.154, pp. 466-473.

Klauber, C.; 2003: Fracture induced reconstruction of a chalcopyrite $CuFeS_2$ surface. Surf. Interf. Analy. Vol. 35, pp. 415-428.

Klauber, C.; Parker, A.; Bronswijk W. V.; and Watling, H.; 2001: Sulphur speciation of leached chalcopyrite surfaces as determined by X-ray photoelectron spectroscopy. Inter. Jour. Miner. Proce. Vol. 62, pp. 65-94.

Kodali, B.; Rao M.B.; Narasu, M.L.; and Pogaku, R.; 2004: Effect of biochemical reactions in enhancement of rate of leaching. Chem. Engin. Sci. Vol.59, pp.5069-5073.

Lacey, D.T.,and Lawson, F.,1970: Kinetics of the liquid-phase oxidation of acid ferrous sulphate by bacterium Thiobacillus ferrooxidans. Biotechnol. Bioeng. Vol.12, pp. 29-50.

Lavalle,L.; Giaveno,A.; Pogliani,C.; and Donati, E.; 2007:Bioleaching of a polymetallic sulphide by native strains of leptospirillum ferrooxidans from Patagonia. Proce. Bioche. doi:10.1016/j.procbio.2007.10.019.

Lazaro I.; and Nicol, M.J.; 2003: The mechanism of the dissolution and passivation of chalcopyrite: an electrochemical study. In: Young, C.A. Alfantazi, A.M. Anderson, C.G. Dreisinger, D.B. Harris B. and James, A. (eds.) Hydrometallurgy 2003, TMS, Warrendale. pp. 405-417.

Leduc, L.G.; and Ferroni, G.D.; 1994:The chemolithotrophic bacterium Thiobacillus ferrooxidans. Microbiol. Rev. Vol.14, pp.103-120.

Liu, Y.G.; Ming, Z.; Zeng, G.M.; Li, X.;, Xu, W.H.; and Fan, T.; 2007: Effect of solids concentration on removal of heavy metals from mine tailings via bioleaching. J. Hazard. Mater. Vol. 141, pp. 202-208.

Lu, Z.Y.; Jeffrey, M.I.; Zhu, Y.; and Lawson, F.; 2000: Studies of pentlandite leaching in mixed oxygenated acidic chloridesulfate solutions. Hydrometall.Vol. 56, pp.63-74.

Mason L.J., and Rice, N.M.,2002: The adaptation of Thiobacillus ferrooxidans for the treatment of nickeliron sulfide concentrates, Miner. Engin. Vol. 15 pp. 759-808.

May, N. Ralph D.E. and Hansford, G.S. (1997) Dynamic redox potential measurement for determining the ferric leach kinetics of pyrite, Miner. Engin. Vol. 10 (11), pp. 1279-1290.

Mousavi, S.M.; Yaghmaei, S.; Vossoughi, M.; Jafari, A.; ,Roostaazad, R.; and Turunen, I.;2007: Bacterial leaching of low-grade ZnS concentrate using indigenous mesophilic and thermophilic strains. Hydrometall.Vol. 85 pp. 59-65.

Natarajan K.A.; and Iwasaki, I.; 1983: Environmental leaching behaviour of coppernickel bearing Duluth gabbro and floatation tailings. Hydrometall. Vol. 10 pp. 329-342.

Nemati, M.; and Webb, C. A.; 1997: Kinetic Model for Biological Oxidation of Ferrous Iron by Thiobacillus ferrooxidans.' Bio-tech. Bio-engin. Vol 53, p. 478-482.

Nicol M.J.; and Lazaro, I. 2003:The role of non-oxidative processes in the leaching of chalcopyrite. In: Riveros, P.A. Dixon, D. Dreisinger, D.B. and Menacho, J.(eds), Copper 2003 Cobre 2003 (Santiago) Volume VI (Book 1), Canadian Institute of Mining, Metallurgy and Petroleum, Montreal. pp. 367-381.

Norris, P.R.; Barr D.W.; and Hinson, D.; 1987: Iron and mineral oxidation by acidophilic bacteria: affinities for iron and attachment to pyrite. In: Norris P.R. and Kelly D.P., (eds.), Biohydrometallurgy: Proc. Intl. Symposium (Warwick, UK, 1987), Science and Technology Letters, Kew. pp. 43-59.

Norris, P.R.; Parrott, L.; and Marsh, R.M.; 1986: Moderately thermophilic mineral-oxidising bacteria. Biotechnol. Bioeng. Symp. No. 16, pp. 253-262.

Okibe, N.; Gericke, M.; Hallberg K.B.; and Johnson, D.B.;2003: Enumeration and characterization of acidophilic microorganisms isolated from a pilot plant stirred tank bioleaching operation, Appl. Environ. Microbi. Vol. 69, pp. 1936-1943.

Parker, A.; Klauber, C.; Stott, M.; Watling H.; and Bronswijk, W.V.; 2004:An X-ray photoelectron spectroscopy study of the mechanism of microbially assisted dissolution of chalcopyrite. In: Tzesos, M. Hatzikioseyan A.and Remoundaki, E. (eds), Biohydrometallurgy: A Sustainable Technology in Evolution (Athens, 2003), National Technical University of Athens, Athens . pp. 1011-1022.

Porro, S.; Raminez, S.; Reche, C.; Curutchet, G.; Alenso-Romanski S.; and Donati, E.; 1997: Bacterial attachment its role in bioleaching process. Process Bioche. Vol. 32 (7), pp. 573-578.

Rao S. R.; and Finch, J.A. 1988: Galvanic interaction studies on sulphide minerals.Canadian Metall. Quarte. Vol. 27, pp. 253-259.

Rawlings, D.E.; and Silver, S.; 1995:. Mining with microbes. Biol.Technol. Vol.13, pp. 773-778.

Rawlings, D.E.; Tributsch H.; and Hansford, G.S.; 1999: Reasons why 'Leptospirillum'-like species rather than Thiobacillus ferrooxidans are the dominant iron-oxidising bacteria in many commercial processes for the biooxidation of pyrite and related ores. Microbi. Vol. 145, pp. 5-13.

Rawlings, D.E.;2002: Heavy metal mining using microbes. Annu. Rev. Microbiol. Vol.56, pp. 65-91.

Reyes, S.A.; 2007: Fungal bioleaching of metals in preservative treated wood. Process. Biochem. Vol.42, pp.798-804.

Rodríguez, L. M.; and Tributsch, H.; 1988: Morphology of bacterial leaching patterns by Thiobacillus ferrooxidans on synthetic pyrite. Arch. Microbiol.Vol. 149, pp. 401-405.

Rodriguez, Y.; Ballester, A.; Blazquez,M.L.; Gonzalez, F.; Munoz, J.A.; 2003: New information on the sphalerite bioleaching mechanism at low and high temperature. Hydrometall. Vol. 71, pp. 57-66.

Romero, J.; Yañez, C.; Vásquez, M., Moore E.R.B.; and Espejo R.T.; 2003: Characterization and identification of an iron-oxidizing, Leptospirillumlike bacterium, present in the high sulfate leaching solution of a commertial bioleaching plant. Res. Microbiol. Vol.154, pp. 353-359.

Rulkens, W.H.; Grotenhuis, J.T.C.; and Tichy, R.; 1995: Heavy Metals. Springer, Berlin, 151 p.

Sand, W.; Gerke, T.; Hallman, R.; and Schippers, A.; 2001: Sulphur chemistry, biofilm, and the (in) direct attack mechanism a critical evaluation of bacterial leaching, Appl. Microbio. Biotech. Vol. 43 (6), pp. 961-966.

Sand,W.; Rohde,K.; Sobotke,B.; and Zenneck,C.; 1992: Evalution of Leptospirillum ferrooxidans for leaching. Appl.Environ.Microbiol.Vol.58, pp. 85-92.

Santos, L.R.G.; Barbosa, A.F.; Souza,A.D.; and Leao,V.A.;2006:.Bioleaching of a complex nickel-iron concentrate by mesophile bacteria.Mine. Engin.Vol.19, pp. 1251-1258.

Schippers, A.; and Sand, W.; 1999: Bacterial leaching of metal sulfides proceeds by two indirect mechanisms via thiosulfate or via polysulfides and sulfur. Appl. Environ. Microbiol.Vol. 65, pp. 319-321.

Schnell, H.A.; 1997: Bioleaching of copper. In: Rawlings, D.E., (ed.), Biomining: Theory, Microbes and Industrial Processes. Springer, pp. 21-43.

Schrenk, M.O.; Edwards, K.J.; Goodman, R.M.; Hamers R.J.; and Banfield, J.F.; 1998: Distribution of Thiobacillus ferrooxidans and Leptospirillum ferrooxidans: implications for generation of acid mine drainage. Sci. Vol. 279, pp. 1519-1522.

Seidel, H.; Ondruschka, J.; Morgenstern, P.; and Stottmeister, U.; 1998: Bioleaching of heavy metals from contaminated aquatic sediments using indigenous sulfur-oxidizing bacteria: a feasibility study. Water Sci. Technol. Vol.37, pp. 387-394.

Sharma I.P.; and Verma A.; 1991: Microbial Reclamination of metals from ores and industrial wastewater. Indi. Jour. . Microbi.Vol.31 (1), p.1.

Shen Y.C.; and Jih, G.L.; 2001: Bioleaching of heavy metals from sediment significance of pH, Chemo. Vol.44 pp. 1093-1102.

Shi, S.; and Fang, Z.;2005: Bioleaching of marmatite

flotation concentrate by adapted mixed mesoacidophilic cultures in an air-lift reactor. Int. Jour. Miner. Process. Vol.76, pp. 3-12.

Shi, S.Y.; Fang, Z.H.; and Ni, J.R.; 2006: Comparative study on the bioleaching of zinc sulphides. Process. Biochem. Vol.41, pp. 438-446.

Shrivatava, K.L.; Tripathi, R.P.; and Jangid, M.L.; 1992: Spontaneous combustion of coal: A mossbauer approach. Fuel. Vol. 71, pp. 377-380.

Silverman M.P.; 1967: Mechanism of bacterial pyrite oxidation.Jour. Bacteriol. Vol. 94, pp. 1046-1051.

Silverman M.P.; and Ehrlich, H.L.; 1964: Microbial formation and degradation of minerals, Adv. Appl. Microbio. Vol. 6, pp. 153-206.

Songrong, Y.; Jiyuan, X.; Guanzhou, Q.; and Yuehua, H.; 2002: Research and application of bioleaching and biooxidation technologies in China. Miner. Engin. Vol.15, pp. 361-363.

Steudel, R.; 1996: Mechanism for the formation of elemental sulfur from aqueous sulfide in chemical and microbiological desulfurization processes. Indu. Engi. Chem. Rese. Vol. 35, pp. 1417-1423.

Sugio,T.; Domatsu,C.; Munakata,O.; Tano,T.; and Imai,K.; 1985: Role of ferric reducing system in sulphur oxidation of Thiobacillus ferrooxidans. Appl. Environ. Microbiol. Vol. 49, pp. 1401-1406.

Sui, C.C.; Brienne, S.H.R.; Rao R. S.; Xu Z.; and Finch, J.A.; 1995: Metal ion production and transfer between sulphide minerals. Miner. Engi. Vol. 8 pp. 1523-1539.

Tichy, R.; Rulkens, W.H.; Grotenhuis, J.T.C.;Nydl, V.; Cuypers, C.; and Fajtl, J.; 1998: Bioleaching of metals from soils or sediments. Water. Sci. Technol. Vol.37, pp. 119-127.

Torma, A.E.; 1977: The role of Thiobacillus ferrooxidans in hydrometallurgical processes. Adv. Biochem. Eng.Vol.6, pp. 1-37.

Torma. A. E.;1971: Microbiological Oxidation of Synthetic Cobalt, Nickel and Zinc Sulphides by Thiobacillus ferrooxidans. Revi. Cana. Biol. Vol. 30, pp. 1971, p. 209.

Tributsch, H.; 2001: Direct vs indirect bioleaching, Hydrometall. Vol. 59, pp. 177-185.

Tripathi, R.P.; and Shrivatava, K.L.; 1990: Characterization of Merta Road lignite and decomposional study of its pyrite using mossbauer spectroscopy. Fuel. Vol. 69, pp. 928-930.

Tuovinen,O.H.; and Kelly,D.P.;1973: Studies on the growth of Thiobacillus ferrooxidans I. Use of membrane filters and ferrous iron agar to determine viable numbers, and comparision with CO_2 fixation and iron oxidation as measures of growth. Arch. Microbiol. Vol. 88, pp. 285-298.

Tuovinen, O.H.; and Kelly, D.P.; 1974: Studies on the growth of Thiobacillus ferrooxidans II. Arch.Microbiol. Vol.95, pp. 153-160.

Umrania, V.V.; 2006: Bioremediation of toxic heavy metals using acidothermophilic autotrophes. Bioresour. Technol. Vol. 97, pp. 1237-1242.

United States Geological Survey (USGS); 1990: Mineral year book.http//minerals. usgs.gov/minerals/pubs/commodity/myb.

Vasquez, M.; and Espejo, R.T.; 1997: Chemolithotrophic bacteria in copper ores leached at high sulfuric acid concentration. Appli. Environ. Microbio. Vol.63, pp. 332-334.

Vishniac, W.; and Santer, M.; 1957: The Thiobacilli. Bacte. Rev. Vol.21, pp. 195-213.

Watling, H.R.; 2006: The bioleaching of sulphide minerals with emphasis on copper sulphides - A review. Hydromet. Vol. 84, pp. 81-108.

Wong, J.W.C.; Xiang, L.; Gu, X.Y.; and Zhou, L.X.; 2004: Bioleaching of heavy metals from anaerobically digested sewage sludge using FeS2 as an energy source. Chemo.Vol. 55, pp. 101-107.

Xie, F.; and Dreisinger, D.B.; 2007: Leaching of silver sulfide with ferricyanidecyanide solution. Hydrometall. Vol. 88, pp. 98-108.

Xu, T.J.; and Ting, Y.P.; 2004: Optimisation on bioleaching of incinerator fly ash by Aspergillus niger use of central composite design. Enzyme Microb. Technol. Vol.35, pp. 444-454.

Zhu, N.W.; Zhang, L.H.; Li, C.J.; and Cai, C.G.; 2003: Recycling of spent nickelcadmium batteries based on bioleaching process. Waste Manag. Vol. 23, pp.703-708.

Hon'ble Reviewers

1. Prof. O.P. Agarwal
   (Former ViceChancellor)
   School of Studies in Zoology
   Jiwaji University
   GWALIOR 474011 INDIA
2. Prof. V.N. Sharma
   Dean
   (Prof. of Biotechnology)
   Meerut Institute of Engineering
   and Technology
   MEERUT 250002 INDIA

Editorial Assistance

Mr. C.P. Dadhich

Economic Mineralization
K.L. Shrivastava (Editor)
2009. pp. 485-509
Scientific Publishers (India), Jodhpur

45

Section-V

Environmental Impact of Coal Mining Activity on Trace Element Concentration of Natural Waters in Pench Valley Coalfield Area, Chhindwara District, Madhya Pradesh

D.C. Gupta

Department of Applied Geology, Barkatullah University, Bhopal 462 026
Email : dcggeology@rediffmail.com

Abstract

The Pench Valley Coalfield (PVC) covers about 239,575 sq. kms. area in Parasia tehsil of Chhindwara district, Madhya Pradesh. The area falls on Survey of India toposheet No. 55 J/12 and 55 J/16 between the latitude 22° 10' - 22° 15' N and longitude 78° 40' - 78° 55' E. The coalseams in PVC area are confined to Barakar formation of Lower Gondwana age, which are affected by network of ENE WSW trending faults. A total of about five coalseams with their split sections have been reported in a column of 50 to 80m thickness within a sequence of Barakar formation. The thickness of coalseams vary from 0.04 to 3.39m. The coalseams attend workable thickness are mined with the help of 13 under ground and 4 opencast mines. The trace elements like B, Ba, Co, Cu, Cr, Ge, Ga, La, Mn, Mo, Nb, Ni, Pb, V, Sr, Sn, Y, and Z are reported by Mukherjee et al. (1982) in Pench Valley coals. These trace elements become mobile with the initiation of coal mining activity, thus produce inverse impacts on environment in general and on public health in particular in PVC area. During last two decades or so, the coal production in Pench Valley area has increased many folds. As a result of increased mining activity, the concentrations of some of the trace elements have increased in natural waters of the area. It is estimated with the help of AAS that concentration of the elements like As, Cd, Cu, Co, Fe, Hg, Mn, Ni, Pb, Se, V and Zn varies between wide limits, and some of the elements like Hg, Se, Cu, Fe, Pb etc. have reached concentration far above the prescribed limits suggested by ISI : 10500 (1983). In the present paper an attempt has been made to discuss the trace elements concentration in natural waters of PVC area. The study reveals that the concentration of trace elements is more intricate in western part rather than in the eastern part of PVC area.

Key words : Mining Environment, Coal Pollution, Pench Valley Coalfield, Medical geology.

INTRODUCTION

"Environment" means the whole complex of physical, chemical, biological, social, cultural, economic and aesthetic factors that affects our growth, reproduction and well-being. The wide spectrum of environment includes the air and water quality, erosion control, natural hazards, land use planning, site selection and design, conservation of animal and plant life, displacement and relocation, resulting from public and private action or natural disaster, noise pollution, preservation of cultural resources, urban design and the quality of constructed environment. The impact of environment affects individual and communities and ultimately determines their form, character, relationship and survival. Both aspects of environment, the natural and the man made are essential to his well being and for the enjoyment of basic human rights - even the right to live itself. The protection and improvement of human environment is therefore, a major issue, which affect the well - being of people and economic development throughout the world.

The continuous stretch of the Satpura-Gondwana Basin, extends over a length of 117 km in east - west direction and retain more or less ENE - WSW Satpura trend. Within the confines of this Master Basin, the Mohpani coalfield forms the northern limit, while Pathakhera defines the southern limit. On the east, the limit is near Haranbhatta and on the west it extends upto Sonada area. The southern portion of the Satpura - Gondwana Basin comprises, Pench Valley Coalfield in the extreme east, followed successively towards west by the "Kanhan Valley Coalfield", the Upper Tawa Valley Coalfield, Pathakhera Coalfield, Dulhara Coalfield and Shahpur Coalfield. The Satpura-Gondwana Basin, occupies a unique position in Peninsular India. It is the westernmost Gondwana Basin and caters the need of coal of western India.

The Pench Valley Coalfield (PVC) area, is a part of Parasia tehsil of Chhindwara district, Madhya Pradesh. The Pench Valley covers a total area of about 239,575 sq. km. and lie between the latitude 22° 10' - 22° 15' N and longitude 78° 40' - 78° 55' E, on Survey of India toposheet Nos. 55 J/12 and 55 J/16 Parasia is the tehsil headquarter of the Chhindwara district and well connected by all weather and fair weather roads to all villages and collieries in the area.

Coals are the natural wealth of any nation and constitute the backbone of all modern industries. Since independence there has been a marked increase in the coal production of the country. The over all impact of such development both in terms of mining and industrial set up has resulted in the deterioration of environment both at macro and micro level. The macro level degradation of environment in the mining areas has been given more emphasis due to its pronounced impact on soil, vegetation, vibration and alteration of the topography and other related changes in the area. However, the micro level changes in all kinds of profile are brought about by chemical processes, which are directly or indirectly related to setup of mining industries as well as other industries. Therefore, environmental geochemistry in relation to water quality becomes an integral part of assessment of environment, which deals with various chemical phenomena taking place in the environment.

Water is one of the most important natural resource for all the living organisms, whether unicellular or multicellular; it is required for their metabolic activities. In addition water is required in various domestic, irrigation, shipping, sanitation, power generation and industrial purposes. Water of all media, is unique in the sense that it has properties of dissolving and carrying in solution, a variety of chemicals and other matter. In this process it easily gets contaminated. If water gets polluted, it becomes a direct threat to both man and other living beings. The stupendous increase in world's population resulting in the spurt in urbanization, industrialization and agriculture etc. and has put tremendous pressure on the limited fresh water resources, thereby threatening the freshwater bodies with pollution. One would agree that most of the problems of water pollution are man-made and resulted due to indiscriminate and unwise use of water bodies and its mismanagement. Since, water is a natural resource, therefore, conservation and keeping up of the good quality of water is of prime importance. However, there still exist data gaps in regard to the natural distribution of trace metals in water bodies of different geological environments. Therefore, it is realized to monitor the level of pollution by various chemical elements as a function of time.

In assessing environmental impacts of coal mining activity in PVC area, it is necessary to estimate the source and extent of pollution, whether it is a natural or artificial.

PREVIOUS WORK

The study of PVC has been carried out fc r the first time by Jerdon and Shanky in 1852. However, the first

Table-1 GEOLOGICAL SUCCESSION OF PENCH VALLEY COALFIELD AREA (After : MECL, 1983)

System	Age	Formation	Lithology	Max. Thickness, encountered in MECL BHs
	Sub-Recent	Alluvium	Sandy and Clayey soils	------
	Upper Cretaceous (?)	Deccan Taprs with Intertrappeans & Intrusives (dykes)	Basalt (massive & vesicular) and basic dykes	187.00 (m)
	----------------	***UNCONFORMITY***	---------------------------------	
Upper Gondwana	Upper Jurassic	Infratrappeans (Jabalpurs (?)	Grit and Red jasper pebble beds with white yellow, grey, clays	30.00 (m)
	----------------	***UNCONFORMITY***	---------------------------------	
	Middle Permian	Moturs	Variegated clays and sandstones with calcareous nodules	340 (m)
Lower Gondwana	Lower Permian	Barakars	Sandstones (fine to coarse grained), shales & Coal seams	140.00 (m)
	Upper Carboniferous	Talchirs	Boulder beds, sandstones (fine to medium grained) and greenish grey shales (splintery on exposure)	10.00 (+)
	----------------	***UNCONFORMITY***	---------------------------------	
	Archaeans		Granite, Hornblend gneises and schists	

geological traverse in Pench Valley was made by Medlicott in 1855. Since then, many workers namely Ashburner (1866), West (1923), Gee (1924), Crookshank (1936) and Chandra, Shrivastava and Ghose Jr. (1958 - 60), have undertaken the detailed study of the area to prove the coal potential in the valley. Many collieries have been located distinctly between the fault blocks. The exploratory mining was started long back in 1976 in Rawanwara Khas area. In recent past many new collieries have come up from time to time by joint venture of Geological Survey of India (GSI), Coal India Limited (CIL), Mineral Exploration Corporation Limited (MECL) and Indian Bureau of Mines (IBM). The total coal production in Pench Valley area as reported by WCL office Parasia is of the order of 181000 tonnes/month from 17 mines. With increasing demand the coal production has enhanced many fold year by year and no environmental appraisal of impact of coal mining in Pench Valley area has been carried so far in any detail.

Environmental Audit Report (1993) submitted by Environmental Planning and Coordination Organization (EPCO), Government of Madhya Pradesh in collaborating with CMRI, Dhanbad have put on record the various environmental problems like acid mine water, water quality of Pench and Kanhan rivers in terms of TDS, SO_4, Fe and BOD, reduced agricultural production, damaged to land, forest property and air quality in Pench Valley area in a general way. This report gives a general mention about presence of different toxic elements in the natural waters of the area. A preliminary study have been performed by Gupta (1999).

Central Mining Research Institute (1986) in collaboration with CMRI, Patna conducted a health survey on more than 7000 coal miners in Jharia coalfield. The study indicate that mine workers have suffered various types of occupational diseases like Pneumoconniosis, Pulmonary T.B., Bronchitis, Eosinophelia and Asthma, Ophthalmic diseases like Cataract, Refractive error, Pterygium, Conjunctivities are very common. In addition to these, other kind of diseases like Intestinal parasitic infection, Anaemia, Cutaneous mycosis, G-6-P-D deficiency, Sickle cell trait and Pb poisoning are very commonly reported in the mine workers due to coal mining activity in Jharia coalfield.

GEOLOGY

The PVC stretches in an east - west direction from Haranbhatta in the east to Datla in the west and represents southern portion of the Satpura - Godwana

Basin. MECL since 1973 has undertaken a detailed exploration in the Pench Valley area to unravel the complex geological structure found in Barakar formation and to prove the total coal reserves for proposed super thermal power plant in the Chhindwara region. So far, detailed exploration in more than 13 blocks was completed and on the basis of exploration data the following geological succession has been suggested by MECL in PVC area. The complete geological succession is given in Table 1 and a detailed geological map of the area is given in Fig. 1.

The Archaean rocks occupy the southern part of the coalfield and consist of granites, hornblende gneisses and schists. Archaean are unconfirmably overlain by Talchir formation which consists of boulder bed, greenish shales and hard compact and fine grained sandstones. The shale are generally olive green in colour and on weathering breaks up into needle-shaped structures.

The coal bearing Barakar are exposed in a linear tract along the southern margin of coalfield and represented by grey feldspathic (often kaolinised) sandstone, coal seams, carbonaceous shales, shalycoal and grey / arenaceous shales. Motur formation comprises variegated and mottled clays, sandy clays, arenaceous shales, hard clays and variegated chloritic sandstones. They overlie Barakar formation and considered equivalent to the Barren Measures.

The infratrappean is represented by a succession of strata comprising jasper conglomerate, followed by

Table-2 Seam nomenclature and thickness of Pench Valley Coals

Coal seams	Thick ness in m.	
	Minimum	Maximum
Coal Seam IA	0.07	2.10
Coal Seam IB	0.21	3.31
Coal Seam IC	0.25	3.19
Parting	0.22	16.59
Coal Seam IIA	0.05	1.60
Coal Seam IIB	0.07	1.42
Parting	1.23	13.21
Coal Seam IIIA	0.06	0.74
Coal Seam IIIB	0.06	2.92
Coal Seam IIIC	0.04	0.92
Parting	9.69	18.23
Coal Seam IVA	0.59	3.39
Coal Seam IVB	0.05	1.11
Parting	0.94	7.78
Coal Seam VA1	0.44	2.80
Coal Seam VA2	0.11	1.31
Coal Seam VB	0.06	3.21
Coal Seam VC	0.11	1.09

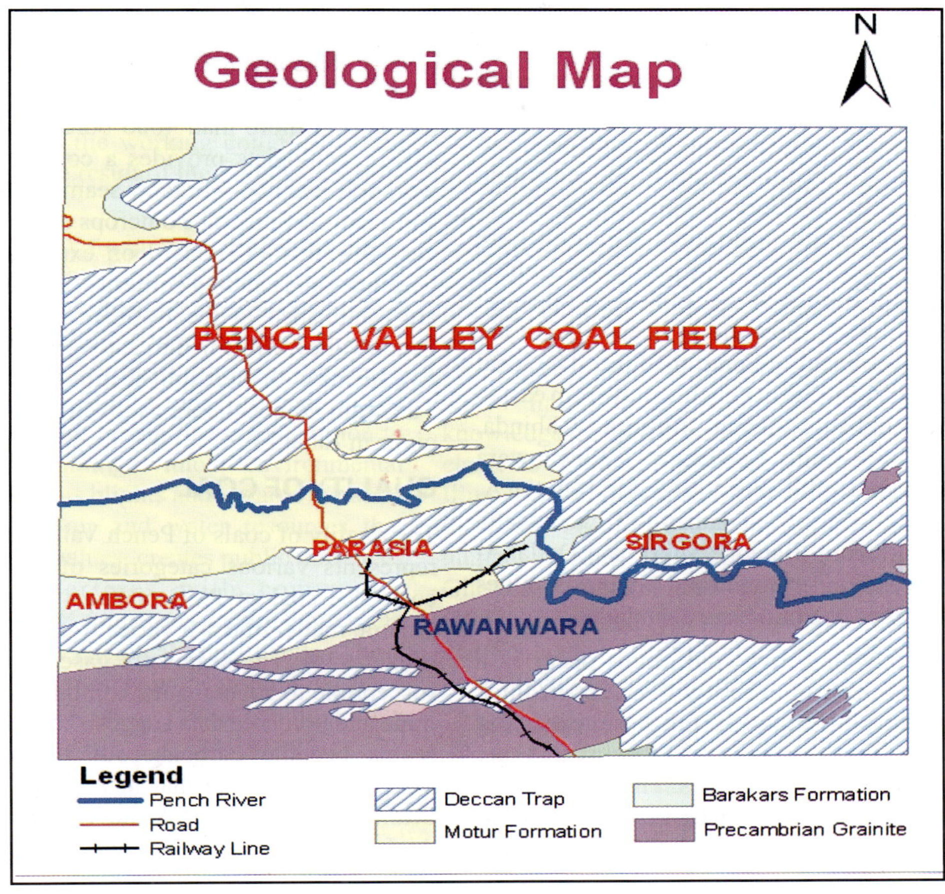

Fig. 1 Geological map of Pench Valley Coalfield area

thick, fine to medium grained argillaceous sandstones and reddish clays which are designated as Jabalpur formation in Pench Valley area by MECL. At many places the jasper conglomerates form the topmost part of the infratrappean stage.

The northern part of the coalfield is mostly covered by the Deccan Traps. A number of flows of basalt with intertrappean beds cover most of the older formations in the PVC area. The flows are also seen covering the Barakar and Talchir formations and in Sirgora area, they overlie the Jabalpur formation.

STRUCTURE

The Pench Valley area to the east is affected by a network of faults, which trends in ENE - WSW direction. These faults show varying amount of dip and throw along the fault plane. The fault zone has prominent topographic expressions in the form of linear ridges and composed of sheared and silicified sandstones and platy quartz etc. Fracturing of rocks and slickenside are quite common. Compared to the western part, the eastern part of the area is more intricately faulted near Khirsadoh and Rawanwara. Though most of the faults have down throw towards north, however, several faults show southerly down throw resulting in the repetition of Barakar strata within the Motur in Eklehra - Bamori and Rawanwara - Sethia areas. One such fault, with down throw towards south, has been observed over a length of about 20 km extending from Ambara in the west to Harrai in the east. It is believed that the throw of the fault increases towards east. Two more faults with southerly throw have been recorded within the Motur formation between Chikli Kalan and Sirgori Khurd and Balagaon villages. The downthrow of these faults are estimated to be between 60 m and 100 m.

In Sialghoghri and Rawanwara collieries, the dip of the formation is towards north - east to east. Rolling dips and horizontal beds are quite common to the west of Ghughri village. To the south of Eklehra and north of Chandametta, the dips are towards SE to SW.

Table-5 SPECTROGRAPHIC ANALYSIS OF COALS OF PENCH VALLEY COALFIELD AREA, MADHYA PRADESH
(After : Mukherjee et al. 1982)

Name of Colliery	Proximate Analysis	Spectrographic analysis (in ppm)																	
	Ash %	B	Ba	Co	Cu	Cr	Ge	Ga	La	Mn	Mo	Nb	Ni	Pb	V	Sr	Sn	Y	Zr
Shivpuri Mine (Top seam)	19.0	35	550	25	50	80	ND	30	20	150	ND	25	70	15	100	800	ND	40	250
	31.3	20	300	10	50	70	ND	10	20	150	ND	25	55	20	70	100	ND	30	509
	20.2	20	350	55	45	50	ND	10	30	3000	ND	20	65	15	60	300	ND	35	350
	30.1	25	800	40	60	50	ND	15	25	1500	ND	20	79	10	60	450	ND	40	400
	24.3	100	500	40	70	60	20	25	20	1000	ND	25	80	20	70	150	10	45	230
	7.5	80	850	70	60	80	80	35	25	2000	ND	20	130	10	60	350	10	50	200
	---	10	250	10	50	70	ND	10	15	150	ND	30	50	20	70	100	ND	30	350
	30.5	15	300	10	55	30	ND	20	15	450	ND	20	50	20	60	160	ND	30	220
	---	10	200	10	50	30	ND	10	15	100	ND	30	50	10	70	60	ND	30	600
	19.0	25	850	70	70	60	55	20	35	300	ND	25	130	25	110	450	15	50	650
	18.2	20	750	85	80	65	ND	15	20	1000	ND	20	70	20	65	200	ND	40	450
	14.3	20	800	40	160	50	10	15	30	2000	ND	25	120	25	70	400	10	40	500
	---	10	700	10	60	50	ND	25	25	150	ND	25	30	25	65	500	ND	30	500
Rawanwara Khas Colliery (Top seam)	13.2	10	170	15	100	60	60	20	20	700	ND	15	70	25	45	70	10	90	200
Eklehra Colliery	20.2	15	250	15	70	50	20	10	20	900	ND	25	60	20	60	70	ND	40	500

Table-6 CHEMICAL ANALYSIS OF COALS OF PENCH VALLEY AREA ON ATOMIC ABSORPTION SPECTROPHOTOMETER
(All values in ppm)

Elements	Eklehra Colliery (Seam-I)	Eklehra Colliery (Seam-II)	Chandametta Colliery (Seam-II)	Chandametta Colliery (Seam-III)	EDC Colliery (Seam-II)	Rawanwara Colliery (Seam-III)	Shivpuri Mine (Seam-V)
As	BDL	0.02	BDL	BDL	BDL	BDL	BDL
Cd	BDL	BDL	0.632	0.04	BDL	---	---
Cu	1.04	0.87	1.05	0.88	0.84	0.764	0.237
Fe	41.22	39.54	72.05	89.50	61.57	33.18	99.466
Ni	0.88	0.45	0.91	0.75	0.68	0.477	0.824
Pb	0.21	0.157	0.08	0.025	0.084	0.186	0.065
V	BDL	0.05	BDL	BDL	BDL	---	---

occurrence of trace elements in coals of different epochs, distributed in different parts of the world, shows the presence of the following elements - Ag, Al As, Au, B, Ba, Be, Bi, Ca, Cd, Cl, Co, Cr, Cs, Cu, F, Fe, Ga, He, Hf, Hg, I, Br, In, K, Na, Li, Mg, Mn, Mo, N, Nb, Ni, P, Pb, Pt, Rh, Pd, Ra, Rb, Re, S, Sb, Sc, Se, Si, Sn, Sr, Ti, Th, Te, Ta, U, V, W, Y, La, Ce, Pr, Nd, Sm, Eu, Gd, Tb, Ho, Dy, Er, Yb, Lu, Zn and Zr. These elements have been grouped according to their geochemical affinity for describing their occurrence.

The trace element study of coal is gaining importance due to its several practical applications. The most important are in understanding the coal genesis, particularly the depositional environment; correlation of coal seams; possibilities of extraction of certain rare metals; and also in the contamination hazards abatement. The inorganic constituents in coal have pronounced effects on just about every aspect of coal utilization. They affect mining, cleaning, combustion, conversion, and disposal. The magnitude of these effects is dependent, in part on the element's concentration and mode of occurrence and on the presence of other elements.

TRACE ELEMENTS IN PENCH VALLEY COALS

The occurrence and concentrations of trace elements in coals of PVC area is not available in full detail. However, Mukherjee et al. (1982) spectrographically analysed the composite coal sample of "Top seam" from Eklehra colliery, Rawanwara Khas colliery and Shivpuri open cast mine. These workers have reported the presence of trace elements like B, Ba, Co, Cu, Cr, Ge, Ga, La, Mn, Mo, Nb, Ni, Pb, V, Sr, Sn, Y and Zr in coal ashes of PVC. The concentration of these elements is given in Table 5.

Recently the author has undertaken the chemical analysis of Seam - I and Seam - II of Eklehra colliery, Seam - II and Seam - III of Chandametta, Seam - II of East Dongar Chikli underground, Seam - III of Rawanwara colliery and Seam - V of Shivpuri mines of Pench Valley area on Atomic Absorption Spectrophotometer (AAS) and determined the trace elements like As, Cd, Cu, Fe, Ni, Pb and V. The chemical analysis of these coals is given in Table 6.

ENVIRONMENTAL QUALITY OF NATURAL WATERS OF PVC AREA

The Pench is the main river that passes through the coalfield. The Ghatamali and Sukri are the tributaries which join the main Pench river at Nawagaon and Hatiyagarh-Dhana village in the eastern part of the area, while in the western part the Mandhan and a part of Sukri are the tributary to main Pench river. These rivers and ground water are the main sources of water for industrial, agricultural and domestic purposes. Due to mining, ground water table has depleted, therefore, the people of this area are entirely depending upon surface water.

Major sources of pollution in this coalfield are due to mine water discharges from different mines, lechate from overburden dumps and domestic effluents from nearby localities. In some cases mine water was found highly acidic in this region, which further deteriorates the ground water and surface water quality.

In PVC area, it is noticed that huge amount of mine water is discharged into Pench river either directly or through some other available drainage. Little quantity of mine water are used for water sprinkling and other industrial purposes. The mine water, is mixed with river water and after slow sand filtration treatment supplied to public for drinking purposes. Mine water of Chandametta and Rawanwara Khas collieries are acidic in nature. Total amount of mine water discharged from these two collieries are approximately 2834.8 kl/day. Total mine water discharged from all other mines is approximately 15050 kl/day, out of which approximately 10826.7 kl/day are discharged on the surface. Thus, approximately 23.7 million liters of water is discharged on surface every day. The rate of acid mine water discharged is 9.3 million liters per day, which is 39.2% of total discharge (EAR,1993). Generally mine water is found too hard to use and in few cases these are acidic, which decreases its suitability and deteriorate the water quality of the Pench river.

The parameters, which constitute the water environment are Biochemical Oxygen Demand (BOD), Dissolved Oxygen (DO), Total Dissolved Solids (TDS), Total Suspended Solids (TSS), Turbidity, Facel Coliforms, Oil and Grease, Toxic metals, pH, Temperature and Nutrients. With the commencement of the mining operations, the water flowing through the dumped overburdens will carry suspended solids and dissolved metals and nutrients, which may affect the water quality of the surface and sub-surface waters. Mine water discharged from underground mines also contains different values of TSS, TDS and Trace metals, which do contaminate surface and ground water resources.

QUALITY OF MINE WATER

The general parameters related to quality of mine water are presented in Table 7. The study of data reveals that water from Chandametta and Rawanwara Khas collieries are highly acidic in nature. The pH value varies from 2.8 - 3.7 and acidity ranges from 56.4 - 130.0 mg/l. In other mines pH ranges from 7.0 - 8.6 and alkalinity varies from 176 - 220 mg/l. The value of Chemical Oxygen Demand (COD) and Biochemical Oxygen Demand (BOD) varies from 147 to 280 mg/l and 9 to 12 mg/l respectively. Hardness values in the mine water vary between 42 - 432 mg/l. Maximum hardness was found in Chhinda and Vishnupuri-II, mine. It is noticed that hardness of water decreases its utilities and the surface water quality. Level of Total Suspended Solids (TSS) in mine waste water ranged from 60 to 195 mg/l, which is higher than the permissible limit as per ISI : 2490 (1982). TDS content in the mine water varies from 755 to 2471 mg/l, and being maximum in the Chandametta mine. Bacterial contamination as coliform organisms has been detected and found varying from 10 - 32 MPN/100 ml, which should be absent in normal course.

QUALITY OF PENCH RIVER WATER

The Pench river receives about 10826.7 kl/day, mine waste water discharge from various mines located in PVC area. This mine water is discharged, directly or indirectly into river together with domestic waste water from nearby localities of the coalfields. The water quality data of the Pench river from upstream of Eklehra mine, downstream of Vishnupuri OPC, upstream of Shivpuri mine discharge and downstream of Shivpuri mine is given in Table 8.

The water quality of the Pench river has been analysed before it enters the mining area, which indicate the base line quality. The study of the water quality of the river water showed 6.7 DO level, 1.1 mg/l BOD level and 8.0 pH level. Sulphate concentration was within 47.33 mg/l, but after traversing through the core zone of mining, the level of TSS, sulphate, iron and BOD increases considerably.

The level of DO is observed to decrease upto 5.3 mg/l at downstream of Shivpuri area. At some of the locations overburden dumps are also placed on the river side, which also contribute to the increased level of TSS, specially during rainy season. The level of TDS increases as it passes through the mining areas. The concentration of sulphate, iron and BOD increases as compared to the values found at upstream of Eklehra mine. It is found that the level of pollution

Table-7 GENERAL QUALITY PARAMETERS RELATED TO MINE WATER DISCHARGED FROM VARIOUS MINES IN PENCH VALLEY COALFIELD AREA. (After : EAR, 1993).

	PARAMETERS												
MINES	**pH**	**Alk.**	**Acidity**	**Hardness**	**Free CO_2**	**COD**	**BOD**	**TDS**	**TSS**	**Ca**	**Mg**	**Cl**	**SO_4**
ISI : 2490 (1974)	**5.5 - 9.0**	**NS**	**---**	**NS**	**---**	**250**	**30**	**2100**	**100**	**NS**	**NS**	**1000**	**1000**
Bamori	8.3	---	---	74.0	---	280.0	---	---	90.0	---	---	7.6	---
Chhinda	7.9	---	---	314	---	120.0	---	---	98.0	---	---	22.8	---
Chandametta	2.8 - 3.7	---	56.4	---	---	---	0.8	2471	200.0	---	---	---	1948.9
East Dongar Chikli	7.0	---	---	282.0	---	164.0	---	---	67.0	---	---	48.0	---
Eklehra	7.1	---	---	---	---	---	---	---	---	---	---	---	540.0
Gajandoh	8.6	---	---	74.0	---	200.0	---	---	60.0	---	---	29.5	---
Ganapati	7.2	---	---	194.0	---	140.0	---	---	83.0	---	---	22.5	---
New Sethia OCP	8.4	---	---	42.0	---	147.0	---	---	88.0	---	---	9.0	---
Newton	7.0	---	---	282.0	---	164.0	---	---	112.0	---	---	48.0	---
Rawanwara Khas	3.5	---	130.0	---	---	---	---	---	---	---	---	---	---
Shivpuri OCP	8.2	---	---	58.0	---	169.0	---	---	94.0	---	---	4.75	---
Vishnupuri	8.3	220.9	---	132.0	2.69	---	---	---	---	104.0	28.0	55.4	35.0
Vishnupuri UG-1	8.1	---	---	224.0	---	160.0	---	---	125.0	---	---	62.0	---
Vishnupuri UG-2	7.9	176.3	---	432.0	3.5	---	---	---	---	284.0	148.0	23.5	130.0
Vishnupuri OCP	7.29	180.0	---	408.0	---	---	0.5	340.0	90.0	---	---	---	222.26

All values are expressed in mg/l NS = Not Specified (---) = Not analysed

Table-8 GENERAL WATER QUALITY PARAMETERS RELATED TO PENCH RIVER
Source : EAR, (1993).

	Up stream of Eklehra Mine	Down stream of Vishnupuri OCP	Up stream of Shivpuri area Mine	Down stream of Shivpuri area Mine
pH	8.0	7.7	7.8	7.4
Colour (Hazen units)	---	---	< 5	< 5
Temperature °C	---	---	34	30.5
Total Suspended Solids	6.0	8.0	24	54.0
Total Dissolved Solids	125.5	176.0	297	301
Alkalinity	172.0	160.0	---	---
Hardness	180.0	224.0	---	---
Dissolved Oxygen	6.9	7.0	6.2	5.3
BOD (5 days at 20°C)	1.1	1.2	9.0	12.0
COD	---	---	23.0	28.0
Oil and Grease	---	---	Nil	Nil
Chloride as (Cl^-)	---	---	63.0	62.0
Phenolic Com. as (C_6H_5OH)	---	---	Nil	Nil
Cyanide as (CN^-)	---	---	ND	ND
Sulphide as (S^{--})	---	---	ND	ND
Sulphate as SO_4^{--}	47.33	98.78	58.0	51.0
Total Nitrogen as (N)	---	---	11.0	12.0
Fluorides as (F^-)	---	---	0.55	0.52
Pesticides	Abstent	Abstent	Abstent	Abstent
Insecticides	Absent	Absent	Absent	Absent
Coliform Organisms (MPN/100 ml)	---	---	11	14
Calcium (Ca^{++})	---	---	88.0	72.0
Magnesium (Mg^{++})	---	---	25.0	24.0

is not significant but it is fact that mining activities are certainly affecting the water quality of the river.

The study of CMRI, indicate that approximately 12500 million litres of mine water, are thrown on the surface annually, which may be the reason for acute scarcity of drinking water in the PVC area. Due to heavy discharge of mine water the assimilative capacity of the water body is severely damaged. The study reveals that on the basis of assimilative capacity, the Pench river water is placed in 'C' category as per CPCB standards. This result indicates that assimilative capacity of the river water still exists, but uncheck discharge of pollutants should be stopped (EAR, 1993). The quantification damage of the quality of Pench river water due to mining activity in the PVC area, has changed from base line 130 units to 89 units. Thus, there is a negative change of - 41 units in quantification damage of Pench river water (EAR, 1993).

TRACE/TOXIC METALS IN SURFACE AND SUB-SURFACE WATERS OF PENCH VALLEY AREA

In order to know the present status of chemical environment of PVC area, the entire valley is divided into two parts namely the eastern and the western by the main road passing through the Parasia town. The eastern part includes the Rawanwara Khas, Vishnupuri, Shivpuri group of collieries with adjoining areas, while western part includes the Chandametta and Eklehra group of mines. From eastern part, a total of thirty five samples have been collected for the determination of concentration of heavy toxic metals. These samples includes eight samples from mine waters, fourteen samples from Pench river waters, five samples from hand pump waters and eight samples from shallow dug well waters. While from the western part, a total of fifty water samples have been collected that includes twenty samples from dug wells, eight samples from tube wells, six from mine water, four from drainage water from mine and twelve from Pench river and their tributaries. These samples have been analysed for As, Cd, Co, Cu, Fe, Hg, Mn, Ni, Pb, Se, V, and Zn elements on Atomic Absorption Spectrophotometer (AAS), model GBC - 902, in the laboratory of Centre of Mining Environment, Indian School of Mines, Dhanbad. The detailed account of analytical concentration of various toxic elements found in the natural waters of eastern and western part of the area is given in Table 9 and 10 respectively.

In the foregoing pages, an attempt has been made to give a detailed account of concentration of individual element. For simplicity purpose, the relative percentage of samples is given, in which the concentration of element is known. This is being followed by the relative percentage of individual type of waters in which the concentration of particular element is known.

Arsenic (As)

In eastern part, out of 35 samples analysed, the As concentration is observed only in six samples, while in rest of the samples, As is found Below Detectable Limits (BDL). Thus, it is estimated that only 17% of the samples are showing As concentration, which is found above the recommended limits.

It is also estimated that only 14% of total river water samples have As contamination. The maximum value of As in Pench river water is of the order of 0.55 ppm (EPRW-1), while the minimum value of 0.18 ppm is recorded in (EPRW-2). These sample points are located near Mayawari village and near junction of Pench river and Sukri nadi respectively. However, in all the water samples obtained from mines, the As content was Below Detectable Limits. The study of As values reveals that 25% of well waters and 40% of handpump waters show considerable concentration of As. The maximum value of As of about 0.400 ppm is found in well (EWW-3) located near Mandla village, while the minimum value of 0.36 ppm is observed in well located at Rawanwara colliery. In handpump water, the maximum value of As is of the order of 0.58 ppm near Lohangi village while the minimum value of 0.067 ppm As is obtained from handpump located at Parasia township. Thus, the As concentration in the water samples of the study area ranges from 0.067 to 0.58 ppm with an average value of 0.356 ppm.

In western part As content in dug well water ranges from 0.05 ppm to 0.62 ppm, in tube well waters it ranges from 0.12 ppm to 0.64 ppm while in river water As concentration is high that varies from 0.458 ppm to 0.95 ppm. In mine water samples the As concentration ranges from 0.03 ppm to 0.55 ppm and in drainage from mine waters, the As content ranges from 0.11 ppm to 0.85 ppm. The study of data reveals that As concentration in western part is generally high in nature.

Cadmium (Cd)

In eastern part, the Cd concentration is found only in one water sample of the study area. This sample (EHPW-2) represent handpump water located near Rawanwara colliery. The concentration of Cd is of order of 0.011 ppm, which is above the recommended limit. Thus, only 2.85% of the total samples of the study area shows Cd concentration, while in rest of the samples, it is below Detectable Limit.

In western part the Cd concentration is found in almost all the water samples of the area. In dug well waters the Cd concentration varies from 0.01 ppm to as high as 0.54 ppm. In tube well waters the Cd content ranges from 0.01 ppm to 0.51 ppm with an average value of 0.11 ppm. In river waters the Cd concentration ranges from 0.046 ppm to 0.14 ppm. In mine water, it concentration ranges from 0.01 ppm to 0.11 ppm while in drainage water from mine the mean Cd content is of the order of 0.079 ppm.

Cobalt (Co)

In eastern part, the Co is found in 29% of the water samples of Pench Valley area, with the value ranging from 0.010 ppm to 0.054 ppm while in rest of the samples the Co concentration is Below Detectable Limit. A detailed study of the data reveals that Co is found in 14% of Pench river waters, 37% of mine waters, 37% of well waters and 40% of hand pump waters.

The maximum value of Co in Pench river waters is 0.014 ppm at EPRW-7, located near Vishnupuri open cast project, while the minimum value of 0.010 ppm Co is found at EPRW-1 located near Mayawari village. On the other hand, the maximum value of 0.054 ppm Co is found in mine water located near Pench east underground incline (EMW-4), while minimum concentration of 0.010 ppm Co is found in mine water from Rawanwara Khas colliery (EMW-3). The maximum value of 0.019 ppm Co is found in well water from Bichhua village (EWW-1), while minimum value of 0.012 ppm Co occurs in well water (EWW-8) from Khirsadoh village. Maximum concentration of 0.052 ppm Co is found in hand pump water (EHPW-3) from Sirgora village while minimum value of 0.013 ppm Co is found in hand pump water (EHPW-1) from Parasia township.

In western part, the Co is found in all the water samples of the area. In dug well and tube well waters the Co concentration varies from 0.11 ppm to 0.52 ppm and 0.014 ppm to 0.27 ppm with an average value of 0.10 ppm in both the cases. In mine waters the Co content slightly depleted and ranges from 0.08 ppm to 0.064 ppm while in drainage water from mine the Co values increases and ranges from 0.015 ppm to 0.054 ppm. In river waters the minimum concentration of the order of 0.015 ppm is observed in WPR-2 while maximum value 0.074 ppm is recorded in sample WPR-6.

Copper (Cu)

In eastern part, Cu is found in 77% of total water samples of the area. Maximum concentration of 1.954 ppm of Cu is observed in water samples from Vishnupuri underground incline (EMW-8), while minimum concentration of 0.092 ppm Cu is recorded in water sample at EPRW-6 collected from the junction of Pench river and the Ghatamali nadi. Further analysis of data reveals that the concentration of Cu is found in 79% of Pench river water samples, 62% of dug well water samples and 60% of hand pump water samples and in all mine water samples, while in rest of the samples the concentration of Cu is below Detectable Limit.

The maximum concentration of 1.425 ppm Cu, is found in water samples (EPRW-11) near Ghaliara village, while the minimum value of 0.092 ppm is found in water sample at EPRW-6, collected at the junction of Pench river and Ghatamali nadi. In mine water samples maximum concentration of 1.954 ppm is found in water sample from Vishnupuri underground incline (EMW-8), while minimum concentration of 0.234 ppm Cu is found in water sample of Shivpuri underground incline (EMW-7). The maximum concentration of 0.721 ppm is found in well water (EWW-4) collected from Dighawani village, while minimum concentration of the order of 0.251 ppm Cu is recorded in well water (WW-2) collected from Mandla village. In hand pump waters, the maximum concentration of 1.185 ppm Cu, is found in hand pump water (EHPW-1) from Parasia township, while minimum concentration of 0.494 ppm Cu is found in hand pump water (EHPW-5) collected from Lohangi village.

In western part, the Cu concentration is found in all the samples which is considerably high. In dug well waters the minimum value o.12 ppm is locked up in WDW-1 while maximum value of the order of 0.96 ppm Cu is recorded in sample no. WDW-7. In tube well waters the Cu concentration ranges from 0.17 ppm (WTW-3) to 1.90 ppm (WTW-6) with an average value of 1.11 ppm Cu. In mine waters the Cu content is high and the minimum concentration 0.991 ppm Cu is reported from WMW-2 while maximum value 1.98 ppm Cu is recorded in sample no. WMW-3. In drainage water from mine the Cu content ranges from 1.05 ppm (WDWM-3) to 1.42 ppm (WDWM-4) with an average value of 1.19 ppm Cu. In river waters the minimum value of the order of 0.98 ppm Cu is reported from Sukri nadi (WSR-1) while maximum concentration of about 1.98 ppm. Cu is recorded from Mandhan river a tributary to main Pench river.

Iron (Fe)

In eastern part, iron is found in 94% of the total water samples of the area, and the maximum value of Fe concentration is 52.402 ppm. In the Pench river waters the maximum concentration of 1.450 ppm Fe, is found in water samples (EPRW-10) collected near Chhinda mine and minimum concentration of 0.119 ppm Fe is seen in water sample (EPRW-4) collected from the east of Hatiyagarh village. In mine water samples, maximum concentration of 52.402 ppm Fe, is found in water sample (EMW-4) from Pench east underground incline, while minimum concentration of 0.115 ppm is seen in water sample (EMW-7) in Shivpuri underground. In 75% well water samples Fe concentration occurs in detectable range, while in 25%, the Fe concentration is Below Detectable Limit. The maximum concentration 0.726 ppm Fe is recorded in well water sample (EWW-4) from Dighawani village and minimum value of 0.156 ppm is found in well water (EWW-5) from Chhinda village. In hand pump waters maximum concentration of 7.108 ppm Fe occurs in (EHPW-3) from Sirgora village, while minimum concentration of 0.242 ppm is found in hand pump water of Lohangi village. Thus, it appears that Fe, is widely distributed in moderate concentration in most of the water samples of the area. It is interesting to mention here that out of total samples analysed about 54.54% of samples shows their concentration above the desirable limits proposed by various environmental agencies.

In western part of Pench Valley, the Fe is recorded in all the water samples of the area. In dug well waters the Fe concentration varies from 0.46 ppm to 1.14 ppm and is recorded in sample nos. WDW-20 and WDW-6 respectively. In tube well waters , the minimum value of 0.41 ppm Fe is recorded in sample no WTW-1, while in sample no. WTW-7 the maximum value of the order of 7.15 ppm Fe is recorded. In mine waters the Fe appears to be very high in its concentration and the Fe content varies from 45.25 ppm (WMW-2) to 53.54 ppm (WMW-5) with an average value of 50.03 ppm. However, in water samples collected from drainage from mine water the Fe content reduces and the value varies from 1.43 ppm to 2.52 ppm, with an average value of 1.79 ppm Fe. In river waters the minimum value 1.02 ppm Fe is recorded in sample no.WSR-1, while the maximum value 1.98 ppm Fe is recorded in sample no. WPR-5 from Pench river.

Mercury (Hg)

Hg in general found in all the water samples of the eastern part of Pench Valley area and it concentration ranges from less than 0.002 ppm to as high as 2.5 ppm. In Pench river waters, the maximum concentration of 2.2 ppm Hg occurs, while minimum value of less than 0.002 ppm Hg is recorded. Maximum value of 2.5 ppm Hg is seen in mine water sample (EMW-7) from Shivpuri underground mine, while minimum value of 0.002 ppm Hg is found in mine water sample (EMW-1) from New Sethia open cast mine. In well water the maximum value, 2.2 ppm of Hg is locked in water sample (EWW-5), located near Chhinda colliery, while the minimum value of 0.002 ppm Hg is found at three sites. In hand pump waters, the maximum concentration of the order of 1.05 ppm is noted in water sample (EHPW-5) near Lohangi village, while minimum value of 0.05 ppm Hg is found in water sample (EHPW-4) near Haranbhatta village. It is interesting to note that the water samples of the area by and large show Hg concentration above the recommended limits.

In western part of the PVC area, the Hg is found in 100% of the samples and its concentration in dug well waters, tube well waters, mine waters, drainage water from mine and in river waters ranges from 0.02 ppm - 1.21 ppm; 0.18 ppm - 1.14 ppm; 0.14 ppm - 2.871 ppm; 1.52 ppm - 2.45 ppm; and 1.047 ppm to 2.64 ppm respectively. The Hg concentration are appear to be high, which is also observed in eastern part of the Pench Valley area.

Manganese (Mn)

In eastern part of the Pench Valley area about 83% of total water samples shows Mn concentration. The highest value of 3.920 ppm Mn is found in water sample (EMW-3) and the lowest value of 0.14 ppm Mn is noted in water sample (EPRW-12). The analytical data as shown in Table 9 reveals that Mn is found in 79% of Pench river waters, 88% of mine waters, 75% of well waters and in all hand pump waters while in rest of the samples, the concentration of Mn is Below Detectable Limit. The maximum value of Mn in Pench river waters ranges from 0.014 ppm in water sample (EPRW-12) to 0.618 ppm in water sample (EPRW-5), while in mine waters the maximum value of Mn locked is 3.920 ppm in water sample (EMW-3) and the minimum value 0.029 ppm in water sample (EMW-6). In the well waters the maximum value of Mn concentration of 0.047 ppm is found in water sample (EWW-3), while minimum value of 0.020 ppm Mn is found in water sample (EWW-2). In hand pump waters the maximum value 0.656 ppm Mn is found in water samples (EHPW-2), while minimum value of 0.026 ppm Mn is found in water sample (EHPW-5).

In western part the Mn is recorded in all the water samples of the area. In dug well waters the Mn concentration varies from 0.01 ppm (WDW-10) to 0.45 ppm (WDW-1) with an average value of 0.092 ppm Mn. In tube well waters the Mn content increases and its value ranges from 0.12 ppm (WTW-3) to 0.98 ppm (WTW-4) with an average value of 0.65 ppm. In mine waters, the Mn concentration again increases and the minimum value of the order of 0.024 ppm Mn is recorded from sample no. WMW-1, while the maximum value 2.97 ppm Mn is observed in sample no. WMW-6. The Mn concentration in drainage water from mine become very high and varies from 0.54 ppm (WDWM-2) to 0.81 ppm (WDWM-3), with an average value of 0.64 ppm Mn. In river waters the Mn concentration decline and minimum value of the order of 0.11 ppm is observed in sample no. WPR-6, while maximum value 1.01 ppm Mn is recorded in sample no. WPR-3.

Nickel (Ni)

It is seen from the Table 9, that Ni concentration is found Below Detectable Limit in all the water samples of the eastern side of the Pench Valley area.

In western part of Pench Valley area the Ni concentration is recorded in all the water samples. In dug well waters the Ni concentration ranges from 0.07

ppm to 0.81 ppm, with an average value of 0.32 ppm. In tube well waters the Ni concentration reduces and the values varies from 0.01 ppm (WTW-5) to 0.71 ppm (WTW-6) with an average value of 0.31 ppm. In mine water the Ni content increases and the minimum value of the order of 0.024 ppm Ni is recorded from sample no. WMW-1, while maximum value of about 2.97 ppm Ni is observed in sample no. WMW-6). In drainage water from mine the Ni values decline and varies from 0.10 ppm (WDW-1) to 1.01 ppm (WDW-3) with an average value of 0.415 ppm. In river water the maximum value of the order of 0.327 ppm Ni is recorded in sample no. WPR-6 while the minimum concentration of the tune of 0.05 ppm Ni is found in sample no.WPR-3.

Lead (Pb)

In eastern part of Pench Valley area the Pb is found to occur only in 40% of the total samples analysed, while in rest of the samples Pb contents are Below Detectable Limit. The concentration of Pb in general varies from 0.014 ppm to 0.681 ppm in water samples from hand pump water (EHPW-3) located near Sirgora village and Pench river water sample (EPRW-10) near Chhinda village respectively.

The analytical data reveals that Pb concentration occurs in 50% samples of Pench river waters in which, it varies from 0.035 ppm in water sample from (EPRW-4), located east of Hatiyagarh village to 0.681 ppm in water sample (EPRW-10), located near Chhinda village. In mine waters about 25% samples have Pb concentration ranging from 0.095 ppm in water sample (EMW-8) to 0.194 ppm in water sample (EMW-2), collected from Rawanwara colliery area. The Pb concentration is found to occur in 37% of well waters and it ranges from 0.131 ppm in water sample (EWW-7) to 0.074 ppm (EWW-8) from well located near Khirsadoh and Khirsadoh Dhana villages respectively. While 40% of hand pump water samples have Pb concentration ranging from 0.014 ppm in water sample (EHPW-3) to 0.155 ppm in water sample (EHPW-2) from hand pump located near Rawanwara colliery area.

In western part of Pench Valley area, the Pb concentration varies from 0.044 ppm (WDW-6) to 1.10 ppm (WDW-1) in dug well waters with an average value of 0.23 ppm. However, in tube well water the minimum value 0.11 ppm Pb is observed in sample no. WTW-3, while maximum concentration of the order of 0.97 ppm is recorded in sample no. WTW-7. In mine water the Pb concentration varies from 0.138 ppm to 0.97 ppm with an average value of 0.49 ppm. In drainage water from mine the Pb content increases considerably that ranges from 0.52 ppm in sample no. WDWM-4 to 0.81 ppm in sample no. WDWM-2, with an average value of 0.645 ppm. In river waters the concentration of Pb reduces and initiate from 0.185 ppm in sample no. WJP-3 to as high as 0.64 ppm in sample no. WJP-1 in the Pench river.

Selenium (Se)

In eastern part of Pench Valley area, the Se contents are found only in 23% of water samples of the study area and its concentration ranges from 0.753 ppm in water sample (EPRW-6) to 1.76 ppm in water sample (EMW-4). The synthesis of data reveals that detectable concentration of Se is found in 29% of Pench river waters, 12% mine waters and well waters and 40% hand pump waters of the area and in rest of the water samples the concentration of Se is below Detectable Limit.

The maximum value of Se in Pench river waters is 1.75 ppm in water sample (EPRW-8) collected from near the bridge, while the minimum value of 0.753 ppm is found in water sample in (EPRW-6) collected at the confluence of Pench river and Ghatamali nadi. The concentration of Se in mine water samples is as high as 1.76 ppm in Pench east underground incline water sample (EMW-4). The value of Se in well water on the other hand is 0.453 ppm in well no. EWW-6 located near Khirsadoh village. The maximum value of Se in hand pump water samples is 1.19 ppm from water sample located in Rawanwara colliery area, while the minimum value of 1.163 ppm is observed in hand pump water at Parasia township (EHPW-1). It is interesting to mention here that Se concentration at all sample points is found to be above desirable limits.

In western part the Se content is present in all the water samples of the area and the concentration ranges from 0.14 ppm in tube well water to 1.94 ppm in drainage water from mine. However, on individual basis the Se content in dug well waters ranges from 0.18 ppm (WDW-17) to 1.51 ppm (WDW-20) with an average value of 0.85 ppm Se. In tube well waters the Se concentration differs from 0.14 ppm (WTW-3) to 1.62 ppm (WTW-7) and the average being 0.91 ppm. In mine water samples the Se concentration increases and the minimum value 0.854 ppm is reported from sample no. WMW-2, while maximum value 1.74 ppm Se is recorded in sample no. WTW-6, with an average value of 1.30 ppm. However, in drainage water from mine the Se content increases considerably and the value varies from 1.22 ppm (WDWM-4) to 1.94 ppm (WDWM-2) with a high average of the order of 1.62

ppm. In river water the Se concentration differs from 0.981 ppm to 1.92 ppm, with an average value of the order of 1.38 ppm. Thus, the river waters also retain high concentration of Se.

Vanadium (V)

The V is found only in 34% of water samples of eastern part of PVC area. The V contents ranges from 0.216 ppm to 0.980 ppm. The study of data reveals that 43% of Pench river waters, 25% of mine water, 37% of well water and 20% of handpump waters show V concentration within detectable limits, while in rest of the samples the concentration of V is below Detectable Limit.

The maximum value of V in Pench river water is 0.870 ppm in sample (EPRW-13) collected near Bijori village, while minimum value of 0.390 ppm is locked in sample (EPRW-12) collected near Pindrai village. In mine water the maximum value of V is 0.746 ppm in water sample (EMW-1) from New Sethia open cast mine, while the minimum value of the order of 0.276 ppm is found in sample from Rawanwara Khas colliery. The maximum concentration of V in well water is 0.980 ppm as observed in well located near Rawanwara colliery (EWW-7), while the minimum concentration of 0.430 ppm is found in water sample from well (EWW-8) located near Khirsadoh Dhana village. In hand pump waters the concentration of V is 0.216 ppm is found in sample (EHPW-1) from hand pump located in Parasia township.

In western part, the V concentration in dug well waters ranges from 0.01 ppm to as high as 1.26 ppm, with an average value of 0.73 ppm. However, in tube well waters the V concentration increases that varies from 0.024 ppm in sample no. WTW-8 to 0.514 ppm in sample no. WTW-6, with an average value of 0.28 ppm. In mine water the minimum value of 0.14 ppm V is reported in sample no. WMW-3, while maximum value of 0.90 ppm V is recorded in WMW-4. The V concentration in drainage water from mine is slightly low than mine waters, and the minimum value of the order of 0.22 ppm and maximum value of about 0.858 ppm is recorded in sample nos. WDWM-4 and WDWM-2 respectively. It is however, interesting to observed that the V concentration in river water become very high and varies from 0.42 ppm (WPR-1) to 1.01 ppm (WPR-5), with an average value of 0.75 ppm.

Zinc (Zn)

The concentration of Zn is found in 89% of the total water samples of eastern part of PVC area and it ranges from as low as 0.658 ppm in Pench river water sample (EPRW-7) collected near Vishnupuri open cast mine to as high as 1.676 ppm in mine water sample (EMW-2) collected from Rawanwara colliery.

The study of data reveals that out of total water samples, the concentration of Zn is found in 86% of Pench river waters, 87% of mine waters, 87% of well waters and in all hand pump water samples of eastern side of Pench Valley area, while in rest the value of Zn is Below Detectable Limit. The maximum value of Zn in Pench river water ranges from 0.998 ppm in sample (EPRW-10) collected from near Chhinda mine to 0.658 ppm in river water sample (EPRW-7) from Vishnupuri open cast project. The maximum value of Zn in mine water is about 1.676 ppm in Rawanwara colliery (EMW-2) while the minimum value of Zn is 0.882 ppm in New Sethia opencast mine (EMW-1). In well waters the Zn content varies from 1.391 ppm in well from Khirsadoh village (EWW-6) to 0.857 ppm in well from Mandla village (EWW-2). In hand pump waters the maximum value of Zn is of the order of 1.561 ppm in hand pump water located near Rawanwara colliery (EHPW-2), while the minimum value of about 0.857 ppm in hand pump water located near Haranbhatta village (EHPW-4). It is interesting to mention here that Zn concentration in all the sampling sites is below recommended limits.

In western part, the Zn concentration is found in all the water samples of the area that ranges from as low as 0.11 ppm to as high as 2.55 ppm. In dug well waters minimum value of 0.14 ppm is locked in sample no. WDW-5, while maximum value 1.981 ppm is recorded in WDW-8, with an average value of 1.08 ppm. In tube well waters the overall minimum and maximum value of Zn is recorded that varies from 0.11 ppm (WTW-4) to 2.55 ppm in WTW-2. The mine water show slightly low value. The minimum value of 0.49 ppm Zn is reported from WMW-3, while maximum value of 1.94 ppm is observed in sample no. WMW-6, with an average value of 1.39 ppm. In drainage water from mine the Zn concentration further decreases that varies from 0.37 ppm (WDWM-2) to 0.88 ppm (WDWM-1), with an average value of 0.72 ppm. In river waters the minimum Zn content of the order of 0.602 ppm is obtained from WSR-2, while maximum concentration of about 0.96 ppm is reported from WSR-1, a tributary to main Pench river.

STUDY OF POLLUTANTS IN WATERS OF PENCH VALLEY AREA

The toxic elements concentration of Pench Valley area

Table-9 TRACE/TOXIC ELEMENT CONCENTRATIONS IN NATURAL WATERS OF EASTERN PART OF PENCH VALLEY AREA

Sam. No.	As	Cd	Co	Cu	Fe	Hg	Mn	Ni	Pb	Se	V	Zn
ISI : 10500 (1983)	0.05	0.01	NS	0.05	0.3	0.001	0.1	NS	0.1	0.01	NS	5.0
1	2	3	4	5	6	7	8	9	10	11	12	13
EMW -1	BDL	BDL	BDL	0.469	2.491	<0.002	0.107	BDL	BDL	BDL	0.746	0.882
EMW -2	BDL	BDL	BDL	1.537	3.187	1.3	2.996	BDL	0.194	BDL	BDL	1.676
EMW -3	BDL	BDL	0.010	1.390	3.887	0.25	3.920	BDL	BDL	BDL	0.276	0.971
EMW -4	BDL	BDL	0.054	0.662	52.402	0.625	3.875	BDL	BDL	1.76	BDL	0.904
EMW -5	BDL	BDL	BDL	1.266	0.859	0.850	0.263	BDL	BDL	BDL	BDL	0.988
EMW -6	BDL	BDL	BDL	0.872	0.319	1.95	0.029	BDL	BDL	BDL	BDL	0.978
EMW -7	BDL	BDL	0.036	0.234	0.115	2.50	BDL	BDL	BDL	BDL	BDL	BDL
EMW -8	BDL	BDL	BDL	1.954	7.117	0.05	0.094	BDL	0.095	BDL	BDL	0.999
EWW -1	BDL	BDL	0.019	BDL	BDL	<0.002	0.028	BDL	BDL	BDL	0.560	0.948
EWW -2	0.400	BDL	BDL	0.251	0.561	<0.002	0.020	BDL	BDL	BDL	BDL	0.857
EWW -3	0.360	BDL	BDL	0.394	BDL	0.550	0.047	BDL	BDL	BDL	0.980	BDL
EWW -4	BDL	BDL	0.015	0.721	0.726	1.850	BDL	BDL	BDL	BDL	BDL	0.950
EWW -5	BDL	BDL	BDL	BDL	0.156	2.200	BDL	BDL	BDL	BDL	BDL	0.959
EWW -6	BDL	BDL	BDL	BDL	0.187	1.750	0.039	BDL	0.121	0.453	BDL	1.391
EWW -7	BDL	BDL	BDL	0.483	0.317	<0.002	0.035	BDL	0.131	BDL	0.430	0.967
EWW -8	BDL	BDL	0.012	0.595	0.200	0.700	0.034	BDL	0.074	BDL	BDL	0.983
EHPW -1	0.067	BDL	0.013	1.185	1.012	0.625	0.035	BDL	BDL	1.163	0.216	0.939

Table-9 Contd.........

Sam.No.	As	Cd	Co	Cu	Fe	Hg	Mn	Ni	Pb	Se	V	Zn
1	2	3	4	5	6	7	8	9	10	11	12	13
EHPW -2	BDL	**0.011**	BDL	**1.028**	**2.890**	**0.400**	**0.656**	BDL	**0.155**	**1.19**	BDL	1.561
EHPW -3	BDL	BDL	0.052	BDL	**7.108**	**0.060**	**0.137**	BDL	0.014	BDL	BDL	0.993
EHPW -4	BDL	BDL	BDL	BDL	**1.281**	**0.050**	0.026	BDL	BDL	BDL	BDL	0.857
EHPW -5	**0.580**	BDL	BDL	**0.494**	0.242	**1.050**	0.022	BDL	BDL	BDL	BDL	0.981
EPR -1	**0.550**	BDL	0.010	**0.335**	**0.356**	**<0.002**	0.016	BDL	**0.134**	**1.186**	BDL	0.928
EPR -2	**0.180**	BDL	BDL	BDL	0.129	**<0.002**	0.023	BDL	BDL	BDL	BDL	0.698
EPR -3	BDL	BDL	BDL	**0.218**	0.129	**<0.002**	0.028	BDL	BDL	BDL	BDL	0.836
EPR -4	BDL	BDL	BDL	BDL	0.199	**<0.002**	BDL	BDL	0.035	BDL	BDL	BDL
EPR -5	BDL	BDL	BDL	**0.159**	**0.620**	**<0.002**	**0.618**	BDL	BDL	BDL	0.463	BDL
EPR -6	BDL	BDL	BDL	**0.092**	**0.326**	**<0.002**	0.016	BDL	BDL	**0.753**	BDL	0.742
EPR -7	BDL	BDL	0.014	**0.469**	0.268	**<0.002**	0.038	BDL	BDL	**1.526**	BDL	0.658
EPR -8	BDL	BDL	BDL	**0.402**	**0.492**	**1.950**	**0.108**	BDL	**0.534**	**1.75**	BDL	0.971
EPR -9	BDL	BDL	BDL	**0.990**	**1.211**	**2.200**	0.024	BDL	**0.130**	BDL	0.463	0.995
EPR -10	BDL	BDL	BDL	**0.664**	**1.450**	**0.450**	0.070	BDL	**0.681**	BDL	0.560	0.998
EPR -11	BDL	BDL	BDL	**1.425**	**0.730**	**0.050**	0.018	BDL	0.089	BDL	0.506	0.998
EPR -12	BDL	BDL	BDL	**0.963**	**0.507**	**0.020**	0.014	BDL	0.072	BDL	0.390	0.974
EPR -13	BDL	BDL	BDL	**0.574**	**0.340**	**0.250**	BDL	BDL	BDL	BDL	0.870	0.882
EPR -14	BDL	BDL	BDL	BDL	0.291	**0.450**	BDL	BDL	BDL	BDL	BDL	0.763

All concentrations are in ppm, NS = Not Specified, BDL = Below Detectable Limit Bold figures indicate the concentrations above the recommended limit as per ISI : 10500 (1983).

Table-10 TRACE/TOXIC METAL CONCENTRATIONS IN WATERS OF WESTERN PART OF PENCH VALLEY COALFIELD AREA

Trace Ele.	As	Cd	Co	Cu	Fe	Hg	Mn	Ni	Pb	Se	V	Zn
1	2	3	4	5	6	7	8	9	10	11	12	13
ISI : 10500 (1983)	0.05	0.01	NS	0.05	0.3	0.001	0.1	NS	0.1	0.01	NS	5.0
SampleNo.												
WDW -1	0.40	0.19	0.52	0.12	0.88	0.45	0.45	0.14	1.10	1.10	0.18	0.84
WDW -2	0.06	0.21	0.11	0.65	0.88	0.17	0.14	0.07	0.14	0.47	0.97	1.21
WDW -3	0.54	0.10	0.04	0.45	0.76	0.21	0.041	0.25	0.34	0.55	1.12	1.28
WDW -4	0.54	0.04	0.017	0.45	0.76	0.21	0.041	0.25	0.34	0.55	1.12	1.28
WDW -5	0.09	0.11	0.022	0.55	0.914	1.02	0.14	0.14	0.114	0.562	0.73	0.14
WDW -6	0.12	0.12	0.014	0.178	1.14	0.74	0.08	0.81	0.044	0.89	0.84	1.34
WDW -7	0.57	0.02	0.054	0.96	0.98	0.247	0.04	0.24	0.147	1.05	0.47	1.01
WDW -8	0.10	0.08	0.024	0.495	0.69	1.01	0.061	0.69	0.245	0.91	0.01	1.981
WDW -9	0.44	0.10	0.074	0.58	0.88	0.22	0.017	0.76	0.18	0.71	0.472	0.981
WDW -10	0.57	0.04	0.15	0.71	0.75	0.02	0.01	0.143	0.07	0.19	1.01	0.947
WDW -11	0.61	0.05	0.08	0.68	0.96	1.06	0.048	0.18	0.124	0.87	0.22	1.96
WDW -12	0.41	0.21	0.017	0.145	1.02	0.421	0.02	0.21	0.512	1.24	0.98	1.97
WDW -13	0.25	0.01	0.21	0.545	0.64	0.54	0.19	0.175	0.26	0.91	0.82	1.44
WDW -14	0.14	0.54	0.24	0.26	0.57	0.17	0.05	0.594	0.154	1.04	0.23	0.984
WDW -15	0.51	0.24	0.014	0.49	0.46	0.14	0.14	0.414	0.094	1.29	0.914	1.54
WDW -16	0.29	0.04	0.029	0.57	0.901	1.21	0.047	0.17	0.184	0.94	1.02	0.41
WDW -17	0.45	0.05	0.071	0.12	0.53	0.164	0.021	0.11	0.135	0.18	0.47	0.155
WDW -18	0.05	0.10	0.024	0.641	0.98	0.97	0.017	0.70	0.09	0.98	0.73	0.978
WDW -19	0.62	0.10	0.017	0.449	0.74	0.02	0.114	0.09	0.141	1.01	1.26	0.147
WDW -20	0.59	0.04	0.011	0.584	0.46	0.182	0.183	0.34	0.21	1.51	0.97	0.981

Table-10 Contd.........

WTW -1	0.59	0.02	0.062	1.55	0.41	0.18	0.52	0.20	0.114	1.09	0.314	0.54
WTW -2	0.54	0.05	0.105	1.05	2.75	1.09	0.87	0.50	0.54	1.11	0.51	2.55
WTW -3	0.12	0.01	0.09	0.17	4.25	0.41	0.12	0.10	0.11	0.14	0.11	0.81
WTW -4	0.47	0.11	0.17	1.11	6.45	1.01	0.98	0.30	0.59	0.54	0.41	0.11
WTW -5	0.12	0.041	0.051	1.51	4.55	0.71	0.41	0.01	0.244	0.91	0.17	2.54
WTW -6	0.21	0.51	0.057	1.90	6.78	1.04	0.74	0.71	0.157	1.00	0.514	0.64
WTW -7	0.64	0.07	0.27	1.187	7.15	1.14	0.73	0.27	0.97	1.62	0.18	1.05
WTW -8	0.25	0.034	0.014	0.43	6.31	0.35	0.81	0.41	0.184	0.87	0.024	0.35
WMW -1	0.03	0.024	0.047	1.84	47.84	0.14	0.024	0.14	0.50	1.74	·0.745	1.57
WMW -2	0.04	0.014	0.024	0.991	45.25	2.31	2.92	0.18	0.138	0.854	0.431	0.954
WMW -3	0.55	0.09	0.008	1.98	53.02	2.34	2.57	0.90	0.184	1.19	0.14	0.49
WMW -4	0.09	0.01	0.064	1.97	49.27	2.14	1.461	0.97	0.90	1.537	0.90	1.47
WMW -5	0.47	0.07	0.058	1.78	53.54	2.09	2.04	0.10	0.97	0.94	0.85	1.93
WMW -6	0.21	0.11	0.015	1.471	51.24	2.871	2.97	0.65	0.54	1.57	0.758	1.94
WDWM -1	0.11	0.05	0.046	1.21	1.54	1.95	0.64	0.10	0.64	1.581	0.742	0.88
WDWM -2	0.58	0.14	0.054	1.09	2.52	2.45	0.54	0.40	0.81	1.94	0.858	0.37
WDWM -3	0.14	0.08	0.042	1.05	1.43	1.52	0.81	1.01	0.61	1.75	0.647	0.83
WDWM -4	0.85	0.046	0.015	1.42	1.68	2.04	0.57	0.45	0.52	1.22	0.22	0.79
WPR -1	0.51	0.05	0.054	1.545	1.08	2.21	0.54	0.21	0.625	1.25	0.42	0.84
WPR -2	0.458	0.14	0.015	1.74	1.45	1.09	0.56	0.14	0.425	1.75	0.954	0.76
WPR -3	0.514	0.08	0.024	1.09	0.90	1.047	1.01	0.05	0.24	1.44	0.821	0.71

Table-10 Contd.........

WPR -4	0.65	0.046	0.018	1.447	1.54	2.25	0.72	0.21	0.51	1.05	0.51	0.658
WPR -5	0.95	0.05	0.019	1.24	1.98	2.18	0.47	0.181	0.485	1.09	1.01	0.902
WPR -6	0.47	0.14	0.074	1.04	1.05	2.21	0.11	0.327	0.35	0.981	0.95	0.824
WMR -1	0.84	0.08	0.017	1.98	1.25	2.64	0.45	0.24	0.612	1.117	0.75	0.93
WSR -1	0.84	0.046	0.021	0.98	1.02	1.98	0.39	0.19	0.447	1.351	0.621	0.96
WSR -2	0.671	0.05	0.039	1.44	1.22	1.48	0.642	0.121	0.541	1.52	0.701	0.602
WJP -1	0.74	0.14	0.041	1.92	1.45	2.01	0.551	0.10	0.64	1.75	0.65	0.881
WJP -2	0.65	0.08	0.055	1.73	1.92	2.42	0.431	0.18	0.42	1.92	0.72	0.73
WJP -3	0.681	0.046	0.07	1.49	1.25	2.08	0.512	0.201	0.185	1.32	0.85	0.912

Table-9 All values are expressed in ppm.; NS = Not Specified,
Bold figure indicate concentration above permissible limit as per ISI : 10500 (1983).
WDW = Dug well water, WTW = Tube well water, WMW = Mine water,
WDWM = Drainage water from mines, WPR = Pench river water, WMR = Mandhan river water,
WSR = Sukri river water, and WJP = Junction point of tributary with Pench river.

are compared with the specifications prescribed by ISI : 10500 (1983). The concentration of different metals above recommended limit in eastern and western part of the area is given in Table Nos. 9 and 10 respectively. The study of data from Table No. 9, reveals that As is found above recommended limit in two samples of well water, hand pump water and Pench river water. The Cd is reported only in one sample of hand pump water of the eastern side and is found above recommended limit. The Cu content is found above prescribed limit in most of the waters except EWW-1, 5, 6, EHPW-3, 4 and EPRW-2, 4 and 14. Similarly the Fe concentration is found above recommended limit in most of the samples except EMW-7, EWW-1, 3, 5, 6, 8, EHPW-5, EPRW-2, 3, 4, 7 and 14. The Hg concentration when compared with ISI : 10500 (1983) specifications it is found above recommended limit in all the samples of the eastern part of the area. The Mn is found above desirable limit in most of the mine water samples but in well water samples it is below the recommended limit. It is only two hand pump water samples (EHPW-2 & 3) and Pench river water sample (EPRW-5 & 8) in which the Mn content are above the desirable limit. The Pb is found above recommended limit in eight samples only that include one sample of mine water (EMW2), two samples of well water (EWW-6 & 7), one sample of hand pump water (EHPW-2) and four samples of Pench river water (EPRW-1, 8, 9, & 10). In the same way, the Se is also found above desirable limit in eight samples of the area that include one sample each of mine water(EMW-4), and well water (EWW-6), two sample of hand pump water(EHPW-1 & 2), and four samples of Pench river water namely EPRW-1, 6, 7, & 8. The Zn is found below recommended limit in all the samples of the eastern part of the Pench Valley area while for Co, Ni, and V the recommended limit is not specified by ISI : 10500 (1983). Thus, it is observed that in the eastern part of the Pench Valley area the surface water and sub-surface waters are contaminated either by two or more than two toxic metals.

In the western part of the Pench Valley coalfield area, the environmental contamination scenario is highly intricate and complex because most of the elements shows their concentration above the recommended limits except Zn. There occur only one sample of dug well water (WDW-18) and two sample (WMW-1 & 2) of mine water in which the As concentration is below the recommended limit. In the same way there is only one sample each of dug well water (WDW-13), tube well water (WTW-3) and mine water (WMW-4) in which the Cd contents are below the recommended limit while in rest of the samples the Cd concentration is far above the desirable limit of drinking water specifications. The concentration of Cu, Fe, Hg and Se is found far above the desirable limit in all fifty water samples of the western part of the area. It is particularly of the interest to record that the Fe concentration in mine waters are extremely high and ranges from 45.25 ppm to 53.54 ppm with a mean value of 50.03 ppm. Similarly the Hg concentration is also very high above the recommended limit (0.001 ppm), in mine water, drainage from mine water and river waters of the area. The Mn concentration in twelve water sample of dug well waters and only one sample of mine water have its value below the recommended limit prescribed by ISI : 10500 (1983). The Pb concentration on the other hand is below desirable limit only in four samples of dug well waters while in rest of the samples it is far above the recommended limit. The study reveals that western part of Pench Valley area shows extreme deterioration of the environment particularly with reference to toxic metals. In the same way this water is also above the recommended limits prescribed by the Ayer and Westcot (1974) for irrigation water and livestock waters.

CONCLUSIONS

On the basis of the data analysed above, it is concluded that natural waters in western part of Pench Valley area is highly contaminated in terms of trace/toxic elements as compared to eastern part. The coal mining activity on large scale in the entire valley is responsible for the mobilization of trace/toxic elements. In addition, the mine water discharges into Pench and its tributaries deteriorate the water quality of surface waters, which in turn affects the ground waters of PVC area. People of the area are using this water for their domestic needs and this is how they are exposed to higher concentration trace/toxic elements therefore, there is every possibility for the proliferation of both carcinogenic as well as non-carcinogenic diseases in their life time exposure.

References

Ashburner (1866) : "Coal Resources of M.P., Jammu and Kashmir", (1983) In Bulletins of GSI, Series - A, No. 45, Coalfields of India, Vol. III, C.S.RajaRao (Ed.), pp. 155-178.

Banerjee, N.N., Rao, H.S. and Lahiri, A. (1972) : Germanium in Indian Coals, Indian Jour. of Tech., Vol. 12 No. 8. pp. 353-358.

Bethell, F.F., (1961) : The distribution and origin of minor elements in coal, Br. Coal Util. Res. Assoc. Mon. Bull., Valley. Vol. 26, 401 p.

Central Fuel Research Institute, (1979) : Indian Coals, Vol. 5, pp. 3-144.

Central Mining and Research Institute, CMRI, (1986) : Findings, Water Pollution, Mine Environment : Status of Research on Environmental Pollution in Mines and Mining areas and miner's health, pp. 18-30.

Chandra, T.K., Shrivastava, A.K. and Ghose, (Jr.) P.K., (1958-60) : A report on the geology and coal resources of the Pench Kanhan Valley coalfiels, District - Chhindwara, M.P., Rep. Geo. Surv. India.

Chou, C.L., Cahill, R.A., Ruch, R.R., and Harvey, R.D., (1982) : Geochemistry of a Bituminous coal seam in Illinois, Transactions of the American Nuclear Society, Vol. 41, pp. 187-188.

Chou, C.L., Cahill, R.A., Steele, J.D., Henderson, L.R., and Vogel, R.S., (1983) : Geochemistry of trace elements and sulfur in the Spring field (No.5) Coal in the Illinois Basin (Abstract), The Geological Siciety of American (96th Annual Meeting), Vol. 15, 543 p.

Crookshank, H., (1936) : The geology of the northern slopes of the Satpuras between Morand and Sher rivers, Mem. Surv. India, Vol. 66, No. 2.

Das, D.K., (1958) : A note on the presence of Germanium in coal of Talchir coalfields, Orissa, Quart. Jour. Geol. Min. Metal. Soc. India, Valley. Vol. 30, No. 4, pp. 235-236.

Dasgupta, H.S., and Kathing, D.T., (1979) : North - East Indian Coal : An assessment of its potential arsenic a Ge source, pp. 1 - 22.

Dey, A.K., (1980) : Trace elements of diamictites used arsenic environment reflector Aabstract), Third Conv. Ind. Assoc. Sed.

Dreher, G.B., and Schleicher, J.A., (1975) : Trace elements in coal by optical emission spectroscopy, In : S.P. Babu (Ed.), Advances in Chemistry Series, Washington D.C., Amer. Chem. Soc., Vol. 141, pp. 35-47.

Environmental Audit Report, EAR, (1993) : An Environmental Audir Report (EAR) of Chhindwara and balaghat districts of M.P. Submitted by Environmental Planning and Coordination Organisation, Government of M.P. in collaboration with CMRI, Dhanbad.

Eskenazi, G., (1977) : On the binding form of tungusten in coal, Chem. Geo., Vol. 19, pp. 153-159.

Eskenazi, G., (1980) : On the geochemistry of Indium in coal forming process, In : Geochim. Cosmochim. Acta, Valley. Vol. 44, No. 7, pp. 1023-1027.

Finkelman, R.B., (1993) : Trace and Minor elements in coal, In : Organic Geochemistry, Michael H. Engel and Stephen A. Macko (Eds.), pp. 593-607, Plenum Press, New York.

Fox, C.S., (1934) : The Lower Gondwana coalfields of India, Mem. Geol. Surv., Vol. 59, 386 p.

Frost, J.K., Santoliquido, P.M., Camp, L.R., and Ruch, R.R., (1975) : Trace elements in coal by neutron activation analysis with radiochemical separations, In : S.P. Babu (Ed.), Trace Elements in Fuel, Washington D.C., Amer. Chem. Soc., Advances in Chemistry Series, Vol. 141, pp. 84-97.

Gee, E.R., (1924) : The geology of the Umaria coalfield, Recommended. Geol. Sur. India, Vol. 60, No. 4, pp. 399-410.

Gluskoter, H.J., and Lindhal, P.C., (1973) : Cd : Mode of occurrence in Illinois Coal, Science, Vol. 181, pp. 264-266.

Gluskoter, H.J., Ruch, R.R., Miller, G.W., Cahill, R.A., Dreher, G.B., and Kuhn, J.K., (1977) : Trace elements in coal : Occurrence and Distribution, III Geol. Surv. Ind., C-499, 154 p.

Goldschmidt, V.M., (1922a) : Stammestypen der Eruptingestgpen videnskapssel ska pets skifter, I. Matnateur, Klasse, No. 10.

Goldschmidt, V.M., (1922b) : Der Stoffuvechsel der, Erde. Ibid, No. 11.

Goldschmidt, V.M., (1930) : Uber das Vorkommen des Germaniums in Steinkohlen und Steinkohlen - Produkten, Nachr. Ges. Wiss. Gottingen, Math. - Phys.. KL., Vol. 3, pp. 398-401.

Goldschmidt, V.M., (1937) : The principles of distribution of chemical elements in minerals and rocks. Jour. Chem. Soc. London, pp. 655-673.

Gupta, D.C., (1999) : "Environmental aspects of selected trace elements associated with coal and natural waters of Pench Valley coalfield of India and their impact on human health". Int. Jour. Coal. Geol. Vol. 40, pp. 133-149.

Hawley, J.E., (1955) : Germanium content of some Nova Scotia coals, Econ. Geol., 50, 517-532, Lancaster.

Indian Standards Institute, 2490 : (1974) "Standards for discharge of waste water effluents". Gazette of India.

Indian Standards Institute, (1983) : Indian Standard Specification for drinking water (28th Feb.1983) IS I : 10500 - 1983. In : Bhu-Jal News (1986) pp. 12-15.

Iyer, V.G., and Sundaram, N., (1955) : Ge in coal, Ind. Min. Jour., Vol. 3, No. 9, pp. 182-183.

Jerdon, and Shanky, (1852) : In Bulletins of GSI, Series - A, No. 45, Coalfields of India, Vol. III, C.S.RajaRao (Ed.) "Coal Resources of M.P., Jammu and Kashmir", (1983), pp. 155-178.

Kuhn, J.E., Harfest, W.F., and Shimp, N.F., (1975) : X-Ray Fluorescence analysis of whole coal, In : S.P. Babu (Ed.), Washington, D.C., Amer. Chem. Soc., Advances in Chemistry Series, Vol. 141, pp. 66-73.

Mallick, A.U., and Ahmed, S., 91979) : Analytical studies on Indian coals with special reference to their ash and mineral contents, Jour. Min. Met. Fuel, Vol. 27, No. 5, pp. 148-152.

Mason, B., (1958) : Principles of Geochemistry, (2nd Ed.), 329p. John Wiley and Sons, New York

Medlicott, H.B., (1855) : In Bulletins of GSI, Series - A, No. 45, Coalfields of India, Vol. III, C.S.RajaRao (Ed.) "Coal Resources of M.P., Jammu and Kashmir", (1983), pp. 155-178.

Mineral Exploration Corporation Limited, MECL, (1983) : A note on Pench Valley coalfield, MECL Pub., pp. 1-13.

Mukherjee, B., and Dutta, R., (1950) : A note on the constituents of the ashes of Indian coals determined spectrographically, Fuel, Vol. 29, pp. 190-192, London.

Mukherjee, K.N., Raja Rao, C.S., Chowdhury, A.N., Pal, J.C., and Das, M., (1982) : Trace elements studies in the major Tertiary and Gondwana coalfields of India, Bull. Geol Surv. India, Series - A, No.49, 115p.

Pareek, H.S., and Bardhan, B., (1985) : Trace elements and their varations along seam profiles of certain coal seams of Middle and Upper Barakar formations (Lower Permian). In : East Bokaro Coalfields (Abstract), pp. 281-314.

Raja Rao, C.S., (1983) : Coal resources of M.P. and Jammu and Kashmir, Bull. Geol. Surv. India, Series-A, No. 45, V. III, 197 p.

Ramage, H., (1927) : Gallium in fly dust, Nature, Vol. 119, pp. 783.

Rankama, K., and Sahama, T.G., (1950) : Geochemistry, The Univ. of Chicago Press, 912p., Chicago.

Rao, B., and Rao, V., (1955) : Minor elements in Indian coals, Jour.Ind.Geol. Sci. Assoc., Vol. 4-5.

Rishi, M.K.,(1970) : Germanium and other minor elements in the coals of Umaria, M.P., Jour. Geol Soc. India, Vol. 11.

Ruch,R.R., Gluskoter,H.J., and Kennedy, E.J., (1971) : Hg content of Illinois coals, Environment. Geol. Notes, 111, U. State Geol Surv., Vol. 43, pp. 1-15, Washington.

Ruch, R.R., Gluskoter, H.J., and Shimp, N.F., (1974) : Occurrence of distribution of potentially volatile trace elements in coal - A final report. Illinois State Geol. Surv., Envirn. Geol. Notes, 72, 96 p.

Singh, R.M., Singh, M.P., and Chandra, D., (1983) : Occurrence, distribution and probable source of the trace elements in Ghugus coals, Wardha Valley, District-Chandrapur and Yeotmal, Maharastra, International Jour. Coal Geol., Vol. 2, pp. 371-381.

Singh, M.P., Singh, R.M., and Chandra, D., (1985) : Environmental and health problems due to geochemical alterations associated with trace elements in coals of Ghugus coalfields, Wardha Valley, Maharastra, Quat. Jour. Geol. Min. Met. Soc. India, Vol. 57, pp. 99-103.

Swaine, D.J., (1962) : Trace elements in coal, their Origin, mode of occurrence and economic importance, Geol. Surv. India, R.O.Div., Coal Respectively. Tech. Commu., 45 p.

Swaine, D.J., (1972) : Trace elements in coal, In : Contribution to Recent geochemistry and analytical chemistry, pp. 482-492.

Swaine, D.J., (1977) : Trace elements in coal, In : Symp. On trace substances in environmental health, D.D. Hemphill (Ed.), xith edition, pp. 107-122.

West, W.D., (1923) : The line of the Narmada and Son Valley, Curr. Sci., Vol. 31, No. 4, pp. 143-144.

Wood, G.H.(Jr.), Kehn, T.M., Carter, M.D., and Culbertson, W.C., (1983) : Coal resource classification system of the U.S. Geological Survey, U.S.Geol. Surv. Circ., 891p.

Hon'ble Reviewers

1. Prof. M.M. Saxena
   Editor in Chief
   'Okoiassay'
   7P3, Pavanpuri South
   BIKANER 334 001 RAJASTHAN
2. Prof. A.S. Sheoran
   Head
   Department of Mining Engineering
   Faculty of Engineering
   JNV University
   JODHPUR 342 011 RAJASTHAN

Editorial Assistance

Dr. Sameer Naval

Economic Mineralization
K.L. Shrivastava (Editor)
2009. pp. 510-513
Scientific Publishers (India), Jodhpur

46

Section-V

Impact of Mining on Biodiversity in India

D.K. KHANDELWAL[1], SHISHIR SHARMA[1], KALPANA KHANDELWAL[2] AND RAJNI SHARMA[3]

[1]Department of Geology, Dungar College, Bikaner University, Bikaner 334 003
[2]Department of Chemistry, M.S. College, Bikaner University, Bikaner 334 003
[3]Department of Economics, Dungar College, Bikaner University, Bikaner 334 003
Email : arnunshandilyabikaner@yahoo.com

Abstract

Biodiversity provides the basic biotic resources that sustain human race. Biodiversity can be divided into five hierarchical levels namely ecosystem, communities, species, population, and genes. The Biodiversity Protection Act, 2002 says that conservation (in situ and ex situ) and sustainable utilization must take into account the complexity as well as relationship between, among, and within each of these levels. As a signatory to the conservation of biodiversity (CBD), it is desirable to have valuation of biodiversity and it's components.

Various legal instruments relevant to biological diversity in India (viz. Environment Protection Act 1986, Wild Life Protection Act 1972, 1991 etc.) make provision for the protection of the environment by controlling and regulating all kinds of pollution caused by prospecting or mining operations, ultimately affecting genetic resources.

Exploration and exploitation of minerals is essential for the economic and industrial development of the developing and developed countries. At present mining activities are being carried out on an extensive scale but mostly irrationally, causing ecological imbalances to the mining sites and ultimately affecting the biodiversity.

Key words : Biodiversity, Mining Environment, Sustainable Development, Mining Law.

INTRODUCTION

Diversity is the 'spice of life' is an old saying. Biodiversity has been the backbone of human food, health, livelihood and security systems, ever since the dawn of the human civilization.

Among the various physical and biological factors, the mining of natural resources is also one of the major causes, which creates considerable environmental pollution (Keller, 1976). Minerals are the basic resources along with air, water, land and biological resources for the sustainance of human race. During the last few decades, activities for mineral exploitation have increased substantially to fulfill the growing demands of the present scientific and industrial world and adversely affect the biological diversity.

BIODIVERSITY

According to the Biological Diversity Act, 2002, the 'Biological Diversity' means the variability among living organisms and the ecological complexes of which they are a part and includes diversity within species or between species and ecosystems.

Biodiversity provides the basic biotic resources that sustain the human race. Biodiversity can be divided into five hierarchical levels namely ecosystem, communities, species, population, and genes (Soule 1991). Conservation and sustainable utilization must take into account the complexity as well as relationships among between and within each of these levels.

The biological diversity is the cumulative resource on this planet in respect of plant and animal genera and species and also the ecosystem's richness. It is the fundamental basis of evolution and the continuing adaptation of human race to changed environment. For Earth, the only planet, where life is known to exist, diversity is the product of over 3000 m.y. of evolution. This also contributes the basis of sustainability and survival of human beings.

Biological diversity of the world is fast depleting, despite all efforts being made to conserve it. This is basically taking place due to habitat destruction, over-harvesting, pollution and the inappropriate introduction of foreign plants and animals. The increasing degradation of natural ecosystems resulting in consequent erosion of native genetic diversity in genetic resources of plants and animals, is assuming an alarming proportions. The loss of biodiversity has to be seen against a greater need to produce food and other commodities under conditions of expanding biotic and abiotic stresses and shrinking per capita availability of arable land and irrigation water. Moreover there are great possibilities for adverse changes in precipitation, temperature, UV radiations and in sea level.

Conservation and proper management of these genetic resources for their sustainable use, therefore, constitute the basis for future insurance of economic growth, development and investment.

It is to prevent continuing genetic erosion and to promote concerted efforts to conserve biodiversity by all nations acting both individually and collectively, that a Global Convention on Biodiversity (CBD) was adopted at United Nations Conference on Environment and Development (UNCED) at Rio De Janeiro on 5th June in 1992, and officially came into force on the 29th December 1993, put it as reference January 2000. Over 170 nations had ratified the CBD.

The CBD is a unique international agreement, since it provides a framework for integrated action in biodiversity, conservation, sustainable use and equitable sharing of benefits. Political, public and media attention to biodiversity conservation has certainly increased since the CBD came into force. Many countries have enacted legislations to give effects to its provisions.

In India, an act to provide for conservation of biological diversity, sustainable use of its components and fair and equitable sharing of the benefits arising out of the use of biological resources knowledge and for matters connected there with or incidental threats has been enacted as "The Biological Diversity Act, 2002 (18 of 2003)". The President assented to the said Act on 5th Feb. 2003.

PRESENT STATUS OF BIODIVERSITY IN INDIA

India currently has 85 National Parks covering 3.6 million sq. kms. And 448 Wildlife Sanctuaries covering about 1,20,000 sq. kms. In the major biogeographic zones. The total extent of protected areas includes 5 World Heritage Sites, 9 Biosphere Reserves and 6 Ramsar Sites. About 7000 endemic species are found in India, that is, they do not occur anywhere else in the world. There are two hot spots in India out of total 25 in the world. These are the western Ghats and the Northeastern region.

Over the last one-century or so, half of our forests have gone, a third of our wetlands drained out, 70% of freshwater polluted, 40% of mangroves destroyed. At least 10% of our 1,35,000 species of plants (over 1500 flowering plants) and animals are threatened with extinction, with a few dozen already have vanished forever. Much of the crop and livestock diversity that once characterized Indian farming, is on the decline, with some areas of the Green Revolution belt losing almost all their seed diversity, and all the 18 indigenous breeds of poultry threatened by breeds introduced from outside India.

MINERAL RESOURCES AND THEIR EXTRACTION

Among the bounties of nature, mineral resources are by for the most valuable treasures to the society for marching ahead on the road of planning to achieve economic development and growth. Mineral products are beyond doubt the backbone of industrialization. The mining may be defined as the removal of minerals from the earth's crust and converting them into consumable products for the welfare and prosperity of mankind. Mining is second only to agriculture as the world's oldest and most important industry. The dependence of primitive societies upon mined products is illustrated by the nomenclature of these epochs : stone age, bronze age, and iron age, a sequence which also shows the increasing complexity of society's relationship with mining. In a sense, the history of mining is the history of civilization itself. There is no doubt that mining is a vital sector of the economy in any country. The minerals, which are extracted from the earth's crust, serve as the raw materials for various industries, on which our standards of living depend. To meet the ever-increasing standards of living, the demand for raw materials keeps on growing and to meet this growing demand, the size of mining must be continued if civilization is to survive. Since mining operations cause irretrievable damage to environment, it must be our endeavour at least to reduce, if not to entirely eliminate, the environmental damage due to mining.

EXTRACTION OF MINERAL RESOURCES AND ENVIRONMENTAL DEGRADATION

Exploration and exploitation of minerals is very essential for the economic and industrial development of any country. Generally in developing countries like India, mining activity is carried out unsystematically and irrationally. This activity reduces or destroys the biological potentials of the land, causing ecological imbalance to the mining sites and ultimately affecting the biodiversity.

In general, mining activities are accompanied by a variety of environmental problems. This process of environmental degradation, which starts with the extraction of minerals resulting in land degradation and addition of pollutants to air and water, continues as the mineral is beneficiated and further processed.

Mining operations can damage the environment and ecology to an unacceptable degree, unless carefully planned and controlled. Environmental damage due to mining and mineral processing can be considered under the following heads :

i. Land degradation due to surface mining.

ii. Land degradation due to underground mining.

iii. Water pollution due to mining and mineral processing.

iv. Air pollution due to mining and mineral processing.

v. Land disposal of solid wastes and refuses dumps.

vi. Land degradation due to deforestation.

vii. Diminishing flora and fauna, and

viii. Aesthetic damage due to abandoned mine sites, etc.

There is a need for specialization to assess the environmental and ecological impact of mining and for mining operations to be a balance between mining, engineering and environmental requirements. With the growing concern for environmental protection, stringent laws have been enacted or are in the offing, in most of the countries of the world. However these stringent environmental laws are making impacts on the economics of the mining and mineral processing industry.

LEGISLATIVE MEASURES FOR ENVIRONMENTAL PROTECTION DUE TO MINING AND MINERAL PROCESSING

Management of mineral resources is the responsibility of the Central Government and the State Governments in terms of Entry 54 of the Union List (List I) and Entry 23 of the State List (List II) of the **Seventh Schedule** of the Constitution of India.

Indian government is much concerned about the preservation and protection of healthy environmental conditions. A number of Acts and Rules have been enacted by the government from time to time inter alia with necessary modification their in. Notifications and guidelines are also being issued. Even our constitution obliges the state and the citizens to protect forests and wildlife and develop compassion towards all living beings.

Legislative control of environmental degradation caused by toxic metals and the protection of workers engaged in mining and processing of ores involves following steps.

1. Laying down a set of rules based on the expert recommendation;
2. Prescribing non offensive levels of air-borne pollutants in the working environment;
3. Adopting the international conventions and recommendations concerning the prevention of occupational risks; and
4. Observing the codes of practice and guidelines in preventions, medical supervision; and improvement of working environment.

State government have adopted a number of acts passed by parliament. For effective enforcement of these Acts, state governments have formulated a number of Rules for the guidance of enforcement agencies. Following are some important legislations that have been made to protect the environment and ecology of the country.

1. Mines and Minerals (Regulation and Development) MMRD Act 1957 (major amendments in 1986, and 1994).
2. The Mineral Concession Rules (MCR) 1960 (amended up to September 1994).
3. The water (Prevention and Control of Pollution) Act 1974 (Amended in 1978, 1988).
4. Forest (conservation) Act, 1980.
5. The Air (Prevention and Control of Pollution) Act 1981 (Amended in 1987).
6. The Environmental (Protection) Act 1986.
7. Mineral Conservation and Development Rules (MCDR) 1988.
8. The Mineral Concession (Amendment) Rule, 2002.

PREVENTIVE MEASURES/ CONCLUDING REMARKS

Of foremost importance is the need to instill in the minds of India's planners and decision-makers, the critical value of biodiversity. It is, first of all, a moral imperative to protect nature, not least of all because we ourselves have evolved from it, and because all species have a right to survive. But, if this does not convince our policy makers, the fact is: we simply cannot do without biodiversity for it is the source of all the essential ecological functions that we all depend on.

Mining operation shall not ordinarily be taken up in identified ecologically fragile and biologically rich areas. Strip mining in such areas should as far as possible be avoided and it should be permitted only when accompanied with comprehensive time-bound reclamation programme.

Cost benefit analysis for ecological and environmental losses and socio-economic distress caused to the people who are displaced be weighed against economic and social gains.

Mining waste dumps cause the maximum damage to the environment. Loss of flora and fauna is a common feature in the mining activity. To conserve the local eco-system and restoration of flora and fauna in that particular area a gene bank should be devised that can be used during reclamation. Reclamation should be done to minimize their harmful potential. Unless we want to convert several million hectares into a wasteland, reclamation has to become an integral part of all mining activity in India. If reclamation is made a part of mining operations from the start, it is easier, cheaper, and less hazardous to carry out than to reclaim abandoned areas. Planning may be made for rehabilitation and restoration of the worked out mine land also. A forestation programme in wasteland or back filled pits can play a vital role in rehabilitation practices.

Biodiversity conservation cannot, however, happen unless it becomes a mass movement and unless the laws, policies, technologies, development projects and demographic trends that are currently threatening it are tackled head-on.

References

Geological Survey of India, 1996. Contributions in Environmental Geology. Spl. Pub. No. 43.

Govt. of India 1994. conservation of Biological Diversity of India : An Approach, Ministry of Environment and Forest (MOEF), New Delhi. McMillan Pub.

Keller, E.A., 1979. Environmental Geology. Charles E. Merrill Publ. Company, Ohio, 454 p.

Mehta, R., 1996. Land Reclamation in opencast coalmines in proceedings of the national seminar on 'Environmental Impact Assessment Methodology Focussing on Mining Environment', pp. 119 135.

Moghe, V.B., et al., 1994. Resource Atlas of Rajasthan, Department of Science and Technology, Govt. of Raj., Pub. Jaipur.

Mulder, Ed, F.J., 1994. Planning and managing the human environment the essential role of the geoscience. COGEOENVIRONMENT (News letter) Publication. The Netherlands, pp. 1-12.

Owen, O.S., 1971. Natural Resources Conservation : An Ecological approach. McMillan Pub.

Roy, S.C., 1984. Final Report on environmental evaluation of impact of mining in Dehradun-Mussoorie areas, UP, GSI Unpublished Report.

Hon'ble Reviewers

1. Prof. R.P. Tiwari
Dean
Faculty of Forestry and Geology
Mizoram University
AIZAWAL 796 009 MIZORAM

2. Prof. S.K. Agarwal
Editor in Chief
'Acta Ecologica'
A594, Talwandi
KOTA 324 005 RAJASTHAN

Editorial Assistance

Dr. Narendra Singh

Economic Mineralization
K.L. Shrivastava (Editor)
2009. pp. 514-525
Scientific Publishers (India), Jodhpur

47

Section-V

Passive Treatment of Acid Mine Drainage from Sulphide Mineral Mining and Processing Industry

A.S. SHEORAN[1] AND V. SHEORAN[2]
[1]Department of Mining Engineering, J.N.V. University, Jodhpur, INDIA 342011
[2]Department of Zoology, J.N.V. University, Jodhpur, INDIA 342005
Email : as_sheoran@yahoo.com

Abstract

Acid Mine Drainage (AMD) is a serious environmental pollution problem of the sulphide mineral mining and processing industry around the world. Active treatment systems commonly used by the Indian mining industry at active mine sites require continual addition of alkaline chemicals like lime, hydrated lime, caustic soda, soda ash etc. These chemicals besides being expensive, lead to production of voluminous sludge. The disposal of this sludge is again an environmental problem to be dealt with. Thus active treatment processes are often expensive both in terms of capital and operating costs.

The main aim of the present paper is to provide information about Passive Treatment technology for acid mine drainage from sulphide mineral processing and mining industry. The need for this technology is because the current approaches are impractical at the sources of AMD at abandoned mine sites. Presently at abandoned mines contaminated water continues to be discharged day after day without any treatment in India.

Keywords : Sulphide mine, Acid Mine Drainage, Mine Environment, Constructed Wetlands.

INTRODUCTION

Mining has been one of the most common activities since ancient times and continues to remain so in the modern world. Minerals extracted from mines are the foundation stones of civilization. Minerals extracted raw from the earth, are processed to yield basic substances such as metals, chemicals, building materials, fuels, fertilizers etc. In fact the development of civilization is linked up with transition from stone implements to those of iron, copper and from wood as fuel to coal, oil and now even nuclear fuel. Industrial society could not exist without these essential commodities. A combined total of about 1150 million tonnes of heavy metals (copper, lead, cobalt, zinc, cadmium and chromium) have been mined by man since the Stone Age. It is further estimated that an annual output of 14 million tonnes of heavy metals are being mined with an annual growth rate of 3.4 percent. While the world population doubled during the period 1959-1990, the production of six major non-ferrous metals (zinc, copper, lead, tin, nickel and aluminium) increased eight fold. One can add to these fuel minerals, industrial minerals and stones. The volume of non-fuel minerals consumed during the five decades since the Second World War has exceeded the total minerals extracted from the earth during all the previous history of mankind (Matagi et al., 1998; Aswathanarayana, 2003).

Mining of the earth resources is the foundation for the development of mankind but mining has never been held in high esteem. Most people concede mining as a necessary evil. Mining causes hazards like deterioration of landscape/aesthetic environment, changes of land use pattern, degradation of soil, air pollution, surface and sub-surface water pollution, vibrations, subsidence and also has socio-economic impacts on the region. Another serious problem facing the mining industry presently is the enormous mass of mine tailings (about 18 billion m^3 per year) and is expected to double in the next 20-30 years. Out of total production, 95% of the ore is rejected as mill tailings after extracting the valuable metal by mineral beneficiation process. Neither the industrialized countries nor more so, the developing countries can afford to avoid mining altogether, as the whole spectrum of industrial activities (including energy generation) is based on minerals. In some African countries mining accounts for 80-90 percent of the gross domestic product (GDP). Mining is an important part of economy in India. Country is rich in many minerals. The overall value of minerals being extracted in India exceeded US $ 11.32 billion during 2001-2002. Contribution of mining to GDP has increased from 0.56 percent (pre-independence period) to 3 percent (Aswathanarayana, 2003; Raju et al., 2003).

The treatment of acidic mine wastewater from mining and processing of sulphide minerals and uranium ores is an environmental challenge to the mining industry. Mining of these ores exposes the sulphide/ pyritic and pyrrhotitic minerals to oxygen and water, which coupled with the microbial activity, lead to the formation of waters that are highly enriched with sulphate and a range of heavy metals commonly known as Acid Mine Drainage (AMD). Other pollutants and toxic metals such as arsenic, cyanide may also be present in acidic mine water at elevated concentrations when these sulphide minerals are processed in minerals processing units. Further, mill tailings amount to about 90% of ore is dumped in tailing impoundments and seepage from these impoundments can be source of AMD for decades. Nordstrom and Alpers (1999) estimated that without preventative measures the Richmond Mine, at California's Iron Mountain, would generate AMD, with pH < 1 and containing several g L- 1 of dissolved metals, for 3000 years. Kalin (2001) estimated, based on oxidation rates derived from tailings pore water samples, that the site of a small zinc/copper mine in northwest Ontario, Canada, would generate AMD for 1000 to 35,000 years. These may be extreme examples but it is not uncommon for base metal mines and their waste products to generate acid for more than 100 years. Since AMD is self-renewing, an ideal solution to it would also be self-perpetuating. A typical AMD is characterized by high acidity (e.g., low pH) high concentration of sulphate and metals such as iron, manganese, aluminum, zinc, copper, nickel, lead, cadmium, arsenic etc.of varying compositions dependent upon the originating minerals deposit types (Fyson et al., 1994; Clarke, 1996; Banks et al., 1997; Kuyucak, 2002; Fillpek et al., 2003; Akcil and Koldas, 2006).

The most evident consequence of AMD contamination to water bodies is, the precipitation along the streambeds, of bright colored orange or yellow secondary iron and sulphur minerals, which are referred to as 'yellow boy'. If left uncontrolled, it results in contamination of ground and surface waters, along with the surrounding land. Because of its leaching capacity, augmented by the activities of the bacteria it attracts, it is self-perpetuating (Kalin et al., 2004). Being associated with both operating and abandoned mining operations, abandoned mine sites may cover very large areas; it is difficult to inhibit the formation of the AMD (Whitehead and Prior, 2005). It has the potential to contaminate ground and local watercourse thus leading to a serious harmful impact on ecosystem (Grande et al., 2005). An insidious feature of AMD is that its sources may remain active for decades or even centuries after mine closure (Modis et al., 1998).

The treatment of AMD is one of the major concerns for the mining industry. Ironically, even under the best reclamation scenario it requires long-term collection and treatment. There are two distinct strategies for treating AMD. The conventional solution is to collect and chemically treat acidified effluents in a centralized treatment plant by continual addition of alkaline chemicals like lime, hydrated lime, caustic soda, soda ash etc. These chemicals besides being expensive, also, lead to production of voluminous sludge (Fiset et al. 2003; Santos et al., 2004; Cohen,

2006). These active treatment processes are often expensive both in terms of capital and operating costs. In Canada, an estimated 6.7 million cubic meters of sludge are produced each year (Zinck et al. 1996). Since the long-term stability and environmental acceptability of these sludges are of great concern.

Alternatively, effluents can be routed through natural and constructed wetlands within which the biotic and abiotic communities treat acidic mine wastewater. In the past decades, therefore, research efforts have been directed towards wetlands as an alternative passive means of treating AMD (Matagi et al., 1998;). Such a passive treatment is economical, non-polluting, and is not a source of secondary wastes. Moreover, a well-engineered passive treatment system is a closed ecological system and hence is self-renewing. (Dunbabin and Bowmer, 1992; Kleinmann and Hedin, 1993; Hedin et al., 1994; Eger, 1994; Bozau and Strauch, 2002; Kalin, 2004; Woulds and Ngwenya, 2004).

In the early 1980's it was observed that the quality of acidic mine water flowing through the natural, sphagnum moss dominated wetlands improved significantly. Further investigations on natural wetland ecosystems have indicated that the wetlands have the significant capability to remove trace metals and improving the low pH values (Dinges, 1982; Wieder and Lang, 1982; 1984). The emphasis, for a decade or more, has been to emulate the natural wetland ecosystem by constructing a substrate based system with emergent vegetation which can be controlled and manipulated to remediate acidic mine water (Mays and Edwards, 2001; Masferrer, 2002; Champagne et al., 2005; Nicomrat et al., 2006).

Wetlands are composite systems where a variety of physical, chemical, microbiological and plant-mediated processes occur that cause significant changes in the water chemistry (Sheoran and Sheoran, 2006). It was accepted that plants probably remove metal into their biomass and that they provided micro aerobic zones for bacteria that may assist in the removal process (Greenway and Simpson, 1996). Subsequently, much work in the development and engineering of wetlands to treat mine water discharge has been carried out by U.S.Bureau of mines (Hallberg and Johnson, 2005). Engineered wetlands have been successfully applied in many locations to treat AMD. The focus of this biological pollution abatement process is to apply bio-geochemical treatment mechanisms at or near the source of mine drainage to immobilize and concentrate metals and raise pH (Eger and Lapakko, 1988; Hiel and Kerins, 1988; Stark et al., 1988; Stillings et al., 1988; Brodie et al., 1989a,b; Hammer and Bastian, 1989; Kleinmann, 1989; Watson et al., 1989; Hedin and Nairn, 1990; Wieder, 1993; Eger, 1994; Stark et al., 1994; Cambridge, 1995; Gusek, 1995; DeBusk et al., 1996; Cole, 1998; McCarthy et al., 1999; Tarutis et al., 1999 ; DeBusk et al., 2000; Ye et al., 2001; Younger et al., 2002; Kalin and Chaves, 2003 ; Karathanasis and Johnson, 2003; Sheoran, 2004; Sheoran and Bhandari 2005; Sheoran2005; Sheoran and Sheoran 2005; Sheoran,2006; Sheoran,2007a,b; Hallberg and Johnson, 2005).

Modern mining faces an environmental challenge to treat the polluting effluents and adequate handling of all the liquid residues, originating in the different stages of extractive mining. The mining sector of India, as in all the countries of great metallurgical production, has had to accept a set of regulation related to emission and discharge of liquid industrial wastes. Part A, effluent, (Schedule-VI, 1993) MOEF, GSR 422(E), 19.05.93 requires the treatment of the polluting effluents and adequate handling of all the liquid residues originating in the different stages of extractive mining (Maiti, 2001). Conventional chemical treatment for the AMD are used at active mine sites in India and no treatment at all is being carried out at abandoned mine sites. The aim of the present paper is to introduce the passive technology in India so that it can be implemented at abandoned as well as active mine sites of sulphide mineral. This technique may be implemented in the field as Indian climate conditions are quite favourable.

COMPONENTS OF CONSTRUCTED WETLAND

A constructed wetland is a shallow basin filled with substrate and planted with vegetation tolerant of saturated condition. Water is introduced at one end and flows on the substrate and discharged at the other end though a weir or other structure, which controls the depth of water. So wetland is a complex assemblage of Water, Basin, Substrate and Plants. Some times a liner is also included beneath the substrate.

POLLUTANT REMOVAL MECHANISM IN WETLANDS

The pollutants removal from acid mine drainage in wetlands is obtained by physical, chemical and various biological processes which involves sedimentation, settling, filtration, adsorption, precipitation, co-precipitation into insoluble compounds. It is, however, difficult to illustrate what actually occurs or which reactions takes place in the wetland. The entire processes are dependent on each other, thus making the whole process of pollutants removal mechanisms in wetlands very complex. More or less, the extent to which these reactions occur depends on the composition of the substrate, sediment pH, nature of wastewater and plant species (Sheoran and Sheoran, 2006).

The Pollutant removal mechanisms of AMD occur in the three main compartments of a wetland, i.e. (1) Soil and substrate, (2) Hydrology, (3) Vegetation. Water is usually present at the surface or within the root for extended periods of time. Hydrology is probably the single most important determinant for the establishment and maintenance of specific types of wetland and wetland processes. It is the permanent or periodic saturation of a wetland area that results in the anaerobic conditions in the soil under which typical wetland biogeo-chemical processes occur. These processes cause the development of characteristic wetland soils, which support a dominant plant community adapted to living in saturated soils. The hydrology compartment also contains heterogeneous polyligands i.e. fulvic, humic and tannic acids, amorphous metal oxyhydroxides, clay, bacterial surfaces, and associated exocopolymers, suspended particles and macro-molecules e.g. polysaccharides, proteins etc.

Soils consist of unconsolidated natural material that supports or is capable of supporting plant life. Under wetland conditions, soils are considered to be hydric, i.e. saturated, flooded, or ponded long enough during the growing season to develop anaerobic conditions in the upper portion of the soil. Hydric soils are developed under conditions sufficiently met to support vegetation typical to wet areas (hydrophytic vegetation). Wetland plants are autotrophic organisms, creating a biomass of reduced carbon compounds that serves as food for a variety of organisms, both micro and macroscopic. Plants also have the ability to remove trace metals from the water through biological uptake and surface adsorption. In the biota, biological conversion occurs through assimilation and metabolism of microorganisms living on and around the macrophyte and plant uptake and metabolism. In permanently anoxic water conditions in wetlands, decomposition of organic matter is by reduction and organic matter accumulates on the sediment surface. The resulting organic sediment surface is responsible for scavenging heavy metals from influent AMD (Sheoran and Sheoran, 2006).

Physico-chemical properties provide many positive attributes for remediating contaminants including heavy metals. The expansive rhizosphere of wetland, herbaceous shrub and tree species provides an enriched culture zone for microbes involved in degradation. Redox conditions in most wetland soil/sediment zones enhance degradation pathways requiring reducing conditions. The concentration of heavy metal ions removed from solution in wetlands is determined by interacting processes of settling, sedimentation, sorption, co-precipitation, cation exchange, photodegradation, phytoaccumulation, biodegradation, microbial activity and plant uptake. It is, however, difficult to illustrate what actually occurs or which reactions takes place in the wetland (Sheoran and Sheoran, 2006).

DESIGN AND DEVELOPMENT OF CONSTRUCTED WETLAND

The design of a constructed wetland for the treatment of acid mine drainage varies upon site characteristics. The most important design considerations are:

Biochemical Processes: As soon as the wastewater enters in the wetland it under goes to number biochemical processes and that have to be carefully considered when planning a construction of a wetland to treat acid mine drainage. Conditions which are favourable to chemical and microbial reducing, adsorption, neutralization and precipitation need to be incorporated into the soil, hydrology, vegetation, and site development plan (Fennessy and Mitsch 1989).

Loading rate: The pollution loading of the wastewater is another important consideration. The maximum ability for a constructed wetland to treat AMD is determined by the loading rate (Girts and Kleinmann 1986).

The Retention Time: The retention period is another important parameter in designing the construed wetland. It is determined by the volume and concentration of AMD, existing effluent standards and the rate of treatment, precipitation, run-off, infiltration, evapotranspiration. All these variables have to be considered when determining the maximum contaminant reduction (Fennessy and Mitsch 1989). A hydrologic holding rate of 200 m3/ha-day and a seven-day holding time are believed to be the optimal conditions for wetlands treating wastewater (Wile et al. 1985).

Slope: Slope is an important factor to consider in wetland design. Brooks (1984) determined a slope of less than 5% maximizes contact by wetland vegetation and substrate, therefore influencing iron and manganese removal. The Environmental Protection Agency recommends a slope of less than 1% will optimize efficiency of wastewater remediation (EPA 1999).

Substrate and Vegetation: Substrate and Vegetation are closely linked, in that the substrate used is dependent upon the desired vegetation (Fennessy and Mitsch 1989). For example, design of Typha wetlands often use composted hay and manure on top of a layer of limestone (Girts and Kleinmann 1986). Increased organic matter content of wetland sediments has been shown to influence increased metal adsorption (Fennessy and Mitsch 1989). Limestone and sewage sludge are considered important additives for soil condition manipulation (Joost et al. 1987; Topper and Sabey 1986).

Vegetation is an important developmental characteristic for wetland construction due to its important role in AMD remediation, and its influence on other wetland characteristics. Perennial vegetation can be established by placing rhizomes directly into a saturated subsurface zone (Fennessy and Mitsch 1989). The use of local stock is favoured because their local adaptability (Fennessy and Mitsch 1989). Two main species used in constructed wetlands are Typha spp. and Shagnum spp. Typha spp. are characteristic of being acid tolerant and thriving under a variety of environmental conditions (Brooks 1984). This versatility and hardiness, along with Typha spp. ability to adsorb iron and manganese, make it and excellent vegetative choice for wetlands constructed to remediate mine drainage (Snyder and Aharrah 1985). Studies on Sphagnum spp. show significant influence in reducing sulfates, iron, and manganese concentrations (Kleinmann 1985). Perry and Kleinmann (1991) found difficulty in establishing Sphagnum spp. due to its sensitivity to the stresses associated with transplanting, abrupt changes in water chemistry, and excessive or insufficient water depth. This establishment problem, along with its having slow generating ability, makes its use in a constructed wetland less viable than Typha spp. This does not preclude its consideration, however, because of Sphagnum spp. excellent ability to aid in AMD remediation. Future research and development of techniques allowing Shagnum spp. use in constructed wetlands could prove critical in this remediation technology.

High Sediment: High Sediment loads in AMD make sediment accumulation an important consideration in wetland design and construction. Ponds designed to accumulate sediment can be constructed into the wetland treatment system (Fennessy and Mitsch 1989). These sediment ponds would be constructed prior to the treatment wetland. Sediments then can be dredged from the pond without disturbing the wetland habitat. Kadlec (1985) has found seasonal harvesting of plant biomass to reduce sediment accumulation, although this does raise a concern of affecting levels of important organic matter accumulation in sediments.

Morphometry or the geometric configuration: Morphometry or the geometric configuration can be extremely influential in a constructed wetlands ability to perform AMD remediation. Designs, which imitate natural wetlands, such as varied shape and slope, are more aesthetically pleasing, provides better wildlife habitat, and enhances macrophyte establishment. Placement of islands within the wetland system increases habitat diversity and act as natural water flow breaks and diversions. a more serpentine shape outperformed a rectangular shape in reducing contaminants of AMD (Brooks 1984; Wile et al., 1985).

Seasonality: Seasonality of the locale in which the wetland is to be constructed is an important factor in wetland design, in particular areas where climate fluctuations are great. The bacteria catalyzing AMD creation is as active during cold seasons as does in warmer seasons. The dormant season and its effect on the vegetation's ability to remediate AMD is therefore important to consider. Typha spp. are subject to a efficiency decline at locales with colder temperatures regime, whereas Sphagnum spp. does not face these same restrictions (McHerron, 1985; Fennessy and Mitsch 1989).

Regulatory: Regulatory issues can hinder the use of constructed wetlands to remediate AMD due to uncertainty of the ability of wetlands to remediate effluent to meet standards on a consistent basis, or whether they can be maintained to retain biological control (Fennessy and Mitsch 1989).

ADVANTAGE OF CONSTRUCTED WETLANDS

Constructed wetland treatment technology have emerged as a viable option for addressing a vide range of water quality problems specially in treating acid mine drainage and effluents loaded with metals. Such high tech treatment systems are quiet efficient in raising pH, reducing the other toxic components. The treatment are often more economic than energy intensive engineered treatment plants. The constructed wetland treatment technologies have the following advantages over the conventional treatment (Chemical or Active treatment plants) (Hedin et. al., 1994; Sheoran, 2004).

Provides a High Quality of Treatment: Properly designed, constructed, maintained and managed wetlands can provide very efficient treatment of acid mine water. Laboratory and field test results show that pollutants were reduced to very acceptable levels. High metal removal rates have been observed in both the constructed wetland and natural wetlands.

Relatively Inexpensive to Construct: Each constructed wetland's design is site specific, taking into consideration such variables as topography, water supply, soil types etc. Selection of a site with accommodating specifications keeps establishment costs low.

Inexpensive to Operate and Maintain: A constructed wetland requires little, if any, energy use and equipment needs are minimal. A well-designed wetland transfers water by gravity through the system. If topography limits the use of gravity, pumps will be necessary which increases the cost of operation. Once established, properly designed and constructed wetlands are largely self-maintaining. Operation and maintenance expenses are low (Energy and supplies are low). Operation and maintenance require only periodic, rather than continuous on site labour. The treatment of acid mine drainage by constructed wetlands is to provide a long term, zero or low-maintenance method of abating acid mine drainage. Although some human intervention may be required.

Tolerance to variability of Loadings: Properly-designed wetlands have shown great tolerance for varying amount of wastewater loading. This is important because varying production levels again especially with mineral processing, changing climatic conditions, and modifications in management can alter loading rates significantly. They are more flexible and less susceptible to loading and able to tolerate fluctuations in flow.

Low Intensity Odor: Odor is a serious problem when handling and treating wastewater especially from mining and mineral processing and if the operation is located in close proximity to residential housing. Research, has shown that odors from wetlands are of very low intensity or are non-existent.

Aesthetically Pleasing: Depending upon design, location and type of vegetation, constructed wetland can enhance the landscape with color, texture, and variety in plant materials. In addition they provide green space and educational areas.

Provide Wildlife Habitat: Wetlands attract some types of wildlife and can add to the usefulness and attractiveness of the area. Passive recreation such as bird watching and photography.

Facilitation of water reuse and recycling:

Environmentally Sensitive: Wetlands are viewed with the favour by the public. They can be built to fit harmoniously into the landscape.

Reliability: The technology offers the added advantage of operational reliability while requiring minimal operational control.

LIMITATIONS OF CONSTRUCTED WETLANDS

Even with the best design, there are some limitations associated with the use of constructed wetlands.

They Generally Require Larger Land Areas : Depending upon the influent load than do conventional (Active) treatment systems. Wetland treatment system may be economical relative to other options only where land is available and affordable.

Requires a Continuous Supply of Water: One of the most important factors of the success of a constructed wetland is the supply of water. In addition to the acid mine drainage to be treated, supplemental water will have to be added if the acid mine water supply is not sufficient to sustain plant population during dry periods. They require a minimum amount of water it they are to survive.

Undesirable land features may make it expensive to construct: Desirable topography and other natural factors such as soil type can make a constructed wetland inexpensive to build; undesirable land features increase construction costs. Changing the lay of the land, adding soil amendments, and/or liners, incorporating pumps, etc, increases the cost of building a wetland significantly.

Weather Limitations: A wetland performance is affected by seasonal weather conditions, which may reduce treatment reliability Seasonal weather conditions, such as cold and drought, reduce the efficacy of the system. Consider fluctuations in weather when designing and operating the wetland. Due to warmer climatic conditions in India this parameter is not going to affect the constructed wetland treatment technology.

Distraction by unusual Overload of Solids: A wetland can fill in with solids or flow pattern can be seriously disrupted when the design does not allow for the removal of solids before the acid mine water enters the constructed wetland.

LONG TERM PERFORMANCE

After almost about two decades, by the growing number of wetlands built around the world for AMD treatment, this technology seems to have firmly established roots. Long-term performance of wetlands has been reported by number of workers. Benning and Otte (1996, 1997) reported about a natural wetland outside of an abandoned lead-zinc mine in Ireland that has apparently been functioning unattended for over 120 years.

The performance of wetland treatment system depends upon the input water quality and the specific mechanism of metal removal processes. If the primary removal mechanism is anaerobic systems, it works efficiently as long as sulphate-reducing bacteria reduce sulphate to sulphide thus precipitating metals as metal sulphides and generating alkalinity. Sulphate reduction would be more likely to offer long-term treatment (Hedin et al., 1994). The problematic mine drainage generally has elevated sulphate concentrations but the limiting reactant is amount of available carbon for sulphate reducing bacteria. For each mole of sulphate that is reduced, two moles of carbon are required. The reaction can be represented as follows (Eqs. 1 and 2):

$$SO_4 + 2CH_2O = H_2S + 2HCO_3 \quad (1)$$

$$H_2S + M^{2+} = MS + 2H^+ \quad (2)$$

Where CH_2O represents a small chain organic compound. Typical average values for sulphate reduction rates range from 200 to 600 millimole/m^3/d in wetlands, but an average rate of 300 millimole/ m^3/d has been recommended for the design of sulphate reduction system (Wildeman et al., 1994; Eger and Wagner, 2001, 2003). Carbon production in natural wetlands has been reported to be about 1kg/ m^2/y (Whittaker and Likens, 1973). For one meter thick substrate this would provide 23 millimole carbon /m^3/d. Since two moles of carbon are required to reduce each mole of sulphate, the reaction rate based on the annual input of carbon would be only about 12 millimole/m^3/d or a factor of 25 less than the recommended typical design rate of 300 millimole/m^3/d. Using this average sulphate reduction rate, and assuming all carbon in the substrate available, the calculated lifetime have been reported to be on the order of at least 20-30 years. But to maintain the acceptable treatment either additional organic material must be added or the total metal and acid load to the system would need to be reduced substantially. In several wetland systems a supplemental carbon source was introduced to increase the rate of sulphate reduction. New material can be added when the removal capacity of the mixture is exhausted (Eger et al., 1994; Benner et al., 2000; Watzlaf et al., 2000).

The volume available to collect and store chemical precipitates or the number of available removal sites governs removal rate in aerobic systems. The primary removal processes in these systems include sedimentation, adsorption, ion exchange, complexation, which are finite; removal will cease unless new removal sites are generated. Although precipitated solids can be periodically removed from wetlands, only systems that are well-vegetated can generate sufficient new removal sites to offer the possibility of long term, low maintenance treatment. (Eger and Wagner, 2003). At a site in Ohio, metal sludge is accumulating at a rate of about 3-4 cm per year (Stark, 1992).Measurements at a number of sites in Pennsylvania show sludge accumulating at 2-3 cm per year (Hellier and Hedin, 1992). These data suggest that dikes that provide one meter of freeboard should provide sufficient volume for 25 to 50 years of treatment. It has been found that water quality at some surface mines improved within a decade after regarding and reclamation were completed. At these sites, 25 to 50 years of treatment may be enough to mitigate the contaminant problem. At surface mine sites with continual contaminant production or at systems constructed to treat drainage from underground mines or coal refuse disposal areas, systems can either be built with greater freeboard or rebuilt when they fill up. Site conditions will determine whether it is more economical to simply bury the wetland system in place and construct a new one, or to excavate away the accumulated solids for proper disposal. Disposal of this excavated sludge is not difficult or unduly expensive because the material is not considered a hazardous waste (Hedin et al., 1994; Stark et al., 1994; Eger and Wagner, 2003). Since the generation of new removal sites is a slow process, long-term treatment will require a large wetland, with a relatively small input load.

CONCLUSIONS

Metal contamination of soils and waters reported around the world has severe impacts on environment and human health. Acid mine drainage is one of the most important sources of heavy metal environmental pollution. The target of paper is the development of methods facilitating the establishment of self-maintaining vegetation covers for the different conditions encountered at acid-generating waste management areas. These methodologies can be applied to promote vegetation covers and wetland communities, which would function as biofiltration systems. Thus an effective mechanism for treating acid mine drainage from mine sties is to filter it through a biological polishing pond before release. The water flows through a swamp of marsh. The plants absorb any heavy metals in the water. With the unwanted material thus removed, clean water can be released into the environment. This method is virtually self-sustaining and can operate unattended for many years. However, the efficiency of constructed wetland is limited in northern countries such as Canada because harsh winters prevent the use of engineered wetland during winter months. But in countries like India the efficiency is activated because of the favourable climatic conditions.

Wetlands help to treat this acid mine water loaded with heavy metals and prevents the spread of metal contamination in surface and subsurface water. High metal removal rates about 100% have been reported in both natural and artificially constructed wetlands. Understanding the basic mechanisms and processes controlling the metal removal increases the probability of success of the treatment wetland application. Heavy metals removal from acid mine drainage in wetlands is removed by physical, chemical and various biological processes which involves sedimentation, settling, filtration, adsorption, precipitation, co-precipitation into insoluble compounds. Further more, a working knowledge of the entire mechanism from substratum to the plants (living and non-living components of the ecosystem) can provide valuable insight into the overall wetland function and structure. This level of understanding is surely going to be very useful for evaluating the metal removal performance of the constructed as well as natural wetlands and for assessing the functional integrity of human impacted, restored and mitigations wetlands.

The use of constructed wetlands to improve water quality is a developing technology. Much has yet to be understood to improve the design of wetland systems for their better adoptability, specially regarding longevity.

References

Akcil, A.; and Koldas, S.; 2006: Acid mine drainage (AMD): causes, treatment and case studies. Jour. Clean. Prod. Vol.14 (12-13), pp.1139-1145.

Aswathanarayana, U.; 2003: Mineral Resources Management and the Environment. A.A. Balkema, Tokyo. 294 p.

Banks, D.; Younger, P. L.; Arnesen, R.T.; Iversen, E. R.; and Banks, S. B.; 1997: Mine-water chemistry: the good, the bad and the ugly. Environ.Geol. Vol.32 (3), pp.157-174.

Beining, B.; and Otte, M.L.; 1996: Retention of metals from an abandoned lead-zinc mine by a wetland at Glendalough, Co. Wicklow, Ireland. Biol. Environ. Vol. 96, pp.117 - 126.

Beining, B.A.; and Otte, M.L.; 1997: Retention of metals and longevity of a wetland receiving mine leachate. In: Brandt J.E., Galevotic J.R., Kost L. and Trouart J. (eds). Proc. 14th Annual National Meeting - Vision 2000: An Environmental Commitment. American Society for Surface Mining and Reclamation, Austin, Texas, pp.43 - 46.

Benner, S.; Blowes, D.; and Ptacek, C.; 2000: Long Term Performance of the Nickel Rim Reactive Barrier: A Summary. In: Proceedings of V International Conference on Acid Rock Drainage, Vol. II, Denver, CO, pp.1221-1225.

Bozau, E.; and Strauch, G.; 2002: Hydrogeological basis for biotechnological remediation of the acidic mining lake 'Rl 111', Lusatia (Germany). Water, Air, Soil Poll.: Focus, Vol. 2(3), pp.15-25.

Brodie, G.A.; Hammer, D.A.; and Tomljanovich, D.A.; 1989a.: Treatment of acid drainage with a constructed wetland at the Tennessee Valley Authority 950 coal mine. Hammer, D.A. (ed). Constructed Wetlands for Waste Water Treatment. Lewis Publishers Inc., Chelsea, MI. pp.201-210.

Brodie, G.A.; Hammer, D.A.; and Tomljanovich, D.A.; 1989b: Constructed wetlands for treatment of an ash pond seepage. Hammer, D.A. (ed). Constructed Wetlands forWaste Water Treatment. Lewis Publishers Inc. Chelsea, MI. pp.211-220.

Brooks, R.P.; 1984: Optimal designs for restored wetlands. In Treatment of Mine Drainage by Wetlands. Contribution #264. Dept. of Biology, Pennsylvania State University. University Park, PA. pp. 19-29.

Cambridge, M.; 1995: Use of passive systems for the treatment and remediation of mine outflows and seepages, Minerals Industry International, pp. 35-42.

Champagne, P.; and Geel, P.; and Parker, W.; 2005: A bench-scale assessment of a combined passive system to reduce concentrations of metals and sulphate in acid mine drainage. Mine Water Environ. Vol.24(3), pp.124-133.

Clarke, L.; 1996: Coal mining and water quality. Jour. Mines Metals Fuels, Vol. 44, pp.181-183.

Cohen, R. R.H.; 2006: Use of microbes for cost reduction of metal removal from metals and mining industry waste streams. Jour. Clean. Prod. Vol.14 (12-13), pp.1146-1157.

Cole, S.; 1998: The emergence of treatment wetlands. Jour. Environ. Sci. Tech.Vol. 32(5), pp.218-223.

Collins, B.S.; Sharitz,R.R.; and Coughlin,D.P.; 2005: Elemental composition of native wetland plants in constructed mesocosm treatment wetlands. Bioreso. Tech.Vol. 96(8), pp.937-948.

Debusk,A.T.; Laughlin, R.B.; and Schwartz, L.N.; 1996: Retention and compartmentalization of lead and cadmium in wetland microcosms. Water Res. Vol. 30 (11), pp. 2707-2716.

DeBusk, T.A.; and DeBusk, W.F.; 2000: The use of wetlands for water treatment. In: Applied Wetlands Science and Technology, Kent, D.M. (Ed.),CRC Press, Boca Raton, FL. pp. 241-279.

Dinges, R.; 1982: Natural systems for water pollution control. Van Nostrand Reinhold Co, New York. 252 p.

Dunbabin, J.S.; and Bowmer, K.H.; 1992: Potential use of constructed wetlands for treatment of industrial waste waters containing metals. Sci. Total Enviro.,Vol. 3, pp. 151-168.

Eger, P.; 1994: Wetland treatment for trace metal removal from mine drainage: the importance of aerobic and anaerobic processes. Water Sci. Tech.,Vol. 29(4) pp. 249-256.

Eger, P.; and Lapakko, K.; 1988: Nickel and copper removal from mine drainage by a natural wetland. Mine Drainage and Surface Mine Reclamation, U.S. Bureau of Mines Information Circular 9183, Pittsburgh, PA.Vol. 1, pp. 301-309.

Eger, P.; and Wagner, J.; 2001: Sulfate reduction- designing systems for long term treatment, Transactions of the Society for Mining, Metallurgy and Exploration, Vol. 3(10), pp.118-124.

Eger, P.; and Wagner, J.; 2003: Wetland treatment systems-How long will they really work? Presented in Mining and the Environment III Conference Sudbury,Ontario,Canada, 14 p.

Eger, P.; Wagner,J.; Kassa, Z.; Melchert, G.; 1994: Metal removal in wetland treatment systems. In: Proceedings of the International Land Reclamation and Mine Drainage Conference/Third International Conference on the Abatement of Acidic Drainage, Pittsburgh, PA, pp. 80-88.

Eger, P.; 1994: Wetland treatment for trace metal removal from mine drainage: the importance of aerobic and anaerobic processes. Water Sci. Tech. Vol. 29(4), pp.249-256.

Ellis, J. B.; Shutes, R. B.; Revitt, D. M.; and Zhang, T. T.; 1994: Use of macrophytes for pollution treatment in urban wetlands. Reso. Conser. Recy.Vol. 11(1-4), pp.1-12.

Environment Protection Agency, 1999: Free water surface wetland for wastewater treatment pp.6-22.

Fennessy, S; and W.J. Mitsch; 1989: Design and use of wetlands for renovation of drainage from coal mines. In : Ecological Engineering: an introduction to ecotechnology. W.J. Mitsch and S.E. Jorgensen (Ed.), John Wiley and Sons, New York, NY. pp. 232-252.

Filipek, L.H.; Hatton, C.; Gusek, J.; and Tsukamoto,T.; 2003: Passive treatment of acid rock drainage(ARD): state of the practice. In: Proceedings of the Tenth International Conference on Tailings and Mine Waste, Colorado,USA pp.293-303

Fiset, J.F.; Zinck, J.M.; and Nkinamubanzi, P.C.; 2003: Chemical stabilization of metal hydroxide sludge.In: Proceedings of the X International Conference on Tailings and Mine Waste, Vail, CO, USA, AA Balkema, pp. 329-332.

Fyson, A.; Kalin, M.; and Adrian, L.W.; 1994: Arsenic and nickel removal by wetland sediments. In : Proceedings of the International Land Reclamation and Mine Drainage Conference and Third International Conference on the Abatement of Acidic Drainage. Pittsburgh, PA. pp. 109-118.

Girts, M.; and Kleinmann. R.; 1986: Construction of wetlands for treatment of mine water. Proc. of Society of Mining Engineers 7-10 Sept. St Louis, MO.

Grande, J. A.; Beltrán, R.; Sáinz, A.; Santos, J.C.; Torre, M. L.; and Borrego, J.; 2005: Acid mine drainage and acid rock drainage processes in the environment of Herrerías Mine (Iberian Pyrite Belt, Huelva-Spain) and impact on the Andevalo Dam. Enviro. Geol. Vol. 47(2), pp. 185-196.

Greenway, M.; and Simpson, J.S.; 1996: Artificial wetlands for wastewater treatment, water reuse and wildlife in Queens Land, Australia. Water Sci. Tech. Vol.33 (10-11), pp. 221-229.

Gusek, J.; 1995: Passive-treatment of acid rock drainage: What is the potential bottom line? Min. Eng. pp. 250-253.

Hallberg, K.B.; and Johnson, D. B.; 2005: Microbiology of a wetland ecosystem constructed to remediate mine drainage from a heavy metal mine. Sci. Total Environ. Vol. 338 (1-2), 53-66.

Hammer, D.A.; and Bastian, R.K.; 1989: Wetland ecosystems : natural water purifiers? Hammer, D.A. (Ed). Constructed Wetlands for Waste Water Treatment. Lewis Publishers Inc., Chelsea,MI. pp.5-19

Hedin, R.S.; and Nairn, R.W.; 1990: Sizing and performance of constructed wetlands: Case studies. Proceedings of the Mining and Reclamation Conference and Exhibition, Vol. II, pp. 385-392.

Hedin, R.S.; Nairn, R.W.; and Kleinmann, R.L.P.; 1994: The passive Treatment of Coal Mine Drainage. U.S. Bureau of Mines Information Circular 9389.35 p.

Hellier, W.W.; and Hedin, R.S.; 1992: The mechanism of iron removal from mine drainages by artificial wetlands at circumneutral pH.In: INTECOL'S IV International Wetlands Conference Abstracts, Columbus, OH, 13 p.

Hiel, M.T.; and Kerins, F.L.; 1988: The Tracy wetlands: A case studies of two passive mine drainage treatment systems in Montana. U.S. Bureau of Mines Information Circular 9183, Pittsburgh, PA. pp. 352-358.

Johnson, D. B.; and Hallberg, K.B.; 2005: Acid mine drainage remediation options: a review. Sci. Total Environ.Vol. 338(1-2), pp. 3-14.

Joost, R.E.; Olsen, F.J.; and Jones, J.H.; 1987: Revegetation and minespoil development of coal refuse amended with sewage sludge and limestone. Jour. Environ. Qual., Vol. 16, pp. 65-68.

Kadlec, R.H.; 1985: Aging phenomena in wastewater wetlands. IN Ecological Considerations in Wetlands Treatment of Municipal Waters. E.R. Kaynor, S. Pelczarski, and J. Benforado (Ed.). Van Nostrand Reinhold, NY. pp. 338-347.

Kalin, M.; 2001: Biogeochemical and ecological consideration in designing wetland treatment systems in post-mining landscapes. Waste Manag. Vol. 21, pp. 191-196.

Kalin, M.; 2004: Passive mine water treatment: the correct approach? Ecol. Eng.,Vol. 22(4-5), pp. 299-304.

Kalin, M.; and Chaves, W. L. C.; 2003: Acid reduction using microbiology: treating AMD effluent emerging from an abandoned mine portal. Hydromet. Vol.71(1-2), pp.217-225.

Kalin, M.; Fyson, A.; and Wheeler, W. N.; 2006: The chemistry of conventional and alternative treatment systems for the neutralization of acid mine drainage. Sci. Total Environ. Vol. 366(2-3), pp. 395-408.

Karathanasis, A. D.; and Johnson, C. M.; 2003: Metal Removal Potential by Three Aquatic Plants in an Acid Mine Drainage Wetland, Mine Water Environ. Vol.22(1), pp. 22 -30.

Kleinman, R.L.P.; 1990; Acid mine drainage, U.S. bureau of mines researches and develops: control methods for both coal and metal mines. Environ. Sci. Techn.Vol. 24(9), pp. 1278-1285.

Kleinmann R.L.P.; and Hedin, R.S.; 1993: Treating mine water using passive method. Jour. Pollu. Engi. Vol.25(13), pp. 20-22.

Kleinmann, R.L.; 1985: Treatment of acid mine water by wetlands. In : Control of Acid Mine Drainage. Proc. of a technology transfer seminar. Bureau of Mines circular #9027. US Dept. of Interior. pp. 48-52.

Kleinmann, R.L.P.; 1989: Biological treatment of mine and mineral processing wastewater. Proceedings of Biohydrometallurgy, Wyoming, CANMET SP89-10, pp. 593-598.

Kuyucak, N.; 2002: Acid mine drainage prevention and control options. CIM Bull. Vol.95(1060), pp. 96-102.

Maiti, S.K.;2001: Handbook of methods in environmental studies Vol. 1. ABD Publishers, Jaipur, 307 p.

Masferrer, J.A.R.; 2002: Passive treatment of acid mine drainage at the La Extranjera Mine (Puertollano, Spain). Mine Water Environ. Vol.21(3), pp. 111-113.

Matagi, S.V.; Swai, D.; and Mugabe, R.; 1998: A review of heavy metal removal mechanisms in wetlands. Afri. Jour. for Trop. Hydrobiol. Fish. Vol. 8, pp. 23-35.

Mays, P. A.; and Edwards, G. S.; 2001: Comparison of heavy metal accumulation in a natural wetland and constructed wetlands receiving acid mine drainage. Ecol.Eng.Vol.16 (4), 487-500.

McCarthy, D.;. Bullock, R.; and Pantano, J.; 1999: Evaluation of wetland treatment for the removal of metals from mining impacted groundwater. In: Proceedings of the IX International Conference on Tailings and Mine Waste, Balkema, Rotterdam, Netherlands. pp. 725-733.

McHerron, L.E.; 1985: The seasonal effectiveness on a Sphagnum wetland in removing iron and manganese from mine drainage. In : Wetlands and Water Management on Mined Lands: Proc. of a Conf. 23-24 Oct. 1985 The Pennsylvania State University. University Park, PA. pp. 365-372.

Modis, K. ; Adam, K. ; Panagopoulos, K. ; and Komtopoulos, A. ; 1998 : Development and Validation of a geostatistical model for prediction of acid mine drainage in underground sulphide mines. J. Trans. Instn. Min. Metall(Sect A: Min. industry) pp. A102-A107.

Nicomrat, D.; Dick, W.A.; and Tuovinen, O. H.; 2006: Assessment of the microbial community in a constructed wetland that receives acid coal mine drainage. Micro. Ecol. Vol. 51(1), pp. 83-89.

Nordstrom, D.K.; and Alpers, C.N.; 1999: Negative pH, efflorescent mineralogy, and consequences for environmental restoration at the Iron Mountain Superfund Site, California. Proc Natl Acad Sci, Vol. 96, pp. 3455-3462.

Perry, A.; and Kleinmann, R.L.P.; 1991: The use of constructed wetlands in the treatment of acid mine drainage. Nat. Reso. Forum ,Vol.15, pp. 178-184.

Raju K.S.; Hasan M.; and Kale Y.G.; 2003: Status of metalliferrous mining in India: An overview and future outlook. 19th World mining Congress, Mining in the 21st Century Quo Vadis?, New Delhi, Nov. 1-5, pp. 403-414.

Santos, S.; Machado, R.; Correia, M. J. N.; and Carvalho, J.R.; 2004: Treatment of acid mining waters. Miner. Eng.Vol. 17(2), pp. 225-232.

Sheoran, A.S.; 2004: Treatment of Acid Mine Drainage by Constructed Wetland: An Ecological Engineering Approach. Ph.D. Thesis, Jai Narain Vyas University, Jodhpur, India. (Unpublished).

Sheoran, A.S.; and Bhandari, S.; 2005: Treatment of mine water by a microbial mat: bench scale experiments. Mine Water Enviro., Vol. 24, pp. 38-42.

Sheoran, A.S.; 2006: Performance of Three Aquatic Plant Species in Bench-scale Acid Mine Drainage Wetland Test Cells. Mine Water Environ. Vol. 25(1), pp. 23-36.

Sheoran, A.S., and Sheoran, V.; 2006: Heavy Metal Removal Mechanism of Acid Mine Drainage in Wetlands: A Critical Review. Miner. Engi. Vol.19, pp.105-116.

Sheoran, A.S.; 2006: A Laboratory Treatment Study of Acid Mine Water of Wetlands with Emergent Macrophyte (Typha angustata). Int. Jour. Surf. Min. Recl. Environ. Vol.20 (3) pp. 209-222.

Sheoran, A.S.; 2007: Acid Mine Drainage Treatment on Bench Scale Wetland with Saccharum Bengalense. Miner. Proce. and Extra. Metall. (Trans. Aus. Inst. Min. Metal. C) Vol.116 (1) pp. 48-53.

Sheoran, A.S.; Management of Acidic Mine Waste Water by Constructed Wetland Treatment Systems: A Bench Scale Study. Europ. Jour. Miner. Proce. and Environ. Prot. (in press).

Snyder, C,D.; and Aharrah, E.C.; 1985: The Typha community: a positive influence on mine drainage and mine restoration. IN Wetlands and Water Management on Mined Lands: Proc. of a Conf. 23-24 Oct. 1985 The Pennsylvania State University. University Park, PA. pp. 187-188.

Stark, L.R.; 1992: Assessing the longevity of a constructed wetland receiving coal mine drainage in eastern Ohio. INTECOL'S IV International Wetlands Conference Abstracts,Columbus, OH, pp. 13.

Stark, L.R.; Kolbash, R.L.; Webster, H.J.; Stevens, S.E.; Dionis, K.A.; and Murphy, E.R..; 1988:The Simco #4 wetland: Biological patterns and performance of a wetland receiving mine drainage. U.S. Bureau of Mines Information Circular 9183, Pittsburgh, PA. pp. 332-344.

Stark, L.R.; Williams, F.M.; Stevens, S.E.; and Eddy, D.P.; 1994. Iron retention and vegetative cover at the Simco constructed wetland: An appraisal through year 8 of operation. In: Proceedings of III International Conference on The Abatement of Acidic Drainage, Bureau of Mines SP 06A-94 Publication, Pittsburgh, PA, pp. 89-98.

Stillings, L.L.; Gryta, J.J.; and Ronning, T.A.; 1988: Iron and manganese removal in a Typha -dominated wetland during ten months following its construction. Mine Drainage and Surface Mine Reclamation, Vol. 1. U.S. Bureau of Mines Information Circular 9183, Pittsburgh, PA. pp. 317-325.

Tarutis Jr, W.J.; Stark L. R.; and Williams, F. M.; 1999: Sizing and performance estimation of coal mine drainage wetlands. Ecol. Engi. Vol.12(3-4), pp.353-372.

Topper, K.F.; and Sabey. B.R.; 1986: Sewage sludge as a coal mine spoil amendment for vegetation in Colorado. Jour. Environ. Qual.Vol. 15, pp. 44-49.

Watson, J.T.; Reed, S.C.; Kadlec, R.H.; Knight, R.L.; and Whitehouse, A.E.; 1989: Performance expectations and loading rates for constructed wetlands. Constructed wetlands for wastewater treatment: municipal, industrial and agricultural, (Ed.) D. A., Hammer.Lewis, pp. 319-352.

Watzlaf, G.; Schroeder, K.; and Kairies,C.; 2000: Long-term performance of alkalinity- producing passive systems for the treatment of mine drainage. Proceedings of XVII Annual Meeting, American Society for Surface Mining and Reclamation, Tampa, FL, pp. 262-274.

Whitehead, P.G.; and Prior, H.; 2005: Bioremediation of Acid Mine Drainage: An Introduction To The Wheal Jane Wetlands Project. Sci. Total Envir. Vol.338(1-2), pp.15-21.

Whittaker,R.H.; and Likens,G.E.; 1973: Carbon in the biota. In: Carbon and the Biosphere. Woodwell, G., Pecan, E., (Eds) CONF 720510. National Technical Information Service, Washington, D.C. pp. 281-302.

Wieder, R. K.; 1993: Ion input/output budgets for five wetlands constructed for acid coalmine drainage treatment. Water, Air, Soil Pollu. Vol.71(3-4), pp. 231 270.

Wieder, R.K.; and Land, G.E.; 1982: Modification of the acid mine drainage in a freshwater wetland. Proceeding of the symposium on wetlands of the Unglaciated Appalachian Region, ed. B.R. McDonald, Morgantown, USA. pp. 43-53.

Wieder, R.K.; and Lang, G.E.; 1984: Influence of wetlands and coal mining on stream water chemistry, Water, Air Pollu. Vol.23, pp. 381-396.

Wildeman, T.; Cevall, J.; Whiting, K.; Gusek, J.; and Scheuering, J.; 1994: Laboratory and pilot scale studies on the treatment of acid rock drainage at a closed gold-mining operationin California. In: Proceedings of the International Land Reclamation and Mine Drainage Conference and III International Conference on Abatement of Acidic Drainage, USBM Pub. SP 06A-94, pp. 379-386.

Wile, I.; Miller, G.; and Black. S.; 1985:Design and use of artificial wetlands. In : Ecological Considerations in

Wetlands Treatment of Municipal Waters. E.R. Kaynor, S. Pelczarski, and J. Benforado (Ed.). Van Nostrand Reinhold, NY. pp. 26-37.

Woulds, C.; and Ngwenya, B.T.; 2004: Geochemical processes governing the performance of a constructed wetland treating acid mine drainage, Central Scotland. Appli. Geochem. Vol. 19(11), pp.1773-1783.

Ye, Z.H.; Whiting, S.N.; Qian, J.H.; Lytle, C.M.; Lin, Z.Q.; and Terry, N.; 2001: Wetlands and aquatic processes, trace elements removal from coal ash leachate by a 10 year old constructed wetland. Jour. Environ. Qual. Vol.30, pp. 1710-1719.

Younger P.L.; Banwart S.A.; and Hedin, R.S.; 2002: Mine Water: Hydrology, Pollution, Remediation, Kluwer Academic, 464 p.

Zinck, J.M.; Wilson, L.J.; Chen, T.T.; Griffith, W.; Mikhail, S.; and Turcotte, A.M.; 1996: Characterization and stability of acid mine drainage treatment sludges, CANMET, Ottawa, Ontario, 374 p.

Hon'ble Reviewers

1. Prof. V.P.Singh
   (Former Head, ISM)
   K-227, Aashiyana Colony
   Kanpur Road
   LUCKNOW-226012, INDIA
2. Dr. Krishan Kumar
   School of Studies in Environmental Sciences
   Jawaherlal Nehru University
   NEW DELHI-110001, INDIA

Editorial Assistance

Dr. Virendra Gaur

Economic Mineralization
K.L. Shrivastava (Editor)
2009. pp. 526-532
Scientific Publishers (India), Jodhpur

48

Section-V

Land Degradation Environmental Impact of Clay Mining in Kolayat, Bikaner

SHISHIR SHARMA[1], D.K. KHANDELWAL[1], A.K. SHANDILYA[1] AND RAJNI SHARMA[2]

[1]Department of Geology, Dungar College, Bikaner University, Bikaner 334 003
[2]Department of Economics, Dungar College, Bikaner University, Bikaner 334 003
Email : arnunshandilyabikaner@yahoo.com

Abstract

Nature has gifted valuable resources and one must learn to exploit them judiciously without causing any imbalance. Clay mining operations and mineral based industries apart from causing alteration in land use usually cause land degradation. Among mining activities, surface mining produces dramatic changes in landscapes due to excavation. Huge voids are created due to the exploitation of clays and overburden. It alters even the topography of the surface, spoils the land and landscape rendering it worthless and unaesthetic. Other potential sources of land degradation due to mining operations include unplanned dumping of waste material from mining and industrial sites.

The land degradation is due to dump nuisance that are mainly caused by the presence of weak rock foundations, height of the dump, groundwater conditions and steep slopes. During rainy season seeping water from these dumps contaminate surface and sub surface water and reduces land's fertility. Dumping of waste causes diversion of streams, deforestation, disappearance of flora, and fauna and thus spoils the eco-system of the area. The paper mainly outlines the causes and effects of land degradation due to clay mining activities and incorporates some remedial measures to be taken into account for better eco friendly mining to ensure maximum resource recovery.

Key words : Land Degradation, Kolayat Clay Deposit, Clay Mining, Environmental Impact.

INTRODUCTION

The rich diversity of the planet wealth is at present in great danger of depletion due to various factors. Nature has provided us with certain resources and one must learn to exploit them intelligently without causing imbalance in nature. Management of natural resources in fragile desert environment is of critical relevance for food, folder, fuel and environmental security in this desert region, which enjoys a special status in Rajasthan because of its mineral wealth. The state of Rajasthan generates almost 50% of the country's wealth of minor minerals. Half of these are mined in the desert region. Clay is one of the most important ceramic non-metallic minerals abundant in Bikaner district.

MINERAL RESOURCES OF BIKANER DISTRICT

The northwestern region of Thar Desert Bikaner, the erstwhile princely state is mostly covered with sand. It has a rich wealth of non-metallic and energy minerals.

Some of the important minerals are Gypsum, Clays, Fuller's Earth, Red and yellow Ochre, Lignite, Bajri, Kankar and Sandstone etc. These resources are the backbone of economy and they are directly concerned with the environment of the area.

DISTRIBUTION OF CLAYS

Clay is the earthy hydrous aluminium silicate formed by the decomposition of preferably felspathic material.

The commercial value of clays depends mainly on their physical properties like plasticity, strength, shrinkage, vitrification range and refractoriness, fired colour, porosity and absorption.

The maximum Indian contribution of clay production comes from Bikaner where this mineral is concentrated in and around Kolayat tehsil of the district. There are four blocks of clay belt of Bikaner viz. Kolayat-Kotri-Gura block, Chandi block, Mudh block and Kotri block and at least 50 mining leases are in operation. The various clays namely ball clay China clay and fire clay are found in Bikaner district.

Ball Clay

It is a plastic sedimentary clay. The name derives from original method of production, which consisted of cutting out the clay in open pits into cubes or balls. The primary mineral is kaolinite. The main impurities are mica, quartz; chlorite etc. Ball clay has no definite chemical composition. It is used in the manufacture of sanitary ware, floor and wall tiles, electrical porcelain and semi vitreous white ware. In Bikaner district it occur at Kotri-Gurha, Chandi, Mudh, Nal, Khari-Charnan etc.

China Clay

It is an altered product derived by weathering or hydrothermal action from rocks rich in feldspars. It may be of primary, secondary or residual in origin. The colour of Kaolin (China Clay) in the green state may vary from white, pale cream to light gray depending on the nature and amount of impurities present in them. It is soft with a soapy touch and consists of small flakes of scales.

It is the basic ingredient in ceramic body composition for the manufacture of porcelain-glazed earthenware and stoneware. It is also used in paper, textile and rubber industries and limited quantity is used in manufacture of pharmaceuticals and insecticides.

Fire Clay

It consists mainly of the mineral Kaolinite with minor amounts of other clay minerals. Mineralogical studies reveal a disorder in the structure of Kaolinite present in fire clays. The term has been commonly applied to refractory clays associated with sedimentary formations. The fire clays used as refractories, are capable of withstanding temperatures around 1700°C. Fire clays have two modes of occurrence

In the Bikaner district fire clay occurs in the lignite bearing areas as an under clay horizon and over clay horizone. Important occurrences are Palana, Barsinghsar and Nal. Other minor occurrences are at Deh, Gurha etc.

Fuller's Earth

Fuller's earth is locally called Multani Mitti. It is a variety of clay, which has high absorptive capacity and rich in exchangeable cations. Mineralogically the fuller's earth is composed or smectite group of clay minerals dominatingly Montmorillonite. It is chiefly used for bleaching, deodourising, cleanifying vegetable, mineral oils and greases. It is also used as a filling and binding materials, and in cosmetics and pharmaceutical preparations. It is reported at Palana, Kesardesar and Mudh areas but it is mined out only at Mudh.

ENVIRONMENTAL IMPACT OF CLAY MINING

The impact of mining operations on environment and socio-economic status of the region commences with exploration activities, extended through extraction and processing of mineral and continue up to post closer of operation. The potential social and environmental issues associated with mining and mineral processing operation are highly significant as well as complex to manage.

Mining, as known, is a land based activity and thus has a great potential to change the landscape, damage the ground vegetation, flora and fauna. It could even disturb the groundwater table if adequate care is not taken. Thus, environmental care in mining has become a specialized area and a discipline by itself demanding well coordinated roles of mining, geology, environmental engineering, chemistry environmental biology and horticulture and forestry specializations.

In recent years environmental concerns have seriously threatened the very existence of mining industry. Environmental concerns have assumed a great deals of significance at the same time ecological awareness and sensitivity has thrown up a series of questions that has spawned a new commitment.

The key questions raised in the environmental analysis of the clay mining of the Bikaner region are includes an environmental damaging activity causing land degradation, destruction of wild life and other natural resources. It creates huge waste and requires space for their dumping and it reflects the number of environmental problems. Huge quantity of air and water pollutions also added to the environment during excavation of clay. The heavy machinery used during mining operations also creates noise pollution.

The environmental impact of clay mining could be classified into the following major categories :

1. Socio-economic impact
2. Land degradation
3. Effect on flora and fauna
4. Air pollution
5. Noise pollution and
6. Water pollution

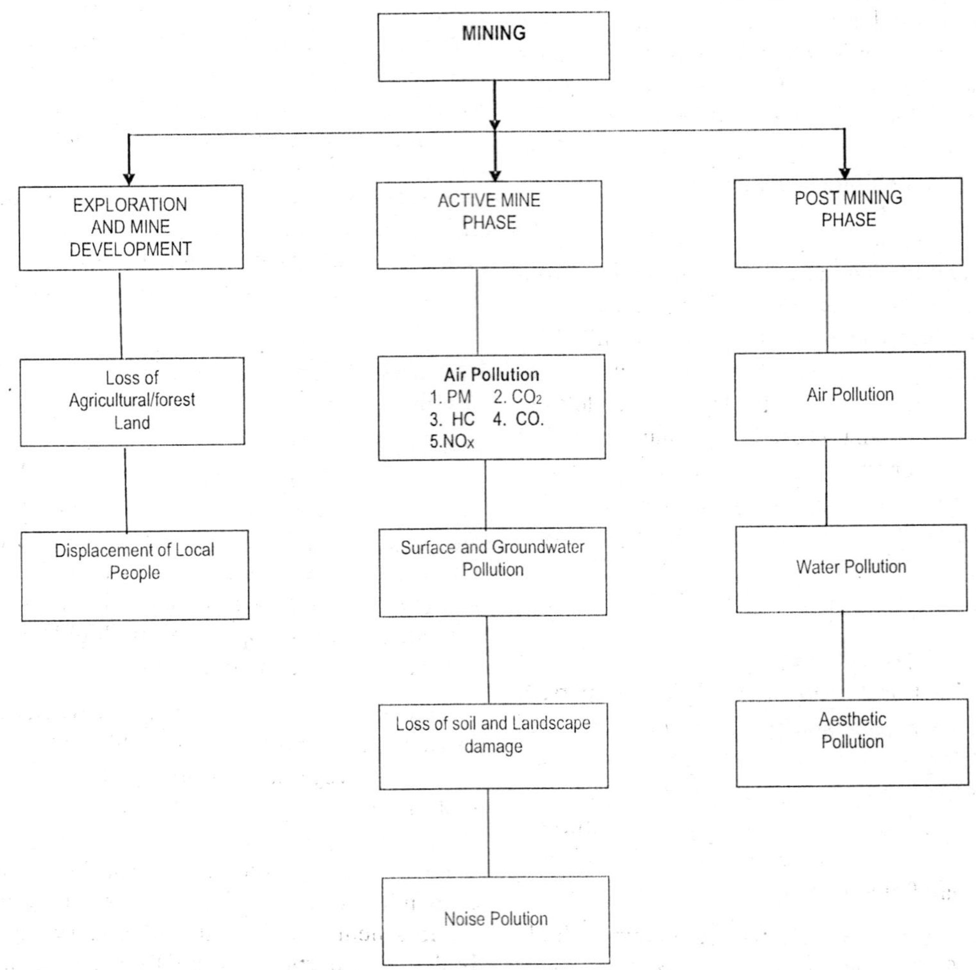

Fig.1 Schematic Environmental Discharge Model of Clay Mining Industries in Bikaner

Mining of clay is divided into three phases

* Pre-mining phase
* Active mining phase and
* Post mining phase

The sources of different component activities of mining environment discharge are listed in the Fig. 1.

Land Degradation due to clay mining

a. Mining wastelands : Lands where large-scale mining operations bring about the degradation of land and resultant mine dumps. Extensive and concentrated surface mining operations are probably one of the great offenders in creating land degradation. This paper is concentrated to the land degradation due to mining operations and some remedial and preventive measures. During the study it has been observed that the unstable nature of waste dumps is the most burning and prominent cause of Land Degradation. The stability of waste dumps is a burning issue because it threatens the working of the mine, degrades land and destroys the soil fertility.

b. Factors affecting stability of the waste dumps

Broadly, the factors that affect the dump stability can be classified as,

1. Dump Geometry
2. Geological and Geo-hydrological conditions
3. Geo-technical properties of the waste dump material
4. Mining equipment and
5. Blasting and other ground movement

c. Sources of land degradation

Land pollution is alteration and spoiling of land and landscape and rendering it worthless for any useful purpose including destruction of nature, flora and fauna etc. This also includes shifting of human habitat, change in the physical characteristics to the topography of the areas, change in the existing service facilities and change in the course of water body and also wind flow. Mining of clay by surface method disturbs the land from which it is mined. Unplanned dumping of solid wastes from mines, indiscriminate disposal of mineral based ceramic industries, unplanned construction of infrastructure facilities are also potential sources of land and landscape pollution due to clay mining based activities.

The damage to the environment within acceptable limits or sustainable development

It is reasonable and relevant to examine the causes of critical effects due to mining activity, not only day to day basis, but on the basis of future impacts also. Nevertheless, Eco-friendly and 'Sustainable Development' mining is one of the solutions for damage to be contained within acceptable limits. The 'Eco-friendly' and 'sustainable Development' Mining consists as following :

a. Scientific Mine Planning-Planning ensures the effective utilization of the deposit and also provides for phase-wise safe development of mining benches. Waste minimization, mineral conservation and judicious land use are the other benefits of efficient mine planning.

b. Pollution Control Measures-This includes installation and maintenance of dust extraction and dust suppression system in mining and processing operations.

c. Environmental Monitoring-Regular monitoring of environmental parameters, including that of performance of pollution control devices, is yet another feature of eco-friendly mining. This helps in EIA in general and statutory compliance in particular.

d. Optimization of Resources Utilization This technology optimizes energy and process water utilization. The three 'R' concepts, namely Recover, Reduce and Recycle Wastes, must also form the part of resources optimization.

IMPACTS OF LAND DEGRADATION

Large-scale surface mining operation has been an important focal point of interest in view of their implications on the significant alterations that they generate in and around a mining area. Briefly they are

1. Complete alteration of the topography of the earth's surface, leaving an area of more or less parallel ridges or of mounds of material, which overlay the excavated mineral deposit. Spreading of waste and overburden materials from industrial and mine plants over large area, make the land unusable and affects the natural harmony, thus contaminating the land.
2. Damage to agricultural land Surface mining of clay results into a constant decline of cultivable

areas, if the mine is situated in such area as it disturbs the land surface constituting different layers of the solid, soils which are in ecological balance with total environment.

3. Deforestation Surface mining of Kolayat destroys all existing vegetation and results into soil erosion, land slips etc. and also changes the existing eco-system some of the important plants may permanently become extinct.
4. Damage to landscape With the development of high technology and improved blasting; surface mining operations have been expanded to more prominent locations and to areas of more valuable land.
5. Appearance After mining, the area consisting of varying percentage of debris and soil materials, becomes devoid of all plants and animal life. These mined out areas remain bare and barren years after years.
6. Socio-economic impacts such as loss of farmlands and displacement of farmers along with loss of forest and forest products also occur due to land degradation.

Other impacts such as population increase in the area, health issues, cost of living and changes in community also occur.

SUGGESTIVE REMEDIES

I. Land Reclamations : Land reclamation in mining areas needs economic, social and environmental elements for consideration.

a. These reclamation areas can be used for bathhouse amenities, plant maintainance workshops, administrative building, ancillary facilities etc.

b. The dumped material can be used for construction of water management structures, site haul roads, dragline errection area etc.

c. It can also be used as the rehabilitation area.

It is suggested to reclaim the degraded land in such a manner to achieve stability and minimize pollution, to return it possibly to productive agricultural and recreational use. To remove all items of mining equipment and rubbish material, to return the top soil and reshape the land to revegetate.

II.Control measures at mine level : Various environmental protective measures are required to be taken at mine level against the land and landscape pollution in the different in different situations like

Measures in Abandoned Mines :

In abandoned mine areas the dumps are most settled and are available for aforestation by suitable species. However, dumps may require some terracing at the top and filling up the eroded portion along the slopes of dumps.

Large open pits could be put to economic usage in many ways. The runoff water may be stored in the open pits which can be used for agriculture, horticulture or afforestation activities.

After proper leveling and seepage prevention the pit may be used for pisciculture by storing Indian and exotic carps. It will not only be a regular source of income but also help in the stabilization process.

Measures in Active Mining Projects :

In this case the measures should be such that these do not either interfere or get disturbed by continued mining operations.

a. A long-term strategy to be adopted for sacking of dumps suitably terraced according to a plan of work.

b. Some pits may become available for filling back the rejects from the development sections, as the mining advances. This will not only minimize the wash offs from the dumps but will also reduce the requirement of space for dumping the waste and overburden.

c. The dead dumps in due course of time should be afforested by suitable vegetation.

Effective implementation of waste disposal system and reclamation of dumps through proper vegetation should be practiced.

Measures in New Mining Projects :

The new areas, which are likely to come up under mining in future, require :

a. Environmental consideration should be an integral part of advance planning as to prevent the rejects from contaminating water courses and silting of rivers, agricultural fields etc. as mining progresses by providing suitable techniques.

b. The proposed reclamation should be implemented and monitored simultaneously with mining operation, as far as possible.

c. The overburden dumped outside the pit should be reclaimed. It is to be decided whether aforestation of such dumps would be sufficient for the purpose or complete shifting of dumps or back filling will be necessary.

III. Methods of stablisation and prevention of waste dumps

It can be achieved by the combination of two methods i.e. debris stabilization and surface stabilization.

Debris stabilization :

Diversion ditches should be constructed around the dump to save the debris from the effect of running water. Equally important is the stacking of debris at a low angle. The debris can be covered with a layer of topsoil, applied with appropriate fertilizers and finally seeded with grasses and deep root trees.

While dumping a waste, a number of benches are formed to break the continuity of the slope. Terraces of approximately 7m widths are constructed at half the wrap-around interval. These terraces will provide access for revegetation as well as provide for drainage and erosion control. The terraces will also provide for wildlife migration.

Surface stabilization :

Surface stability of waste dumps is of paramount importance in controlling environment. Vegetation is the only method of achieving long-term stability. Other methods are.

a. Physical stabilization : Respreading topsoil, water spraying and mulching.

b. Chemical stabilization : Application of appropriate reagents to form and erosion resistant crest. Spreading of 5 cm of topsoil over the waste increases infiltration, reduces run-off and erosion and enhances plane survival and growth.

CONCLUSIONS

Without exception, no mining activity in the area is eco-friendly. The environmental degradation brought about by unscientific, unplanned and ruthless mining is common. There is no provision to deal with the overburden from mines, for instance, the site area approach and capacity of dumping sites in these mining areas are ill-conceived, improper and inadequate leading to a variety of problems in the area. There is absolutely no effort to utilize mine waste by back filling to help in reclamation, restoration and rehabilitation of the terrain without affecting the drainage and water regimes. No attempt has ever been made to use the rubble in road construction and mine pits for waste disposal. Instead, the overburden is left out on productive land.

The problem of land degradation and landscape pollution posed by clay mining activity could assume serious environmental threat in near future. Therefore, proper waste dumping plans in or outside the pits, the reclamation measures or retrieval of dumps, aforestation, and revegetation has to be integral with the project plan.

This is the right time to formulate land use plans to cover pre-operational, operational and post operational phases of mine. Poor or no land use planning, with no respect for desert flora and fauna, has eroded the productive top soil, which is essential for good agriculture and perennial grasses, herbs and shrubs on which a variety of animal taxa survive in nature. As a result, the eco-system is becoming barren and there is greater desertification and poorer biomass turnover. The above measures will definitely improve the environmental conditions. A multi disciplinary approach and involvement of related research organization is essential for effective and pragmatic land utilization in the clay mining industry of Bikaner region.

References

Atkinson, K. and Mitchell, P.B., 1994. The environmental impact of mining and some novel methods of reclamation, Proc. of the Int. Symp. The impact of mining on environment-Problems and solutions, pp. 305-310.

Bhattacharya, B.C., 1988. Bikaner Clays, Raj. Min. Bull., Vol. 12, No. 1, pp. 1-14.

Chaulya, S.K. singh, T.N. and Singh, J., 1993. Effect of mine waste disposal on environment and its protection, EUROCK 1993, Portugal, Balkana Ed, Netherland.

Geological Survey of India, 1980. Desert Tracts of Rajasthan. A Decade of Environmental Geo Scientific Studies by Geological Survey of India. Sp. Pub. No. 9, pp. 107-110.

Ghosh, R.N., 1983. Tertiary Clay Deposits of Kolayat and adjacent in Bikaner dist. Rajasthan. In Indian Minerals, Vol. 37, Plate 4, pp. 56 -68.

Ghosh, R.N., 1971. Geological mapping and traverses with special reference to the economic Mineral resources of the area around Kolayat, district Bikaner, Rajasthan (F.S. 1969 70). Unpublished Report. Geological Survey of India.

Gupta, R.K. and Prakash, I., 1975. Environmental Analysis of the Thar Desert. Eng. Book Depot. Dehra Dun, pp. 22 -201.

Keller, E.A., 1979. Environmental Geology. Charles E,

Merrill Pub. Company, Ohio, 454 p.

Khandelwal, N.K. and Mozumdar, B.K., 1987. Stability of overburden dumps, J. of Mines, Metals and Fuels, 11(6).

Pareek, H.S., 1984. Pre-Quarternary geology and mineral resources of northwestern Rajasthan, Geol. Surv. Ind. Mem. Vol. 115.

Radhakrishana, B.P. and Ramchandran, K.K., 1986. India's Environment Problems and Perspectives. Geol. Soc. Ind. Mem. No. 5.

Sharma, S. and Shandilya, A.K., 2001. Environmental impact of Bajri Mining around Bikaner City. Proc. of the National Seminar on Small Scale Mining, Deptt. Of Min. Engg., Faculty of Engg. JNV Univ. Jodhpur, pp. 110-115.

Singh, T.N. and Premkrishan, R., 2000. Stability of waste dumps and its environmental impacts, Proc. of the National Seminar on People and the Mining Industry The Future, Deptt. of Min. Engg., Faculty of Engg., JNV Univ. Jodhpur, pp. 141-155.

Soni, A.K. and Mohnot, S.M., 2001. Environmental concerns of small-scale mining. Proc. of the National Seminar on Small Scale Mining, Deptt. of Min. Engg., Faculty of Engg., JNV Univ. Jodhpur, pp. 93-94.

Wasteland Atlas of India, 2000. Deptt of rural Dev., Deptt. of Land Resources, Govt. of India and NRSA, Hydrabad Region, Mining of Water Resources GOI.

Hon'ble Reviewers

1. Prof. Madhumas Chandra Khare
   School of Studies in Earth Science
   Jiwaji University
   GWALIOR 474011 INDIA
2. Prof. D. Mohan
   Professor of Ecology
   Department of Zoology
   JNV University
   JODHPUR 342001 INDIA

Editorial Assistance

Mrs. Veena Chaudhary

Economic Mineralization
K.L. Shrivastava (Editor)
2009. pp. 533-540
Scientific Publishers (India), Jodhpur

A Remote Sensing Study on Environment Impact of Mining : A Case Study of Rampura-Agucha Mine

N.K.KALRA, T.S.SHARMA, ASHOK GEHLOT AND RAJEEV JAIN

Remote Sensing Application Centre, Department of Science and Technology,Pal Road, Jodhpur 342 001
Email : trilok_01@sify.com

Abstract

The mining of economic mineral resources is essential for the development and progress of the country. But unscientific and ruthless mining brings environmental problems. Remote sensing by virtue of synoptic view of large area, multi date and multi resolution capability in different wave length regions have proved to be of great assistance in providing latest information on landscape manifestations and environmental degradation over the years. Development and Planning of landscape always require accurate information on spatial distribution of various land use / land cover units. Conventional methods to collect such information have been tedious and time consuming.

Present study has been taken to assess land use / land cover details and respective change detection in and around Rampura Agucha Mine area in Bhilwara district of Rajasthan which is being worked by M/s Hindustan Zinc Limited, Udaipur. Indian Remote Sensing Satellite LISS -III and PAN sensors data in digital mode of Feb.,2003 have been used to assess spatial extent of various land use units. Based on mapping of 15 Km radius of mine site, statistics of various units have been generated. A comparison has also been made with earlier information of 1998 and accordingly some observations have also been made.

Key words : Remote Sensing, Agucha Mine, Mining Environment

INTRODUCTION

The main aim of the study has been to prepare land use / land cover thematic map and determine spatial distribution of various units in the study area and their changes over the period of time and (1998-2003) by using remote sensing data.

The study area covers 15 Km. radius around the Rampura Agucha Mines, District Bhilwara, Rajasthan covering geographically 25°42'26" to 25°58'40" North latitudes and 74°27'35" to 74°45'28" East longitudes. The area falls in 45-K/9 and 13 Survey of India toposheets.

Physiographically the area is almost plain with few hillocks in north portion with moderate deep soil. Two important river namely Khari near Gulabpura and Hurda town flowing west to south east and Mansi, flowing south to north east near village Rupaheli and Agucha, join each other near village Phuliya which is about 20 Km. from the site. Both rivers are seasonal in nature. Tanks have been constructed in several villages of the study area, the water from which is utilized for irrigation as well as for domestic use. The excess of soil moisture is used for in-situ cultivation in rabi season to harvest Gram crop. Among these tanks the Arwar dam is bigger one which is built on Mansi river. The dug wells have moderate to poor yield due to which most of the agriculture depends on rain fall. The climate of the area is moderate type with hot dry summer and cold winter season. A meterological station has been set up at environmental deptt. in HZL Campus it self at Agucha.The normal average annual rainfall of the area 400 mm.. The heaviest rain recorded is 806 mm in the year 1990 and lowest is 153 mm in the year 2002.

Geologically in general the study area is underlined by Pre-cambrian rocks which consist mainly banded gneissic complex. This rock unit is light gray in color and medium to fine grained. The present average depth of the mine pit is about 90 meters. At this depth the under ground water seepage problem has not been observed although the Agucha village pond is an up stream side.

15 Km. radius from mine area was selected for the eco-environment study of the area for its land use land cover change and environmental change detection of the area over a period from 1998 to 2003. The data used for the study include IRS-1D Satellite data (LISS-III + PAN) merged product dated 6th Feb.,2003, earlier report of study area, 1998 and other collateral data. (Fig.1)

Administratively, about 15 % of the part of the area comes under Ajmer district and remaining 85 % of the area comes under Hurda and Shahpura tehsil of Bhilwara district. The study reveals that area under forest which falls totally under scrub type forest or dry tropical forest covered randomly by small bushes, and grasses only. This area is fenced with Jouliflora and Kair plants.

METHODOLOGY ADOPTED

For land use analysis the methodology is based on image analysis of satellite data through image analysis software adopted with a 2,3 and 4 band combination of standard False color composite, generated on 1:25,000 scale for identification of different land use / land cover classes based on the image characteristics like tone, size, shape, pattern, texture, location, association, etc. A broad image interpretation key have been developed based on such characteristics to enable interpretation of satellite image for land use / land cover analysis.

LANDUSE/LANDCOVER CLASSES

Land use refers to man's activities and the various use which are carried on land. Land cover refers to natural vegetation water bodies, river course, rocks, Hillocks and soil etc.

Built-up land

It is defined as an area of human habitation developed due to non-agricultural use and that which has a cover of buildings ,transport network. and other places of public utility.

Agricultural land

Land which is primarily used for farming and for production of various crops. It also includes land under crop (Irrigated and un-irrigated, fallow and plantation, etc.)

Crop Land

It includes those lands with standing crop as on the date of the satellite imagery. The crop may be of either kharif (July to Sept., it coincides with south-west monsoon season and is associated with rain fed crops) or Rabi (It includes standing crops during season October to March and associated with area under good irrigation or kharif + Rabi seasons (it includes standing crops during both kharif and rabi seasons), such areas are associated with double crop.

Fallow Land

It is described as agricultural land which is taken up for cultivation but is temporarily allowed to rest and is uncropped for one or more seasons, but less than one year.

Plantation

It is the area under agricultural/ Horticulture tree crop plants, comprising tea, coffee. rubber. citrus, orchards and other horticulture nurseries. Generally plantation are well separated from crop lands during rabi season by their relatively low reflectance response, particularly in red and infra-red regions.

Forest

It is an area within the notified forest boundaries having various trees and grasses where the vegetative density is less than 20 % of the canopy cover. It is the result of both biotic and abiotic influence.

Waste Lands

It is described as, degraded land which loses the land productivity qualitatively or quantitatively, through a variety of process, and can be brought under some use after reclamation. and is currently under -utilized and lands. It is deteriorating due to lack of appropriate water and soil management.

Saline/Alkaline Land

Land that has adverse effects on the growth of most plants due to the action or presence of excess soluble or high exchangeable sodium salt. The predominant salts are carbonates and bicarbonates of sodium.

Land with/without scrub

These occupy at relatively higher topography like uplands or high grounds with or without scrub. These lands are generally prone to deterioration or erosion. These exclude hilly and mountainous terrain.

Grass Land

These are areas covered with natural grass along with other vegetation, often grown for fodder to feed cattle and other animals, some of such land may be wire fenced.

Barren rocky/stony Land

This area may be defined as the rock exposures of varying lithology often barren and devoid of soil cover and vegetation.

Discussion

After the generation and analysis of the satellite data for the various land use / land cover classes, the statistics have been generated for the different classes using GIS Software. (Fig. 2 and Table-1)

LAND USE/LAND COVER CHANGE ANALYSIS

Mapping of various landscape units have been done using remote sensing inputs as per details mentioned above. Their aerial extent have been calculated. A comparison of present land use (2003) and of earlier available (1998) has also been shown in Table-2. Salient observations are :-

Fig.1 Rampura-Agucha Mine area as viewed through IRSID (LISS III + PAN) satellite data

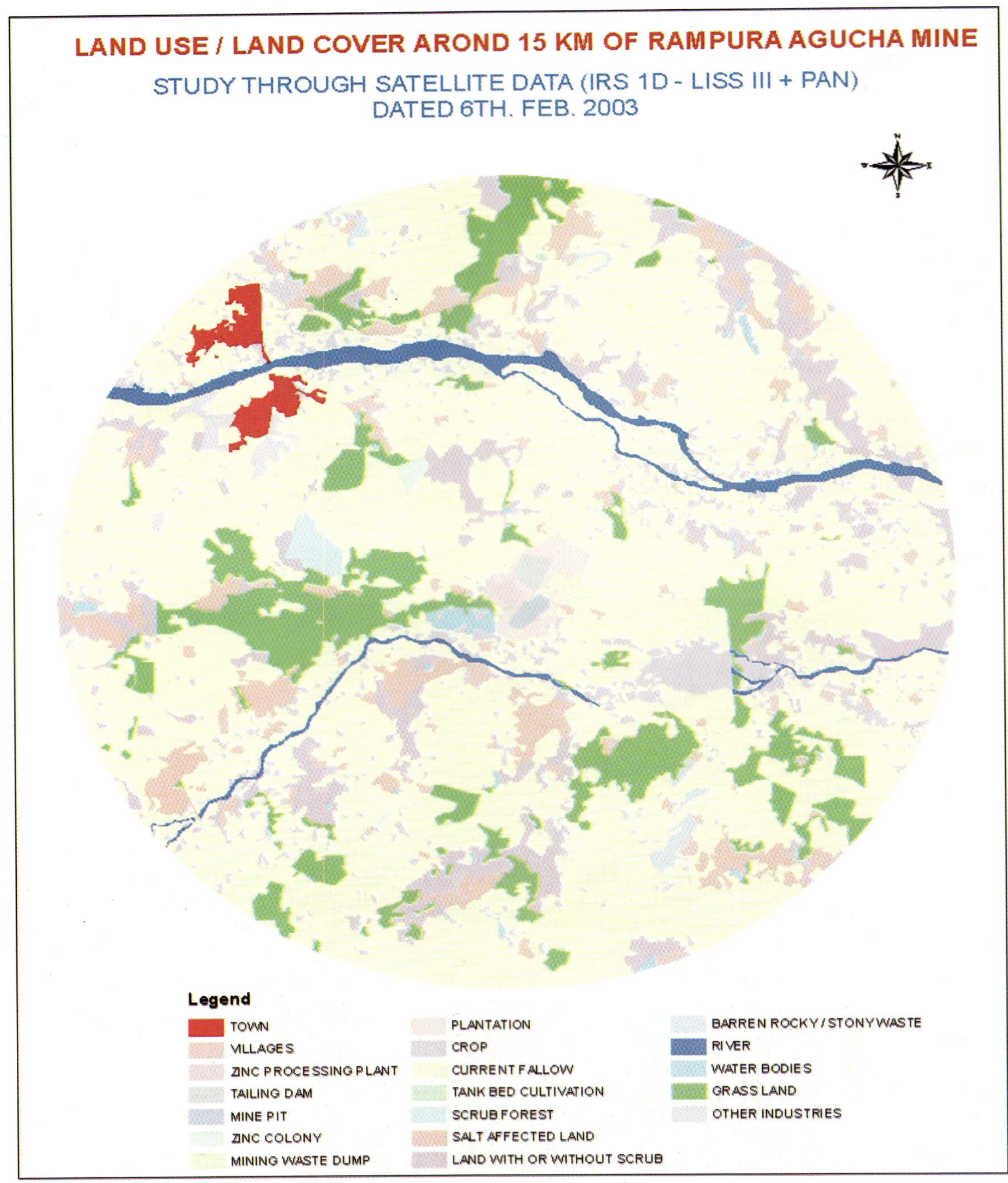

Fig.2 Landuse and Landcover Map

Fig.3 Plantation near mine pit

Fig.4 Plantation near waste dump

Fig.5 Plantation near tailing dam of Rampura Agucha mine

1. Built up land in 15 Km. Radius :

The study reveals an increase of 125 ha. in built up area as compared to total built up area 502 ha. in 1998 which indicates expansion of residential area 24 % at Gulabpura and Bijaiinagar town. Similarly over all increase in rural habitation area in villages, in the 15 Km. radius, is 1264 ha. which is 39 % indicating rise of population.

(a) Built up land in Rampura-Agucha Mine premises :

Various activities in Rampura Agucha Mine Campus take place for different purposes among these are- Zinc processing plant and other administrative block ,Tailing Dam, Mine pit, Mining waste dump, plantation in premises.Processing plant and other building occupy about 101.00 ha. Area.

(b) Tailing Dam :

It is space used for discharge of slurry mixed water. It is 18 meter high embankment area, surrounded wall is made up of stones, inner side residual solid wastes are collected along the walls such that inner lining is thick and solidified with pace of time. The circumference of it has also increased. Waste water is recycled for using in mining operation. As compared to 1998 occupancy, there has been increase of 9 ha. (Fig.5)

(c) Mine Pit :

The Mineral excavation area has observed an increase in aerial extent from 66 ha. in 1998 to 89 ha., an increase of 35 % indicating increase in mining operations. (Fig.3)

(d) Waste dump :

It is being accumulated with in premises in south east direction of Plant at a distance of about 0.5 Km. There is increase in spatial extent of dump to the extent of 37 ha., an increase of 52 % in between 1998 to 2003. However, the dumping ground are being treated with plantation of *Acacia Juliflora* and local species.(Fig.4)

(e) Industrial Plantation :

The dumping area is being intensively planted having more than 60 % plantation a dense growth, mostly acacia species . In other area plants are also being planted. There is increase of 45 ha.under plantation which shows about 23 % rise. The Agriculture in the area is mostly rainfed, tankfed and occasionally Irrigated. There is no remarkable change in overall agriculture area from 1998 to 2003. However at the time of satellite imagery taken high proportion of the agriculture area was fallow probably due to low rainfall in the last three years.

2. Scrub Forest :

In 15 Km. radius area a large plot of scrub forest exist under Panchayat ownership. The vegetation is of

Table-1 LAND USE / LAND COVER AREA STATISTICS OF RAMPURA AGUCHA MINE IN 15 Km. RADIUS

S.No.	Land use/ Land cover category	Area in Ha.	% of total area
1	**BUILT UP LAND**		
	Town & area under urbanization	627.00	0.88
	Villages	1264.04	1.79
	TOTAL	**1891.04**	**2.67**
	RELATED TO RAMPURA MINE		
	Zinc processing plant & Adm. Building	101.02	0.14
	Tailing Dam	109.15	0.15
	Mine pit	89.10	0.12
	Waste Dump	108.49	0.15
	Industrial Plantation	241.22	0.34
	Future proposed mining waste dump site	220.23	0.31
	TOTAL	**869.21**	**1.21**
2	**AGRICULTURE LAND**		
	Crop (Winter season Crop)	3459.16	04.90
	Current fallow land	46230.63	65.47
	Plantation	627.49	00.88
	Tank bed cultivation	33.56	00.05
	TOTAL	**50350.84**	**71.30**
3	**FOREST**		
	Scrub forest	89.50	0.12
	TOTAL	**89.50**	**0.12**
4	**WASTELAND**		
	Saline/Alkaline land	2945.97	4.17
	Land with or without scrub	5174.85	7.32
	Grass land	6784.00	9.60
	Barren rocky/Stony waste	295.69	0.42
	TOTAL	**15200.51**	**21.51**
5	**WATER BODIES**		
	River / Stream	1615.69	2.28
	Reservoirs/Pond/ Tank	464.68	0.65
	TOTAL	**2080.37**	**2.94**
6	**OTHERS**		
	Other industries in the area	123.18	0.17
	TOTAL	**123.18**	**0.17**
	GRAND TOTAL	**70604.65**	**99.92**

moderate density (10 to 40 %) but significantly no changed is observed, showing no adverse impact of mining on the natural vegetation in adjoining areas.

3. Wasteland :

Wastelands are mostly eroded, saline /alkaline affected land and barren rocky. There has been an increase of about 15 % in vegetation cover which indicate decrease in wasteland area.

4. Water bodies :

River streams are seasonal but aerial extent has slightly increased. Ponds and tank have witnessed decrease of submergence area to the extent of 72 % mainly due to severe drought conditions and failure of monsoon in last three years.

5. Other industrial growth :

In 15 Km.radius area there has been establishment of other industries also like Mayur textile, Bhilwara etc. covering about 36 ha. increase. Showing industrial growth of the area.

CONCLUSION

From the study of satellite data of Rampura Agucha mine area following conclusions can be drawn:--

1. There is considerable increase in habilitation part area due to new and extension of existing habilitation in 15 Km. radius.
2. Cropping pattern and natural vegetation of the area is in healthy condition but due to scarcity of ground water and rainfall pattern only kharif crop is prominent amongst all three crops rabi, kharif and zaid.
3. In some areas saline land is also present because of natural depression and saline water.
4. Tailing dam system is well managed at the mine area for discharge of slurry mixed water.
5. Waste dump of the mine has increase simultaneously dense vegetation cover has also increased.
6. Good plantation has developed around the mine area.
7. Looking to increase in vegetation cover from 1998 to 2003, there is no adverse affect on natural vegetation and environment due to mining activities in adjoining villages and in the mining area itself.

ACNOWLEDGEMENT

The author are grateful to Director, Department of Science and Technology, Govt. of Rajasthan for administrative support and General Manager, Hindustan Zinc Ltd. for sponsoring the project.

Hon'ble Reviewers

1. Prof. Damodar Sharma
(Professor of Remote Sensing)
Vice -Chancellor
Rajasthan Technical University
Rawatbhata Road
KOTA 324010 RAJASTHAN

2. Dr. S.S. Rathore
Head
Department of Mining Engineering
MPA Technical University
UDAIPUR 313 001 RAJASTHAN

Editorial Assistance

Ms. Tina Prajapat

Economic Mineralization
K.L. Shrivastava (Editor)
2009. pp. 541-545
Scientific Publishers (India), Jodhpur

Demand and Supply for Geoscientists in India for the 21st Century

MIKE KATZ
KCM International School of Mining Engineering
University of New South Wales Sydney, NSW, 2052 AUSTRALIA

Abstract

The demand and supply of geoscientists in countries like India has increased due to the recent unprecedented rapid economic and mineral development. Mineral and energy resources contribute significantly to the economies of many countries including India. This mining development has also been stimulated due to the establishment of favorable for investment mining laws and the accompanying needs to strengthen government geological surveys and mine departments. The growth of these economies has led to overall expansion of industries and infrastructure that also require increasing geotechnical and environmental inputs. Geoscientists with the appropriate training are now in demand in exploration, mine geology, mine environment and in the geotechnical and overall environmental areas. These trends indicate increasing future demands also in government and industry research and development. On the supply side there are some problems with the available training facilities and of attracting the numbers and quality of students, especially on the post-graduate level. The greatest demand is for technologically capable graduates that can meet the needs of an increasingly sophisticated, challenging and changing workplace, given that this demand is often cyclical characterised by alternating periods of shortage and oversupply of geoscientists. A guide that can be utilised to estimate the demand and supply for geoscientists is based on a simple Australian based model which correlates the mining in GDP (MGDP) with the professional geoscientific stock (GEO) and the manpower projected increase with the MGDP predicted growth rate. Applying this model correlation to India results in some reasonable preliminary estimates. Although the numbers of geoscientists necessary to meet the needs of the 21st century are speculative projections the increasingly technological quality required is without question. The current globalisation and economic competitiveness of the rapidly changing mining sector also places pressure on the quantity and especially the quality of professional geoscientist demand and supply.

Key words : Mineral Mining Industry, Geoscientists, Demand and Supply Employment Opportunities, Globalisation

INTRODUCTION

For geoscientists the growth and development of the mineral sector has created many employment opportunities. The growth in the mineral sector, however cyclical, has been further advanced by the introduction of favourable for investment exploration and mining policies, mineral prospects and the availability of quality geological and geophysical maps and data. In the last few years the emphasis has been to set up mining laws and policies that would attract more foreign investment in line with the liberalization and increasingly competitive nature of the global economy and mining industry. Past mineral and energy development of countries like India in coal and iron ore has more recently opened up to foreign investments in gold and base metal exploration. The introduction of favourable mining laws, codes and contracts in highly prospective countries like India has led to an increased foreign involvement in mineral exploration.

EDUCATION AND TRAINING INSTITUTES

McDivitt's (1993) survey of geoscience education and training has provided much useful information which may be common to countries like India. Although there has been progress in the strengthening of geoscience education and training, many problems still exist in training for middle and high level posts in universities, research institutes and in the mineral industry. In most cases, the basic programs in these countries are adequate. Facilities for university level geoscientific training are relatively new, modern, and have good equipment and facilities. However, the development has not been long enough to permit the necessary balance of facilities, training programs, research programs and staff training. The facilities and programs for geoscientific training are limited in mineral poor countries. Mineral rich countries like India usually have full structures for training of nationals however, in most cases, the rapidly changing industry requirements for practical graduates trained in new technological applications and management methods are not being adequately addressed.

SUPPLY AND DEMAND FACTORS

In the ideal situation, supply and demand of geoscientists should be in balance, given that the mining industry is characterised by cyclical and fluctuating demand periods. Indicators are that in various countries supply exceeds demand, demand exceeds supply and/or there is some local equilibrium perhaps disguising oversupply or shortfalls. In all cases there is a lack of appropriate R and D experts for the development of national technical self reliance. Most, if not all, exploration, mining and geotechnical products, goods and services are from overseas which results in the avoidance of employment or underemployment of significant numbers on R and D functions. In general, there exists persistent short falls in the supply of geoscientists covered by a propensity to rely on R and D work done off shore. This is not in the national short or long term interest and unless the supply of geoscientific expertise increases, the nationally-essential demand may never arise.

DEMAND FOR GEOSCIENTISTS

The demand for geoscientists is divided between the public and private sectors. The government tends to employ many of the graduates as administrators, educators and engineers in the civil service, and in statutory boards. Geoscientists who enter the private sector are not exclusively employed in the exploration, mining or geotechnical fields. Some are employed as managerial or sales personnel and only a few professionals are self-employed. The demand has increased where economies have experienced high mineral sector growth rate that will depend on economic factors such as cyclical mineral and energy commodities world market conditions. In addition to continued but slower quantitative growth, there will be a substantial qualitative change in the composition of employment in the 21st century. Industrial upgrading means a higher ratio of indirect to direct labour, of skilled to unskilled workers, and of engineers and technicians to production operators. Economic conditions may lead to short - long term investment problems that may result in a slowdown in the mineral industry as well as the geotechnical/ construction industry's demand for geoscientists. The demand may also accelerate if major technological or functional shifts occur. Moving into R and D activities, which have considerable economies of scale, will also multiply demand. The demand will also shift towards more management and planning and fewer technical jobs, though the numbers of both may grow. Development of a new comparative advantage in a more skilled, physical, and human capital-intensive exploration, mining and geotechnical sectors, that will create a strong demand for geoscientists also depends on the availability, cost and quality of the professionals themselves.

SUPPLY OF GEOSCIENTISTS

Manpower forecasts project, that if countries like India are to achieve a reasonable annual GDP growth in the 21st century its tertiary and post-secondary institutions must together take in substantial numbers of new students each year. The problem of finding enough suitable applicants for geoscience courses will be particularly severe. Institutes are expected to raise their intakes and present indications suggest that only a small percentage of students are enrolled in geology.

As science and engineering enrolments are expected to grow faster than those of non-engineering courses the proportion of students studying geology may also rise.

Currently, a few percent of secondary school-leavers proceed to the university and a similar percent to technical and vocational colleges. Only a fraction of these students have the ability and inclination to enroll in geology programs which are variable and often restricted to only a few university departments. Female students, for example, have traditionally preferred other fields of study. Even if the quantitative targets are reached, they may be inadequate in terms of the quality and type of geoscientist required for upgrading and intensive activities like R and D. The geoscientists required for these activities must be highly intelligent, well trained, imaginative, and versatile and an improvement in quality over the ones now being produced.

RESEARCH AND DEVELOPMENT DEMAND AND SUPPLY

The quality of geoscientists is crucial in the development of research and development capabilities. Government organisations and the exploration, mining and geotechnical industries are now becoming more sophisticated. In order to be internationally competitive there is a need for greater investment in R and D and a greater demand for professionals with postgraduate degrees and research training and experience. In all cases research trends are away from the fundamental to the applied, especially in regard to possible local, unique exploration, mining and geotechnical problems which cannot be solved by using outside R and D results. R and D expertise is absolutely necessary for national technical self-reliance.

The expectation is that the R and D demand will increase during the next few decades which will require more quality student recruitment and higher retention rates as follows:

Recruitment

* more women
* quality of teaching in schools
* encouraging young students
* encouraging post graduate studies and continuing education
* equal opportunity and greater access
* awareness of challenging career
* availability of scholarships retention
* ensuring degree completion - grants
* preventing the loss of qualified persons - fellowships
* short term to long term appointments
* retraining and conversion

There will also be pressure on strengthening existing research facilities and developing new ones in academia, government and industry organisations.

GEOSCIENTIST MANPOWER PLANNING - AUSTRALIAN MODEL

The provisional supply-demand geoscientific manpower prediction method used here considers a simple economic model based on 1988 - 1989 Australian data when supply and demand was in equilibrium (Katz, 1994). Since then there has been fluctuations in this workforce due to the mining industry sharp ups and downs which is currently the case. This model guide assumes an approximate balanced labour market and that the mining (and energy) component of the GDP (MGDP) and its growth rate directly correlates with the number and stock of geoscientists (GEO).

In Australia in 1988-89 there were estimated to be about 5000 geoscientists out of a total mining industry workforce of about 100,000. Figures for that year indicate that the Australian MGDP accounted for about $16,000 million, so it can be shown that for every $100 million MGDP, the workforce requires about 30 geoscientists (GEO). Given an estimated MGDP growth rate of about five per cent, 250 geoscientists graduates (GG) would be required each year to meet this demand. As there is a necessary shift toward R and D, the supply of more highly qualified personnel will have to be increased to keep up with the progressive demand from academia, government and industry. Based on Australian earth science R and D data there were about 800 researchers (RS) in the estimated geoscientific stock of 5000, about 15% per cent of the total. Earth science research expenditure (RE) amounts to about $160 million which is about 8% of the national R and D and over 1% of the MGDP. Using these figures, as an estimate of the R and D supply/demand component, for every $1 million spent on R and D requires about four geoscientific researchers (RS) and about one per cent of MGDP should be expended on earth science R and D. Of the total 250 annual geoscientific graduates (GG), about 15% (40) should be postgraduate (PG) to Ph.D level.

INDIAN MODEL

The Australian model based on a simple (MGDP) - (GEO) correlation estimate (every $100 million of MGDP requires about 30 geoscientists) will have to be modified in regard to various national and international conditions including education, technology, labour factors and fluctuating economic conditions.

Table 1 shows some current important Australian and Indian mineral economic parameters that must be taken into consideration. The gross domestic product (GDP) is an indication of the size of the economy and the mining in GDP (MGDP) and the mining in export (ME) reflects the importance of the mining sector. Applying this MGDP-GEO model correlation in Table 2 estimates of geoscientists (GEO) in India required by this level of activity at the start of the 21st century would be in the order of 16,600 with an ideal percentage (~ 15%) of 2,500 R and D experts. Research expenditure (RE) estimates for India should be in the order of one per cent of MGDP or 1,660 million dollars. Assuming a conservative MGDP growth rate in the order of five per cent, some 850 geoscience graduates (GG) are required including 125 postgraduate students (PG). These figures are only first order approximations, based on Australian model data, and must be treated as very speculative. How these results match actual stock and graduate figures can only be examined when detailed data becomes available.

ENROLMENTS AND GRADUATES

A major problem of supply of geoscientists is the rate of enrolments and graduation. Enrolment rates are dependent on the numbers of secondary school students who opt for science and engineering and those that choose geosciences as their specialty. Enrolment rates in science and engineering are increasing with technology development and those that select geology courses may be streamed into that direction or show interest because of employment opportunities on graduation and/or take these courses as a 'soft' science option. In Australia about 13 per cent of all university students enrol in natural sciences. Of these, about five per cent of the science students are in the geosciences and the graduates amount to

Table-1 Current Mining in GDP and Export Estimates (US$ millions rounded off)

Country	GDP	Mining in GDP (MGDP)	Mining in Export (ME)
Australia	611,700	9%	35%
India	3,319,000	5%	20%

Table-2 Geoscientific Professional Manpower and Research Expenditure Guess Estimates

MGDP and RE (in US$ million)						
COUNTRY	MGDP	RE	GEO	RS	GG	PG
Australia	**55,000**	**550**	**5,500**	**825**	**275**	**45**
India	**166,000**	**1,660**	**16,600**	**2,500**	**850**	**125**
MGDP	Mining in GDP (GDP X % Mining)					
RE	Research Expenditure (1% of MGDP)					
GEO	Geoscientific Stock (10/$100 million MGDP)					
RS	Research Stock (15% of GEO)					
GG	Geosciences Graduates (5% of GEO)					
PG	Geosciences Postgraduates (15% of GG)					

about 20 percent of those enrolled. Of those graduates about three per cent are geoscience graduates. It is interesting to compare these figures from Australia on enrolments and graduate numbers to the data and estimates for the countries in the region (UNESCO). India may reflect most third world countries and indicate relatively low enrolments of about 5% in the natural sciences, with graduation rates of about 10%. Of these enrollees and graduates only a small proportion would be in the geosciences (<5%).

CONCLUSIONS

In all cases, economic and structural changes must be taken into account in demand and supply of geoscientists in India in the 21st century. Given that prediction of supply is as reliable as weather forecasting, and demand projections even more difficult given the cyclical nature of the demand. Table 2 gives a guess estimate of numbers and research expenditure at the start of the 21st century based on an assumed and speculative demand/supply balance and some positive economic growth rates. The changes indicate that there will be increasing and pressing emphasis on high-level, technical and management skills to meet these future needs. The speculative projected stock of the geoscientists of the 21st century will be required to have a qualitative edge in order to deal with new technology; in mineral evaluation, exploration, geological data processing, geostatistics, mine geology and environmental geology as well as new and modern exploration, geotechnical, environmental and impact management methods and techniques. Increasing R and D expenditures will be necessary. The geoscientist with technical research, computer and management skills should be in high demand in the future developing public and private mining sector. In response to these new changes all geoscience education and training institutes in India will have to develop new and innovative programs with the support of government and industry that will not only serve the modern needs of undergraduate and postgraduate students but also be able to set up continuing education, retraining and upgrading opportunities for geoscientists already in the workplace.

References

Katz, M.; (1994) : The Demand and Supply for Mining Professionals in the Asia / Pacific Region, Asian. Jour. Mining, Sept. 1994, pp. 59 62.

McDivitt, J.F.; (1993) : University Training Programs for Mineral and Energy in the Asia/Pacific Region. MEF, PECC Publication, 188 p.

Hon'ble Reviewers

1. Prof. U. Aswathnarayana
   (Hon. Director, Mahadevan International
   Centre for Water Management)
   B-16, Santishikha Apartments
   Somajiguda
   HYDERABAD 500 082 INDIA
2. Prof. S.P. Tandon
   Pro-Vice Chancellor
   Delhi University
   DELHI 110 007 INDIA

Editorial Assistance

Prof. Srikant Ojha